MATHEMATICS OF CLASSICAL AND QUANTUM PHYSICS

FREDERICK W. BYRON, JR.

AND

ROBERT W. FULLER

TWO VOLUMES BOUND AS ONE

Dover Publications, Inc., New York

This Dover edition, first published in 1992, is an unabridged, corrected republication of the work first published in two volumes by the Addison-Wesley Publishing Company, Reading, Mass., 1969 (Vol. One) and 1970 (Vol. Two). It was originally published in the "Addison-Wesley Series in Advanced Physics."

Manufactured in the United States of America
Dover Publications, Inc., 31 East 2nd Street, Mineola, N.Y. 11501

Library of Congress Cataloging-in-Publication Data

Byron, Frederick W.
Mathematics of classical and quantum physics / Frederick W. Byron, Jr., Robert W. Fuller.
p. cm.
"Unabridged, corrected republication of the work first published in two volumes by the Addison-Wesley Publishing Company, Reading, Mass., 1969 (Vol. One) and 1970 (Vol. Two) . . . in the 'Addison-Wesley series in advanced physics' "—T.p. verso.
Includes bibliographical references and index.
ISBN 0-486-67164-X (pbk.)
1. Mathematical physics. 2. Quantum theory. I. Fuller, Robert W. II. Title.
QC20.B9 1992
530.1'5—dc20

92-11943
CIP

To Edith and Ann

PREFACE

This book is designed as a companion to the graduate level physics texts on classical mechanics, electricity, magnetism, and quantum mechanics. It grows out of a course given at Columbia University and taken by virtually all first year graduate students as a fourth basic course, thereby eliminating the need to cover this mathematical material in a piecemeal fashion within the physics courses. The two volumes into which the book is divided correspond roughly to the two semesters of the full-year course. The consolidation of the mathematics needed for graduate physics into a single course permits a unified treatment applicable to many branches of physics. At the same time the fragments of mathematical knowledge possesed by the student can be pulled together and organized in a way that is especially relevant to physics. The central unifying theme about which this book is organized is the concept of a *vector space*. To demonstrate the role of mathematics in physics, we have included numerous physical applications in the body of the text, as well as many problems of a physical nature.

Although the book is designed as a textbook to complement the basic physics courses, it aims at something more than just equipping the physicist with the mathematical techniques he needs in courses. The mathematics used in physics has changed greatly in the last forty years. It is certain to change even more rapidly during the working lifetime of physicists being educated today. Thus, the physicist must have an acquaintance with abstract mathematics if he is to keep up with his own field as the mathematical language in which it is expressed changes. It is one of the purposes of this book to introduce the physicist to the language and the style of mathematics as well as the content of those particular subjects which have contemporary relevance in physics.

The book is essentially self-contained, assuming only the standard undergraduate preparation in physics and mathematics; that is, intermediate mechanics, electricity and magnetism, introductory quantum mechanics, advanced calculus and differential equations. The level of mathematical rigor is generally comparable to that typical of mathematical texts, but not uniformly so. The degree of rigor and abstraction varies with the subject. The topics treated are of varied subtlety and mathematical sophistication, and a logical completeness that is illuminating in one topic would be tedious in another.

While it is certainly true that one does not need to be able to follow the proof of Weierstrass's theorem or the Cauchy–Goursat theorem in order to be able to compute Fourier coefficients or perform residue integrals, we feel that the student who has studied these proofs will stand a better chance of growing mathematically *after* his formal coursework has ended. No reference work, let alone a text, can cover all the mathematical results that a student will need. What *is* perhaps possible, is to generate in the student the confidence that he can find what he needs in the mathematical literature, and that he can understand it and use it. It is our aim to treat the limited number of subjects we do treat in enough detail so that after reading this book physics students will not hesitate to make direct use of the mathematical literature in their research.

The backbone of the book—the theory of vector spaces—is in Chapters 3, 4, and 5. Our presentation of this material has been greatly influenced by P. R. Halmos's text, *Finite-Dimensional Vector Spaces.* A generation of theoretical physicists has learned its vector space theory from this book. Halmos's organization of the theory of vector spaces has become so second-nature that it is impossible to acknowledge adequately his influence.

Chapters 1 and 2 are devoted primarily to the mathematics of classical physics. Chapter 1 is designed both as a review of well-known things and as an introduction of things to come. Vectors are treated in their familiar three-dimensional setting, while notation and terminology are introduced, preparing the way for subsequent generalization to abstract vectors in a vector space. In Chapter 2 we detour slightly in order to cover the mathematics of classical mechanics and develop the variational concepts which we shall use later. Chapters 3 and 4 cover the theory of finite dimensional vector spaces and operators in a way that leads, without need for subsequent revision, to infinite dimensional vector spaces (Hilbert space)—the mathematical setting of quantum mechanics. Hilbert space, the subject of Chapter 5, also provides a very convenient and unifying framework for the discussion of many of the special functions of mathematical physics. Chapter 6 on analytic function theory marks an interlude in which we establish techniques and results that are required in all branches of mathematical physics. The theme of vector spaces is interrupted in this chapter, but the relevance to physics does not diminish. Then in Chapters 7, 8, and 9 we introduce the student to several of the most important techniques of theoretical physics—the Green's function method of solving differential and partial differential equations and the theory of integral equations. Finally, in Chapter 10 we give an introduction to a subject of ever increasing importance in physics—the theory of groups.

A special effort has been made to make the problems a useful adjunct to the text. We believe that only through a concerted attack on interesting problems can a student really "learn" any subject, so we have tried to provide a large selection of problems at the end of each chapter, some illustrating or extending mathematical points, others stressing physical applications of techniques developed in the text. In the later chapters of the book, some rather significant results are left as problems or even as a programmed series of problems, on the theory that as the student develops confidence and sophistication in the early chapters he will be able, with a few hints, to obtain some nontrivial results for himself.

The text may easily be adapted for a one-semester course at the graduate (or advanced undergraduate) level by omitting certain chapters of the instructor's choosing. For example, a one-semester course could be based on Volume 1. Another possibility, and one essentially used by one of the authors at the University of California at Berkeley, is to give a semester course based on the material in Chapters 3, 4, 5, and 10. On the other hand, a one-semester course in advanced mathematical methods in physics could be constructed from Volume II.

Certain sections within a chapter which are difficult and inessential to most of the rest of the book are marked with an asterisk.

In writing a book of this kind one's debts proliferate in all directions. In addition to the book of Halmos, we have been influenced by Courant–Hilbert's treatment of, and T. D. Lee's lecture notes on, Hilbert space, Riesz and Nagy's treatment of integral equations, and M. Hamermesh's book, *Group Theory*.

A special debt of gratitude is owed to R. Friedberg whose comments on the material have been extremely helpful. In particular, the presentation of Section 5.10 is based on his lecture notes.

Parts of the manuscript have also been read and taught by Ann L. Fuller, and her comments have improved it greatly. Richard Haglund and Steven Lundeen read and commented on the manuscript. Their painstaking work has removed many blemishes, and we thank them most sincerely.

Much of this book appeared in the form of lecture notes at Columbia University. Thanks are owed to the many students there, and elsewhere, who pointed out errors, or otherwise helped to improve the manuscript. Also, the enthusiasm of the students studying this material at Berkeley provided important encouragement.

While all the above named people have helped us to improve the manuscript, we alone are responsible for the errors and inadequacies that remain. We will be grateful if readers will bring errors to our attention so corrections can be made in subsequent printings.

One of us (FWB) held an Alfred P. Sloan Fellowship during much of the period of the writing; he gratefully thanks Professors M. Demeur and C. J. Joachain for their hospitality at the Université Libre de Bruxelles. The other author (RWF) would like to thank R. A. Rosenbaum of Wesleyan University, the University's Center for Advanced Studies, and its director, Paul Horgan, for their hospitality during the course of much of the work. We would also like to thank F. J. Milford and Battelle Memorial Institute's Seattle Research Center for providing support that facilitated the completion of the work.

Many of the practical problems of producing the manuscript were alleviated by the valued assistance of Rae Figliolina, Cheryl Gruger, Barbara Hollisi, and Barbara Satton.

Amherst, Mass. F.W.B., Jr.
Hartford, Conn. R.W.F.
January 1969

CONTENTS

VOLUME ONE

VOLUME ONE

CHAPTER 1

VECTORS IN CLASSICAL PHYSICS

INTRODUCTION

In this chapter we shall review informally the properties of the vectors and vector fields that occur in classical physics. But we shall do so in a way, and in a notation, that leads to the more abstract discussion of vectors in later chapters. The aim here is to bridge the gap between classical three-dimensional vector analysis and the formulation of abstract vector spaces, which is the mathematical language of quantum physics. Many of the ideas that will be developed more abstractly and thoroughly in later chapters will be anticipated in the familiar three-dimensional setting here. This should provide the subsequent treatment with more intuitive content. This chapter will also provide a brief recapitulation of classical physics, much of which can be elegantly stated in the language of vector analysis—which was, of course, devised expressly for this purpose. Our purpose here is one of informal introduction and review; accordingly, the mathematical development will not be as rigorous as in subsequent chapters.

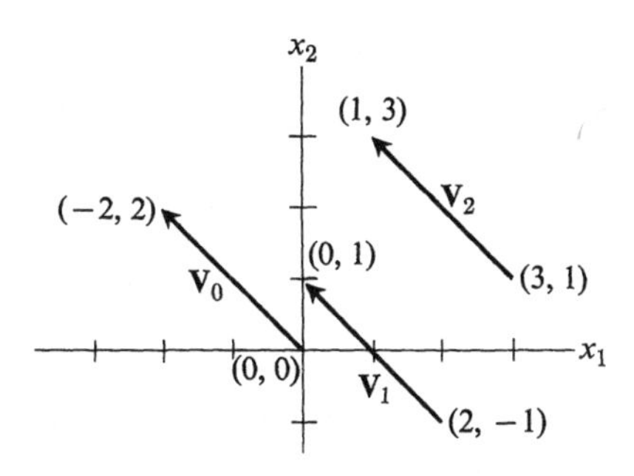

Fig. 1.1 Three equivalent vectors in a two-dimensional space.

1.1 GEOMETRIC AND ALGEBRAIC DEFINITIONS OF A VECTOR

In elementary physics courses the geometric aspect of vectors is emphasized. A vector, **x**, is first conceived as a directed line segment, or a quantity with both a magnitude and a direction, such as a velocity or a force. A vector is thus distinguished from a scalar, a quantity which has only magnitude such as temperature, entropy, or mass. In the two-dimensional space depicted in Fig. 1.1, three vectors of equal magnitude and direction are shown. They form an equivalence class which may be represented by $\mathbf{V}_0$, the unique vector whose initial point is at the origin. We shall gradually replace this elementary characterization of vectors and scalars with a more fundamental one. But first we must develop another language with which to discuss vectors.

An algebraic aspect of a vector is suggested by the one-to-one correspondence between the unique vectors (issuing from the origin) that represent equivalence classes of vectors, and the coordinates of their terminal points, the ordered pairs of real numbers (x_1, x_2). Similarly, in three-dimensional space we associate a geometrical vector with an ordered triple of real numbers, (x_1, x_2, x_3), which are called the *components* of the vector. We may write this vector more briefly as x_i, where it is understood that i extends from 1 to 3. In spaces of dimension greater than three we rely increasingly on the algebraic notion of a vector, as an ordered n-tuple of real numbers, $(x_1, x_2, \cdots, x_n)$. But even though we can no longer construct physical vectors for n greater than three, we retain the geometrical language for these n-dimensional generalizations. A formal treatment of the properties of such abstract vectors, which are important in the theory of relativity and quantum mechanics, will be the subject of Chapters 3 and 4. In this chapter we shall restrict our attention to the three-dimensional case.

There are then these two complementary aspects of a vector: the geometric, or physical, and the algebraic. These correspond to plane (or solid) geometry and analytic geometry. The geometric aspect was discovered first and stood alone for centuries until Descartes discovered algebraic or analytic geometry. Anything that can be proved geometrically can be proved algebraically and vice-versa, but the proof of a given proposition may be far easier in one language than in the other.

Thus the algebraic language is more than a simple alternative to the geometric language. It allows us to formulate certain questions more easily than we could in the geometric language. For example, the tangent to a curve at a point can be defined very simply in the algebraic language, thus facilitating further study of the whole range of problems surrounding this important concept. It is from just this formulation of the problem of tangents that the calculus arose.

It is said of Niels Bohr that he never felt he understood philosophical ideas until he had discussed them with himself in German, French, and English as well as in his native Danish. Similarly, one's understanding of geometry is strengthened when one can view the basic theorems from both the geometric and the algebraic points of view. The same is true of the study of vectors. It is all too easy to rely on the algebraic language to carry one through vector analysis, skipping blithely over the physical, geometric interpretation of the differential operators. We shall try to bring out the physical meanings of these operators as well as review their algebraic manipulation. The basic operators of vector analysis crop up everywhere in physics, so it pays to develop a physical picture of what these operators do—that is, what features they "measure" of the scalar or vector fields on which they operate.

1.2 THE RESOLUTION OF A VECTOR INTO COMPONENTS

One of the most important aspects of the study of vectors is the resolution of vectors into components. In fact, this will remain a central feature in Chapter 5, where we deal with Hilbert space, the infinite-dimensional generalization of a vector space. In three dimensions, any vector $\mathbf{x}$ can be expressed as a linear combination of any three noncoplanar vectors. Thus $\mathbf{x} = \alpha\mathbf{V}_1 + \beta\mathbf{V}_2 + \gamma\mathbf{V}_3$, where α, β, and γ are scalars. If we denote the length of a vector $\mathbf{x}$ by $|\mathbf{x}|$, then $\alpha|\mathbf{V}_1|$, $\beta|\mathbf{V}_2|$, and $\gamma|\mathbf{V}_3|$ are the components of $\mathbf{x}$ in the $\mathbf{V}_1$, $\mathbf{V}_2$, and $\mathbf{V}_3$ directions. The three vectors $\mathbf{V}_1$, $\mathbf{V}_2$, and $\mathbf{V}_3$ need not be perpendicular to each other—any three noncoplanar vectors form a base, or *basis*, in terms of which an arbitrary vector may be decomposed or expanded. But it is often most convenient to choose the basis vectors perpendicular to each other. In this case the basis is called *orthogonal*; otherwise it is called oblique. We shall deal almost exclusively with sets of orthogonal basis vectors.

A particularly useful set of basis vectors is the *Cartesian basis*, consisting of three mutually orthogonal vectors of unit length which have the same direction at all points in space. We shall denote unit vectors by the letter $\mathbf{e}$ in this chapter; accordingly, the Cartesian basis is the set $(\mathbf{e}_1, \mathbf{e}_2, \mathbf{e}_3)$ shown in Fig. 1.2. Such a set of base vectors is called *orthonormal*, because the vectors are orthogonal to each other and are normalized (have unit length). We shall not distinguish between a "basis" and a "coordinate system" in this treatment.

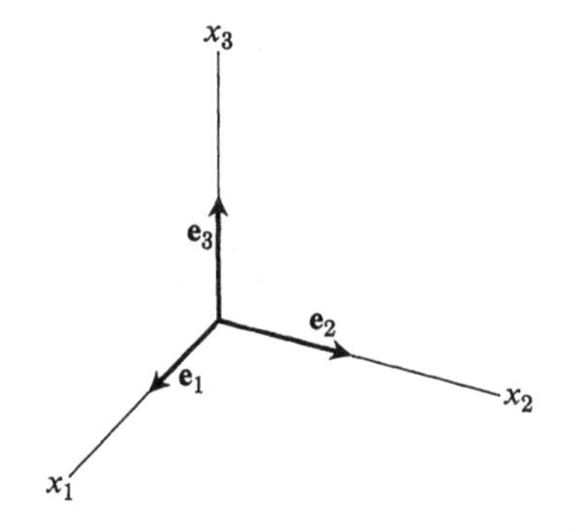

Fig. 1.2 Right-handed Cartesian basis or coordinate system.

The basis (or coordinate system) in Fig. 1.2 is *right-handed*, i.e., if the fingers of the right hand are extended along the positive x_1-axis and then curled toward the positive x_2-axis, the thumb will point in the positive x_3-direction. If any one of the basis vectors is reversed, we have a *left-handed* orthogonal basis. A mathematical definition of "handedness" will be given in Section 1.5.

An arbitrary vector may be expressed in terms of this Cartesian basis as $\mathbf{x} = x_1\mathbf{e}_1 + x_2\mathbf{e}_2 + x_3\mathbf{e}_3$. The $\mathbf{e}_i$, or ith, component of $\mathbf{x}$ with respect to this basis is x_i, for $i = 1, 2, 3$. There are a great many other orthonormal bases, such as those of cylindrical and spherical and other curvilinear coordinate systems, which can greatly simplify the treatment of problems with special symmetry features. We shall deal with these in Section 1.7.

1.3 THE SCALAR PRODUCT

The scalar ("inner" or "dot") product of two vectors $\mathbf{x}$ and $\mathbf{y}$ is the real number defined in geometrical language by the equation

$$\mathbf{x}\cdot\mathbf{y} = |\mathbf{x}|\,|\mathbf{y}|\cos\theta\,,$$

where θ is the angle between the two vectors, measured from $\mathbf{x}$ to $\mathbf{y}$. Since $\cos\theta$ is an even function, the scalar product is commutative:

$$\mathbf{x}\cdot\mathbf{y} = \mathbf{y}\cdot\mathbf{x}\,.$$

Moreover, the scalar product is distributive with respect to addition:

$$\mathbf{x}\cdot(\mathbf{y}+\mathbf{z}) = \mathbf{x}\cdot\mathbf{y} + \mathbf{x}\cdot\mathbf{z}\,.$$

This equation has a familiar and reasonable appearance, but that is only because we automatically interpret it algebraically, where we usually take distributivity for granted. The reader will find it instructive to prove this by geometrical construction.

If $\mathbf{x}\cdot\mathbf{y} = 0$, it does not follow that one or both of the vectors are zero. It may be that they are perpendicular. Note that the length of a vector $\mathbf{x}$ is given by

$$|\mathbf{x}| \equiv x = (\mathbf{x}\cdot\mathbf{x})^{1/2}\,,$$

since $\cos\theta = 1$ for $\theta = 0$. In particular, for Cartesian basis vectors, we have

$$\mathbf{e}_i\cdot\mathbf{e}_j = \delta_{ij}\,, \tag{1.1}$$

where δ_{ij} is the Kronecker delta defined by

$$\delta_{ij} = \begin{Bmatrix} 1 & \text{if} & i = j \\ 0 & \text{if} & i \neq j \end{Bmatrix}. \tag{1.2}$$

If we expand two arbitrary vectors, $\mathbf{x}$ and $\mathbf{y}$, in terms of the Cartesian basis,

$$\mathbf{x} = x_1\mathbf{e}_1 + x_2\mathbf{e}_2 + x_3\mathbf{e}_3 = \sum_{i=1}^{3} x_i\mathbf{e}_i\,,$$

$$\mathbf{y} = y_1\mathbf{e}_1 + y_2\mathbf{e}_2 + y_3\mathbf{e}_3 = \sum_{i=1}^{3} y_i\mathbf{e}_i\,,$$

then

$$\begin{aligned}\mathbf{x}\cdot\mathbf{y} &= \Big(\sum_i x_i\mathbf{e}_i\Big)\cdot\Big(\sum_j y_j\mathbf{e}_j\Big)\\ &= \sum_{i,j} x_iy_j\,\mathbf{e}_i\cdot\mathbf{e}_j = \sum_{i,j} x_iy_j\,\delta_{ij}\\ &= \sum_i x_iy_i\,.\end{aligned} \tag{1.3}$$

Here we have used the distributivity of the scalar product;

$$\sum_{i,j} \quad \text{stands for} \quad \sum_i\sum_j\,.$$

This last expression may be taken as the algebraic definition of the scalar product.

It follows that the length of a vector is given in terms of the scalar product by $|\mathbf{x}| = (\mathbf{x}\cdot\mathbf{x})^{1/2} = (\sum_i x_i^2)^{1/2}$. This equation provides an independent way of associating with any vector, a number called its length. We see that the notion of length need not be taken as inherent in the notion of vector, but is rather a consequence of defining a scalar product in a space of abstract vectors. Thus in Chapter 3 we shall study abstract vector spaces in which no notion of length has been defined. Then, in Chapter 4 we shall add an inner (or scalar) product to this vector space and focus on the enriched structure that results from this addition.

We shall now introduce a notational shorthand known as the *Einstein summation convention*. Einstein, in working with vectors and tensors, noticed that whenever there was a summation over a given subscript (or superscript), that subscript appeared *twice* in the summed expression, and vice versa. Thus one could simply omit the redundant summation signs, interpreting an expression like $x_i y_i$ to mean summation over the repeated subscript from 1 to, in our case, 3. If there are two distinct repeated subscripts, two summations are implied, and so on. In a letter, Einstein refers with tongue in cheek to this observation as "a great discovery in mathematics," but if you don't believe it is, just try getting along without it! (Another story in this connection—probably apocryphal—has it that the printer who was setting type for one of Einstein's papers noticed the redundancy and suggested omitting the summation signs.)

We shall adopt Einstein's summation convention throughout this chapter. In terms of this convention we have, for example,

$$\mathbf{x} = x_i \mathbf{e}_i \,,$$

$$\mathbf{x}\cdot\mathbf{y} = x_i y_j \delta_{ij} = x_i y_i = x_j y_j \,,$$

$$\mathbf{x}\cdot\mathbf{e}_i = x_j \mathbf{e}_j \cdot \mathbf{e}_i = x_j \delta_{ij} = x_i \,.$$

The last equation defines x_i, the component of $\mathbf{x}$ in the $\mathbf{e}_i$ direction, $\mathbf{x}\cdot\mathbf{e}_i$ is also called the *projection* of $\mathbf{x}$ on the $\mathbf{e}_i$ axis. The set of numbers $\{x_i\}$ is called the representation (or the *coordinates*) of the vector $\mathbf{x}$ in the basis (or the coordinate system) $\{\mathbf{e}_i\}$.

1.4 ROTATION OF THE COORDINATE SYSTEM: ORTHOGONAL TRANSFORMATIONS

We shall now consider the relationship between the components of a vector expressed with respect to two different Cartesian bases with the same origin, as shown in Fig. 1.3. Any vector $\mathbf{x}$ can be resolved into components with respect to either the K or the K' system. For example, in K we have

$$\mathbf{x} = (\mathbf{x}\cdot\mathbf{e}_j)\mathbf{e}_j = x_j \mathbf{e}_j \,, \tag{1.4}$$

where we are using the summation convention. In particular, if we take $\mathbf{x} = \mathbf{e}'_i$ $(i = 1, 2, 3)$, we can express the primed set of basis vectors in terms of the unprimed set:

$$\mathbf{e}'_i = (\mathbf{e}'_i \cdot \mathbf{e}_j)\mathbf{e}_j \equiv a_{ij}\mathbf{e}_j \qquad (j = 1, 2, 3)\ . \tag{1.5}$$

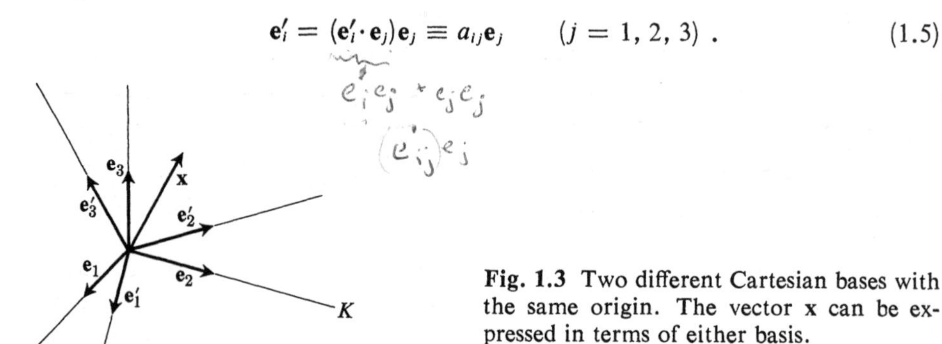

Fig. 1.3 Two different Cartesian bases with the same origin. The vector $\mathbf{x}$ can be expressed in terms of either basis.

The nine terms a_{ij} defined by Eq. (1.5), are the directional cosines of the angles between the six axes. These numbers may be written as the square array,

$$R \equiv (a_{ij}) \equiv \begin{bmatrix} a_{11} & a_{12} & a_{13} \\ a_{21} & a_{22} & a_{23} \\ a_{31} & a_{32} & a_{33} \end{bmatrix}; \tag{1.6}$$

R is known as the *rotation matrix* in three dimensions, since it describes the consequences of a change from one basis to another (rotated) basis.

Note that in defining the matrix elements by the equation

$$a_{ij} \equiv \mathbf{e}'_i \cdot \mathbf{e}_j\ , \tag{1.7a}$$

we have adopted a certain convention. We could just as well have defined matrix elements a'_{ij} by

$$a'_{ij} \equiv \mathbf{e}'_j \cdot \mathbf{e}_i = a_{ji}\ . \tag{1.7b}$$

Almost all authors use the convention of Eq. (1.7a), but this is an arbitrary choice; a completely consistent development of the theory is possible based on the definition (1.7b). In fact, in abstract vector space theory, matrices are usually defined by a convention consistent with Eq. (1.7b) rather than by our rule, Eq. (1.7a). However, by replacing a_{ij} by a_{ji} in Eq. (1.6) and the equations we shall derive presently—thus interchanging rows and columns or *transposing* the matrix R—we may shuttle back and forth between conventions. In chapter 4 we shall reconsider these issues in a more general setting that permits an easy and complete systematization.

It is apparent that the elements of the rotation matrix are not independent. Since the basis vectors form an orthonormal set, it follows from Eq. (1.5) that

$$\delta_{ij} = \mathbf{e}'_i \cdot \mathbf{e}'_j = a_{ik}(\mathbf{e}_k \cdot \mathbf{e}'_j) = a_{ik}a_{jk}\ . \tag{1.8}$$

Equation (1.8) stands for a set of nine equations (of which only six are distinct), each involving a sum of three quadratic terms. It is left to the reader to show (by expanding the unprimed vectors in terms of the primed basis and taking scalar products) that we also have the relation

$$a_{ki}a_{kj} = \delta_{ij} \, . \tag{1.9}$$

The expressions (1.8) and (1.9) are referred to as orthogonality relations; the corresponding transformations (Eq. 1.5) are called *orthogonal transformations.*

In an n-dimensional space, the rotation matrix will have n^2 elements, upon which the orthogonality relations place $\frac{1}{2}(n^2 + n)$ conditions, as the reader can verify. Thus

$$n^2 - \tfrac{1}{2}(n^2 + n) = \tfrac{1}{2}n(n - 1)$$

of the a_{ij} are left undetermined. In a two-dimensional space this leaves one free parameter, which we may take as the angle of rotation. In a three-dimensional space there are three degrees of freedom, corresponding to the three so-called *Euler angles* used to describe the orientation of a rigid body.

Equation (1.5), together with the orthogonality relations, tells us how one set of orthogonal basis vectors is expressed in terms of another rotated set. Now we ask: How are the *components* of a vector in K related to the components of that vectors in K', and vice versa?

Any vector $\mathbf{x}$ may be expressed either in the K system as $\mathbf{x} = x_j\mathbf{e}_j$, or in the K' system as $\mathbf{x} = x'_i\mathbf{e}'_i$. Let us first express the x_j, the components of $\mathbf{x}$ with respect to the basis $\mathbf{e}_j$, in terms of the x'_i, the components of $\mathbf{x}$ with respect to the basis $\mathbf{e}'_i$. Using Eq. (1.5), we have

$$\mathbf{x} = x'_i\mathbf{e}'_i = x'_i a_{ij}\mathbf{e}_j = x_j\mathbf{e}_j \, . \tag{1.10}$$

Now, since the basis vectors are orthogonal, we may identify their coefficients in Eq. (1.10):

$$x_j = a_{ij}x'_i \, . \tag{1.11}$$

(More formally, if $\alpha_i\mathbf{e}_i = \beta_i\mathbf{e}_i$, then $(\alpha_i - \beta_i)\mathbf{e}_i = 0$. Since this is a sum, it does not follow automatically that $\alpha_i = \beta_i$. However, taking the scalar product with $\mathbf{e}_j$ gives $(\alpha_i - \beta_i)\delta_{ij} = 0$, whence $\alpha_j = \beta_j$.)

To derive the inverse transformation, one could, of course, repeat the above procedure, substituting for the unprimed vectors. That is, instead of using Eq. (1.5), one could use the corresponding relation for the unprimed basis vectors in terms of the primed:

$$\mathbf{e}_i = (\mathbf{e}_i \cdot \mathbf{e}'_j)\mathbf{e}'_j = a_{ji}\mathbf{e}'_j \, .$$

However, using the orthogonality relations, we can derive this result directly from Eq. (1.11). Multiplying it by a_{kj}, summing over j, and using Eq. (1.8), we have

$$a_{kj}x_j = a_{kj}a_{ij}x'_i = \delta_{ki}x'_i = x'_k \, , \tag{1.12}$$

which gives the primed components in terms of the unprimed components.

In summary, we have

$$\begin{aligned}
\mathbf{x} &= x'_i \mathbf{e}'_i = x_j \mathbf{e}_j\,, \\
\mathbf{e}'_i &= a_{ij}\mathbf{e}_j\,, \qquad \mathbf{e}_i = a_{ji}\mathbf{e}'_j \\
x'_i &= a_{ij}x_j\,, \qquad x_i = a_{ji}x'_j \\
a_{ik}a_{jk} &= a_{ki}a_{kj} = \delta_{ij}\,.
\end{aligned} \tag{1.13}$$

It should be understood that these equations refer to the components of *one* vector, $\mathbf{x}$, as expressed with respect to *two* different sets of basis vectors, $\mathbf{e}_i$ and $\mathbf{e}'_i$. Thus unprimed basis vectors can be expressed in terms of the other (primed) basis. Thus a_{ij} is the jth component of $\mathbf{e}'_i$, expressed with respect to the unprimed basis, and a_{ji} is the jth component of $\mathbf{e}_i$ expressed with respect to the primed basis.

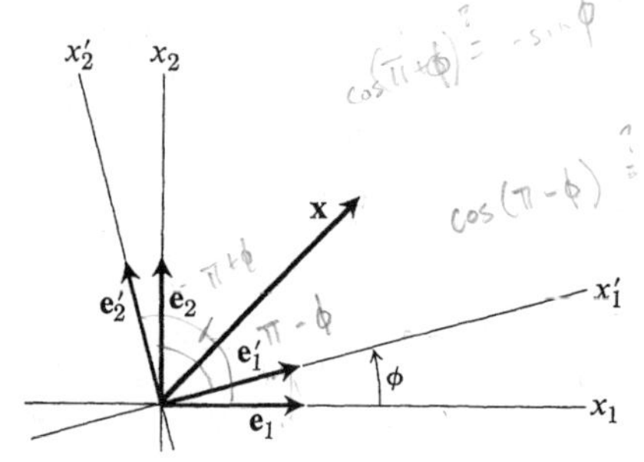

Fig. 1.4 Rotation in two dimensions, or rotation in three dimensions about an axis, x_3, orthogonal to the x_1, x_2, x'_1, x'_2 axes.

Example 1.1. *The two-dimensional rotation matrix.* We have defined the elements of the rotation matrix in Eqs. (1.5) and (1.6). For the two-dimensional case we have four coefficients: $a_{ij} \equiv (\mathbf{e}'_i \cdot \mathbf{e}_j)$, for $i, j = 1, 2$. From Fig. 1.4 it is clear that

$$(a_{ij}) = \begin{bmatrix} \cos\varphi & \sin\varphi \\ -\sin\varphi & \cos\varphi \end{bmatrix}. \tag{1.14a}$$

The first subscript of a_{ij} labels the row and the second subscript labels the column of the element a_{ij}. This rotation matrix tells us what happens to the components of a single vector $\mathbf{x}$ when we go from one basis, $\mathbf{e}_i$, to a new basis, $\mathbf{e}'_i$, by rotating the basis counterclockwise through an angle $(+\varphi)$. From Eq. (1.13), the components x_i of a vector $\mathbf{x}$ relative to the $\mathbf{e}_i$ basis and the components x'_i of that same vector relative to the $\mathbf{e}'_i$ basis are related by $x'_i = a_{ij}x_j$, or written out in full,

$$\begin{aligned}
x'_1 &= \cos\varphi x_1 + \sin\varphi x_2\,, \\
x'_2 &= -\sin\varphi x_1 + \cos\varphi x_2\,.
\end{aligned} \tag{1.15a}$$

Here x_i and x'_i refer to the components of a single vector with respect to two bases. The vector $\mathbf{x}$ sits *passively* as the basis with respect to which it is expressed rotates beneath it.

But there is another way to interpret these equations. We may regard the x_i as the components of one vector, $\mathbf{x}$, and the x'_i as the component of *another* vector, $\mathbf{x}'$, obtained from $\mathbf{x}$ by rotating $\mathbf{x}$ through the angle $(-\varphi)$. Let $X \equiv \{\mathbf{e}_i\}$ be the original set of basis vectors, and let $X' \equiv \{\mathbf{e}'_i\}$ denote the basis obtained by rotating X through the angle $(+\varphi)$. Then the components of $\mathbf{x}'$ referred to X are numerically equal to the components of $\mathbf{x}$ referred to X'. The idea behind this is actually quite simple; the difficulties are largely notational and can best be bypassed by drawing a diagram. The rotation of a vector through an angle $(-\varphi)$ produces a new vector with components in the original fixed basis equal to the components of the original vector, viewed as fixed, with respect to a new basis obtained from the original basis by rotating the original basis through the angle $(+\varphi)$.

Thus the equations that describe the *active* transformation of one vector into a new vector, rotated with respect to the original vector through an angle $(+\varphi)$, are obtained by substituting the angle $(-\varphi)$ into Eq. (1.15a), which gives

$$\begin{aligned} x'_1 &= \cos\varphi x_1 - \sin\varphi x_2 \\ x'_2 &= \sin\varphi x_1 + \cos\varphi x_2 \,. \end{aligned} \tag{1.16a}$$

Here the x_i and x'_i refer to the components of two vectors with respect to a single basis. Note that Eqs. (1.16a) may be written as

$$x'_i = a_{ji} x_j \,, \tag{1.16b}$$

where the a_{ij} are defined in Eq. (1.14a).

If we had defined the rotation matrix as the *transpose* of (a_{ij}), that is, if we had used Eq. (1.7b), then we would have

$$(a'_{ij}) = \begin{bmatrix} \cos\varphi & -\sin\varphi \\ \sin\varphi & \cos\varphi \end{bmatrix} \tag{1.14b}$$

replacing Eq. (1.14a);

$$x'_i = a_{ji} x_j \qquad \text{or} \qquad x_i = a_{ij} x'_j \,, \tag{1.15b}$$

replacing Eq. (1.15a) for the components of a single vector in two different bases, and

$$x'_i = a_{ij} x_j \qquad \text{or} \qquad x_i = a_{ji} x'_j \,, \tag{1.16c}$$

replacing Eq. (1.16b), for the components of a transformed vector with respect to a single basis.

We may extend the rotation matrix (Eq. 1.14a) to the three-dimensional case of a rotation of basis vectors about the x_3-axis. Denoting this rotation matrix by $R(\varphi)$, we have

$$R(\varphi) \equiv (a_{ij}) = \begin{bmatrix} \cos\varphi & \sin\varphi & 0 \\ -\sin\varphi & \cos\varphi & 0 \\ 0 & 0 & 1 \end{bmatrix}. \tag{1.17}$$

Example 1.2. *The three-dimensional rotation matrix* $R(\varphi, \theta, \psi)$. Suppose that we want to transform to a coordinate system in which the new z-axis, x_3', is in an arbitrarily specified direction, say along the vector **V** in Fig. 1.5. Such a rotation may be compounded of two three-dimensional rotations about an axis, such as those discussed in Example 1.1. First we rotate the coordinate system counterclockwise about the common x_3-x_3' axis through an angle φ. This gives

$$\mathbf{e}_i' = a_{ij}\mathbf{e}_j\,, \tag{1.18}$$

where the a_{ij} are given by Eq. (1.17). Now we rotate clockwise through an angle θ, that is, counterclockwise through the angle $(-\theta)$, in the $x_3'x_1'$-plane about the x_2'-axis. (We could as well have rotated about the x_1' axis but the sequence we have chosen is the conventional one.) The appropriate rotation matrix for this rotation of base about the x_2'-axis is

$$(b_{ij}) \equiv \begin{bmatrix} \cos\theta & 0 & -\sin\theta \\ 0 & 1 & 0 \\ \sin\theta & 0 & \cos\theta \end{bmatrix}, \tag{1.19}$$

and the new basis vectors, $\mathbf{e}_i''$, are given in terms of the primed ones by

$$\mathbf{e}_i'' = b_{ij}\mathbf{e}_j'\,.$$

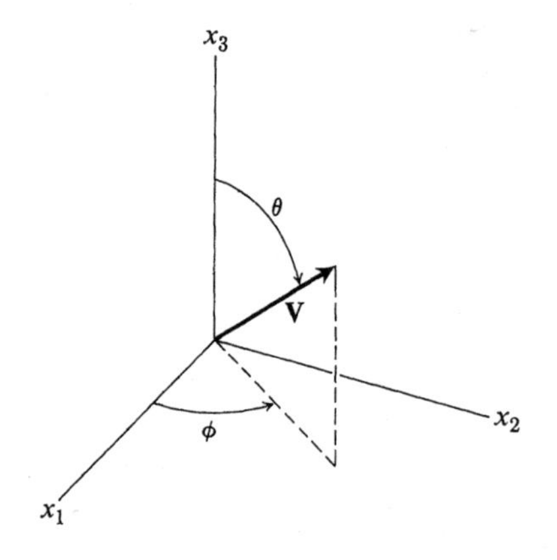

Fig. 1.5 The vector, **V**, which determines the z-axis of a rotated coordinate system.

Therefore, using Eq. (1.18), we have

$$\mathbf{e}_i'' = b_{ij}a_{jk}\mathbf{e}_k\,. \tag{1.20}$$

To go directly from the unprimed system to the doubly primed system we must know the coefficients c_{ik} in the equation

$$\mathbf{e}_i'' = c_{ik}\mathbf{e}_k\,. \tag{1.21}$$

Knowing these coefficients is equivalent to knowing the three-dimensional rotation matrix. From Eqs. (1.20) and (1.21), we see that

$$c_{ik} = b_{ij}a_{jk}\,. \tag{1.22}$$

Using this result and the matrices (1.17) and (1.19), we may compute the elements c_{ik}. The resulting rotation matrix is

$$R(\varphi, \theta) \equiv (c_{ik}) = \begin{bmatrix} \cos\varphi\cos\theta & \sin\varphi\cos\theta & -\sin\theta \\ -\sin\varphi & \cos\varphi & 0 \\ \cos\varphi\sin\theta & \sin\varphi\sin\theta & \cos\theta \end{bmatrix}. \tag{1.23}$$

This rotation matrix contains the matrix $R(\varphi)$ [Eq. (1.17)] as the special case $\theta = 0$; thus $R(\varphi, 0) = R(\varphi)$. Equation (1.22) is a special case as the general operation of matrix multiplication which will be treated fully in Chapter 3. The components of a vector $\mathbf{x}$ relative to the $\mathbf{e}_i$-basis and the components of that same vector $\mathbf{x}$ relative to the $\mathbf{e}_i''$-basis are related according to

$$x_i'' = c_{ij}x_j. \tag{1.24}$$

The rotation matrix $R(\varphi, \theta)$ does not represent the most general possible rotation. One more rotation is possible, a counterclockwise rotation through an angle ψ in the $x_1''x_2''$-plane about the x_3''-axis. This third rotation about an axis is described by the rotation matrix

$$(d_{ij}) = \begin{bmatrix} \cos\psi & \sin\psi & 0 \\ -\sin\psi & \cos\psi & 0 \\ 0 & 0 & 1 \end{bmatrix}.$$

And the grand "rotation of rotations," that may be achieved by compounding the three rotations about axes, through φ, θ, and ψ, is described by the rotation matrix whose elements are

$$[R(\varphi, \theta, \psi)]_{ij} = d_{ik}b_{kl}a_{lj}].$$

The reader may verify that the resulting matrix is

$$R(\varphi, \theta, \psi) = \begin{bmatrix} \cos\varphi\cos\theta\cos\psi - \sin\varphi\sin\psi & \sin\varphi\cos\theta\cos\psi + \cos\varphi\sin\psi & -\sin\theta\cos\psi \\ -\cos\varphi\cos\theta\sin\psi - \sin\varphi\cos\psi & -\sin\varphi\cos\theta\sin\psi + \cos\varphi\cos\psi & \sin\theta\sin\psi \\ \cos\varphi\sin\theta & \sin\varphi\sin\theta & \cos\theta \end{bmatrix}. \tag{1.25}$$

The angles (φ, θ, ψ) are called the *Euler angles*. Their definition varies widely—the probability is small that two distinct authors' general rotation matrix will be the same. Note that $R(\varphi, \theta, 0) = R(\varphi, \theta)$ and $R(\varphi, 0, 0) = R(\varphi)$. The reader might note in passing that the determinant of $R(\varphi, \theta, \psi)$ (and all the other rotation matrices), has the value one. We shall prove this in Chapter 4.

Originally we introduced a vector as an ordered triple of numbers. The rule for expressing the components of a vector in one coordinate system in terms of its components in another system tells us that if we fix our attention on a physical vector and then rotate the coordinate system $(K \to K')$, the vector will have different numerical components in the rotated coordinate system. So we are led to realize that a vector is really more than an ordered triple. Rather, it is many sets of ordered triples which are related in a definite way. One still specifies a vector by giving three ordered numbers, but these three numbers are distinguished from an arbitrary collection of three numbers by including the law of transformation under rotation of the coordinate frame as part of the definition. This law tells how all vectors change if the coordinate system changes. Thus one physical vector may be represented by infinitely many ordered triples. The particular triple depends on the orientation of the coordinate system of the observer. This is important because physical results must be the same regardless of one's vantage point, that is, regardless of the orientation of one's coordinate system. This will be the case if a given physical law involves vectors on both sides of the equation. Now, from this point of view, the transformation rule of Eq. (1.11) and the orthogonality relations, Eq. (1.9), may be used *to define* vectors. This is the natural starting point for a generalization to tensor analysis.

Since the orthogonal transformations are linear and homogeneous, it follows that the sum of two vectors is a vector and will transform acccording to Eq. (1.11) under orthogonal transformations. Also, if the equation $\mathbf{x} = \alpha\mathbf{y}$ (for example, $\mathbf{F} = m\mathbf{a}$), with α a scalar, holds in one coordinate system, it holds in any other which is related to the first by an orthogonal transformation. The reader may want to carry out the proofs of these statements formally.

We now prove a simple, but very important theorem.

Theorem. The scalar product is invariant under orthogonal transformations.

Proof. We see that this statement is obviously true when we consider the geometrical definition of the scalar product, for the lengths of vectors and the angle between them do not change as the axes are rotated. The algebraic proof is less transparent, but it allows some important generalizations. We have

$$\mathbf{x}' \cdot \mathbf{y}' = x_i' y_i' = a_{ij} x_j a_{ik} y_k = a_{ij} a_{ik} x_j y_k = \delta_{jk} x_j y_k = x_j y_j = \mathbf{x} \cdot \mathbf{y} ,$$

which completes the proof of the theorem.

Now scalars, ϕ, are invariant under rotations:

$$\phi' = \phi , \tag{1.26}$$

and the components of a vector transform according to

$$x_i' = a_{ij} x_j . \tag{1.27}$$

It is easy to generalize these notions and write down what is called a *Cartesian tensor of the second rank*. In three dimensions, this is a set of nine components, T_{ij}, which under orthogonal transformations behave according to the rule

$$T_{ij}' = a_{im} a_{jn} T_{mn} .$$

A vector is a tensor of the first rank and a scalar is a tensor of zeroth rank. Generalization to tensors of higher rank is clearly possible, but we shall defer further discussion of tensors to Section 1.8.

The importance of thinking of these quantities in terms of their transformation properties lies in the requirement that physical theories must be invariant under rotation of the coordinate system. The inclination of the coordinate axes that we superimpose on a physical situation must not affect the physical answers we get. Or, to put it another way, observers who study a situation in different coordinate systems must agree on all physical results. For example, we may view the flight of a projectile (Fig. 1.6) from either the K or the K' system. In K, Newton's second law is $\mathbf{F} = m\mathbf{a}$, and the equations of motion are

$$m\ddot{x}_1 = 0, \qquad m\ddot{x}_2 = -mg .$$

Letting the initial $(t = 0)$ conditions be

$$x_i(0) = 0 , \qquad i = 1, 2,$$
$$\dot{x}_1(0) = v_0 \cos\theta ,$$
$$\dot{x}_2(0) = v_0 \sin\theta ,$$

we find for the trajectory in K,

$$x_2 = \frac{-g}{2v_0^2 \cos^2\theta} x_1^2 + \tan\theta x_1 .$$

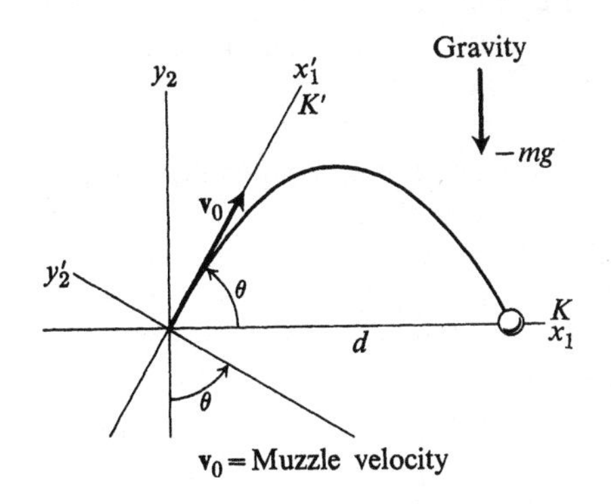

Fig. 1.6 The parabolic trajectory of a projectile viewed in coordinate systems K and K'.

In K', Newton's law is $F' = ma'$, and the equations of motion become

$$m\ddot{x}_1' = -mg\sin\theta , \qquad m\ddot{x}_2' = -mg\cos\theta .$$

The initial conditions become

$$x_i(0) = 0 , \qquad i = 1, 2,$$
$$\dot{x}_1(0) = v_0 ,$$
$$\dot{x}_2(0) = 0 ,$$

and the trajectory is

$$x_1' = x_2' \tan\theta + v_0\left(\frac{2x_2'}{-g\cos\theta}\right)^{1/2} .$$

Do the trajectories as expressed in the primed and unprimed variables describe the same physical path? They had better! Otherwise Newton's law does not provide a frame-independent description. It is left to the reader to reassure himself that all is well.

The important question is why this all works out. Just where is frame-independence for *rotated* coordinate systems built into Newton's laws? The answer is that two *vectors* which are equal in one frame, say K, are equal in a rotated frame, K'. The linear homogeneous character of the transformation law for vector components *guarantees* this. Instead of deriving the equations of motion in K' by looking at the physical situation in the frame, we could have derived them from the equations of motion as stated for K by applying the rotation matrix directly to the relevant vectors—forces and accelerations. It is instructive to carry this out once in one's life.

The key point is that on both sides of the equation there are *vector* quantities; hence under rotation of the basis vectors, both sides transform the same way. If on one side there were two numbers that remained constant under rotation (two such numbers would *not* be the components of a vector), while the other side was transforming like a vector, the equation would have a different form after transformation, and it would give different predictions. The world goes on independent of the inclination of our coordinate system, and we incorporate this isotropy of space into our theories from the start in the requirement that all terms in an equation be tensors of the same rank: all tensors of second rank, all vectors, or all scalars.

Another point worth noting is that since we get the same physical results in any frame, we can solve the problem in the frame where it is solved most easily—in our example, frame K. In general, we can establish a tensor equation in any particular frame and know immediately that it holds for every frame.

In summary then, the invariance of a physical law under orthogonal transformation of the spatial coordinate system requires that all the terms of the equation be tensors of the same rank. We say then that the terms are *covariant* under orthogonal transformations, i.e., they "vary together."

Later we shall view the Lorentz transformation of special relativity as an orthogonal transformation in four-dimensional space ("space-time" or Minkowski space), and again, we shall insist that all the terms of an equation be tensors (in this case, "four-tensors") of the same rank. This will ensure that the laws of physics are invariant under Lorentz transformations; that is, for all observers moving with uniform relative velocity.

1.5 THE VECTOR PRODUCT

The vector (or "cross") product of two vectors $\mathbf{x}$ and $\mathbf{y}$ is a vector—as we might expect—and is written $\mathbf{z} \equiv \mathbf{x} \times \mathbf{y}$. In geometrical language, we define the magnitude of the vector $\mathbf{z}$ by

$$|\mathbf{z}| = |\mathbf{x} \times \mathbf{y}| = |\mathbf{x}||\mathbf{y}| \sin \theta ,$$

where θ is the angle measured from $\mathbf{x}$ to $\mathbf{y}$ in such a way that $\theta \leq \pi$. $\mathbf{z}$ is defined to be perpendicular to the plane containing $\mathbf{x}$ and $\mathbf{y}$, and to point in a direction given by the right-hand rule applied to $\mathbf{x}$ and $\mathbf{y}$, with fingers swinging in the direction of θ from $\mathbf{x}$ to $\mathbf{y}$, and the thumb giving the direction of $\mathbf{z}$. If one's thumb points "up" as one swings one's fingers from $\mathbf{x}$ to $\mathbf{y}$, then it will point "down" as one swings one's fingers from $\mathbf{y}$ to $\mathbf{x}$ (remember that $\theta \leq \pi$). Therefore, the vector product is anticommutative:

$$(\mathbf{x} \times \mathbf{y}) = -(\mathbf{y} \times \mathbf{x}) .$$

Therefore, $\mathbf{x} \times \mathbf{x} = \mathbf{0}$ (which is obvious geometrically, since $\sin 0 = 0$).

As an example of the vector product, consider the set of orthonormal basis vectors in the right-handed coordinate system of Fig. 1.2. It follows from the definition of the vector product that these basis vectors obey the relations

$$\mathbf{e}_i \times \mathbf{e}_j = \mathbf{e}_k , \tag{1.28}$$

where i, j, and k are any even permutation of the subscripts 1, 2, and 3. [Oddness and evenness of a permutation of (1, 2, 3) refer to the oddness or evenness of the number of interchanges of adjacent numbers needed to get to the order (1, 2, 3). Thus (3, 1, 2) is an even permutation and (3, 2, 1) is an odd permutation.] In this connection we define the symbol ε_{ijk} which will be useful in much the same way as the Kronecker delta:

$$\varepsilon_{ijk} = \begin{cases} +1 & \text{if } (i, j, k) \text{ is an even permutation of } (1, 2, 3) , \\ -1 & \text{if } (i, j, k) \text{ is an odd permutation of } (1, 2, 3) , \\ 0 & \text{otherwise (e.g., if 2 or more indices are equal)} . \end{cases} \tag{1.29}$$

There is, in fact, a very useful identity relating the ε_{ijk} symbol and the Kronecker delta. It is

$$\varepsilon_{ijk}\varepsilon_{klm} = \delta_{il}\delta_{jm} - \delta_{jl}\delta_{im} . \tag{1.30}$$

We leave the verification to the reader. It also follows immediately from Eq. (1.29) that

$$\varepsilon_{ijk} = \varepsilon_{kij} = \varepsilon_{jki} = -\varepsilon_{jik} = -\varepsilon_{kji} = -\varepsilon_{ikj} .$$

The handedness of a coordinate system may now be defined mathematically: a set of basis vectors $\mathbf{e}_i$ is said to form a right-handed Cartesian coordinate system if

$$\mathbf{e}_i \times \mathbf{e}_j = \varepsilon_{ijk}\mathbf{e}_k . \tag{1.31}$$

The coordinate system is left-handed if $\mathbf{e}_i \times \mathbf{e}_j = -\varepsilon_{ijk}\mathbf{e}_k$. Clearly, the replacement of any basis vector by its negative simply reverses the handedness of the coordinate system. We shall use right-handed coordinate systems throughout the book.

The algebraic definition of the vector product is

$$\mathbf{z} = \mathbf{x} \times \mathbf{y} = x_j\mathbf{e}_j \times y_k\mathbf{e}_k = x_jy_k\mathbf{e}_j \times \mathbf{e}_k = x_jy_k\varepsilon_{jki}\mathbf{e}_i = \varepsilon_{ijk}x_jy_k\mathbf{e}_i . \tag{1.32}$$

(Again, we have assumed that the vector product is distributive.) Thus the ith component of $\mathbf{z}$ is

$$z_i = \mathbf{z} \cdot \mathbf{e}_i = \varepsilon_{ijk}x_jy_k . \tag{1.33}$$

The proofs of the familiar vector identities become very simple in this notation.

Example 1.3. In order to establish the vector identity,

$$\mathbf{x} \times (\mathbf{y} \times \mathbf{z}) = \mathbf{y}(\mathbf{x} \cdot \mathbf{z}) - \mathbf{z}(\mathbf{x} \cdot \mathbf{y}) \,, \tag{1.34}$$

we must prove that the ith components of the two expressions are equal:

$$\begin{aligned} [\mathbf{x} \times (\mathbf{y} \times \mathbf{z})]_i &= \varepsilon_{ijk} x_j \varepsilon_{klm} y_l z_m \\ &= (\delta_{il}\delta_{jm} - \delta_{jl}\delta_{im}) x_j y_l z_m \\ &= y_i x_j z_j - z_i x_j y_j \\ &= [\mathbf{y}(\mathbf{x} \cdot \mathbf{z}) - \mathbf{z}(\mathbf{x} \cdot \mathbf{y})]_i \,. \end{aligned}$$

Even easier is the proof of the identities

$$\mathbf{x} \cdot (\mathbf{y} \times \mathbf{z}) = \mathbf{z} \cdot (\mathbf{x} \times \mathbf{y}) = \mathbf{y} \cdot (\mathbf{z} \times \mathbf{x}) \,, \tag{1.35}$$

which we leave to the reader.

In Section 1.4, we derived the orthogonality relations

$$a_{ik} a_{jk} = a_{ki} a_{kj} = \delta_{ij} \tag{1.36}$$

by considering the scalar products of the basis vectors. As we noted, these relations among the elements of the rotation matrix mean that the a_{ij} are not independent of each other. A simple way to solve the system of Eqs. (1.36) is to take components of the vector product relations among the basis vectors. The lth component of

$$\mathbf{e}_i \times \mathbf{e}_j = \varepsilon_{ijk} \mathbf{e}_k$$

in the primed frame is

$$\varepsilon_{lmn} (\mathbf{e}_i \cdot \mathbf{e}'_m)\, (\mathbf{e}_j \cdot \mathbf{e}'_n) = \varepsilon_{ijk} (\mathbf{e}_k \cdot \mathbf{e}'_l) \;;$$

here we are using Eq. (1.33) in the primed frame. Recalling that $a_{ij} = \mathbf{e}'_i \cdot \mathbf{e}_j$, we have immediately

$$\varepsilon_{ijk} a_{lk} = \varepsilon_{lmn} a_{mi} a_{nj} \,, \tag{1.37}$$

which gives any one of the a_{ij}'s in terms of the others, since upon choosing i and j there is only one term in the sum on the left.

We may use this result to prove that the vector product of two vectors does indeed transform like a vector. Suppose that two objects $\mathbf{x}$ and $\mathbf{y}$ transform as vectors under orthogonal transformations:

$$x'_j = a_{jn} x_n \qquad \text{and} \qquad y'_k = a_{km} y_m \,.$$

We must show that $\mathbf{z} = \mathbf{x} \times \mathbf{y}$ also transforms as a vector, i.e., that

$$z'_i \equiv (\mathbf{x}' \times \mathbf{y}')_i = a_{ij} z_j \,.$$

Using Eqs. (1.37), with suitably chosen dummy subscripts, one easily carries out the proof. We have

$$z'_i = \varepsilon_{ijk} x'_j y'_k = \varepsilon_{ijk} a_{jn} x_n a_{km} y_m = \varepsilon_{nml} a_{il} x_n y_m = a_{ij} \varepsilon_{jnm} x_n y_m = a_{ij} z_j \,.$$

Although the vector product transforms like a vector under rotations, it does not behave in all respects like an "ordinary" vector. After all, it is the *product* of two vectors. Each of the vectors that enter into the vector product changes sign when we replace all the basis vectors $\mathbf{e}_i$ by their negatives, $-\mathbf{e}_i$. This is called *inversion* of the coordinate system.

In two-dimensional space, inversion is just rotation by 180°. In three dimensions, however, inversion is equivalent to changing from a right-handed to a left-handed coordinate system. Analogous results hold for spaces of higher dimension: In even-dimensional spaces, inversion can be achieved by a rotation of basis vectors, whereas in odd-dimensional spaces, inversion requires the *reflection* of the basis vectors in the origin.

Vectors like the position $\mathbf{r}$ and the momentum $\mathbf{p}$ change sign under inversion. They are called *polar*, or ordinary, vectors. But a vector product of two polar vectors, such as $\mathbf{L} = \mathbf{r} \times \mathbf{p}$, will not change sign under inversion. Such vectors are called *axial* vectors or *pseudovectors*. The scalar product of a polar vector and a pseudovector is a *pseudoscalar*; it changes sign under an inversion, whereas a scalar does not.

Just as equations in classical physics cannot equate tensors of different rank because they have different transformation properties under *rotations*, we cannot equate pseudovectors to vectors or pseudoscalars to scalars, because they transform differently under *inversions*. For example, the equation

$$\frac{d\mathbf{L}}{dt} = \mathbf{r} \times \mathbf{F} = \mathbf{N}$$

relates the *two pseudovectors*, angular momentum and torque.

1.6 A VECTOR TREATMENT OF CLASSICAL ORBIT THEORY

In order to illustrate the power and use of vector methods, we shall employ them to work out the Keplerian orbits. This treatment is to be contrasted with that based on solving the differential equations of motion found, for example, in Chapter 3 of Goldstein.*

We first prove Kepler's second law which states that angular momentum is constant in a central force field. Differentiating the angular momentum

$$\mathbf{L} = \mathbf{r} \times \mathbf{p}$$

with respect to time, we obtain

$$\dot{\mathbf{L}} = \dot{\mathbf{r}} \times \mathbf{p} + \mathbf{r} \times \dot{\mathbf{p}}\,.$$

The first vector product vanishes because $\mathbf{p} = m\dot{\mathbf{r}}$ so that $\dot{\mathbf{r}}$ and $\mathbf{p}$ are parallel. The second vector product is simply $\mathbf{r} \times \mathbf{F}$ by Newton's second law, and hence vanishes for all forces directed along the position vector, that is, for all central forces. Thus $\mathbf{L} = \mathbf{r} \times m\mathbf{v}$ is a constant vector in central force motion. This implies that the position vector $\mathbf{r}$, and therefore the entire orbit, lies in a fixed

* H. Goldstein, *Classical Mechanics*. Reading, Mass.: Addison-Wesley, 1950.

plane in three-dimensional space. This result is essentially Kepler's second law, which is often stated in terms of the conservation of areal velocity, $|\mathbf{L}|/2m$.

We now turn to a familiar, but, in fact, very special central force, the inverse-square force of gravitation and electrostatics. For such a force, Newton's second law becomes

$$m\frac{d\mathbf{v}}{dt} = -\frac{k}{r^2}\mathbf{n}, \tag{1.38}$$

where $\mathbf{n} = \mathbf{r}/r$ is a unit vector in the $\mathbf{r}$-direction, and k is a constant; for the gravitational case it is Gm_1m_2, and for the electrostatic case, it is q_1q_2 in cgs units.

First we note that

$$\mathbf{v} = \dot{\mathbf{r}} = \dot{r}\mathbf{n} + r\dot{\mathbf{n}},$$

and using this in the definition of angular momentum, we find that

$$\mathbf{L} = mr^2(\mathbf{n} \times \dot{\mathbf{n}}). \tag{1.39}$$

Now we consider

$$\begin{aligned}\frac{d}{dt}(\mathbf{v} \times \mathbf{L}) &= \dot{\mathbf{v}} \times \mathbf{L} = \frac{-k}{mr^2}(\mathbf{n} \times \mathbf{L}) \\ &= -k[\mathbf{n}(\dot{\mathbf{n}}\cdot\mathbf{n}) - \dot{\mathbf{n}}(\mathbf{n}\cdot\mathbf{n})].\end{aligned}$$

Here we have used the constancy of $\mathbf{L}$, Eqs. (1.38) and (1.39), and the vector identity, Eq. (1.34). Since $\mathbf{n}\cdot\mathbf{n} = 1$, it follows by differentiation that $\dot{\mathbf{n}}\cdot\mathbf{n} = 0$. Thus we obtain

$$\frac{d}{dt}(\mathbf{v} \times \mathbf{L}) = k\dot{\mathbf{n}},$$

and integrating, we have

$$\mathbf{v} \times \mathbf{L} = k\mathbf{n} + \mathbf{C}, \tag{1.40}$$

where $\mathbf{C}$ is a constant vector. Thus $\mathbf{C} = \mathbf{v} \times \mathbf{L} - k\mathbf{n}$ is a vector constant of the motion. It lies along, and fixes the position of, the major axis of the orbit, as we shall see after we complete the derivation of the orbit.

Now we form the scalar quantity

$$L^2 = \mathbf{L}\cdot(\mathbf{r} \times m\mathbf{v}) = m\mathbf{r}\cdot(\mathbf{v} \times \mathbf{L}) = mr(k + C\cos\theta), \tag{1.41}$$

using Eqs. (1.35) and (1.40), where θ is the angle measured from $\mathbf{C}$ to $\mathbf{r}$. Solving for r, we obtain

$$r = \frac{L^2/km}{1 + C/k\cos\theta} = \frac{A}{1 + \varepsilon\cos\theta}. \tag{1.42}$$

This equation is the polar form for a conic section with one focus at the origin, where θ is measured from the constant vector $\mathbf{C}$ (which we may take to be the x-axis) to $\mathbf{r}$, and ε represents the *eccentricity* of the conic section; depending on its values, the conic section is a hyperbola, parabola, ellipse, or circle. The eccentricity is easily determined in terms of the constants of motion:

$$\begin{aligned}\varepsilon &= \frac{|\mathbf{C}|}{k} = \frac{1}{k}\,|(\mathbf{v}\times\mathbf{L}) - k\mathbf{n}| \\ &= \frac{1}{k}[|\mathbf{v}\times\mathbf{L}|^2 + k^2 - 2k\mathbf{n}\cdot(\mathbf{v}\times\mathbf{L})]^{1/2}\,.\end{aligned}$$

Now $|\mathbf{v}\times\mathbf{L}|^2 = v^2L^2$ because $\mathbf{v}$ is perpendicular to $\mathbf{L}$. Using Eq. (1.41), we obtain

$$\varepsilon = \frac{1}{k}\left[v^2L^2 + k^2 - \frac{2kL^2}{mr}\right]^{1/2} = \left[1 + \frac{2L^2}{mk^2}\left(\frac{m}{2}v^2 - \frac{k}{r}\right)\right]^{1/2} = \left[1 + \frac{2L^2E}{mk^2}\right]^{1/2},$$

where E is the constant energy of the system. We can now write out in full the orbit equation known as Kepler's first law:

$$r = \frac{L^2/km}{1 + (1 + 2L^2E/mk^2)^{1/2}\cos\theta}\,. \tag{1.43}$$

The vector $\mathbf{C}$ lies along the major axis in a direction from the focus to the perihelion, and its magnitude is

$$|\mathbf{C}| = k\varepsilon = k\,(1 + 2L^2E/mk^2)^{1/2}\,.$$

This is most easily seen by considering its value at the perihelion. Note that this vector vanishes for a circle ($\varepsilon = 0$) since there is no unique major axis. The existence of a further integral of the motion (besides energy and angular momentum) is due to degeneracy of the motion associated with potentials of the Coulomb form, k/r.* This degeneracy is perhaps more familiar in the analogous quantum-mechanical problem, the hydrogen atom, than in classical orbit theory. The above vector treatment is based on the existence of this special integral of the motion and is therefore not a general substitute for the usual methods based on the differential equations of motion.

1.7 DIFFERENTIAL OPERATIONS ON SCALAR AND VECTOR FIELDS

If to each point x_i ($i = 1, 2, 3$) in some region of space there corresponds a scalar, $\varphi(x_i)$, or a vector, $\mathbf{V}(x_i)$, we have a scalar or a vector field. Typical scalar fields are the temperature or density distribution in an object, or the electrostatic potential. Typical vector fields are the gravitational force, the velocity at each point in a moving fluid (e.g., a hurricane), or the magnetic-field intensity. Fields are functions defined at points of physical space, and may be time-dependent or time-independent.

---

* There is an excellent discussion of this by David F. Greenberg in *Accidental Degeneracy*, *Am. J. Phys.* **34**, 1101 (1966). Our vector **C** is therein called by its historical name, the Runge-Lenz vector, and it is studied in a more general setting.

Table 1.1.

Name of coordinate system:	Cartesian	Cylindrical	Spherical	Parabolic*
q_1	x	r	r	ξ
q_2	y	θ	θ	η
q_3	z	z	φ	φ
h_1	1	1	1	$\sqrt{\xi^2+\eta^2}$
h_2	1	r	r	$\sqrt{\xi^2+\eta^2}$
h_3	1	1	$r\sin\theta$	$\xi\eta$

In this section we shall define various differential operations on scalar and vector fields—the gradient, divergence, curl, and Laplacian—and examine the form these operations take in general curvilinear coordinate systems. We shall also consider the aspect of the field that each of these operators "measures." If we know what information the different operations extract from given fields, we shall be able to understand more intuitively the physical content of the basic equations of classical physics.

For the solution of certain physical problems, spherical, cylindrical, or still other coordinates are often superior to Cartesian coordinates. For example, the problem of finding the electrostatic potential between two charged concentric spherical shells is much simpler in spherical coordinates than in Cartesian or cylindrical coordinates, because in spherical coordinates the potential can be expressed as a function of one of the three coordinates alone, namely, r.

It is possible to proceed formally from the definitions of the various differential operators in Cartesian coordinates and the coordinate transformation equations to expressions for these operators in the other coordinate systems. Alternatively, one can derive the required expressions from coordinate-free, geometrical definitions. We shall take the latter course, giving general derivations valid for any set of orthogonal curvilinear coordinates q_i related to the distance ds by the line element or "metric"

$$ds^2 = h_1^2\,dq_1^2 + h_2^2\,dq_2^2 + h_3^2\,dq_3^2\,. \tag{1.44}$$

If only one of the three orthogonal coordinates q_1, q_2, q_3 is varied, the corresponding line element may be written

$$ds_i = h_i\,dq_i\,, \qquad i = 1, 2, \text{ or } 3 \qquad \text{(no summation)}\,. \tag{1.45}$$

The three most common (and one less common) systems are shown in Table 1.1.

The Gradient

The gradient is a differential operator defined on a scalar field φ. The gradient of a scalar field, written grad φ, is a vector field defined by the requirement

* Parabolic coordinates are useful in the study of the quantum-mechanical hydrogen atom, and in investigations of the Stark effect.

that

$$\text{grad}\, \varphi \cdot d\mathbf{s} = d\varphi\,, \tag{1.46}$$

where $d\varphi$ is the differential change in φ corresponding to the arbitrary space displacement $d\mathbf{s}$. From this we see that

$$d\varphi = |\text{grad}\, \varphi|\, |d\mathbf{s}| \cos\theta\,, \tag{1.47}$$

where θ is the angle between the vector grad φ and the displacement vector $d\mathbf{s}$. Thus it is clear that the rate of change of φ is greatest if the differential displacement is in the direction of grad φ, when $\theta = 0$ and $\cos\theta$ has its maximum value, 1. This defines the direction of the vector grad φ at a point in space: It is the direction of maximum rate of change of φ from that point, i.e., the direction in which $d\varphi/ds$ is greatest. It also follows from Eq. (1.47) that the magnitude of the vector grad φ is simply this maximal rate of change, i.e. the rate of change $|d\varphi/ds|$ in the direction in which φ is changing most rapidly.

We may summarize this by saying that the gradient of φ is the directional derivative in the direction of the maximum rate of change of φ. For example, let φ be the elevation from sea level of points on the surface of a mountain, and therefore proportional to gravitational potential. The equipotentials are the lines of constant φ or constant altitude. Displacements along equipotentials produce no change in φ ($d\varphi = 0$), but displacements perpendicular to equipotentials are in the direction of most rapid change of altitude, and $d\varphi$ takes on its maximum value. The vector grad φ is always perpendicular to equipotential lines or surfaces.

The general form of the ith component of the gradient vector follows from the definition of the gradient, Eq. (1.46), and Eq. (1.45). We have

$$(\text{grad}\, \varphi)_i = \lim_{ds_i \to 0} \frac{\varphi(q_i + dq_i) - \varphi(q_i)}{ds_i} = \frac{1}{h_i} \frac{\partial\varphi}{\partial q_i}\,. \tag{1.48}$$

We shall use the abbreviation

$$\frac{\partial}{\partial q_i} \equiv \partial_i\,. \tag{1.49}$$

Therefore

$$(\text{grad}\, \varphi)_i = \frac{1}{h_i} \partial_i \varphi\,. \tag{1.50}$$

In Cartesian coordinates, where $h_i = 1$, this becomes

$$(\text{grad}\, \varphi)_i = \partial_i \varphi\,, \tag{1.51}$$

and consequently

$$\text{grad}\, \varphi = \mathbf{e}_i \partial_i \varphi\,, \tag{1.52}$$

where summation from 1 to 3 is implied. This equation is often written in terms of the symbolic vector operator, ∇, called the "del" operator; it is defined in

Cartesian coordinate systems by

$$\nabla_{\text{Cartesian}} \equiv \mathbf{e}_i \partial_i \,. \tag{1.53}$$

In terms of the del operator, we have, in Cartesian coordinate systems,

$$\text{grad}\,\varphi = \nabla_{\text{Cartesian}}\,\varphi \,. \tag{1.54}$$

The ∇ symbol is often used to denote the grad operator irrespective of the type of coordinates, in which case we have

$$\text{grad}\,\varphi \equiv \nabla\varphi = \frac{1}{h_1}(\partial_1\varphi)\mathbf{e}_1 + \frac{1}{h_2}(\partial_2\varphi)\mathbf{e}_2 + \frac{1}{h_3}(\partial_3\varphi)\mathbf{e}_3 \,, \tag{1.55}$$

where $\mathbf{e}_i$ is the unit vector corresponding to q_i in the positive q_i-direction; ∇ is an operator and should not be thought of as a vector. It differs from a vector just as d/dx differs from a number; ∇ acquires meaning only when operating on a scalar or vector function. Taken alone, it has no magnitude; however, it does transform properly under rotations, so it is sometimes treated formally as a vector.

The Divergence

The divergence is a differential operation defined on a vector field $\mathbf{V}$. The divergence of a vector field, written div $\mathbf{V}$, is a scalar field. It is defined at a point x_i in coordinate-free form as

$$\text{div}\,\mathbf{V} = \lim_{\Delta\tau\to 0} \frac{1}{\Delta\tau} \int_\sigma \mathbf{V}\cdot d\boldsymbol{\sigma} \,, \tag{1.56}$$

where $\Delta\tau$ is the volume enclosed by the surface σ, and $d\boldsymbol{\sigma} = \mathbf{n}\,d\sigma$, where $\mathbf{n}$ is the outward-directed unit vector normal to the "infinitesimal" element of surface $d\sigma$ surrounding the point x_i. We omit the demonstration that this definition is independent of the shape of $\Delta\tau$.

The physical meaning of the divergence of a vector field can be determined straight from this definition. The divergence at a point measures the flux of the vector field through an infinitesimal surface per unit volume, or the source strength per unit volume at a point.

We shall give some specific examples as soon as we informally prove Gauss's theorem. This result is almost immediately apparent from the definition, Eq. (1.56). We form the integral of div $\mathbf{V}$ over a finite volume τ. This region can be subdivided into arbitrarily many, arbitrarily small subregions, $\Delta\tau_i$. Each infinitesimal surface element $d\sigma$ belonging to a subregion $\Delta\tau_i$ in the interior of τ appears in two integrals (over two adjoining subregions). But since $\mathbf{n}$ is oppositely directed in these two internal surface elements, the integrals over them cancel. The only surface elements over which the integrals are not canceled by integrals over adjoining subregions lie on the surface σ of the finite volume τ. Thus we obtain Gauss's theorem:

$$\int_\tau \text{div}\,\mathbf{V}\,d\tau = \int_\sigma \mathbf{V}\cdot d\boldsymbol{\sigma} \,. \tag{1.57}$$

We now discuss an important example of the divergence operator which illustrates its physical meaning.

Example 1.4. Let the stationary volume τ, with surface σ, enclose a fluid of density ρ moving with velocity $\mathbf{v}$. The mass in τ will be

$$\int_\tau \rho \, d\tau ,$$

so the time rate of change of the mass in τ will be

$$\frac{\partial}{\partial t}\int_\tau \rho \, d\tau .$$

The mass of the fluid leaving τ per unit time can be calculated as

$$\int_\sigma \rho\mathbf{v}\cdot d\boldsymbol{\sigma} = \int_\tau \operatorname{div}(\rho\mathbf{v}) \, d\tau ,$$

where we have used the divergence theorem, Eq. (1.57). Now conservation of mass requires that

$$\frac{\partial}{\partial t}\int_\tau \rho \, d\tau + \int_\tau \operatorname{div}(\rho\mathbf{v}) \, d\tau = 0 ,$$

or

$$\int_\tau \left[\frac{\partial\rho}{\partial t} + \operatorname{div}(\rho\mathbf{v})\right] d\tau = 0 .$$

For this to hold for an arbitrary volume τ, it must be true that

$$\frac{\partial\rho}{\partial t} + \operatorname{div}(\rho\mathbf{v}) = 0 . \tag{1.58}$$

This is the equation of continuity, or conservation of mass. It is just a restatement of conservation of mass, which was assumed in the derivation. A similar equation holds for charge in electromagnetism: ρ is then the charge density and $\rho\mathbf{v} = \mathbf{J}$ is the current density.

We now obtain an expression for the divergence of a vector field in orthogonal curvilinear coordinates from the integral definition, Eq. (1.56). We must calculate the integral of the outward normal component of $\mathbf{V}$, $\mathbf{V}\cdot\mathbf{n}$, over the surface of an infinitesimal volume $\Delta\tau = ds_1\, ds_2\, ds_3$. Let the volume be bounded by the six surfaces

$$q_i = a_i \qquad \text{and} \qquad q_i = a_i + \delta q_i , \qquad \text{for } i = 1, 2, \text{ and } 3 .$$

The integral of $\mathbf{V}\cdot\mathbf{n}$ over the surface of constant q_1 at the point $q_1 = a_1 + \delta q_1$ is given by

$$\int_{a_2}^{a_2+\delta q_2} dq_2 \int_{a_3}^{a_3+\delta q_3} dq_3 h_2(a_1 + \delta q_1, q_2, q_3) h_3(a_1 + \delta q_1, q_2, q_3)\, V_1(a_1 + \delta q_1, q_2, q_3)$$

(remember that, in this case, $\mathbf{n}$ points in the positive q_1-direction), while for the integral of $\mathbf{V}\cdot\mathbf{n}$ over the surface of constant q_1 at the point $q_1 = a_1$, we have

$$-\int_{a_2}^{a_2+\delta q_2} dq_2 \int_{a_3}^{a_3+\delta q_3} dq_3 h_2(a_1, q_2, q_3) h_3(a_1, q_2, q_3)\, V_1(a_1, q_2, q_3)$$

(here $\mathbf{n}$ points in the *negative* q_1 direction). Combining these two terms, we get

$$\int_{a_2}^{a_2+\delta q_2} dq_2 \int_{a_3}^{a_3+\delta q_3} dq_3 [V_1(a_1+\delta q_1, q_2, q_3) h_2(a_1+\delta q_1, q_2, q_3) h_3(a_1+\delta q_1, q_2, q_3) - V_1(a_1, q_2, q_3) h_2(a_1, q_2, q_3) h_3(a_1, q_2, q_3)]\,.$$

Now, as δq_1 becomes small, this can be written as

$$\int_{a_2}^{a_2+\delta q_2} dq_2 \int_{a_3}^{a_3+\delta q_3} dq_3 \left[\frac{\partial}{\partial q_1}(V_1 h_2 h_3)\right]_{q_1=a_1} \delta q_1\,.$$

Finally, as we let δq_2 and δq_3 become small, we can replace this expression by

$$\left[\frac{\partial}{\partial q_1}(V_1 h_2 h_3)\right]_{q_i=a_i} \delta q_1 \delta q_2 \delta q_3\,.$$

Analogous expressions hold for the other two pairs of surfaces. Adding these three expressions and dividing both sides of Eq. (1.57) by

$$d\tau = h_1 h_2 h_3 \delta q_1 \delta q_2 \delta q_3\,,$$

we obtain a general formula for the divergence:

$$\operatorname{div}\mathbf{V} = \frac{1}{h_1 h_2 h_3}\left[\frac{\partial(h_2 h_3 V_1)}{\partial q_1} + \frac{\partial(h_3 h_1 V_2)}{\partial q_2} + \frac{\partial(h_1 h_2 V_3)}{\partial q_3}\right]. \tag{1.59}$$

As must be the case, the divergence of a vector field is a scalar field. In Cartesian coordinates, Eq. (1.59) reduces to

$$\operatorname{div}\mathbf{V} = \partial_i V_i = \nabla_{\text{Cartesian}}\cdot\mathbf{V}\,. \tag{1.60}$$

The scalar product of the ∇ symbol and a vector field is often used to denote div $\mathbf{V}$ regardless of the type of coordinates, in which case,

$$\operatorname{div}\mathbf{V} = \nabla\cdot\mathbf{V} = \frac{1}{h_1 h_2 h_3}\left[\frac{\partial(h_2 h_3 V_1)}{\partial q_1} + \frac{\partial(h_3 h_1 V_2)}{\partial q_2} + \frac{\partial(h_1 h_2 V_3)}{\partial q_3}\right]. \tag{1.61}$$

Note that the components of grad and div considered as vector operators are not equal in curvilinear coordinates, even though the symbol ∇ is often used in writing both operations.

The Curl

The curl is a differential operator defined on a vector field $\mathbf{V}$, and is written curl $\mathbf{V}$. The curl of a vector field is itself a vector field, and is defined in coordinate-free form as

$$(\operatorname{curl}\mathbf{V})\cdot\mathbf{n} = \lim_{\Delta\sigma\to 0}\frac{1}{\Delta\sigma}\oint_\lambda \mathbf{V}\cdot d\boldsymbol{\lambda}\,, \tag{1.62}$$

where $\Delta\sigma$ is the area of the surface bounded by the closed path λ, and $d\boldsymbol{\lambda} = \mathbf{t}\,d\lambda$, where $\mathbf{t}$ is a unit tangent vector along λ; $(\text{curl } \mathbf{V}) \cdot \mathbf{n}$ is the component of curl $\mathbf{V}$ in the $\mathbf{n}$-direction, i.e., normal to the surface $\Delta\sigma$ in a direction given by the right-hand rule applied to the path of integration about λ. We omit the demonstration that surfaces of different shapes and orientations give equivalent results. The direction of curl $\mathbf{V}$ is given by the orientation of a plane surface, like the direction of the vector product. It will turn out that curl $\mathbf{V}$ is a pseudovector like the vector product.

Stokes's theorem follows immediately from the definition of the curl. We simply form the integral of curl $\mathbf{V}$ over a finite surface σ bounded by the finite curve λ. We subdivide the surface into arbitrarily many, arbitrarily small parts, each of which can be considered an infinitesimal plane surface. For each such surface we form the integrals in Eq. (1.62), and then add them. All contributions to line integrals arising from arcs of curves interior to the bounding perimeter λ are canceled by contributions from line integrals along arcs of adjoining infinitesimal plane surfaces, since $\mathbf{t}$ is oppositely directed along these arcs. The only arcs along which the integrals are not canceled by pairs lie on the bounding curve λ of the finite surface σ. Thus we obtain Stokes's theorem:

$$\int_\sigma (\text{curl } \mathbf{V}) \cdot d\boldsymbol{\sigma} = \oint_\lambda \mathbf{V} \cdot d\boldsymbol{\lambda}\,, \tag{1.63}$$

where $d\boldsymbol{\sigma} = \mathbf{n}\,d\sigma$.

We shall now illustrate the physical meaning of the curl of a vector field with several examples.

Example 1.5. Consider a bucket of water rotating with constant angular velocity $\boldsymbol{\omega}$ about a central vertical axis $\mathbf{k}$ until the water surface stabilizes. We assume that the velocity of the water (in cylindrical coordinates) is $\mathbf{v} = \omega r\boldsymbol{\theta}$, where $\boldsymbol{\theta}$ is a unit vector in the θ-direction. Let us calculate curl $\mathbf{v}$ from Eq. (1.62), taking a circle of radius r for the curve λ.

$$\begin{aligned}\mathbf{k} \cdot \text{curl } \mathbf{v} &= \lim_{\Delta\sigma \to 0} \frac{1}{\Delta\sigma} \oint_\lambda \mathbf{v} \cdot d\boldsymbol{\lambda} = \lim_{r \to 0} \frac{1}{\pi r^2} \oint_\lambda \omega r\boldsymbol{\theta} \cdot \boldsymbol{\theta}\,d\lambda \\ &= \lim_{r \to 0} \frac{\omega r}{\pi r^2} \oint_\lambda d\lambda = \lim_{r \to 0} \frac{\omega r}{\pi r^2}(2\pi r) = 2\omega\,.\end{aligned}$$

The component of the curl in the $\mathbf{k}$-direction is twice the angular velocity. The curl of a velocity field is then a measure of the *angular* velocity of the fluid. If a miniature paddle wheel is placed anywhere in a velocity field, its angular velocity measures the curl of the field. If it does not rotate, the component of the curl in the direction perpendicular to the plane of the paddle wheel is zero. However, there may be curl despite straight-line flow. For example, water moving in a pipe will move more slowly along the wall of the pipe than water in the center, due to friction effects. Therefore a paddle wheel will be turned by the water even though the water always moves in straight lines. On the other hand, there need not necessarily be curl in curved-line flow. If a paddle wheel is placed in an appropriately designed "whirlpool," in which the fluid moves with a velocity given by $\mathbf{v} = \boldsymbol{\theta} v_0(a/r)$ in cylindrical coordinates (where a, and v_0 are constants) the velocity field will have zero curl at all points $r \neq 0$, and the paddle wheel will not rotate. To show this we compute explicitly curl $\mathbf{v}$ at any point in the whirlpool different from $r = 0$; $\mathbf{v}$ is the velocity field of the fluid flow. The component of curl $\mathbf{v}$ in the $\mathbf{k}$-direction is

$$\mathbf{k}\cdot\text{curl }\mathbf{v} = \lim_{A\to 0}\frac{1}{A}\oint_\lambda \mathbf{v}\cdot d\boldsymbol{\lambda},$$

where A is the area of the cork-shaped region (that contains the paddle wheel in Fig. 1.7) bounded by the four arcs: l_1, l_2, l_3, and l_4. The integral about the perimeter of A may be broken up into line integrals over these four arcs. The integrals over l_3 and l_4 vanish because $\mathbf{v}$ is perpendicular to $d\boldsymbol{\lambda}$ at every point on these arcs. Thus

$$\begin{aligned}\mathbf{k}\cdot\text{curl }\mathbf{v} &= \lim_{A\to 0}\frac{1}{A}\left[\int_{l_1}\mathbf{v}\cdot d\boldsymbol{\lambda} + \int_{l_2}\mathbf{v}\cdot d\boldsymbol{\lambda}\right] \\ &= \lim_{A\to 0}\frac{1}{A}\left[\int_{\theta_2}^{\theta_1}\frac{av_0}{r_1}\boldsymbol{\theta}\cdot\boldsymbol{\theta} r_1\, d\theta + \int_{\theta_1}^{\theta_2}\frac{av_0}{r_2}\boldsymbol{\theta}\cdot\boldsymbol{\theta} r_2\, d\theta\right] = 0,\end{aligned}$$

because the integrals are equal in magnitude and opposite in sign, the contributions along arcs l_1 and l_2 canceling. The x- and y-components of curl $\mathbf{v}$ are clearly zero.

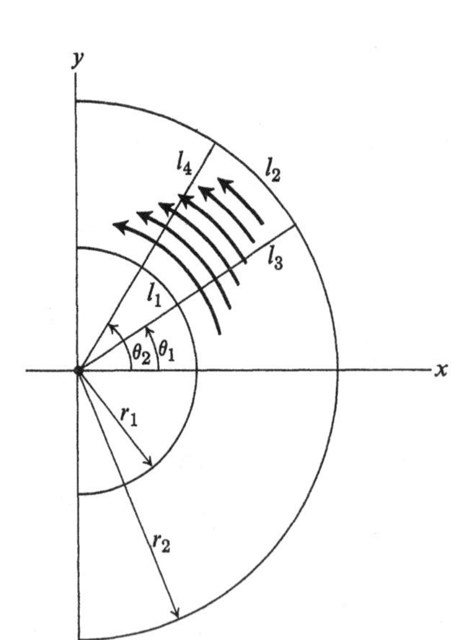

Fig. 1.7 The curl of a whirlpool. The flow lines are shown in boldface; the length of the lines indicates the speed of flow.

It is the purpose of streamlining to provide a surface past which air or water will flow with a minimum of curl, for motion with curl develops eddies that waste energy in heating up the air or water.

Of course a paddle wheel is appropriate only for fluid velocity fields. For electric fields, gravitational fields, etc., a "generalized paddle wheel," or curl-meter, must be visualized—one which responds to the forces associated with the particular vector field in question.

We complete our discussion of the curl operator, which up to now has been a coordinate-free treatment, by obtaining an expression for the components of the curl of a vector field in an orthogonal curvilinear coordinate system. To obtain the ith component of curl $\mathbf{V}$, we must calculate the line integral of the tangential component of the vector field $\mathbf{V}$, $\mathbf{V}\cdot\mathbf{t}$, along the boundary of an infinitesimal area $d\boldsymbol{\sigma}$ perpendicular to the ith basis vector. Consider an element of area in the q_2q_3-plane defined by $d\sigma = ds_2\, ds_3$ and oriented in the direction of increasing q_1. The 1-component of curl $\mathbf{V}$ is given by the sum of the line integrals, $\int \mathbf{V}\cdot d\boldsymbol{\lambda}$, along the four arcs that lead around $d\sigma$ from the point (q_2, q_3) to $(q_2 + dq_2, dq_3)$ to $(q_2 + dq_2, q_3 + dq_3)$ to $(q_2, q_3 + dq_3)$ back to (q_2, q_3). Thus, in brief, we have

$$(\text{curl } \mathbf{V})_1 = \lim_{d\sigma\to 0} \frac{1}{ds_2\, ds_3}[(V_2\, ds_2)_{q_3} + (V_3\, ds_3)_{q_2+dq_2} - (V_2\, ds_2)_{q_3+dq_3} - (V_3\, ds_3)_{q_2}]$$

$$= \lim_{ds_2\, ds_3\to 0} \left[\frac{(V_3\, ds_3)_{q_2+dq_2} - (V_3\, ds_3)_{q_2}}{ds_2\, ds_3} - \frac{(V_2\, ds_2)_{q_3+dq_3} - (V_2\, ds_2)_{q_3}}{ds_2\, ds_3}\right].$$

Therefore

$$(\text{curl } \mathbf{V})_1 = \frac{1}{h_2h_3}\left[\frac{\partial(h_3V_3)}{\partial q_2} - \frac{\partial(h_2V_2)}{\partial q_3}\right]. \tag{1.64}$$

The reader may want to amplify this argument in the fashion of the discussion that led to the expression for the divergence. The other two components of curl $\mathbf{V}$ may be obtained by cyclic permutation of the coordinate indices.

In Cartesian coordinates, Eq. (1.64) and the two additional equations for the other components reduce to

$$(\text{curl } \mathbf{V})_i = \varepsilon_{ijk}\partial_j V_k\,, \tag{1.65}$$

so in Cartesian coordinates,

$$\text{curl } \mathbf{V} = \varepsilon_{ijk}\mathbf{e}_i\partial_j V_k = \boldsymbol{\nabla}_{\text{Cartesian}} \times \mathbf{V}\,. \tag{1.66}$$

The vector product of the $\boldsymbol{\nabla}$ symbol and a vector field is often used to denote curl $\mathbf{V}$, irrespective of the type of coordinates, so we may write

$$(\text{curl } \mathbf{V}) \equiv (\boldsymbol{\nabla} \times \mathbf{V}) = \frac{1}{h_2h_3}[\partial_2(h_3V_3) - \partial_3(h_2V_2)]\mathbf{e}_1 + \frac{1}{h_3h_1}[\partial_3(h_1V_1) - \partial_1(h_3V_3)]\mathbf{e}_2 + \frac{1}{h_1h_2}[\partial_1(h_2V_2) - \partial_2(h_1V_1)]\mathbf{e}_3\,, \tag{1.67}$$

where $\mathbf{e}_i$ is the unit vector corresponding to q_i.

The Laplacian

The Laplacian of a scalar field is defined as the divergence of the gradient of that scalar field. From Eqs. (1.50) and (1.61), we have the following expression for the Laplacian in orthogonal curvilinear coordinates:

$$\text{Laplacian } \varphi \equiv \text{div (grad } \varphi) \equiv \nabla \cdot (\nabla \varphi) \equiv \nabla^2 \varphi$$
$$= \frac{1}{h_1 h_2 h_3}\left[\partial_1\left(\frac{h_2 h_3}{h_1}\partial_1 \varphi\right) + \partial_2\left(\frac{h_3 h_1}{h_2}\partial_2 \varphi\right) + \partial_3\left(\frac{h_1 h_2}{h_3}\partial_3 \varphi\right)\right]. \quad (1.68)$$

In Cartesian coordinates, this reduces to

$$\nabla^2_{\text{Cartesian}} \varphi = \partial_i \partial_i \varphi \, , \quad (1.69)$$

where summation is implied.

Several of the most important partial differential equations of classical physics involve the Laplacian, e.g., Laplace's equation ($\nabla^2 \varphi = 0$), Poisson's equation ($\nabla^2 \varphi = \rho$), the heat and diffusion equations [$\nabla^2 \varphi = K(\partial \varphi / \partial t)$], the wave equation [$\nabla^2 \varphi + (1/c^2)(\partial^2 \varphi / \partial t^2) = 0$], Helmholtz's equation ($\nabla^2 \varphi + k^2 \varphi = 0$), and others. Once we know what information the Laplacian extracts from a scalar field, we shall be able to intuit directly the physical meaning of these equations, which comprise a good part of classical physics.

To simplify the problem, we may use Cartesian coordinates without loss of generality. Let the scalar field φ have the value φ_0 at a certain point which we take as the origin of the Cartesian coordinate system. Consider a cube of side a, centred at the origin. The average value of φ in the cube is

$$\bar{\varphi} = \frac{1}{a^3} \iiint_{-a/2}^{a/2} \varphi \, dx_1 \, dx_2 \, dx_3 \, , \quad (1.70)$$

where we integrate over the interior of the cube. In order to evaluate the volume integral, we expand $\varphi(x_i)$ in a triple Taylor series in the three coordinates x_i, which we write as

$$\varphi(x_i) = \varphi_0 + (\partial_i \varphi)_0 x_i + \tfrac{1}{2}(\partial_i \partial_j \varphi)_0 x_i x_j + \cdots \, .$$

Substituting the Taylor expansion of $\varphi(x_i)$ into Eq. (1.70) and integrating from $-a/2$ to $a/2$, we see that the integral of any term that is linear in any x_i (in fact, any term containing an odd function of x_i) will vanish. The integrals of the quadratic (even) terms are easily carried out, and collecting them we obtain

$$\bar{\varphi} \cong \varphi_0 + \frac{a^2}{24}(\partial_i \partial_i \varphi)_0$$

or

$$(\nabla^2 \varphi)_0 \cong \frac{24}{a^2}(\bar{\varphi} - \varphi_0) \, , \quad (1.71)$$

with the approximation becoming better as the size of the cube (a) approaches zero. The value of the Laplacian of a scalar field at a point is therefore a measure of the difference between the average value of the field in an infinitesimal neighborhood of that point and the value of the field at the point itself.

This result provides us with intuitive physical interpretations of the equations that contain the Laplacian operator. The simplest of these is Laplace's equation,

$$\nabla^2\varphi = 0\,, \tag{1.72}$$

which governs the scalar potential of the gravitational and electrostatic fields in the absence of mass or charge. We see that the average value of such a potential in a neighborhood of a point P must be equal to its value at P. A function with this property is called a *harmonic function*. We shall see in Chapter 6 that the real and imaginary parts of an analytic function are harmonic functions in two-dimensional space. Laplace's equation tells us immediately that a harmonic function cannot increase or decrease in all directions from a given point P, for then $\bar{\varphi}$ could not equal φ_P. If φ is a local maximum at P along a given line, it must be a local minimum along some other line through P; hence P must be a saddle point.

There also exist situations in which the value of a scalar at a point differs from the average value in the neighborhood of that point by an amount which is a function of space. This is the situation described by Poisson's equation:

$$\nabla^2\varphi = \rho(x_i)\,. \tag{1.73}$$

The scalar function $\rho(x_i)$ measures the density of mass or charge at the point x_i. Poisson's equation says that the departure of $\bar{\varphi}$ about x_i from $\varphi(x_i)$ is proportional to the density at x_i.

There are several physical processes in which the greater the difference between the value of φ_P and the average, $\bar{\varphi}$, in the neighborhood of P, the greater is the time rate of change of φ needed to equalize this difference. This is expressed mathematically by an equation known as the heat equation:

$$\nabla^2\varphi = K\frac{\partial\varphi}{\partial t}\,, \tag{1.74}$$

where K is a constant > 0. This equation governs heat-conduction processes (then φ is the temperature) and diffusion processes (then φ is the density of the diffusing material). The temperature or density changes with time at a rate proportional to the difference between its local and average values. The heat equation is a quantitative expression of Newton's law of cooling. Knowing what it is that the Laplacian operator does, we could go straight to an equation of the form of Eq. (1.74) if we were trying to express Newton's law of cooling mathematically.

Finally, the deviation of a local value from the average value can be proportional to the second time derivative, as in the wave equation:

$$\nabla^2\varphi = \frac{1}{c^2}\frac{\partial^2\varphi}{\partial t^2}\,. \tag{1.75}$$

The second derivative corresponds to accelerating the value of φ towards $\bar{\varphi}$, but in such a way that $\partial\varphi/\partial t \neq 0$ at equilibrium (i.e., at $\varphi_P = \bar{\varphi}$). The system moves through equilibrium and repeats its motion, like a simple harmonic oscillator.

We leave the interpretation of Helmholtz's equation

$$\nabla^2\varphi + k^2\varphi = 0 \tag{1.76}$$

to the reader (Problem 10).

Thus far we have regarded the Laplacian as operating only on a scalar field. it is useful to extend the application of this operator to vector fields in the following natural way. In Cartesian coordinates we define

$$\nabla^2_{\text{Cartesian}}\mathbf{V} \equiv \mathbf{e}_i\nabla^2 V_i \,. \tag{1.77}$$

Thus in Cartesian coordinates, the ith component of the vector $\nabla^2_{\text{Cartesian}}\mathbf{V}$ is the Laplacian of the ith component of $\mathbf{V}$, a scalar, defined in Eq. (1.69); that is

$$(\nabla^2_{\text{Cartesian}}\mathbf{V})_i = \partial_j\partial_j V_i \,. \tag{1.78}$$

To generalize the definition of the Laplacian of a vector to orthogonal curvilinear coordinates, we prove in Cartesian coordinates the vector identity

$$\nabla \times (\nabla \times \mathbf{V}) = -\nabla^2_{\text{Cartesian}}\mathbf{V} + \nabla(\nabla\cdot\mathbf{V}) \,, \tag{1.79}$$

which gives an expression for the curl of the curl of a vector field. Then we *define* the Laplacian of a vector in *curvilinear* coordinates, using Eq. (1.79), as

$$\nabla^2\mathbf{V} \equiv -\nabla \times (\nabla \times \mathbf{V}) + \nabla(\nabla\cdot\mathbf{V}) \,. \tag{1.80}$$

Since the curl, gradient, and divergence operators are all defined in curvilinear coordinates, and Eq. (1.80) holds in Cartesian coordinates, it is the natural definition of $\nabla^2\mathbf{V}$ in curvilinear coordinates. The proof of Eq. (1.79) parallels very closely that for the vector identity of Eq. (1.34), using the fact that in Cartesian coordinates,

$$[\nabla \times (\nabla \times \mathbf{V})]_i = \varepsilon_{ijk}\partial_j\varepsilon_{klm}\partial_l V_m \,.$$

In general, when dealing with the various differential operators, it is essential to keep firmly in mind the scalar or vector character of the field that results from the operation, and the scalar or vector character of the operand. The Laplacian is the only operator that is defined on both scalar and vector fields and it produces a scalar or vector field accordingly. The gradient operates on a scalar field and produces a vector field; the divergence operates on a vector field and produces a scalar field; and the curl operates on a vector field and produces a vector field.

There is a natural generalization of the Laplacian operator to four-dimensional space-time (Minkowski space) called the D'Alembertian operator, which is written $\Box^2$. It is defined in terms of the Laplacian as

$$\Box^2 \equiv \nabla^2 - \frac{1}{c^2}\frac{\partial^2}{\partial t^2} \,. \tag{1.81}$$

The significance of the minus sign will be discussed in Section 4.6. The operand of the D'Alembertian, as of the Laplacian, may be either a scalar or a vector field. In terms of the D'Alembertian, the scalar wave equation becomes simply

$$\Box^2\varphi = 0\,. \tag{1.82}$$

The divergence of the vector grad φ is the Laplacian of φ. This suggests the following questions:

1. What is the curl of grad φ?
2. What is the divergence of the curl **V**?

A direct calculation in Cartesian coordinates gives:

$$1. \qquad [\text{curl } (\text{grad } \varphi)]_i = \varepsilon_{ijk}\partial_j\partial_k\varphi = 0\,; \tag{1.83}$$

$$2. \qquad \text{div } (\text{curl } \mathbf{V}) = \partial_i\varepsilon_{ijk}\partial_j V_k = \varepsilon_{ijk}\partial_i\partial_j V_k = 0\,. \tag{1.84}$$

These two proofs need a little explaining. In both we have the product of a term, ε_{ijk}, that is antisymmetric in any pair of subscripts, and a term of the form $\partial_m\partial_n$ which is symmetric in the subscripts. (A sufficient condition for the equality of cross derivatives, $\partial_m\partial_n\psi = \partial_n\partial_m\psi$, $m \neq n$, is their continuity, as is shown in any calculus text. We are assuming that the fields in physics possess this property.) Now the sum over the product of symmetric and antisymmetric terms vanishes. The reader can easily see this either by writing it out or by noting that if

$$a_{ij} = a_{ji} \qquad \text{and} \qquad b_{ij} = -b_{ji}\,,$$

then

$$s \equiv a_{ij}b_{ij} = -a_{ji}b_{ji} = -s\,,$$

since the subscripts are dummies; therefore $s = 0$.

These theorems hold in any curvilinear coordinate system. In fact, one way to show that div (curl **V**) vanishes is to apply the coordinate-free statement of Gauss's theorem to a vector field derived from the curl of a vector field:

$$\int_\tau \text{div } (\text{curl } \mathbf{V})\, d\tau = \int_\sigma (\text{curl } \mathbf{V}) \cdot d\boldsymbol{\sigma}\,.$$

To evaluate the integral of curl **V** over the closed surface σ bounding τ, we draw a closed curve λ of any size and shape on the surface σ. This separates it into two surfaces, both bounded by λ. Now we use Stokes's theorem (Eq. 1.63) to evaluate the integral of curl **V** over these two surfaces. Since the two integrals have the common boundary λ which is traversed in a clockwise direction for one and a counterclockwise direction for the other, their values are equal in magnitude and opposite in sign. Hence their sum, which equals the surface integral of curl **V**, is zero. Since this holds for any volume τ, it must be the case that div (curl **V**) = 0.

The two theorems expressed in Eqs. (1.83) and (1.84) have very important converses. Namely, if curl $\mathbf{V} = \mathbf{0}$, then **V** may be expressed as the gradient of a scalar field: $\mathbf{V} = \nabla\varphi$; and if div $\mathbf{V} = 0$, then $\mathbf{V}$ may be expressed as the curl of a vector field: $\mathbf{V} = \text{curl}\,\mathbf{A}$, where φ and $\mathbf{A}$ are referred to as the scalar and vector potential, respectively. Since our concern here is to gain an intuitive, physical picture of the differential operators and their role in classical physics, we shall not go into the proofs of these theorems, which are found in any book on vector analysis. A closely related result of potential theory is that one can determine a vector field uniquely from a knowledge of its curl and divergence, and suitable boundary conditions. Since Maxwell's equations are expressed in terms of the curl and divergence of the electromagnetic fields, it is essential to the program of classical electromagnetic field theory that this information be sufficient to solve for the fields themselves. This theorem is stated and proved in detail in Panofsky and Phillips.*

We conclude this section with a brief discussion of the self-consistency of Maxwell's equations in the light of the theorem that div (curl $\mathbf{V}$) $= 0$. This example also indicates what a compact and elegant notation the vector differential operators provide.

Most of the equations that bear Maxwell's name were already known when he came on the scene. What Maxwell did was to make a vital addition to one equation in the set of four empirical equations that described electrical and magnetic phenomena. This finishing touch filled in a missing symmetry in the equations and laid the basis for the prediction of wave phenomena. Prior to Maxwell, the following four laws were known. (We express them in differential form, in cgs units and standard notation.)

Coulomb's law: $$\nabla\cdot\mathbf{D} = 4\pi\rho \tag{1.85}$$

Nonexistence of magnetic monopoles: $$\nabla\cdot\mathbf{B} = 0 \tag{1.86}$$

Faraday's law (of electromagnetic induction): $$\nabla\times\mathbf{E} = -\frac{1}{c}\frac{\partial\mathbf{B}}{\partial t} \tag{1.87}$$

Ampere's law (for steady-state circuits): $$\nabla\times\mathbf{H} = \frac{4\pi}{c}\mathbf{J} \tag{1.88}$$

In Maxwell's day these equation were written out as cumbersome partial differential equations, which, of course, made it much harder to analyze them. In their present form any physics major could note, as Maxwell did, that Eq. (1.88) necessarily implies that div $\mathbf{J} = 0$. This equation holds only in the static case, with closed circuits; but if the charge is changing with time, and we assume charge conservation, then the continuity equation (1.58) holds. The continuity equation and Eq. (1.85) imply that

$$\nabla\cdot\mathbf{J} = -\frac{\partial\rho}{\partial t} = -\frac{\partial}{\partial t}\left(\frac{1}{4\pi}\nabla\cdot\mathbf{D}\right) = -\nabla\cdot\frac{1}{4\pi}\frac{\partial\mathbf{D}}{\partial t}.$$

* W. H. K. Panofsky and M. Phillips, *Classical Electricity and Magnetism.* Reading, Mass.: Addison-Welsey, 1955.

Thus if the term $(1/4\pi)(\partial\mathbf{D}/\partial t)$ is added to $\mathbf{J}$ on the right-hand side of Eq. (1.88), we would have

$$\nabla \times \mathbf{H} = \frac{4\pi}{c}\mathbf{J} + \frac{1}{c}\frac{\partial \mathbf{D}}{\partial t}, \tag{1.89}$$

this choice would ensure that $\nabla\cdot(\nabla \times \mathbf{H}) \equiv 0$; then the equations need no longer be restricted to the static case. The addition of this term, the so-called displacement current, was Maxwell's contribution. It fills in a missing symmetry in the equations, for they now express the fact that a time-dependent electric field generates a magnetic field, just as Faraday's law stated that a magnetic field changing in time generates an electric field. Therefore, the possibility of self-sustaining electromagnetic waves is suggested once Maxwell's new term is added to Eq. (1.88). Without the displacement-current term, the equations do not predict wave phenomena. The term is therefore of such tremendous significance that the whole set of equations is named for Maxwell.

1.8 CARTESIAN-TENSORS

Many physically significant quantities are neither scalars nor vectors. That is, the transformation law obeyed by these quantities under orthogonal transformations is neither

$$\phi' = \phi, \tag{1.90}$$

which defines a scalar [Eq. (1.26)], nor

$$x_i' = a_{ij}x_j, \tag{1.91}$$

which defines a vector [Eq. (1.27)].

If under an orthogonal transformation of coordinates in three dimensions, the 3^N quantities $T_{i_1i_2\cdots i_N}$ (where each of the N indices extends from 1 to 3) transform according to the rule

$$T'_{i_1i_2\cdots i_N} = a_{i_1j_1}a_{i_2j_2}\cdots a_{i_Nj_N}T_{j_1j_2\cdots j_N}, \tag{1.92}$$

then $T_{i_1i_2\cdots i_N}$ are the components of an Nth-rank *tensor* (in three dimensions). Note that in Eq. (1.92), N summations are implied. Clearly a tensor of the first rank has three components that satisfy Eq. (1.91), so it is a vector. And a tensor of the zeroth rank has one invariant component and so it is a scalar. The most useful other case is the tensor of second rank. It has nine components, T_{ij}, which obey the transformation law

$$T'_{ij} = a_{ik}a_{jl}T_{kl}. \tag{1.93}$$

We shall discuss several applications of second-rank tensors in physics later in this section. But first it must be noted that what we are doing here really does not go under the name of tensor analysis. We are concerned here only with *Cartesian-tensors*, which are defined in terms of their transformation properties under a very special class of coordinate transformations, namely, orthogonal transformations. Cartesian-tensors are sets of quantities with the transformation law, Eq. (1.92), under *orthogonal* transformations. As a general rule, *tensors* are sets of quantities with a certain law of transformation under *arbitrary* coordinate transformations, which is a much more stringent condition. Thus it is "easier" to qualify as a Cartesian-tensor than a tensor. Any tensor is a Cartesian-tensor, but a Cartesian-tensor need not be a tensor, because although a Cartesian-tensor obeys the proper transformation rule for the special orthogonal coordinate transformations, it may fail to do so for other coordinate transformations. This statement sounds paradoxical—grammatically it is, since on the surface it is a self-contradictory statement of the form: "Any tree is a maple tree, but a maple tree need not be a tree." We have tried to lessen this grammatical annoyance by writing "Cartesian-tensor" instead of "Cartesian tensor." The dash means that Cartesian-tensors are not a subclass of tensors, but new objects; the adjective does not select a subclass out of the class of objects called tensors.

We now discuss some notions of tensor analysis restricting our attention to three-dimensional Cartesian-tensors. However, we shall drop the qualifier "Cartesian," assuming that it is understood that all the tensors we refer to henceforth are, in fact, Cartesian-tensors.

First we consider the *contraction* of tensors, which is one of the most important operations defined on tensors. Contraction is an operation that produces a new tensor from an object whose tensor character is known. (In German the word for contraction is *Verjüngung*, which is translated as "rejuvenation.")

We shall need the result (whose proof we leave to the reader) that if $S_{i_1 i_2 \cdots i_N}$ is an Nth rank tensor and $T_{j_1 j_2 \cdots j_M}$ is an Mth rank tensor, then $S_{i_1 \cdots j_N} T_{j_1 \cdots j_M}$ has 3^{N+M} components and is an $(N+M)$th rank tensor.

Contraction of a tensor $T_{i_1 i_2 \cdots i_N}$ is defined as multiplication by, and thereby automatic summation over the indices of, the Kronecker delta, $\delta_{i_l i_m}$, where $l \neq m$, and l and $m \leq N$. Thus if $T_{i_1 \cdots i_N}$ is a tensor of Nth rank, then

$$H_{i_1 \cdots i_{j-1} i_{j+1} \cdots i_{m-1} i_{m+1} \cdots i_N} = \delta_{i_j i_m} T_{i_1 \cdots i_j \cdots i_m \cdots i_N} \tag{1.94}$$

is the contraction of T. We now prove that H is a tensor of rank $N-2$. First note that it has 3^{N-2} components, so that if it is a tensor, it is of rank $N-2$. Now

$$\begin{aligned} H'_{i_1 \cdots i_{j-1} i_{j+1} \cdots i_{m-1} i_{m+1} \cdots i_N} &= \delta'_{i_j i_m} T'_{i_1 \cdots i_j \cdots i_m \cdots i_N} \\ &= \delta_{i_j i_m} a_{i_1 n_1} \cdots a_{i_j n_j} \cdots a_{i_m n_m} \cdots a_{i_N n_N} T_{n_1 \cdots n_j \cdots n_m \cdots n_N} \\ &= a_{i_1 n_1} \cdots a_{i_m n_j} \cdots a_{i_m n_m} \cdots a_{i_N n_N} T_{n_1 \cdots n_j \cdots n_m \cdots n_N} . \end{aligned}$$

Since $a_{i_m n_j} a_{i_m n_m} = \delta_{n_j n_m}$, the last expression becomes, by Eq. (1.94),

$$a_{i_1 n_1} \cdots a_{i_{j-1} n_{j-1}} a_{i_{j+1} n_{j+1}} \cdots a_{i_{m-1} n_{m-1}} a_{i_{m+1} n_{m+1}} \cdots a_{i_N n_N} H_{n_i \cdots n_{j-1} n_{j+1} \cdots n_{m-1} n_{m+1} \cdots n_N} ,$$

which proves that H is a tensor (of rank $N-2$).

These general proofs for Nth rank tensors demand a certain virtuosity in subscript manipulation. Having indicated the general method in this one instance, we shall now retreat to what is fortunately the most interesting physical case, $N = 2$.

By way of illustration, consider the product of two vectors, x_i and y_j. We know that the product $x_i y_j$ is a second-rank tensor, and we can contract it to get a quantity which, according to the theorem, must be a tensor of rank $2 - 2 = 0$ (i.e., a scalar). Thus

$$\delta_{ij} x_i y_j = x_i y_i = \mathbf{x} \cdot \mathbf{y} , \tag{1.95}$$

the scalar product.

Two important special types of second-rank tensors are the symmetric and antisymmetric tensors. Unlike vectors, tensors are not, in general, easily visualized geometrically, but for symmetric second-rank tensors ($T_{ij} = T_{ji}$) and antisymmetric second-rank tensors ($T_{ij} = -T_{ji}$) a geometrical representation is possible.

For an antisymmetric tensor of second rank, the three diagonal elements must be zero since $T_{ii} = -T_{ii}$. The other six elements occur in pairs of equal magnitude and opposite sign. Thus a tensor of this type is characterized by only three independent quantities (say T_{12}, T_{13}, and T_{23}), and a vector A_k can be associated with it according to the equation

$$A_i = \tfrac{1}{2} \varepsilon_{ijk} T_{jk} . \tag{1.96}$$

By comparing Eq. (1.96) with Eq. (1.33), we see that the vector associated with the second-rank antisymmetric tensor is in fact a pseudovector.

For a symmetric tensor of second rank, there are six independent components: the three diagonal elements, and the three pairs of equal off-diagonal elements. Identifying a second-rank symmetric tensor with a vector is therefore out of the question. However, it takes six independent parameters to specify a quadratic surface:

$$Ax_1^2 + Bx_2^2 + Cx_3^2 + Dx_1x_2 + Ex_1x_3 + Fx_2x_3 = 1 .$$

It is not surprising, therefore, that any symmetric second-rank tensor can be uniquely represented by the surface

$$T_{ij} x_i x_j = \pm 1 ,$$

where the sign is that of the determinant of T_{ij}. The sign convention is needed, for otherwise, for the symmetric tensor $T_{ij} = -\delta_{ij}$ we would get the surface

$$-x_1^2 - x_2^2 - x_3^2 = 1 ,$$

which does not exist in real space. With the sign convention, we get a sphere of radius one about the origin.

An arbitrary second-rank tensor, R_{ij}, can be written as the sum of a symmetric and an antisymmetric tensor:

$$R_{ij} = \underbrace{\tfrac{1}{2}(R_{ij} + R_{ji})}_{\text{symmetric}} + \underbrace{\tfrac{1}{2}(R_{ij} - R_{ji})}_{\text{antisymmetric}} . \tag{1.97}$$

It is not uncommon that after reading for the first time about tensors the student will say, "Yes, fine, but what *is* a tensor?" Providing him with a list of examples of tensors, and reiterating the definition in terms of transformation properties leaves him feeling a bit cheated. The problem here is really one of semantics. It would be clearer not to speak of "tensors in general," but instead to define "tensor character" and speak of given quantities as having or lacking a certain tensor character. All one means when one speaks of a tensor is a quantity which possesses tensor character. We conclude this section by discussing three physical quantities which have tensor character: the moment-of-interia tensor, the electric-polarization tensor, and the multipole tensor.

Example 1.6. *The moment of inertia tensor.* If a rigid body with continuous distribution of mass having density $\rho(\mathbf{r})$ rotates about one fixed point in the body, the total angular momentum about that point is

$$\mathbf{L} = \int \rho(\mathbf{r} \times \mathbf{v})\, d\tau ,$$

where $\mathbf{r}$ is the radius vector from the stationary point to the mass element $dm = \rho\, d\tau$, and $\mathbf{v} =$ the velocity of dm relative to the stationary point. It follows from kinematical considerations that $\mathbf{v} = \boldsymbol{\omega} \times \mathbf{r}$, where $\boldsymbol{\omega}$ is the angular velocity of the rigid body. Therefore

$$\begin{aligned} \mathbf{L} &= \int \rho[\mathbf{r} \times (\boldsymbol{\omega} \times \mathbf{r})]\, d\tau \\ &= \int \rho[\boldsymbol{\omega} r^2 - \mathbf{r}(\boldsymbol{\omega} \cdot \mathbf{r})]\, d\tau . \end{aligned}$$

Note that $\mathbf{L}$ is not parallel to $\boldsymbol{\omega}$ unless $\boldsymbol{\omega}$ is perpendicular to $\mathbf{r}$.

The inertia tensor, I_{ij}, is defined as

$$I_{ij} = \int \rho[r^2\delta_{ij} - r_i r_j]\, d\tau . \tag{1.98}$$

Note that it really is a tensor, as it is the sum of products of vectors. In fact, it is a symmetric second-rank tensor. As we shall see in chapter 4, this means that this tensor can be diagonalized—i.e., principal axes and principal moments of inertia can be found. In terms of the inertia tensor I_{ij}, we can rewrite the earlier expression for $\mathbf{L}$ as

$$\begin{aligned} I_{ij}\omega_j &= \int \rho[r^2\delta_{ij}\omega_j - r_i r_j \omega_j]\, d\tau \\ &= \int \rho[r^2\omega_i - r_i(\mathbf{r} \cdot \boldsymbol{\omega})]\, d\tau = L_i . \end{aligned}$$

Goldstein writes this equation in the symbolic form $\mathbf{L} = \mathbf{I} \cdot \boldsymbol{\omega}$, where $\mathbf{I}$ is an operator on $\boldsymbol{\omega}$. $\mathbf{I}$ is not a scalar that merely changes the magnitude of $\boldsymbol{\omega}$; it is a tensor, and thus this tensor may be viewed as changing both the magnitude and the direction of $\boldsymbol{\omega}$ in relating $\boldsymbol{\omega}$ to $\mathbf{L}$.

Example 1.7. *The electric polarization tensor*. The polarization vector $\mathbf{P}$ is related to the electric displacement vector $\mathbf{D}$ by

$$\mathbf{D} = \varepsilon_0 \mathbf{E} + \mathbf{P}\,.$$

If $\mathbf{P}$ is parallel to $\mathbf{E}$, we write $\mathbf{P} = \varepsilon_0 \chi \mathbf{E}$ (where χ is the electric susceptibility), so $\mathbf{D} = \varepsilon_0(1 + \chi)\,\mathbf{E} = \varepsilon_0 k\mathbf{E}$, where k is the dielectric constant. But $\mathbf{P}$ will only be parallel to $\mathbf{E}$ if the dielectric material polarizes isotropically. This can hardly be expected in crystals with molecular structures of low symmetry. In crystals of symmetry lower than cubic, the relation between each of the components of $\mathbf{P}$ and of the $\mathbf{E}$-vector is still linear, but the constants of proportionality in the various directions are not necessarily the same. The special case of $\mathbf{P}$ parallel to $\mathbf{E}$ will occur in gases, liquids, amorphous solids, Jello ® puddings, and cubic crystals. If this is not the case, the vector proportionality between $\mathbf{P}$ and $\mathbf{E}$ must be replaced by a tensor equation that allows for different constants of proportionality in different directions of $\mathbf{E}$. We have

$$P_i = \sum_{j=1}^{3} \varepsilon_0 \chi_{ij} E_j\,,$$

where χ_{ij} is the susceptibility tensor.

Example 1.8. *The multipole-moment tensor*. The multipole tensor arises in the multipole expansion of the potential $\varphi(\mathbf{r})$ due to an arbitrary localized charge distribution $\rho(\mathbf{r}')$. From potential theory (e.g., Panofsky and Phillips, Chapter 1), we know that if div $\mathbf{E} = \rho$, and if curl $\mathbf{E} = 0$ (electrostatics in free space), then $\mathbf{E} = -\nabla\varphi$, where

$$\varphi(\mathbf{r}) = \int_{\tau'} \frac{\rho(\mathbf{r}')}{|\mathbf{r} - \mathbf{r}'|}\, d\tau'\,. \tag{1.99}$$

The charge of density $\rho(\mathbf{r}')$ is localized in volume τ'. To calculate the field at points $\mathbf{r}$, remote from the source—i.e., for $|\mathbf{r}| > |\mathbf{r}'|$—we expand $1/R$ in powers of r'/r:

$$\begin{aligned}
\frac{1}{R} &\equiv |\mathbf{r} - \mathbf{r}'|^{-1} = (r^2 - 2\mathbf{r}\cdot\mathbf{r}' + r'^2)^{-1/2} \\
&= \frac{1}{r}\left[1 + \left(-\frac{2\mathbf{r}\cdot\mathbf{r}'}{r^2} + \frac{r'^2}{r^2}\right)\right]^{-1/2} \\
&= \frac{1}{r}\left[1 - \frac{1}{2}\left(-\frac{2\mathbf{r}\cdot\mathbf{r}'}{r^2} + \frac{r'^2}{r^2}\right) + \frac{1}{2}\cdot\frac{3}{4}\left(-\frac{2\mathbf{r}\cdot\mathbf{r}'}{r^2} + \frac{r'^2}{r^2}\right)^2 + \cdots\right] \\
&= \frac{1}{r}\left[1 + \frac{\mathbf{r}\cdot\mathbf{r}'}{r^2} - \frac{1}{2}\frac{r'^2}{r^2} + \frac{3}{8}\left(\frac{4(\mathbf{r}\cdot\mathbf{r}')^2}{r^4}\right) + \cdots\right] \\
&= \frac{1}{r}\left[1 + \frac{\mathbf{r}\cdot\mathbf{r}'}{r^2} + \frac{1}{2}\left(\frac{3(\mathbf{r}\cdot\mathbf{r}')^2}{r^4} - \frac{r'^2}{r^2}\right)\right. \\
&\qquad \left. + \text{ terms of order } \left(\frac{r'}{r}\right)^3 \text{ and higher.}\right]
\end{aligned}$$

Table 1.2.

Symbol	Definition	Name	Tensor character
q	$\int_{\tau'} \rho(r')\, d\tau'$	monopole (charge)	scalar
$\mathbf{p}$	$\int_{\tau'} \mathbf{r}'\rho(\mathbf{r}')\, d\tau'$	dipole	vector
Q_{ij}	$\int_{\tau'} (3x_i'x_j' - \delta_{ij}r'^2)\rho(\mathbf{r}')\, d\tau'$	quadrupole	second-rank symmetric tensor

Note at this point that if we put $\mathbf{r}\cdot\mathbf{r}' = rr' \cos\theta$, the multipole expansion can be written as

$$\frac{1}{R} = \frac{1}{r}\left[1 + \left(\frac{r'}{r}\right)\cos\theta + \left(\frac{r'}{r}\right)^2 \frac{3\cos^2\theta - 1}{2} + \cdots\right]. \tag{1.100}$$

In Chapter 6 we shall prove that, for $r' < r$,

$$\frac{1}{R} \equiv \frac{1}{|\mathbf{r} - \mathbf{r}'|} = \frac{1}{r}\sum_{l=0}^{\infty}\left(\frac{r'}{r}\right)^l P_l\,(\cos\theta)\ , \tag{1.101}$$

where the $P_l\,(\cos\theta)$ are the Legendre polynominals:

$$\begin{aligned} P_0\,(\cos\theta) &= 1\ , \\ P_1\,(\cos\theta) &= \cos\theta\ , \\ P_2\,(\cos\theta) &= \tfrac{1}{2}\,(3\cos^2\theta - 1)\ , \\ P_3\,(\cos\theta) &= \tfrac{1}{2}\,(5\cos^3\theta - 3\cos\theta)\ , \text{ etc.} \end{aligned} \tag{1.102}$$

We have proved by direct expansion that the first three terms of Eq. (1.101) are correct. If we now substitute the expansion of R^{-1} into the expression for $\varphi(r)$, we have

$$\begin{aligned} \varphi(r) &= \frac{1}{r}\int_{\tau'} \rho(\mathbf{r}')\, d\tau' + \frac{\mathbf{r}}{r^3}\cdot\int_{\tau'} \mathbf{r}'\rho(\mathbf{r}')\, d\tau' \\ &\quad + \frac{1}{2}\frac{x_i x_j}{r^5}\int_{\tau'} [3x_i'x_j' - \delta_{ij}r'^2]\,\rho(\mathbf{r}')\, d\tau' + \cdots . \end{aligned} \tag{1.103}$$

The first three multipoles are defined in Table 1.2.

Higher-order poles are defined by expanding the potential to higher order. It is clear from the definition of the quadrupole-moment tensor that it is symmetric. The *trace* of a tensor T_{ij} is defined as $\text{tr}\,(T_{ij}) \equiv \sum_i T_{ii}$ = the sum of the diagonal elements. Note that

$$\begin{aligned} \text{tr}\, Q_{ij} &= \sum_i Q_{ii} = \sum_i \int_{\tau'} (3x_i'x_i' - \delta_{ii}r'^2)\,\rho(\mathbf{r}')\, d\tau' \\ &= \int_{\tau'} (3r'^2 - 3r'^2)\,\rho\, d\tau' = 0\ . \end{aligned}$$

So the trace of the quadrupole-moment tensor is identically zero. Knowledge of the quadrupole moment is useful in determining the shape of nuclei and thus obtaining information on nuclear forces. In the multipole expansion, the first term is just the total nuclear charge and the second term is the dipole moment, which can be shown to vanish for quantum-mechanical systems in stationary states. The third term is a quadrupole moment. By measuring it and then fitting the result to a model which predicts such a quadrupole moment, we learn the shape of the nucleus in question. I. I. Rabi and co-workers showed that the deuteron possesses a nonzero quadrupole moment such that it must have the shape of a prolate ellipsoid, 50% longer than it is wide. This implies a *non-central nuclear force* between the neutron and the proton, a fact of great importance.

PROBLEMS

1. Prove by vector methods that the diagonals of a rhombus are orthogonal.
2. A particle moves in the $x_1 x_2$-plane so that its position vector is given by
$$\mathbf{r} = a \cos \omega t \mathbf{e}_1 + a \sin \omega t \mathbf{e}_2 \qquad (\omega = \text{const}) .$$
Show that
a) the velocity $\mathbf{v}$ is perpendicular to $\mathbf{r}$;
b) the acceleration $\mathbf{a}$ is directed toward the origin; find its magnitude;
c) $\mathbf{L} = \mathbf{r} \times m\mathbf{v}$ is a constant vector; find its magnitude.
3. Prove from simple vector arguments and the conservation laws of mechanics that in an elastic collision between two particles of equal mass, one initially at rest, the velocity vectors after collision are perpendicular (if nonzero).
4. A charged particle moves in a region of space where there is a uniform magnetic field $\mathbf{B}$ (in the z-direction say,) and a uniform electric field $\mathbf{E}$ in the yz-plane. Write the equations of motion for the particle. Then solve them, taking for initial conditions (at $t = 0$) $x = y = z = 0$; $\dot{x} = \dot{x}_0$, $\dot{y} = \dot{z} = 0$. Show that there is a drift velocity equal to E_y/B in the positive x-direction. If $E_y = 0$, then the orbit becomes a spiral about a line parallel to the z-axis.
5. Prove that $\varepsilon_{ijk}\varepsilon_{ijk} = 6$.
6. Prove the vector identities:
a) $\text{grad}\,(\mathbf{A}\cdot\mathbf{B}) = (\mathbf{A}\cdot \text{grad})\,\mathbf{B} + (\mathbf{B}\cdot \text{grad})\,\mathbf{A} + \mathbf{A}\,\text{curl}\,\mathbf{B} + \mathbf{B}\,\text{curl}\,\mathbf{A}$;
b) $\nabla\cdot(\mathbf{U} \times \mathbf{V}) = \mathbf{V}\cdot(\nabla \times \mathbf{U}) - \mathbf{U}\cdot(\nabla \times \mathbf{V})$.
7. Prove Green's theorem, that is,
$$\int_\tau (\varphi\nabla^2\psi - \psi\nabla^2\varphi)\, d\tau = \int_\sigma (\varphi\nabla\psi - \psi\nabla\varphi)\cdot d\boldsymbol{\sigma} .$$
Hint: Apply Gauss's divergence theorem to the vectors $\mathbf{A} = \varphi\nabla\psi$ and $\mathbf{B} = \psi\nabla\varphi$.
8. If the electric potential of a point charge were
$$\varphi(r) = \frac{q}{r^{1-\varepsilon}}, \qquad \varepsilon \ll 1 \qquad \text{(positive or negative)}$$
instead of q/r, many of the results of physics would be altered. Calculate for $r \neq 0$:
a) $\mathbf{E} = -\nabla\varphi$, b) $\nabla\cdot\mathbf{E} = -\nabla^2\varphi$, c) $\nabla \times \mathbf{E}$,
d) Derive the analog of Gauss's law for a spherical region of radius R.

Examine the limits as $\varepsilon \to 0$, and compare with the usual Coulomb potential. One of the consequences of such a potential is discussed in Problem 8, Panofsky and Phillips, Chapter 1.

9. A physical system known as a vector meson is described by the three vector fields **E**, **H**, **A** and a real scalar field V, which satisfy the following equations:

$$\nabla \cdot \mathbf{E} = -\mu^2 V$$

$$\nabla \times \mathbf{H} = \frac{\partial \mathbf{E}}{\partial t} - \mu^2 \mathbf{A} \qquad (\mu^2 \text{ is a positive constant})$$

$$\mathbf{H} = \nabla \times \mathbf{A}\,, \qquad \mathbf{E} = \frac{\partial \mathbf{A}}{\partial t} - \nabla V$$

a) Show that $\nabla \cdot \mathbf{A} + \dfrac{\partial V}{\partial t} = 0$

b) Show that V satisfies the equation:

$$\nabla^2 V - \mu^2 V - \frac{\partial^2 V}{\partial t^2} = 0\,.$$

c) When V is time-independent, it satisfies $\nabla^2 V - \mu^2 V = 0$. If this is the case inside a closed surface and if V vanishes on the closed surface, show that $V = 0$ inside.

10. a) What class of physical phenomena are described by Helmholtz's equation,

$$\nabla^2 \varphi + k^2 \varphi = 0\,?$$

b) Interpret this equation physically in the light of the discussion of the physical meaning of the Laplacian, and square your interpretation with the answer to part (a).

11. a) Derive the law of conservation of charge from Maxwell's equations.

b) Express the electric field in terms of the scalar potential φ and the vector potential **A** where $\mathbf{B} = \text{curl}\, \mathbf{A}$. (*Note*: For the general time-dependent case, **E** is not equal to grad φ alone.)

12. A second-rank antisymmetric tensor F_{ij} is characterized by three independent quantities. Consequently, a vector H_k can be associated with such a tensor. Prove that $H_k = \frac{1}{2}\varepsilon_{ijk}F_{ij}$ *if and only if* $F_{ij} = \varepsilon_{ijk}H_k$.

13. *The electromagnetic field tensor in free space.* Problems 11 and 12 have laid the groundwork for the four-dimensional theory of the electromagnetic field that will unfold as this problem is done. We shall use Gaussian units. To begin, if we identify the vector H_k of Problem 11 with the magnetic field, then problem 12 relates H_k to a second-rank antisymmetric tensor, F_{ij}.

a) Prove

$$F_{ij} = \frac{\partial A_j}{\partial x_i} - \frac{\partial A_i}{\partial x_j} \equiv \partial_i A_j - \partial_j A_i$$

where curl $\mathbf{A} = \mathbf{H}$.

b) Show directly from (a) that F_{ij} is antisymmetric.

So far, F_{ij} is three-dimensional and has $3^2 = 9$ components. We now extend it to four dimensions.

c) Make the identifications $x_4 = ict$ and $A_4 = i\varphi$, where $i = \sqrt{-1}$ and φ is the electrostatic potential. Prove that $F_{4j} = -F_{j4} = iE_j$ $(j = 1, 2, 3)$, using F_{ij} as expressed in (a) and problem 11(b).

d) Verify that $F_{ii} = 0$ $(i = 1, 2, 3, 4)$.
Thus F_{ij} is a four-dimensional antisymmetric second-rank tensor with $4^2 = 16$ components. The four diagonal elements are zero. Only six of the remaining 12 elements are independent, expressing the fact that **E** and **H** have a total of six components.

e) Write out F_{ij} as a 4 by 4 array explicitly in terms of the components of **E** and **H**. At this point you might want to see how you are doing. You should obtain

$$F_{ij} = \begin{bmatrix} 0 & H_z & -H_y & -iE_x \\ -H_z & 0 & H_x & -iE_y \\ H_y & -H_x & 0 & -iE_z \\ iE_x & iE_y & iE_z & 0 \end{bmatrix}.$$

f) All four of Maxwell's equations can be expressed in two equations written in terms of the electromagnetic field tensor:

i. $\partial_l F_{ik} + \partial_i F_{kl} + \partial_k F_{li} = 0,$
ii. $\partial_k F_{ik} = (4\pi/c)\, J_i\,,$

where $J_i = (\mathbf{J})_i$ for $i = 1, 2, 3$ and $J_4 = ic\rho$ (ρ = electric-charge density). In these two equations i, k, and l extend from 1 to 4. Prove that these equations subsume Maxwell's equations, and give nothing more. (This will involve considering different combinations of subscripts, etc.)

g) Derive the four-dimensional form of the equation of conservation of charge directly from Maxwell's equation (ii) in part (f). Interpret it in four-dimensional form.

h) Show that the Lorentz force law is

$$f_i = \frac{1}{c} F_{ik} J_k\,, \qquad i = 1, 2, 3\,,$$

where f_i are the three components of the Lorentz force. Interpret the quantity icf_4.

14. Prove that the trace of the electromagnetic energy-momentum tensor, T_{ik}, is zero, where

$$T_{ik} = \frac{1}{4\pi}\left(F_{il}F_{kl} - \frac{1}{4}F_{lm}^2\delta_{ik}\right);$$

F_{ij} is the electromagnetic field tensor of Problem 13.

15. Prove that the dipole moment of a charge distribution is independent of the origin of coordinates if the total charge is zero.

16. a) A charge q is uniformly distributed upon a circle that has radius R, is concentric with the z-axis, and lies in the plane $z = 0$. A charge $(-q)$ is located at the center of the circle. Find (i) the electric monopole moment (ii) the electric-dipole moment vector and (iii) the electric-quadrupole moment tensor, expressing the latter in an appropriate 3 by 3 array. Using the symmetry of the quadrupole tensor and the fact that its trace vanishes, as well as the inherent symmetry of the physical system under consideration, one can reduce the labor enormously.

b) Determine for this charge distribution the multipole expansion for the potential at distances r from the origin for which $(r/R) \gg 1$; carry the expansion through the quadrupole contribution, and express the spatial dependence in the final result in terms of the spherical coordinates r, θ, φ.

17. Find the first three multipole moments for a single point charge q sitting at the point $\mathbf{r} = R(\mathbf{i} + \mathbf{j} - \mathbf{k})$. Determine, through three terms in the multipole expansion, the potential (in spherical coordinates) at a point very far away from the charge.

FURTHER READINGS

ADLER, R., M. BAZIN, and M. SCHIFFER, *Introduction to General Relativity.* New York: McGraw-Hill 1965. An introduction to tensor analysis and Riemannian spaces as well as a most readable account of general relativity.

HOPF, L., *Introduction to the Differential Equations of Physics.* New York: Dover, 1948. A short discussion of the equations of classical physics, emphasizing their intuitive content.

LANDAU, L., and E. LIFSHITZ, *The Classical Theorn of Fields.* Reading, Mass.: Addison-Wesley, 1951.

SKILLING, H. H. *Fundamentals of Electric Waves.* New York: Wiley, 1942. This book contains nice examples of grad, div, and curl operations.

SPIEGEL, M. R. *Theory and Problems of Vector Analysis.* New York: Schaum, 1959. An excellent set of solved review problems.

SPRINGER, C. E. *Tensor and Vector Analysis.* New York: Ronald Press Co., 1962.

SYNGE J. L., and A. SCHILD, *Tensor Calculus.* Toronto: University of Toronto Press, 1949.

CHAPTER 2

CALCULUS OF VARIATIONS

INTRODUCTION

One of the oldest problems in mathematical physics is that of trying to minimize expressions which do not depend on some simple continuous variable, but rather on a function. That is, the problem of interest is not just to determine the point at which some particular function takes on its smallest value, but to determine the entire functional dependence of some function in such a way that it will minimize an integral involving that function and its derivatives. Such problems played a crucial role in the development of classical physics during the eighteenth and nineteenth centuries, and their importance did not diminish with the arrival of quantum concepts in the twentieth century. This subject is called the calculus of variations, and in this chapter we will discuss briefly the ideas and applications of the calculus of variations to classical physics in order to familiarize the reader with the concepts and techniques involved. In later chapters we will have occasion to return to these ideas in a more modern context.

2.1 SOME FAMOUS PROBLEMS

The calculus of variations is an old subject. The study of calculus itself had been in existence for only a few years before questions were being asked that we now view as part of the subject of the calculus of variations. Three such questions are:

1) What plane curve connecting two given points has the shortest length?

We can formulate this question mathematically as follows. Let (x_A, y_A) and (x_B, y_B) be two points in the xy-plane with $x_A < x_B$. Let $y = y(x)$ connect them—thus, $y(x_A) = y_A$ and $y(x_B) = y_B$. The length of the arc connecting them will be

$$I = \int_{(x_A,\, y_A)}^{(x_B,\, y_B)} ds\,.$$

Since $ds^2 = dx^2 + dy^2$, we have

$$ds = \sqrt{1 + y'^2}\, dx\,,$$

where y' denotes the derivative of y with respect to x. The problem then is to choose the function $y(x)$, which is buried in the *integrand*, in such a way as to make the *integral* I a minimum.

2) In 1696 Johann Bernoulli of the University of Basel posed the following problem: Given two points A and B in a vertical plane, find the path AMB which the movable particle M will traverse in the shortest time, assuming that its acceleration is due only to gravity. This is the famous *brachistochrone* problem (from the Greek, *brachistos* = shortest, *chronos* = time), and it marks the beginning of general interest in the calculus of variations.

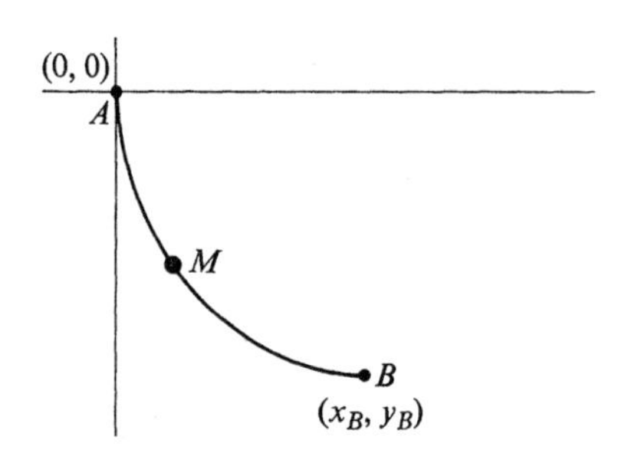

Fig. 2.1 Bernoulli's brachistochrone.

Let us take A as the origin of our coordinate system, assume that the particle of mass m initially has zero velocity (at A) and, as did Bernoulli, assume that there is no friction. We take the y-axis directed vertically downward for convenience. The speed along the curve AMB of Fig. 2.1 is $v = ds/dt$, so the total time of descent is

$$I = \int_A^B \frac{ds}{v} = \int_{x_A=0}^{x_B} \frac{\sqrt{1 + y'^2}}{v}\, dx .$$

To complete the specification of the problem, we must compute v as a function of position. We know that the increase of kinetic energy is equal to the decrease of potential energy if there is no friction. Hence

$$\tfrac{1}{2}mv^2 = mgy$$

so that

$$I = \frac{1}{\sqrt{2g}} \int_0^{x_B} \sqrt{\frac{1 + y'^2}{y}}\, dx .$$

The problem again is to find the function $y(x)$ which minimizes I.

3) A third problem is that of finding the minimum surface of revolution passing through two given fixed points, (x_A, y_A) and (x_B, y_B). Let us suppose that the curve $y(x)$ is to be rotated about the x-axis as illustrated in Fig. 2.2. We want to minimize the area of the surface so generated. We assume that $y_A > 0$, $y_B > 0$ and $y(x) \geq 0$ for $x_A \leq x \leq x_B$ as shown in the diagram. The integral we seek to minimize is the surface area:

$$I = 2\pi \int_A^B y\, ds = 2\pi \int_{x_A}^{x_B} y\sqrt{1 + y'^2}\, dx ,$$

where $y(x_A) = y_A$ and $y(x_B) = y_B$.

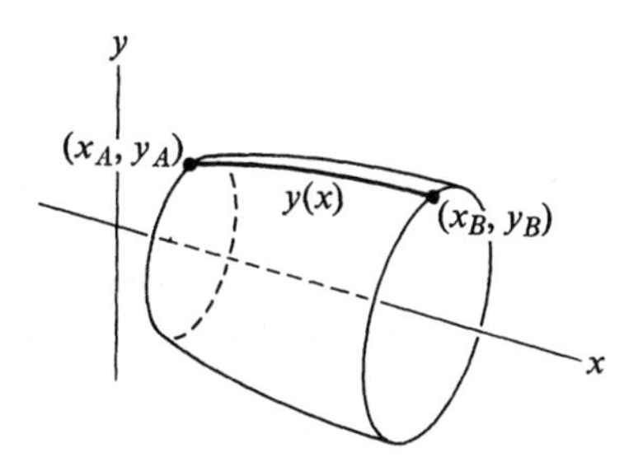

Fig. 2.2 The minimum surface of revolution.

All three of these problems are special cases of the more general question: What is the function $y(x)$, with $y(x_A) = y_A$ and $y(x_B) = y_B$ which minimizes the integral

$$I = \int_{x_A}^{x_B} f(x, y, y')\, dx\,,$$

where f is a *given* function of x, y, and y'? In geometrical language we are looking for the *path of integration* from x_A to x_B which minimizes the integral I.

2.2 THE EULER-LAGRANGE EQUATION

In this section we derive the differential equation that $y(x)$ must obey in order to minimize the integral

$$I = \int_{x_A}^{x_B} f(x, y, y')\, dx\,.$$

x_A, x_B, $y(x_A) = y_A$, $y(x_B) = y_B$ and f are all given, and f is assumed to be a twice-differentiable function of all its arguments. We denote the function which minimizes I by $y(x)$ (see Fig. 2.3) and consider the one-parameter family of comparison functions (or test functions), $\bar{y}(x, \epsilon)$, which satisfy the conditions:

a) $\bar{y}(x_A, \epsilon) = y_A$, $\bar{y}(x_B, \epsilon) = y_B$ for all ϵ;

b) $\bar{y}(x, 0) = y(x)$, the desired minimizing function;

c) $\bar{y}(x, \epsilon)$ and all its derivatives through second order are continuous functions of x and ϵ.

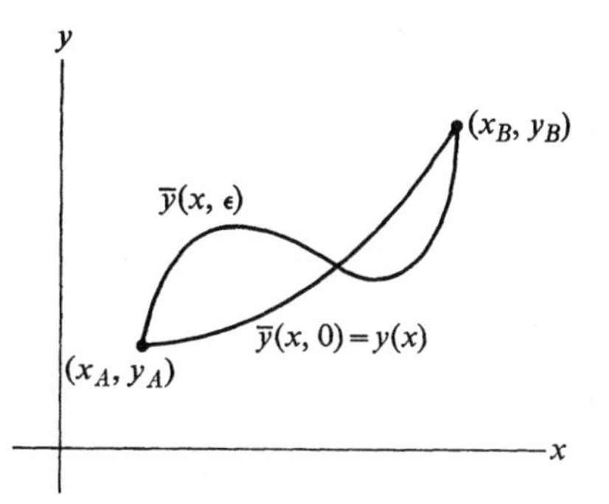

Fig. 2.3 A one-parameter family of comparison functions.

For a given comparison function, the integral

$$I(\epsilon) = \int_{x_A}^{x_B} f(x, \bar{y}, \bar{y}')\, dx$$

is clearly a function of ϵ. Also, since setting $\epsilon = 0$ corresponds, by condition (b), to replacing $\bar{y}$ by $y(x)$ and $\bar{y}'$ by $y'(x)$, we see that $I(\epsilon)$ should be a minimum with respect to ϵ for the value $\epsilon = 0$ according to the designation that $y(x)$ is the actual minimizing function. This is true for *any* $\bar{y}(x, \epsilon)$.

What we are doing here is considering a family of "smooth" curves passing through the points (x_A, y_A) and (x_B, y_B) which are labeled by the variable ϵ. The curve which makes the number I as small as possible has the label $\epsilon = 0$. We should emphasize that the choice of $\epsilon = 0$ is entirely arbitrary. We could equally well designate the minimizing function by the label $\epsilon = 1.67$ and change condition (b) accordingly. Since this curve by assumption minimizes I, and since I depends continuously on ϵ and is a differentiable function of ϵ [this follows from condition (c)], we can apply the standard methods of elementary calculus to determine the minimum.

A necessary, *but not sufficient*, condition for a minimum is the vanishing of the first derivative. Thus we have

$$\left[\frac{dI}{d\epsilon}\right]_{\epsilon=0} = 0 \tag{2.1}$$

as a necessary condition for the integral to take on a minimum value at $\epsilon = 0$. We now differentiate $I(\epsilon)$ with respect to ϵ (remember, x is not a function of ϵ, only $\bar{y}$ and $\bar{y}'$ are):

$$\frac{dI}{d\epsilon} = \int_{x_A}^{x_B} \left[\frac{\partial f}{\partial \bar{y}} \frac{d\bar{y}}{d\epsilon} + \frac{\partial f}{\partial \bar{y}'} \frac{d\bar{y}'}{d\epsilon}\right] dx$$

which we may write according to condition (c) as

$$\frac{dI}{d\epsilon} = \int_{x_A}^{x_B} \left[\frac{\partial f}{\partial \bar{y}} \frac{d\bar{y}}{d\epsilon} + \frac{\partial f}{\partial \bar{y}'} \frac{d}{dx}\left(\frac{d\bar{y}}{d\epsilon}\right)\right] dx\,.$$

Now, if we integrate the second term by parts, we will obtain

$$\frac{dI}{d\epsilon} = \int_{x_A}^{x_B} \frac{\partial f}{\partial \bar{y}} \frac{d\bar{y}}{d\epsilon}\, dx + \left[\frac{d\bar{y}}{d\epsilon} \frac{\partial f}{\partial \bar{y}'}\right]_{x_A}^{x_B} - \int_{x_A}^{x_B} \frac{d\bar{y}}{d\epsilon} \frac{d}{dx}\left(\frac{\partial f}{\partial \bar{y}'}\right) dx\,.$$

But by condition (a), $\bar{y}(x_A, \epsilon) = y_A$ and $\bar{y}(x_B, \epsilon) = y_B$ for *all* ϵ. Hence

$$\left.\frac{d\bar{y}}{d\epsilon}\right|_{x=x_A} = 0 = \left.\frac{d\bar{y}}{d\epsilon}\right|_{x=x_B}$$

so we get, finally,

$$\frac{dI}{d\epsilon} = \int_{x_A}^{x_B} \left[\frac{\partial f}{\partial \bar{y}} - \frac{d}{dx}\left(\frac{\partial f}{\partial \bar{y}'}\right)\right] \frac{d\bar{y}}{d\epsilon}\, dx\,.$$

We now require that $I(\epsilon)$ have a minimum at $\epsilon = 0$, that is,

$$\left[\frac{dI}{d\epsilon}\right]_{\epsilon=0} = 0 = \int_{x_A}^{x_B} \left[\frac{\partial f}{\partial \bar{y}} - \frac{d}{dx}\left(\frac{\partial f}{\partial \bar{y}'}\right)\right]_{\epsilon=0} \left[\frac{d\bar{y}}{d\epsilon}\right]_{\epsilon=0} dx \,.$$

Now setting $\epsilon = 0$ is equivalent to setting $\bar{y}(x, \epsilon) = y(x)$, $\bar{y}'(x, \epsilon) = y'(x)$, and $\bar{y}''(x, \epsilon) = y''(x)$. [Note that the integrand does depend on $\bar{y}''$, and in taking the limit $\epsilon = 0$, we need to know that the second derivative $y''(x, \epsilon)$ is a continuous function of its two variables. This is guaranteed by condition (c).] Calling

$$\left[\frac{d\bar{y}}{d\epsilon}\right]_{\epsilon=0} = \eta(x) \,,$$

we obtain

$$\int_{x_A}^{x_B} \left[\frac{\partial f}{\partial y} - \frac{d}{dx}\left(\frac{\partial f}{\partial y'}\right)\right] \eta(x)\, dx = 0 \,. \tag{2.2}$$

But $\eta(x)$, apart from the fact that it vanishes at x_A and x_B by condition (a) and is continuous and differentiable by condition (c), is completely arbitrary. Thus for the integral in Eq. (2.2) to vanish, we must have

$$\frac{\partial f}{\partial y} - \frac{d}{dx}\left(\frac{\partial f}{\partial y'}\right) = 0 \,. \tag{2.3}$$

This equation was derived by Euler in 1744. It is known as the Euler-Lagrange equation because it is also the basis for Lagrange's formulation of classical mechanics. If we expand the total derivative with respect to x, Eq. (2.3) becomes

$$\frac{\partial f}{\partial y} - \frac{\partial^2 f}{\partial x\, \partial y'} - \frac{\partial^2 f}{\partial y\, \partial y'} y' - \frac{\partial^2 f}{\partial^2 y'} y'' = 0 \,,$$

a second-order differential equation, whose solution supplies the twice-differentiable minimizing function $y(x)$, provided the minimum exists. The condition of Eq. (2.1) is only a *necessary* condition for a minimum; the solution $y(x)$ could also produce a maximum or even a point of inflection of the function $I(\epsilon)$ at $\epsilon = 0$. Thus what we have really found is an *extremizing* function $y(x)$. From the nature of the problem, we can usually tell the nature of the extremum without investigating the sign of higher derivatives of $I(\epsilon)$, which is in general a very complicated procedure. For example, in the three problems discussed in Section 2.1, it is clear on physical grounds that maxima do not exist. Also, there is the question whether we have found a relative or absolute minimum. In general, we shall not investigate these problems in detail mathematically, but shall rely on a physical interpretation of the situation to tell us the type of extremum involved.

In many cases of physical interest, the Euler-Lagrange equation can be greatly simplified. In particular, note that in all the examples considered in

Section 2.1, the function $f(x, y, y')$ does not depend *explicitly* on x, only implicitly through the x-dependence of y and y'. With this in mind, consider the quantity

$$\frac{d}{dx}\left[y'\frac{\partial f}{\partial y'} - f\right].$$

Performing the differentiations, we get

$$\begin{aligned}\frac{d}{dx}\left[y'\frac{\partial f}{\partial y'} - f\right] &= y''\frac{\partial f}{\partial y'} + y'\frac{d}{dx}\frac{\partial f}{\partial y'} - \frac{\partial f}{\partial x} - \frac{\partial f}{\partial y}y' - \frac{\partial f}{\partial y'}y'' \\ &= -y'\left[\frac{\partial f}{\partial y} - \frac{d}{dx}\left(\frac{\partial f}{\partial y'}\right)\right] - \frac{\partial f}{\partial x}.\end{aligned}$$

But if y satisfies the Euler-Lagrange equation, that is, Eq. 2.3, and if f does not depend explicitly on x so that

$$\frac{\partial f}{\partial x} = 0,$$

then we conclude that

$$\frac{d}{dx}\left[y'\frac{\partial f}{\partial y'} - f\right] = 0.$$

We can integrate this equation to get

$$y'\frac{\partial f}{\partial y'} - f = \text{const.} \tag{2.4}$$

Thus we are left with a simple *first-order* differential equation for $y(x)$.

The techniques used in deriving the one-variable Euler-Lagrange equation can readily be extended to integrals of the form

$$I = \int_{x_A}^{x_B} f(x, y_1, y_2, \cdots, y_n; y_1', y_2', \cdots, y_n')\, dx.$$

We want to determine the functions $y_1(x), y_2(x), \cdots, y_n(x)$ which extremize I, subject to the conditions

$$\begin{aligned} y_1(x_A) &= y_{1A}, & y_1(x_B) &= y_{1B} \\ y_2(x_A) &= y_{2A}, & y_2(x_B) &= y_{2B} \\ &\vdots & &\vdots \\ y_n(x_A) &= y_{nA}, & y_n(x_B) &= y_{nB}. \end{aligned}$$

This problem is virtually identical to the one we have just solved. In fact, if we imagine that $(n - 1)$ of the functions $\bar{y}_j$ are given their extremizing form, $y_j(x)$, and if we then vary the one remaining function, say $\bar{y}_i$, in such a way as to extremize I, then the problem does indeed coincide with the extremal problem which we have just solved. The equation satisfied by $y_i(x)$, the ith extremizing function, is then according to Eq. 2.3, just

$$\frac{\partial f}{\partial y_i} - \frac{d}{dx}\left(\frac{\partial f}{\partial y_i'}\right) = 0\,, \qquad i = 1, 2, \cdots, n\,. \tag{2.5}$$

Again, just as in the case of one dependent variable, if f does not depend explicitly on x, then we get the first integral

$$\sum_{i=1}^{n} y_i' \frac{\partial f}{\partial y_i'} - f = \text{const} \tag{2.6}$$

which follows immediately by evaluating

$$\frac{d}{dx}\left[\sum_{i=1}^{n} y_i' \frac{\partial f}{\partial y_i'} - f\right]$$

and assuming that the $y_i(x)$ satisfy Eq. (2.5).

Note that as a problem-solving device Eq. (2.6), for the case of several dependent variables, is not nearly as useful as Eq. (2.4), for the case of one dependent variable. Equation (2.4) reduces the entire problem to solving a single first-order differential equation while Eq. (2.6) provides only an equation linking the first derivatives of *all* the y_i. In this case one still must come to grips with some second-order differential equations.

2.3 SOME FAMOUS SOLUTIONS

In this section we shall discuss the solutions to the three classic problems posed in Section 2.1 in the order in which they were presented there.

Problem 1. The shortest line. In this case, $f = \sqrt{1 + y'^2}$. Applying the Euler-Lagrange equation directly, we get, since $\partial f/\partial y = 0$,

$$\frac{d}{dx}\left(\frac{\partial f}{\partial y'}\right) = 0 \qquad \text{or} \qquad \frac{\partial f}{\partial y'} = \frac{y'}{\sqrt{1 + y'^2}} = \text{const.}$$

But this just tells us that y' equals some constant, say $y' = A$, so that

$$y = Ax + B\,,$$

where A and B are constants which can be chosen so that the curve, a straight line, passes through the fixed endpoints.

In this case f does not depend explicitly on x so we can equally well use Eq. (2.4),

$$y' \frac{\partial f}{\partial y'} - f = \text{const.}$$

Using the given form of f, we get immediately

$$\frac{1}{\sqrt{1 + y'^2}} = \text{const,}$$

from which we conclude that y' is a constant, so again we get $y = Ax + B$.

Problem 2. The brachistochrone. For this problem, we saw in Section 2.1 that

$$f = \sqrt{\frac{1 + y'^2}{y}}$$

(we have dropped an irrelevant factor of $1/\sqrt{2g}$). Since f does not depend explicitly on x, we use Eq. (2.4) to get

$$\frac{y'^2}{\sqrt{y(1 + y'^2)}} - \sqrt{\frac{1 + y'^2}{y}} = C \qquad \text{or} \qquad \frac{1}{y(1 + y'^2)} = C^2 .$$

Letting $1/C^2 = 2a$, we get

$$y' = \sqrt{\frac{2a - y}{y}} ,$$

and integration yields

$$x - x_0 = \int \sqrt{\frac{y}{2a - y}}\, dy .$$

If we make the change of variable $y = a(1 - \cos\theta)$, then the integral becomes

$$x - x_0 = 2a \int \sin^2 \frac{\theta}{2}\, d\theta = a(\theta - \sin\theta) .$$

Thus the solution to the brachistochrone problem is, in parametric form,

$$x = a(\theta - \sin\theta) + x_0 \qquad y = a(1 - \cos\theta) .$$

These are the equations of a cycloid generated by the motion of a fixed point on the circumference of a circle of radius a which rolls on the positive side of the line $y = 0$, that is, on the underside of the x-axis of Fig. 2.1. There exists one and only one cycloid through the origin and the point (x_B, y_B); a suitable choice of a and x_0 will give this cycloid.

When Bernoulli first solved the brachistochrone problem, he did not do it in this way. Although he boasted of having discovered a wonderful solution, he did not publish it immediately, but instead challenged other mathematicians —especially his elder brother, Jacob, with whom he had been carrying on a bitter feud, and had publicly characterized as an incompetent—to find the answer. He finally published his result in 1697.* Perhaps it was Bernoulli who first said, "It is not enough for you to succeed; your colleagues must fail."

The cycloid had been discovered only a short time before Bernoulli's work in connection with mechanical problems involving the ideal pendulum. (An ideal pendulum is one whose period is exactly independent of the amplitude of oscillation. For a pendulum whose bob describes a circular path—the only one that is easy to design—the period is strictly independent of amplitude only in the limit of zero amplitude.) Huygens had discovered that a mass point

---

* See Courant and Robbins, *What is Mathematics?*, pp. 379–384.

oscillating without friction under the influence of gravity on a vertical cycloid has a period independent of amplitude; thus the cycloid had been called the tautochrone. With Bernoulli's discovery, this curve was rechristened the brachistochrone.

Problem 3. The minimum surface. In this case, we found in Section 2.1 that $f = y\sqrt{1 + y'^2}$. Again f does not depend explicitly on x, so

$$y' \frac{\partial f}{\partial y'} - f = -b,$$

where we write the constant as $-b$ for convenience. Using our particular f, we get immediately

$$\frac{y}{\sqrt{1 + y'^2}} = b$$

from which we obtain

$$x - x_0 = \int \frac{dy}{\sqrt{\frac{y^2}{b^2} - 1}} = b \cosh^{-1}\left(\frac{y}{b}\right).$$

Thus

$$y = b \cosh \frac{x - x_0}{b}.$$

Since the curve is supposed to lie entirely above the x-axis, we see that we must have $b > 0$.

The curve is a catenary (from the Latin, *catena* = chain); the surface it generates is a catenoid of revolution. The constants b and x_0 must be picked so that the catenary passes through the points (x_A, y_A) and $x(_B, y_B)$ without dropping below the x-axis. This cannot always be done! The theory developed here is deficient, primarily in that it will only determine for us the *twice-differentiable* minimizing curves. Let us examine the problem of connecting the points (x_A, y_A) and (x_B, y_B) in more detail. If the catenary is to pass through (x_A, y_A), then b and x_0 are related by

$$y_A = b \cosh \frac{x_A - x_0}{b}.$$

This tells us that

$$x_0 = x_A - b \cosh^{-1} \frac{y_A}{b},$$

so the equation of the catenary becomes

$$y = b \cosh\left[\frac{x - x_A}{b} + \cosh^{-1}\left(\frac{y_A}{b}\right)\right].$$

This equation describes a one-parameter family of catenaries which pass through the point (x_A, y_A), as shown in Fig. 2.4. In the last equation $\cosh^{-1} y_A/b$ must be chosen negative if the catenaries are to take on minima to the right of x_A.

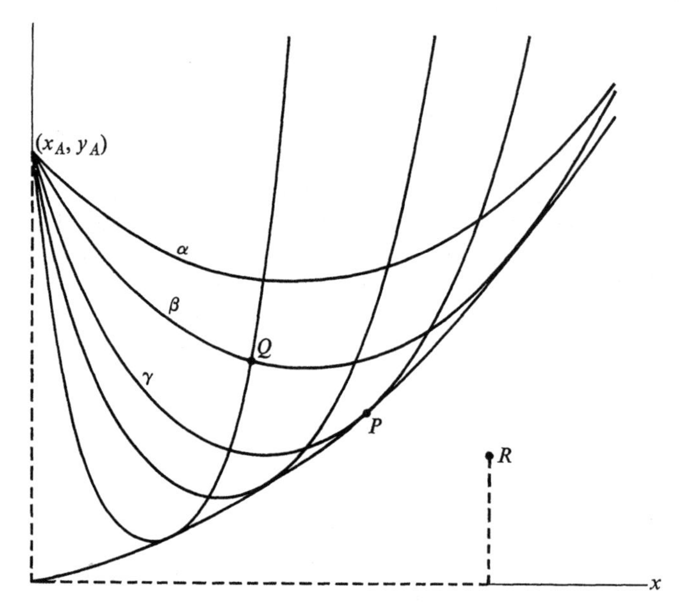

Fig. 2.4

One can show that every catenary $(\alpha, \beta, \gamma, \cdots)$ through (x_A, y_A) which does not intersect the x-axis is tangent to the envelope curve in Fig. 2.4. The envelope curve is a parabola with its focus at (x_A, y_A). No such catenary can ever pass through a point such as R, which is separated from (x_A, y_A) by the envelope. Also, only one catenary passes through a point like P lying on the envelope, but two catenaries pass through a point like Q lying on the same side of the envelope as (x_A, y_A). Thus we see that for a point like R, there is no twice-differentiable miminizing curve. It can be shown that the minimum area is generated in this case by the broken line whose segments are

$$\begin{aligned} x &= x_A \quad &&\text{for} \quad 0 \leq y \leq y_A \,, \\ y &= 0 \quad &&\text{for} \quad x_A \leq x \leq x_R \,, \\ x &= x_R \quad &&\text{for} \quad 0 \leq y \leq y_R \,. \end{aligned}$$

This solution is called the Goldschmidt discontinuous solution. It generates a surface of area equal to $\pi(y_A^2 + y_R^2)$, a spool-shaped surface. This surface is always a relative minimum, but it turns out that there exists a range of points (x_B, y_B) for which the catenary provides the absolute minimum. We enter the range of such points if we go sufficiently far above (or to the left of) the envelope. Then the upper catenary provides an absolute minimum surface of revolution. The upper catenary through a point such as Q always provides at least a relative minimum. The lower catenary through such a point as Q does not even generate a relative minimum, nor does the (unique) catenary through a point such as P.

We have analyzed in detail this defective "solution" to point out the limitations and the subtleties of the theory if studied rigorously.

2.4 ISOPERIMETRIC PROBLEMS—CONSTRAINTS

It was once the practice to reward exceptional military or civil service in the following way. The king would give to the one being honored all the land that he could encompass by a ploughed furrow in a specified period of time. The problem thus arose of finding the plane curve of given length which encloses the greatest possible area—the isoperimetric problem ("same perimeter").

In mathematical terms, we are trying to derive the differential equation that must be satisfied by the function $y(x)$ which makes the integral

$$I = \int_{x_A}^{x_B} f(x, y, y')\, dx$$

an extremum and also satisfies a second condition, such as giving the integral

$$J = \int_{x_A}^{x_B} g(x, y, y')\, dx$$

a given value with $y(x_A) = y_A$ and $y(x_B) = y_B$ both satisfied. Of course, one could easily imagine a more general class of problems. For example, we could allow for the possibility of more than one dependent variable, say $y_1(x)$, $y_2(x)$, $\cdots$, $y_n(x)$, and we might envision several integrals which have assigned values, $J_1, J_2, \cdots, J_m$. However, for the purposes of a proof, it suffices to illustrate the method on the simple case stated above. We shall then write the more general result, the proof of which we will leave to the reader as an exercise in developing a notation suitable to the increased generality.

The procedure is essentially the same as in Section 2.2. We consider a *two-parameter* family of functions $\bar{y}(x, \epsilon_1, \epsilon_2)$ with the following conditions:

(α) $\bar{y}(x_A, \epsilon_1, \epsilon_2) = y_A$, $\bar{y}(x_B, \epsilon_1, \epsilon_2) = y_B$ for all ϵ_1 and ϵ_2;

(β) $\bar{y}(x, 0, 0) = y(x)$, the desired extremizing function;

(γ) $\bar{y}(x, \epsilon_1, \epsilon_2)$ has continuous derivatives through second order in all variables.

We form

$$I(\epsilon_1, \epsilon_2) = \int_{x_A}^{x_B} f(x, \bar{y}, \bar{y}')\, dx\,,$$

$$J(\epsilon_1, \epsilon_2) = \int_{x_A}^{x_B} g(x, \bar{y}, \bar{y}')\, dx = \text{const.}$$

Just as in Section 2.2, we want to extremize $I(\epsilon_1, \epsilon_2)$ with respect to ϵ_1 and ϵ_2, requiring the extremum to be at $\epsilon_1 = 0$, $\epsilon_2 = 0$, but now subject to the condition that $J(\epsilon_1, \epsilon_2)$ have a fixed value. Clearly, ϵ_1 and ϵ_2 are not independent if

$J(\epsilon_1, \epsilon_2)$ has a fixed value, and it is for this reason that we elected to work with a two-parameter family of functions, $\bar{y}(\epsilon_1, \epsilon_2)$. If we had just $J(\epsilon)$, then the constant value of J would determine ϵ, which would, in turn, determine $I(\epsilon)$.

Now we use the techniques of elementary calculus to form the quantity

$$K(\epsilon_1, \epsilon_2) = I(\epsilon_1, \epsilon_2) + \lambda J(\epsilon_1, \epsilon_2),$$

where λ is a constant—as yet undetermined—referred to as a *Lagrange multiplier*. The condition for an extremum of $K(\epsilon_1, \epsilon_2)$ [and hence of $I(\epsilon_1, \epsilon_2)$] to occur at $\epsilon_1 = 0, \epsilon_2 = 0$ is

$$\left[\frac{\partial K}{\partial \epsilon_1}\right]_{\substack{\epsilon_1=0\\ \epsilon_2=0}} = 0 = \left[\frac{\partial K}{\partial \epsilon_2}\right]_{\substack{\epsilon_1=0\\ \epsilon_2=0}}$$

where K is given simply by

$$K(\epsilon_1, \epsilon_2) = \int_{x_A}^{x_B} h(x, \bar{y}, \bar{y}')\, dx\,,$$

and $h = f + \lambda g$. We now must calculate the two first partial derivatives of K:

$$\frac{\partial K}{\partial \epsilon_j} = \int_{x_A}^{x_B}\left[\frac{\partial h}{\partial \bar{y}}\frac{\partial \bar{y}}{\partial \epsilon_j} + \frac{\partial h}{\partial \bar{y}'}\frac{\partial \bar{y}'}{\partial \epsilon_j}\right] dx\,, \qquad j = 1, 2\,.$$

Integrating by parts exactly as in Section 2.2, we get

$$\frac{\partial K}{\partial \epsilon_j} = \int_{x_A}^{x_B}\left[\frac{\partial h}{\partial \bar{y}} - \frac{d}{dx}\left(\frac{\partial h}{\partial \bar{y}'}\right)\right]\frac{\partial \bar{y}}{\partial \epsilon_j}\, dx\,, \qquad j = 1, 2\,,$$

where we have used condition (α) to drop the boundary term. Now we set $\epsilon_1 = 0$, $\epsilon_2 = 0$ and require that the two first partial derivatives of K vanish. Thus

$$\int_{x_A}^{x_B}\left[\frac{\partial h}{\partial y} - \frac{d}{dx}\left(\frac{\partial h}{\partial y'}\right)\right]\eta_j(x)\, dx = 0\,, \qquad j = 1, 2\,,$$

where we have written

$$\left[\frac{\partial \bar{y}}{\partial \epsilon_j}\right]_{\substack{\epsilon_1=0\\ \epsilon_2=0}} = \eta_j(x)\,.$$

But the functions $\eta_1(x)$ and $\eta_2(x)$ are arbitrary, so we conclude that

$$\frac{\partial h}{\partial y} - \frac{d}{dx}\left(\frac{\partial h}{\partial y'}\right) = 0$$

is the equation for the extremizing function $y(x)$.

This looks just like the Euler-Lagrange equation of Section 2.2, except that the quantity $h = f + \lambda g$ replaces f. Note that the solution of the Euler-Lagrange equation involves two constants of integration *plus* the multiplier λ. These three numbers are just enough to ensure that $y(x)$ passes through the two endpoints, (x_A, y_A) and (x_B, y_B), *and* that the integral J has the prescribed value.

The generalizations of this result are not difficult. First, if we have m constraint integrals, $J_1, J_2, \cdots, J_m$, then the function h which obeys the Euler-Lagrange equation is just

$$h = f + \sum_{i=1}^{m} \lambda_i g_i \, ,$$

where

$$J_i = \int_{x_A}^{x_B} g_i(x, y, y')\, dx = \text{const.}$$

If we want to allow for the possibility of n dependent variables, a simple extension of notation shows that (as indicated in Section 2.2) we obtain the set of Euler-Lagrange equations

$$\frac{\partial h}{\partial y_i} - \frac{d}{dx}\left(\frac{\partial h}{\partial y_i'}\right) = 0 \, , \qquad i = 1, 2, \cdots, n \, , \tag{2.7}$$

where again

$$h = f + \sum_{i=1}^{m} \lambda_i g_i \, . \tag{2.8}$$

The reader can easily go through the few steps necessary to obtain Eqs. (2.7) and (2.8). Note that if there are m constraint integrals, it will be necessary to use an $(m + 1)$-parameter collection of functions, $\bar{y}(x, \epsilon_1, \epsilon_2, \cdots, \epsilon_{m+1})$, in the general case.

We now set about solving the original isoperimetric problem. To find the curve of fixed length L and maximum area, we want to find the functions $x(t)$ and $y(t)$ which maximize

$$A = \tfrac{1}{2} \int_{t_A}^{t_B} (xy' - x'y)\, dt \, .$$

This is the formula for the area inside a closed curve in the xy-plane, when we parametrize the coordinates x and y by the variable t. The reader should show that in polar coordinates the expression above is equivalent to

$$A = \tfrac{1}{2} \int_{0}^{2\pi} r^2(\theta)\, d\theta \, .$$

The length is given by

$$L = \int_{A}^{B} ds = \int_{t_A}^{t_B} \sqrt{x'^2 + y'^2}\, dt \, .$$

Thus our function $h(t, x, y, x', y')$ is

$$h = \tfrac{1}{2}(xy' - x'y) + \lambda\sqrt{x'^2 + y'^2} \, .$$

We have one independent variable, t, two dependent variables, $x(t)$ and $y(t)$, and one constraint integral (and hence one Lagrange multiplier). Using Eq. (2.7), we get

$$\frac{\partial h}{\partial x} - \frac{d}{dt}\left(\frac{\partial h}{\partial x'}\right) = 0 = \tfrac{1}{2}y' - \frac{d}{dt}\left(-\tfrac{1}{2}y + \lambda\frac{x'}{\sqrt{x'^2 + y'^2}}\right),$$

so that

$$y' - \lambda\frac{d}{dt}\left(\frac{x'}{\sqrt{x'^2 + y'^2}}\right) = 0 .$$

Similarly, the y equation yields

$$x' + \lambda\frac{d}{dt}\left(\frac{y'}{\sqrt{x'^2 + y'^2}}\right) = 0 .$$

Now these two equations are exact differentials so we may immediately integrate them to obtain

$$y - \lambda\frac{x'}{\sqrt{x'^2 + y'^2}} = y_0 , \qquad x + \lambda\frac{y'}{\sqrt{x'^2 + y'^2}} = x_0 ,$$

where x_0 and y_0 are constants to be determined by the fixed endpoint condition. We may rewrite these equations as

$$x - x_0 = -\lambda\frac{y'}{\sqrt{x'^2 + y'^2}}, \qquad y - y_0 = \lambda\frac{x'}{\sqrt{x'^2 + y'^2}} .$$

If we square both sides and add the two equations, we get immediately

$$(x - x_0)^2 + (y - y_0)^2 = \lambda^2 ,$$

so the extremal curve is a circle of radius λ about the point (x_0, y_0). Since the fixed length of the curve is L, we see that

$$\lambda = \frac{L}{2\pi} .$$

The points x_0 and y_0 are to be chosen so that the circle of radius $L/2\pi$ passes through the desired endpoint (which is the same as the starting point since the curve is closed). It is clear on physical grounds that the extremum here is a maximum, not a minimum.

The solution to this isoperimetric problem is particularly simple; it turns out to be just a circle. However, the reader should note that the equations obtained from the Euler-Lagrange formalism are very complicated, coupled, nonlinear differential equations. If one were to choose the functions $f(t, x, y, x', y')$ and $g(t, x, y, x', y')$ at random, the chances of being able to solve the resulting Euler-Lagrange equations in closed form would be small indeed.

Another famous isoperimetric problem is the determination of the shape of a hanging perfectly flexible rope of given length, fastened at its two endpoints. We assume the rope takes on a shape, consistent with the constraints, that minimizes its gravitational potential energy. Let the rope be held at its endpoints (x_A, y_A) and (x_B, y_B) as shown in Fig. 2.5. We want to determine the curve $y(x)$ passing through these fixed endpoints that minimizes the potential energy.

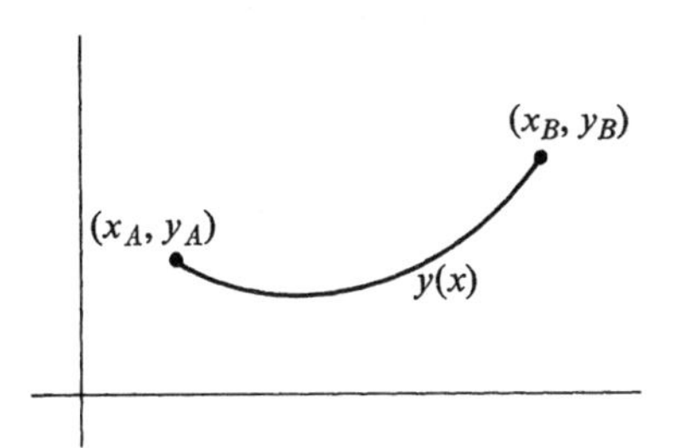

Fig. 2.5 A freely hanging rope held at its end points.

If ρ denotes the constant mass per unit length of the rope, then the potential energy relative to the x-axis is given by

$$I = \rho g \int_A^B y\, ds = \rho g \int_{x_A}^{x_B} y \sqrt{1 + y'^2}\, dx \,.$$

The constraint is that the length L of the rope is fixed:

$$L = \int_{x_A}^{x_B} ds = \int_{x_A}^{x_B} \sqrt{1 + y'^2}\, dx \,.$$

Thus

$$h = \rho g y \sqrt{1 + y'^2} - \rho g y_0 \sqrt{1 + y'^2} \,,$$

where we have written our Lagrange multiplier as $\lambda = -\rho g y_0$, y_0 to be determined. If we now change variables from y to $\eta = y - y_0$, we see that

$$h = \rho g \eta \sqrt{1 + \eta'^2} \,, \qquad \eta = \eta(x) \,.$$

But this has the same form as did the function f which we used to determine the minimum surface of revolution. Thus, just as in that case, the Euler-Lagrange equation will lead to

$$\eta = b \cosh\left(\frac{x - x_0}{b}\right),$$

where b and x_0 are constants to be determined. Since $\eta = y - y_0$, the equation of the curve having the shape of a freely hanging rope with fixed endpoints turns out to be

$$y = y_0 + b \cosh\left(\frac{x - x_0}{b}\right).$$

The three constants x_0, y_0, and b are chosen so that the curve passes through (x_A, y_A) and (x_B, y_B) and so that the length is L. This presents numerical problems, but is always possible in principle.

It may have occurred to the reader that there are other types of constraint equations which could arise in practice. We close this section by discussing a particularly important class of constraints which make their appearance primarily in classical mechanics. Suppose that we have an integral

$$I = \int_{x_A}^{x_B} f(x, y_1, y_2, \cdots, y_n; y_1', y_2', \cdots, y_n')\, dx$$

which we want to extremize subject to the conditions

$$\begin{aligned} g_1(x, y_1, y_2, \cdots, y_n) &= 0\,, \\ g_2(x, y_1, y_2, \cdots, y_n) &= 0\,, \\ &\vdots \\ g_m(x, y_1, y_2, \cdots, y_n) &= 0\,, \end{aligned}$$

where $m < n$. (Otherwise we would have constrained ourselves out of business.) Now, one possible way to proceed would be to solve these m equations for m of the y's in terms of the remaining $(n - m)y$'s and insert these values into the integrand of I. However, in many cases the functions g_i are sufficiently complicated to make this an undesirable approach. Also, for reasons of symmetry it is often not desirable to single out some variables and write all the other variables in terms of these special ones. Finally, even if it is a fairly simple matter to eliminate variables, we might not want to do so for physical reasons. We shall see in the next section that this is often the case in classical mechanics.

We now outline an alternative procedure, treating the case in which we have just one constraint equation,

$$g(x, y_1, y_2 \cdots, y_n) = 0\,,$$

and want to minimize

$$I = \int_{x_A}^{x_B} f(x, y_1, y_2, \cdots, y_n; y_1', y_2', \cdots, y_n')\, dx\,.$$

As usual, we introduce a collection of test functions $\bar{y}_i(x, \epsilon_1, \epsilon_2)$ satisfying

α') $\bar{y}_i(x_A, \epsilon_1, \epsilon_2) = y_{iA}$, $\bar{y}_i(x_B, \epsilon_1, \epsilon_2) = y_{iB}$ for all ϵ_1, ϵ_2, that is, fixed endpoints again;

β') $\bar{y}_i(x, 0, 0) = y_i(x)$, where $y_i(x)$ are the desired extremizing functions; and

γ') $\bar{y}_i(x, \epsilon_1, \epsilon_2)$ has continuous derivatives through second order.

We now form

$$I(\epsilon_1, \epsilon_2) = \int_{x_A}^{x_B} f(x, \bar{y}_1, \cdots, \bar{y}_n; \bar{y}_1', \cdots, \bar{y}_n')\, dx\,,$$

and as before we require that $I(\epsilon_1, \epsilon_2)$ have an extremal value at $\epsilon_1 = 0 = \epsilon_2$, subject to

$$g(x, \bar{y}_1, \cdots, \bar{y}_n) = 0\,.$$

An equation of this form yields a *local* constraint, holding at all points, whereas constraints of the integral type encountered thus far are *global* constraints, whose implications are weaker than those of the local type. [The local equation actually implies that one of the $\bar{y}_i$ can be written in terms of the other $(n-1)$ $\bar{y}_i$, whereas the global constraint only gives a relation between the ϵ variables.] It would, of course, be possible to convert this equation into a global constraint just by integrating it from x_A to x_B, but in doing so we would lose most of the equation's content. However, we can retain the equation's generality and at the same time cast it into the familiar integral form if we first multiply it by an *arbitrary* function of x, say $\phi(x)$, and *then* integrate from x_A to x_B. Doing this, we obtain

$$J(\epsilon_1, \epsilon_2) = \int_{x_A}^{x_B} \phi(x) g(x, \bar{y}_1, \cdots, \bar{y}_n)\, dx = 0 ,$$

where $\phi(x)$ is some completely arbitrary function.

Just as we did in the first part of this section, we set

$$K(\epsilon_1, \epsilon_2) = I(\epsilon_1, \epsilon_2) + \lambda J(\epsilon_1, \epsilon_2)$$

and obtain a necessary condition for an extremum by requiring that

$$\left[\frac{\partial K}{\partial \epsilon_1}\right]_{\substack{\epsilon_1=0\\ \epsilon_2=0}} = 0 = \left[\frac{\partial K}{\partial \epsilon_2}\right]_{\substack{\epsilon_1=0\\ \epsilon_2=0}} , \tag{2.9}$$

where K is given by

$$K(\epsilon_1, \epsilon_2) = \int_{x_A}^{x_B} h(x, \bar{y}_1, \cdots, \bar{y}_n; \bar{y}_1', \cdots, \bar{y}_n')\, dx ,$$

and

$$h = f + \lambda \phi(x) g .$$

For notational convenience, we define $\lambda(x) \equiv \lambda\phi(x)$; $\lambda(x)$ is a completely undetermined function at present. Using our (by now) standard method, we compute the two first partial derivatives of $K(\epsilon_1, \epsilon_2)$ and find

$$\frac{\partial K}{\partial \epsilon_j} = \sum_{i=1}^{n} \int_{x_A}^{x_B} \left[\frac{\partial h}{\partial \bar{y}_i}\frac{\partial \bar{y}_i}{\partial \epsilon_j} + \frac{\partial h}{\partial \bar{y}_i'}\frac{\partial \bar{y}_i'}{\partial \epsilon_j}\right] dx , \qquad j = 1, 2 .$$

Integrating the second term by parts, we find that

$$\frac{\partial K}{\partial \epsilon_j} = \sum_{i=1}^{n} \int_{x_A}^{x_B} \left[\frac{\partial h}{\partial \bar{y}_i} - \frac{d}{dx}\left(\frac{\partial h}{\partial \bar{y}_i'}\right)\right]\frac{\partial \bar{y}_i}{\partial \epsilon_j}\, dx , \qquad j = 1, 2 ,$$

where condition (α') has enabled us to drop the terms coming from the endpoints. Now we impose Eq. (2.9) as a necessary condition for an extremum. Setting

$$\left[\frac{\partial \bar{y}_i}{\partial \epsilon_j}\right]_{\substack{\epsilon_1=0\\ \epsilon_2=0}} = \eta_j^{(i)}(x) , \tag{2.10}$$

we get

$$\sum_{i=1}^{n} \int_{x_A}^{x_B} \left[\frac{\partial h}{\partial y_i} - \frac{d}{dx}\left(\frac{\partial h}{\partial y_i'}\right)\right]\eta_j^{(i)}(x)\, dx = 0 \tag{2.11}$$

for $j = 1, 2$. It is at this point that we must be careful. One is tempted to use the familiar argument to the effect that the functions $\eta_j^{(i)}(x)$ are completely arbitrary except for the fact that they are smooth and vanish at x_A and x_B. However, in this case, the argument is not correct! Let us take $g(x, y_1 \cdots, y_n) = 0$ and differentiate with respect to ϵ_1. We obtain

$$\frac{dg}{d\epsilon_1} = 0 = \frac{\partial g}{\partial \bar{y}_1}\frac{\partial \bar{y}_1}{\partial \epsilon_1} + \frac{\partial g}{\partial \bar{y}_2}\frac{\partial \bar{y}_2}{\partial \epsilon_1} + \cdots + \frac{\partial g}{\partial \bar{y}_n}\frac{\partial \bar{y}_n}{\partial \epsilon_1}\,.$$

If we now set $\epsilon_1 = 0 = \epsilon_2$ and use the definition of Eq. (2.10), we get

$$\sum_{i=1}^{n} \frac{\partial g}{\partial y_i}\eta_1^{(i)}(x) = 0\,.$$

Thus there is a linear relation between the functions $\eta_1^{(i)}(x)$ and similarly between the functions $\eta_2^{(i)}(x)$, so we certainly cannot claim that the $\eta_j^{(i)}(x)$ are completely arbitrary. However, we still have the function $\lambda(x)$ at our disposal. Let us focus our attention on Eq. (2.11) with $j = 1$; the case $j = 2$ is handled analogously. Since g does not depend on the derivatives of the $y_i(x)$, we may write Eq. (2.11) in more detail as

$$\sum_{i=1}^{n} \int_{x_A}^{x_B} \left[\frac{\partial f}{\partial y_i} - \frac{d}{dx}\left(\frac{\partial f}{\partial y_i'}\right) + \lambda(x)\frac{\partial g}{\partial y_i}\right]\eta_1^{(i)}(x)\,dx = 0\,. \qquad (2.12)$$

Now let us assume that $\eta_1^{(1)}(x)$ is given in terms of the remaining $\eta_1^{(i)}(x)$. Rather than substitute this value of $\eta_1^{(1)}(x)$ into Eq. (2.12), let us choose $\lambda(x)$ so that

$$\frac{\partial f}{\partial y_1} - \frac{d}{dx}\left(\frac{\partial f}{\partial y_1'}\right) + \lambda(x)\frac{\partial g}{\partial y_1} = 0\,. \qquad (2.13)$$

Thus we need not know the expression for $\eta_1^{(1)}(x)$, since its coefficient in our basic equation is zero. With $\lambda(x)$ determined in this way, Eq. (2.12) becomes

$$\sum_{i=2}^{n} \int_{x_A}^{x_B} \left[\frac{\partial f}{\partial y_i} - \frac{d}{dx}\left(\frac{\partial f}{\partial y_i'}\right) + \lambda(x)\frac{\partial g}{\partial y_i}\right]\eta_1^{(i)}(x)\,dx = 0\,,$$

where the term $i = 1$ no longer appears. Now the last $n - 1$ of the $\eta_1^{(i)}(x)$ are completely independent, so we may safely conclude that

$$\frac{\partial f}{\partial y_i} - \frac{d}{dx}\left(\frac{\partial f}{\partial y_i'}\right) + \lambda(x)\frac{\partial g}{\partial y_i} = 0\,, \qquad i = 2, 3, \cdots, n\,. \qquad (2.14)$$

These $n - 1$ equations are identical in form with Eq. (2.13), but we obtained them in a rather different manner. We can now combine Eqs. (2.13) and (2.14) into the single relation

$$\frac{\partial h}{\partial y_i} - \frac{d}{dx}\left(\frac{\partial h}{\partial y_i'}\right) = 0\,, \qquad i = 1, 2, \cdots, n\,, \qquad (2.15)$$

where $h = f + \lambda(x)g$. Note that we need not have singled out $\eta^{(1)}(x)$ in obtaining Eq. (2.15); any $\eta^{(i)}(x)$ would have led to the same result in which all the original dependent variables appear on an equal footing.

Thus we have n second-order differential equations for the n unknown functions $y_n(x)$ and the function $\lambda(x)$. [Note that no derivatives of $\lambda(x)$ appear. This is because we have assumed that g does *not* depend on the derivatives of the $y_i(x)$. Hence these equations are not differential equations as far as $\lambda(x)$ is concerned.] The additional equation necessary to make the problem well-posed is just

$$g(x, y_1, y_2, \cdots, y_n) = 0 ,$$

where g is a *given* function. The solutions $y_1(x)$, $y_2(x)$, $\cdots$, $y_n(x)$ are determined uniquely by the boundary conditions

$$y_i(x_A) = y_{iA} , \qquad y_i(x_B) = y_{iB} , \qquad i = 1, 2, \cdots, n .$$

Finally, the reader should check that if we have m constraint equations of the form

$$g_i(x, y_1, y_2, \cdots, y_n) = 0 , \qquad i = 1, 2, \cdots, m ,$$

then we again obtain the system of equations given by Eq. (2.15), but in this case the function h is given by

$$h = f + \sum_{i=1}^{m} \lambda_i(x) g_i ,$$

where the $\lambda_i(x)$ are to be determined as part of the solution to the system.

2.5 APPLICATION TO CLASSICAL MECHANICS

With the mathematical equipment of the last four sections, a fairly large part of classical mechanics can be conveniently expressed. If a classical-mechanical system, specified by generalized coordinates $q_1(t)$, $q_2(t)$, $\cdots$, $q_n(t)$, has a potential function $V(q_1, q_2, \cdots, q_n, t)$, then its Lagrangian can be constructed:

$$L(t, q_i, \dot{q}_i) = T(q_i, \dot{q}_i) - V(q_i, t) ,$$

where T is the kinetic energy, and we use a dot to denote differentiation with respect to time. It can be shown that the motion of the system now follows from Hamilton's principle.

Hamilton's principle. The motion of system whose Lagrangian is L, from time t_A to time $t_B (t_A < t_B)$, is such as to render the integral

$$I = \int_{t_A}^{t_B} L \, dt$$

an extremum with respect to the functions $q_i(t)$ $(i = 1, 2, \cdots, n)$ for which $q_i(t_A)$ and $q_i(t_B)$ are prescribed; t_A and t_B are arbitrary, but fixed, times.

It follows immediately from our work that I is extremized if and only if

$$\frac{\partial L}{\partial q_i} - \frac{d}{dt}\left(\frac{\partial L}{\partial \dot{q}_i}\right) = 0 \qquad i = 1, 2, \cdots, n . \tag{2.16}$$

These are Lagrange's equations of motion, a set of n simultaneous second-order differential equations whose solution gives the functions $q_i(t)$. The $2n$ constants of integration are evaluated by specifying the state of the system at some of time, t_A, that is by giving the values of $q_i(t_A)$ and $\dot{q}_i(t_A)$ for all i.

For example, suppose that

$$T = \tfrac{1}{2} m(\dot{x}_1^2 + \dot{x}_2^2 + \dot{x}_3^2) \,, \qquad V = V(x_1, x_2, x_3) \,,$$

that is, suppose that a particle of mass m is moving in an arbitrary potential field. Then

$$L = \tfrac{1}{2} m(\dot{x}_1^2 + \dot{x}_2^2 + \dot{x}_3^2) - V(x_1, x_2, x_3) \,,$$

and Lagrange's equations give us

$$-\frac{\partial V}{\partial x_i} - \frac{d}{dt}(m\dot{x}_i) = 0 \,, \qquad i = 1, 2, 3 \,,$$

or

$$m\ddot{x}_i = -\frac{\partial V}{\partial x_i} \,.$$

Since $-\partial V/\partial x_i$ is the force on the particle in the x_i-direction, this is simply Newton's second law which may be written in vector form as

$$m\ddot{\mathbf{r}} = -\nabla V = \mathbf{F} \,.$$

We should point out that it is possible to extend the Lagrangian formulation to more general potentials, such as the velocity-dependent potentials of electromagnetic theory, if the forces can be written as

$$F_i = -\frac{\partial V}{\partial q_i} + \frac{d}{dt}\left(\frac{\partial V}{\partial \dot{q}_i}\right).$$

In this case, we again find that

$$L = T - V \,,$$

and Hamilton's principle holds. Thus Eq. (2.16) will still govern the motion of such a system.

If L does not depend explicitly on t, we have immediately from Eq. (2.6)

$$\sum_{i=1}^{n} \frac{\partial L}{\partial \dot{q}_i} \dot{q}_i - L = E \,, \tag{2.17}$$

where E is a constant. The significance of E may be seen as follows. If V does not depend on the $\dot{q}_i$, then

$$\sum_{i=1}^{n} \frac{\partial L}{\partial \dot{q}_i} \dot{q}_i = \sum_{i=1}^{n} \frac{\partial T}{\partial \dot{q}_i} \dot{q}_i \,.$$

Under many circumstances T will take the form

$$T = \sum_{l,m} a_{lm}(q_i) \dot{q}_l \dot{q}_m$$

so that

$$\sum_{i=1}^{n} \frac{\partial T}{\partial \dot{q}_i} \dot{q}_i = 2T .$$

Thus Eq. (2.17) can be written as

$$2T - L = 2T - (T - V) = T + V = E ,$$

that is, we may identify E with the total ("conserved") energy.

The quantity

$$p_i \equiv \frac{\partial L}{\partial \dot{q}_i}$$

is called the *generalized momentum* corresponding to (or *conjugate* to) the generalized coordinate q_i. If some particular $q_k(t)$ does not appear in L, $q_k(t)$ is said to be an *ignorable* coordinate. For the generalized momentum p_k conjugate to an ignorable coordinate, we have

$$\frac{d}{dt}\left(\frac{\partial L}{\partial \dot{q}_k}\right) = \dot{p}_k = 0$$

so that p_k is a constant. This is just a statement of the conservation of momentum.

As an example, let us consider the motion of a particle in a plane, when the potential (and hence the force) depends only on the radial coordinate. Using polar coordinates in the plane, we have

$$L = \tfrac{1}{2} m(\dot{r}^2 + r^2\dot{\theta}^2) - V(r) ;$$

θ is an ignorable coordinate, and therefore

$$\frac{\partial L}{\partial \dot{\theta}} = mr^2\dot{\theta} = p_\theta , \tag{2.18}$$

where p_θ is a constant which is seen to be the angular momentum of the particle. In this case, since the Lagrangian does not depend explicitly on t, we also have from Eq. (2.17),

$$\tfrac{1}{2} m(\dot{r}^2 + r^2\dot{\theta}) + V(r) = E . \tag{2.19}$$

Combining Eqs. (2.18) and (2.19), we obtain

$$\dot{\theta} = \frac{p_\theta}{mr^2} , \qquad \tfrac{1}{2} m\left(\dot{r}^2 + \frac{p_\theta^2}{m^2r^2}\right) + V(r) = E ,$$

so in this case our constants of the motion, p_θ and E, reduce the problem to one involving only first derivatives, i.e., the problem is reduced to evaluating some integrals. Of course, for a complicated $V(r)$ these integrals might have to be done numerically.

Finally, let us consider a mechanical problem with a constraint. In particular we ask: What is the motion of a particle moving under gravity if it is constrained to move always on the surface of a fixed solid sphere of radius

a? It will clearly be to our advantage to use spherical coordinates. Therefore we write

$$T = \frac{1}{2} m(\dot{r}^2 + r^2\dot{\theta}^2 + r^2 \sin^2 \theta \dot{\phi}^2) ,$$
$$V = mgz = mgr \cos \theta ,$$

and

$$L = \frac{1}{2} m(\dot{r}^2 + r^2\dot{\theta}^2 + r^2 \sin^2 \theta \dot{\phi}^2) - mgr \cos \theta .$$

We want to extremize

$$I = \int_{t_A}^{t_B} L \, dt ,$$

subject to the constraint

$$g(t, r, \theta, \phi) = r - a = 0 .$$

According to our discussion in Section 2.4, we form the function

$$\begin{aligned} L' &= L + \lambda g \\ &= \frac{1}{2} m(\dot{r}^2 + r^2\dot{\theta}^2 + r^2 \sin^2 \theta \dot{\phi}^2) - mgr \cos \theta + \lambda r - \lambda a \end{aligned}$$

and use this function in the Euler-Lagrange equations. For the r-coordinate, we find that

$$\lambda - mg \cos \theta + mr \sin^2 \theta \dot{\phi}^2 + mr\dot{\theta}^2 - m\ddot{r} = 0 , \tag{2.20a}$$

and for the θ-coordinate,

$$mgr \sin \theta + mr^2 \sin \theta \cos \theta \dot{\phi}^2 - m\frac{d}{dt}(r^2\dot{\theta}) = 0 . \tag{2.20b}$$

Now ϕ is an ignorable coordinate, so we get immediately

$$mr^2 \sin^2 \theta \dot{\phi} = p_\phi$$

from which we obtain

$$\dot{\phi} = \frac{p_\phi}{mr^2 \sin^2 \theta} . \tag{2.20c}$$

Thus we have three coupled differential equations to solve, together with the relation

$$r - a = 0 . \tag{2.20d}$$

Putting Eqs. (2.20c) and (2.20d) into Eqs. (2.20a) and (2.20b), we get

$$\lambda = mg \cos \theta - ma\dot{\theta}^2 - \frac{p_\phi^2}{ma^3 \sin^2 \theta} \tag{2.21a}$$

and

$$\ddot{\theta} = \frac{g}{a} \sin \theta + \frac{p_\phi^2 \cos \theta}{m^2 a^4 \sin^3 \theta} . \tag{2.21b}$$

Eq. (2.20c) becomes simply

$$\dot{\phi} = \frac{p_\phi}{ma^2 \sin^2 \theta} . \tag{2.21c}$$

Thus we see that everything hinges on the solution to Eq. (2.21b); once $\theta(t)$ is determined, Eq. (2.21a) gives us $\lambda(t)$ directly, and upon integrating Eq. (2.21c) we will get $\phi(t)$. In fact, Eq. (2.21b) is not as complicated as it looks. If we multiply both sides of the equation by $\dot{\theta}(t)$ they both become exact differentials, and we can integrate once to obtain

$$\dot{\theta}^2 = C - 2\frac{g}{a} \cos\theta - \frac{p_\phi^2}{m^2 a^4 \sin^2 \theta} . \tag{2.22}$$

Such a relation is not surprising; the reader can show that this follows directly from the conservation of energy. The constant C is just equal to $2E/ma^2$, where E is the total conserved energy. Hence the problem is reduced to evaluating the integral implied by Eq. (2.22). This integral cannot, in general, be evaluated in closed form, although in certain limiting cases general features of the solution can be determined without much difficulty.

It is interesting to consider $\lambda(t)$ in some detail. The first term in Eq. (2.21a) for $\lambda(t)$ is the negative of the radial component of the gravitational force on the particle. The second term is just the negative of the centrifugal force due to the θ-motion (a radial force), and the last term is the radial component of the centrifugal force due to the ϕ-motion. Thus $\lambda(t)$ is precisely the force which the sphere must exert to keep the particle on the sphere. One can show that in general the Lagrange multipliers are related in a simple way to the forces required to maintain the constraints in question. Thus they provide us with direct physical information about the system under study.

2.6 EXTREMIZATION OF MULTIPLE INTEGRALS

As our final extension of the variational problem, we now want to consider the case when the integral to be extremized has more than one independent variable. We shall deal directly with the case of two independent variables and leave the more general case as an exercise.

Consider, then, the integral

$$I = \iint_D f\left(x_1, x_2, y, \frac{\partial y}{\partial x_1}, \frac{\partial y}{\partial x_2}\right) dx_1\, dx_2 ,$$

where x_1 and x_2 are the two independent variables and D is some fixed domain in the x_1x_2-plane. We shall denote the boundary of D by C. Just as before, we shall prescribe the values of $y(x_1, x_2)$ at all points of the boundary C, that is,

$$[y(x_1, x_2)]_{(x_1,x_2)\in C} = g(C) ,$$

where $g(C)$ is a given function, defined on the one-dimensional curve, C.

Therefore we want our one-parameter family of test functions to satisfy

a) $[\bar{y}(x_1, x_2, \epsilon)]_{(x_1, x_2)\epsilon C} = g(C)$ for all values of ϵ;
b) $\bar{y}(x_1, x_2, 0) = y(x_1, x_2)$, the desired extremizing function; and
c) $\bar{y}(x_1, x_2, \epsilon)$ has continuous first and second derivatives.

We now form

$$I(\epsilon) = \iint_D f\left(x_1, x_2, \bar{y}, \frac{\partial \bar{y}}{\partial x_1}, \frac{\partial \bar{y}}{\partial x_2}\right) dx_1\, dx_2$$

and require that

$$\left[\frac{dI}{d\epsilon}\right]_{\epsilon=0} = 0 .$$

Proceeding in the usual manner, we find that

$$\begin{aligned} \frac{dI}{d\epsilon} &= \iint_D \left[\frac{\partial f}{\partial \bar{y}}\frac{d\bar{y}}{d\epsilon} + \frac{\partial f}{\partial\left(\frac{\partial \bar{y}}{\partial x_1}\right)}\frac{d}{d\epsilon}\left(\frac{\partial \bar{y}}{\partial x_1}\right) + \frac{\partial f}{\partial\left(\frac{\partial \bar{y}}{\partial x_2}\right)}\frac{d}{d\epsilon}\left(\frac{\partial \bar{y}}{\partial x_2}\right)\right] dx_1\, dx_2 \\ &= \iint_D \left[\frac{\partial f}{\partial \bar{y}}\frac{d\bar{y}}{d\epsilon} + \frac{\partial f}{\partial\left(\frac{\partial \bar{y}}{\partial x_1}\right)}\frac{\partial}{\partial x_1}\left(\frac{d\bar{y}}{d\epsilon}\right) + \frac{\partial f}{\partial\left(\frac{\partial \bar{y}}{\partial x_2}\right)}\frac{\partial}{\partial x_2}\left(\frac{d\bar{y}}{d\epsilon}\right)\right] dx_1\, dx_2 . \end{aligned}$$

We now need to make use of the two-dimensional generalization of integration by parts—Green's Theorem—which reads

$$\iint_D \left[\frac{\partial P}{\partial x_1} + \frac{\partial Q}{\partial x_2}\right] dx_1\, dx_2 = \int_C (P\, dx_2 - Q\, dx_1) .$$

This is valid if $P(x_1, x_2)$ and $Q(x_1, x_2)$ and their first partial derivatives are continuous in D. If we write

$$P = \phi(x_1, x_2) A(x_1, x_2) \quad \text{and} \quad Q = \phi(x_1, x_2) B(x_1, x_2),$$

then we get

$$\iint_D \left[A\frac{\partial \phi}{\partial x_1} + B\frac{\partial \phi}{\partial x_2}\right] dx_1\, dx_2 = -\iint_D \left[\frac{\partial A}{\partial x_1} + \frac{\partial B}{\partial x_2}\right]\phi\, dx_1\, dx_2 + \int_C [A\, dx_2 - B\, dx_1]\,\phi .$$

By making the identification

$$\phi = \frac{d\bar{y}}{d\epsilon}, \qquad A = \frac{\partial f}{\partial\left(\frac{\partial \bar{y}}{\partial x_1}\right)}, \qquad B = \frac{\partial f}{\partial\left(\frac{\partial \bar{y}}{\partial x_2}\right)},$$

we obtain

$$\frac{dI}{d\epsilon} = \iint_D \left[\frac{\partial f}{\partial \bar{y}} - \frac{\partial}{\partial x_1} \frac{\partial f}{\partial\left(\dfrac{\partial \bar{y}}{\partial x_1}\right)} - \frac{\partial}{\partial x_2} \frac{\partial f}{\partial\left(\dfrac{\partial \bar{y}}{\partial x_2}\right)} \right] \frac{d\bar{y}}{d\epsilon}\, dx_1\, dx_2$$

$$+ \int_C \frac{d\bar{y}}{d\epsilon} \left[\frac{\partial f}{\partial\left(\dfrac{\partial \bar{y}}{\partial x_1}\right)} dx_2 - \frac{\partial f}{\partial\left(\dfrac{\partial \bar{y}}{\partial x_2}\right)} dx_1 \right].$$

But by condition (a), $d\bar{y}/d\epsilon$ vanishes on C, so we get, using our extremum requirement,

$$\iint_D \left[\frac{\partial f}{\partial \bar{y}} - \frac{\partial}{\partial x_1} \frac{\partial f}{\partial\left(\dfrac{\partial \bar{y}}{\partial x_1}\right)} - \frac{\partial}{\partial x_2} \frac{\partial f}{\partial\left(\dfrac{\partial \bar{y}}{\partial x_2}\right)} \right] \eta(x_1, x_2)\, dx_1\, dx_2 = 0\,,$$

where

$$\left[\frac{d\bar{y}}{d\epsilon}\right]_{\epsilon=0} = \eta(x_1, x_2)\,.$$

Since $\eta(x_1, x_2)$ is an *arbitrary* smooth function inside D, the equation

$$\frac{\partial f}{\partial y} - \frac{\partial}{\partial x_1} \frac{\partial f}{\partial\left(\dfrac{\partial y}{\partial x_1}\right)} - \frac{\partial}{\partial x_2} \frac{\partial f}{\partial\left(\dfrac{\partial y}{\partial x_2}\right)} = 0 \tag{2.23}$$

must hold everywhere in D. This is the Euler-Lagrange equation for two independent variables; the partial derivatives reflect the presence of more than one independent variable.

As an example of a variational problem with two independent variables, we consider the vibrating string. Let a perfectly flexible elastic string be stretched under constant tension τ along the x-axis with its endpoints fixed at $x = 0$ and $x = l$. We consider small, transverse, frictionless (undamped) vibrations—the simplest possible case—and we neglect gravity. The amplitude of the vibration, at time t and position x $(0 \leq x \leq l)$, is $y(x, t)$. The slope of the string is given by $\partial y/\partial x$ and assumed to be small:

$$\left|\frac{\partial y}{\partial x}\right| \ll 1\,.$$

This will be the case in most situations that arise in practice. The velocity is $\partial y/\partial t$. The fixed endpoint condition says that $y(0, t) = y(l, t) = 0$ for all t.

We now want to compute the Lagrangian for this mechanical system and then determine the motion by Hamilton's principle. First we find the potential energy V. The work done on the string to effect a given distorted configuration all goes into changing the length of the string because we assume perfect flexibility. Since the stretching force is equal to the (constant) tension τ, the potential energy is simply

$$V = \tau\left[\int_0^l \sqrt{1 + \left(\frac{\partial y}{\partial x}\right)^2}\, dx - l\right].$$

The term in brackets is the difference between the instantaneous length of the string and its equilibrium length l. The zero of V thus occurs when the length of the string is equal to l. Since we are assuming

$$\left|\frac{\partial y}{\partial x}\right| \ll 1,$$

we may write

$$\sqrt{1 + \left(\frac{\partial y}{\partial x}\right)^2} \cong 1 + \frac{1}{2}\left(\frac{\partial y}{\partial x}\right)^2.$$

To compute the kinetic energy, we assume a mass density along the string $\rho(x) > 0$. Thus the mass contained in an element of length dx is $\rho(x)\, dx$, with the associated kinetic energy

$$dT = \tfrac{1}{2}\rho(x)\, dx\left(\frac{\partial y}{\partial t}\right)^2.$$

The total kinetic energy is therefore

$$T = \tfrac{1}{2}\int_0^l \rho(x)\left(\frac{\partial y}{\partial t}\right)^2 dx.$$

With T and V determined, we now form the Lagrangian:

$$L = T - V = \int_0^l \left[\tfrac{1}{2}\rho(x)\left(\frac{\partial y}{\partial t}\right)^2 - \tfrac{1}{2}\tau\left(\frac{\partial y}{\partial x}\right)^2\right] dx.$$

The quantity in brackets,

$$\mathscr{L} = \tfrac{1}{2}\rho(x)\left(\frac{\partial y}{\partial t}\right)^2 - \tfrac{1}{2}\tau\left(\frac{\partial y}{\partial x}\right)^2,$$

is called the *Lagrangian density*, since its space integral yields the Lagrangian. Now, Hamilton's principle says that the function which describes the actual motion of the string is the one which renders

$$I = \int_{t_1}^{t_2} L\, dt = \int_{t_1}^{t_2}\int_0^l \mathscr{L}\, dx\, dt$$

an extremum with respect to those functions $y(x, t)$ which satisfy $y(0, t) = y(l, t) = 0$ for all t and for which $y(x, t_1)$ and $y(x, t_2)$ are given functions of x.

We now have available the mathematical tools necessary for determining the equation which $y(x, t)$ must satisfy. The domain D in the xt-plane is a rectangle defined by $0 \leq x \leq l$ and $t_1 \leq t \leq t_2$. Using Eq. (2.23), we have

$$\frac{\partial \mathscr{L}}{\partial y} - \frac{\partial}{\partial x}\frac{\partial \mathscr{L}}{\partial\left(\frac{\partial y}{\partial x}\right)} - \frac{\partial}{\partial t}\frac{\partial \mathscr{L}}{\partial\left(\frac{\partial y}{\partial t}\right)} = 0,$$

and inserting our expression for $\mathscr{L}$, we get

$$\frac{\partial^2 y}{\partial x^2} = \frac{\rho(x)}{\tau} \frac{\partial^2 y}{\partial t^2} .$$

This is the partial differential equation which describes the motion of the string —the wave equation.

We now generalize to the case of n independent variables. We want to extremize the multiple integral

$$I = \int_D \mathscr{L}\left(x_1, x_2, \cdots, x_n, \phi; \frac{\partial \phi}{\partial x_1}, \frac{\partial \phi}{\partial x_2}, \cdots, \frac{\partial \phi}{\partial x_n}\right) dx_1\, dx_2 \cdots dx_n ,$$

where $\phi = \phi(x_1, x_2 \cdots, x_n)$, D is some domain in the n-dimensional space, and ϕ is some prescribed function defined on the $(n-1)$-dimensional boundary of D. The Euler-Lagrange equation which gives the extremizing function ϕ is

$$\frac{\delta \mathscr{L}}{\delta \phi} \equiv \frac{\partial \mathscr{L}}{\partial \phi} - \sum_{i=1}^{n} \frac{\partial}{\partial x_i} \frac{\partial \mathscr{L}}{\partial\left(\frac{\partial \phi}{\partial x_i}\right)} = 0 , \tag{2.24}$$

where we have introduced $\delta \mathscr{L}/\delta\phi$, the so-called functional derivative. We have already seen that the Lagrangian density

$$\mathscr{L} = \frac{1}{2}\left[\rho(x)\left(\frac{\partial \phi}{\partial t}\right)^2 - \tau\left(\frac{\partial \phi}{\partial x}\right)^2\right]$$

leads to the one-dimensional wave equation.

As another example we consider the Lagrangian density

$$\begin{aligned}\mathscr{L} &= (\nabla\phi)^2 = \nabla\phi\cdot\nabla\phi \\ &= \left[\left(\frac{\partial \phi}{\partial x}\right)^2 + \left(\frac{\partial \phi}{\partial y}\right)^2 + \left(\frac{\partial \phi}{\partial z}\right)^2\right] .\end{aligned}$$

Note that from the basic properties of the gradient, $\mathscr{L}$ is invariant under rotations. The Euler-Lagrange equation for the case of three independent variables and one dependent variable is

$$\frac{\partial \mathscr{L}}{\partial \phi} - \sum_{i=1}^{3} \frac{\partial}{\partial x_i} \frac{\partial \mathscr{L}}{\partial\left(\frac{\partial \phi}{\partial x_i}\right)} = 0 ,$$

where $x_1 = x$, $x_2 = y$, $x_3 = z$. Substituting $\mathscr{L} = (\nabla\phi)^2$, we get

$$\frac{\partial^2 \phi}{\partial x^2} + \frac{\partial^2 \phi}{\partial y^2} + \frac{\partial^2 \phi}{\partial z^2} = \nabla^2\phi = 0 .$$

Thus Laplace's equation is the necessary condition for extremizing

$$I = \int_D (\nabla\phi)^2\, dx\, dy\, dz .$$

If $\phi(x, y, z)$ is fixed on the boundary of D. We see that a rotationally invariant $\mathscr{L}$ has led to a rotationally invariant equation for ϕ.

These results may be further generalized to the case where there are several, say m, dependent variables, $\phi_j(x_1, x_2, \cdots, x_n)$, involved in $\mathscr{L}$. For example, if we extremize

$$I = \int_D \mathscr{L}\left(x_i, \phi_j, \frac{\partial \phi_j}{\partial x_i}\right) dx_1\, dx_2 \cdots dx_n ,$$

we obtain the set of m coupled second-order partial differential equations

$$\frac{\partial \mathscr{L}}{\partial \phi_j} - \sum_{i=1}^{n} \frac{\partial}{\partial x_i} \frac{\partial \mathscr{L}}{\partial\left(\dfrac{\partial \phi_j}{\partial x_i}\right)} = 0 , \qquad j = 1, 2, \cdots, m . \tag{2.25}$$

These versions of the Euler-Lagrange equation are used in field theory where the partial differential equations governing certain physical situations are derived by constructing a "reasonable" Lagrangian density $\mathscr{L}$ and then postulating a generalized Hamilton's principle which leads to the vanishing of the functional derivative, in other words, to the Euler-Lagrange equation. Reasonable Lagrangians may be partially determined from invariance considerations and symmetry properties: e.g., the Lagrangian density may be required to be Lorentz invariant, or invariant under certain gauge transformations.

As yet nothing has been said in this section about constraints that lead to isoperimetric problems. The generalization of this procedure to isoperimetric problems goes through exactly as before. We mention the result here for the general case of m dependent variables, n independent variables, and p integral constraints of the form

$$J_k = \int_D g_k\left(x_i, \phi_j, \frac{\partial \phi_j}{\partial x_i}\right) dx_1\, dx_2 \cdots dx_n ,$$

where $i = 1, \cdots, n$; $j = 1, \cdots, m$; $k = 1, \cdots, p$. Here the J_k are p fixed constants. Subject to these p constraints we want to extremize the multiple integral

$$I = \int_D \mathscr{L}\left(x_i, \phi_j, \frac{\partial \phi_j}{\partial x_i}\right) dx_1\, dx_2 \cdots dx_n .$$

In strict analogy with the isoperimetric problem discussed previously, we do this by forming the new function

$$h = \mathscr{L} + \sum_{k=1}^{p} \lambda_k g_k , \tag{2.26}$$

where the λ_k are p constant Lagrange multipliers. Now h satisfies the appropriate system of Euler-Lagrange equations, namely,

$$\frac{\partial h}{\partial \phi_i} - \sum_{i=1}^{n} \frac{\partial}{\partial x_i} \frac{\partial h}{\partial\left(\dfrac{\partial \phi_j}{\partial x_i}\right)} = 0, \qquad j = 1, \cdots, m . \tag{2.27}$$

As an illustration of this general formalism, let us consider the Lagrange density

$$\mathscr{L} = \frac{\hbar^2}{2m}\left[\frac{\partial\psi^*}{\partial x}\frac{\partial\psi}{\partial x} + \frac{\partial\psi^*}{\partial y}\frac{\partial\psi}{\partial y} + \frac{\partial\psi^*}{\partial z}\frac{\partial\psi}{\partial z}\right] + V(x, y, z)\psi^*\psi ,$$

where $\psi^*(x, y, z)$ and $\psi(x, y, z)$ are complex-valued functions and V is real-valued. (An asterisk denotes complex conjugation.) We want to extremize

$$I = \int_D \mathscr{L}\, dx\, dy\, dz ,$$

subject to the constraint

$$J = \int_D \psi^*\psi\, dx\, dy\, dz = 1 .$$

This is a variational problem with three independent variables (x, y, z), two dependent variables (the real and imaginary parts of ψ, ψ_R and ψ_I respectively), and one constraint. According to Eq. (2.26), we now form

$$\begin{aligned} h &= \mathscr{L} + \lambda\psi^*\psi \\ &= \frac{\hbar^2}{2m}\left[\left(\frac{\partial\psi_R}{\partial x}\right)^2 + \left(\frac{\partial\psi_R}{\partial y}\right)^2 + \left(\frac{\partial\psi_R}{\partial z}\right)^2 + \left(\frac{\partial\psi_I}{\partial x}\right)^2 + \left(\frac{\partial\psi_I}{\partial y}\right)^2 + \left(\frac{\partial\psi_I}{\partial z}\right)^2\right] \\ &\quad + [V(x, y, z) + \lambda][\psi_R^2 + \psi_I^2] . \end{aligned}$$

Substituting into Eq. (2.27) and making the identification $\phi_1 = \psi_R$, $\phi_2 = \psi_I$, we get for $j = 1$,

$$\lambda\psi_R + V\psi_R - \frac{\hbar^2}{2m}\sum_{i=1}^{3}\frac{\partial}{\partial x_i}\left(\frac{\partial\psi_R}{\partial x_i}\right) = 0$$

and for $j = 2$,

$$\lambda\psi_I + V\psi_I - \frac{\hbar^2}{2m}\sum_{i=1}^{3}\frac{\partial}{\partial x_i}\left(\frac{\partial\psi_I}{\partial x_i}\right) = 0 ,$$

where as usual we call $x = x_1$, $y = x_2$, $z = x_3$. Calling $\lambda = -E$, we may write these as a single equation,

$$-\frac{\hbar^2}{2m}\nabla^2\psi + V\psi = E\psi .$$

This equation is the time-independent Schrödinger equation for the motion of a nonrelativistic particle in a potential $V(x, y, z)$. In this case, the Lagrange multiplier plays the role of the energy eigenvalue in the Schrödinger equation. We shall see this in a more general setting in Chapter 4, where we shall also consider the interesting question of whether or not minima occur at the extreme values.

*2.7 INVARIANCE PRINCIPLES AND NOETHER'S THEOREM

Having introduced the basic concepts of the calculus of variations, we now look back to the starting point of our investigations and ask why it was, that in considering the trial function $\bar{y}(x, \epsilon)$ in previous sections, we always demanded that $\bar{y}(x, \epsilon)$ obey fixed endpoint conditions. Clearly, there is a large class of problems which falls into this category, but there are many others which lie outside it; for example, the problem of finding the path connecting a fixed point P to a vertical line L such that a particle moving under gravity will go from P to L in minimum time. Here one endpoint is fixed and one is free to lie anywhere on L.

From the mathematical point of view, the reason for starting with fixed endpoint restrictions is to look at the simplest things first. From the point of view of the physicist, the reason is clearer, namely, Hamilton's principle refers to variations of the fixed endpoint variety. Since this principle has played a crucial role in the development of physics, it is not surprising that the physicist's attention is first focussed on this special variational problem.

However, now that we have familiarized ourselves with the simplest problems, let us look at the problem of varying the integral

$$I = \int_{x_A}^{x_B} f(x, y, y')\, dx \tag{2.28}$$

from a more general point of view. For simplicity, we will consider the case of infinitesimal transformations, parametrized by a small number ϵ. If we were working in the context of previous sections, we would consider a transformation

$$y(x) \to \bar{y}(x) \equiv y(x) + \epsilon\eta(x) ,$$

which is just a special case of the transformation $y(x) \to \bar{y}(x, \epsilon)$ considered in Section 2.2, *if* we require that $\eta(x)$ vanish at x_A and x_B so that $\bar{y}(x_A) = y(x_A)$ and $\bar{y}(x_B) = y(x_B)$ for *all* ϵ. The simplest generalization of this type of variation consists of dropping the restrictions on $\eta(x)$. A further generalization would be to allow η to have a more complicated functional form. For example, η might depend explicitly on the functions y and y'; then we would write

$$y(x) \to \bar{y}(x, y, y') \equiv y(x) + \epsilon\eta(x, y, y') . \tag{2.29}$$

The primary reason for considering such elaborate transformations, particularly those for which the endpoints are not fixed, is that there are many transformations—often very simple ones—in the physical world which are of this type. For example, we might want to consider a translated trajectory of a classical particle $[x(t) \to x(t) + \epsilon]$ or perhaps a rotated trajectory. We will see that by allowing for the possibility of variations such as these we will be able to obtain very general information about quantities which are conserved in physical problems.

Pursuing the analogy suggested by classical physics (via Hamilton's principle) we can also imagine the possibility of translating a particle trajectory in time, for example, by the transformation $t \to t + \epsilon$. In order to allow for variations of this type, we consider another extension of the variational approach, which is to *shift* (not necessarily in a uniform manner) the curve along which we integrate to some other region of space. For example, we could transform each point x into a point $\bar{x}$ via the transformation

$$x \to \bar{x} = x + \epsilon\xi(x, y, y') , \tag{2.30}$$

that is, this transformation depends not only on the point x in question, but also on the function y occurring in the integrand of I. With the transformations of Eqs. (2.29) and (2.30), we can form a quantity $\bar{I}(\epsilon)$ analogous to the I of Eq. (2.28):

$$\bar{I}(\epsilon) = \int_{\bar{x}_A}^{\bar{x}_B} f[\bar{x}, \bar{y}(\bar{x}), \bar{y}'(\bar{x})]\, d\bar{x} . \tag{2.31}$$

Note that $\bar{I}(\epsilon = 0) = I$. In Eq. (2.31), $\bar{x}_A$ and $\bar{x}_B$ are obtained by applying the transformation of Eq. (2.30) to x_A and x_B, respectively. In Eq. (2.31), the significant quantity is $\bar{y}(\bar{x})$. This symbol does *not* mean that we just replace x by $\bar{x}$ in Eq. (2.29). Rather, it means that we consider the *transformed curve* $\bar{y}$, obtained by writing x in terms of $\bar{x}$ [using Eq. (2.30)] and substituting that value of x into Eq. (2.29). Since we will be interested only in evaluating $\bar{I}(\epsilon)$ *through first order in* ϵ, we can use x or $\bar{x}$ interchangeably in the functions ξ and η, because in zeroth order, x and $\bar{x}$ are equal. In Eq. (2.31), the prime on $\bar{y}'$ denotes differentiation with respect to $\bar{x}$, while in Eq. (2.28) the prime on y' denotes differentiation with respect to x, that is, a prime will always denote differentiation with respect to the *independent* variable, whatever it may be.

Since ϵ is small, let us try to compute $\bar{I}(\epsilon)$ in power series form. For reasons that will soon become clear, we will concentrate on the first two terms in the series. Using Eqs. (2.29) and (2.30), we have

$$\bar{y}(\bar{x}) = y[(\bar{x} - \epsilon\xi(\bar{x}, y, y')] + \epsilon\eta(\bar{x}, y, y') ,$$

where y and y' are functions of $\bar{x}$. Expanding through terms in ϵ, we find that

$$\bar{y}(\bar{x}) = y(\bar{x}) - \epsilon\frac{\partial y}{\partial \bar{x}}\xi(\bar{x}, y, y') + \epsilon\eta(\bar{x}, y, y')$$

or

$$\bar{y}(\bar{x}) = y(\bar{x}) + \epsilon\rho(\bar{x}, y, y') , \tag{2.32}$$

where we have made the definition

$$\rho(\bar{x}, y, y') \equiv \eta(\bar{x}, y, y') - \frac{\partial y}{\partial \bar{x}}\xi(\bar{x}, y, y') \tag{2.33}$$

in order to simplify subsequent calculations. Using Eqs. (2.31) and (2.32), we can write $\bar{I}(\epsilon)$ as

$$\bar{I}(\epsilon) = \int_{\bar{x}_A}^{\bar{x}_B} f[\bar{x}, y(\bar{x}) + \epsilon\rho(\bar{x}, y, y'), y'(\bar{x}) + \epsilon\rho'(\bar{x}, y, y')]\, d\bar{x} . \tag{2.34}$$

Now, according to Eq. (2.30),

$$\bar{x}_A = x_A + \epsilon\xi(x, y, y')\,|_{x=x_A} \equiv x_A + \delta_A\,, \tag{2.35a}$$

$$\bar{x}_B = x_B + \epsilon\xi(x, y, y')\,|_{x=x_B} \equiv x_B + \delta_B\,, \tag{2.35b}$$

so let us break up the integral in Eq. (2.34) as follows:

$$\int_{\bar{x}_A}^{\bar{x}_B} = \int_{x_A+\delta_A}^{x_B+\delta_B} = \int_{x_A+\delta_A}^{x_B} + \int_{x_B}^{x_B+\delta_B} = \int_{x_A}^{x_B} - \int_{x_A}^{x_A+\delta_A} + \int_{x_B}^{x_B+\delta_B}$$

Thus

$$\begin{aligned}\bar{I}(\epsilon) &= \int_{x_A}^{x_B} f[\bar{x}, y(\bar{x}) + \epsilon\rho(\bar{x}, y, y'), y'(\bar{x}) + \epsilon\rho'(\bar{x}, y, y')]\, d\bar{x} \\ &\quad + \int_{x_B}^{x_B+\delta_B} f[\bar{x}, y(\bar{x}), y'(\bar{x})]\, d\bar{x} - \int_{x_A}^{x_A+\delta_A} f[\bar{x}, y(\bar{x}), y'(\bar{x})]\, d\bar{x} + \mathcal{O}(\epsilon^2)\end{aligned}$$

where $\mathcal{O}(\epsilon^2)$ refers collectively to all terms of order ϵ^2 and higher. In this expression, we have dropped terms in ϵ in the integrands of the last two integrals since these integrals are already proportional to ϵ (because the length of the integration range is proportional to ϵ). Thus terms in ϵ in the integrand will contribute only terms of order ϵ^2. We now expand the first integral through order ϵ and evaluate the last two integrals. We find that

$$\begin{aligned}\bar{I}(\epsilon) &= \int_{x_A}^{x_B} f[\bar{x}, y(\bar{x}), y'(\bar{x})]\, d\bar{x} \\ &\quad + \epsilon \int_{x_A}^{x_B} \left[\rho(\bar{x}, y, y') \frac{\partial}{\partial y} f(\bar{x}, y, y') + \rho'(\bar{x}, y, y') \frac{\partial}{\partial y'} f(\bar{x}, y, y') \right] d\bar{x} \\ &\quad + \delta_B[f(\bar{x}, y, y')]_{\bar{x}=x_B} - \delta_B[f(\bar{x}, y, y')]_{\bar{x}=x_A} + \mathcal{O}(\epsilon^2)\,.\end{aligned}$$

Integrating the last term in the second integral by parts and using Eqs. (2.33), (2.35a), and (2.35b), we get

$$\begin{aligned}\bar{I}(\epsilon) &= \int_{x_A}^{x_B} f(\bar{x}, y, y')\, d\bar{x} \\ &\quad + \epsilon \int_{x_A}^{x_B} \left[\frac{\partial f}{\partial y} - \frac{d}{d\bar{x}}\frac{\partial f}{\partial y'}\right]\left[\eta(\bar{x}, y, y') - \frac{dy}{d\bar{x}}\xi(\bar{x}, y, y') \right] d\bar{x} \\ &\quad + \epsilon \left\{\frac{\partial f}{\partial y'}\eta(\bar{x}, y, y') + \left[f(\bar{x}, y, y') - y'\frac{\partial f}{\partial y'}\right]\xi(\bar{x}, y, y')\right\}_{x_A}^{x_B} + \mathcal{O}(\epsilon^2)\,.\end{aligned}$$

But the first integral is just I (remember that the variable of integration can be called either x or $\bar{x}$ since it is a dummy variable), so we may write

$$\begin{aligned}\bar{I}(\epsilon) - I &= \epsilon \int_{x_A}^{x_B} \left[\frac{\partial f}{\partial y} - \frac{d}{dx}\frac{\partial f}{\partial y'}\right](\eta - y'\xi)\, dx \\ &\quad + \epsilon\left[\frac{\partial f}{\partial y'}\eta + \left(f - y'\frac{\partial f}{\partial y'}\right)\xi\right]_{x_A}^{x_B} + \mathcal{O}(\epsilon^2)\,,\end{aligned} \tag{2.36}$$

where we have written the dummy independent variable as x throughout.

Note that if in Eq. (2.36) we set $\xi \equiv 0$ (that is, $x = \bar{x}$) and require η to vanish at both x_A and x_B, then we find

$$\bar{I}(\epsilon) - \bar{I}(\epsilon = 0) = \epsilon \int_{x_A}^{x_B} \left[\frac{\partial f}{\partial y} - \frac{d}{dx} \frac{\partial f}{\partial y'} \right] \eta \, dx + \mathcal{O}(\epsilon^2) , \tag{2.37}$$

since $\bar{I}(\epsilon = 0) = I$. If we require that

$$\frac{\bar{I}(\epsilon) - \bar{I}(\epsilon = 0)}{\epsilon} \to 0$$

as $\epsilon \to 0$, then the integral on the right-hand side of Eq. (2.37) must vanish. From this we conclude that

$$\frac{\partial f}{\partial y} - \frac{d}{dx}\left(\frac{\partial f}{\partial y'}\right) = 0 .$$

This is just the basic Euler-Lagrange equation [Eq. (2.3)] derived in Section 2.2. Thus we see that if we require that the rate of change of $\bar{I}$ with respect to ϵ vanish for variations of the type $\xi \equiv 0$, $\eta\,|_{x_A} = 0 = \eta\,|_{x_B}$, then we obtain the same differential equation for y as we did in Section 2.2. The reader should be able to see the reason for this result.

The generalization of Eq. (2.36) to the case of several dependent variables, $y_1, y_2, \cdots, y_n$, is straightforward, just as it was in Section 2.2. The reader can show that in this case, if

$$I = \int_{x_A}^{x_B} f(x, y_1, y_2, \cdots, y_n; y'_1, y'_2, \cdots, y'_n) \, dx , \tag{2.38}$$

we have

$$\begin{aligned} \bar{I}(\epsilon) - \bar{I}(\epsilon = 0) &= \epsilon \int_{x_A}^{x_B} \sum_{i=1}^{n} \left[\frac{\partial f}{\partial y_i} - \frac{d}{dx}\frac{\partial f}{\partial y'_i} \right] (\eta_i - y'_i \xi) \, dx \\ &\quad + \epsilon \left[\sum_{i=1}^{n} \frac{\partial f}{\partial y_i} \eta_i + \left(f - \sum_{i=1}^{n} y'_i \frac{\partial f}{\partial y'_i} \right) \xi \right]_{x_A}^{x_B} + \mathcal{O}(\epsilon^2) , \end{aligned} \tag{2.39}$$

where we have written

$$\bar{y}_i(x) = y_i(x) + \epsilon \eta_i(x, y_1, \cdots, y_n; y'_1, \cdots, y'_n) ,$$

$$\bar{x} = x + \epsilon \xi(x, y_1, \cdots, y_n; y'_1, \cdots, y'_n) .$$

Again, if we assume $\xi \equiv 0$, $\eta_i\,|_{x_A} = 0 = \eta_i\,|_{x_B}$ for all i, then we obtain the Euler-Lagrange equations for n dependent variables [Eq. (2.5)],

$$\frac{\partial f}{\partial y_i} - \frac{d}{dx}\left(\frac{\partial f}{\partial y'_i}\right) = 0 \qquad i = 1, 2, \cdots, n .$$

A particularly interesting situation arises when the functions $y_i(x)$ which we are varying are solutions to the Euler-Lagrange equations. Then, according to Eq. (2.5), the integral in Eq. (2.39) vanishes, and we have simply

$$\bar{I}(\epsilon) - \bar{I}(\epsilon = 0) = \epsilon\left[\sum_{i=1}^{n} \frac{\partial f}{\partial y_i'}\eta_i + \left(f - \sum_{i=1}^{n} y_i' \frac{\partial f}{\partial y_i'}\right)\xi\right]_{x_A}^{x_B}. \tag{2.40}$$

Now we have already discovered that at least in one special case the Euler-Lagrange equations can be manipulated in such a way as to produce a constant of the motion or first integral. This was the case when $f(x, y, y')$ did not depend *explicitly* on x, but only implicitly, through y and y'. In that case we found by direct calculation that

$$y' \frac{\partial f}{\partial y'} - f = \text{const}, \tag{2.41a}$$

and, generally, in the case of several dependent variables,

$$\sum_{i=1}^{n} y_i' \frac{\partial f}{\partial y_i'} - f = \text{const}. \tag{2.41b}$$

If f is equal to the Lagrangian function L and if the independent variable is the time t, then when we denote differentiation with respect to time by a dot, this last equation becomes

$$\sum_{i=1}^{n} \dot{q}_i \frac{\partial L}{\partial \dot{q}_i} - L = \text{const}, \tag{2.42}$$

where the q_i are the generalized coordinates. As was pointed out in Section 2.5, in most cases of physical interest the constant in Eq. (2.42) is the conserved total energy of the system.

The essential element in the derivation of Eq. (2.41a) [and also Eq. (2.41b)] was the fact that f did not depend explicitly on x. Another way of obtaining Eq. (2.41a) is to show that the integral I is *invariant* under the transformation

$$\bar{x} = x + \epsilon\xi, \qquad \bar{y} = y(x) + \epsilon\eta \tag{2.43}$$

with

$$\xi = 1, \qquad \eta = 0. \tag{2.44}$$

By invariance, we mean simply that

$$\bar{I}(\epsilon) = \bar{I}(\epsilon = 0) = I. \tag{2.45}$$

Eq. (2.45) can be derived as follows. If $f(x, y, y') = F(y, y')$, then

$$I = \int_{x_A}^{x_B} F(y, y')\, dx$$

and

$$\bar{I}(\epsilon) = \int_{\bar{x}_A}^{\bar{x}_B} F[\bar{y}(\bar{x}), \bar{y}'(\bar{x})]\, d\bar{x}.$$

Following the prescription for $\bar{x}$ and $\bar{y}$, we have

$$\bar{x}_A = x_A + \epsilon\,, \qquad \bar{x}_B = x_B + \epsilon\,,$$

and

$$\bar{y}(\bar{x}) = y(\bar{x} - \epsilon)\,.$$

Thus

$$\bar{I}(\epsilon) = \int_{x_A+\epsilon}^{x_B+\epsilon} F\left[y(\bar{x} - \epsilon), \frac{d}{d\bar{x}}y(\bar{x} - \epsilon)\right] d\bar{x}\,.$$

But

$$\frac{d}{d\bar{x}}y(\bar{x} - \epsilon) = \frac{dy(\bar{x} - \epsilon)}{d(\bar{x} - \epsilon)}\,,$$

so that

$$\bar{I}(\epsilon) = \int_{x_A+\epsilon}^{x_B+\epsilon} F\left[y(\bar{x} - \epsilon), \frac{dy(\bar{x} - \epsilon)}{d(\bar{x} - \epsilon)}\right] d\bar{x}\,.$$

If we make the change of variable $x = \bar{x} - \epsilon$, then

$$\bar{I}(\epsilon) = \int_{x_A}^{x_B} F\left[y(x), \frac{dy(x)}{dx}\right] dx = \bar{I}(\epsilon = 0) = I\,.$$

Thus I is *invariant* under the set of transformations defined by Eqs. (2.43) and (2.44). But this means that the difference $\bar{I}(\epsilon) - I$ must vanish, so according to Eq. (2.40), we obtain

$$\epsilon\left[\frac{\partial f}{\partial y'}\eta + \left(f - y'\frac{\partial f}{\partial y'}\right)\xi\right]_{x_A}^{x_B} = 0\,.$$

Using Eq. (2.44), we can reduce this to

$$\left(f - y'\frac{\partial f}{\partial y'}\right)_{x=x_A} = \left(f - y'\frac{\partial f}{\partial y'}\right)_{x=x_B}\,.$$

Since x_A and x_B are arbitrary, we must have

$$f - y'\frac{\partial f}{\partial y'} = \text{const},$$

which is exactly the same first integral derived in Section 2.2 by a very different method. Thus we see that the existence of a first integral (or constant of the motion if we have in mind a physical problem) is related to the invariance of I under a transformation of the type discussed in this section. This important result is a special case of a powerful theorem due to Emmy Noether.

Theorem 2.1. If

$$I = \int_{x_A}^{x_B} f(x, y_1, \cdots, y_n;\, y_1', \cdots, y_n')\, dx$$

is invariant under the infinitesimal transformation generated by

$$\bar{x} = x + \epsilon\xi(x, y_1, \cdots, y_n; y_1', \cdots, y_n') ,$$

$$\bar{y}_i(x) = y_i(x) + \epsilon\eta_i(x, y_1, \cdots, y_n; y_1', \cdots, y_n') ,$$

in the sense that

$$I(\epsilon) = \int_{x_A}^{x_B} f[\bar{x}, \bar{y}_1(\bar{x}), \cdots, \bar{y}_n(\bar{x}); \bar{y}_1'(\bar{x}), \cdots, \bar{y}_n'(\bar{x})]\, d\bar{x} = I ,$$

then there exists a first integral of the related Euler-Lagrange equations which is given by

$$\sum_{i=1}^{n} \frac{\partial f}{\partial y_i'}\eta_i + \left(f - \sum_{i=1}^{n} y_i'\frac{\partial f}{\partial y_i'}\right)\xi = \text{const.} \tag{2.46}$$

Proof. Since $y_1, \cdots, y_n$ satisfy the Euler-Lagrange equations, by Eq. (2.40) we have

$$\bar{I}(\epsilon) - I = \epsilon\left[\sum_{i=1}^{n} \frac{\partial f}{\partial y_i'}\eta_i + \left(f - \sum_{i=1}^{n} y_i'\frac{\partial f}{\partial y_i'}\right)\xi\right]_{x_A}^{x_B} + \mathcal{O}(\epsilon^2) ;$$

since I is assumed to be invariant, we have $\bar{I}(\epsilon) - I = 0$, so that

$$\left[\sum_{i=1}^{n} \frac{\partial f}{\partial y_i'}\eta_i + \left(f - \sum_{i=1}^{n} y_i'\frac{\partial f}{\partial y_i'}\right)\xi\right]_{x=x_A} = \left[\sum_{i=1}^{n} \frac{\partial f}{\partial y_i'}\eta_i + \left(f - \sum_{i=1}^{n} y_i'\frac{\partial f}{\partial y_i'}\right)\xi\right]_{x=x_B} .$$

Since x_A and x_B are arbitrary, we conclude that

$$\sum_{i=1}^{n} \frac{\partial f}{\partial y_i'}\eta_i + \left(f - \sum_{i=1}^{n} y_i'\frac{\partial f}{\partial y_i'}\right)\xi = \text{const},$$

which is the required result.

The first integral obtained in the case that f does not depend explicitly on x is an example of this result. Note that the use of the term first integral is justified since Eq. (2.46) contains only *first* derivatives of the y_i, whereas the Euler-Lagrange equations are *second-order* differential equations. We have already mentioned that if f is the Lagrangian of a dynamical system,

$$f = L(t, q_1, \cdots, q_n; \dot{q}_1, \cdots, \dot{q}_n) \equiv T - V ,$$

then this first integral corresponds to the conservation of energy in the case that the Lagrangian does not depend *explicitly* on time.

Another interesting example which is furnished by a classical mechanical Lagrangian is the case where V, the potential energy, depends only on the vectors joining pairs of particles. For example, in a two-particle system, this would mean that

$$L = \frac{m_1}{2}(\dot{x}_1^2 + \dot{y}_1^2 + \dot{z}_1^2) + \frac{m_2}{2}(\dot{x}_2^2 + \dot{y}_2^2 + \dot{z}_2^2) - V(\mathbf{r}_1 - \mathbf{r}_2) .$$

Such a Lagrangian is invariant under the transformation

$$\bar{t} = t + \epsilon\tau\,, \qquad \bar{x}_1 = x_1 + \epsilon\xi_1\,, \qquad \bar{x}_2 = x_2 + \epsilon\xi_2\,, \qquad \bar{y}_1 = y_1 + \epsilon\eta_1\,,$$
$$\bar{y}_2 = y_2 + \epsilon\eta_2\,, \qquad \bar{z}_1 = z_1 + \epsilon\zeta_1\,, \qquad \bar{z}_2 = z_2 + \epsilon\zeta_2\,,$$

where

$$\tau = 0\,, \qquad \xi_1 = 1\,, \qquad \xi_2 = 1\,,$$
$$\eta_1 = \eta_2 = \zeta_1 = \zeta_2 = 0\,.$$

Using Eq. (2.46), we have immediately

$$\sum_{i=1}^{2} \frac{\partial L}{\partial \dot{x}_i}\xi_i = \text{const}, \qquad \text{or} \qquad \frac{\partial L}{\partial \dot{x}_1} + \frac{\partial L}{\partial \dot{x}_2} = m_1\dot{x}_1 + m_2\dot{x}_2 = \text{const.}$$

This simply states that the x-component of the total linear momentum of the system is a constant of the motion. Similarly, invariance under translation of the y-coordinates leads to the same result for the y-component of the total linear momentum, and invariance under translation of the z-coordinates leads to conservation of the z-component of the total linear momentum. Thus we see that conservation of *linear* momentum is an immediate consequence of the *translational* invariance of the system, defined in terms of the integral I, appearing in Hamilton's principle.

As a final example, let us consider the Lagrangian of a particle moving in a central potential $V(r)$. We have for L,

$$L = \frac{m}{2}(\dot{x}^2 + \dot{y}^2 + \dot{z}^2) - V(r)\,, \qquad r = (x^2 + y^2 + z^2)^{1/2}\,. \tag{2.47}$$

Consider now an arbitrary rotation about the z-axis. This corresponds to the transformation

$$\bar{t} = t\,, \qquad \bar{x} = x\cos\epsilon + y\sin\epsilon\,, \qquad \bar{y} = -x\sin\epsilon + y\cos\epsilon\,, \qquad \bar{z} = z\,, \tag{2.48}$$

as may be seen by looking back at Eq. (1.15). The corresponding values of τ, ξ, η and ζ are

$$\tau = 0\,, \qquad \xi = y\,, \qquad \eta = -x\,, \qquad \zeta = 0\,,$$

since the infinitesimal transformation corresponding to $\bar{x}$ and $\bar{y}$ is [letting ϵ become small in Eq. (2.48)]

$$\bar{x} = x + \epsilon y\,, \qquad \bar{y} = y - \epsilon x.$$

Substituting Eqs. (2.48) into the Lagrangian of Eq. (2.47), we see that it is invariant under rotations about the z-axis. Equation (2.46) then gives

$$\frac{\partial L}{\partial \dot{x}}\xi + \frac{\partial L}{\partial \dot{y}}\eta = \text{const}\,, \qquad \text{or} \qquad x\frac{\partial L}{\partial \dot{y}} - y\frac{\partial L}{\partial \dot{x}} = \text{const}\,.$$

But

$$\frac{\partial L}{\partial \dot{x}} = p_x , \qquad \frac{\partial L}{\partial \dot{y}} = p_y ,$$

so we get

$$xp_y - yp_x = (\mathbf{r} \times \mathbf{p})_z = L_z = \text{const},$$

which is a statement of the conservation of angular momentum about the z-axis. In an identical manner, it follows from the invariance of the system under rotations about the x and y axis that L_x and L_y are also conserved in the motion of the system; i.e., the *total* angular momentum of the system is conserved.

Thus, from the examples considered, we see that one of the great advantages of being able to formulate classical mechanics in terms of Hamilton's principle is that from this viewpoint we are easily able to relate conservation laws to the existence of underlying symmetries under which the quantity

$$I = \int_{t_A}^{t_B} L\, dt$$

is invariant.

PROBLEMS

1. *The shortest line (polar coordinates).* Find the shortest distance between two points using polar coordinates, i. e., using them as a line element:
$$ds^2 = dr^2 + r^2 d\theta^2 .$$
2. *The shortest line (three dimensions).* Prove that the shortest distance between two points in three-dimensional space is a straight line. Take the origin and (a, b, c) as the two points and actually determine the equation of the line.
3. *Geodesics on a sphere.* On a sphere of fixed radius a, the line element is given by
$$ds^2 = a^2 d\theta^2 + a^2 \sin^2 \theta d\varphi^2 .$$
Determine the equation of the curve that extremizes the distance between two points. Is the extremum a minimum, a maximum, or what? [*Hint*: Let θ be the independent variable so the integrand of the integral to be extremized is independent of the *dependent* variable φ.] Now prove: $\varphi = \alpha - \sin^{-1} (k \cot \theta)$, where α and k are constants. Then change from the spherical coordinates to Cartesian coordinates to see that this is the equation of a plane through the origin; its intersection with the sphere $r = a$ is the geodesic—a great circle—on the surface of the sphere.
4. *Inverse isoperimetric problem* ("isoareametric" problem). Prove that of all simple closed curves enclosing a given area, the least perimeter is possessed by the circle.
5. Demonstrate the following reciprocity relationship for the simple isoperimetric problem: The particular function which renders
$$I = \int_{x_1}^{x_2} f(x, y, y')\, dx$$

an extremum with respect to functions which give

$$J = \int_{x_1}^{x_2} g(x, y, y')\, dx$$

a prescribed value, also renders J an extremum with respect to functions which give I a prescribed value if the Lagrange multiplier $\lambda \neq 0$.

6. In all our discussions so far on finding the function f for which $I = \int_{x_1}^{x_2} f\, dx$ is an extremum, it has been assumed that f depends only on x, y, and y':

$$f = f(x, y, y')\ .$$

Show that if f also involves the second derivative of y with respect to x, $f = f(x, y, y', y'')$, then for fixed endpoints and prescribed y' at the endpoints, the Euler-Lagrange equation is

$$\frac{\partial f}{\partial y} - \frac{d}{dx}\left(\frac{\partial f}{\partial y'}\right) + \frac{d^2}{dx^2}\left(\frac{\partial f}{\partial y''}\right) = 0\,.$$

What continuity assumptions must be made about derivatives of y?

7. Fermat's principle states: *If the velocity of light is given by the continuous function $u = u(y)$, the actual light path connecting the points (x_1, y_1) and (x_2, y_2) in a plane is one which extremizes the time integral*

$$I = \int_{(x_1,\, y_1)}^{(x_2,\, y_2)} \frac{ds}{u}\,.$$

(Actually, refinements are needed to make this formulation of Fermat's principle hold for all cases.)

a) Derive Snell's law from Fermat's principle; that is, prove that $\sin\phi/u = \text{const}$, where ϕ is the angle shown in Fig. 2.6.

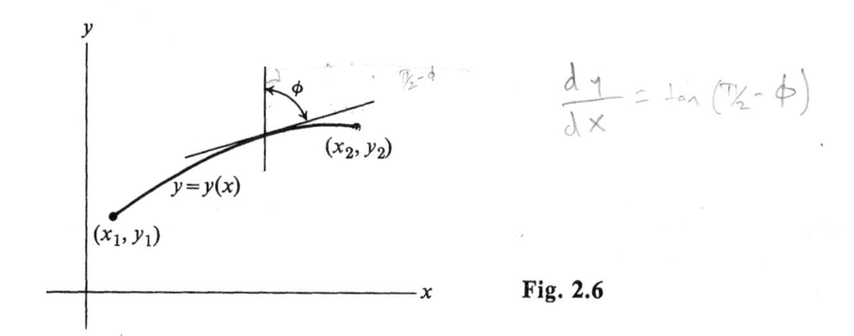

Fig. 2.6

b) Suppose that light travels in the xy-plane in such a way that its speed is proportional to y; then prove that the light rays emitted from any point are circles with their centers on the x-axis.

8. *Brachistochrone problem for a central gravitational force.* A particle of mass m is to move under the gravitational attraction of another stationary spherical mass M from a point at a finite distance $r = R_0$ from the center of M to a point on the surface of M. We want to determine the path such that the particle moving along

it will travel between the two fixed endpoints in a time less than the time of transit along any other path. Taking for the two endpoints ($\theta = 0$, $r = R_0$) and ($\theta = \alpha$,

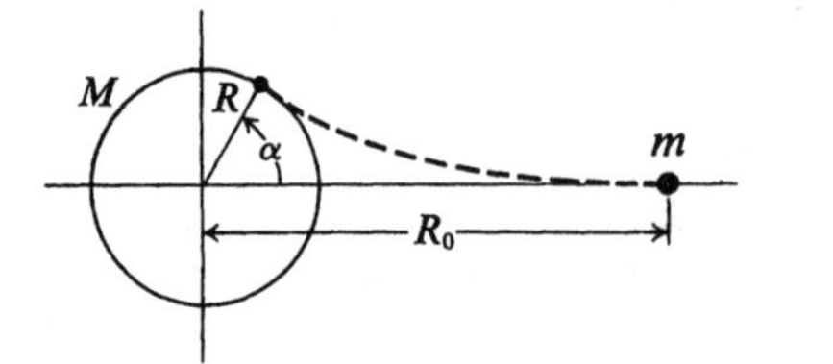

Fig. 2.7

$r = R$) as shown in Fig. 2.7, set up in polar coordinates the equation that determines the path and solve for θ as a function of r. The best one can do analytically is to express θ as an elliptic integral with a variable upper limit and lower limit chosen to satisfy one of the boundary conditions. The determination of this path tells us how best to design interplanetary mail chutes.

9. Show that if ψ and ψ^* are taken as two independent field variables (dependent variables) the real Lagrangian density

$$\mathscr{L} = \frac{\hbar^2}{2m}\nabla\psi\cdot\nabla\psi^* + V\psi\psi^* - \frac{i\hbar}{2}\left(\psi^*\dot{\psi} - \psi\dot{\psi}^*\right),$$

where $\dot{\psi} = \partial\psi/\partial t$ leads to the time-dependent Schrödinger equation

$$H\psi = \left(\frac{-\hbar^2}{2m}\nabla^2 + V\right)\psi = i\hbar\frac{\partial\psi}{\partial t}$$

and its complex conjugate.

10. a) A charged scalar meson is described by two real fields $\varphi_1(x_\mu)$, $\varphi_2(x_\mu)$, where $x_\mu = (x, y, z, it)$, (natural units: $c = \hbar = 1$). The Lagrange density for such a meson in an external electromagnetic field, described by the vector potential $A_\mu(x)$, is

$$\mathscr{L} = -\frac{1}{2}\left(\frac{\partial\varphi_1}{\partial x_\mu}\frac{\partial\varphi_1}{\partial x_\mu} + \frac{\partial\varphi_2}{\partial x_\mu}\frac{\partial\varphi_2}{\partial x_\mu}\right) - \frac{m^2}{2}\left(\varphi_1^2 + \varphi_2^2\right)$$
$$+ eA_\mu\left[\varphi_2\frac{\partial\varphi_1}{\partial x_\mu} - \varphi_1\frac{\partial\varphi_2}{\partial x_\mu}\right] + \frac{1}{2}e^2A_\mu A_\mu(\varphi_1^2 + \varphi_2^2)\,.$$

Here we use the Einstein summation convention (summation over repeated indices). Find the Euler-Lagrange equations for φ_1 and φ_2 corresponding to this Lagrangian; e and m are constants; A_μ is a given function of x.

b) Sometimes the complex fields φ and φ^* are introduced in place of the two real fields φ_1 and φ_2:

$$\varphi(x_\mu) = \frac{1}{\sqrt{2}}\left[\varphi_1(x_\mu) - i\varphi_2(x_\mu)\right],$$
$$\varphi^*(x_\mu) = \frac{1}{\sqrt{2}}\left[\varphi_1(x_\mu) + i\varphi_2(x_\mu)\right].$$

Show that the Lagrangian density in terms of these two new complex dependent field variables in free space ($A_\mu = 0$) is

$$\mathscr{L}_0 = -\left(\frac{\partial\varphi^*}{\partial x_\mu}\frac{\partial\varphi}{\partial x_\mu} + m^2\varphi^*\varphi\right).$$

Now determine the Euler-Lagrange equations for this Lagrangian density (either independently or from part (a) of the problem by transformation of φ_1 and φ_2 into φ and φ^*). You will find that the Euler-Lagrange equations are

$$(\Box^2 - m^2)\varphi = 0\,, \qquad \text{and} \qquad (\Box^2 - m^2)\varphi^* = 0\,,$$

the Klein-Gordon equation which describes a spinless particle in free space. Here the two equations describe the two charged π-mesons.

11. The Lagrangian density $\mathscr{L}$, which generates a given set of Euler-Lagrange equations, is not unique. Prove this result by showing that adding a divergence to $\mathscr{L}$ does not alter the Euler-Lagrange equations. That is, let

$$\mathscr{L}' = \mathscr{L} + \sum_k \frac{\partial f_k}{\partial x_k}\,,$$

where

$$\mathscr{L} = \mathscr{L}\left(x_k,\, \varphi_j,\, \frac{\partial \varphi_j}{\partial x_k}\right);$$

$$f_k = f_k(\varphi_j)\;;$$

$j = 1, \cdots, m$ indexes the dependent field variables, and
$k = 1, \cdots, n$ indexes the independent variables.

Now prove that

$$\frac{\delta \mathscr{L}'}{\delta \varphi_j} = \frac{\delta \mathscr{L}}{\delta \varphi_j}\,,$$

where we define

$$\frac{\delta \mathscr{L}}{\delta \varphi_j} \equiv \frac{\partial \mathscr{L}}{\partial \varphi_j} - \sum_i \frac{\partial}{\partial x_i} \frac{\partial \mathscr{L}}{\partial(\partial \varphi_j/\partial x_i)}\,.$$

12. A Lagrangian density for classical electrodynamics is

$$\mathscr{L} = -\frac{1}{16\pi} \sum_{i,j} F_{ij}F_{ij} + \frac{1}{c} \sum_i J_i A_i$$

with

$$F_{ij} = \frac{\partial A_j}{\partial x_i} - \frac{\partial A_i}{\partial x_j}\,.$$

The four-vectors $\{A_i\}$ and $\{J_i\}$ are defined by

$$\{A_i\} \equiv \{A_1, A_2, A_3, i\varphi\} \equiv \{\mathbf{A}, i\varphi\}\,,$$
$$\{J_i\} \equiv \{j_1, j_2, j_3, ic\rho\} \equiv \{\mathbf{j}, ic\rho\}\,,$$

where $\mathbf{A}$ is the vector potential ($\mathbf{H} = \nabla \times \mathbf{A}$), φ is the scalar potential

$$\left(\mathbf{E} = -\nabla\varphi - \frac{1}{c}\frac{\partial \mathbf{A}}{dt}\right),$$

$\mathbf{j}$ is the current vector, and ρ is the charge density (see Problem 1–13). Show that this Lagrangian density can be written in a more conventional form as

$$\mathscr{L} = \frac{1}{8\pi}(\mathbf{E}^2 - \mathbf{H}^2) + \frac{1}{c}\mathbf{j}\cdot\mathbf{A} - \rho\varphi\,.$$

Using Lagrange's equations with **A** and φ as dependent variables, show that this Lagrangian density leads to the familiar Maxwell equations.

13. Prove that a particle moving under gravity in a plane from a fixed point P to a vertical line L will reach the line in minimum time by following the cycloid from P to L that intersects L at right angles.

FURTHER READINGS

Akhiezer, N. I., *The Calculus of Variations*. Waltham, Mass.: Blaisdell, 1962. This advanced book explores sufficient conditions for extrema, and distinguishes various types of extrema.

Bliss, *Calculus of Variations*. La Salle, Ill.: Open Court, 1925.

Lanczos, C., *The Variational Principles of Mechanics*. Toronto: University of Toronto Press, 1949. This book on classical mechanics emphasizes general formulations and underlying principles, presented in the spirit of natural philosophy.

Landau, L., and E. Lifshitz, *The Classical Theory of Fields*. Reading, Mass.: Addison-Wesley, 1951. The variational formulation of electricity and magnetism.

Weinstock, R., *Calculus of Variations*. New York: McGraw-Hill, 1952. An introductory account with many applications to physics.

Wigner, E. P., *Symmetries and Reflections*. Bloomington, Ind.: Indiana University Press, 1967. Essays on invariance principles, symmetry, and conservation laws, etc.

Yourgrau, W., and S. Mandelstam, *Variational Principles in Dynamics and Quantum Theory*. New York: Pitman, 1962.

CHAPTER 3

VECTORS AND MATRICES

INTRODUCTION

The theory of vector spaces is to quantum mechanics what calculus is to classical mechanics. In this chapter we shall deal principally with finite-dimensional vector space, but in such a way that the generalization to infinite-dimensional spaces (Chapters 5 and 8), which forms the mathematical language of quantum theory, is natural and entails little revision. In fact, almost all the theorems proved in this chapter for finite-dimensional spaces will hold in the infinite-dimensional generalization.

Our discussion of vector spaces will be a good deal more abstract and formal than the discussion of vectors in Chapter 1. We now concentrate primarily on the algebraic aspects of vectors. However, much of the geometric language is retained and given a precise algebraic interpretation.

3.1 GROUPS, FIELDS, AND VECTOR SPACES

The definition of an abstract vector to be given in this chapter depends very much on the properties of scalars, even though the vectors will turn out to obey quite different combinatorial rules than the scalars. For example, the components of a vector are scalars. And a scalar may operate on a vector: e.g., a scalar may multiply a vector and thus change its magnitude.

It is customary to speak of the vectors, or the vector space, as being defined "over a field of scalars." The word *field* is just shorthand for the list of axioms that the *scalars* obey. We usually take these rules for granted, but we shall state them explicitly here. In the process, it will be convenient first to introduce another algebraic structure called a *group*. Like a field, a group is a system consisting of a set of elements and rules for combining the members of this set.

Suppose, then, that G is a set of elements, and let $\cdot$ be an operation defined on the elements of G such that, if a and $b \in G$, then $(a \cdot b) \in G$ as well.* Such an operation is said to be *closed*. Then we define a group as follows.

Definition 3.1. A group is a system $\{G, \cdot\}$, consisting of the set G and a single *closed* operation $\cdot$, which satisfies the following three axioms:

1. If $a, b, c \in G$, then $a \cdot (b \cdot c) = (a \cdot b) \cdot c$ (associativity).
2. There exists an identity element $e \in G$ such that for all $a \in G$, $a \cdot e = e \cdot a = a$.

* The symbol $\in$ means "belongs to" or "is a member of."

Abelian group: $(a \cdot b) = (b \cdot a)$

3. For every $a \in G$, there exists an inverse element in G, denoted a^{-1}, such that $a \cdot a^{-1} = a^{-1} \cdot a = e$.*

A group is said to be *abelian* if it is commutative, that is, if $a \cdot b = b \cdot a$, where $a, b \in G$. Often the dot $(\cdot)$ is omitted if it is interpreted as the usual multiplication, but for some sets the dot will be interpreted as $+$ (addition). As used in our definition, it denotes any operation that combines two elements of the set to give a third element of the set, just so the rule of combination satisfies the three axioms.

The theory of groups is very important in physics. For instance, in the study of quantum mechanics one often hears, "It can be shown by group theory that . . ." We shall discuss this subject in its own right in Chapter 10.

Examples of Groups.

1. The set of all real numbers excluding 0 forms a group with respect to the usual multiplication $(\cdot)$. The identity is 1, and for each x the inverse is $x^{-1} \equiv 1/x$. This group is abelian.
2. The integers with addition as operation compose another abelian group; 0 is the identity element, and the inverse is $n^{-1} \equiv -n$.
3. The set of all two-dimensional rotation matrices [see Eq. (1.14)]

$$r(\phi) \equiv \begin{pmatrix} \cos\phi & \sin\phi \\ -\sin\phi & \cos\phi \end{pmatrix}, \tag{3.1}$$

   is an abelian, *continuous* group—continuous because there is an infinite number of elements defined by the continuous variation of the parameter φ. Multiplication of group elements (matrices) was defined briefly in Eq. (1.22). Matrix multiplication will be discussed in detail in this Chapter. The unit element of this group is $r(0)$ and the inverse of $r(\phi)$ is $r(-\phi)$.
4. The set of three-dimensional rotations is a noncommutative continuous group. The reader may demonstrate this by noting that the net result of two consecutive rotations in three dimensions is dependent on their order (see Figs. 3.1 and 3.2).
5. The set of permutations of n objects is a group. For example, the $3! = 6$ possible permutations of 3 elements may be written as

$$\begin{bmatrix} 1 & 2 & 3 \\ 1 & 2 & 3 \end{bmatrix} = I, \quad \begin{bmatrix} 1 & 2 & 3 \\ 3 & 1 & 2 \end{bmatrix} = F, \quad \begin{bmatrix} 1 & 2 & 3 \\ 2 & 3 & 1 \end{bmatrix} = D$$
$$\begin{bmatrix} 1 & 2 & 3 \\ 2 & 1 & 3 \end{bmatrix} = A, \quad \begin{bmatrix} 1 & 2 & 3 \\ 1 & 3 & 2 \end{bmatrix} = C, \quad \begin{bmatrix} 1 & 2 & 3 \\ 3 & 2 & 1 \end{bmatrix} = B. \tag{3.2}$$

   For example, the permutation F means that 1 is replaced by 3, 2 is replaced by 1, and 3 is replaced by 2. The product of two permutations, denoted by $*$, is the single permutation that accomplishes what two applied successively, from right to left would accomplish, e.g.,

$$A * C = \begin{bmatrix} 1 & 2 & 3 \\ 2 & 1 & 3 \end{bmatrix} * \begin{bmatrix} 1 & 2 & 3 \\ 1 & 3 & 2 \end{bmatrix} = \begin{bmatrix} 1 & 2 & 3 \\ 2 & 3 & 1 \end{bmatrix} = D.$$

---

* This is not a minimal definition from a mathematical point of view. Such a definition will be given in Chapter 10.

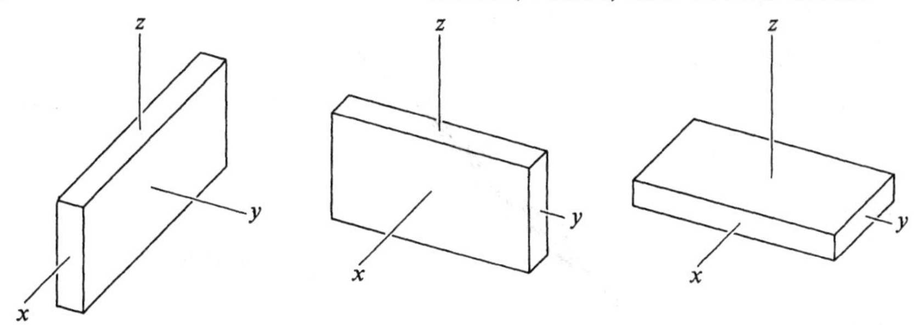

Fig. 3.1 A rotation about the z-axis followed by one about the y-axis.

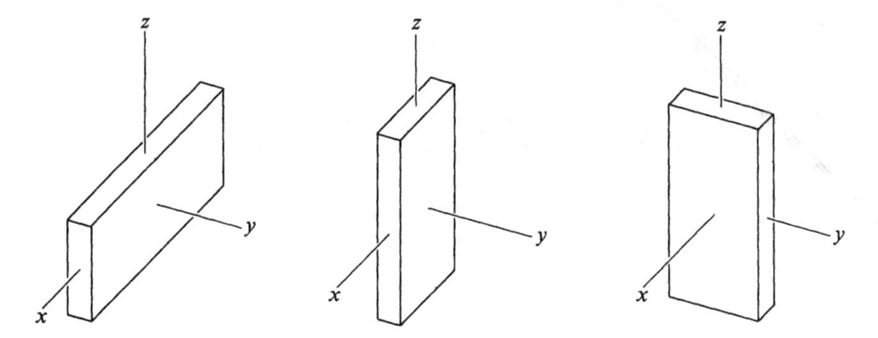

Fig. 3.2 A rotation about the y-axis followed by one about the z-axis.

This is obtained as follows: First the permutation C taken $1 \to 1$, $2 \to 3$, $3 \to 2$. Then A taken $1 \to 2$, $2 \to 1$, $3 \to 3$. The combined effect of first C and then A may be written:

$$1 \to 1 \to 2\,, \qquad 2 \to 3 \to 3\,, \qquad 3 \to 2 \to 1\,.$$

The net result of the two permutations acting successively, from right to left, is thus: $1 \to 2$, $2 \to 3$, $3 \to 1$; that is, the single permutation D. Similarly,

$$C * A = \begin{bmatrix} 1 & 2 & 3 \\ 1 & 3 & 2 \end{bmatrix} * \begin{bmatrix} 1 & 2 & 3 \\ 2 & 1 & 3 \end{bmatrix} = \begin{bmatrix} 1 & 2 & 3 \\ 3 & 1 & 2 \end{bmatrix} = F\,. \tag{3.3}$$

Thus this group is not abelian. The 36 possible multiplications for the permutation group of three objects are shown in Table 3.1.

Table 3.1

	I	A	B	C	D	F
I	I	A	B	C	D	F
A	A	I	F	D	C	B
B	B	D	I	F	A	C
C	C	F	D	I	B	A
D	D	B	C	A	F	I
F	F	C	A	B	I	D

(3.4)

Now we define a field in terms of a group.

Definition 3.2. A *field* is a system $\{F, +, \cdot\}$ satisfying the three axioms:

1. The system $\{F, +\}$ is an abelian group with the identity element 0.

2. Let F' be the set of all $x \in F$ except $x = 0$. Then system $\{F', \cdot\}$ is an abelian group with the identity element e.

3. Let $a, b, c \in F$; then $a \cdot (b + c) = a \cdot b + a \cdot c$ (distributivity of $\cdot$ with respect to $+$).

Examples of a Field. The set of all rational numbers, the set of all real numbers and the set of all complex numbers with the usual multiplication and addition are the most familiar examples of a field.

Now we can define a vector space V. We assume we are given a field F; the scalars to be used are all elements of F.

Definition 3.3. A *vector space* over a field F is a set of elements, $V(F)$, called vectors, which satisfy the following axioms:

1. There exists an operation $(+)$ on the vectors such that $\{V(F), +\}$ forms an abelian group; the identity element is denoted by 0.

2. For every $\alpha \in F$ and $x \in V(F)$, there exists a vector $\alpha x \in V(F)$; furthermore,

   a) $\alpha(\beta x) = (\alpha\beta)x$, b) $1(x) = x$, for all x.

   c) $\alpha(x + y) = \alpha x + \alpha y$, d) $(\alpha + \beta)x = \alpha x + \beta x$.

A vector space $V(F)$ is called a real or a complex vector space according to whether F is the field of real numbers or of complex numbers.

We shall adopt the following notational conventions:

a) Vector spaces will be written $V_n(F)$, where the subscript denotes the dimension of the space (to be defined shortly) and the quantity in parentheses is the field over which the vector space is defined. Where the dimensionality and/or the field are apparent from the context, we may write just V_n or V.

b) Vectors in the space will be denoted by letters at the end of the alphabet, e.g., x, y, z; they will *not* be written in bold-face type as in Chapter 1. The reader will be able to distinguish the scalar and the vector zero from the context.

c) The scalar components of vectors will be written with subscripted Greek letters, thus: $(\alpha_1, \alpha_2, \cdots, \alpha_n)$. The elements of matrices will be written with doubly subscripted Roman letters: a_{12}, etc.

d) Scalars which operate on vectors will be denoted by unsubscripted Roman or Greek letters: a, b, c, or α, β, γ. These may be either real or complex.

e) Sets of basis vectors will be denoted $\{x_i\}$, or $\{x_1, x_2, \cdots, x_n\}$. In accordance with convention b), such expressions are *not* to be interpreted as the scalar components of a single vector. For example, in this notation we write,

$$x = \sum_i \alpha_i x_i ,$$

where x is a vector whose ith component is α_i in the basis $\{x_i\}$.

Examples of Vector Spaces.

1. For a fixed positive integer n, the set of all n-tuples of complex numbers $(x = (\xi_1, \xi_2, \cdots, \xi_n))$ forms a vector space if addition of vectors and multiplication of a vector by a scalar are defined by

$$\begin{aligned} x + y &= (\xi_1, \xi_2, \cdots, \xi_n) + (\eta_1, \eta_2, \cdots, \eta_n) \\ &= (\xi_1 + \eta_1, \xi_2 + \eta_2, \cdots, \xi_n + \eta_n); \\ \alpha x &= \alpha(\xi_1, \cdots, \xi_n) = (\alpha\xi_1, \cdots, \alpha\xi_n); \end{aligned}$$

furthermore,

$$0 = (0, 0, \cdots, 0);$$

and

$$-x = (-\xi_1, -\xi_2, \cdots, -\xi_n) .$$

It is easy to show that all the axioms for a vector space are satisfied. (In fact, this vector space is the prototype for all finite-dimensional vector spaces, as we shall see later.)

2. The set of all complex numbers, where we interpret $x + y$ and αx as ordinary complex numerical addition and multiplication, is a complex vector space.

3. The set P of all polynomials in a real variable t with complex coefficients is a complex vector space, provided that we interpret vector addition and scalar multiplication as ordinary addition of two polynomials and the multiplication of a polynomial by a complex number. The zero vector, 0, in P is the polynomial identically zero; P is not a finite-dimensional vector space.

As special cases of (2) and (3), we have the real vector spaces composed of the sets of all real numbers and of all polynomials in t with real coefficients, where the operations $+$ and $\cdot$ are those of ordinary arithmetic.

3.2 LINEAR INDEPENDENCE

We already have an idea of what it means to say that any two noncollinear vectors in the plane are linearly independent, and we also know that if you add any third vector in the plane, the resulting set of three vectors is linearly dependent. That is, we know that any one of the three can be expanded in terms of the other two: one cannot find three vectors in a two-dimensional space that are linearly independent. In these simple statements we have hinted at the content of several theorems that we shall prove in this section. The definition of linear independence given below will tie in neatly with these intuitive notions.

A word on summation notation: in general, we shall not use the Einstein summation convention in this chapter because it is often necessary to indicate the *range* of the summation. Thus we write $\sum_{i=1}^{n}$ to mean the sum from 1 to n. Sometimes, however, when the range of summation is perfectly clear from the context, we may omit the range, writing simply $\sum_i$.

Definition 3.4. A finite set $\{x_i\}$ of n vectors is *linearly independent* if and only if

$$\sum_{i=1}^{n} \alpha_i x_i = 0 \Longrightarrow \alpha_i = 0$$

for all i. If there exists a set of scalars $\{\alpha_i\}$ that are not all zero, such that

$$\sum_{i=1}^{n} \alpha_i x_i = 0 ,$$

then the set of vectors $\{x_i\}$ is *linearly dependent*.

The definition says that if the $\{x_i\}$ are linearly independent—as are, for example, the $\mathbf{e}_i$ unit vectors in Cartesian three-space—the zero vector can be obtained as a linear combination of them only in a trivial way, i.e., all $\alpha_i = 0$. But if the zero vector is a nontrivial linear combination of the x_i, so that not all $\alpha_i = 0$, then

$$\sum_{i=1}^{n} \alpha_i x_i = 0 \Longrightarrow x_j = -\frac{1}{\alpha_j} \sum_{i \neq j} \alpha_i x_i ,$$

where α_j is a nonzero α_i. (The symbol $\sum_{i \neq j}$ means that the sum is over all i except $i = j$.) In this case the vector x_j *depends* linearly on the others in the sense that x_j is a linear combination of the others. Note that any set of vectors containing the zero vector is linearly dependent.

We have already mentioned that any two nonzero, noncollinear vectors in the plane are linearly independent, while any three vectors in the plane are linearly dependent. Another illustration of Definition 3.4 may be found in the vector space P of polynomials mentioned earlier. Let

$$x_1(t) = 2 - t + t^2 , \qquad x_2(t) = t - t^2 + 2t^3 , \qquad x_3(t) = 1 + t^3 ,$$

be three vectors in P. If we set $\alpha_1 = \alpha_2 = 1$, and $\alpha_3 = -2$, then

$$\sum_{i=1}^{3} \alpha_i x_i = x_1 + x_2 - 2x_3 = 0 .$$

This means that x_1, x_2, and x_3 are linearly dependent, for we have found a set of α_i which are not all zero such that $\sum_i \alpha_i x_i$ vanishes. However, the infinite set of vectors

$$\begin{aligned} x_0(t) &= 1 \\ x_1(t) &= t \\ &\vdots \\ x_n(t) &= t^n \\ &\vdots \end{aligned}$$

is a linearly independent set because if for any n

$$\sum_{i=0}^{n} \alpha_i x_i = 0 ,$$

then we have the polynomial identity:

$$\alpha_0 + \alpha_1 t + \cdots + \alpha_n t^n = 0 ,$$

which implies

$$\alpha_i = 0 \qquad \text{for} \quad 0 \leq i \leq n ,$$

since this must hold for all t.

The most important result involving the notions of linear dependence and independence is contained in the following theorem.

Theorem 3.1. The set of nonzero vectors $\{x_i\}$ is linearly dependent if and only if some x_i, $2 \leq i \leq n$, is a linear combination of the preceding ones.

Proof. We first prove the necessity of the condition. We assume that the x_i are linearly dependent. This means that for some i (and let us pick the smallest such i and call it k) we know that

$$\sum_{i=1}^{k} \alpha_i x_i = 0 ,$$

where $k \leq n$ and not all the $\alpha_i = 0$. Now we know $\alpha_k \neq 0$, because if $\alpha_k = 0$, the set $\{x_i, \cdots, x_{k-1}\}$ would be linearly dependent, contrary to the definition of k as the *smallest* value of i for which the vectors $x_1, \cdots, x_k$ are linearly dependent. Therefore

$$x_k = -\frac{1}{\alpha_k}\left[\sum_{i=1}^{k-1} \alpha_i x_i\right] .$$

Thus x_k has been explicitly expressed as a linear combination of preceding vectors. The proof of the sufficiency of the condition is simple. If we know that

$$x_j = \sum_{i=1}^{j-1} \alpha_i x_i \qquad \text{for some } j ,$$

then letting $\alpha_j = -1$, we have

$$\sum_{i=1}^{j} \alpha_i x_i = 0 .$$

For the remaining vectors $(x_{j+1}, \cdots, x_n)$, we set $\alpha_{j+1} = \alpha_{j+2} = \cdots = \alpha_n = 0$, so that

$$\sum_{i=1}^{n} \alpha_i x_i = 0 ,$$

which implies linear dependence, since not all the α_i's are zero. QED

In the next section we relate the concept of linear independence to that of dimensionality.

3.3 BASES AND DIMENSIONALITY

Definition 3.5. A *basis* in a vector space V is a set $\{x_i\}$ of linearly independent vectors such that every vector in V is a linear combination of elements of $\{x_i\}$. A vector space is finite-dimensional if it has a finite set of basis vectors; otherwise it is infinite-dimensional.

The reader may want to prove that an equivalent alternative definition is that a basis is a maximal linearly independent subset of V.

If every vector in V is a linear combination of elements of $\{x_i\}$, we say that the set $\{x_i\}$ *spans* V, so a basis is a linearly independent set that spans V.

Examples of Bases.

1. In real three-dimensional space, the three vectors

$$\mathbf{e}_1 = (1, 0, 0)\,, \qquad \mathbf{e}_2 = (0, 1, 0)\,, \qquad \mathbf{e}_3 = (0, 0, 1)$$

form a basis. It is easy to show they are linearly independent, and that every vector, x, can be expanded as a linear combination of these three basis vectors, as

$$x = (\xi_1, \xi_2, \xi_3) = \sum_{i=1}^{3} \xi_i \mathbf{e}_i\,.$$

2. For the vector space of polynomials, P, the set $\{x_n\}$, where $x_n(t) = t^n$, $n = 0, 1, 2, \cdots$, is a basis, but not a finite one, so P is infinite-dimensional.

Theorem 3.2. Every vector in V has a unique representation as a linear combination of vectors of a fixed basis of V.

Proof. Let $\{x_i\}$ be a fixed basis in V. Then for every $x \in V$, x can be written as $x = \sum_i \alpha_i x_i$. The theorem says that the α_i are uniquely determined. To prove this, we assume the opposite, namely that there exists another representation of x, $x = \sum_i \beta_i x_i$. If this is the case, then by subtracting, we find that

$$0 = x - x = \sum_i (\alpha_i - \beta_i) x_i\,.$$

But since the x_i form a basis, they are linearly independent, and by the definition of linear independence, $(\alpha_i - \beta_i) = 0$ for all i. Thus $\alpha_i = \beta_i$ and the representation is unique. QED

Suppose that we are given a collection of linearly independent vectors $\{y_i\}$ where $y_i \in V$ for all i. If we consider the set of all linear combinations of the y_i, we see that this set of vectors is itself a vector space, every element of which lies in V. We say that this collection is the *subspace* of V *spanned* by the set $\{y_i\}$. In particular, V is a subspace of itself. Now it may happen that the set

$\{y_i\}$ is not a basis for V. We would expect that it is possible to add some vectors to the set $\{y_i\}$ and turn it into a basis for V. The following theorem shows that this is indeed the case.

Theorem 3.3. If V is a finite-dimensional vector space and $\{y_1, y_2, \cdots, y_m\}$ is a set of linearly independent vectors, then unless the y's are a basis, we may extend $\{y_1, y_2, \cdots, y_m\}$ by adding $y_{m+1}, \cdots, y_n$ so that the set $\{y_1, y_2, \cdots, y_n\}$ is a basis for V.

Proof. Let $\{x_1, x_2, \cdots, x_l\}$ be a basis for V. Then the set of vectors

$$S_1 = \{y_1, y_2, \cdots, y_m, x_1, x_2, \cdots, x_l\}$$

is linearly dependent. Hence, by Theorem 3.1, some member of S_1 is a linear combination of the vectors which precede it. It must be one of the x's since the y's are linearly independent. Call this element x_i and remove it from S_1 to form S_2:

$$S_2 = \{y_1, y_2, \cdots, y_m, x_1, x_2, \cdots, x_{i-1}, x_{i+1}, \cdots, x_l\} .$$

We claim that every vector in V can be expressed as a linear combination of elements of S_2. To see this, we note that the set S_1 spanned V, and the only element of S_1 missing from S_2 is x_i, which is a linear combination of elements of S_2; in fact, x_i is a linear combination of the vectors $\{y_1, y_2, \cdots, y_n, x_1, \cdots, x_{i-1}\}$.

Now, if the set S_2 is linearly independent, then we have found a basis for V; if not, there must be some x which is a linear combination of the elements of S_2 which precede it (again it must be an x, since the y's are linearly independent). We eliminate this x and form S_3. For the same reasons as before, every vector in V can be written as a linear combination of the elements of S_3. If S_3 is linearly independent, then it is a basis for V; if not, we continue this process until we find some S_n which is linearly independent. Since the S's have only a finite number of elements, this search can take only a finite number of steps. In this way, we find a basis which contains the set $\{y_1, y_2, \cdots, y_m\}$, as required. QED

Now the subspace spanned by the set $\{y_i\}$ is contained in V and the basis for this subspace, is contained in the basis for V. We would like to say that the subspace is of smaller dimension than V and to define the dimensionality of a space to be the number of elements in its basis. However, we do not yet know that this number is unique. The reader may think it obvious that all is well, but the demonstration of the uniqueness of the number of elements in a basis actually involves a rather subtle line of reasoning. Since the concept of dimensionality is an absolutely fundamental one, we shall go through this proof in detail.

Theorem 3.4. The number of elements in any basis of a finite-dimensional vector space is the same as in any other basis.

Proof. Let $\{x_1, x_2, \cdots, x_n\}$ and $\{y_1, y_2, \cdots, y_m\}$ be bases for V. Since n must either be equal to m or be different from m, we may assume without loss of generality that $n \leq m$. The idea of the proof will be to show that $n \geq m$, which combined with the assumption $n \leq m$ leads to the conclusion $n = m$, which is just what we want to prove.

Now a set of basis vectors has two essential properties: (1) the vectors must be linearly independent; and (2) any vector in V must be a linear combination of the elements of the basis. We shall consider the set $\{y_i\}$ to have property (1) and the set $\{x_i\}$ to have property (2). Of course, each set has both the properties in question, but for the purposes of this proof we shall focus only on one property per set.

Consider the set

$$S_1 = \{y_m, x_1, x_2, \cdots, x_n\} .$$

Since $\{x_i\}$ has property (2), every vector in V is a linear combination of elements of S_1, and S_1 is a linearly dependent set. Hence by the argument of Theorem 3.3, there is some x, say x_i, which is a linear combination of the elements of S_1 which precede it. Let us remove this element and form T_1:

$$T_1 = \{y_m, x_1, \cdots, x_{i-1}, x_{i+1}, \cdots, x_n\}$$

with every vector in V a linear combination of elements of T_1, for the reasons given in Theorem 3.3. Now form

$$S_2 = \{y_{m-1}, y_m, x_1, \cdots, x_{i-1}, x_{i+1}, \cdots, x_n\} .$$

Clearly, S_2 is linearly dependent, and every vector in V is a linear combination of elements of S_2. Thus some x must be a linear combination of the preceding elements of S_2 (it must be an x rather than a y because the y's are linearly independent). Call this element x_j, and subtract it from S_2 to form T_2:

$$T_2 = \{y_{m-1}, y_m, x_1, \cdots, x_{i-1}, x_{i+1}, \cdots, x_{j-1}, x_{j+1}, \cdots, x_n\} .$$

Now we continue this process, in each step eliminating an x and inserting a y. After n steps, all the x's are gone and we have

$$T_n = \{y_{m-n+1}, \cdots, y_m\} .$$

But all the T_n have property (2); that is, all the T_n span V. Therefore if $m > n$, y_1 (or any y) is a linear combination of the other y's. But this contradicts the fact that the basis $\{y_i\}$ is linearly independent. The only way to avoid this contradiction is to have $n \geq m$. But we have assumed $n \leq m$, so we have the desired result. QED

If we had assumed $n \geq m$, then we would have reversed the roles of $\{y_i\}$ and $\{x_i\}$ in our proof and concluded that $m \geq n$, thereby obtaining the same result. Now we are free to make the definition that we are after:

Definition 3.6. The *dimension* of a finite-dimensional vector space V is the unique number of elements in a basis of V.

Note that in an n-dimensional vector space, any set containing more than n vectors is linearly dependent.

3.4 ISOMORPHISMS

Let F^n be the set of n-tuples of elements in a field F. In the first example given of a vector space, it was stated that F^n may serve as a prototype of the n-dimensional vector space V over F. We want to prove that every finite-dimensional vector space over a field F is "essentially the same as" some F^n.

Definition 3.7. Two vector spaces U and V (over the same field) are *isomorphic* if there exists a one-to-one correspondence between the vectors $x^{(i)}$ in U and $y^{(i)}$ in V, say $y^{(i)} = f(x^{(i)})$, such that

$$f(\alpha_1 x^{(1)} + \alpha_2 x^{(2)}) = \alpha_1 f(x^{(1)}) + \alpha_2 f(x^{(2)}) ,$$

and also a one-to-one correspondence between the vectors in V and the vectors in U.

Thus an isomorphism is a one-to-one correspondence preserving all linear relations, or, what is the same, preserving "algebraic structure." We shall examine this definition to see what is behind it, and in so doing generalize it.

Consider two closed systems $S = \{E, *\}$ and $S' = \{E', \cdot\}$, each with one operation. Assume that there is a one-to-one mapping (or function) f which assigns to each element $a, b, c, \cdots, \in E$ an element $f(a), f(b), f(c), \cdots, \in E'$, depicted thus:

$$\begin{array}{ccc} E & & E' \\ \hline a & \xrightarrow{f} & f(a) \\ b & \xrightarrow{f} & f(b) \, . \end{array}$$

Assume that $a*b = c$; then $c \rightarrow f(c) = f(a*b)$. If the algebraic structure of S and S' is to be identical, then it is necessary that a multiplication carried out on the left in S be paralleled by a multiplication of the corresponding elements on the right in S'. Since $a*b = c$, this would require the following equality to hold:

$$f(a) \cdot f(b) = f(c) = f(a*b) \, .$$

If this equation is satisfied for every a, b, then the systems $S = \{E, *\}$ and $S' = \{E', \cdot\}$ are said to be *isomorphic*.

As an example of two systems that are isomorphic—although at first glance they look very different—consider the six permutations [Eqs. (3.2)] and the six 2×2 matrices

$$\begin{bmatrix} 1 & 0 \\ 0 & 1 \end{bmatrix} = I' , \quad \begin{bmatrix} -1/2 & \sqrt{3}/2 \\ -\sqrt{3}/2 & -1/2 \end{bmatrix} = F' , \quad \begin{bmatrix} -1/2 & -\sqrt{3}/2 \\ \sqrt{3}/2 & -1/2 \end{bmatrix} = D' ,$$
$$\begin{bmatrix} 1 & 0 \\ 0 & -1 \end{bmatrix} = A' , \quad \begin{bmatrix} -1/2 & -\sqrt{3}/2 \\ -\sqrt{3}/2 & 1/2 \end{bmatrix} = C' , \quad \begin{bmatrix} -1/2 & \sqrt{3}/2 \\ \sqrt{3}/2 & 1/2 \end{bmatrix} = B' . \tag{3.5}$$

We have not as yet dealt extensively with the multiplication of matrices, although the reader will recall that it was briefly considered in Section 1.4,

Eq. (1.22). There we encountered the multiplication of two arrays of numbers, (a_{ij}) and (b_{ij}), to give a third, (c_{ij}), using the rule

$$c_{ij} = \sum_{k=1}^{2} a_{ik} b_{kj} \,, \tag{3.6}$$

where the first subscript labels the row, and the second the column, of an element.

It is easy to verify that the six matrices of Eq. (3.5) form a non-abelian group with respect to matrix multiplication (Eq. 3.6) which we shall later prove is associative, and that the group multiplication table (3.4), is identical with that of the permutation group of three objects, if we prime all the elements. Thus the multiplicative group structure is maintained if we associate each element of the permutation group with the corresponding primed element of the group of matrices. Then, for example, under this mapping f, we have

$$B \xrightarrow{f} B'$$
$$D \xrightarrow{f} D'$$
$$B*D = A \xrightarrow{f} A' = B' \cdot D'$$

so that

$$f(B*D) = B' \cdot D' = f(B) \cdot f(D) \,.$$

A relation of the same form holds for any of the other elements, so these two systems are isomorphic.

In general, if two systems have the same multiplication table, they are isomorphic, or *structurally identical*. We can find out anything we want to know about one system by studying the other and then translating the results back into the language of the first. Suppose, for example, that we want to study a group of rotations. We look for a set of matrices which is isomorphic to the group of rotations; we can then study the matrix group to learn about the rotations. Such a group of matrices is called a *representation* of the original group. Thus the set of matrices $\{I', A', B', C', D', F'\}$ above is a matrix representation of the permutation group (sometimes called a *symmetric group*) for a set of three objects. We shall study group representations in Chapter 10.

In the Definition (3.7) of isomorphism given originally, as between two vector spaces, the operation $(+)$ is the same in both systems, U and V, which are both vector spaces. The example above makes it clear that the operation could be different in the two systems. Also, the linear operation of the multiplication of a vector by a scalar has been included at the same time as the linear operation of the addition of two vectors. The one condition could be broken down into two as follows:

$$f(x^{(1)} + x^{(2)}) = f(x^{(1)}) + f(x^{(2)}) \,, \qquad f(\alpha x) = \alpha f(x) \,.$$

We now prove the fundamental result regarding the algebraic structure of a vector space.

Theorem 3.5. Every n-dimensional vector space $V_n(F)$ over F is isomorphic to the space F^n.

Proof. Let $\{x_1, \cdots, x_n\}$ be any basis in V. Each vector $x \in V$ can therefore be written as

$$x = \sum_{i=1}^{n} \alpha_i x_i .$$

We know that the scalar coordinates α_i $(i = 1, \cdots, n)$ are uniquely determined (Theorem 3.2.). Now we propose the one-to-one correspondence $x \rightleftarrows (\alpha_1, \cdots, \alpha_n)$ between V and F^n; that is,

$$f(x) = (\alpha_1, \cdots, \alpha_n) ,$$

as an isomorphism of V and F^n. If

$$y = \sum_{i=1}^{n} \beta_i x_i ,$$

then

$$cx + dy = c\sum_{i=1}^{n} \alpha_i x_i + d\sum_{i=1}^{n} \beta_i x_i = \sum_{i=1}^{n} (c\alpha_i + d\beta_i)x_i ,$$

so that

$$f(cx + dy) = (c\alpha_1 + d\beta_1, c\alpha_2 + d\beta_2, \cdots, c\alpha_n + d\beta_n) .$$

Now we must compute $cf(x) + df(y)$ and compare this with $f(cx + dy)$:

$$cf(x) = c(\alpha_1, \alpha_2, \cdots, \alpha_n) = (c\alpha_1, c\alpha_2, \cdots, c\alpha_n) ,$$
$$df(y) = d(\beta_1, \beta_2, \cdots, \beta_n) = (d\beta_1, d\beta_2, \cdots, d\beta_n) ,$$

so that

$$cf(x) + df(x) = (c\alpha_1 + d\beta_1, \cdots, c\alpha_n + d\beta_n) = f(cx + dy) .$$

Therefore the one-to-one correspondence f is an isomorphism between $V(F)$ and F^n. Thus $V(F)$ and F^n are isomorphic. QED

It follows from this theorem that any two vector spaces of the same dimension over the same field are isomorphic, because both are isomorphic to F^n, and therefore isomorphic to each other. This theorem also means that from here on, we could restrict our attention to n-tuples, the prototypes of vectors, without loss of generality. But this necessitates the choice of a representation (or basis) and working with coordinates, and besides being tedious, it obscures important properties of vectors and vector spaces that are independent of the choice of basis. For the most part, therefore, we shall treat vector spaces without reference to a specific basis.

3.5 LINEAR TRANSFORMATIONS

Definition 3.8. A *linear transformation* or *linear operator* A on a vector space V is a correspondence that assigns to every $x, y \in V$, a vector Ax, $Ay \in V$ in such a way that

$$A(ax + by) = aAx + bAy ,$$

where a and b are scalars. The product C of two linear transformations A and B is defined by the equation $Cx \equiv A(Bx)$, and written $C = AB$. The sum of two linear transformations is defined by the equation $(A + B)x \equiv Ax + Bx$.

Examples of linear transformations.

1. Let $V(F)$ be the vector space of n-tuples of real numbers, and let a_{ij} $(i, j = 1, 2, \cdots, n)$ be a set of n^2 real numbers. If we define $y = Ax$ by

$$y_i = \sum_{j=1}^{n} a_{ij}x_j ,$$

then A is a linear transformation. If for the case $n = 2$ we take $a_{11} = a_{22} = \cos\phi$, and $a_{12} = -a_{21} = \sin\phi$, then $A \equiv (a_{ij})$ becomes the rotation matrix of Eq. (1.14). This linear transformation corresponds to rotating the vector x through an angle ϕ to produce the new vector y.

2. Two special linear transformations O and I are defined by:

$$Ox \equiv 0 , \qquad Ix \equiv x .$$

3. The differentiation and multiplication linear transformations, D_x and X, are defined by:

$$D_x f(x) \equiv \frac{df(x)}{dx} \qquad \text{and} \qquad Xf(x) \equiv xf(x) .$$

Strictly speaking, D_x is an operator in the sense of Definition 3.8 only if, given a function in V, the derivative of this function is also in V. For example, if V is the space of continuous functions, then $|x| \in V$, but $d/dx\,|x|$ is not in V. Thus D_x is not defined everywhere in the space.

The product of D_x and X produces the following result when applied to a vector $f(x)$:

$$(D_xX)f(x) = D_x(Xf(x)) = \frac{d}{dx}(xf(x)) = f(x) + x\frac{df(x)}{dx} .$$

On the other hand, the product of X and D_x gives

$$(XD_x)\,f(x) = X(D_x f(x)) = x\frac{df(x)}{dx} .$$

Therefore

$$(D_xX)\,f(x) \neq (XD_x)\,f(x) ,$$

or for short, omitting the operand,

$$D_x X \neq X D_x .$$

In fact,

$$(D_x X - X D_x) f(x) = f(x) , \qquad \text{or} \qquad D_x X - X D_x = I .$$

This particular pair of noncommuting operators plays a central role in quantum mechanics, where the operator associated with the x-component of momentum is $p_x = -i\hbar D_x$, and X is the position operator. (There is a slight embarrassment with the notation here. Although we have written the multiplication operator as X above, it is customary among physicists to write this operator simply as x, since it has the effect of multiplying the operand by x. We shall henceforth abide by this convention.) Thus we write

$$[p_x x - x p_x] f(x) = -i\hbar f(x) , \qquad \text{or} \qquad p_x x - x p_x = -i\hbar .$$

The operator $(p_x x - x p_x)$ is the called the *commutator* of p_x and x. It is written as $[p_x, x]$. The rule $[p_x, x] = -i\hbar$ is taken as an axiom of quantum mechanics.

4. In the space P_m of polynomials of degree $\leq m$, the linear transformation D is given by

$$D p_m(x) = \frac{d}{dx} p_m(x) .$$

Let

$$D^1 = D , \qquad D^2 = DD , \qquad D^n = \underbrace{DD \cdots D}_{n \text{ times}} .$$

Then if $n > m$, $D^n p_m = 0$ for all polynomials in P_m, so that $D^n = 0$. Thus even if an operator is not the zero operator, some power of it may be the zero operator!

Most of the formal algebraic properties of numerical multiplication, with the notable exception of commutativity, are valid in the algebra of linear transformations. Thus we have:

a) $AO = OA = O$,

b) $AI = IA = A$,

c) $A(B + C) = AB + AC$,

d) $A(BC) = (AB)C$,

e) $(aA) = a(A)$, where a is a scalar.

We prove (c):

$$\begin{aligned}[A(B + C)]x &= A[(B + C)x] = A(Bx) + A(Cx) \\ &= (AB)x + (AC)x = (AB + AC)x ,\end{aligned}$$

where we have used the definitions of product and sum in Definition (3.8). Thus the *sum* of two linear transformations is commutative and associative because addition in V is so. Property (d) above says that the product is also associative. The reader may now prove that the set of all linear transformations on a vector space $V(F)$ is itself a vector space over F.

3.6 THE INVERSE OF A LINEAR TRANSFORMATION

The following definition is designed to cover the infinite dimensional case as well as the finite dimensional case.

Definition 3.9. If a linear transformation A has both the following two properties, then there exists an *inverse* to A, denoted by A^{-1}, and we say that A is *invertible*.

a) $x \neq y \Longrightarrow Ax \neq Ay$ (or equivalently, $Ax = Ay \Longrightarrow x = y$).

b) For every $y \in V$ there exists an $x \in V$ such that $Ax = y$.

We construct A^{-1} as follows: If y_0 is any vector, we may [by (b)] find an x_0 such that $y_0 = Ax_0$. This x_0 is in fact uniquely determined, since if $Ax_0 = y_0$ and $Ax = y_0$, then $Ax_0 = Ax \Longrightarrow x_0 = x$ [by (a)]. We define $A^{-1}y_0 \equiv x_0$. Now, is A^{-1} a *linear* transformation? To establish linearity, we must prove

$$A^{-1}(\alpha x + \beta y) = \alpha A^{-1}x + \beta A^{-1}y .$$

If $Ax = u$ and $Ay = v$, then linearity of A implies

$$A(\alpha x + \beta y) = \alpha u + \beta v ,$$

so that

$$A^{-1}(\alpha u + \beta v) = \alpha x + \beta y = \alpha A^{-1}u + \beta A^{-1}v ,$$

and the inverse is linear. It follows immediately from the definition of the inverse that if A is invertible,

$$AA^{-1} = A^{-1}A = I .$$

The following theorem shows that whenever a linear transformation plays the role of A^{-1} in the above equation, then it must be the inverse of A.

Theorem 3.6. If A, B, and C are linear transformations such that $AB = CA = I$, then A is invertible and $A^{-1} = B = C$.

Proof. To prove A is invertible, we must prove the two conditions, (a) and (b), of Definition (3.9).

a) If $Ax = Ay$ then $CAx = CAy$, which implies $x = y$ (since $CA = I$).

b) If y is any vector and $x = By$, then $y = ABy = Ax$ (since $AB = I$).

Therefore A is invertible.

Now multiply $AB = I$ on the left and $CA = I$ on the right by A^{-1} to get $A^{-1} = B = C$. QED

Neither of the two conditions, $AB = I$ or $CA = I$, is alone sufficient to ensure the invertibility of A, unless we are in a finite-dimensional vector space, or unless we also know that B or C is unique (see Problem 10); in these cases, one such condition suffices on any vector space.

To illustrate these statements, consider the two operators D_x and S operating on vectors of the infinite-dimensional vector space P, where

$$D_x p(x) \equiv \frac{d}{dx} p(x) , \qquad Sp(x) \equiv \int_0^x p(t)\, dt .$$

Although $D_x S = I$, neither D_x nor S is invertible. D_x violates condition (a), since, for instance, $D_x(x^2 + 3) = D_x(x^2 + 1)$; S violates condition (b), for if $p_0 = x^2 + 1$, there exists no $p(t)$ such that $\int_0^x p(t)\, dt = p_0$.

Theorem 3.7. Let A be a linear transformation defined on a finite-dimensional vector space V. If $Ax = 0$ implies that $x = 0$ then A is invertible.

Proof. If $Ax = Ay$, then $Ax - Ay = A(x - y) = 0$, which, by hypothesis, implies $(x - y) = 0$, or $x = y$, hence condition (a) for invertibility is satisfied. To prove condition (b), let $\{x_1, \cdots, x_n\}$ be a basis in V. (Note that we can pick a *finite* basis.) If it can be shown that $\{Ax_1, \cdots, Ax_n\}$ is also a basis, then every vector $y \in V$ can be written in the form

$$y = \sum_{i=1}^{n} \alpha_i A x_i = A\left(\sum_{i=1}^{n} \alpha_i x_i\right) ,$$

fulfilling condition (b). Thus we ask: Is $\{Ax_1, \cdots, Ax_n\}$ also a basis? It is a set of n vectors in an n-dimensional vector space, so it will be a basis if the set is linearly independent. To prove linear independence, we assume that

$$\sum_i \alpha_i A x_i = 0 .$$

It follows that

$$A\left(\sum_{i=1}^{n} \alpha_i x_i\right) = 0 \Longrightarrow \sum_{i=1}^{n} \alpha_i x_i = 0 ,$$

by hypothesis. The linear independence of the x_i now implies $\alpha_i = 0$ for all i, so the set $\{Ax_i\}$ is linearly independent and hence does form a basis. QED

The converse also holds: If A is invertible, then (obviously) $Ax = 0$ implies $x = 0$.

It should be noted that the condition of Theorem 3.7, that $Ax = 0 \Longrightarrow x = 0$, is equivalent to condition (a) of Definition 3.9. Thus Theorem 3.7 shows that on a *finite* dimensional space, condition (a) alone suffices for invertibility.

With the help of Theorem 3.6 the reader should have no trouble in supplying the simple proofs of the statements comprising the following theorem.

Theorem 3.8. If A and B are invertible, then AB is invertible and $(AB)^{-1} = B^{-1}A^{-1}$. If A is invertible and $\alpha \neq 0$, then αA is invertible and $(\alpha A)^{-1} = (1/\alpha)A^{-1}$. If A is invertible, then A^{-1} is invertible and $(A^{-1})^{-1} = A$.

3.7 MATRICES

Just as abstract vectors can be represented in terms of coordinates relative to a particular basis, so abstract linear transformations can be represented as matrices in terms of a given basis. We now develop the connection between linear transformations and matrices.

Definition 3.10. Let V be an n-dimensional vector space, with $X = \{x_1, \cdots, x_n\}$ a basis in V, and let A be a linear transformation mapping V into itself.

Since every vector is a linear combination of the $\{x_i\}$, we have in particular

$$Ax_j = \sum_{i=1}^{n} a_{ij} x_i , \qquad \text{for } j = 1, \cdots, n . \tag{3.7}$$

The set of n^2 scalars (a_{ij}) is the *matrix* of A relative to the basis X (or in the "coordinate system" X).

$$(a_{ij}) \equiv \begin{bmatrix} a_{11} & a_{12} & \cdots & a_{1n} \\ a_{21} & a_{22} & \cdots & a_{2n} \\ \cdot & & & \cdot \\ \cdot & & & \cdot \\ \cdot & & & \cdot \\ a_{n1} & a_{n2} & \cdots & a_{nn} \end{bmatrix} \equiv [A] \equiv A \qquad \text{(by sloppy convention)} .$$

That is, every transformed vector Ax_j (for all j) can be reexpressed in terms of the basis vectors x_i by n coefficients, as previously. But there are n basis vectors, hence n transformed vectors Ax_j ($j = 1, \cdots, n$), and hence n^2 coefficients, a_{ij}.

Note carefully the order of indices in the equation that defines the a_{ij}. The rule is: write Ax_j as a linear combination of $(x_1, \cdots, x_n)$, and write the coefficients so obtained as the jth *column* of the matrix A. The first index on $a_{ij} \equiv [A]_{ij}$ is the row index, the second the column index. The location of the elements in the matrix representation of A will depend on the ordering of the vectors in the basis, but we are free to adopt a fixed ordering of basis vectors if it becomes desirable to associate a uniquely-arranged array of elements with a linear transformation. We say that two matrices (a_{ij}) and (b_{ij}) are equal if and only if $a_{ij} = b_{ij}$, for all i and j.

In Section 1.4, Eq. (1.13), we expanded one set of basis vectors, $\mathbf{e}'_i$, in terms of another set, $\mathbf{e}_i$, to define a matrix. There we had the equation

$$\mathbf{e}'_j = \sum_i a_{ji} \mathbf{e}_i .$$

The $\mathbf{e}_i$ played the role that the x_i play now. The x_i in Chapter 1 referred to components of the vectors $\mathbf{x}$, whereas now the x_i are themselves vectors in an abstract vector space. However, the convention of Chapter 1 that results in the above equation is not our present one. In terms of the convention of this chapter, we should have written in Chapter 1,

$$\mathbf{e}'_j = \sum_i a_{ij} \mathbf{e}_i .$$

This notational inconsistency, and the reasons for retaining it, were discussed in Chapter 1. All the equations of Chapter 1 may be converted to our present convention by simply replacing a_{ij} by a_{ji}.

Now let us see how the application of a linear transformation A to a vector x looks in terms of the representation of these abstract objects with respect to a basis, $X = \{x_1, \cdots, x_n\}$. For any $x \in V$,

$$x = \sum_{j=1}^{n} \alpha_j x_j .$$

And the matrix of the linear transformation is defined by

$$Ax_j = \sum_{i=1}^{n} a_{ij} x_i .$$

Therefore

$$y = Ax = A \sum_{j=1}^{n} \alpha_j x_j = \sum_{j=1}^{n} \alpha_j A x_j = \sum_{j=1}^{n} \alpha_j \sum_{i=1}^{n} a_{ij} x_i = \sum_{i,j=1}^{n} a_{ij} \alpha_j x_i . \tag{3.8}$$

But the vector y, produced by A operating on x, may also be represented in terms of the basis X:

$$y = \sum_{i=1}^{n} \beta_i x_i .$$

Subtracting this from Eq. (3.8), we have

$$\sum_{i=1}^{n} \left[\beta_i - \sum_{j=1}^{n} a_{ij} \alpha_j \right] x_i = 0 ,$$

which, from the linear independence of the basis, implies

$$\beta_i = \sum_{j=1}^{n} a_{ij} \alpha_j .$$

This conforms exactly with our earlier specialized result, Eq. (1.16c), the β_i corresponding to the x'_i. This equation is often written symbolically as

$$y = Ax , \tag{3.9}$$

where y and x stand for the coordinate representatives of the vectors y and x, and A stands for the matrix representing the linear transformation A, all referred to the same basis X. A more accurate notation than that in Eq. (3.9) is

$$[\beta] = [A][\alpha] , \tag{3.10}$$

where $[\beta]$ and $[\alpha]$ are matrices of n rows and one column, called *column matrices*, or *column vectors*, defined by

$$[\beta]_{i1} = \beta_i , \qquad i = 1, \cdots, n ,$$

and similarly for α, and $[A]$ is the matrix of A. In practice, however, one usually falls back on the sloppier, simple notation of Eq. (3.9).

Row vectors are matrices of one row and n columns. Some authors distinguish carefully between row and column vectors but we shall not; the context always enables one to do so if necessary.

Thus far, we have discussed only square matrices, i.e., those having the same number of rows and columns. The reason for this is that the operators, A, in question are defined to take vectors in V into vectors in V, that is, to map V into itself. However, we could equally well imagine a linear operator B which maps V into some other vector space U. If V and U happen to be of different dimensionalities, the matrix which represents B will not be square.

Let $\{x_i\}$ be a basis for V and $\{y_i\}$ be a basis for U; then Bx_j is a vector in U so we may write

$$Bx_j = \sum_{i=1}^{m} b_{ij} y_i \qquad j = 1, 2, \cdots, n\,,$$

where we have taken V to be n-dimensional and U to be m-dimensional, so that (b_{ij}) is an $m \times n$ matrix, having m rows and n columns. We leave it to the reader to show that if x is a vector in V and y is a vector in U such that $y = Bx$, and if

$$y = \sum_{i=1}^{m} \beta_i y_i\,, \qquad x = \sum_{i=1}^{n} \alpha_i x_i\,,$$

then

$$\beta_i = \sum_{j=1}^{n} b_{ij}\alpha_j\,, \qquad i = 1, 2, \cdots, m\,,$$

which we again write as

$$[\beta] = [B][\alpha]$$

in analogy with Eq. (3.10).

Example. We shall examine the form of the matrices of the differentiation and multiplication operators, D and X. Earlier we remarked that in talking about linear operators, one had to be sure that they were defined on the space in question. D, for example, is certainly defined on the space P_n of polynomials of degree $\leq n$. However, it turns every element of P_n into an element of P_{n-1}; this is why D is not invertible when considered as an operator mapping P_n into P_n.

Since P_{n-1} is a subspace of P_n, we can consider D as either an operator which maps P_n into P_{n-1}, or P_n into P_n. In the former case, D will be represented by an $n \times (n+1)$ matrix, in the latter by an $(n+1) \times (n+1)$ matrix, [By an unfortunate convention, P_n is an $(n+1)$-dimensional space.]

We use the obvious bases

$$y_j(t) = t^{j-1}\,, \qquad j = 1, 2, \cdots, n \qquad \text{for } P_{n-1};$$

$$x_j(t) = t^{j-1}\,, \qquad j = 1, 2, \cdots, n+1 \qquad \text{for } P_n\,.$$

Then

$$Dx_j = (j-1)t^{j-2} = (j-1)y_{j-1} = \sum_{i=1}^{n} d_{ij}y_i\,, \qquad j = 1, 2, \cdots, n+1\,.$$

Therefore $d_{ij} = j - 1$ if $i = j - 1$, and $d_{ij} = 0$ otherwise. Thus

$$[D] = (d_{ij}) = \begin{bmatrix} 0 & 1 & 0 & 0 & \cdots & 0 & 0 \\ 0 & 0 & 2 & 0 & & 0 & 0 \\ 0 & 0 & 0 & 3 & & 0 & 0 \\ \vdots & & & & & & \vdots \\ 0 & 0 & 0 & 0 & & n-1 & 0 \\ 0 & 0 & 0 & 0 & \cdots & 0 & n \end{bmatrix}. \tag{3.11}$$

Thus $[D]$ is an $n \times (n+1)$ matrix. If we wanted to think of D as mapping P_n into P_n, we would represent D by a matrix obtained from the one above by adding an $(n+1)$th row of zeroes.

The operator X can only be represented as a nonsquare matrix, since it maps P_{n-1} into P_n, and P_n is not a subspace of P_{n-1}. Thus X will be represented by an $(n+1) \times n$ matrix. We leave it to the reader to show that

$$[X] \equiv (x_{ij}) = \begin{bmatrix} 0 & 0 & 0 & \cdots & & 0 \\ 1 & 0 & 0 & & & 0 \\ 0 & 1 & 0 & & & 0 \\ 0 & 0 & 1 & & & 0 \\ \vdots & & & & & \vdots \\ 0 & 0 & 0 & & 1 & 0 \\ 0 & 0 & 0 & \cdots & 0 & 1 \end{bmatrix}. \tag{3.12}$$

We shall return to the matrix representation of the commutator $[D, X]$ after first discussing the addition and multiplication of the matrices of linear transformations.

The problem is to find the matrices associated with the sum and product of two linear transformations whose matrices are known in some fixed basis, $X = \{x_1, \cdots, x_n\}$. Let the matrices of A and B with respect to X be (a_{ij}) and (b_{ij}), respectively. We want to know the matrices of

1. $S \equiv \alpha A + \beta B\,,$

where α and β are scalars, and

2. $C = AB\,,$

with respect to the basis X.

$$1.\quad Sx_j = (\alpha A + \beta B)x_j = \alpha A x_j + \beta B x_j = \alpha \sum_{i=1}^{n} a_{ij}x_i + \beta \sum_{i=1}^{n} b_{ij}x_i$$

$$= \sum_{i=1}^{n} (\alpha a_{ij} + \beta b_{ij})x_i = \sum_{i=1}^{n} s_{ij}x_i\,,$$

where the final expression simply defines the matrix of S, (s_{ij}). The linear independence of the basis vectors implies that

$$[S]_{ij} = s_{ij} = \alpha a_{ij} + \beta b_{ij} \, .$$

2. $$Cx_j = ABx_j = A(Bx_j) = A\Big(\sum_{k=1}^{n} b_{kj} x_k\Big) = \sum_{k=1}^{n} b_{kj} A x_k$$

$$= \sum_{k=1}^{n} b_{kj} \sum_{i=1}^{n} a_{ik} x_i = \sum_{i,k=1}^{n} a_{ik} b_{kj} x_i = \sum_{i=1}^{n} c_{ij} x_i \, .$$

Therefore, by the same reasoning as in (1) above

$$[C]_{ij} = c_{ij} = \sum_{k=1}^{n} a_{ik} b_{kj} \, . \tag{3.13}$$

This equation defines the multiplication of two $n \times n$ matrices. We derived a special instance of such matrix multiplication in Eq. (1.22).

We can use Eq. (3.13) to define the multiplication of two arrays of numbers. Therefore, the product, AB, of two matrices will be defined if and only if the number of columns of A equals the number of rows of B. For example,

$$\begin{bmatrix} 1 & 2 \\ 3 & 4 \\ 5 & 6 \end{bmatrix} \begin{bmatrix} 1 & 3 & 5 & 7 \\ 2 & 4 & 6 & 8 \end{bmatrix} = \begin{bmatrix} 5 & 11 & 17 & 23 \\ 11 & 25 & 39 & 53 \\ 17 & 39 & 61 & 83 \end{bmatrix} .$$

The product of an $(l \times m)$ matrix (l rows and m columns) and an $(m \times n)$ matrix is an $(l \times n)$ matrix. Thus Eq. (3.10) represents the multiplication of an $(n \times n)$ matrix, A, by the $(n \times 1)$ column matrix, α, to yield the $(n \times 1)$ column matrix, β.

Clearly, the matrices of the special linear transformation, O and I, are

$$[O]_{ij} = 0 \, , \qquad [I]_{ij} = \delta_{ij} \, . \tag{3.14}$$

Note that for the zero and identity matrices:

$$OA = AO = O \, , \qquad IA = AI = A \, ,$$

just as for the corresponding linear transformations.

At this juncture, one could prove that linear transformations and matrices are related isomorphically. The proof is simple and is omitted; the intuitive content of the theorem is only a slight generalization of Theorem 3.5, which demonstrated that vectors are isomorphic to n-tuples. Henceforth we can be sure that any theorem proved for a linear transformation automatically holds for the matrix associated with it relative to a given basis, and vice versa. Thus, the sloppy notation which fails to distinguish between linear transformations and matrices has a measure of theoretical justification.

The most unusual of the properties of linear transformations when contrasted with the properties of numbers is that they need not commute. The matrices of linear transformations will therefore exhibit this same property. From Eq.

(3.13) we see that, in general, $(AB)_{ij} = \sum_{k=1}^{n} a_{ik}b_{jk}$, will not be equal to $(BA)_{ij} = \sum_{k=1}^{n} b_{ik}a_{kj}$.

Example. Consider the products of the matrices of the differentiation and multiplication operators defined in Eqs. (3.11) and (3.12). Since D is an $n \times (n+1)$ matrix and X is an $(n+1) \times n$ matrix, DX will be an $n \times n$ matrix. We have

$$(DX)_{ij} = \sum_{k=1}^{n+1} d_{ik}x_{kj} = \begin{bmatrix} 1 & 0 & 0 & \cdots & 0 \\ 0 & 2 & 0 & & 0 \\ 0 & 0 & 3 & & 0 \\ \cdot & & & & \cdot \\ \cdot & & & & \cdot \\ \cdot & & & n-1 & 0 \\ 0 & 0 & 0 \cdots & 0 & n \end{bmatrix}.$$

It is evident from Eqs. (3.11) and (3.12) for D and X that XD will be $(n+1)$-dimensional, whereas DX is n-dimensional. This is sensible, since X maps P_{n-1} into P_n, and D maps P_n back into P_{n-1}; thus DX maps P_{n-1} into P_{n-1}, and DX is n-dimensional. However, XD maps P_n into P_n, because D first maps P_n into P_{n-1} and then X maps P_{n-1} into P_n; therefore XD is $(n+1)$-dimensional. A simple calculation gives

$$(XD)_{ij} = \sum_{k=1}^{n} x_{ik}d_{kj} = \begin{bmatrix} 0 & 0 & 0 & \cdots & 0 & 0 \\ 0 & 1 & 0 & & 0 & 0 \\ 0 & 0 & 2 & & 0 & 0 \\ \vdots & & & & & \vdots \\ 0 & 0 & 0 & & n-1 & 0 \\ 0 & 0 & 0 & & 0 & n \end{bmatrix}.$$

Suppose that we restrict XD to mapping P_{n-1} into P_{n-1}, so that D is $(n-1) \times n$ and X is $n \times (n-1)$. Then we get an $n \times n$ matrix obtained from the above matrix by lopping off the $(n+1)$ row and $(n+1)$ column. If we now compare DX with XD, both operating on P_{n-1}, we see that they do not commute and that in fact

$$DX - XD = I,$$

in agreement with the discussion in Section 3.5.

Even though the matrices D and X are not square for any finite-dimensional space, they may, in a rather imprecise way, be thought of as square infinite-dimensional matrices if n increases without limit, since the difference between n and $n+1$ becomes negligible when n is arbitrarily large. They are often dealt with as such in field theory.

We close this section by introducing a number of special type of matrices.

1. *Diagonal matrices.* The special property of a *diagonal* matrix is that its only nonzero elements lie on its northwest-southeast diagonal, that is, $a_{ij} = 0$ if

$i \neq j$, or $a_{ij} = d_i\delta_{ij}$. Clearly, any two $n \times n$ diagonal matrices commute, and their product is also a diagonal matrix.

2. *Idempotent matrices.* A matrix A is said to be *idempotent* if $A^2 = A$. Thus the matrices I, O, and

$$\begin{bmatrix} 1 & -1 \\ 0 & 0 \end{bmatrix}$$

are idempotent.

3. *The transpose of a matrix.* Given $A = (a_{ij})$, we define the *transpose* of A, written $\tilde{A}$, as

$$[\tilde{A}]_{ij} = a_{ji} .$$

Theorem 3.9.

a) If $\tilde{A} = B$, then $\tilde{B} = A$ (consequently, $\tilde{\tilde{A}} = A$),

b) $\widetilde{A + B} = \tilde{A} + \tilde{B}$,

c) $\widetilde{AB} = \tilde{B}\tilde{A}$.

Proof. The proofs of (a) and (b) are trivial. For (c), let $C = AB$. Then

$$[\tilde{C}]_{ij} \equiv [\widetilde{AB}]_{ij} = [AB]_{ji} = \sum_k [A]_{jk}[B]_{ki} = \sum_k [\tilde{B}]_{ik}[\tilde{A}]_{kj} = [\tilde{B}\tilde{A}]_{ij} ,$$

that is,

$$\tilde{C} \equiv \widetilde{AB} = \tilde{B}\tilde{A} . \qquad \text{QED}$$

4. *The adjoint of a matrix.* We define the adjoint of a matrix A to be the matrix $A^\dagger$ whose elements are given by

$$[A^\dagger]_{ij} = a_{ji}^* .$$

Clearly, this can also be written as $A^\dagger = \tilde{A}^*$. As a simple corollary of Theorem 3.9, we have the following results:

a) If $A^\dagger = B$, then $B^\dagger = A$, that is, $(A^\dagger)^\dagger = A$.

b) $(A + B)^\dagger = A^\dagger + B^\dagger$.

c) $(AB)^\dagger = B^\dagger A^\dagger$.

5. *Symmetric and antisymmetric matrices.* An $n \times n$ matrix A is symmetric if $A = \tilde{A}$, that is, if $a_{ij} = a_{ji}$. A is antisymmetric if $A = -\tilde{A}$; that is, if $a_{ij} = -a_{ji}$. The diagonal elements of an antisymmetric matrix must all be zero. Any matrix A may be expressed as the sum of a symmetric and an antisymmetric matrix, for we may write identically

$$A = \underbrace{\tfrac{1}{2}(A + \tilde{A})}_{\text{symmetric}} + \underbrace{\tfrac{1}{2}(A - \tilde{A})}_{\text{antisymmetric}} ,$$

exactly as for tensors [Eq. (1.97)].

6. *Singular and nonsingular matrices.* An $n \times n$ matrix A is said to singular if there exists an $n \times n$ matrix B such that $AB = I$. Otherwise A is singular. We shall see that if the matrix of a linear transformation is nonsingular, then the linear transformation is invertible, and conversely; in other words, a necessary and sufficient condition for invertibility, of either A or B, on a *finite*-dimensional vector space is $AB = I$. $BA = I$ is not needed on a finite-dimensional space, as it is in Theorem 3.6.

3.8 DETERMINANTS

Aside from being useful in solving systems of linear algebraic equations, the determinant of a matrix is important because it can give information about the associated linear transformation. For example, we shall prove the general result (anticipated in Section 1.4) that the determinant of a matrix corresponding to an orthogonal transformation is always ± 1. Also, we shall see that the determinant of a matrix A is zero if and only if A is singular.

The determinant of a matrix A, written det A, is a function that acts on a matrix to yield a number.

Definition 3.11. For a given linear transformation, A, which is represented by the $n \times n$ matrix (A_{ij}),

$$\det A \equiv \epsilon_{abc\cdots n} A_{1a} A_{2b} \cdots A_{Nn}\,, \tag{3.16a}$$

where we are using the Einstein summation convention, as we shall do throughout this section. $\epsilon_{abc\cdots n}$ is the antisymmetric symbol on N objects, defined by

$$\epsilon_{abc\cdots n} \equiv \begin{cases} +1 \text{ if } (abc\cdots n) \text{ is an even permutation of } (1, 2, 3, \cdots, N)\,, \\ -1 \text{ if } (abc\cdots n) \text{ is an odd permutation of } (1, 2, 3, \cdots, N)\,, \\ \;\;0, \text{ otherwise.} \end{cases}$$

The reader will note that this notation is not without flaws. The subscripts over which we are summing are denoted by $a, b, c, \cdots$, etc. Clearly, if we have a 100-dimensional vector space, there will not be enough letters in the English alphabet to serve as subscripts in Definition 3.11. However, the reader should have no difficulty in imagining an alphabet with 100 letters which we could use in such an emergency; e.g., English + Greek + Arabic + Hebrew + $\cdots$ + Navaho. In discussing determinants, we shall content ourselves with the familiar alphabet. Thus in this section, when we write A_{Nn}, by n we understand the Nth letter of the alphabet. We could, of course, always resort to using subscripted subscripts, such as $i_1, i_2, \cdots$, etc.; however, since the main problem with determinants is a notational one, it is perhaps useful to pare down the symbols as much as possible.

If $(abc\cdots n)$ is some permutation of $(1, 2, 3, \cdots, N)$ we may write $\epsilon_{abc\cdots n} = (-1)^P$, where P is the number of transpositions of subscripts needed to produce the normal order $(1, 2, 3, \cdots, N)$. If P is even or odd, we speak of the permutation as being of even or odd parity. This ϵ symbol is the natural generalization of the symbol ϵ_{ijk} which proved so useful in the treatment of vectors and tensors in Chapter 1.

Det A is seen to consist of a sum of $N!$ products, each product consisting of one element from every row and every column. Products containing more than a single term from a given column do not contribute to det A because the corresponding ϵ symbol vanishes. From this it is immediately clear that an equivalent definition for det A is

$$\det A = \epsilon_{abc\cdots n} A_{a1} A_{b2} \cdots A_{nN} . \tag{3.16b}$$

The roles of the rows and columns have simply been reversed, or what is the same thing, the rows have become the columns and vice versa. Thus Eq. (3.16b) tells us that

$$\det A = \det \tilde{A} = (\det A^{\dagger})^* . \tag{3.17}$$

This alternative definition of det A means that for any theorem that can be proved for rows, there will be a corresponding theorem that holds for columns.

If at any time in this treatment of the determinants of $n \times n$ matrices the reader cannot see the "forest" (the meaning) for the "trees" (the subscripts), he should simply write out the equations for the 2×2 or 3×3 case. In fact, when any theorem concerning $n \times n$ matrices is stated, it is often very helpful in proving (or disproving!) it to consider its meaning in the 2×2 case. One should not feel that any "loss of face" is involved in trying out these general theorems on the simplest nontrivial special case. After all, the general theorems were originally derived by abstraction from just such special cases.

We now list a number of the properties of determinants.

1. A common factor of each element of a row (or column) may be factored out as a multiplicative constant.
2. If any row (or column) is all zeros, the determinant is zero.
3. det $I = 1$.
4. Interchanging any two rows (or columns) changes the sign of the determinant.
5. If any two rows (or columns) are equal, the determinant is zero.
6. det $(AB) = \det A \det B$.
7. A scalar multiple of a row (or column) may be added to another row (or column) without changing the value of the determinant.
8. det $A = 0$ if and only if the row (or column) vectors of A are linearly dependent.

The proofs of (1) through (3) are simple. We shall prove properties (4), (5), and (6). It will be clear how to prove (7), and then (8), after the proofs of (4), (5), and (6) have been studied; so we shall leave these for the reader to practice on.

The first step in proving (4) is to show that if two *adjacent* rows of a matrix are interchanged, its determinant merely changes sign. Let A' be the matrix obtained from A by interchanging the Kth and Lth rows, which we take to be adjacent. Then

$$\begin{aligned}\det A' &= \epsilon_{a\cdots kl\cdots n}A_{1a}A_{2b}\cdots A_{Lk}A_{Kl}\cdots A_{Nn}\\ &= \epsilon_{a\cdots lk\cdots n}A_{1a}\cdots A_{Kk}A_{Ll}\cdots A_{Nn} = -\det A\,,\end{aligned}$$

where we have interchanged the order of the factors in the product $A_{Kk}A_{Ll}$, interchanged the summed over, dummy subscripts, k and l, and, finally, interchanged the order of lk in the ϵ-symbol which introduces the minus sign. Property (4) now follows if we note that the interchange of *any* two rows can be effected by an odd number of transpositions of adjacent rows. Property (5) is an immediate consequence of (4).

We now derive property (6): The determinant of the product of two matrices is the product of their determinants. We first note that

$$\epsilon_{abc\cdots n}\det A = \epsilon_{a'b'c'\cdots n'}A_{aa'}A_{bb'}\cdots A_{nn'}\,, \tag{3.18a}$$

for if $(abc\cdots n)$ are set equal to $(1, 2, 3, \cdots, N)$, Eq. (3.18a) reduces to the definition of $\det A$ (Eq. 3.16a). If two (or more) of the unprimed subscripts are equal, then both sides of Eq. (3.18a) vanish since the right-hand side represents determinant of a matrix with two (or more) equal rows; and if $(abc\cdots n)$ is a permutation of $(1, 2, 3, \cdots, N)$, then the expressions on the right and left of Eq. (3.18a) will give $(-1)^P \det A$ where P is the parity of the permutation. In exactly the same manner, it follows that

$$\epsilon_{abc\cdots n}\det A = \epsilon_{a'b'c'\cdots n'}A_{a'a}A_{b'b}\cdots A_{n'n}\,. \tag{3.18b}$$

Now consider the product $(\det A)(\det B)$. Applying Eq. (3.16b) to B, we have

$$(\det A)(\det B) = (\det A)\,\epsilon_{abc\cdots n}B_{a1}B_{b2}\cdots B_{nN}\,;$$

using Eq. (3.18b), we may write this as

$$\begin{aligned}(\det A)(\det B) &= \epsilon_{a'b'c'\cdots n'}A_{a'a}A_{b'b}\cdots A_{n'n}B_{a1}B_{b2}\cdots B_{nN}\\ &= \epsilon_{a'b'c'\cdots n'}(A_{a'a}B_{a1})(A_{b'b}B_{b2})\cdots(A_{n'n}B_{nN})\\ &= \epsilon_{a'b'c'\cdots n'}(AB)_{a'1}(AB)_{b'2}\cdots(AB)_{n'N}\\ &= \det(AB)\,. \qquad \text{QED}\end{aligned}$$

There is a way to picture this result geometrically that is considerably more intuitive than the proof we have given, and also suggests the relevance of determinants to physics. In three dimensions, the volume of a parallelepiped defined by three vectors **A**, **B**, and **C** [Fig. (3.3)] is

$$\mathbf{A}\cdot(\mathbf{B}\times\mathbf{C}) = \epsilon_{ijk}A_iB_jC_k = \det\begin{bmatrix}A_1 & A_2 & A_3\\ B_1 & B_2 & B_3\\ C_1 & C_2 & C_3\end{bmatrix}.$$

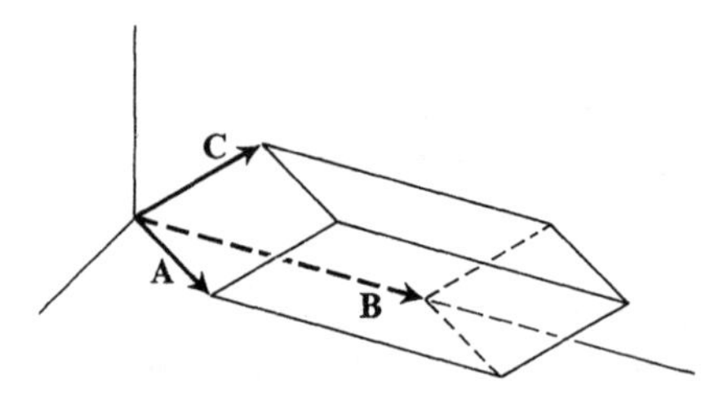

Fig. 3.3 Parallelepiped defined by three vectors, **A**, **B**, and **C**.

Given this interpretation of the quantity $\mathbf{A}\cdot(\mathbf{B}\times\mathbf{C})$ as a *volume*, it is not surprising that it is invariant under rotations of the coordinate system, and hence is a scalar. Actually, it is a pseudoscalar; it changes sign under inversion of the coordinate system. Hence the physical volume is really $|\mathbf{A}\cdot(\mathbf{B}\times\mathbf{C})|$.

This association of a determinant with a volume may be generalized to more or fewer dimensions. In two dimensions, the "volume" becomes the area of a parallelogram determined by two vectors in a plane. Let I denote the matrix which corresponds to the unit square:

$$I = \begin{bmatrix} 1 & 0 \\ 0 & 1 \end{bmatrix}.$$

The row vectors of I determine the endpoints of the sides of the unit square with one corner at the origin. Now a matrix T multiplying I and giving T may be viewed as transforming the unit square to a parallelogram defined by the row vectors of T. The area of this parallelogram is det T.

We can decompose an arbitrary parallelogram represented by a matrix A, of area det A, into little *squares* of arbitrarily small area s_i where $\sum_i s_i =$ det A. Then a matrix T multiplying A, and changing it into another parallelogram whose matrix is TA, can be viewed as operating on each of the little squares of which A consists. Under the transformation, each little of area s_i becomes a little parallelogram of area $a_i = (\det T)s_i$. Therefore the new parallelogram whose matrix is TA has an area of

$$\sum_i a_i = \sum_i (\det T)s_i = (\det T)(\sum_i s_i) = \det T \det A\,.$$

But the area of the parallelogram represented by the matrix TA is also given by det (TA). Therefore det TA = det T det A.

The key ideas of this proof expressed in three-dimensional language are:

1. The unit cube is carried by a linear transformation T into a parallelepiped of volume det T. The determinant of the linear transformation gives the ratio of the final volume to that of the original cube.

2. The determinant of the matrix of a linear transformation gives the ratio of the volumes of two arbitrary parallelepipeds whose corresponding matrices are connected by the matrix of the linear transformation. The determinant is thus a scale factor for measuring volume changes.

3. If we go from one volume to another to a third, and so on, we accumulate these scale factors, and the total factor relating the initial and the final volume is the product of all the determinants of all the matrices of the linear transformations which relate the different parallelepipeds to each other. On the other hand, the first parallelepiped is related to the last by the product of all the matrices; the volume of the parallelepiped that results from the successive transformations is just the determinant of this product of matrices. Thus the product of the determinants of n matrices equals the determinant of the product of the n matrices. This proof holds for spaces of any dimension.

A very important combinatorial result that reduces the evaluation of the determinant of an $n \times n$ matrix to the evaluation of the determinants of $(n-1) \times (n-1)$ matrices is the following:

Theorem 3.10. $\det A = \sum_j A_{Jj} \operatorname{cof}(A_{Jj})$, for any row J, that is, J is not summed on. We include the summation on j explicitly to remind the reader that J is *not* summed. By $\operatorname{cof}(A_{Jj})$ we mean the cofactor of the element A_{Jj}, which equals $(-1)^{J+j}$ times the determinant of the $(n-1) \times (n-1)$ matrix obtained from A by striking out the Jth row and the jth column.

Proof.
$$\det A = \epsilon_{abc\cdots n} A_{1a} A_{2b} \cdots A_{Nn}$$
$$= \sum_j A_{Jj} [\epsilon_{a\cdots ijk\cdots n} A_{1a} \cdots A_{Ii} A_{Kk} \cdots A_{Nn}] , \tag{3.19}$$

where we have put the cofactor of A_{Jj} in brackets. Now

$$\operatorname{cof}(A_{Jj}) = (-1)^{J-1} \epsilon_{ja\cdots ik\cdots n} A_{1a} \cdots A_{Ii} A_{Kk} \cdots A_{Nn} , \tag{3.20}$$

where we have moved the subscript j in the symbol ahead of the others by $(J-1)$ transpositions. (Remember that j means the Jth letter of the alphabet!) These transpositions produce a factor $(-1)^{J-1}$. Now if the reader will stare at Eq. (3.20) for a moment, he will see that $\operatorname{cof}(A_{Jj})$ is nearly equal to what we asserted in the statement of the theorem: we have removed the Jth row, so there is no factor A_{Jj}, and in summing on the set $(a \cdots ik \cdots n)$ no subscript can take on the numerical value of j (remember that in Eq. (3.20) j is fixed at some integer between 1 and N), because we have excluded the jth column. Only the j in the ϵ-symbol stands between us and our result. But

$$\epsilon_{ja\cdots ik\cdots n} = (-1)^{j-1} \tilde{\epsilon}_{a\cdots ik\cdots n} , \tag{3.21}$$

where on the right-hand side of Eq. (3.21) the subscripts $a, \cdots, i, k, \cdots, n$ take on all values between 1 and N, *excluding* the (fixed) value taken by j. The normal ordering in this $\tilde{\epsilon}$ symbol will be the usual chronological order, $1, 2, \cdots, j-1, j+1, \cdots, N$. Thus we obtain

$$\operatorname{cof}(A_{Jj}) = (-1)^{J+j} \tilde{\epsilon}_{a\cdots ik\cdots n} A_{1a} \cdots A_{Ii} A_{Kk} \cdots A_{Nn} ,$$

where in the $N-1$ summations implied by the Einstein convention, the indices *may not take on the value* j. Therefore the cofactor is just $(-1)^{J+j}$ times the determinant of the $(n-1) \times (n-1)$ matrix obtained from A by striking out the Jth row and the jth column. QED

An analog to Theorem 3.10 for columns is obtained in exactly the same way. It is

$$\det A = \sum_j A_{jJ} \operatorname{cof}(A_{jJ}) , \tag{3.22}$$

for any column J, that is, no summation on J.

A simple and useful generalization of this basic result is

Theorem 3.11.

$$\delta_{KJ} \det A = \sum_j A_{Kj} \operatorname{cof}(A_{Jj}) , \tag{3.23a}$$

for all J and K.

Proof. For $J = K$, this reduces to Theorem 3.10. For $J \neq K$, the left-hand side vanishes as does the right-hand side since the expansion on the right-hand side is equivalent to the expression for the determinant of a matrix with two equal rows. To see this, we write

$$\begin{aligned}\sum_j A_{Kj} \operatorname{cof}(A_{Jj}) &= \sum_j A_{Kj}(-1)^{J+j}\tilde{\epsilon}_{a\cdots ik\cdots n}A_{1a}\cdots A_{Ii}A_{Kk}\cdots A_{Nn} ,\\ &= \sum_j \epsilon_{ja\cdots ik\cdots n}A_{1a}\cdots A_{Ii}A_{Kj}A_{Kk}\cdots A_{Nn}(-1)^{J-1}\\ &= \sum_j \epsilon_{a\cdots ijk\cdots n}A_{1a}\cdots A_{Ii}A_{Kj}A_{Kk}\cdots A_{Nn} .\end{aligned}$$

This last expression is just the determinant of a matrix containing two equal rows (the Kth) and hence vanishes. QED

Here we have just reversed the reasoning which led in Theorem 3.10 to the expression for $\operatorname{cof}(A_{Jj})$. Similarly, it follows that

$$\delta_{JK} \det A = \sum_j A_{jK} \operatorname{cof}(A_{jJ}) , \tag{3.23b}$$

for all J and K.

Before ending our discussion of determinants, we should perhaps say a final word about the $\tilde{\epsilon}$ symbol we defined in proving Theorem 3.10. Any ϵ-type symbol is just a way of ordering a set of integers. The ϵ-symbol orders the first N integers, while $\tilde{\epsilon}$ orders the first N integers with *one* integer missing, i.e., it has only $N-1$ subscripts (if it is to be nontrivial). Clearly, we could also define $\tilde{\tilde{\epsilon}}$, which would order the first N integers with *two* integers missing. In all such cases the defining properties of the symbol are completely analogous to those which we listed for the ϵ-symbol. The relation, Eq. (3.21),

$$\epsilon_{ja\cdots ik\cdots n} = (-1)^{j-1}\tilde{\epsilon}_{a\cdots ik\cdots n}$$

is evident when one recognizes that the same number of transpositions is required to order $(a\cdots ik\cdots n)$ numerically in ϵ as in $\tilde{\epsilon}$. Then an additional $j-1$

transpositions are required to get j to the correct position in ϵ. Hence the factor $(-1)^{j-1}$ multiplying $\tilde{\epsilon}$.

We now proceed to apply these results to the theory of linear operators.

Definition 3.11. The *classical adjoint* of an $n \times n$ matrix A is the $n \times n$ matrix, adj A, where $(\text{adj } A)_{ij} = \text{cof } (A_{ji})$.

It follows directly from Eqs. (3.23a) and (3.23b) that

$$A(\text{adj } A) = (\text{adj } A)A = (\det A)I \,. \tag{3.24}$$

This is a matrix equation. The matrix $[(\det A)I]$ is diagonal, with $\det A$ for every diagonal element. (A word of caution; the *classical adjoint* is not the same as the *adjoint operator* to be defined in Section 4.4). Now we prove an extremely useful theorem.

Theorem 3.12. The matrix A has an inverse, A^{-1}, if and only if $\det A \neq 0$.

Proof. If $\det A \neq 0$, we may divide Eq. (3.24) by $\det A$ to get

$$A\left(\frac{\text{adj } A}{\det A}\right) = \left(\frac{\text{adj } A}{\det A}\right)A = I \,.$$

By theorem 3.6, this implies that there exists a unique inverse

$$A^{-1} = \left(\frac{\text{adj } A}{\det A}\right) .$$

Conversely, if there exists an inverse, A^{-1}, then

$$AA^{-1} = I = A^{-1}A \,.$$

Now take the determinant of both sides to get

$$\det (AA^{-1}) = \det A \cdot \det A^{-1} = \det I = 1 \,, \tag{3.25}$$

which implies that $\det A \neq 0$. QED

Corollary. A matrix A is nonsingular if and only if $\det A \neq 0$.

Proof. If A is nonsingular, then by definition there exists a matrix B such that $AB = I$. It follows that $\det A \det B = 1$, so that $\det A \neq 0$. If $\det A \neq 0$, then from theorem (3.12) A has a unique inverse, $A^{-1} = \text{adj } A/\det A$, so there exists a matrix $B = A^{-1}$ such that $AB = I$. Thus, as mentioned previously, having an inverse is the same thing as being nonsingular (in a finite-dimensional vector space). Some authors define a matrix A as nonsingular if $\det A \neq 0$ (and singular if $\det A = 0$), and then follow a slightly different logical route to what is the same key result: that a nonsingular matrix has a unique inverse given by

$$A^{-1} = \frac{1}{\det A} (\text{adj } A) \,. \tag{3.26}$$

It follows from Eq. (3.25) that

$$\det(A^{-1}) = \frac{1}{\det A}. \tag{3.27}$$

The expression $1/A$ is meaningless because A is a matrix; "$1/banana$" is equally meaningless. But $1/\det A \equiv [\det A]^{-1}$ is a scalar because $\det A$ is a scalar-valued function of a matrix A.

Once we have a way of computing the inverse of a matrix, we can solve systems of linear equations, such as $Ax = y$. In the three-dimensional case this shorthand notation [introduced in Eqs. (3.9) and (3.10)] stands for

$$\begin{bmatrix} a_{11} & a_{12} & a_{13} \\ a_{21} & a_{22} & a_{23} \\ a_{31} & a_{32} & a_{33} \end{bmatrix} \begin{bmatrix} \xi_1 \\ \xi_2 \\ \xi_3 \end{bmatrix} = \begin{bmatrix} \eta_1 \\ \eta_2 \\ \eta_3 \end{bmatrix}.$$

By carrying out the indicated multiplication, we get the system of three equations in three unknowns:

$$\begin{aligned} a_{11}\xi_1 + a_{12}\xi_2 + a_{13}\xi_3 &= \eta_1 \\ a_{21}\xi_1 + a_{22}\xi_2 + a_{23}\xi_3 &= \eta_2 \\ a_{31}\xi_1 + a_{32}\xi_2 + a_{33}\xi_3 &= \eta_3 . \end{aligned} \tag{3.28}$$

Now if $\det A \neq 0$, A^{-1} exists and is given by

$$A^{-1} = [\det A]^{-1} \operatorname{adj} A .$$

Then

$$A^{-1}Ax = Ix = A^{-1}y = [\det A]^{-1} (\operatorname{adj} A)\, y . \tag{3.29}$$

This solves the system for the unknown ξ's in terms of the given η's. Writing out Eq. (3.29) explicitly gives

$$\xi_i = (\det A)^{-1} \sum_{j=1}^{n} [\operatorname{adj} A]_{ij}\eta_j = (\det A)^{-1} \sum_{j=1}^{n} \operatorname{cof}(A_{ji})\eta_j . \tag{3.30}$$

The quantity

$$\sum_{j=1}^{n} \eta_j \operatorname{cof}(A_{ji})$$

is just the determinant of the matrix obtained by replacing the ith column of A by the η's. Equation (3.30) is known as *Cramer's rule*. It is usually *not* the most efficient way to solve large linear systems; there are numerical techniques that are much faster. These are discussed in books on numerical analysis.

For a homogeneous system of equations, where η_j vanishes for all j, Cramer's rule implies that we can only have a nontrivial solution (i.e., some ξ's $\neq 0$) if $\det A = 0$. We can also see this directly from Theorem 3.7, which says that if A is nonsingular, then $Ax = 0$ implies that $x = 0$. Clearly, then, a necessary condition for a nontrivial solution ($x \neq 0$) will be the singularity of A. It is shown in books on linear equations that this is also a sufficient condition for a nontrivial solution. However, the reader should *not* deduce that if $\det A = 0$, then the equation $Ax = y$ has no solution unless $y = 0$. This is false, as Theorem 3.17 will show.

Example. We calculate the inverse of the matrix

$$A = (a_{ij}) = \begin{bmatrix} 1 & 3 & 4 \\ 2 & 1 & 0 \\ 1 & 2 & 3 \end{bmatrix}.$$

First, we note that $\det A = -3$, so that A is nonsingular and has an inverse. From Eq. (3.26), we find that

$$[A^{-1}]_{ij} = \frac{1}{\det A}[\text{adj } A]_{ij} = \frac{1}{\det A}\text{cof }(A_{ji});$$

$$A^{-1} = -\frac{1}{3}\begin{bmatrix} (-1)^{1+1}\det\begin{pmatrix} 1 & 0 \\ 2 & 3 \end{pmatrix} & (-1)^{2+1}\det\begin{pmatrix} 3 & 4 \\ 2 & 3 \end{pmatrix} & (-1)^{3+1}\det\begin{pmatrix} 3 & 4 \\ 1 & 0 \end{pmatrix} \\ (-1)^{1+2}\det\begin{pmatrix} 2 & 0 \\ 1 & 3 \end{pmatrix} & (-1)^{2+2}\det\begin{pmatrix} 1 & 4 \\ 1 & 3 \end{pmatrix} & (-1)^{3+2}\det\begin{pmatrix} 1 & 4 \\ 2 & 0 \end{pmatrix} \\ (-1)^{1+3}\det\begin{pmatrix} 2 & 1 \\ 1 & 2 \end{pmatrix} & (-1)^{2+3}\det\begin{pmatrix} 1 & 3 \\ 1 & 2 \end{pmatrix} & (-1)^{3+3}\det\begin{pmatrix} 1 & 3 \\ 2 & 1 \end{pmatrix} \end{bmatrix}$$

$$= -\frac{1}{3}\begin{bmatrix} 3 & -1 & -4 \\ -6 & -1 & 8 \\ 3 & 1 & -5 \end{bmatrix} = \begin{bmatrix} -1 & \frac{1}{3} & \frac{4}{3} \\ 2 & \frac{1}{3} & -\frac{8}{3} \\ -1 & -\frac{1}{3} & \frac{5}{3} \end{bmatrix}.$$

The reader may check the result by showing that

$$AA^{-1} = A^{-1}A = I.$$

Again it should be understood that there are other numerical methods for inverting matrices which are far more efficient for many purposes.

3.9 SIMILARITY TRANSFORMATIONS

We now investigate the changes induced in the matrix representative of a linear transformation when the set of basis vectors with respect to which it is expressed is changed. Let

$$X = \{x_1, \cdots, x_n\}$$

and

$$Y = \{y_1, \cdots, y_n\}$$

be two given bases in an n-dimensional vector space. We assume that they are related to each other by the linear transformation A:

$$y_i = Ax_i.$$

Now any linear transformation that transforms one basis into another must be invertible, because for any vector

$$x = \sum_i \alpha_i x_i ,$$

if $Ax = 0$, then

$$A\Big(\sum_i \alpha_i x_i\Big) = \sum_i \alpha_i A x_i = \sum_i \alpha_i y_i = 0 ,$$

which implies that $\alpha_i = 0$, since the y_i are linearly independent; thus $x = 0$. Therefore, by theorem 3.7, A^{-1} exists.

Suppose that we expand a given vector x in the two bases X and Y:

$$x = \sum_i \alpha_i x_i \qquad \text{and} \qquad x = \sum_i \beta_i y_i .$$

Now the question is, how are α_i and β_i related? Or, to put it another way, how do the coordinates transform under a change of basis? We have

$$x = \sum_i \beta_i y_i = \sum_i \beta_i A x_i = \sum_i \beta_i \sum_j a_{ji} x_j = \sum_{i,j} a_{ji} \beta_i x_j .$$

But

$$x = \sum_j \alpha_j x_j ;$$

therefore, with a change of dummy subscripts, we get

$$\sum (\alpha_j - \sum_i a_{ji}\beta_i) x_j = 0 \Longrightarrow \alpha_i = \sum_j a_{ij}\beta_j .$$

This equation tells us how the coordinates of a given vector change when we go from one basis to another. It is exactly of the form of Eq. (1.15b), in which the same question is answered, under the same convention for defining matrix elements; there the x_i correspond to the present α_i. We may write it in the abbreviated form

$$\alpha_X = [A]_X \beta_Y ,$$

where the subscripts serve to remind us that the column vectors α and β represent x in the X and Y bases, respectively, and that the matrix representative of A is in the X basis.

We now turn to the analogous question for the matrix representation of a linear transformation, B. What is the relation between the matrix representation (b_{ij}) of B with respect to the basis X, and its representation with respect to the basis Y, (b'_{ij})? We have

$$Bx_j = \sum_i b_{ij} x_i \qquad \text{and} \qquad By_j = \sum_i b'_{ij} y_i .$$

We know that

$$By_j = BAx_j = B \sum_k a_{kj} x_k ,$$

where (a_{kj}) is the matrix representation of A with respect to X. Thus

$$By_j = \sum_k a_{kj} Bx_k = \sum_k a_{kj} \sum_i b_{ik} x_i = \sum_i (\sum_k b_{ik} a_{kj}) x_i . \tag{3.31}$$

But we also have

$$By_j = \sum_k b'_{kj} y_k = \sum_k b'_{kj} A x_k = \sum_k b'_{kj} \sum_i a_{ik} x_i = \sum_i (\sum_k a_{ik} b'_{kj}) x_i . \tag{3.32}$$

The linear independence of the basis vectors x_i now allows us to deduce from Eqs. (3.31) and (3.32) that

$$\sum_k b_{ik} a_{kj} = \sum_k a_{ik} b'_{kj} , \tag{3.33}$$

or, in matrix notation,

$$[B]_X [A]_X = [A]_X [B]_Y . \tag{3.34}$$

The subscripts indicate the basis with respect to which the matrix representative is determined. Since A has an inverse, we may write Eq. (3.34) as

$$[B]_Y = [A]_X^{-1} [B]_X [A]_X . \tag{3.35}$$

Definition 3.12. Two matrices, B and C, are said to be *similar* if there exists an invertible matrix, A, such that $C = A^{-1}BA$; we say that C is related to B by a *similarity transformation*.

Thus the previous result may be stated as "similar matrices represent the same linear transformation with respect to different bases."

Now we prove two important theorems about similar matrices.

Theorem 3.13. If B and C are similar matrices,

$$\det B = \det C .$$

Proof.

$$\det C = \det (A^{-1}BA) = \det A^{-1} \det B \det A = \det B ,$$

from Eq. (3.27). QED

Definition 3.13. The *trace* of a matrix A, written tr A, is the sum of the diagonal elements:

$$\operatorname{tr} A \equiv \sum_i a_{ii} .$$

Theorem 3.14. If B and C are similar matrices, tr B = tr C.

Proof.

$$\begin{aligned} \operatorname{tr} C &= \sum_i c_{ii} = \sum_i [A^{-1}BA]_{ii} = \sum_{i,j,k} [A^{-1}]_{ij} [B]_{jk} [A]_{ki} \\ &= \sum_{i,j,k} [A]_{ki} [A^{-1}]_{ij} [B]_{jk} = \sum_{j,k} [AA^{-1}]_{kj} [B]_{jk} \\ &= \sum_{j,k} \delta_{kj} [B]_{jk} = \sum_k [B]_{kk} = \operatorname{tr} B . \qquad \text{QED} \end{aligned}$$

Therefore both the determinant and the trace are invariant under similarity transformations. The reader may prove as a simple exercise that $\operatorname{tr}(AB) = \operatorname{tr}(BA)$ for arbitrary $n \times n$ matrices A, B. Theorem (3.14) may be derived without recourse to summations and subscripts if one uses this result.

3.10 EIGENVALUES AND EIGENVECTORS

Definition 3.14. A scalar λ is an *eigenvalue* and a *nonzero* vector x is an *eigenvector* of a linear transformation A if

$$Ax = \lambda x .$$

When λ and x are eigenvalues and eigenvectors of A, all that A does in operating on x is to multiply x by a constant, thus possibly changing its length, but not changing its "direction."

In quantum mechanics, it is assumed that any observable that exists in nature can be represented by a linear operator A and that the result of any physical measurement of the observable must be one of the eigenvalues of A. For example, the allowed values of the energy of a system with a Hamiltonian H are the values of E which satisfy $Hu = Eu$, where u denotes the eigenvectors describing the physical states associated with the allowed energies E. This is Schrödinger's time-independent equation.

We now investigate the problem of determining the eigenvalues and eigenvectors of an operator. The matrix equation $Ax = \lambda x$ is the same as $Ax - \lambda Ix = 0$ or $(A - \lambda I)x = 0$. We know that if $(A - \lambda I)$ is nonsingular, then this equation has only the trivial solution, $x = 0$. So in order for there to be an eigenvector, the matrix $(A - \lambda I)$ must be singular. That is,

$$\det(A - \lambda I) = \det \begin{bmatrix} a_{11} - \lambda & a_{12} & \cdots & a_{1n} \\ a_{21} & a_{22} - \lambda & \cdots & a_{2n} \\ \vdots & \vdots & & \vdots \\ a_{n1} & a_{n2} & \cdots & a_{nn} - \lambda \end{bmatrix} = 0 . \tag{3.36}$$

Equation (3.36) is called the *characteristic equation* for A. The polynomial $\det(A - \lambda I)$ is called the *characteristic polynomial*. The eigenvalues of A, which are usually labeled λ_i, are the n roots (not necessarily distinct) of the characteristic polynomial. The set of eigenvalues $\{\lambda_i\}$ is called the *spectrum* of A. If a given eigenvalue is a multiple root of the characteristic polynomial, the spectrum is said to be *degenerate*.

We know from algebra that every polynomial of degree n has n (not necessarily distinct) roots in the complex field. In fact, we shall prove this fundamental result in Chapter 6. Of course, it is not true that every polynomial has a root in the *real* field (e.g., the polynomial $x^2 + 1$ has only the roots $\pm i$). In order to ensure that the operators we deal with have eigenvalues and eigenvectors, we shall henceforth work in the complex field, unless we explicitly restrict the discussion to the real field.

By putting the eigenvalues back into the equation $Ax_i = \lambda_i x_i$ one at a time, we can solve for the eigenvector x_i that belongs to each λ_i. Note that any linear combination of eigenvectors belonging to the same eigenvalue is itself an eigenvector belonging to that eigenvalue. If we let x_i and x_j be eigenvectors belonging to the same eigenvalue λ, then

$$Ax_i = \lambda x_i \qquad \text{and} \qquad Ax_j = \lambda x_j\,,$$

and

$$A(\alpha x_i + \beta x_j) = \alpha A x_i + \beta A x_j = \alpha\lambda x_i + \beta\lambda x_j = \lambda(\alpha x_i + \beta x_j)\,.$$

Definition 3.15. The *geometric multiplicity* of an eigenvalue λ is the number of linearly independent eigenvectors belonging to λ.

Another type of multiplicity, the algebraic multiplicity, may also be defined. The *algebraic multiplicity* of a given root is the number of times it is repeated in the solution to the characteristic equation. It must be carefully distinguished from the notion of geometric multiplicity defined here.

Theorem 3.15. If A and B are similar, then A and B have the same eigenvalues and these eigenvalues have the same geometric multiplicities.

Proof. Consider $Ax_i = \lambda_i x_i$ and suppose that $P^{-1}AP = B$. Then

$$P^{-1}Ax_i = \lambda_i P^{-1}x_i\,,$$

so

$$P^{-1}APP^{-1}x_i = \lambda_i P^{-1}x_i\,,$$

where we have simply inserted $I = PP^{-1}$ into the left-hand side of the equation. Thus

$$By_i = \lambda_i y_i\,, \qquad \text{where} \qquad y_i = P^{-1}x_i\,,$$

and it follows that B has the same eigenvalues, λ_i, as A.

The eigenvector of B belonging to λ_i is $y_i = P^{-1}x_i$, and we must prove that if there are k linearly independent eigenvectors x_i satisfying $Ax_i = \bar{\lambda}x_i$—so that the geometric multiplicitly of $\bar{\lambda}$ is k—then there are k linearly independent eigenvectors y_i satisfying $By_i = \bar{\lambda}y_i$. Consider a linear combination of the eigenvectors y_i belonging to $\bar{\lambda}$:

$$\sum_{i=1}^{k} \alpha_i y_i = \sum_{i=1}^{k} \alpha_i P^{-1}x_i = P^{-1}\Big(\sum_{i=1}^{k} \alpha_i x_i\Big)\,.$$

Thus if we assume

$$\sum_{i=1}^{k} \alpha_i y_i = 0\,,$$

it follows that

$$P^{-1}\Big(\sum_{i=1}^{k} \alpha_i x_i\Big) = 0\,.$$

Since P exists, this implies that

$$\sum_{i=1}^{k} \alpha_i x_i = 0 .$$

Hence $\alpha_i = 0$, by the assumed linear independence of the k eigenvectors x_i. Therefore

$$\sum_{i=1}^{k} \alpha_i y_i = 0 \Longrightarrow \alpha_i = 0 ,$$

so the k eigenvectors, y_i, must also be linearly independent. QED

A somewhat cryptic alternative proof that similar matrices have the same eigenvalues consists of the observation that

$$(B - \lambda I) = P^{-1}AP - \lambda I = P^{-1}(A - \lambda I)P ,$$

and therefore

$$\det (B - \lambda I) = \det (A - \lambda I) .$$

Thus, since the two similar matrices A and B have the same characteristic polynomial, they must have the same eigenvalues, and these eigenvalues must have the same *algebraic* multiplicities. Note that this result is *not* equivalent to that of Theorem 3.15. The fact that similar matrices have the same characteristic polynomial implies that the coefficients of each power of λ must be equal. Examination of the characteristic polynomial will show that the coefficient of λ^0 (the constant term in the polynomial) is $\det A$; thus $\det A = \det B$. And since the coefficient of λ^{n-1} is $\operatorname{tr} A$, it follows that $\operatorname{tr} A = \operatorname{tr} B$. This is, in fact, an alternative proof of the invariance of the determinant and trace under similarity transformations; indeed, it points to the existence of a whole class of such invariants, namely, the n coefficients of the various powers of λ ($\lambda^0, \lambda^1, \lambda^2, \cdots, \lambda^{n-1}$) in the characteristic polynomial of an nth order matrix. We shall soon see why the determinant and trace are singled out as having special significance.

If A were a diagonal matrix of the form

$$D = \begin{bmatrix} d_{11} & 0 & 0 & \cdots & 0 \\ 0 & d_{22} & 0 & & 0 \\ 0 & 0 & d_{33} & & 0 \\ \vdots & & & & \vdots \\ 0 & 0 & 0 & \cdots & d_{nn} \end{bmatrix} ,$$

then the equation $Dx = \lambda x$ would be, in effect, already solved. For in this case the characteristic equation is

$$\det (D - \lambda I) = \prod_{i=1}^{n} (d_{ii} - \lambda) = 0 ,$$

which has the solutions

$$\lambda_i = d_{ii} , \qquad i = 1, 2, \cdots, n ;$$

that is, the eigenvalues are simply the diagonal elements. Thus if we can find a similarity transformation that diagonalizes a matrix A (and as we have seen, such transformations do preserve eigenvalues), we shall have simultaneously solved for the eigenvalues of A.

An important result in this connection is

Theorem 3.16. Let $X_i = (p_{1i}, \cdots, p_{ni})$ be an eigenvector of A (defined on V_n) belonging to λ_i, $(i = 1, \cdots, n)$, so that $AX_i = \lambda_i X_i$. If the vectors X_i span V_n, then the matrix $P = (p_{ij})$ is a diagonalizing matrix for A:

$$P^{-1}AP = \begin{bmatrix} \lambda_1 & & & 0 \\ & \lambda_2 & & \\ & & \ddots & \\ 0 & & & \lambda_n \end{bmatrix} = D = (\lambda_i \delta_{ij}) = (d_{ij}) \,.$$

Conversely, if A can be diagonalized by a similarity transformation, then the set of eigenvectors of A spans V.

Proof. The *column* vectors of P are the eigenvectors, X_i, of A. If these n vectors span V, they are linearly independent, so $\det P \neq 0$ and P^{-1} exists.

We have

$$AX_j = \lambda_j X_j \,,$$

or

$$\sum_k a_{ik} p_{kj} = \lambda_j p_{ij} = \sum_k p_{ik} \delta_{kj} \lambda_j = \sum_k p_{ik} d_{kj} \,,$$

for $i = 1, \cdots, n$; $j = 1, \cdots, n$. Thus

$$AP = PD \Longrightarrow D = P^{-1}AP \,,$$

since P is invertible. The converse follows by simply reversing the steps. QED

Thus a spanning set of eigenvectors is both a necessary and sufficient condition for diagonalizability by a similarity transformation. An example (see Problem 18) of a matrix whose eigenvectors do not span the space is

$$A = \begin{bmatrix} 1 & a \\ 0 & 1 \end{bmatrix} .$$

Note that one can always find a P such that $AP = PD$ even if there is no P such that $P^{-1}AP = D$.

Since any matrix A has at least one eigenvalue (call it λ) and one eigenvector (call it p), then the matrix P, built with n columns, each of which is p, will satisfy $AP = PD$. In fact, $AP = \lambda P$, so $D = \lambda I$. But this P is not invertible, since its columns are equal.

Theorem 3.16 is not really very useful in computations because one can only construct the diagonalizing matrix P if one knows the eigenvectors, and usually one can find the eigenvectors only after one knows the eigenvalues. Thus to get the eigenvalues of A by constructing a matrix P that diagonalizes

A, one must already know the eigenvalues of A. The main significance of this theorem is that it provides a necessary and sufficient condition for diagonalizability.

Example 1. We shall compute the eigenvalues, eigenvectors, and a diagonalizing matrix P for the Pauli spin matrix

$$\sigma_x = \begin{bmatrix} 0 & 1 \\ 1 & 0 \end{bmatrix}.$$

In quantum mechanics the operator for the x-component of spin is

$$s_x = \frac{\hbar}{2}\sigma_x .$$

The characteristic equation for σ_x is

$$\det(\sigma_x - \lambda I) = [(-\lambda)^2 - 1] = 0 \Longrightarrow \lambda = \pm 1 ;$$

thus the spin, s_x, has the eigenvalues $\pm\hbar/2$. To determine the eigenvectors that correspond to these eigenvalues, we put them back one at a time, into the equation $\sigma_x x = \lambda x$. For $\lambda = +1$,

$$\begin{bmatrix} 0 & 1 \\ 1 & 0 \end{bmatrix}\begin{bmatrix} \xi_1 \\ \xi_2 \end{bmatrix} = \begin{bmatrix} \xi_1 \\ \xi_2 \end{bmatrix} \Longrightarrow \xi_1 = \xi_2 = a ,$$

so an eigenvector belonging to $\lambda = +1$ has the form

$$\begin{bmatrix} a \\ a \end{bmatrix}, \qquad a \neq 0 .$$

Similarly, an eigenvector belonging to $\lambda = -1$ has the form

$$\begin{bmatrix} b \\ -b \end{bmatrix}, \qquad b \neq 0 .$$

Thus

$$P = \begin{bmatrix} a & b \\ a & -b \end{bmatrix}, \qquad \text{and} \qquad \det P = -2ab \neq 0 .$$

Therefore P^{-1} exists; it is given explicitly by

$$P^{-1} = \frac{1}{\det P}(\operatorname{adj} P) = \frac{-1}{2ab}\begin{bmatrix} -b & -b \\ -a & a \end{bmatrix}.$$

Now

$$P^{-1}\sigma_x P = \frac{1}{2ab}\begin{bmatrix} b & b \\ a & -a \end{bmatrix}\begin{bmatrix} 0 & 1 \\ 1 & 0 \end{bmatrix}\begin{bmatrix} a & b \\ a & -b \end{bmatrix} = \begin{bmatrix} 1 & 0 \\ 0 & -1 \end{bmatrix} = \begin{bmatrix} \lambda_1 & 0 \\ 0 & \lambda_2 \end{bmatrix}.$$

Another way to solve for the eigenvalues in this particular case is to note that since the square of any Pauli spin matrix is the identity matrix (Problem 12), that, in particular

$$x = Ix = \sigma_x^2 x = \sigma_x\sigma_x x = \sigma_x \lambda x = \lambda\sigma_x x = \lambda^2 x \Longrightarrow \lambda^2 = 1 ;$$

since $x \neq 0$, we see that $\lambda = \pm 1$. This calculation shows that the eigenvalues of any matrix whose square is I are ± 1.

Because equations of the type $Ax = 0$ are so important in physics, one often gets the impression that since $Ax = 0$ has a solution only if $\det A = 0$, then if $\det A = 0$, an equation of the type $Ax = b$ has no solution unless $b = 0$. This is false. Clearly, $Ax = b$ cannot have a solution for all vectors b; there is, nevertheless, a large class of vectors for which a solution exists. In fact, we can prove the following result, which will be useful when we develop the perturbation theory in Chapter 4.

Theorem 3.17. If $\det A = 0$, and if the eigenvectors of A span the space V on which A acts, then $Ax = b$ has a solution if and only if b may be written as a linear combination of those eigenvectors of A which belong to nonzero eigenvalues. We shall sometimes speak of such a vector b as *lying in the subspace* of V spanned by those eigenvectors of A which belong to nonzero eigenvalues.

Proof. We begin by noting that the determinant of A is equal to the product of the eigenvalues of A (see Problem 3.19). Since $\det A = 0$, it follows that A must have *at least* one zero eigenvalue. We label the eigenvalues by $\lambda_1, \lambda_2, \cdots, \lambda_n$ and the corresponding eigenvectors by $x_1, x_2, \cdots, x_n$. We will include degenerate eigenvalues in our labeling as many times as they occur and make the labeling convention that the zero eigenvalues come first, i.e., if we have m eigenvalues equal to zero, then $\lambda_1 = 0, \lambda_2 = 0, \cdots, \lambda_m = 0$. Since the x_i span the space, we may write

$$b = \sum_{i=1}^{n} \beta_i x_i \, .$$

Similarly,

$$x = \sum_{i=1}^{n} \xi_i x_i \, ,$$

so the equation $Ax = b$ may be written

$$A \sum_{i=1}^{n} \xi_i x_i = \sum_{i=1}^{n} \beta_i x_i \Longrightarrow \sum_{i=1}^{n} (\lambda_i \xi_i - \beta_i) x_i = 0 \, .$$

Since the x_i span V, we have $\lambda_i \xi_i - \beta_i = 0$, for all i. Thus, if a solution exists, then $\beta_i = 0$ for $i = 1, 2, \cdots, m$, since $\lambda_i = 0$ for $i = 1, 2, \cdots, m$. This means that if a solution exists, b may be written as a linear combination of just those x_i belonging to nonzero eigenvalues, and so half the theorem is proved.

On the other hand, if $\beta_i = 0$ for $i = 1, 2, \cdots, m$, then we can solve for the remaining ξ_i:

$$\xi_i = \beta_i / \lambda_i \, , \qquad i = m + 1, \cdots, n \, ,$$

while the first m of the ξ_i are arbitrary. Thus any vector of the form

$$x = \xi_1 x_1 + \xi_2 x_2 + \cdots + \xi_m x_m + \frac{\beta_{m+1}}{\lambda_{m+1}} x_{m+1} + \cdots + \frac{\beta_n}{\lambda_n} x_n$$

is a solution to $Ax = b$, so we have an m-fold infinite family of solutions for any b obeying the conditions of the theorem! QED

This arbitrariness in x should not be surprising. Given a solution vector x and an arbitrary multiple of any solution of $Ay = 0$, the sum $x + ay$ is a solution. For if $Ax = b$, then

$$A(x + ay) = Ax + aAy = Ax = b\,.$$

But if A has m eigenvalues equal to zero and m associated linearly independent eigenvectors, $x_1, x_2, \cdots, x_m$, then of course

$$Ax_i = 0x_i = 0\,, \qquad i = 1, \cdots, m\,,$$

so the eigenvectors $x_1, x_2, \cdots, x_m$ are solutions of the homogeneous equation $Ay = 0$. Thus their coefficients in the expansion of x will be arbitrary. This also illustrates the fact that an equation of the type $Ay = 0$ may have more than one linearly independent solution.

The set of all vectors y satisfying $Ay = 0$ is called the *null space* or *kernel* of A; the dimension of this space is called the *nullity* of A. The dimension of the space V on which A acts, minus the nullity of A, is called the *rank* of A.

We now turn to a second example of the eigenvalue-eigenvector problem which is of paramount importance in modern physics.

Example 2. *Energy Eigenvalues and Eigenvectors of the Simple Harmonic Oscillator by Operator Algebra.* The eigenvalue-eigenvector problem for the one-dimensional simple harmonic oscillator is to find the allowed energies, E_n, and the corresponding eigenstates for the quantum-mechanical system whose Hamiltonian is

$$H = (p^2/2m) + \tfrac{1}{2}\, m\omega^2 x^2\,. \tag{3.37}$$

The spring constant k has been replaced using the relation $\omega = (k/m)^{1/2}$. We must solve the equation

$$Hu_n = E_n u_n\,, \tag{3.38}$$

called Schrödinger's equation, for the energy eigenvalues E_n, and the corresponding eigenvectors u_n.

One way to solve this eigenvalue-eigenvector problem is to make the usual Schrödinger substitutions

$$p \rightarrow -i\hbar\frac{\partial}{\partial x}\,, \qquad x \rightarrow x\,, \tag{3.39}$$

which satisfy the commutation relation,

$$[p, x] = -i\hbar\,, \tag{3.40}$$

and then solve the resulting ordinary differential equation, insisting on solutions which decrease properly at large distances. The eigenvalues can be determined in a more elegant way, however, by operator or matrix algebra techniques.

This method derives great importance from its central role in quantum field theory. Consider the operators:

$$a \equiv (xm\omega + ip)\,(2m\omega\hbar)^{-1/2} \tag{3.41}$$

$$a^\dagger \equiv (xm\omega - ip)\,(2m\omega\hbar)^{-1/2}\,. \tag{3.42}$$

The definitions of these operators differ widely in the literature. Their product is related to the Hamiltonian, so there is considerable freedom in the disposition of constant multipliers and phase factors. The use of the dagger notation is in accord with later definitions. The product of $a^\dagger$ and a is

$$\begin{aligned} a^\dagger a &= (2m\omega\hbar)^{-1}(xm\omega - ip)(xm\omega + ip) \\ &= (2m\omega\hbar)^{-1}[m^2\omega^2x^2 + p^2 - im\omega(px - xp)] \\ &= (\hbar\omega)^{-1}(H - (i\omega/2)[p, x]) = (\hbar\omega)^{-1}(H - \hbar\omega/2)\,, \end{aligned}$$

where the commutation relation (3.40) has been used in the last step. Therefore, the Hamiltonian may be expressed in terms of $a^\dagger$ and a by

$$H = \hbar\omega(a^\dagger a + \tfrac{1}{2})\,. \tag{3.43}$$

A parallel calculation shows that

$$aa^\dagger = (\hbar\omega)^{-1}(H + \hbar\omega/2)\,. \tag{3.44}$$

Consequently,

$$[a, a^\dagger] \equiv aa^\dagger - a^\dagger a = I\,. \tag{3.45}$$

We shall also need to know the commutator of H with $a^\dagger$ and a.

$$[H, a^\dagger] = \hbar\omega[(a^\dagger aa^\dagger + \tfrac{1}{2}a^\dagger) - (a^\dagger a^\dagger a + \tfrac{1}{2}a^\dagger)] = \hbar\omega a^\dagger[a, a^\dagger] = \hbar\omega a^\dagger\,. \tag{3.46}$$

Similarly,

$$[H, a] = -\hbar\omega a\,. \tag{3.47}$$

We can see on physical grounds that the simple harmonic oscillator only has energies greater than or equal to zero. This can also be shown from the operator algebra. In Section 4.4, we shall use the fact that both the operators p and x are Hermitian and therefore have real eigenvalues to prove that their squares can have only nonnegative (real) eigenvalues. Consequently, the Hamiltonian can have only positive (or zero) real eigenvalues, since $m, \omega > 0$. Assuming this result, let $E_0 \geq 0$ denote the lowest energy which the harmonic oscillator can have, the so-called *ground state*. Let u_0 be the eigenvector belonging to E_0*. Then

$$Hu_0 = E_0 u_0\,.$$

Now

$$aHu_0 = E_0 au_0\,.$$

* Note that we are assuming that u_0 is the *only* eigenvector belonging to E_0. For the particular operator, a, that we are using here, this assumption is correct, as we will see below. However, this fact does *not* follow just from the commutation relations.

Using the commutation relation (3.47) for H and a, we obtain

$$aHu_0 = (Ha + \hbar\omega a)u_0 = E_0 a u_0\,,$$

or

$$Hau_0 = (E_0 - \hbar\omega)au_0\,.$$

This is in the form of an eigenvalue equation with a new eigenvalue, $E_0 - \hbar\omega$, and a new eigenvector, (au_0). But since by assumption E_0 is the lowest allowed eigenvalue, this can only be a trivial solution to the eigenvalue equation; that is, $au_0 \equiv 0$. If we operate on this equation with $a^\dagger$, we obtain

$$a^\dagger a u_0 = (\hbar\omega)^{-1}(H - \hbar\omega/2)u_0 = 0\,,$$

or

$$Hu_0 = (\hbar\omega/2)u_0\,.$$

Therefore $E_0 = \hbar\omega/2$ is the energy of the ground state, the lowest allowed energy level. To get the other energy levels, we operate on $Hu_0 = E_0 u_0$ with $a^\dagger$.

$$a^\dagger H u_0 = (Ha^\dagger - \hbar\omega a^\dagger)u_0 = E_0 a^\dagger u_0\,,$$

or

$$H(a^\dagger u_0) = (E_0 + \hbar\omega)a^\dagger u_0\,.$$

That is, $(E_0 + \hbar\omega)$ is an energy eigenvalue of H corresponding to the eigenvector $a^\dagger u_0$. If we again operate on this equation with $a^\dagger$, we get

$$a^\dagger H a^\dagger u_0 = (E_0 + \hbar\omega)(a^\dagger)^2 u_0\,.$$

Using the commutation relations between $a^\dagger$ and H, we get

$$(Ha^\dagger - \hbar\omega a^\dagger)(a^\dagger u_0) = (E_0 + \hbar\omega)(a^\dagger)^2 u_0\,,$$

or

$$H(a^\dagger)^2 u_0 = (E_0 + 2\hbar\omega)(a^\dagger)^2 u_0\,.$$

Thus $(E_0 + 2\hbar\omega)$ is an energy eigenvalue of H corresponding to the eigenvector $a^{\dagger 2}u_0$. By operating n times with $a^\dagger$, one obtains

$$H(a^\dagger)^n u_0 = (E_0 + n\hbar\omega)(a^\dagger)^n u_0\,.$$

Successive application of $a^\dagger$ creates *all* the eigenvalues and eigenvectors (Problem 23). For this reason, $a^\dagger$ is called the *creation operator*. Calling the nth eigenvalue E_n, and the nth eigenvector u_n, we have

$$E_n = (n + \tfrac{1}{2})\hbar\omega\,, \qquad \text{and} \qquad u_n = c_n(a^\dagger)^n u_0\,, \tag{3.48}$$

where c_n is an arbitrary constant. Note that $c_0 = 1$.

We can solve for u_0 explicitly by using $au_0 = 0$, and making the usual substitutions (3.39) for p and x. Thus

$$au_0 = (2m\omega\hbar)^{-1/2}[xm\omega + \hbar(d/dx)]u_0 = 0\,,$$

or

$$\frac{du_0}{dx} + \frac{m\omega}{\hbar}xu_0 = 0\,.$$

This equation is easily solved, by one integration, to give

$$u_0 = N_0 e^{-(m\omega/2\hbar)x^2},$$

where the constant of integration, N_0, is determined by the normalization condition:

$$1 = \int_{-\infty}^{\infty} |u_0|^2\, dx = |N_0|^2 (\hbar\pi/m\omega)^{1/2}.$$

Therefore

$$N_0 = (m\omega/\hbar\pi)^{1/4},$$

and u_0 is completely determined:

$$u_0 = (m\omega/\hbar\pi)^{1/4} e^{-(m\omega/2\hbar)x^2}.$$

The other eigenfunctions can be derived from this by applying Eq. (3.48). Even the constants c_n may be found by a purely algebraic technique; we shall prove in Section 4.4 that

$$c_n = (1/n!)^{1/2}. \tag{3.49}$$

For completeness, we mention that the eigenfunctions generated by applying Eq. (3.48) are proportional to the Hermite polynomials, $H_n(y)$, times $\exp(-y^2/2)$. That is,

$$u_n(x) = N_n H_n(\alpha x) e^{-\alpha^2 x^2/2}, \tag{3.50}$$

where

$$\alpha^2 = m\omega/\hbar, \qquad \text{and} \qquad N_n = \left(\frac{\alpha}{\pi^{1/2} 2^n n!}\right)^{1/2}.$$

The first three Hermite polynomials are

$$H_0(y) = 1, \qquad H_1(y) = 2y, \qquad H_2(y) = 4y^2 - 2.$$

We have not *proved* here that the solution can be expressed in terms of Hermite polynomials. In fact, one may more easily arrive at this result by putting Eq. (3.38) in the form of a second-order differential equation and solving it using the standard method of expansion in a power series.

To complete this example, we discuss the effect of the operator, a, on the eigenfunction u_n. Remembering that $au_0 = 0$, and writing $(a^\dagger)^n$ as $a^{\dagger n}$, we have

$$\begin{aligned}
au_n &= c_n a a^{\dagger n} u_0 \\
&= c_n\{[a, a^\dagger] a^{\dagger n-1} u_0 + a^\dagger a a^{\dagger n-1} u_0\} \\
&= c_n\{a^{\dagger n-1} u_0 + a^\dagger a a^{\dagger n-1} u_0\} \\
&= c_n\{a^{\dagger n-1} u_0 + a^\dagger [a, a^\dagger] a^{\dagger n-2} u_0 + a^\dagger (a^\dagger a) a^{\dagger n-2} u_0\} \\
&= c_n\{a^{\dagger n-1} u_0 + a^{\dagger n-1} u_0 + a^{\dagger 2} a a^{\dagger n-2} u_0\} \\
&= c_n\{2a^{\dagger n-1} + a^{\dagger 2} a a^{\dagger n-2}\} u_0 .
\end{aligned}$$

By continuing this process n times, we finally get

$$au_n = c_n (n a^{\dagger n-1} u_0 + a^{\dagger n} a u_0) = c_n n a^{\dagger n-1} u_0,$$

so that

$$au_n = n\frac{c_n}{c_{n-1}}(c_{n-1}a^{\dagger n-1}u_0) = n\frac{c_n}{c_{n-1}}u_{n-1} = n^{1/2}u_{n-1}\,. \tag{3.51}$$

We may also calculate the effect of $a^\dagger$ on u_n. Using Eqs. (3.43) and (3.48), we get

$$Hu_n = \hbar\omega(a^\dagger a + \tfrac{1}{2})u_n = \hbar\omega(n + \tfrac{1}{2})u_n\,.$$

Therefore

$$a^\dagger a u_n = n u_n\,,$$

so using Eq. (3.51), we obtain

$$a^\dagger(n^{1/2}u_{n-1}) = n u_n\,,$$

which, letting $n \rightarrow (n+1)$, becomes

$$a^\dagger u_n = (n+1)^{1/2}u_{n+1}\,. \tag{3.52}$$

Because of Eqs. (3.51) and (3.52), it is customary to call $a^\dagger$ and a, *raising* and *lowering* operators, or *creation* and *annihilation* operators, respectively.

3.11 THE KRONECKER PRODUCT

In this section, we want to discuss a method of combining linear operators which act on different vector spaces. We have already seen how to form the "product" of two linear operators, B_1 and B_2, which act on the same vector space U. Suppose now that we have a linear operator B acting on an n-dimensional vector space U, which takes vectors in U into vectors in U. In other words, B maps U into U, or symbolically,

$$B\colon\ U \rightarrow U\,.$$

Similarly, let A act on the m-dimensional vector space V:

$$A\colon\ V \rightarrow V\,.$$

Another important class of operators is the collection of linear operators which map U into V in a manner which preserves linear operations. That is, if C is some such operator, we want

$$C(\alpha x_1 + \beta x_2) = \alpha C x_1 + \beta C x_2\,.$$

Note that since $x_1, x_2 \in U$, and $Cx_1, Cx_2 \in V$, U and V must be vector spaces over the same field of scalars for this equation to make sense.

Just as we did for operators which map U into itself, we can represent operators like C by an array of scalars in the following manner. Let $\{u_i\}$ be a basis in U and $\{v_i\}$ be a basis in V. Then Cu_j is in V and must be a linear combination of the v_i. We write therefore

$$Cu_j = \sum_{k=1}^{m} c_{kj}v_k\,, \qquad j = 1, 2, \cdots, n\,.$$

Note that the array c_{kj} is in general not square, but rather has m rows and n columns. Thus if x is a vector in U such that

$$x = \sum_{j=1}^{n} \alpha_j u_j \, ,$$

and y is a vector in V such that

$$y = \sum_{i=1}^{m} \beta_i v_i \, ;$$

then, if $y = Cx$, the β's and α's are related by

$$\beta_i = \sum_{j=1}^{n} c_{ij}\alpha_j \, , \qquad i = 1, 2, \cdots, m \, .$$

We now want to consider this collection of linear operators which maps U into V as a *vector space*. The vectors in this new vector space are operators. Now C is represented by an $m \times n$ matrix, having (mn) elements. Therefore, our vector space will be mn-dimensional, since every $m \times n$ matrix can be written as a linear combination of the mn matrices

$$P_{11} = \begin{bmatrix} 1 & 0 & 0 & \cdots \\ 0 & 0 & 0 & \\ \vdots & & & \end{bmatrix}, \qquad P_{12} = \begin{bmatrix} 0 & 1 & 0 & \cdots \\ 0 & 0 & 0 & \\ \vdots & & & \end{bmatrix}, \quad \text{etc.},$$

or, in other words, the mn matrices P_{ij} defined by

$$(P_{ij})_{kl} = \delta_{ik}\delta_{jl} \, . \tag{3.53}$$

These are linearly independent, and span the space. Thus they form a basis for the vector space of $m \times n$ matrices. Since there are mn of them, we have an mn-dimensional vector space, which we shall denote by $\mathscr{L}$. [In forming this vector space we make the logical linearity requirements, such as $(A + B)x = Ax + Bx$ for any $x \in U$, etc.]

For example, consider the following 2×3 matrix which maps from a three-dimensional vector space into a two-dimensional one:

$$C = \begin{bmatrix} 1 & 2 & 3 \\ 4 & 5 & 6 \end{bmatrix} .$$

In this case, $\mathscr{L}$ is six-dimensional, and if we use the basis suggested above, we may write

$$\{C\} = 1 \times \begin{bmatrix} 1 & 0 & 0 \\ 0 & 0 & 0 \end{bmatrix} + 2 \times \begin{bmatrix} 0 & 1 & 0 \\ 0 & 0 & 0 \end{bmatrix} + 3 \times \begin{bmatrix} 0 & 0 & 1 \\ 0 & 0 & 0 \end{bmatrix}$$
$$+ 4 \times \begin{bmatrix} 0 & 0 & 0 \\ 1 & 0 & 0 \end{bmatrix} + 5 \times \begin{bmatrix} 0 & 0 & 0 \\ 0 & 1 & 0 \end{bmatrix} + 6 \times \begin{bmatrix} 0 & 0 & 0 \\ 0 & 0 & 1 \end{bmatrix},$$

where we adopt the convention of bracketing an element of $\mathscr{L}$ when considered as a "vector" rather than as an "operator." In this case, $\{C\}$ can be represented in this particular (but very natural) basis by

$$\{C\} \to \begin{bmatrix} 1 \\ 2 \\ 3 \\ 4 \\ 5 \\ 6 \end{bmatrix}.$$

With the preliminary definitions and discussion out of the way, it is a simple matter to define the *Kronecker* (or *direct*) *product*. Let A and B be operators defined on V and U, which respectively map V and U into themselves, and let C be some operator which maps U into V. For our given A and B, there is another operator that maps U into V which one would want to associate naturally with C, namely

$$C' = AC\tilde{B}.$$

That is, $\tilde{B}$ maps U into U, C maps U into V, and finally, A maps V into V. The reason we use $\tilde{B}$ rather than B is to make the properties of the Kronecker product as symmetrical as possible.

Now consider $\{C\}$ and $\{C'\}$ as elements of $\mathscr{L}$. We define the Kronecker product of A and B, denoted by $A \otimes B$, to be that operator on $\mathscr{L}$ which takes $\{C\}$ into $\{C'\}$, that is,

$$(A \otimes B)\{C\} = \{C'\} = \{AC\tilde{B}\}, \tag{3.54}$$

for all $C \in \mathscr{L}$.

The properties of the Kronecker product are easy to derive from Eq. (3.54). We have immediately

$$I \otimes I = I, \tag{3.55}$$

where by I we mean the unit operator in whatever dimensional space is appropriate. Other immediate consequences of Eq. (3.54) are

$$(A_1 + A_2) \otimes B = A_1 \otimes B + A_2 \otimes B \tag{3.56}$$

and

$$A \otimes (B_1 + B_2) = A \otimes B_1 + A \otimes B_2. \tag{3.57}$$

Now we consider

$$\begin{aligned}(A_1 \otimes B_1)(A_2 \otimes B_2)\{C\} &= (A_1 \otimes B_1)\{A_2C\tilde{B}_2\} \\ &= \{A_1A_2C\tilde{B}_2\tilde{B}_1\} \\ &= (A_1A_2C\widetilde{B_1B_2}) \\ &= (A_1A_2 \otimes B_1B_2)\{C\},\end{aligned}$$

so we have established the important property

$$(A_1 \otimes B_1)(A_2 \otimes B_2) = (A_1A_2 \otimes B_1B_2), \tag{3.58}$$

where $A_1, A_2 \in V$, and $B_1, B_2 \in U$. The fact that the right-hand side of (3.58) is $A_1A_2 \otimes B_1B_2$ rather than $A_1A_2 \otimes B_2B_1$ is the reason for our use of $\tilde{B}$ rather than B in our definition of the Kronecker product. We leave it to the reader to show that if both A^{-1} and B^{-1} exist, then

$$(A \otimes B)^{-1} = A^{-1} \otimes B^{-1}. \tag{3.59}$$

Finally, let us see what $A \otimes B$ looks like when represented as a matrix in the basis of Eq. (3.53). There is a minor notational annoyance which we should first point out: Our basis "vectors," the P_{ij}, are indexed by two subscripts instead of the usual single subscript that we are used to seeing. This is because we are perversely regarding operators (matrices) as vectors. Thus we now have to talk about the ij-component of a *vector*! Likewise, we shall have to refer to the (ij), (kl)-component of a matrix of an operator like $A \otimes B$.

The (ij), (kl) element of $A \otimes B$ is found in the usual way. We write

$$(A \otimes B)\{P_{kl}\} = \sum_{i,j} [A \otimes B]_{ij,kl}\{P_{ij}\},$$

where we denote the (ij), (kl) element of $A \otimes B$ by the symbol $[A \otimes B]_{ij,kl}$. But

$$(A \otimes B)\{P_{kl}\} = \{AP_{kl}\tilde{B}\},$$

by Eq. (3.54), so we obtain the equation

$$\{AP_{kl}\tilde{B}\} = \sum_{i,j} [A \otimes B]_{ij,kl}\{P_{ij}\}.$$

In other words, the number $[A \otimes B]_{ij,kl}$ is just the ijth element of the *matrix* $AP_{kl}\tilde{B}$, that is,

$$[A \otimes B]_{ij,kl} = \sum_{\mu,\nu} A_{i\mu}(P_{kl})_{\mu\nu}\tilde{B}_{\nu j}.$$

But by Eq. (3.53), we get

$$\begin{aligned} [A \otimes B]_{ij,kl} &= \sum_{\mu,\nu} A_{i\mu}\delta_{k\mu}\delta_{l\nu}\tilde{B}_{\nu j} \\ &= A_{ik}\tilde{B}_{lj}. \end{aligned}$$

Since $\tilde{B}_{lj} = B_{jl}$, we get finally

$$[A \otimes B]_{ij,kl} = A_{ik}B_{jl}, \tag{3.60}$$

so the elements of $A \otimes B$ are obtained easily from the elements of A and the elements of B.

Note that the trace of $A \otimes B$ is just

$$\operatorname{tr}(A \otimes B) = \sum_{i,j} [A \otimes B]_{ij,ij} = \sum_{i,j} A_{ii}B_{jj} = (\operatorname{tr} A)(\operatorname{tr} B),$$

or

$$\operatorname{tr}(A \otimes B) = (\operatorname{tr} A)(\operatorname{tr} B). \tag{3.61}$$

This is a relation which we shall find very useful when we discuss group theory.

It is instructive to see how these things look written out in matrix form. To do this we have to adopt an ordering for the subscripts (ij). We shall write the column vectors, which represent elements of $\mathscr{L}$, in the following order:

$$\begin{bmatrix} C_{11} \\ C_{12} \\ \vdots \\ C_{1n} \\ C_{21} \\ C_{22} \\ \vdots \\ C_{2n} \\ \vdots \\ C_{m1} \\ C_{m2} \\ \vdots \\ C_{mn} \end{bmatrix}$$

Hence our matrices, which represent objects like $A \otimes B$, will take the form

$$\begin{bmatrix} [A\otimes B]_{11,11} & [A\otimes B]_{11,12} & \cdots & [A\otimes B]_{11,1n} & [A\otimes B]_{11,21} & \cdots & [A\otimes B]_{11,mn} \\ [A\otimes B]_{12,11} & [A\otimes B]_{12,12} & \cdots & [A\otimes B]_{12,1n} & [A\otimes B]_{12,21} & \cdots & [A\otimes B]_{12,mn} \\ \vdots & \vdots & & \vdots & \vdots & & \vdots \\ [A\otimes B]_{1n,11} & [A\otimes B]_{1n,12} & \cdots & [A\otimes B]_{1n,1n} & [A\otimes B]_{1n,21} & \cdots & [A\otimes B]_{1n,mn} \\ [A\otimes B]_{21,11} & [A\otimes B]_{21,12} & \cdots & [A\otimes B]_{21,1n} & [A\otimes B]_{21,21} & \cdots & [A\otimes B]_{21,mn} \\ \vdots & \vdots & & \vdots & \vdots & & \vdots \\ [A\otimes B]_{mn,11} & [A\otimes B]_{mn,12} & \cdots & [A\otimes B]_{mn,1n} & [A\otimes B]_{mn,21} & \cdots & [A\otimes B]_{mn,mn} \end{bmatrix}.$$

Using Eq. (3.60), we obtain

$$\begin{bmatrix} A_{11}B_{11} & A_{11}B_{12} & \cdots & A_{11}B_{1n} & A_{12}B_{11} & \cdots & A_{12}B_{1n} & \cdots & A_{1m}B_{1n} \\ A_{11}B_{21} & A_{11}B_{22} & \cdots & A_{11}B_{2n} & A_{12}B_{21} & \cdots & A_{12}B_{2n} & \cdots & A_{1m}B_{2n} \\ \vdots & \vdots & & \vdots & \vdots & & \vdots & & \vdots \\ A_{11}B_{n1} & A_{11}B_{n2} & \cdots & A_{11}B_{nn} & A_{12}B_{n1} & \cdots & A_{12}B_{nn} & \cdots & A_{1m}B_{nn} \\ A_{21}B_{11} & A_{21}B_{12} & \cdots & A_{21}B_{1n} & A_{22}B_{11} & \cdots & A_{22}B_{1n} & \cdots & A_{2m}B_{1n} \\ \vdots & \vdots & & \vdots & \vdots & & \vdots & & \vdots \\ A_{21}B_{n1} & A_{21}B_{n2} & \cdots & A_{21}B_{nn} & A_{22}B_{n1} & \cdots & A_{22}B_{nn} & \cdots & A_{2m}B_{nn} \\ \vdots & \vdots & & \vdots & \vdots & & \vdots & & \vdots \\ A_{m1}B_{n1} & A_{m1}B_{n2} & \cdots & A_{m1}B_{nn} & A_{m2}B_{n1} & \cdots & A_{m2}B_{nn} & \cdots & A_{mm}B_{nn} \end{bmatrix}.$$

Inspecting this result, we see that if we break the matrix representative of $A \otimes B$

into $n \times n$ blocks (there will be m^2 of them) starting in the upper left corner, then the matrix of $A \otimes B$ can be written symbolically as

$$A \otimes B \to \begin{bmatrix} A_{11}B & A_{12}B & \cdots & A_{1n}B \\ \vdots & & & \vdots \\ A_{m1}B & A_{m2}B & \cdots & A_{mm}B \end{bmatrix}.$$

That is, to form the matrix of $A \otimes B$, we replace the ij-element of the matrix of A, A_{ij}, by $A_{ij}B$, that is, by an $n \times n$ array. Note that in this form the theorem on traces [Eq. (3.61)] is particularly easy to see.

As an example of this kind of matrix, we consider the 2×3 matrix

$$C = \begin{bmatrix} 1 & 2 & 3 \\ 4 & 5 & 6 \end{bmatrix},$$

which, when considered as a vector in $\mathscr{L}$ may be represented by

$$\{C\} \to \begin{bmatrix} 1 \\ 2 \\ 3 \\ 4 \\ 5 \\ 6 \end{bmatrix}.$$

Now suppose that

$$B = \begin{bmatrix} 1 & 0 & 1 \\ 2 & 1 & 1 \\ 1 & 3 & 1 \end{bmatrix}, \qquad \text{so} \qquad \tilde{B} = \begin{bmatrix} 1 & 2 & 1 \\ 0 & 1 & 3 \\ 1 & 1 & 1 \end{bmatrix},$$

and

$$A = \begin{bmatrix} 1 & 0 \\ 1 & 2 \end{bmatrix}.$$

B maps U into itself and A maps V into itself. Thus

$$C' = AC\tilde{B} = \begin{bmatrix} 4 & 7 & 10 \\ 24 & 45 & 60 \end{bmatrix}$$

so that

$$\{C'\} \to \begin{bmatrix} 4 \\ 7 \\ 10 \\ 24 \\ 45 \\ 60 \end{bmatrix}.$$

The matrix of $A \otimes B$ is easily computed, either from Eq. (3.60) or by the picture given above. We get

$$A \otimes B \rightarrow \begin{bmatrix} 1 & 0 & 1 & 0 & 0 & 0 \\ 2 & 1 & 1 & 0 & 0 & 0 \\ 1 & 3 & 1 & 0 & 0 & 0 \\ 1 & 0 & 1 & 2 & 0 & 2 \\ 2 & 1 & 1 & 4 & 2 & 2 \\ 1 & 3 & 1 & 2 & 6 & 2 \end{bmatrix}$$

and we leave it to the reader to check that, indeed,

$$(A \otimes B)\{C\} = \{C'\} .$$

PROBLEMS

1. Prove that the n complex nth roots of unity together with multiplication of complex numbers form an abelian group.
   [*Hint*: From De Moivre's formula, the n complex nth roots of unity are given by
   $$e_k = \cos\frac{2k\pi}{n} + i\sin\frac{2k\pi}{n} = e^{2ik\pi/n}, \qquad k = 0, 1, \cdots, n-1,$$
   where $i^2 = -1$.]
2. Show that 3 vectors in three-dimensional space are linearly dependent if and only if $\mathbf{A}\cdot(\mathbf{B} \times \mathbf{C}) = 0$.
3. In the vector space of real n-tuples, prove that the n vectors
   $$\begin{aligned} x_1 &= (1, 1, \cdots, 1, 1), \\ x_2 &= (0, 1, \cdots, 1, 1), \\ x_3 &= (0, 0, 1, \cdots, 1, 1), \\ &\;\;\vdots \\ x_n &= (0, 0, \cdots, 0, 1) \end{aligned}$$
   are linearly independent.
4. Prove that any set of $n + 1$ vectors in an n-dimensional vector space is linearly dependent.
5. Prove that if the set of n-vectors $\{x_i\}$ is a basis for a vector space V, then the set $\{c_i x_i\}$ is also a basis for V, where the c_i are arbitrary nonzero scalars. Interpret this geometrically.
6. Prove that a basis is a maximal linearly independent set of vectors.
7. a) Prove that any two $n \times n$ diagonal matrices commute and that their product is also a diagonal matrix.
   b) If A commutes with all diagonal matrices B, then A is diagonal.
8. Prove that if the matrix A is idempotent and A is not the unit matrix, then A is singular.
9. Define the commutator $[A, B]$ of two matrices by $[A, B] = AB - BA$, and the anticommutator $\{A, B\}$ by $\{A, B\} = AB + BA$. Show that

a) $[A, B] = -[B, A]$, b) $[AB, C] = A[B, C] + [A, C]B$, c) $[AB, C] = A\{B, C\} - \{A, C\}B$, d) $[A, [B, C]] + [B, [C, A]] + [C, [A, B]] = 0$.

10. Show that if a linear transformation A is known to have a *unique* left inverse (that is $BA = I$), then B is also a right inverse (that is $AB = I$).
[*Hint*: Consider the linear transformation $C = AB - I + B$, which you suspect (but must prove) equals B if the theorem is true, etc.]

11. Prove that any antisymmetric matrix of *odd order* is singular. The "odd order" qualification is essential; the Pauli spin matrix

$$\sigma_y = \begin{bmatrix} 0 & -i \\ i & 0 \end{bmatrix}$$

is a second-order antisymmetric matrix, but since $\sigma_y^2 = I$, σ_y *does have* an inverse, namely $\sigma_y^{-1} = \sigma_y$.

12. The Pauli spin matrices are defined as follows:

$$\sigma_1 = \begin{bmatrix} 0 & 1 \\ 1 & 0 \end{bmatrix}, \qquad \sigma_2 = \begin{bmatrix} 0 & -i \\ i & 0 \end{bmatrix}, \qquad \sigma_3 = \begin{bmatrix} 1 & 0 \\ 0 & -1 \end{bmatrix},$$

where $i = \sqrt{-1}$. Prove that

$$\sigma_i\sigma_j + \sigma_j\sigma_i = 2\delta_{ij} \qquad \text{and} \qquad \sigma_i\sigma_j - \sigma_j\sigma_i = 2i\sum_{k=1}^{3} \epsilon_{ijk}\sigma_k .$$

Thus the Pauli matrices anticommute.

13. Prove that $\text{tr}\,(AB) = \text{tr}\,(BA)$ for arbitrary $n \times n$ matrices, A and B. Use this result to prove that if C and D are similar, then $\text{tr}\,(C) = \text{tr}\,(D)$.

14. Let A be an $n \times n$ matrix with real entries. Prove that $A = 0$ if and only if $\text{tr}\,(\tilde{A}A) = 0$.

15. Let A be an $n \times n$ idempotent matrix (that is, $A^2 = A$).
    a) Prove that each eigenvalue of A is either 0 or 1.
    b) It can be shown that an idempotent matrix can always be diagonalized by a similarity transformation. Using this, prove that $\text{tr}\,A$ is integral and $0 \leq \text{tr}\,A \leq n$.

16. Show that the eigenvalues of

$$M = \begin{bmatrix} 3 & 5 & 8 \\ -6 & -10 & -16 \\ 4 & 7 & 11 \end{bmatrix}$$

are $\lambda = 0, 1, 3$, and that the corresponding eigenvectors are

$$\begin{bmatrix} 1 \\ 1 \\ -1 \end{bmatrix}, \quad \begin{bmatrix} 1 \\ -2 \\ 1 \end{bmatrix}, \quad \text{and} \quad \begin{bmatrix} 4 \\ -8 \\ 5 \end{bmatrix}.$$

Construct a diagonalizing matrix P, prove that its inverse exists, compute its inverse, and verify that $P^{-1}MP$ is diagonal with the eigenvalues on the diagonal. Note that $\det M = 0$. Is it true that any diagonalizable matrix with an eigenvalue equal to zero is singular?

17. Denote the two-dimensional rotation matrix for a rotation of the coordinate axes through an angle θ by

$$A = \begin{bmatrix} \cos\theta & \sin\theta \\ -\sin\theta & \cos\theta \end{bmatrix}.$$

Find the eigenvalues and eigenvectors of A. Find a diagonalizing matrix for A, that is, a matrix P such that $P^{-1}AP$ is a diagonal matrix. Demonstrate that P is such a matrix by inverting P and forming the product $P^{-1}AP$.

18. Show that if

$$A = \begin{bmatrix} 1 & a \\ 0 & 1 \end{bmatrix}$$

there exists no nonsingular matrix P of order 2 such that $P^{-1}AP = D$, where D is a diagonal matrix. This example shows that not every matrix can be diagonalized by a similarity transformation.

19. Show that the determinant of *any* matrix is equal to the product of its eigenvalues and that the trace of a matrix is equal to the sum of its eigenvalues. Thus show that if a matrix has a zero eigenvalue it is not invertible.

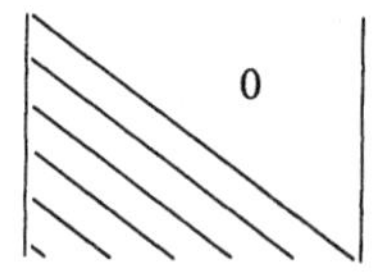

Fig. 3.4

20. Show that by adding and subtracting rows (multiples of rows) systematically, any determinant can be reduced to evaluating a determinant of the type shown in Fig. 3.4, where there are terms on the diagonal and below the diagonal, and only zeros above the diagonal. Clearly, this same process can now be applied to the columns to leave us with the trivial problem of evaluating the determinant of a diagonal matrix. However, the evaluation is already trivial after the first step. Why?

21. Derive an expression for the inverse of a matrix A in terms of the eigenvalues and eigenvectors of A. Assume that the inverse of A exists, and that the eigenvectors of A span the space in which A acts.

22. Consider an $n \times n$ matrix Q whose elements are given by $Q_{ij} = v_i v_j$, where v is a vector in n-dimensional space. Find the eigenvalues and eigenvectors of Q. Are the eigenvectors all unique?

23. Show that all eigenvalues of the quantum oscillator have the form $E_n = \hbar\omega(n + \frac{1}{2})$. [*Hint*: Assume that there is an eigenvector y and associated eigenvalue λ such that $Hy = \lambda y$ and using $[a, H] = \hbar\omega a$, create eigenvalues $\lambda - \hbar\omega$, $\lambda - 2\hbar\omega$, etc., and arrive at a contradiction.]

24. Show that if A^{-1} and B^{-1} exist, then $A^{-1} \otimes B^{-1} = (A \otimes B)^{-1}$.

25. If A and B are similar, prove that A^n and B^n are similar.

26. Show that $(C + D)^{-1} = C^{-1} - C^{-1}D(C + D)^{-1}$. Thus for D "small" we can iterate to get $(C + D)^{-1} = C^{-1} - C^{-1}DC^{-1} + C^{-1}DC^{-1}DC^{-1} \cdots$ which is the operator analogue of the familiar expansion of $(x + y)^{-1}$ when y is "small."

27. *The Dirac matrices.* Assume that for some n there exist *four* $n \times n$ matrices which satisfy the anticommutation relations $\gamma_i\gamma_j + \gamma_j\gamma_i = 2I\delta_{ij}$, where $i, j = 1, 2, 3,$ or 4.
 a) Prove that $\gamma_i^2 = I$ (the unit matrix), for $i = 1, 2, 3,$ or 4.
 b) Prove that det $\gamma_i = \pm 1$, for $i = 1, 2, 3,$ or 4.
 c) If λ is an eigenvalue of any one of the γ_i, show that $\lambda = \pm 1$.
 d) Show that tr $\gamma_i = 0$, for $i = 1, 2, 3,$ or 4.
 The final part of this problem consists in showing that there exists no set of four $n \times n$ matrices of order less than 4 which satisfies the above anticommutation relations. You may, of course, do this in any way you like. One way is to prove the statement via the following two steps.
 e) Show that there exists no set of four $n \times n$ matrices of odd order which satisfies the anticommutation relations.
 f) Show that any *four* 2×2 matrices cannot satisfy the anticommutation relations.
 [*Hint*: The three 2×2 Pauli matrices do satisfy the relations

$$\sigma_i\sigma_j + \sigma_j\sigma_i = 2I\delta_{ij}, \qquad i, j = 1, 2, 3,$$

 and any 2×2 matrices can be expanded as a linear combination of the three Pauli matrices and the 2×2 unit matrix.]

 (*Note*: That there exist four nonunique 4×4 matrices satisfying these relations is shown in any standard treatment of the Dirac equation. This problem shows that one need not bother looking for four matrices of lower than fourth order which obey the anticommutation relations, a point of great theoretical significance.)

28. Show that the operator

$$T = I + \frac{xD}{1!} + \frac{(xD)^2}{2!} + \cdots + \frac{(xD)^n}{n!},$$

 where $D = d/dt$, acts as a translation operator on the space of polynomials (in the variable t) of degree $\leqq n$, that is,

$$Tf(t) = f(t + x),$$

 if $f(t)$ is in the space of polynomials of degree n. This is the end of the problem; an application of this result follows, which is typical of various symbolic operator solutions of problems in physics (e. g. Blankenbecler's method for obtaining integrals over the Fermi-Dirac distribution function*). The present application is due largely to L. N. Hand. The result was obtained by other methods by Jacob Bernoulli (1654–1705).

 We first note that in the space of continuous functions which are infinitely differentiable, the translation operator would be given by

$$T = e^{xD}.$$

 Some interesting results can be obtained using this formula. For example, a general expression for $\sum_{k=0}^{k=n} k^m$ can be derived for any integers $n \geqslant 0$ and $m > 0$. Let $F_m(n) = \sum_{k=0}^{n} k^m$. Now $F_m(n) - F_m(n-1) = n^m$. Since $F_m(n)$ is a polynomial,†

---

* See Kittel, *Elementary Statistic Physics*, New York, Wiley 1958, § 20.
† See Stewart, *Theory of Numbers*, New York, Macmillan 1952, p. 19.

we can write $e^{-D}F_m(n) = F_m(n-1)$, and, consequently,

$$F_m(n) - F_m(n-1) = (1 - e^{-D})F_m(n) = n^m$$

or, symbolically,

$$F_m(n) = \left(\frac{1}{1-e^{-D}}\right)n^m .$$

Now,

$$\frac{1}{1-e^{-x}} = x^{-1} + \frac{1}{2} + \sum_{j=0}^{\infty} \frac{(-1)^j B_{j+1}}{[2(j+1)]!} x^{2j+1} ,$$

where the B_i's are the Bernoulli numbers

$$B_1 = \tfrac{1}{6}, \qquad B_2 = \tfrac{1}{30}, \qquad B_3 = \tfrac{1}{42}, \qquad B_4 = \tfrac{1}{30}, \quad \text{etc.}$$

(See any mathematical tables.) If we blithely use this series for $x = D$, interpreting D^{-1} as integration, and note that $D^p n^m = [m!/(m-p)!]n^{m-p}$, then we can calculate $F_m(n)$.

$$F_m(n) = \frac{n^{m+1}}{m+1} + C_m + \frac{n^m}{2} + \sum_{j=0}^{J} \frac{(-1)^j B_{j+1} m!}{[2(j+1)]!\,[m-(2j+1)]!} n^{[m-(2j+1)]},$$

where J equals $(m-1)/2$ or $(m-2)/2$, whichever is integral. The constant of integration, C_m, can be determined by using $F_m(0) = 0$. We see immediately that $C_m = 0$ if m is even and $C_m = (-1)^{(m+1)/2}B_{(m+1)/2}/(m+1)$ if m is odd; in other words C_m is chosen so that no constant term appears in $F_m(n)$. For example,

$$F_1(n) = \sum_{k=0}^{n} k = \frac{n^2}{2} + C_1 + \frac{n}{2} + \frac{B_1}{2} .$$

In order to satisfy $F_1(0) = 0$ we must choose the constant equal to $-B_1/2$, giving the usual result:

$$F_1(n) = \tfrac{1}{2}(n^2 + n) .$$

Similarly, we find the formulas:

$$F_2(n) \equiv \sum_{k=0}^{n} k^2 = \frac{n^3}{3} + \frac{n^2}{2} + \frac{n}{6} = \frac{1}{6}n(n+1)(2n+1) ,$$

$$F_3(n) \equiv \sum_{k=0}^{n} k^3 = \frac{n^4}{4} + \frac{n^3}{2} + \frac{n^2}{4} = \left[\frac{n(n+1)}{2}\right]^2 ,$$

$$F_4(n) \equiv \sum_{k=0}^{n} k^4 = \frac{n^5}{5} + \frac{n^4}{2} + \frac{n^3}{3} - \frac{n}{30} ,$$

$$F_5(n) \equiv \sum_{k=0}^{n} k^5 = \frac{n^6}{6} + \frac{n^5}{2} + \frac{5}{12}n^4 - \frac{n^2}{12} .$$

Of course, the level of rigor in this derivation is simply appalling. However, the fact that we get away with it makes it plausible that this derivation could be made rigorous. Very often new mathematics results from the formalization of just such sloppy forays into the unknown. The theory of distributions which formalizes the symbolic functions, such as delta functions, used by physicists for fifty years prior to their formalization, is a case in point. The present example, however, is a treatment of results long known and rigorously established. It is presented partly as an antidote to the more rigorous material covered in this and subsequent chapters.

FURTHER READINGS

HALMOS, P. R., *Finite-Dimensional Vector Spaces*, Second edition. Princeton, N. J.: D. Van Nostrand, 1958. The classic.

HILDEBRAND, F. B., *Introduction to Numerical Analysis*. New York: McGraw-Hill, 1956. The bridge from theory to thesis.

LANG, S., *Linear Algebra*. Reading, Mass.: Addison-Wesley, 1965.

MARCUS, M., and H. MINC, *A Survey of Matrix Theory and Matrix Inequalities*. Boston: Allyn and Bacon, 1964. Exhaustive reference to properties of, and theorems about, matrices.

CHAPTER 4

VECTOR SPACES IN PHYSICS

INTRODUCTION

So far, the concepts of length and angle have been missing from our discussion of vector spaces. Thus there have been few applications to physics. In this chapter the structure of the vector space will be enormously enriched by the addition of a numerical function, the *inner* (or scalar) product, in terms of which the length of a vector and the angle between two vectors can be defined. Vector spaces in which an inner product has been defined are called *inner-product spaces.* The study of inner-product spaces will enable us to make a real juncture with physics. Accordingly, in the latter sections of this chapter we shall discuss several examples from classical and quantum physics.

4.1 THE INNER PRODUCT

In Section 1.3 we defined the scalar product of two ordinary vectors in real three-dimensional space. In the notation used there, we wrote the scalar product of the vectors $\mathbf{x} = (x_1, x_2, x_3)$ and $\mathbf{y} = (y_1, y_2, y_3)$ as

$$\mathbf{x}\cdot\mathbf{y} \equiv \sum_{i=1}^{3} x_i y_i \,.$$

In this notation the length of the vector $\mathbf{x}$ is denoted by $|\,\mathbf{x}\,|$ and is given by the quantity

$$|\,\mathbf{x}\,| \equiv (\mathbf{x}\cdot\mathbf{x})^{1/2} = \left(\sum_{i=1}^{3} x_i^2\right)^{1/2}.$$

The angle θ between two vectors $\mathbf{x}$ and $\mathbf{y}$ is given by

$$\theta = \cos^{-1}\left[\frac{\mathbf{x}\cdot\mathbf{y}}{|\,\mathbf{x}\,|\;|\,\mathbf{y}\,|}\right].$$

In abstract vector spaces we shall denote the length, or magnitude, of a vector x by the symbol $||\,x\,||$, in order to reserve single vertical bars for the absolute value of a complex number. It is customary in vector-space theory to refer to the scalar product as the *inner* product, and to write the inner product of two vectors, x and y, as (x, y). In this notation, the length of the vector x is written

$$||\,x\,|| \equiv (x, x)^{1/2}\,, \tag{4.1}$$

and the angle θ between two vectors is written

$$\theta = \cos^{-1}\left[\frac{(x, y)}{||x||\,||y||}\right]. \tag{4.2}$$

As long as we refer to real three-dimensional vector spaces, we can define the quantity (x, y) just as we defined $\mathbf{x}\cdot\mathbf{y}$. That is, if $x = (\xi_1, \xi_2, \xi_3)$ and $y = (\eta_1, \eta_2, \eta_3)$, then

$$(x, y) \equiv \sum_{i=1}^{3} \xi_i \eta_i . \tag{4.3}$$

The generalization to real n-dimensional vector spaces is also straightforward—we merely extend the sum over i to n. However, if we try to retain this definition of the inner product for a *complex* vector space, we immediately encounter difficulties. For example, consider the "length" of the one-dimensional vector ix. The above definition tells us that its length is

$$||ix|| = (ix, ix)^{1/2} = (-x^2)^{1/2} ,$$

which is imaginary if x is real. But a distance function must be real-valued. We could skirt the difficulty by restricting our attention to real vector spaces, but only at the price of losing relevance to much of physics. In particular, in quantum theory the underlying field is not the real field, but the field of complex numbers.

There is a simple, but seemingly artificial, remedy for this situation. Let x and y be vectors with *complex* components $(\xi_1, \xi_2, \cdots, \xi_n)$ and $(\eta_1, \eta_2, \cdots, \eta_n)$ referred to some basis. We shall not define the inner product of these vectors by Eq. (4.3), but rather by

$$(x, y) \equiv \sum_{i=1}^{n} \xi_i^* \eta_i . \tag{4.4}$$

The asterisk (*) denotes complex conjugation of ξ_i, which is a number in the complex field.* With this definition of the inner product, the length of the vector x is

$$||x|| \equiv (x, x)^{1/2} = \left(\sum_{i=1}^{n} \xi_i^* \xi_i\right)^{1/2} = \left(\sum_{i=1}^{n} |\xi_i|^2\right)^{1/2}, \tag{4.5}$$

where $|\xi_i|$ is the absolute value of ξ_i. Since $|\xi_i|^2 \geq 0$, $||x||$ is always real. Moreover, if the vector space is real, $\xi_i = \xi_i^*$, and this definition reduces to the familiar one of Eq. (4.3).

For a complex vector space, the inner product is not symmetrical as it is in a real vector space. That is,

$$(x, y) \neq (y, x) ,$$

* If $z = a + ib$, where $i \equiv \sqrt{-1}$, then $z^* \equiv a - ib$. The absolute value of z is the quantity $|z| \equiv (zz^*)^{1/2} = (a^2 + b^2)^{1/2}$.

but rather

$$(x, y) = (y, x)^* . \tag{4.6}$$

Nevertheless, this definition will work out satisfactorily.

We now give the formal definition of an inner-product space.

Definition 4.1. An *inner product* in a (real or complex) vector space is a scalar-valued function of the ordered pair of vectors x and y, such that

1. $(x, y) = (y, x)^*$;
2. $(\alpha x + \beta y, z) = \alpha^*(x, z) + \beta^*(y, z)$, where α and β are scalars in the field underlying vector space;
3. $(x, x) \geq 0$ for any x; $(x, x) = 0$ if and only if $x = 0$.

An *inner-product space* is a vector space with an inner product. A real vector space with an inner product, that is, a real inner-product space, is called a *Euclidean* space; a complex inner-product space is called a *unitary* space.

In a real vector space, the complex conjugate in (1) and (2) adds nothing and may be ignored. In either case, real or complex, (1) implies that (x, x) is real, so that the inequality in (3) makes sense. The quantity $(x, x)^{1/2} \equiv ||x||$ is often referred to as the *norm* of the vector x. Note that

$$||ax|| = (ax, ax)^{1/2} = [a^*a(x, x)]^{1/2} = |a| \cdot ||x|| .$$

It follows from (1) and (2) that $(x, \alpha y) = \alpha(x, y)$.

Requirement (3) says that the inner product is a *positive-definite* form—it is *positive* unless x is *definitely* the zero vector, in which case it is *definitely* zero. As we proceed, the reader will see that certain theorems can be proved without reference to this property.

Examples of an Inner Product.

1. As we have seen, if

$$x = (\xi_1, \cdots, \xi_n) \qquad \text{and} \qquad y = (\eta_1, \cdots, \eta_n) ,$$

then

$$(x, y) \equiv \sum_{i=1}^{n} \xi_i^* \eta_i$$

satisfies all the axioms of an inner product.

2. If $x(t)$ and $y(t)$ are polynomials in the complex vector space P, defined on the closed interval $[0, 1]$, then an inner product is

$$(x, y) \equiv \int_0^1 x(t)^* y(t) \, dt;$$

3. If $x = (\xi_1, \cdots, \xi_n)$ and $y = (\eta_1, \cdots, \eta_n)$ are column vectors over a real field, and P is a diagonal matrix whose diagonal elements are real positive numbers,

then we may define an inner product as

$$(x, y) \equiv \tilde{x}Py = (\xi_1, \cdots, \xi_n)\begin{bmatrix} P_{11} & & 0 \\ & \ddots & \\ 0 & & P_{nn} \end{bmatrix}\begin{pmatrix} \eta_1 \\ \vdots \\ \eta_n \end{pmatrix} = \sum_{i=1}^{n} \xi_i P_{ii} \eta_i .$$

If x is a column vector, $\tilde{x}$ denotes its transpose, a row vector.

4. If $x = (\xi_1, \xi_2, \xi_3, \xi_4)$ and $y = (\eta_1, \eta_2, \eta_3, \eta_4)$ are column four-vectors over a real field, and

$$M = \begin{bmatrix} 1 & 0 & 0 & 0 \\ 0 & 1 & 0 & 0 \\ 0 & 0 & 1 & 0 \\ 0 & 0 & 0 & -1 \end{bmatrix},$$

then the quantity $(x, y) \equiv \tilde{x}My = \xi_1 n_1 + \xi_2 n_2 + \xi_3 n_3 - \xi_4 n_4$ satisfies requirements (1) and (2) for an inner product, but not (3)—it is not positive-definite. Thus the entity $\tilde{x}My$ is not an inner product. The quantity $\tilde{x}My$ plays an important role in the theory of special relativity; M is called the *Lorentz metric*.

4.2 ORTHOGONALITY AND COMPLETENESS

Now that the concepts of length and angle are defined, the theory of vector spaces becomes more immediately relevant to physics. One of the central notions of mathematical physics is that of complete orthogonal sets of vectors or functions. In this section we lay the groundwork for the theory of such sets.

Definition 4.2. Two vectors x and y are *orthogonal* if and only if $(x, y) = 0$.

Two vectors in a three-dimensional space are orthogonal if the angle between them is 90°; our definition is simply a generalization of this idea for n-dimensional—or even infinite-dimensional—spaces. As we might expect, if $(x, y) = 0$, then $(x, y) = (y, x)^* = 0$, so that $(y, x) = 0$ as well. Thus, although the inner product is unsymmetric, orthogonality is a symmetric relation. Note that the zero vector is orthogonal to every vector.

Definition 4.3. A set of vectors $\{x_1, x_2, \cdots\}$ is *orthonormal* if $(x_i, x_j) = \delta_{ij}$, for all i and j.

That is, a set of vectors is orthonormal if each vector is *orthogonal* to all others in the set and is *normalized* to unit length. A vector x may be normalized by dividing by its length to form the new vector $x/||x||$. One orthonormal set of vectors is the set of three unit vectors, $\{\mathbf{e}_i\}$, for the Cartesian space introduced in Section 1.3.

Definition 4.4. In a finite-dimensional vector space, an orthonormal set is *complete* if it is not contained in any larger orthonormal set.

We now prove several basic theorems involving these concepts.

Theorem 4.1. An orthonormal set is linearly independent. (Hence any orthonormal set of n vectors is a basis for an n-dimensional vector space.)

Proof. If $\{x_1, \cdots, x_n\}$ is any finite orthonormal set, then for any j,

$$\sum_i \alpha_i x_i = 0 \Longrightarrow 0 = \left(x_j, \sum_i \alpha_i x_i\right)$$
$$= \sum_i \alpha_i (x_j, x_i) = \sum_i \alpha_i \delta_{ij} = \alpha_j. \quad \text{QED}$$

Theorem 4.2. *Bessel's inequality.* If $X = \{x_1, \cdots, x_n\}$ is any finite orthonormal set in an inner product space, and x is any vector, then

$$||x||^2 \geq \sum_i |\alpha_i|^2, \tag{4.7}$$

where $\alpha_i = (x_i, x)$. Furthermore, the vector $x' \equiv x - \sum_i \alpha_i x_i$ is orthogonal to each x_j.

Proof. We first prove the inequality

$$0 \leq ||x'||^2 = (x', x') = \left(x - \sum_i \alpha_i x_i, x - \sum_j \alpha_j x_j\right)$$
$$= (x, x) - \sum_i \alpha_i^* (x_i, x) - \sum_j \alpha_j (x, x_j) + \sum_{i,j} \alpha_i^* \alpha_j (x_i, x_j)$$
$$= ||x||^2 - \sum_i |\alpha_i|^2 - \sum_j |\alpha_j|^2 + \sum_j |\alpha_j|^2 = ||x||^2 - \sum_i |\alpha_i|^2,$$

where we have used $(x_i, x_j) = \delta_{ij}$. Thus

$$||x||^2 \geq \sum_i |\alpha_i|^2.$$

To prove the second part of the theorem, we simply calculate

$$(x', x_j) = (x, x_j) - \sum_i \alpha_i^* (x_i, x_j) = \alpha_j^* - \alpha_j^* = 0. \quad \text{QED}$$

There are a number of ways to characterize a complete orthonormal set. Halmos has collected these and presented the proof of their equivalence in the form of a theorem.

Because it illustrates a very important logical technique, we adopt essentially his statement of the theorem and his logical route in its proof.*

Theorem 4.3. If $X = \{x_i\}$ is a finite orthonormal set of m vectors in an inner-product space V, the following six conditions on X are equivalent to each other:

1. X is complete;
2. if $(x_i, x) = 0$ for $i = 1, \cdots, m$, then $x = 0$;
3. X spans V,
4. if $x \in V$, then $x = \sum_{i=1}^m (x_i, x) x_i$,
5. if $x, y \in V$, then $(y, x) = \sum_{i=1}^m (y, x_i)(x_i, x)$—Parseval's equation;
6. if $x \in V$, then $||x||^2 = \sum_{i=1}^m |(x_i, x)|^2 = \sum_{i=1}^m |\alpha_i|^2$, in the notation of Theorem 4.2; i.e., the equal sign holds in Bessel's inequality.

* See Halmos, P. R., *Finite-Dimensional Vector Spaces*, Second Edition. Princeton, N.J.: D. Van Nostrand, 1958. p. 124.

Proof. In this proof, we shall establish the implications

$$(1) \Longrightarrow (2) \Longrightarrow (3) \Longrightarrow (4) \Longrightarrow (5) \Longrightarrow (6) \Longrightarrow (1).$$

Thus, given any one of the six conditions, any of the other five can be reached by a logical chain, which establishes their logical equivalence. To begin, we assume that (1) is true and derive (2).

(1) $\Longrightarrow$ (2). We actually show this in the equivalent form: not (2) $\Longrightarrow$ not (1). If $(x_i, x) = 0$ for all i and $x \neq 0$, then we may adjoin $x/||x||$ to X and thus obtain an orthonormal set larger than X. Hence x is not complete according to Definition (4.4) of a complete set.

(2) $\Longrightarrow$ (3). Again, we actually show: not (3) $\Longrightarrow$ not (2). If X does not span V, then there is some $x \in V$ which cannot be written as a linear combination of the x_i, and therefore there exist no constants α_i such that

$$x = \sum_i \alpha_i x_i \,.$$

In particular, the set of constants $\alpha_i = (x_i, x)$ will be such that

$$x \neq \sum_i \alpha_i x_i \,.$$

Hence

$$x' = x - \sum_i (x_i, x) x_i \neq 0 \,,$$

and yet, by the second part of Theorem 4.2, x' is orthogonal to each x_i. This statement is equivalent to "not (2)."

(3) $\Longrightarrow$ (4). If every x may be written in the form

$$x = \sum_j \alpha_j x_j \,,$$

then

$$(x_i, x) = \sum_j \alpha_j (x_i, x_j) = \sum_j \alpha_j \delta_{ij} = \alpha_i \,,$$

so that

$$x = \sum_i (x_i, x) x_i \,.$$

(4) $\Longrightarrow$ (5). If

$$x = \sum_i \alpha_i x_i \qquad \text{and} \qquad y = \sum_j \beta_j x_j \,,$$

with $\alpha_i = (x_i, x)$ and $\beta_j = (x_j, y)$, then

$$\begin{aligned}(y, x) &= \left(\sum_j \beta_j x_j, \sum_i \alpha_i x_i\right) \\ &= \sum_{i,j} \beta_j^* \alpha_i (x_j, x_i) = \sum_i \beta_i^* \alpha_i = \sum_i (y, x_i)(x_i, x) \,.\end{aligned}$$

(5) $\Longrightarrow$ (6). Set $x = y$ in (5).

(6) $\Longrightarrow$ (1). [Actually, not (1) $\Longrightarrow$ not (6).] Assume that X is not complete. It is therefore contained in a larger orthogonal set, i.e., there exists a nonzero vector, x_0, orthogonal to each x_i in X; that is, $(x_i, x_0) = 0$, for all i. Thus

$$||x_0||^2 \neq \sum_i |(x_i, x_0)|^2 ,$$

since the right-hand side vanishes and the left does not. This statement is precisely "not (6)." This completes the proof of the equivalence of the six conditions.

Theorem 4.4. *Schwarz's inequality.* If x, y are vectors in an inner-product space, then

$$|(x, y)| \leq ||x|| \cdot ||y|| . \tag{4.8}$$

Proof. If $y = 0$, both sides vanish. If $y \neq 0$, then the set consisting of the single vector $y/||y||$ is orthonormal, and therefore by Bessel's inequality

$$\left|\left(x, \frac{y}{||y||}\right)\right|^2 \leq ||x||^2 \Longrightarrow |(x, y)| \leq ||x|| \cdot ||y|| . \quad \text{QED}$$

Examples of Schwarz's inequality.

1. In two-dimensional Euclidean space, the Schwarz inequality just says that $|\cos\theta| \leq 1$.

2. In n-dimensional unitary space, the Schwarz inequality is known as the Cauchy inequality. It asserts that for any two sequences of complex numbers, $(\alpha_1, \cdots, \alpha_n)$ and $(\beta_1, \cdots, \beta_n)$,

$$\left|\sum_{i=1}^n \alpha_i^* \beta_i\right|^2 \leq \sum_{i=1}^n |\alpha_i|^2 \cdot \sum_{i=1}^n |\beta_i|^2 . \tag{4.9}$$

3. In the space P, Schwarz's inequality becomes

$$\left|\int_0^1 x(t)^* y(t)\, dt\right|^2 \leq \int_0^1 |x(t)|^2\, dt \cdot \int_0^1 |y(t)|^2\, dt . \tag{4.10}$$

4.3 COMPLETE ORTHONORMAL SETS

It is very handy to have an orthonormal set of vectors as a basis for a vector space. Now we shall show by induction how one may construct an orthonormal basis from an arbitrary basis using a technique known as the *Gram-Schmidt orthogonalization process*. From Theorem 4.3 [(3) $\Longrightarrow$ (1)], this orthonormal basis will be a complete orthonormal set.

Let $X = \{x_1, \cdots, x_n\}$ be any basis in V. We shall construct a complete orthonormal set $Y = \{y_1, \cdots, y_n\}$ such that each y_j is a linear combination of

$x_1, \cdots, x_j$. Now $x_1 \neq 0$ because X is linearly independent; thus we may take $y_1 = x_1/||x_1||$. Now we put $y_2 = (x_2 - \alpha_1 y_1)/||x_2 - \alpha_1 y_1||$ and determine α_1 so that $(y_1, y_2) = 0$. Thus

$$(y_1, y_2) = \frac{(y_1, x_2) - \alpha_1}{||x_2 - \alpha_1 y_1||} = 0 \Longrightarrow \alpha_1 = (y_1, x_2) .$$

Therefore

$$y_2 = \frac{x_2 - (y_1, x_2)\, y_1}{||x_2 - (y_1, x_2)\, y_1||} .$$

Note that

$$x_2 - (y_1, x_2)\, y_1 = x_2 - \frac{(y_1, x_2)}{||x_1||} x_1 \neq 0 ,$$

since X is linearly independent. Thus we have found two vectors, y_1 and y_2, such that $(y_i, y_j) = \delta_{ij}$, $i, j = 1, 2$.

Now we discuss the induction step: Suppose that $y_1, \cdots, y_m$, $(m < n)$, have been found so that they form an orthonormal set and so that each y_j $(j = 1, \cdots, m)$ is a linear combination of $x_1, \cdots, x_j$. Let

$$y_{m+1} = \frac{x_{m+1} - (\alpha_1 y_1 + \cdots + \alpha_m y_m)}{||x_{m+1} - (\alpha_1 y_1 + \cdots + \alpha_m y_m)||} .$$

Then, for $j \leq m$,

$$(y_j, y_{m+1}) = \frac{(y_j, x_{m+1}) - \alpha_j}{||x_{m+1} - (\alpha_1 y_1 + \cdots + \alpha_m y_m)||} = 0 \Longrightarrow \alpha_j = (y_j, x_{m+1}) .$$

which implies that $\alpha_j = (y_j, x_{m+1})$. Thus

$$y_{m+1} = \frac{x_{m+1} - [(y_1, x_{m+1})y_1 + \cdots + (y_m, x_{m+1})y_m]}{||x_{m+1} - [(y_1, x_{m+1})y_1 + \cdots + (y_m, x_{m+1})y_m]||} . \tag{4.11}$$

The linear independence of the basis X ensures us that $y_{m+1} \neq 0$. When this process has been repeated n times, we shall have generated an orthonormal set of n basis vectors.

A matrix A is defined as a representation of the linear transformation A with respect to some basis $X = \{x_1, \cdots, x_n\}$. If we take the basis X to be an orthonormal set (or make it so by the Gram-Schmidt process), then we have a simple relationship between matrix elements and inner products which is used constantly in quantum mechanics. From Eq. 3.7, we have

$$Ax_j = \sum_k a_{kj} x_k .$$

If we now assume the x_i's to be an *orthonormal* basis and form the inner product (x_i, Ax_j), we have

$$(x_i, Ax_j) = \left(x_i, \sum_k a_{kj} x_k\right) = \sum_k a_{kj}(x_i, x_k) = \sum_k a_{kj}\delta_{ik} = a_{ij} .$$

Therefore the matrix elements of a linear transformation with respect to an orthonormal basis are given by

$$a_{ij} = (x_i, Ax_j) \, . \tag{4.12}$$

All that is being done in the Gram-Schmidt process is to start with some vector x_1, normalize it, and then subtract from the next vector, x_2, any component of the normalized x_1 which it may contain. After normalizing the modified x_2, one goes on to x_3 and removes all components of the previous normalized vectors. The process can be continued indefinitely, although clearly in a finite-dimensional space it must eventually terminate. At each step of the operation, one *decreases* (or leaves unchanged) the norm of the vector in question, since one subtracts some of its components. This fact is the basis for a very simple application of the Gram-Schmidt process to the theory of determinants. Let us denote a matrix A by $\{a_1, a_2, \cdots, a_n\}$, where a_1 is the first row, a_2 is the second row, etc. Now if the row vectors a_i form an orthogonal set of vectors, clearly $\det (AA^\dagger) = ||a_1||^2||a_2||^2 \cdots ||a_n||^2$. Since, according to Eq. (3.17) $\det A^\dagger = (\det A)^*$, it follows that

$$|\det A| = ||a_1|| \, ||a_2|| \cdots ||a_n|| \, .$$

Here the norm of the row vector a_i has its usual meaning, that is,

$$||a_i|| = \left(\sum_{j=1}^{n} a_{ij}^* a_{ij} \right)^{1/2}$$

[see Eq. (4.5)].

Now consider an arbitrary matrix $B = \{b_1, b_2, \cdots, b_n\}$. We know that to any row of B one can add an arbitrary multiple of any other row and leave its determinant unchanged. Thus, by the Gram-Schmidt process, we can create from the row vectors of B a new matrix C which we denote by

$$\{\beta_1 c_1, \beta_2 c_2, \cdots \beta_n c_n\} \, ,$$

where $\{c_1, c_2, \cdots, c_n\}$ is a set of orthonormal vectors and $\{\beta_1, \beta_2, \cdots, \beta_n\}$ is a set of scalars, each of which is respectively less than or equal to $||b_1||$, $||b_2||$, $\cdots$, $||b_n||$. This follows from Eq. (4.11). Now since we have added multiples of rows to other rows, $\det C = \det B$. Moreover, the c_i are normalized to 1, so we find (just as for $\det A$, above) that

$$|\det B| = |\det C| = |\beta_1| \, |\beta_2| \cdots |\beta_n| \, .$$

Since $|\beta_i| \leq ||b_i||$, we have

Hadamard's inequality. If b_i denotes the ith row of the matrix B, then

$$|\det B| \leq ||b_1|| \, ||b_2|| \cdots ||b_n|| \, .$$

This result is useful in the theory of integral equations.

We close this section with a simple but important result that will be required continually in the proofs of subsequent theorems.

Theorem 4.5. A linear transformation A on an inner-product space is the zero transformation if and only if $(x, Ay) = 0$, for all x and y.

Proof. If $A = 0$, then $(x, Ay) = (x, 0) = 0$ (by property 2 in the definition of the inner product). Conversely, if $(x, Ay) = 0$ for all x and y, we take $x = Ay$. Then $(Ay, Ay) = 0 \Longrightarrow Ay = 0$, for all y (by positive definiteness of the inner product) and therefore $A = 0$. QED

4.4 SELF-ADJOINT (HERMITIAN AND SYMMETRIC) TRANSFORMATIONS

The most important linear transformations in quantum physics are the so-called *self-adjoint operators*, which will be defined presently. First, however, we must define the adjoint transformation.

Definition 4.5. Let A be a linear transformation defined on a vector space V. For every A, the operator $A^\dagger$, which satisfies the equation

$$(Ax, y) = (x, A^\dagger y), \qquad \text{for all } x, y \tag{4.13}$$

is called the *adjoint* of A.

Since

$$(x, Ay) = (Ay, x)^* = (y, A^\dagger x)^* = (A^\dagger x, y),$$

the adjoint may be defined equivalently by

$$(x, Ay) = (A^\dagger x, y), \qquad \text{for all } x, y. \tag{4.14}$$

To be strictly rigorous, we ought at this point to prove that the adjoint of a given linear transformation always exists. We leave the proof to the reader (Problem 18), however, since it is of little interest for the purposes we have in mind. The interested reader may find a coordinate-independent proof in a text on vector spaces.*

Moreover, the adjoint $A^\dagger$ is unique. Suppose that

$$(x, By) = (Ax, y) = (x, Cy)$$

for all x, y. Then $(x, [B - C]y) = 0$ for all x, y which implies $B = C$ by Theorem 4.5.

To show that $A^\dagger$ is in fact a *linear* transformation, we compute

$$\begin{aligned}(x, A^\dagger[\alpha y + \beta z]) &= (Ax, [\alpha y + \beta z]) = (Ax, \alpha y) + (Ax, \beta z) \\ &= \alpha(Ax, y) + \beta(Ax, z) = \alpha(x, A^\dagger y) + \beta(x, A^\dagger z) \\ &= (x, \alpha A^\dagger y) + (x, \beta A^\dagger z).\end{aligned}$$

Therefore

$$(x, A^\dagger[\alpha y + \beta z] - \alpha A^\dagger y - \beta A^\dagger z) = 0 \Longrightarrow A^\dagger(\alpha y + \beta z) = \alpha A^\dagger y + \beta A^\dagger z,$$

so $A^\dagger$ is a linear operator. We now prove a few simple properties of the adjoint.

---

* For example, Hoffman and Kunze, p. 237.

Theorem 4.6.

1. $(A + B)^\dagger = A^\dagger + B^\dagger$.
2. $(AB)^\dagger = B^\dagger A^\dagger$.
3. $(\alpha A)^\dagger = \alpha^* A^\dagger$, where α is a scalar.
4. $(A^\dagger)^\dagger = A$.

Proof.

1. For all x and y,

$$\begin{aligned}(x, [A + B]^\dagger y) &= ([A + B]x, y) \\ &= (Ax + Bx, y) = (Ax, y) + (Bx, y) \\ &= (x, A^\dagger y) + (x, B^\dagger y) \\ &= (x, [A^\dagger + B^\dagger]y) .\end{aligned}$$

The result now follows from Theorem 4.5.

2. $(x, [AB]^\dagger y) = (ABx, y) = (Bx, A^\dagger y) = (x, B^\dagger A^\dagger y)$, and again, using Theorem 4.5, the result follows. QED

We leave the proofs of properties (3) and (4), which are very similar to those we have done, to the reader.

Theorem 4.7. Let the matrix of A be (a_{ij}) with respect to an orthonormal basis X. Then the matrix of $A^\dagger$ with respect to X is $[A^\dagger]_{ij} = a_{ji}^*$, that is, $[A^\dagger] = [\tilde{A}]^*$.

Proof. From Eq. (4.12), we have

$$a_{ij} = (x_i, Ax_j) .$$

Therefore

$$a_{ij} = (A^\dagger x_i, x_j) = (x_j, A^\dagger x_i)^* = [A^\dagger]_{ji}^* \Longrightarrow [A^\dagger]_{ij} = a_{ji}^* . \quad \text{QED}$$

Thus the definition of the adjoint of an operator reduces to that of the adjoint of a matrix, given in Section 3.7.

Definition 4.6. If $A = A^\dagger$, A is its own adjoint, and A is called a *self-adjoint* operator. In real inner-product spaces, a self-adjoint operator is called a *symmetric* operator, and in a complex inner-product space, a *Hermitian* operator.

The operators that correspond to physical observables in quantum mechanics are all Hermitian operators.

It follows immediately from Theorem 4.6 that if A and B are self-adjoint, then so is their sum. Similarly, if A is self-adjoint, then the operator αA will be self-adjoint if and only if α is real. We also derive the following simple result:

Theorem 4.8. If A and B are self-adjoint, then AB is self-adjoint if and only if A and B commute (i.e., if and only if their commutator, $[A, B]$, vanishes).

Proof. If $AB = BA$, then

$$(AB)^\dagger = B^\dagger A^\dagger = BA = AB\,,$$

so AB is self-adjoint. If $(AB)^\dagger = AB$, then

$$AB = (AB)^\dagger = B^\dagger A^\dagger = BA\,, \qquad \text{so } [A, B] = 0\,. \text{ QED}$$

The reason why a self-adjoint operator on a real inner-product space is called a symmetric operator is apparent from Theorem 4.7. For if $A = A^\dagger$, the matrix representations satisfy

$$a_{ij} = [A^\dagger]_{ij} = a_{ji}^* = a_{ji}\,,$$

because the matrix elements belong to the field of real numbers. Thus $A = \tilde{A}$, and A is what we have previously called a symmetric matrix.

We now prove two theorems that are variations of Theorem 4.5 and will play a role much like the role which that theorem has played in many subsequent proofs. Basically, we are concerned with strengthening that theorem for more restricted applications to self-adjoint operators.

Theorem 4.9. Let A be a self-adjoint operator in a real inner-product space. Then $A = 0$ if and only if $(x, Ax) = 0$, for all x.

Proof. If $A = 0$, then $(x, Ax) = 0$. Thus the proof of the necessity of the condition is trivial. However, its sufficiency is not. To establish sufficiency consider the expression

$$(x + y, A[x + y]) = (x, Ax) + (y, Ay) + (x, Ay) + (y, Ax)\,.$$

From this we get

$$(x, Ay) + (y, Ax) = (x + y, A[x + y]) - (x, Ax) - (y, Ay)$$

for all x and y. Since each term on the right-hand side is of the form (x, Ax), which is zero by hypothesis, it follows that

$$(x, Ay) + (y, Ax) = 0\,.$$

But

$$(y, Ax) = (A^\dagger y, x) = (Ay, x) = (x, Ay)^* = (x, Ay)\,,$$

since $A^\dagger = A$ and the inner product is a real number. Therefore $(x, Ay) = 0$ for all x and y, which, by Theorem 4.5, implies that $A = 0$. QED

If we know that the inner-product space is complex, as is the case in most of the applications to physics, we may strengthen this result still further.

Theorem 4.10. If A is any linear transformation on a unitary space, then $A = 0$ if and only if $(x, Ax) = 0$ for all x.

Proof. As in the previous theorem, necessity is obvious. The proof of sufficiency follows easily from a result we obtained in proving the last theorem. We showed that

$$(x, Ay) + (y, Ax) = 0$$

for all x and y, if $(x, Ax) = 0$ for all x. No further conditions on A were used in establishing this result, so we are free to use it here. Since it holds for all y, we may replace y by iy to obtain

$$(x, Aiy) + (iy, Ax) = i[(x, Ay) - (y, Ax)] = 0 .$$

It follows that

$$(x, Ay) - (y, Ax) = 0 ,$$

for all x and y. Adding this to the result above, we have $(x, Ay) = 0$ for all x and y. Theorem 4.5 now implies that $A = 0$. QED

As a consequence of Theorems 4.9 and 4.10, we have the following result.

Corollary. If A is a self-adjoint linear transformation in an inner-product space (real *or* complex), then $A = 0$ if and only if $(x, Ax) = 0$, for all x.

A fundamental result for quantum mechanics is the following:

Theorem 4.11. A linear transformation A, on a unitary space, is Hermitian if and only if (x, Ax) is real for all x.

Proof. If $A = A^\dagger$, then $(x, Ax) = (A^\dagger x, x) = (x, Ax)^*$, so (x, Ax) is real. Conversely, if (x, Ax) is always real, then

$$(x, Ax) = (x, Ax)^* = (Ax, x) = (x, A^\dagger x) .$$

Therefore $(x, [A - A^\dagger]x) = 0$ for all x, and from Theorem 4.10 it follows that $A = A^\dagger$. QED

Note that this theorem is not true in a real inner-product space; i.e., if (x, Ax) is real it does not follow that A is symmetric. Clearly, (x, Ax) is always real in a real inner-product space; but in a complex inner-product space its reality for all x has the important consequence that A is Hermitian.

The quantity (x, Ax) is called the *expectation value* of the operator A in quantum mechanics. For normalized wave functions, it is the average value of many measurements of the observable A on systems in the state described by x. Naturally, this number must be real. Therefore the operators of quantum mechanics corresponding to physical observable must be Hermitian.

A closely related result is that the eigenvalues of a Hermitian matrix are all real numbers. Since in quantum theory the only possible results of a physical measurement are the eigenvalues of the corresponding Hermitian operator, this result, which we prove next, is also fundamental to the theory.

Theorem 4.12. The eigenvalues of a Hermitian operator are real.

Proof.

$$Ax = \lambda x \Longrightarrow (x, Ax) = \lambda ||x||^2 \Longrightarrow \lambda = \frac{(x, Ax)}{||x||^2} ,$$

which, by Theorem 4.11, is real if A is Hermitian. QED

Example. The Hamiltonian for the simple harmonic oscillator is

$$H = (p^2/2m) + \tfrac{1}{2}\, m\omega^2 x^2 .$$

Earlier [Example 2, Section 3.10] we claimed that its eigenvalues were all positive. We can now give a simple demonstration of this fact. Let $Hu = Eu$, where the eigenvectors u are normalized, that is, $||u|| = 1$. Then

$$\begin{aligned} E = (u, Hu) &= \left(u, \frac{p^2}{2m} u\right) + (u, \tfrac{1}{2}\, m\omega^2 x^2 u) \\ &= \frac{1}{2m}\,(p^\dagger u, pu) + \tfrac{1}{2}\, m\omega^2 (x^\dagger u, xu) \\ &= \frac{1}{2m}\,(pu, pu) + \tfrac{1}{2}\, m\omega^2 (xu, xu)\ , \end{aligned}$$

since p and x are both Hermitian. Thus

$$E = (1/2m)||pu||^2 + \tfrac{1}{2}\, m\omega^2 ||xu||^2 \geq 0\ ,$$

since $m, \omega > 0$.

It is clear from this example that the eigenvalues of the square of *any* Hermitian operator will always be positive or zero. This also follows from Theorem 4.12.

Example. In our earlier treatment of the eigenvalue-eigenvector for the harmonic-oscillator problem, we stated that the nth normalized eigenvector was

$$u_n = c_n (a^\dagger)^n u_0\ , \tag{4.15}$$

where $c_n = (1/n!)^{1/2}$ [Eqs. (3.48) and (3.49)]. We can now provide a simple proof of this fact by the same operator algebra techniques.

The creation and annihilation operators

$$a^\dagger = (xm\omega - ip)(2m\omega\hbar)^{-1/2}\ , \qquad a = (xm\omega + ip)(2m\omega\hbar)^{-1/2}$$

are adjoints of each other, as the notation has anticipated; p and x are themselves Hermitian.

Normalization of the eigenvectors u_n requires that $(u_n, u_n) = 1$. Since

$$u_n = \frac{c_n}{c_{n-1}}\, c_{n-1} a^\dagger (a^\dagger)^{n-1} u_0 = \frac{c_n}{c_{n-1}}\, a^\dagger u_{n-1}\ ,$$

we may write this as

$$\begin{aligned} 1 &= \left(\frac{c_n}{c_{n-1}}\, a^\dagger u_{n-1}, \frac{c_n}{c_{n-1}}\, a^\dagger u_{n-1}\right) = \left|\frac{c_n}{c_{n-1}}\right|^2 (a^\dagger u_{n-1}, a^\dagger u_{n-1}) \\ &= \left|\frac{c_n}{c_{n-1}}\right|^2 (u_{n-1}, aa^\dagger u_{n-1}) = \left|\frac{c_n}{c_{n-1}}\right|^2 (\hbar\omega)^{-1} \left(u_{n-1}, \left(H + \frac{\hbar\omega}{2}\right) u_{n-1}\right)\ , \end{aligned}$$

where we have used Eq. (3.44). Now using Eq. (3.48), we obtain

$$1 = \left|\frac{c_n}{c_{n-1}}\right|^2 n\,(u_{n-1}, u_{n-1}) = \left|\frac{c_n}{c_{n-1}}\right|^2 n\,.$$

Therefore

$$|c_n|^2 = \frac{1}{n}\,|c_{n-1}|^2\,.$$

By iterating this, we obtain

$$|c_n|^2 = \frac{1}{n}\frac{1}{n-1}\,|c_{n-2}|^2 = \frac{1}{n}\frac{1}{n-1}\frac{1}{n-2}\,|c_{n-3}|^2 = \cdots = \frac{1}{n!}\,|c_0|^2 = \frac{1}{n!}\,,$$

since $c_0 = 1$. Therefore, the choice

$$c_n = \left(\frac{1}{n!}\right)^{1/2}$$

will ensure that all the eigenfunctions generated by (4.15) are normalized.

4.5 ISOMETRIES—UNITARY AND ORTHOGONAL TRANSFORMATIONS

We now introduce another very important class of linear transformations.

> **Definition 4.7.** If a linear transformation U on an inner-product space is such that $U^\dagger U = I$, U is called an isometry (or *isometric* transformation). If, in addition, $UU^\dagger = I$, then the isometry U is called a *unitary* transformation. A linear transformation A satisfying $\tilde{A}A = I$ is called an *orthogonal* transformation.

In finite-dimensional spaces, the existence of a left inverse, $U^\dagger$ $(U^\dagger U = I)$, implies that $U^\dagger$ is also a right inverse $(UU^\dagger = I)$. Thus our definition need make no mention of left and right in finite-dimensional spaces. It follows that if U is an isometry in a finite-dimensional space, then U is invertible and hence unitary. Moreover, in this case we have:

$$(U^{-1})^\dagger = (U^\dagger)^\dagger = U = (U^{-1})^{-1}\,.$$

We discuss several alternative characterizations of isometric transformations in the following theorem, which is applicable to any vector space, whether finite- or infinite-dimensional.

> **Theorem 4.13.** The following three conditions on a linear transformation U on an inner-product space are equivalent to each other.
>
> 1. $U^\dagger U = I$;
> 2. $(Ux, Uy) = (x, y)$ for all x and y;
> 3. $||Ux|| = ||x||$ for all x.

Proof. If $U^\dagger U = I$, then $(Ux, Uy) = (x, U^\dagger Uy) = (x, y)$ for all x and y, and, in particular, if $y = x$, $||Ux||^2 = ||x||^2$ for all x. This proves that

$$(1) \Longrightarrow (2) \Longrightarrow (3)\,.$$

To show (3) $\Longrightarrow$ (1) and thus establish the equivalence of the three conditions, we note that

$$||Ux||^2 = (Ux, Ux) = (U^\dagger Ux, x) = (x, x) \Longrightarrow ([U^\dagger U - I]x, x) = 0 ,$$

for all x. Now, if the operator $[U^\dagger U - I]$ is self-adjoint, we may use the corollary to Theorem 4.10 to deduce $U^\dagger U = I$, which would complete the proof. Since

$$[U^\dagger U - I]^\dagger = (U^\dagger U)^\dagger - I^\dagger = U^\dagger (U^\dagger)^\dagger - I = [U^\dagger U - I] ,$$

$[U^\dagger U - I]$ is self-adjoint. QED

This theorem explains why an isometry is so called. Property 3 tells us that these transformations preserve lengths. Recalling that the cosine of angle θ between the vectors x and y is given by

$$\cos\theta = \frac{(x, y)}{||x||\,||y||} ,$$

we see that isometric linear transformations also preserve angles between vectors because they preserve scalar products (property 2) and lengths (property 3). This holds in n-dimensional space, too, where the definition of the "angle" between vectors is formally identical to this one.

Theorem 4.14. If $\{x_i\}$ is a complete orthonormal set, then so is $\{Ux_i\}$ for any isometry U.

Proof.

$$(Ux_i, Ux_j) = (x_i, U^\dagger Ux_j) = (x_i, x_j) = \delta_{ij} ,$$

so the set $\{Ux_i\}$ is orthonormal.

To prove the completeness of $\{Ux_i\}$, we show that the set $\{Ux_i\}$ satisfies Parseval's identity (condition 5 of Theorem 4.3). We have

$$\sum_i (y, Ux_i)(Ux_i, x) = \sum_i (U^\dagger y, x_i)(x_i, U^\dagger x) = (U^\dagger y, U^\dagger x) ,$$

since the set $\{x_i\}$ is complete. But

$$(U^\dagger y, U^\dagger x) = (y, UU^\dagger x) = (y, x) .$$

Therefore, Parseval's identity is satisfied, and the set $\{Ux_i\}$ is complete. QED

Isometries, therefore, are linear transformations that convert one orthonormal basis into another. We have previously seen an example of such transformations in Section 1.4, where we considered rotations of basis vectors in three-dimensional Cartesian space. There we defined a vector as a triple of numbers, x_i, which upon rotation of the basis vectors transform according to the rule

$$x_j' = \sum_i a_{ij} x_i ,$$

where the coefficients a_{ij} satisfy the relations

$$\sum_k a_{ik}a_{jk} = \sum_k a_{ki}a_{kj} = \delta_{ij} .$$

[See Eqs. (1.13).] Now let $[U]_{ij} = u_{ij}$, so that $[U^\dagger]_{ij} = u_{ji}^*$. The matrix representation of the operator equation, $UU^\dagger = U^\dagger U = I$, is

$$[UU^\dagger]_{ij} = \sum_k u_{ik}u_{jk}^* = [U^\dagger U]_{ij} = \sum_k u_{ki}^* u_{kj} = \delta_{ij} . \tag{4.16}$$

In a real inner-product space, $u_{ij} = u_{ij}^*$, and Eq. (4.16) reduces to the orthogonality relations above. We anticipated this result when we called rotations of vectors in three dimensions *orthogonal transformations.*

It was pointed out that the determinants of the various rotation matrices of Section 1.4 all had the value 1. We now prove the related general result.

Theorem 4.15. If A is an orthogonal matrix, then $\det A = \pm 1$.

Proof. If A is an orthogonal matrix then by definition $\tilde{A}A = I$. Thus

$$\det(\tilde{A}A) = \det \tilde{A} \det A = (\det A)^2 = 1 \Longrightarrow \det A = \pm 1 ,$$

where we have used Eq. (3.17). QED

The $+1$ corresponds to a rotation, and the -1 corresponds to a rotation plus an *inversion*—that is, a switch from a right-handed to a left-handed coordinate system. If U is a unitary matrix, then a similar proof shows that $|\det U| = 1$, or $\det U = e^{i\phi}$, where ϕ is any real number (Problem 1).

4.6 THE EIGENVALUES AND EIGENVECTORS OF SELF-ADJOINT AND ISOMETRIC TRANSFORMATIONS

We are now in a position to prove some general theorems about the spectra of certain linear operators. These theorems are of fundamental importance in quantum mechanics, where we are concerned with finding the eigenvalues and eigenvectors of Hermitian operators.

Theorem 4.16. If A is a self-adjoint linear transformation, then every eigenvalue of A is real.

Proof. The eigenvalues of A are the λ such that $Ax = \lambda x$, where $x \neq 0$. Now $(x, Ax) = (x, \lambda x) = \lambda(x, x)$, and since A is self-adjoint,

$$(x, Ax) = (Ax, x) = (\lambda x, x) = \lambda^*(x, x).$$

Since x is an eigenvector, $x \neq 0$, and hence $(x, x) \neq 0$. Therefore $\lambda = \lambda^*$, so λ is real. QED

This result is only a trivial extension of Theorem 4.12. We include it both to demonstrate this alternative proof—which is the one usually given in the context of quantum mechanics—and to collect together in one section all the main results on the spectra of the operators we have studied. We turn now to the spectra of isometries.

Theorem 4.17. Every eigenvalue of an isometry has absolute value 1.

Proof. If U is an isometry and $Ux = \lambda x$ $(x \neq 0)$ then

$$||x|| = ||Ux|| = |\lambda| \, ||x|| \Longrightarrow |\lambda| = 1$$

since $||x|| \neq 0$. QED

In the proof of this theorem, we know that $||x|| = (x, x)^{1/2} \neq 0$ because $x \neq 0$, and our inner product is such that $(x, x) = 0$ if and only if $x = 0$.

We now prove the central theorem regarding the spectral properties of the transformations we have been considering.

Theorem 4.18. If A is either self-adjoint or isometric, then eigenvectors of A belonging to *distinct* eigenvalues are orthogonal.

Proof. Suppose that $Ax_1 = \lambda_1 x_1$ and $Ax_2 = \lambda_2 x_2$, $\lambda_1 \neq \lambda_2$.

a) If A is self-adjoint, then

$$(x_1, Ax_2) = \lambda_2 (x_1, x_2)$$

and also

$$(x_1, Ax_2) = (Ax_1, x_2) = \lambda_1^* (x_1, x_2) = \lambda_1 (x_1, x_2) \, ,$$

since $A = A^\dagger$ and therefore the eigenvalues are real (Theorem 4.16). Therefore, $(\lambda_1 - \lambda_2)(x_1, x_2) = 0$, and hence $(x_1, x_2) = 0$, since $\lambda_1 \neq \lambda_2$.

b) If A is an isometry, then

$$(x_1, x_2) = (Ax_1, Ax_2) = \lambda_1^* \lambda_2 (x_1, x_2) = (\lambda_2/\lambda_1)(x_1, x_2) \, ,$$

since by Theorem 4.17 $|\lambda|^2 = \lambda\lambda^* = 1$ for an isometry. Therefore, since $\lambda_1 \neq 0$ we have

$$(\lambda_1 - \lambda_2)(x_1, x_2) = 0 \, .$$

But $\lambda_1 \neq \lambda_2$, so we get $(x_1, x_2) = 0$. QED

When we solve the characteristic polynomial for an $n \times n$ self-adjoint or isometric transformation we get n eigenvalues. If any two are different, the corresponding eigenvectors are orthogonal, and if they are all distinct, then the eigenvectors are all orthogonal and span the space on which the matrix is defined. If k, $(k \leq n)$ of them are the same $(= \lambda_0)$, it can and will be shown (Theorem 4.20) that there always exist k linearly independent (but not necessarily orthogonal) eigenvectors belonging to λ_0. Of course all these k eigenvectors are orthogonal to all the other $(n - k)$ which belong to eigenvalues unequal to λ_0. And these k linearly independent eigenvectors belonging to λ_0 may be replaced by an orthogonal set of k eigenvectors by applying the Gram-Schmidt orthogonalization process to them. This procedure is often followed in quantum mechanics for degenerate eigenvalues. We can summarize by saying that the *algebraic* multiplicity (the number of times a given root is repeated in the solution to the characteristic equation) equals the *geometric* multiplicity (the number of linearly independent eigenvectors corresponding to that root) for self-adjoint and isometric linear transformations. Therefore *the eigenvectors of self-adjoint and isometric linear transformations span the space and can be chosen to be a complete orthonormal set.* We shall prove this fundamental result in the next section.

Now we shall look at an example of the use of the preceding theorems.

Example. *The rotation matrix and the Lorentz matrix.* We want to examine the rotation matrix and the Lorentz matrix from the point of view of the previous theorems on eigenvalues and eigenvectors. The two-dimensional rotation matrix, which we denote here by R, is given by

$$R = \begin{bmatrix} \cos\phi & \sin\phi \\ -\sin\phi & \cos\phi \end{bmatrix}.$$

As we saw in Section 1.4, it describes the effect of rotating the basis of a vector space. The Lorentz matrix describes the effect of a Lorentz transformation. The reader is no doubt familiar with the equations of the Lorentz transformation which relate the space-time coordinates of two observers moving relative to each other with uniform velocity v along their common x_3x_3'-axis. For this configuration we have

$$\begin{aligned} x_1' &= x_1, & x_2' &= x_2 \\ x_3' &= \gamma(x_3 - vt), & t' &= \gamma(t - vx_3/c^2)\,, \end{aligned} \tag{4.17}$$

where $\gamma \equiv (1 - \beta^2)^{-1/2}$ and $\beta \equiv v/c$.

The transformation described by Eqs. (4.17) can also be expressed in matrix form. One device for doing so is to write $x_4 = ict$. This is motivated by the fact that then

$$\sum_{i=1}^{4} x_i^2 = x_1^2 + x_2^2 + x_3^2 - c^2t^2\,, \tag{4.18}$$

and this quantity is an invariant in special relativity expressing the invariance of the speed of light in different frames of reference. The coordinates $(x_1, x_2, x_3, x_4 = ict)$ are referred to as the coordinates of Minkowski space. If we now define a Lorentz transformation by

$$u' = Lu\,,$$

where $u = (u_1, u_2, u_3, u_4)$ is any four vector in Minkowski space and L is the Lorentz matrix,

$$L = \begin{bmatrix} 1 & 0 & 0 & 0 \\ 0 & 1 & 0 & 0 \\ 0 & 0 & \gamma & i\beta\gamma \\ 0 & 0 & -i\beta\gamma & \gamma \end{bmatrix}, \tag{4.19}$$

then the reader can easily check that Lx gives precisely the transformed vector x' whose components are given by Eqs. (4.17).

A simple and useful parametric form for the Lorentz matrix results if we let

$$\gamma \equiv \frac{1}{\sqrt{1-\beta^2}} \equiv \cosh\alpha\,. \tag{4.20}$$

Both γ and $\cosh\alpha$ range from a minimum of $1(\beta = 0)$ to a maximum of ∞ $(v = c,\ \beta = 1)$. It follows that

$$\sinh\alpha = (\cosh^2\alpha - 1)^{1/2} = \frac{\beta}{\sqrt{1-\beta^2}}\,.$$

and

$$\alpha = \tanh^{-1}\beta\,.$$

The matrix L becomes

$$L = L(\alpha) = \begin{bmatrix} 1 & 0 & 0 & 0 \\ 0 & 1 & 0 & 0 \\ 0 & 0 & \cosh\alpha & i\sinh\alpha \\ 0 & 0 & -i\sinh\alpha & \cosh\alpha \end{bmatrix}. \tag{4.21}$$

Now let us examine the general properties of R and L. We first note that $\tilde{R}R = I$ and $\tilde{L}L = I$, that is, the inverse of each matrix is equal to its transpose. For the rotation matrix, this expresses the fact that the inverse of a rotation R through an angle ϕ is obtained by replacing ϕ by $-\phi$ in R, which is equivalent to transposing R. In the case of a Lorentz transformation, the inverse is seen on physical grounds to correspond to replacing v by $-v$ (that is, α by $-\alpha$) in the matrix, which is equivalent to transposing L. We may summarize by saying that both R and L are *orthogonal* matrices. However, virtually all the theorems proved above refer to unitary matrices, so if we want to draw any general conclusions about R and L, we had best see if R and L have the more powerful property of unitarity. Since the elements of R are real, $\tilde{R} = R^\dagger$, so we have immediately $R^\dagger R = I$, that is, R is indeed a unitary matrix. However, $\tilde{L} \neq L^\dagger$, because the elements of L are complex. A straightfoward multiplication shows that $L^\dagger L \neq I$, so L is *not* a unitary matrix. Thus our theorems on unitary transformations tell us nothing about L. They do, however, tell us that as far as rotations are concerned, R must have eigenvalues whose modulus is unity and eigenvectors which are orthogonal to each other (if the eigenvalues are distinct). A simple calculation shows that the two eigenvalues of R are given by $\lambda_+ = e^{i\phi}$ and $\lambda_- = e^{-i\phi}$. The corresponding eigenvectors are $x_+ = (1, i)$ and $x_- = (1, -i)$. Thus

$$|\lambda_\pm| = |e^{\pm i\phi}| = 1\,,$$

which fulfills the prediction of Theorem 4.17, and

$$(x_+, x_-) = 0\,,$$

which is in accordance with Theorem 4.18.

If we want to obtain similar information about L, we will have to rely on theorems which do not depend on unitarity. In this case we do not have far to look, since it is clear that $L^\dagger = L$, that is, L is a self-adjoint matrix. Thus we conclude immediately that the eigenvalues of L are real (Theorem 4.12) and that the eigenvectors corresponding to distinct eigenvalues are orthogonal (Theorem 4.18). A simple calculation (Problem 16) shows that the eigenvalues of L are $\lambda_1 = 1$, $\lambda_2 = 1$, $\lambda_+ = \cosh\alpha + \sinh\alpha$ and $\lambda_- = \cosh\alpha - \sinh\alpha$. They are real as expected from Theorem 4.12. The corresponding eigenvectors are $x_1 = (1, 0, 0, 0)$, $x_2 = (0, 1, 0, 0)$, $x_+ = (0, 0, 1, -i)$ and $x_- = (0, 0, 1, i)$. They are orthogonal in accordance with Theorem 4.18. Thus, only by considering L as a Hermitian matrix can we obtain any information about the spectral properties of L. The *orthogonality* of L has so far told us nothing. The only useful fact that we can derive from orthogonality is $\det L = \pm 1$ (Theorem 4.15). This result is also true for R (that is, $\det R = \pm 1$), since R is orthogonal as well as unitary. Note that for both the rotation matrix and the Lorentz matrix the determinant is equal to the product of the eigenvalues as we should expect from the invariance of the determinant under similarity transformations (Theorem 3.13). The reader can easily check that in this case the trace is also invariant under similarity transformations (Theorem 3.14).

Since one is perhaps inclined to think that there should be little difference between orthogonal matrices and unitary matrices, it may be disturbing to some readers that we have been unable to deduce any important consequences from the *orthogonality* of L. One might imagine that the difficulties with complex orthogonal matrices such as L come from the form of the inner product (with its unsymmetric complex conjugation) which we use, and that perhaps if we used a simple Euclidean type inner product such as

$$[x, y] = \sum_{i=1}^{n} \xi_i \eta_i \tag{4.22}$$

—which after all motivated the introduction to *real* orthogonal matrices in the first place (Chapter 1)—then we could save some theorems. Because of the missing complex conjugation, we are giving up several properties of the usual inner product, in particular, positive-definiteness. However, the *pseudo inner product* defined by Eq. (4.22) has certain attractive properties. It is linear:

$$[x + y, z] = [x, z] + [y, z]\,; \tag{4.23}$$

it is symmetric:

$$[x, y] = [y, x]\,; \tag{4.24}$$

and it is homogeneous:

$$[\alpha x, y] = \alpha[x, y]\,. \tag{4.25}$$

Note that in a *real* vector space, these properties are the same as those given by axioms (1) and (2) of Definition 4.1. Using these general properties, the reader can investigate the subject of *pseudo inner-product spaces* along the lines which we have followed for inner-product spaces. Many of the results in this section and in the previous ones have close analogues in a pseudo inner-product space.

The main geometric problem is to elucidate the role of the *null-vector*, that is, a vector $x \neq 0$ for which $[x, x] = 0$. As far as operator theory goes, the key definition is that of the *transpose operator*, $\tilde{A}$, which satisfies $[x, Ay] = [\tilde{A}x, y]$ for all x and y. When represented by a matrix, $\tilde{A}$ has just the form one should expect, and the study of *symmetric matrices* ($\tilde{S} = S$) and orthogonal matrices ($\tilde{O}O = I$) proceeds in much the same manner as in this section, except that certain difficulties associated with the existence of null-vectors must be dealt with. Among the results which one obtains is the fact that eigenvectors of a symmetric matrix belonging to distinct eigenvalues are pseudo orthogonal, i.e., if x_1 and x_2 are the eigenvectors in question, $[x_1, x_2] = 0$. Note that the eigenvalues of a symmetric operator need not be real in a general (complex) vector space. Also, if O is an orthogonal matrix (such as the Lorentz matrix), then every eigenvalue of O belonging to a *nonnull* eigenvector is equal to 1 or -1. The demonstration of these results and others we leave as a problem (Problem 4.27). It will be found that the theorems which one can obtain for orthogonal operators in a pseudo inner-product space are far weaker than the theorems for unitary operators in a true inner-product space.

We close this section on spectral theory with an interesting generalization that contains many of the previous theorems as special cases. It shows that these results really depended only on the fact that the linear transformations in question commuted with their adjoints.

Definition 4.8. A linear transformation is *normal* if and only if $AA^\dagger = A^\dagger A$.

All self-adjoint (Hermitian and symmetric) and isometric (unitary and real orthogonal) linear transformations are normal. In addition, anti-self-adjoint operators [anti-symmetric ($A = -\tilde{A}$) and anti-Hermitian ($A = -A^\dagger$) operators] are normal.

Theorem 4.19. If A is normal, then eigenvectors belonging to distinct eigenvalues are orthogonal.

Proof. We first prove the preliminary result, that if A is normal, then

$$Ax = \lambda x \Longleftrightarrow A^\dagger x = \lambda^* x \, .$$

Normality of A implies that

$$||Ax||^2 \equiv (Ax, Ax) = (A^\dagger Ax, x) = (AA^\dagger x, x) = (A^\dagger x, A^\dagger x) = ||A^\dagger x||^2 \, .$$

Now, if A is normal, $(A - \lambda)$ is normal, and since $(A - \lambda)^\dagger = A^\dagger - \lambda^*$, using $(A - \lambda)$ in place of A in the above, we obtain

$$||Ax - \lambda x||^2 = ||A^\dagger x - \lambda^* x||^2 \, .$$

It follows immediately that

$$Ax = \lambda x \Longleftrightarrow A^\dagger x = \lambda^* x \, .$$

Now we may prove the theorem itself. If $Ax_1 = \lambda_1 x_1$ and $Ax_2 = \lambda_2 x_2$, then

$$(x_1, Ax_2) = \lambda_2(x_1, x_2) ,$$

and

$$(x_1, Ax_2) = (A^\dagger x_1, x_2) = (\lambda_1^* x_1, x_2) = \lambda_1(x_1, x_2) .$$

Therefore

$$(\lambda_1 - \lambda_2)(x_1, x_2) = 0 \Longrightarrow (x_1, x_2) = 0 ,$$

if $\lambda_1 \neq \lambda_2$. QED

Any matrix A may be decomposed into the sum of a self-adjoint and an anti-self-adjoint matrix, for we have the identity

$$A \equiv \tfrac{1}{2}(A + A^\dagger) + \tfrac{1}{2}(A - A^\dagger) ;$$

here $\frac{1}{2}(A + A^\dagger)$ is self-adjoint and $\frac{1}{2}(A - A^\dagger)$ is anti-self-adjoint. We may also decompose A as follows:

$$A \equiv [\tfrac{1}{2}(A + A^\dagger)] + i[1/2i(A - A^\dagger)] . \tag{4.26}$$

Then both $B \equiv \frac{1}{2}(A + A^\dagger)$ and $C \equiv 1/2i(A - A^\dagger)$ are *Hermitian*. We note that B and C commute if and only if A and $A^\dagger$ commute, that is, A is normal. This result for operators is analogous to the decomposition of a complex number into the sum of a real and an imaginary part.

4.7 DIAGONALIZATION

Diagonalizing a matrix is equivalent to solving for its eigenvalues—they will simply be the diagonal elements. Thus the solution to a physical problem often involves diagonalizing a matrix. In classical mechanics, we may want to diagonalize the moment-of-inertia tensor, Eq. (1.98), and in quantum theory if we can diagonalize the matrix representation of an operator corresponding to an observable, then the diagonal elements will be the possible results of a measurement of the observable.

We have seen that the n-eigenvectors of an $n \times n$ normal matrix A with a nondegenerate spectrum (i.e., n with distinct eigenvalues) are orthogonal. Thus they are linearly independent and span the space, so by Theorem 3.16 we may construct a diagonalizing matrix P by taking as the ith column of P the ith eigenvector. Then

$$(P^{-1}AP)_{ij} = \lambda_i \delta_{ij} .$$

It is easy to see that the diagonalizing matrix P is in fact a unitary matrix. To show this, let

$$P = \begin{bmatrix} p_{11} & p_{12} & \cdots & p_{1n} \\ p_{21} & p_{22} & & p_{2n} \\ \vdots & & & \vdots \\ p_{n1} & p_{n2} & \cdots & p_{nn} \end{bmatrix} .$$

In the notation of Theorem 3.16, $X_i = (p_{1i}, \cdots p_{ni})$ is the ith eigenvector. We are free to assume it has been normalized. Since we know the eigenvectors are orthogonal, we have

$$(X_i, X_j) = \sum_{k=1}^{n} p_{ki}^* p_{kj} = \delta_{ij} ,$$

which, written in matrix notation, is just $P^\dagger P = I$, so P is unitary.

The fact that a normal matrix with distinct eigenvalues can be diagonalized by a unitary similarity transformation, is a very important result. But the restriction to nondegenerate spectra is a serious one, because most physical situations involve degenerate spectra. It would therefore be useful to remove that restriction; and there are good reasons to suspect that it can in fact be removed altogether, leaving the rest of the statement intact.

Consider the situation for Hermitian matrices. If the eigenvalues of a Hermitian matrix A are not distinct, we can perturb the elements of A by adding a Hermitian matrix δA so that the resulting Hermitian matrix $A + \delta A$ has distinct eigenvalues. This is a familiar situation in quantum theory, where the inclusion of a previously neglected perturbation, such as a magnetic field, removes the degeneracy of the spectrum. Once the degeneracy is removed, all the eigenvalues are distinct, the previous reasoning applies, and we know there exists a unitary similarity transformation that diagonalizes the matrix $A + \delta A$. As long as δA is nonzero and the eigenvalues are distinct, we know that the corresponding eigenvectors remain orthogonal, since $A + \delta A$ is *Hermitian.* (For an arbitrary matrix, this argument would break down.) Since the eigenvectors must be orthogonal for arbitrarily small but nonzero δA, it is very plausible that when δA is finally reduced to zero, they do not suddenly become linearly dependent. It is therefore reasonable to expect that as $\delta A \to 0$ we continue to have a full slate of k linearly independent eigenvectors for a k-fold root of a characteristic polynomial, and hence that there exists a unitary similarity transformation that diagonalizes A.*

We now prove that a unitary diagonalizing matrix can be constructed for any Hermitian matrix, repeated eigenvalues or not.

Theorem 4.20. Any Hermitian matrix A may be diagonalized by a unitary similarity transformation; i.e., there exists a unitary matrix U such that $U^{-1}AU$ is a diagonal matrix (the diagonal elements of which are the eigenvalues of A).

* From a physical point of view, we know that if the perturbation δA is turned off slowly and gently, that is, *adiabatically*, the quantum numbers will not change; they are so-called *adiabatic invariants*. (In fact, it was precisely this observation that led Bohr and Sommerfeld to quantize the classical action integrals, which were known to be adiabatic invariants.) The fact that the quantum numbers do not change means that the eigenvectors remain unchanged as the perturbation diminishes adiabatically in strength. Thus there is no danger of the eigenvectors, which are orthogonal as long as $\delta A \neq 0$, becoming linearly dependent as $\delta A \to 0$.

Proof. We shall prove the theorem by induction from the 2×2 case. That is, we prove it explicitly for the 2×2 case, and then assuming it for the $(n-1) \times (n-1)$ case we show that it must hold for the $n \times n$ case. This procedure proves the theorem for any n. We could, of course, prove the theorem by induction from the 1×1 Hermitian matrix (i.e., a matrix consisting of a single real element). We use the 2×2 matrix because this is the smallest matrix for which an eigenvalue can have multiplicity greater than 1, and thus this case is more instructive.

We know the theorem is true if the eigenvalues are distinct and, in particular, it is true in this case for 2×2 matrices. In the degenerate 2×2 case, where the eigenvalues are not distinct, they must be equal. Clearly, this can be the case only if A is (already) a diagonal matrix of the form

$$A = \begin{bmatrix} a & 0 \\ 0 & a \end{bmatrix}.$$

(If the reader wants to verify this explicitly, he may do so by solving the characteristic polynomial of the most general 2×2 Hermitian matrix,

$$A = \begin{bmatrix} a & b \\ b^* & c \end{bmatrix},$$

when a and c are real.) In the degenerate 2×2 case, any vector is an eigenvector because

$$\begin{bmatrix} a & 0 \\ 0 & a \end{bmatrix} \begin{bmatrix} x_1 \\ x_2 \end{bmatrix} = a \begin{bmatrix} x_1 \\ x_2 \end{bmatrix}.$$

Thus we are free to choose two orthonormal eigenvectors, say

$$u_1 = \begin{bmatrix} 1 \\ 0 \end{bmatrix} \quad \text{and} \quad u_2 = \begin{bmatrix} 0 \\ 1 \end{bmatrix},$$

and therefore

$$U = \begin{bmatrix} 1 & 0 \\ 0 & 1 \end{bmatrix}$$

is a unitary operator which "brings" A to diagonal form by a similarity transformation, (Of course, A is in diagonal form to start with.) This exhausts all the possibilities for the 2×2 case. Now we prove the $n \times n$ case by induction.

We assume that any $(n-1) \times (n-1)$ Hermitian matrix can be diagonalized by a unitary similarity transformation. Let λ_1 be any eigenvalue of A, and denote a corresponding normalized eigenvector by

$$v_1 = (v_{11}, v_{21}, \cdots, v_{n1}).$$

We now form a unitary operator V by choosing the eigenvector v_1 as the first column of V, and putting in any other set of $n-1$ orthonormal vectors which are all orthogonal to v_1 as the remaining $n-1$ columns. Thus

$$V = (v_{i1} \mid v_{i2} \mid \cdots \mid v_{in}),$$

where the v_{ij} form the jth column vector, where $Av_1 = \lambda_1 v_1$ (that is, $\sum_l a_{kl} v_{l1} = \lambda_1 v_{k1}$), and where $(v_i, v_j) = \delta_{ij}$.

Now let us calculate the matrix $V^{-1} AV$. The general element is given by

$$(V^{-1}AV)_{ij} = (V^\dagger AV)_{ij} = \sum_k v_{ki}^* \sum_l a_{kl} v_{lj} , \tag{4.27}$$

and, in particular,

$$(V^{-1}AV)_{i1} = \sum_k v_{ki}^* \sum_l a_{kl} v_{l1} = \sum_k v_{ki}^* \lambda_1 v_{k1} = \lambda_1 \delta_{i1} .$$

Since

$$(V^{-1}AV)^\dagger = V^\dagger A^\dagger (V^{-1})^\dagger = V^{-1}AV ,$$

$V^{-1}AV$ is Hermitian, so

$$(V^{-1}AV)_{1i} = (\lambda_1 \delta_{1i})^* = \lambda_1 \delta_{1i} ,$$

since the eigenvalues are real. Consequently, the matrix $V^{-1}AV$ has the form

$$V^{-1}AV = \begin{bmatrix} \lambda_1 & 0 & 0 \cdots 0 \\ 0 & & \\ 0 & & \\ \vdots & & A' \\ 0 & & \end{bmatrix} \equiv B ,$$

where A' is the $(n-1) \times (n-1)$ Hermitian matrix whose elements are given by Eq. (4.27). Now by our induction hypothesis, A' *can* be diagonalized by an $(n-1) \times (n-1)$ unitary matrix, V', that is, $(V')^{-1}A'V'$ is diagonal. Therefore, the $n \times n$ matrix $B = V^{-1}AV$ can be diagonalized.

The only remaining question is whether there is a single unitary diagonalizing matrix for A. To complete the construction explicitly, we define the unitary matrix

$$W = \begin{bmatrix} 1 & 0 \cdots 0 \\ 0 & \\ \vdots & V' \\ 0 & \end{bmatrix} .$$

Then, because $(V')^{-1}A'V'$ is diagonal (by hypothesis),

$$W^{-1}BW = W^\dagger BW$$

will be diagonal. But

$$W^{-1}BW = W^{-1}V^{-1}AVW = (VW)^{-1}A(VW) = U^{-1}AU ,$$

where $U = VW$. The unitarity of U follows from the unitarity of V and W, as

$$U^\dagger = (VW)^\dagger = W^\dagger V^\dagger = W^{-1}V^{-1} = (VW)^{-1} = U^{-1} .$$

Thus there exists a unitary matrix U, which we have constructed explicitly, which diagonalizes any Hermitian matrix by a similarity transformation. The proof is complete.

This means that one can always find a complete set of orthonormal basis vectors with respect to which the matrix of a Hermitian linear transformation A is diagonal. This proof also shows that for every degenerate eigenvalue λ_0, of algebraic multiplicity $k(k \leq n)$, it is always possible to find k linearly independent eigenvectors belonging to λ_0. This follows either by directly investigating the process of construction used in the theorem, or by simply noting that the determinant of a unitary matrix always has absolute value one and so is never zero. Therefore its column vectors, which are the eigenvectors of A, must be linearly independent by property (8) of determinants (Section 3.8). Thus the algebraic and geometric multiplicities are equal for Hermitian operators; the eigenvectors of a Hermitian operator form a complete set.

This theorem has been stated for Hermitian matrices and *unitary* similarity transformations. However, as inspection will verify, the proof may be modified (by merely omitting the complex conjugates and replacing daggers by tildes) to give the following theorem.

Theorem 4.21. Any (real) symmetric matrix may be diagonalized by a real orthogonal similarity transformation.

It has been pointed out that any matrix A can be written as the sum of two Hermitian matrices: $A = B + iC$ (Eq. 4.26). We have just proved that any Hermitian matrix can be diagonalized by a unitary similarity transformation. Applying this result to B and C, we may be inclined to ask: Why can't we prove that any matrix (A), Hermitian or not, can be so diagonalized? The answer is that diagonalizing B involves a specific set of basis vectors—eigenvectors of B—and there is no reason to believe that that the basis in which B assumes diagonal form will simultaneously cast C into diagonal form, or, in other words, that the eigenvectors of B will also be eigenvectors of C. However, if B and C do possess a common set of eigenvectors, then it will be possible to diagonalize B and C simultaneously by a single similarity transformation, and therefore A will be diagonalizable. This prompts us to look for a condition on two Hermitian matrices which ensures that they can be simultaneously diagonalized. The following theorem, which is also of great importance in quantum theory, provides such a condition.

Theorem 4.22. Two Hermitian matrices, B and C, commute if and only if there exists a complete orthonormal set of common eigenvectors.

Proof.

1. Assume that there exists a common complete orthonormal set of eigenvectors, $\{x_i\}$. Then $Bx_i = b_i x_i$ and $Cx_i = c_i x_i$. Therefore, $BCx_i = B(c_i x_i) = c_i Bx_i = c_i b_i x_i$, and similarly, $CBx_i = c_i b_i x_i$. Therefore $(BC - CB)x_i = 0$, for any x_i. But the x_i span the vector space, so $(BC - CB)x = 0$ for any x, $BC = CB$.

2. Now for the more difficult "half" of the theorem: Assume that $BC = CB$ and $Bx_i = b_i x_i$, where the $\{x_i\}$ are orthonormal. Then

$$CBx_i = b_i Cx_i \Longrightarrow B(Cx_i) = b_i(Cx_i)$$

which implies that Cx_i is an eigenvector of B corresponding to the eigenvalue b_i. There are now two possibilities:

i) The spectrum is nondegenerate, so the multiplicity of all the b_i is one. Then for all i, Cx_i must equal some constant times x_i: $Cx_i = c_i x_i$, that is, x_i is an eigenvector of C with eigenvalue c_i so the proof is complete in the nondegenerate case.

ii) The spectrum is degenerate, i.e., there exist eigenvalues b_i of multiplicity $m > 1$. Then there must exist m eigenvectors, $(Cx_i^{(1)}, Cx_i^{(2)}, \cdots, Cx_i^{(k)}, \cdots, Cx_i^{(m)})$, belonging to the eigenvalue b_i, each of which must be a linear combination of the m linearly independent eigenvectors $(x_i^{(1)}, \cdots, x_i^{(k)}, \cdots, x_i^{(m)})$ which also belong to the eigenvalue b_i. We shall choose the $x_i^{(k)}$ to be an orthonormal set, and write these linear combinations as

$$\begin{aligned} Cx_i^{(1)} &= \sum_{j=1}^{m} \alpha_{j1} x_i^{(j)} , \\ &\vdots \\ Cx_i^{(k)} &= \sum_{j=1}^{m} \alpha_{jk} x_i^{(j)} , \\ &\vdots \\ Cx_i^{(m)} &= \sum_{j=1}^{m} \alpha_{jm} x_i^{(j)} . \end{aligned} \tag{4.28}$$

The coefficient matrix (α_{ij}) is Hermitian, as the reader may show.

For the nondegenerate case, $m = 1$ for all i. Thus the treatment of the degenerate case includes the nondegenerate case. Now we want to find linear combinations of the eigenvectors, $x_i^{(k)}$, of B belonging to b_i that are simultaneously eigenvectors of C. If the linear combination

$$\sum_{k=1}^{m} \gamma_k x_i^{(k)} ,$$

were to be an eigenvector of C (as it is of B), it would be necessary that the expression

$$C\left(\sum_{k=1}^{m} \gamma_k x_i^{(k)}\right) \equiv \sum_{j,k=1}^{m} \gamma_k \alpha_{jk} x_i^{(j)}$$

be of the form

$$c\left(\sum_{j=1}^{m} \gamma_j x_i^{(j)}\right) .$$

This will be possible if there exists γ_k's such that

$$\sum_{k=1}^{m} \alpha_{jk}\gamma_k = c\gamma_j . \tag{4.29}$$

But this is just the eigenvalue-eigenvector problem for the $m \times m$ Hermitian matrix (α_{jk}). Hence we know there exist m linearly independent solutions (the eigenvectors) which can be chosen to be orthonormal. Thus we find m eigenvalues, c_μ, some of which may be equal, and m orthonormal eigenvectors, $\gamma^{(\mu)}$, and we have the m linear combinations

$$y_i^{(\mu)} = \sum_{k=1}^{m} \gamma_k^{(\mu)} x_i^{(k)} ,$$

which are simultaneous orthonormal eigenvectors of B and C, as the reader may verify. QED

In quantum mechanics, this theorem (really the infinite-dimensional generalization of this theorem) shows that the operators H (Hamiltonian), L^2 (total angular momentum squared) and L_z (z-component of angular momentum), which all commute, possess a mutual set of eigenvectors and therefore the corresponding observables can be measured simultaneously. It also follows that p_x and x cannot be measured simultaneously.

Although we have proved this theorem for two *Hermitian* matrices, because of their greater relevance to quantum theory, it obviously holds for (real) symmetric matrices as well. We simply omit the complex conjugates and substitute "symmetric" for "Hermitian."

We now generalize the earlier result on the diagonalizability of Hermitian matrices (Theorem 4.20), and at the same time characterize precisely the class of matrices that can be diagonalized by a unitary similarity transformation.

Theorem 4.23. A matrix A can be diagonalized by a unitary similarity transformation if and only if it is normal.

Proof. Any matrix A can be decomposed into the sum of two Hermitian matrices; according to Eq. (4.26)

$$A = B + iC ,$$

where B and C are Hermitian. Now A can be diagonalized by a unitary similarity transformation if and only if the Hermitian matrices B and C can be simultaneously so diagonalized; and a necessary and sufficient condition for this is that B and C commute (Theorem 4.22). But, as was mentioned in connection with Eq. (4.26), B and C commute if and only if A and $A^\dagger$ commute, i.e., if and only if A is normal. Therefore a necessary and sufficient condition for a matrix to be diagonalizable by a unitary similarity transformation is that it be a normal matrix. QED

This theorem tells us that any unitary matrix can be diagonalized by a unitary similarity transformation. A real orthogonal matrix is, of course, a unitary matrix and therefore it can be diagonalized by a unitary similarity transformation. Note, however, that the diagonalizing matrix will, in general, be a *unitary*, not a real orthogonal matrix. For example, the rotation matrix is a real orthogonal matrix, but it can only be diagonalized by a unitary matrix.

This is to be contrasted with the situation for symmetric matrices. Since a symmetric matrix is Hermitian, it can be diagonalized by a unitary matrix. But since the eigenvalues of a symmetric matrix are real, the diagonalizing unitary matrix is in fact real orthogonal. However, the eigenvalues of a real orthogonal matrix will, in general, be complex and therefore the unitary matrix that diagonalizes it cannot usually be chosen to be real orthogonal.

4.8 ON THE SOLVABILITY OF LINEAR EQUATIONS*

In the past few sections we discussed self-adjoint operators in considerable detail. In this section, we want to obtain some results which relate the properties of a linear transformation A to the properties of the adjoint operator $A^\dagger$. These results will be used in the theory of integral equations (Chapter 8). We begin by noting the simple result:

Theorem 4.24. If A is a linear operator on a finite-dimensional vector space with eigenvalues λ_i, then $A^\dagger$ has eigenvalues λ_i^*.

Proof. If λ belongs to the spectrum of A, then $\det(A - \lambda I) = 0$. From the general relation

$$\det[(A - \lambda I)^\dagger] = [\det(A - \lambda I)]^* ,$$

it follows immediately that

$$\det(A^\dagger - \lambda^* I) = 0 ,$$

which is the statement that λ^* is in the spectrum of $A^\dagger$. QED

A more interesting result concerns the orthogonality properties of the eigenvectors of A and $A^\dagger$. The following theorem gives us an important generalization of Theorem 4.18.

Theorem 4.25. If $\{\lambda_i\}$ is the spectrum of A (and therefore $\{\lambda_i^*\}$ is the spectrum of $A^\dagger$), then if ϕ_j is any eigenfunction of A belonging to λ_j and if ψ_i is any eigenfunction of $A^\dagger$ belonging to λ_i^*, we have $(\psi_i, \phi_j) = 0$ whenever $\lambda_i \neq \lambda_j$.

Proof. Consider $A\phi_j = \lambda_j\phi_j$. Taking the inner product from the left by ψ_i, we get

$$(\psi_i, A\phi_j) = \lambda_j(\psi_i, \phi_j) ,$$

and hence

$$(A^\dagger\psi_i, \phi_j) - \lambda_j(\psi_i, \phi_j) = 0 .$$

* The material in this section will not be required until we reach Chapter 8.

But $A^\dagger \psi_i = \lambda_i^* \psi_i$, so

$$(\lambda_i - \lambda_j)(\psi_i, \phi_j) = 0 .$$

If $\lambda_i \neq \lambda_j$, we conclude that $(\psi_i, \phi_j) = 0$, as stated in the theorem. QED

Now, for a self-adjoint matrix A, we were able to show that every eigenvalue of the spectrum of A which had algebraic multiplicity $m \geq 1$ had precisely m associated linearly independent eigenvectors. For a general matrix, this result is not true, but there is a related theorem.

Theorem 4.26. Let A be a linear operator on a finite-dimensional vector space,and let λ be an eigenvalue of A with λ^* the corresponding eigenvalue of $A^\dagger$. Then if there are μ linearly independent eigenvectors, ϕ_i belonging to λ, there are also μ linearly independent eigenvectors ψ_i belonging to λ^*. Note that μ may be less than m, the algebraic multiplicity of λ (and of λ^*).

Proof. First of all, the set $\{\phi_i\}$ may be chosen to be orthonormal; if it had not been so, we could have applied the Gram-Schmidt process to it, using the fact that if ϕ_1 and ϕ_2 are both eigenvectors belonging to λ, then $\alpha\phi_1 + \beta\phi_2$ also belongs to λ. The same comment applies to the set $\{\psi_i\}$. We therefore assume that we are dealing with *orthonormal* sets. Call μ the number of elements of $\{\phi_i\}$ and ν the number of elements of $\{\psi_i\}$. We want to show that $\mu = \nu$.

We assume first that $\mu < \nu$, and define an operator B by

$$Bx \equiv Ax + \sum_{j=1}^{\mu} \psi_j(\phi_j, x) .$$

B is clearly a linear operator. We now show that λ is not an eigenvalue of B. Consider

$$Bx = \lambda x ,$$

that is,

$$Ax + \sum_{j=1}^{\mu} \psi_j(\phi_j, x) = \lambda x . \tag{4.30}$$

Taking the inner product from the left by ψ_k, we get

$$(\psi_k, Ax) + \sum_{j=1}^{\mu} (\psi_k, \psi_j)(\phi_j, x) = \lambda(\psi_k, x) .$$

But $(\psi_k, \psi_j) = \delta_{kj}$, so

$$(\psi_k, Ax) + (\phi_k, x) = \lambda(\psi_k, x) , \qquad k = 1, 2, \cdots, \mu$$

or

$$(\phi_k, x) = \lambda(\psi_k, x) - (A^\dagger\psi_k, x) .$$

However, $A^\dagger\psi_k = \lambda^*\psi_k$, so

$$(\phi_k, x) = \lambda(\psi_k, x) - (\lambda^*\psi_k, x) = 0\,, \qquad k = 1, 2, \cdots, \mu\,. \tag{4.31}$$

Returning to Eq. (4.30), we see that

$$Ax = \lambda x\,.$$

But since x is an eigenvector of A with eigenvalue λ, it can be written as

$$x = \sum_{i=1}^{\mu} c_i\phi_i\,.$$

Using Eq. (4.31) together with the orthonormality of ϕ_i, we get

$$c_k = (\phi_k, x) = 0\,, \qquad k = 1, 2, \cdots, \mu\,.$$

Thus $x = 0$. Since we have shown that $(B - \lambda I)x = 0$ implies that $x = 0$, then by Theorem 3.7 the inverse of $(B - \lambda I)$ exists and, in particular, we can solve the inhomogeneous equation

$$(B - \lambda I)x = \psi_\nu\,,$$

which can be rewritten as

$$Ax + \sum_{j=1}^{\mu} \psi_j(\phi_j, x) - \lambda x = \psi_\nu\,.$$

We now take the inner product from the left by ψ_ν and find

$$(\psi_\nu, Ax) - \lambda(\psi_\nu, x) = 1\,,$$

where the term involving the summation vanishes because of assumption $\nu > \mu$. Since $A^\dagger\psi_\nu = \lambda^*\psi_\nu$, we conclude that

$$(\lambda^*\psi_\nu, x) - \lambda(\psi_\nu, x) = 1$$

or $0 = 1$, a contradiction. This impossibility leads us to the conclusion that our original assumption ($\nu > \mu$) is inconsistent. Hence $\mu \geq \nu$. The same argument, using the operator C defined by

$$Cx \equiv A^\dagger x + \sum_{j=1}^{\nu} \phi_j(\psi_j, x)\,,$$

shows that $\mu > \nu$ is also impossible, so we are left with just $\mu = \nu$. QED

We emphasize again that even though $\mu = \nu$, μ may nevertheless be smaller (but not larger) than m, the algebraic multiplicity of λ and λ^*. With this result, we can obtain an important generalization of Theorem 3.17.

Theorem 4.27. In a finite-dimensional vector space, the equation

$$(A - \lambda I)x = y\,,$$

has a solution if and only if $(\psi_i, y) = 0$ for all ψ_i satisfying

$$(A^\dagger - \lambda^* I)\psi_i = 0.$$

Proof. The desired result follows readily from the previous theorem. There we established that if

$$A\phi_i = \lambda\phi_i \qquad (i = 1, 2, \cdots, \mu)$$

and

$$A^\dagger\psi_i = \lambda^*\psi_i \qquad (i = 1, 2, \cdots, \mu) ,$$

then λ is not an eigenvalue of B, where B is defined by

$$B \equiv Ax + \sum_{j=1}^{\mu} \psi_j(\phi_j, x) .$$

Hence the equation

$$(B - \lambda I)x = \left[Ax + \sum_{j=1}^{\mu} \psi_j(\phi_j, x) - \lambda x\right] = y \tag{4.32}$$

has a solution because det $(B - \lambda I) \neq 0$ and $(B - \lambda I)^{-1}$ therefore exists. Let us first assume that $(\psi_i, y) = 0$ for all i. Then taking the inner product from the left with ψ_i in Eq. (4.32), we find that

$$(\psi_i, Ax) + (\phi_i, x) - \lambda(\psi_i, x) = (\psi_i, y) , \qquad i = 1, 2, \cdots, \mu .$$

Since $(\psi_i, Ax) = (A^\dagger\psi_i, x) = (\lambda^*\psi_i, x) = \lambda(\psi_i, x)$, it follows that

$$(\phi_i, x) = (\psi_i, y) = 0 \qquad \text{for all } i.$$

Hence Eq. (4.32) reduces to

$$(A - \lambda I)x = y . \tag{4.33}$$

Since Eq. (4.32) has a solution, so does Eq. (4.33), which is what was to be shown.

The converse is trivial to prove. If we assume that $(A - \lambda I)x = y$ has a solution x, then taking the inner product on the left by ψ_i, we get

$$(\psi_i, Ax) - \lambda(\psi_i, x) = (\psi_i, y) , \qquad i = 1, 2, \cdots, \mu .$$

Since we just saw that $(\psi_i, Ax) = \lambda(\psi_i, x)$, we have immediately $(\psi_i, y) = 0$, for all i. QED

Note that if we assume that A has a complete set of eigenvectors, then the above theorem reduces to the results of Theorem 3.17. For in this case, calling $\{x_i\}$ the eigenvectors of A, we write

$$x = \sum_{i=1}^{n} \xi_i x_i , \qquad y = \sum_{i=1}^{n} \eta_i x_i ,$$

so

$$(A - \lambda I)x = \sum_{i=1}^{n} (\lambda_i - \lambda)\xi_i x_i = y = \sum_{i=1}^{n} \eta_i x_i ,$$

where λ_i is the eigenvalue corresponding to x_i. Thus

$$\sum_{i=1}^{n} [(\lambda_i - \lambda)\xi_i - \eta_i]x_i = 0,$$

and since the x_i are linearly independent,

$$(\lambda_i - \lambda)\xi_i = \eta_i,$$

for all i. If $\lambda_\nu = \lambda$, then a solution can exist only if $\eta_\nu = 0$. Similarly, if a solution does exist when $\lambda_\nu = \lambda$, then $\eta_\nu = 0$. Clearly, if $\lambda_\nu = \lambda$ and $\eta_\nu = 0$, then coefficient ξ_ν is completely arbitrary. Assuming for notational convenience that only one λ_ν gives trouble, the solution for x is

$$x = \sum_{\substack{i=1 \\ i\neq\nu}}^{n} \frac{\eta_i}{\lambda_i - \lambda} x_i + a x_\nu,$$

where a is *any* constant. Thus, under the assumption of a complete set of eigenvectors belonging to A, we can obtain an explicit solution to the equations, as we did in Theorem 3.17, where we took $\lambda = 0$ for convenience. Note that the relation $\eta_\nu = 0$ means that y has the form

$$y = \sum_{\substack{i=1 \\ i\neq\nu}}^{n} \eta_i x_i .$$

But, denoting by y_ν the eigenvector of $A^\dagger$ belonging to λ_ν^*, we have by Theorem 4.25,

$$(y_\nu, y) = 0,$$

so that the statement $\eta_\nu = 0$ is equivalent to $(y_\nu, y) = 0$. Again we are assuming a nondegenerate spectrum. The wording can easily be changed slightly to cover the case of a multiple eigenvalue.

Thus with virtually no assumptions about the operator A, we have been able to obtain very general conditions for the solvability of a system of linear equations. These results will be of great importance when we discuss integral equations in Chapter 8.

4.9 MINIMUM PRINCIPLES

In this section we want to show how the concepts of the calculus of variations can be used to obtain some striking results when applied to inner products in a vector space.

In Chapter 2, we considered the general problem of extremizing the value of an integral with some fairly smooth function in the integrand. The Euler-Lagrange equations that we found there are second-order differential equations which are, in general, nonlinear and therefore very complicated. Consider, for instance, a basic problem of classical physics—Newton's second law for an arbitrary potential:

$$m\ddot{\mathbf{r}} = -\nabla V(\mathbf{r}) .$$

As we saw in Section 2.5, this equation can be derived readily from the Euler-Lagrange equations for the system. But it is, in general, a system of nonlinear equations which is extremely difficult to solve. This difficulty is often concealed by the fact that special forms for $V(\mathbf{r})$ lead to the existence of conserved quantities. In such cases, the solution of this system of second-order differential equations is reduced to doing the first integrals.

Now there is a special, but nevertheless important, class of problems for which the variational technique is useful and can also be put in a much more general mathematical setting than we have heretofore considered. Let us suppose that we have a finite- or infinite-dimensional inner product space V, and let A be some *linear* operator mapping V into itself. Now consider the quantity

$$I = (x, Ax) .$$

The vector x depends continuously on some parameter ϵ in such a way that the inner product is a function of ϵ. Then we write

$$I(\epsilon) = (x(\epsilon), Ax(\epsilon)) .$$

We are interested in extremizing $I(\epsilon)$ under certain conditions. Therefore we shall want to take derivatives of $I(\epsilon)$. If we assume that A is a linear operator and that $x(\epsilon)$ is a differentiable function of ϵ, then it is easy to establish the "chain rule":

$$\frac{dI}{d\epsilon} = \left(\frac{dx}{d\epsilon}, Ax\right) + \left(x, A\frac{dx}{d\epsilon}\right) . \tag{4.34}$$

We obtain this simple result because A is a *linear* operator. Note that x is a very general object. It could be an n-tuple of numbers, or it could be a function of several variables (besides ϵ) in a space of functions with an inner product given, for example, by an integral over the space.

Equation (4.34) suggests that the techniques of Chapter 2 might be applicable to the problem of extremizing inner products on a vector space. This is indeed true if A is *Hermitian*. Let us now consider this case in detail. We want to extremize the expression

$$I = (x, Ax) ,$$

where A is self-adjoint and $x \in V$, an inner-product space. Now the reader will see immediately that we have not posed the problem in quite the right way. For, given any x which leads to, say, a negative value of I, we can produce an arbitrarily large negative value of I just by increasing the norm of x. Similarly, if we have an x which gives some positive value for I, then we can produce an arbitrarily large positive value of I by increasing the norm of x. Thus $I = 0$ is the only possible stationary value for the problem as we have stated it. To prevent this artificial increase in magnitude of I, let us *constrain* the vectors x

to have a fixed norm. For convenience, we shall take $||x|| = 1$. Then rephrased, the problem is to extremize I with respect to the class of vectors x satisfying the constraint equation

$$J = (x, x) = 1 .$$

To handle this problem, we introduce a two-parameter family of vectors,* $\bar{x}(\epsilon_1, \epsilon_2)$, satisfying

a) $\bar{x}(\epsilon_1, \epsilon_2) \in V$ for all ϵ_1, ϵ_2;

b) $\bar{x}(0, 0) = x$, the extremizing vector;

c) $\partial\bar{x}/\partial\epsilon_1$ and $\partial\bar{x}/\partial\epsilon_2$ exist and are in V.

Then we form

$$I(\epsilon_1, \epsilon_2) = (\bar{x}(\epsilon_1, \epsilon_2), A\bar{x}(\epsilon_1, \epsilon_2)) , \tag{4.35a}$$

$$J(\epsilon_1, \epsilon_2) = (\bar{x}(\epsilon_1, \epsilon_2), \bar{x}(\epsilon_1, \epsilon_2)) = 1 , \tag{4.35b}$$

and consider

$$K(\epsilon_1, \epsilon_2) = I(\epsilon_1, \epsilon_2) - \lambda J(\epsilon_1, \epsilon_2) , \tag{4.35c}$$

where λ is a Lagrange multiplier. According to condition (b) above,

$$\left.\frac{\partial K}{\partial \epsilon_1}\right]_{\epsilon_1=\epsilon_2=0} = 0 = \left.\frac{\partial K}{\partial \epsilon_2}\right]_{\epsilon_1=\epsilon_2=0} \tag{4.36}$$

is the condition for an extremum. Note that Eqs. (4.35 a, b, c) tell us that

$$K(\epsilon_1, \epsilon_2) = (\bar{x}(\epsilon_1, \epsilon_2), [A - \lambda]\bar{x}(\epsilon_1, \epsilon_2)) .$$

We write $(A - \lambda I)$ as simply $(A - \lambda)$ to economize. Also, since A is Hermitian, $I(\epsilon_1, \epsilon_2)$ is real, so λ is real. We now use Eq. (4.34) to calculate the two first partial derivatives of K.

$$\frac{\partial K}{\partial \epsilon_1} = \left(\frac{\partial \bar{x}}{\partial \epsilon_1}, [A - \lambda]\bar{x}\right) + \left(\bar{x}, [A - \lambda]\frac{\partial \bar{x}}{\partial \epsilon_1}\right) ,$$

$$\frac{\partial K}{\partial \epsilon_2} = \left(\frac{\partial \bar{x}}{\partial \epsilon_2}, [A - \lambda]\bar{x}\right) + \left(\bar{x}, [A - \lambda]\frac{\partial \bar{x}}{\partial \epsilon_2}\right) .$$

Setting $\epsilon_1 = 0 = \epsilon_2$ and using Eq. (4.34), we find that

$$(\eta_1, [A - \lambda]x) + (x, [A - \lambda]\eta_1) = 0 , \tag{4.37a}$$

$$(\eta_2, [A - \lambda]x) + (x, [A - \lambda]\eta_2) = 0 , \tag{4.37b}$$

where

$$\eta_1 = \left.\frac{\partial \bar{x}}{\partial \epsilon_1}\right]_{\epsilon_1=\epsilon_2=0} , \qquad \eta_2 = \left.\frac{\partial \bar{x}}{\partial \epsilon_2}\right]_{\epsilon_1=\epsilon_2=0} ;$$

---

* Note that the argument does not depend on the number of parameters (except that *at least* two are needed). In practice one can use more than two parameters if necessary.

$(A - \lambda)$ is self-adjoint, because λ is real. Hence, from Eqs. (4.37a, b),

$$([A - \lambda]x, \eta_i) + (\eta_i, [A - \lambda]x) = 0 , \qquad i = 1, 2.$$

Now η_i is arbitrary, so we may set $\eta_i = (A - \lambda)x$. We conclude that

$$2([A - \lambda]x, [A - \lambda]x) = 0 , \qquad \text{so} \qquad (A - \lambda)x = 0 .$$

We have, therefore, a remarkable link between the eigenvalue problem for self-adjoint operators and the calculus of variations, which can be stated as follows:

Theorem 4.28. If A is a linear self-adjoint operator, $Ax = \lambda x$ if and only if $I = (x, Ax)$ is extremized with respect to the constraint $(x, x) = 1$. The extremized value of I is just λ, the Lagrange multiplier.

It is interesting to see under what circumstances I actually takes on a minimum value. First, it is clear that if λ is any eigenvalue other than the lowest one, the extremum of I will not be a minimum. Therefore only the lowest eigenvalue and its corresponding eigenvector can yield a minimum for I.

In a finite-dimensional vector space, there is always a lowest eigenvalue. In this case, we have the theorems which establish completeness for self-adjoint operators (Theorems 4.20 and 4.21) at our disposal. Thus, any $x \in V$ can be written as

$$x = \sum_{i=1}^{n} \alpha_i x_i ,$$

where the x_i are the eigenfunctions of A and where we take the α_i to be real for convenience. (The reader can easily go through the discussion with α_i complex, handling the real and imaginary parts as independent variables). Thus calling λ_i the eigenvalues of A and ordering them according to $\lambda_1 \leq \lambda_2 \leq \cdots \leq \lambda_n$, we find that

$$(x, Ax) = \sum_{i=1}^{n} \alpha_i^2 \lambda_i , \qquad (x, x) = \sum_{i=1}^{n} \alpha_i^2 = 1 .$$

We have already proved that in any inner-product space, the eigenvalues provide us with an extremum of (x, Ax). Therefore, we need not show this again in the case of a finite-dimensional space. However, in order to illustrate the ideas involved, we derive this result once more for this simple case. In trying to extremize (x, Ax) with respect to α_i, holding $(x, x) = 1$, we set the first derivatives of

$$K(\alpha_i) = \sum_{i=1}^{n} \alpha_i^2(\lambda_i - \lambda)$$

equal to zero, where λ is a Lagrange multiplier. We have

$$\frac{dK}{d\alpha_i} = 2\alpha_i(\lambda_i - \lambda) , \qquad i = 1, \cdots, n .$$

Thus the extrema are determined by

$$\alpha_i(\lambda_i - \lambda) = 0\,, \qquad i = 1, \cdots, n\,.$$

If λ is not equal to any of the λ_i, then the only possible solution is $\alpha_i = 0$ for all i. This is not admissible, since then

$$\sum_{i=1}^{n} \alpha_i^2 \neq 1\,.$$

Therefore, for some j, with $1 \leq j \leq n$, $\lambda = \lambda_j$; also, $\alpha_j = 1$ and $\alpha_i = 0$ for $i \neq j$.

Now if $\lambda = \lambda_1$, $\alpha_1 = 1$, and $\alpha_i = 0$ for $i \geq 2$. Then all the second derivatives of K with respect to the α_i, for $i \geq 2$, are positive and I takes on a minimum in this case. The value of I at this minimum is precisely λ_1. For an extremum at $\lambda = \lambda_k$, $1 < k < n$, $\alpha_k = 1$, and $\alpha_i = 0$ for $i \neq k$. Now some of the second partial derivatives will be positive and some negative, so we have a saddle point of some type. For the case $\lambda = \lambda_n$, all the second derivatives are negative and we have a maximum; this maximum value of I is just $I = \lambda_n$. Hence we obtain

Theorem 4.29. Let A be a self-adjoint operator in an n-dimensional inner-product space. If λ_1 is the least eigenvalue of A and λ_n the largest eigenvalue then the minimum and maximum values of (x, Ax) are λ_1 and λ_n, respectively, for all vectors $x \in V$ such that $(x, x) = 1$.

We shall return to the infinite-dimensional extension of this theorem in later chapters. The reader may have noticed that we also assumed tacitly above that the spectrum of A was nondegenerate. A little work will show that this assumption is unnecessary.

Now consider λ_2, the second smallest eigenvalue of A. If we take the class of all x such that $(x, x) = 1$, then λ_2 is a saddle point of (x, Ax), not a true minimum. However, if we restrict ourselves to the smaller class of vectors satisfying both $(x, x) = 1$ and $(x, x_1) = 0$ where x_1 is the eigenvector belonging to λ_1, then (x, Ax) takes on its smallest value when $x = x_2$, and this smallest value is λ_2. If we restrict the normalized vector x to lie entirely within the subspace of V spanned by $\{x_2, x_3, \cdots, x_n\}$, or better, entirely outside the subspace of V spanned by x_1, then (x, Ax) takes on its minimum value for $x = x_2$ and this minimum value is just λ_2.

A definition will be useful in formulating the final generalization of these considerations.

Definition 4.9. If V is a vector space and U is a subspace of V, the set of all vectors *not* lying in U is called the *complement of U in V*; we denote this set by U_c.

Then we have

Theorem 4.30. Let A be a self-adjoint operator in an n-dimensional inner-product space V, whose eigenvalues are ordered by $\lambda_1 \leq \lambda_2 \leq \cdots \leq \lambda_n$.

Further, let V_i be the subspace of V spanned by the eigenvectors of A belonging to the first i eigenvalues. Then for the class of vectors $\{x\}$ which are normalized to unity and lie in the complement of V_i, the minimum value of (x, Ax) with respect to members of this class is assumed when $x = x_{i+1}$, and the minimum value is λ_{i+1}.

Proof. The result follows immediately if we note that all x in the class can be written in the form

$$x = \sum_{j=i+1}^{n} \alpha_j x_j$$

so that

$$(x, Ax) = \sum_{j=i+1}^{n} \alpha_j^2 \lambda_j \qquad \text{and} \qquad (x, x) = \sum_{j=i+1}^{n} \alpha_j^2 = 1 \, .$$

From this point on, the proof proceeds in the same fashion as the proof of the previous theorem. QED

As we have derived them, Theorems 4.29 and 4.30 are valid only for finite-dimensional spaces. However, Theorem 4.28 is valid for a space of any dimension, and this suggests that, since the eigenfunctions of A produce extrema of (x, Ax), it may be possible to find minimum theorems similar to 4.29 and 4.30 for infinite-dimensional spaces. We shall find in Chapter 9 that this is indeed the case.

Now the difficulty with which we must come to grips in the infinite-dimensional case is twofold:

1. there may be an infinite number of eigenvectors;
2. there may be a continuous spectrum.

A glance at the proofs of Theorems 4.29 and 4.30 suggests that difficulty (1) is not serious. However, difficulty (2) is very serious indeed. In fact, it will force us to revise our concept of completeness, upon which the proofs of Theorems 4.29 and 4.30 are based.

It turns out that for most Hermitian operators of interest in physics, (x, Ax) takes on a minimum value equal to the lowest eigenvalue of A when x is equal to the corresponding eigenvector. In most physical problems, it is intuitively clear that a lowest eigenvalue exists. For example, in the nonrelativistic Schrodinger equation for a potential of given strength, it is obvious that the binding energy due to that potential cannot be an arbitrarily large negative number. Thus we suspect that there will be a lower bound for the energy eigenvalues. However, all this discussion of minimum principles will be irrelevant if there is no such lowest eigenvalue.

On the other hand, a particle not in a bound state can have an arbitrarily large positive energy, so we might guess that the spectrum resulting from the Schrodinger equation in this case will not be bounded above. Moreover, we know on physical grounds that these so-called *scattering states* can have any energy whatever, as opposed to bound states, which are restricted to certain discrete energy levels, as in the case of the hydrogen atom. Therefore, this part of the spectrum corresponds to states which can lie arbitrarily close together. This is roughly what we mean by the term *continuous spectrum*.

A slightly more mathematical way of looking at the continuous spectrum is to say that it corresponds to functions which "solve" a given operator equation but do not lie in the vector space. We shall go into such matters in more detail in later chapters, but we can suggest some of the difficulties of dealing with infinite-dimensional spaces by considering the quantum-mechanical kinetic-energy operator

$$\frac{p^2}{2m} = -\frac{\hbar^2}{2m}\frac{d^2}{dx^2} \tag{4.38}$$

acting on the vector space of functions f which satisfy the condition

$$(f, f) = \int_{-\infty}^{\infty} |f|^2\, dx < \infty .$$

Such functions are said to be *square-integrable*; we shall show later that they do indeed form a vector space.

Now consider the eigenvalue equation

$$\frac{p^2}{2m}\phi = E\phi . \tag{4.39}$$

Clearly, $\phi = e^{ikx}$, $E = \hbar^2k^2/2m$ are solutions to the differential equation implied by Eqs. (4.38) and (4.39), for *all* k, and indeed they are the only solutions if E is positive. However,

$$\int_{-\infty}^{\infty} |\phi|^2\, dx = \infty$$

for every value of k, so the ϕ's do not lie in the vector space in question. Nevertheless, it turns out that every function in the space can be written as a linear combination of the ϕ's in the generalized sense that, if $f(x)$ is in the space of square-integrable functions, there exists a function $\tilde{f}(k)$ such that

$$f(x) = \int_{-\infty}^{\infty} \tilde{f}(k)e^{ikx}\, dk.$$

This is the famous Fourier integral theorem. It assures us that every vector in the space of square-integrable functions can be written as a *continuous* linear combination of the "eigenvector" of Eq. (4.39).

Thus the eigenvectors for a continuous eigenvalue spectrum look as though they can be handled very much like vectors in finite-dimensional vector spaces —provided that the notion of completeness can be generalized in a suitable way. The study of such problems is central in the theory of infinite-dimensional vector spaces.

Some of the most interesting and relevant applications of the theorems of this section are in the solution of eigenvalue problems for infinite-dimensional vector spaces. Ignoring for now the fact that we have not rigorously established these results for such spaces, let us consider some examples from quantum mechanics.

Let A be a given linear operator. One possible technique for finding the lowest eigenvalue is to take a trial vector x with a number of variable parameters, calculate (x, Ax), and then minimize this inner product with respect to the variable parameters in x. By adding more and more parameters, we can make the value of (x, Ax) smaller and smaller; however, it can never drop below λ_1, the lowest eigenvalue of A. Clearly, we must be guided by common sense in this process; for example, if we were inept enough to pick a trial vector in such a way that it was always orthogonal to the true eigenvector belonging to the lowest eigenvalue, then by Theorem 4.30, we would actually be closing in on the second eigenvalue, λ_2, as we increased the number of parameters. A consideration of the physics of the problem will usually enable us to avoid such pitfalls.

Examples. The Hamiltonian of the quantum harmonic oscillator is

$$H = \frac{p^2}{2m} + \frac{1}{2} m\omega^2 x^2 .$$

If we choose p and x as in Eq. (3.39), this becomes

$$H = -\frac{\hbar^2}{2m}\frac{d^2}{dx^2} + \frac{1}{2} m\omega^2 x^2 .$$

Let us call our trial vector ϕ and consider

$$I = (\phi, H\phi) = \int_{-\infty}^{\infty} \phi^*(x) H\phi(x)\, dx .$$

Now, on physical grounds alone we would expect the true eigenfunction to fall off rather rapidly with distance because of the strong harmonic attraction. Hence, a reasonable guess for ϕ might be

$$\phi = Ce^{-\alpha|x|/2} , \qquad \alpha > 0 .$$

The constraint $(\phi, \phi) = 1 = C^2 \int_{-\infty}^{\infty} e^{-\alpha|x|}\, dx$ can be met by choosing $C = (\alpha/2)^{1/2}$. Thus

$$\phi(x) = (\sqrt{\alpha/2})e^{-\alpha|x|/2} , \qquad \alpha > 0 ,$$

and putting this value of ϕ into our expression for I, we find that

$$I = (\hbar^2\alpha^2/8m) + (m\omega^2/\alpha^2) .$$

It is evident that the extremum for I is a minimum; also, a Lagrange multiplier is not necessary since we have normalized ϕ to 1. Thus, setting $dI/d\alpha = 0$, we find that the minimum occurs at

$$\alpha = \alpha_0 = \left(\frac{8m^2\omega^2}{\hbar^2}\right)^{1/4} .$$

The value which I takes on at this minimum is $I(\alpha_0) = (1/\sqrt{2})\hbar\omega$, which is greater than the eigenvalue, $\lambda_1 = \frac{1}{2}\hbar\omega$, by about 40 percent. We have seen in Section 3.10 that the correct eigenvector is

$$u_0 = \left(\frac{m\omega}{\pi\hbar}\right)^{1/4} e^{-(m\omega/2\hbar)x^2},$$

which does not resemble our result,

$$\phi = \left(\frac{m^2\omega^2}{2\hbar^2}\right)^{1/8} \exp\left[-\left(\frac{m^2\omega^2}{2\hbar^2}\right)^{1/4} |x|\right],$$

too closely. Nevertheless, the value which we find for energy is only 40 percent larger than the true lowest eigenvalue. An interesting problem is to guess a trial function of the form

$$\phi' = Ce^{-\alpha x^2/2}$$

and to show that upon minimizing $(\phi', H\phi')$ one obtains the true eigenvalue and eigenvector.

As another example, consider the calculation of the ground-state energy of helium. The Hamiltonian is

$$H = -\frac{\hbar^2}{2m}(\nabla_1^2 + \nabla_2^2) - 2e^2\left(\frac{1}{r_1} + \frac{1}{r_2}\right) + \frac{e^2}{|\mathbf{r}_1 - \mathbf{r}_2|},$$

where 1 and 2 index the two electrons moving about the nucleus of charge $Z = +2e$. The final term in this Hamiltonian representing the Coulomb repulsion between the electrons makes the problem impossible to solve exactly. If we neglect it the problem can be solved, giving for the ground-state eigenfunction the product of two ground-state hydrogen atom eigenfunctions:

$$\phi_0 = \left(\frac{Z^3}{\pi a^3}\right) \exp\left(-\frac{Z(r_1 + r_2)}{a}\right),$$

where $a = \hbar^2/me^2 =$ the Bohr radius. The corresponding energy is -54.4 eV, which is not in particularly good agreement with the experimental value of -78.60 eV. However, if we use a trial function of this *form* and replace Z by $Z_{\text{eff}} = Z - \sigma$, we can do much better. This is physically reasonable because each electron is partially screened from the full charge of the nucleus by the presence of the other electron. We then want to minimize

$$I = (\phi_0, H\phi_0),$$

where H is the complete Hamiltonian for the helium atom, including the interaction between electrons, and

$$\phi_0 = \left[\frac{(Z - \sigma)^3}{\pi a^3}\right] \exp\left[-\frac{(Z - \sigma)(r_1 + r_2)}{a}\right].$$

We know that $I(\sigma) \geq E_0$, the ground state energy of the helium atom. By direct integration, we find that

$$I(\sigma) = -(e^2/a)\left(\tfrac{11}{4} + \tfrac{5}{8}\sigma - \sigma^2\right),$$

$I(\sigma)$ has a minimum for $\sigma = \frac{5}{16}$. Therefore the best estimate of E_0 with a trial function of this form is

$$E_0 = I(\tfrac{5}{16}) = -\tfrac{729}{256}(e^2/a) = -77.46\,\text{eV}\,,$$

since $e^2/a = 27.20$ eV. This is quite close to (and greater than) the measured value of -78.60 eV.

Trial functions with many adjustable parameters can be used. They lead to phenomenally close agreement between theory and experiment.*

4.10 NORMAL MODES

Now that we know how to bring a linear operator to diagonal form, it is a simple matter to define powers of a fairly large class of linear operators. Let A be a linear operator on an n-dimensional vector space, and suppose it can be diagonalized by a similarity transformation

$$S^{-1}AS = D\,,$$

where D is defined by

$$D_{ij} = \lambda_i \delta_{ij}\,,$$

and $\lambda_1, \lambda_2, \cdots, \lambda_n$ are the eigenvalues of A. Then A^α, where α is any real number, may be defined by

$$A^\alpha = SD^\alpha S^{-1}\,, \qquad \text{where} \qquad [D^\alpha]_{ij} = \lambda_i^\alpha \delta_{ij}\,.$$

Note that if α is negative and A has a zero eigenvalue, then A^α is not well defined. Since $SS^{-1} = I$, it is easily seen that $A^\alpha A^\beta = A^{(\alpha+\beta)}$.

Self-adjoint operators can always be diagonalized, so we focus our attention on them. A particular class of self-adjoint operators for which negative powers always exist is the positive-definite operator.

Definition 4.10. An operator H on a vector space V is said to be *positive-definite* if and only if $(x, Hx) > 0$ for all nonzero x in V. If we have just $(x, Hx) \geq 0$, H is called a *positive* operator.

Clearly, H is self-adjoint since (x, Hx) is real for all x. Also, a positive-definite operator must have strictly positive eigenvalues since

$$Hx = \lambda x$$

implies that

$$\lambda = \frac{(x, Hx)}{(x, x)}\,,$$

so if $(x, Hx) > 0$ for all nonzero x, then $\lambda > 0$, since $x = 0$ can never be an eigenvector.

---

* See C. L. Pekeris, *Phys. Rev.* **115**, 1216 (1959).

For positive-definite H, $H^{1/2}$ is Hermitian, which may be seen as follows:

$$H^{1/2} = UD^{1/2}U^{-1} = UD^{1/2}U^{\dagger} ,$$

where U is the *unitary* operator which, by Theorem 4.20, diagonalizes H. Since the eigenvalues of H are positive, $D^{1/2}$ is a real diagonal matrix, and therefore $D^{1/2}$ is Hermitian. Thus

$$(H^{1/2})^{\dagger} = (UD^{1/2}U^{\dagger})^{\dagger} = (U^{\dagger})^{\dagger}(D^{1/2})^{\dagger}U^{\dagger} = UD^{1/2}U^{\dagger} = H^{1/2} ,$$

so $H^{1/2}$ is self-adjoint, as is also $H^{-1/2}$. In a similar way, it follows that if H is (real) symmetric, then so are $H^{1/2}$ and $H^{-1/2}$.

A particularly interesting example of a positive-definite operator occurs in classical mechanics. Suppose that $(x_1, \cdots, x_n)$ is a system of generalized coordinates—not necessarily orthogonal—which may be considered as components of a vector x.* Under many circumstances, the kinetic energy of the system will take the form

$$K = \tfrac{1}{2} \sum_{i,j} T_{ij}\dot{x}_i\dot{x}_j ,$$

where (T_{ij}) is an array of real constants, e.g., masses, which are independent of the coordinates and of the time. They may be chosen so that $T_{ij} = T_{ji}$. The kinetic energy K can be written very compactly in our notation as

$$K = \tfrac{1}{2}(\dot{x}, T\dot{x}) ,$$

where the components of the vector $\dot{x}$ are $\dot{x}_1, \dot{x}_2, \cdots, \dot{x}_n$. Since K is the kinetic energy, it must be strictly positive for any nonzero vector x; thus T is positive-definite as well as (real) symmetric.

The potential energy will, in general, be a complicated function of the coordinates. But if we confine our attention to small bounded motions about the equilibrium point, which we can take without loss of generality to be located at $x = 0$, then we may write

$$V(x) = [V]_{x=0} + \sum_i \left[\frac{\partial V}{\partial x_i}\right]_{x=0} x_i + \tfrac{1}{2} \sum_{i,j} \left[\frac{\partial^2 V}{\partial x_i \partial x_j}\right]_{x=0} x_i x_j + \cdots$$

The subscript on each bracket indicates that the quantity inside the bracket is evaluated at $x = 0$. Equilibrium corresponds to an extremum of the potential energy, so

$$\left[\frac{\partial V}{\partial x_i}\right]_{x=0} = 0, \qquad \text{for all } i.$$

Since the constant term can only shift the zero of energy we drop it; also, we neglect terms in x_i^3 and higher, since we are interested in small amplitude

* For reasons of convenience, we are going to suspend the notational conventions for vectors and their components throughout this section.

motion. Thus

$$V(x) = \tfrac{1}{2} \sum_{i,j} U_{ij} x_i x_j , \qquad \text{where} \qquad U_{ij} = \left[\frac{\partial^2 V}{\partial x_i \partial x_j} \right]_{x=0} = U_{ji} ,$$

so U is (real) symmetric. In our more compact notation,

$$V = \tfrac{1}{2} (x, Ux) .$$

The matrix U is an array of *constants*, as was the matrix T in the expression for the kinetic energy. For bounded motion, we want a local minimum of the potential energy, so that if we move away from the equilibrium point in any direction the potential energy will increase. This may be described mathematically by saying

$$\sum_{i,j} \left[\frac{\partial^2 V}{\partial x_i \partial x_j} \right]_{x=0} \epsilon_i \epsilon_j > 0 ,$$

for all possible choices of the vector ϵ. For many physical purposes, it is convenient to weaken this condition slightly and require only that

$$\sum_{i,j} \left[\frac{\partial^2 V}{\partial x_i \partial x_j} \right]_{x=0} \epsilon_i \epsilon_j \geq 0 ,$$

for all possible ϵ. Thus U is a *positive* operator, so that

$$(x, Ux) \geq 0 .$$

We shall now study the general dynamical solutions for the motion of such a system, making use of the powerful machinery developed in this chapter. This will give us the main results of the theory of coupled systems without recourse to the messy algebra often encountered in traditional discussions.

For such a system, the Lagrangian may be written as

$$L = K - V = \tfrac{1}{2} (\dot{x}, T\dot{x}) - \tfrac{1}{2} (x, Ux) .$$

From Lagrange's equations,

$$\frac{\partial L}{\partial x_i} - \frac{d}{dt} \left(\frac{\partial L}{\partial \dot{x}_i} \right) = 0 \qquad \text{for all } i,$$

we get immediately the set of equations

$$\sum_j T_{ij} \ddot{x}_j + \sum_j U_{ij} x_j = 0 \qquad \text{for all } i.$$

This can be written simply as

$$T\ddot{x} + Ux = 0 . \tag{4.40}$$

By analogy with the simple harmonic-oscillator problem, let us look for a solution of the form

$$x = \xi e^{\pm i\omega t} ,$$

where the scalar ω and the vector ξ are to be determined. Substituting this into Eq. (4.40), we get

$$U\xi - \omega^2 T\xi = 0\,, \tag{4.41}$$

which is an eigenvalue problem, but of a type we have not encountered before. Our notation suggests that ω^2 should not be negative, and this is indeed the case since

$$\omega^2 = \frac{(\xi,\, U\xi)}{(\xi,\, T\xi)}\,,$$

and $(\xi, U\xi) \geq 0$, $(\xi, T\xi) > 0$.

The properties of the solutions to Eq. (4.41) are easily found by using the existence of $T^{1/2}$ and $T^{-1/2}$, although we shall see that in solving an actual problem it will not be necessary to compute $T^{1/2}$ or $T^{-1/2}$. Equation (4.41) implies that

$$T^{-1/2}U\xi = \omega^2 T^{1/2}\xi\,,$$

or since $T^{-1/2}T^{1/2} = I$,

$$T^{-1/2}UT^{-1/2}T^{1/2}\xi = \omega^2 T^{1/2}\xi\,.$$

Thus if we call

$$\eta = T^{1/2}\xi \tag{4.42}$$

and

$$W = T^{-1/2}UT^{-1/2}\,,$$

we have

$$W\eta = \omega^2\eta\,, \tag{4.43}$$

which is a simple eigenvalue problem of the usual form. It is clear that $T^{-1/2}$ will be (real) symmetric if T is real and positive-definite, as well as symmetric. Therefore since U is (real) symmetric,

$$\tilde{W} = \widetilde{T^{-1/2}UT^{-1/2}} = (\widetilde{T^{-1/2}})\,\tilde{U}\,(\widetilde{T^{-1/2}}) = T^{-1/2}UT^{-1/2} = W\,,$$

and thus W is (real) symmetric.

Thus, by Theorem 4.21, W has n eigenvalues ω_i^2, some of which may have multiplicity greater than one, and n real orthonormal eigenvectors $\eta^{(i)}$, where

$$(\eta^{(i)}, \eta^{(j)}) = \delta_{ij}\,.$$

With the $\eta^{(i)}$ determined, we find the corresponding $\xi^{(i)}$ which are solutions to Eq. (4.41). Equation (4.42) yields immediately

$$\xi^{(i)} = T^{-1/2}\eta^{(i)}\,.$$

For the $\xi^{(i)}$, Eq. (4.42) yields

$$(\eta^{(i)}, \eta^{(j)}) = (T^{1/2}\xi^{(i)}, T^{1/2}\xi^{(j)}) = (\xi^{(i)}, T\xi^{(j)})\,,$$

so that

$$(\xi^{(i)}, T\xi^{(j)}) = \delta_{ij} . \tag{4.44}$$

Since the $\eta^{(i)}$ are real, so are the $\xi^{(i)}$. From the $\xi^{(i)}$, we get as solutions to Eq. (4.40) the vectors $x_{\pm}^{(i)}(t)$:

$$x_{\pm}^{(i)}(t) = \xi^{(i)} e^{\pm i\omega_i t} .$$

The magnitude of these vectors is as yet undetermined. In a physical problem, we would have a set of boundary conditions to determine these magnitudes, for example, $x_i(0) = \alpha_i$ and $\dot{x}_i(0) = \beta_i$ for all i, which in our notation looks like $x(0) = \alpha$, $\dot{x}(0) = \beta$. Our general solution can be written as a linear combination of the $x_{\pm}^{(i)}(t)$. Thus

$$\begin{aligned} x(t) &= \sum_k [C_k x_{\dagger}^{(k)}(t) + D_k x_{-}^{(k)}(t) \\ &= \sum_k \xi^{(k)} [C_k e^{i\omega_k t} + D_k e^{-i\omega_k t}] . \end{aligned}$$

In order to emphasize the fact that $x(t)$ is real, we can use a linear combination of $\sin \omega_k t$ and $\cos \omega_k t$ instead of $e^{+i\omega_k t}$ and $e^{-i\omega_k t}$. We have then

$$x(t) = \sum_k \xi^{(k)} [A_k \cos \omega_k t + B_k \sin \omega_k t] . \tag{4.45}$$

Applying the boundary conditions, we obtain

$$\alpha = \sum_k \xi^{(k)} A_k , \qquad \beta = \sum_k \omega_k \xi^{(k)} B_k .$$

Hence

$$\begin{aligned} (\xi^{(j)}, T\alpha) &= \sum_k (\xi^{(j)}, T\xi^{(k)}) A_k , \\ (\xi^{(j)}, T\beta) &= \sum_k \omega_k (\xi^{(j)}, T\xi^{(k)}) B_k , \end{aligned}$$

and using the orthogonality relation of Eq. (4.44), we get

$$A_j = (\xi^{(j)}, T\alpha) , \qquad B_j = (\xi^{(j)}, T\beta)/\omega_j . \tag{4.46}$$

If the boundary conditions are such that all the A_j and B_j vanish except for one value of j, then we have simple harmonic motion with only one frequency, ω_j, occurring. The relative amplitude with which each coordinate varies under these circumstances is given by the various components of $\xi^{(j)}$. It is called the jth *normal mode*, and ω_j is called the *normal frequency* of the jth mode. The most general motion is a linear superposition of normal modes.

The meaning of those normal modes can be made clearer by looking at the Lagrangian, L, once more:

$$L = \tfrac{1}{2} (\dot{x}, T\dot{x}) - \tfrac{1}{2} (x, Ux) .$$

Writing, in analogy with Eq. (4.42),

$$y = T^{1/2} x , \qquad x = T^{-1/2} y ,$$

we get

$$L = \tfrac{1}{2}(\dot{y}, \dot{y}) - \tfrac{1}{2}(y, T^{-1/2}UT^{-1/2}y) ,$$

where we have made use of the fact that $T^{-1/2}$ is self-adjoint. Using our definition W in Eq. (4.42), we get

$$L = \tfrac{1}{2}(\dot{y}, \dot{y}) - \tfrac{1}{2}(y, Wy) .$$

However, our previous work tells us that W is real symmetric and therefore can be diagonalized by an orthogonal transformation O. The diagonal matrix obtained by the similarity transformation of W with O is a matrix Ω with the square of the normal frequencies down the diagonal; O is just an orthogonal matrix whose columns are the $\eta^{(j)}$. Since $\tilde{O}O = I$, we get

$$\begin{aligned} L &= \tfrac{1}{2}(\dot{y}, O\tilde{O}\dot{y}) - \tfrac{1}{2}(y, O\tilde{O}WO\tilde{O}y) \\ &= \tfrac{1}{2}(\tilde{O}\dot{y}, \tilde{O}\dot{y}) - \tfrac{1}{2}(\tilde{O}y, \Omega\tilde{O}y) . \end{aligned}$$

If we call $z = \tilde{O}y$, then

$$L = \tfrac{1}{2}(\dot{z}, \dot{z}) - \tfrac{1}{2}(z, \Omega z) . \tag{4.47}$$

Thus, in terms of this specially constructed vector z, the Lagrangian assumes a simple form which is most obvious if L is written in terms of components:

$$L = \tfrac{1}{2}\sum_{i=1}^{n}(\dot{z}_i^2 - \omega_i^2 z_i^2) ,$$

whose equations of motion are just

$$\ddot{z}_i + \omega_i^2 z_i = 0 \qquad \text{for all } i.$$

Therefore, by going to this coordinate system, we find that the coupled oscillatory motion decouples into a set of simple harmonic motions whose solutions are

$$z_i = A_i \cos\omega_i t + B_i \sin\omega_i t \qquad \text{for all } i.$$

These various z_i correspond to the normal modes discussed above. This may be seen as follows. With z determined, we can invert our previous steps to get $x = T^{-1/2}Oz$. We leave it to the reader to show that the matrix O' defined by $O' = T^{-1/2}O$ is just the matrix whose columns are composed of the vectors $\xi^{(j)}$:

$$O'_{ik} = \xi_i^{(k)} . \tag{4.48}$$

O' is not an orthogonal matrix, although it is easy to show that $\tilde{O}'TO' = I$. Thus we have $x = O'z$, and if we write this in components, we get

$$x_i = \sum_k O'_{ik}z_k \qquad \text{for all } i.$$

Using Eq. (4.48), we obtain

$$x_i = \sum_k \xi_i^{(k)}(A_k \cos\omega_k t + B_k \sin\omega_k t) ,$$

where we have made use of our general solution for z_k. In vector notation this is just

$$x = \sum_k \xi^{(k)}(A_k \cos \omega_k t + B_k \sin \omega_k t) ,$$

which is our result of Eq. (4.45). Thus we see that our definition of normal modes corresponds exactly to the solution of the transformed Lagrangian of Eq. (4.47): simple harmonic motion of one particular coordinate in a special coordinate system!

As we promised earlier, we never actually need to obtain $T^{\pm 1/2}$; we used this only to discover the general properties of the solutions. The procedure involving $T^{\pm 1/2}$ is *not* a practical algorithm. In practice, we work directly with Eq. (4.41) and solve the equation

$$\det (U - \omega^2 T) = 0$$

to determine the possible values of ω^2. Then the eigenvectors $\xi^{(i)}$ are found from Eq. (4.41). The boundary conditions are applied by using Eq. (4.46).

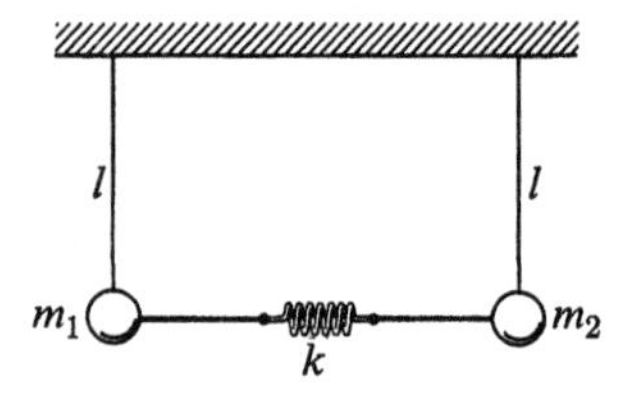

Fig. 4.1. Two pendula of length l coupled by a spring with spring constant k. m_1 and m_2 are the masses of the two bobs.

Consider, for example, the system shown in Fig. 4.1. The kinetic energy is

$$K = \tfrac{1}{2} m_1 l^2 \dot{\theta}_1^2 + \tfrac{1}{2} m_2 l^2 \dot{\theta}_2^2 ,$$

where θ_1 and θ_2 are the angular displacements of m_1 and m_2 from the vertical. Thus

$$T = \begin{bmatrix} m_1 l^2 & 0 \\ 0 & m_2 l^2 \end{bmatrix} .$$

For small oscillations,

$$V = \tfrac{1}{2} m_1 g l \theta_1^2 + \tfrac{1}{2} m_2 g l \theta_2^2 + \tfrac{1}{2} k l^2 (\theta_1 - \theta_2)^2 ,$$

so

$$U = \begin{bmatrix} m_1 g l + k l^2 & -k l^2 \\ -k l^2 & m_2 g l + k l^2 \end{bmatrix} .$$

Now

$$\det (U - \lambda T) = 0$$

implies that

$$\det\begin{bmatrix} m_1gl + kl^2 - m_1l^2\lambda & -kl^2 \\ -kl^2 & m_2gl + kl^2 - m_2l^2\lambda \end{bmatrix} = 0\,,$$

which leads to the secular equation

$$[m_1l^2(\lambda - g/l) - kl^2][m_2l^2(\lambda - g/l) - kl^2] - k^2l^4 = 0\,.$$

The solutions are

$$\lambda_1 = \frac{g}{l}\,, \qquad \lambda_2 = \frac{g}{l} + \frac{k}{\mu}\,,$$

where $\mu = m_1m_2/(m_1 + m_2)$. Thus

$$\omega_1 = \left(\frac{g}{l}\right)^{1/2}, \qquad \omega_2 = \left(\frac{g}{l} + \frac{k}{\mu}\right)^{1/2}.$$

The corresponding eigenvectors are

$$\xi^{(1)} = N\begin{bmatrix} 1 \\ 1 \end{bmatrix}, \qquad \xi^{(2)} = M\begin{bmatrix} -m_2 \\ m_1 \end{bmatrix},$$

where N and M must be determined so that $\xi^{(1)}$ and $\xi^{(2)}$ are properly normalized. The first mode has just the frequency which one simple pendulum would have and corresponds to the two pendula swinging in phase with each other without compressing the spring. The second mode corresponds to the pendula beating against each other, thus compressing the spring. Equation (4.44) gives the normalization as

$$\xi^{(1)} = \frac{1}{l(m_1 + m_2)^{1/2}}\begin{bmatrix} 1 \\ 1 \end{bmatrix},$$

$$\xi^{(2)} = \frac{1}{l(m_1 + m_2)^{1/2}}\begin{bmatrix} -(m_2/m_1)^{1/2} \\ +(m_1/m_2)^{1/2} \end{bmatrix}.$$

One can easily check that $(\xi^{(1)}, T\xi^{(2)}) = 0$ as required.

To find the complete solution, we need to know the initial conditions. Suppose that for the time $t = 0$, $\theta_1 = \alpha_1$, and $\theta_2 = \alpha_2$ give the positions of m_1 and m_2. Then in the notation of Eq. (4.46),

$$\theta(0) = \begin{bmatrix} \theta_1(0) \\ \theta_2(0) \end{bmatrix} = \begin{bmatrix} \alpha_1 \\ \alpha_2 \end{bmatrix} = \alpha \qquad \dot{\theta}(0) = \begin{bmatrix} \dot{\theta}_1(0) \\ \dot{\theta}_2(0) \end{bmatrix} = \begin{bmatrix} 0 \\ 0 \end{bmatrix} = \beta\,.$$

Thus Eq. (4.45) tells us that $B_1 = B_2 = 0$ and

$$A_1 = (\xi^{(1)}, T\alpha) = \frac{1}{l(m_1 + m_2)^{1/2}}(1, 1)\begin{bmatrix} m_1l^2 & 0 \\ 0 & m_2l^2 \end{bmatrix}\begin{bmatrix} \alpha_1 \\ \alpha_2 \end{bmatrix} = \frac{\alpha_1m_1 + \alpha_2m_2}{(m_1 + m_2)^{1/2}}\,l;$$

$$A_2 = (\xi^{(2)}, T\alpha)$$

$$= \frac{1}{l(m_1 + m_2)^{1/2}}\left[-\left(\frac{m_2}{m_1}\right)^{1/2}, \left(\frac{m_1}{m_2}\right)^{1/2}\right]\begin{bmatrix} m_1l^2 & 0 \\ 0 & m_2l^2 \end{bmatrix}\begin{bmatrix} \alpha_1 \\ \alpha_2 \end{bmatrix} = \mu^{1/2}(\alpha_2 - \alpha_1)l\,.$$

Equation (4.45) for this case is

$$\theta(t) = \sum_{k=1}^{2} \xi^{(k)} A_k \cos \omega_k t ,$$

so upon writing out the components of $\theta(t)$, we get

$$\theta_1(t) = \frac{\alpha_1 m_1 + \alpha_2 m_2}{m_1 + m_2} \cos \omega_1 t + \frac{m_2(\alpha_1 - \alpha_2)}{m_1 + m_2} \cos \omega_2 t ,$$

$$\theta_2(t) = \frac{\alpha_1 m_1 + \alpha_2 m_2}{m_1 + m_2} \cos \omega_1 t - \frac{m_1(\alpha_1 - \alpha_2)}{m_1 + m_2} \cos \omega_2 t .$$

Note that if $\alpha_1 = \alpha_2 = \alpha$, then we have motion with only ω_1 present; if $\alpha_1 = (\mu/m_1)\alpha$, $\alpha_2 = -(\mu/m_2)\alpha$, then we have motion with only ω_2 present. These then are two sets of initial conditions which excite the two possible normal modes.

4.11 PERTURBATION THEORY—NONDEGENERATE CASE

Perturbation theory is a very useful technique by which one can obtain approximate solutions to eigenvalue-eigenvector problems that are too complicated to be solved exactly. The following situation is a common one. We want to find the eigenvalues and eigenvectors of a self-adjoint operator A which may be split up into the sum of two self-adjoint operators:

$$A = A_0 + \epsilon A_1 . \tag{4.49}$$

We suppose that we know something about A_0. For example, we might know some eigenvalue and its associated eigenvector. We would like to find the corresponding eigenvalue and eigenvector of A, given that the influence of A_0 is known to be predominant. We have already anticipated this situation by writing the "perturbing" operator in the form ϵA_1, where ϵ is some small parameter.

Thus we assume that the equation

$$A_0 x_n^{(0)} = \lambda_n^{(0)} x_n^{(0)} \tag{4.50}$$

has been solved at least for some n. We now want to look for the corresponding solution of the original problem

$$A x_n = \lambda_n x_n \tag{4.51}$$

which has the form

$$\begin{aligned} \lambda_n &= \lambda_n^{(0)} + \epsilon \lambda_n^{(1)} + \cdots = \sum_{i=0}^{\infty} \epsilon^i \lambda_n^{(i)} , \\ x_n &= x_n^{(0)} + \epsilon x_n^{(1)} + \cdots = \sum_{i=0}^{\infty} \epsilon^i x_n^{(i)} . \end{aligned} \tag{4.52}$$

We shall not at this point discuss the convergence of these power series expansions in ϵ, except to observe that it would be astonishing if they converged for all ϵ.

By substituting the expansions into Eq. (4.51) and dropping for the moment the subscript n on the eigenvalues and eigenvectors, we obtain

$$(A_0 + \epsilon A_1) \sum_{i=0}^{\infty} \epsilon^i x^{(i)} = \left(\sum_{i=0}^{\infty} \epsilon^i \lambda^{(i)}\right)\left(\sum_{j=0}^{\infty} \epsilon^j x^{(j)}\right)$$

or

$$\sum_{i=0}^{\infty} \epsilon^i A_0 x^{(i)} + \sum_{i=0}^{\infty} \epsilon^{i+1} A_1 x^{(i)} = \sum_{i,j=0}^{\infty} \epsilon^{i+j} \lambda^{(i)} x^{(j)} . \tag{4.53}$$

To solve this, we equate coefficients of like powers of ϵ on the left- and right-hand sides of (4.53). The right-hand side is not in a good form for comparison, but if we reshuffle a little, we get

$$\sum_{i=0}^{\infty} \sum_{j=0}^{\infty} \epsilon^{i+j} \lambda^{(i)} x^{(j)} = \sum_{\nu=0}^{\infty} \sum_{\mu=0}^{\nu} \epsilon^{\nu} \lambda^{(\mu)} x^{(\nu-\mu)} .$$

If we set $x^{(\alpha)} \equiv 0$ when $\alpha \leq -1$, then Eqs. (4.53) may be rewritten as

$$\sum_{\nu=0}^{\infty} \epsilon^{\nu} A_0 x^{(\nu)} + \sum_{\nu=0}^{\infty} \epsilon^{\nu} A_1 x^{(\nu-1)} = \sum_{\nu=0}^{\infty} \sum_{\mu=0}^{\nu} \epsilon^{\nu} \lambda^{(\mu)} x^{(\nu-\mu)} .$$

Now we equate powers of ϵ^{ν} to get

$$A_0 x^{(\nu)} + A_1 x^{(\nu-1)} = \sum_{\mu=0}^{\nu} \lambda^{(\mu)} x^{(\nu-\mu)} ,$$

or

$$[A_0 - \lambda^{(0)}] x^{(\nu)} = \sum_{\mu=1}^{\nu} \lambda^{(\mu)} x^{(\nu-\mu)} - A_1 x^{(\nu-1)} , \qquad \nu = 0, 1, \cdots \tag{4.54}$$

Here we write $[A - \lambda^{(0)} I]$ as $[A - \lambda^{(0)}]$ to simplify the notation. Thus we have an equation for $x^{(\nu)}$ in terms of $x^{(0)}, x^{(1)}, \cdots, x^{(\nu-1)}$, the lower-order functions. However, the right-hand side of the equation for $x^{(\nu)}$ also contains $\lambda^{(\nu)}$ which might seem to involve some difficulty, since $x^{(\nu)}$ cannot be found until we know $\lambda^{(\nu)}$. However, if we take the inner product of both sides of Eq. (4.54) with $x^{(0)}$, we get

$$(x^{(0)}, [A_0 - \lambda^{(0)}] x^{(\nu)}) = \sum_{\mu=1}^{\nu} \lambda^{(\mu)} (x^{(0)}, x^{(\nu-\mu)}) - (x^{(0)}, A_1 x^{(\nu-1)}) .$$

Using Eq. (4.50) and the fact that A_0 is Hermitian, we see that the left-hand side vanishes, so

$$\lambda^{(\nu)} = (x^{(0)}, A_1 x^{(\nu-1)}) - \sum_{\mu=1}^{\nu-1} \lambda^{(\mu)} (x^{(0)}, x^{(\nu-\mu)})$$

for $\nu = 1, 2, \cdots$. Hence $\lambda^{(\nu)}$ is determined in terms of the $\lambda^{(i)}$ and $x^{(i)}$ of orders *lower* than ν.

Thus, in principle, we should be able to solve the hierarchy of equations implied by Eq. (4.54), with the first $(n - 1)$ equations providing the input for the nth equation. The first few equations of the sequence are

$$[A_0 - \lambda^{(0)}]x^{(0)} = 0\,,$$

$$[A_0 - \lambda^{(0)}]x^{(1)} = -[A_1 - \lambda^{(1)}]x^{(0)}\,, \tag{4.55a}$$

$$[A_0 - \lambda^{(0)}]x^{(2)} = -[A_1 - \lambda^{(1)}]x^{(1)} + \lambda^{(2)}x^{(0)}\,, \tag{4.55b}$$

$$[A_0 - \lambda^{(0)}]x^{(3)} = -[A_1 - \lambda^{(1)}]x^{(2)} + \lambda^{(2)}x^{(1)} + \lambda^{(3)}x^{(0)}\,. \tag{4.55c}$$

The first equation is just Eq. (4.50), and provides the starting point for solving all the others, order by order.

All of the Eqs. (4.55) except for the first are of the type $Hx = h$, discussed in Theorem 3.17. Since $\lambda^{(0)}$ is an eigenvalue of A_0, $\det H = 0$. Therefore h must lie entirely within the subspace spanned by the eigenvectors of H belonging to nonzero eigenvalues, that is, h must be orthogonal to all the eigenvectors of H belonging to a zero eigenvalue. In this case, h must be orthogonal to all eigenvectors of A_0 belonging to $\lambda^{(0)}$.

It is here that we see the first practical difference between the case of degenerate eigenvalues and nondegenerate ones. If we have only nondegenerate eigenvalues in mind, then there is only *one* eigenvector to which h must be orthogonal. We shall discuss this case in some detail first, since its bookkeeping is simplest.

Let us assume, then, that the spectrum of A_0 is nondegenerate, and imagine that we have determined a particular $x^{(0)}$ from Eq. (4.50). The discussion of the preceding paragraph tells us that Eq. (4.55a) can give a solution for $x^{(1)}$ only if

$$(x^{(0)}, [A_1 - \lambda^{(1)}]x^{(0)}) = 0\,.$$

Since we may choose $x^{(0)}$ to be normalized to 1, this yields

$$\lambda^{(1)} = (x^{(0)}, A_1 x^{(0)})\,. \tag{4.56a}$$

If we apply the same requirement to Eq. (4.55b), we have

$$\lambda^{(2)} = (x^{(0)}, [A_1 - \lambda^{(1)}]x^{(1)})\,. \tag{4.56b}$$

This gives us $\lambda^{(2)}$ in terms of quantities which are, in principle, known. Similarly, Eq. (4.55c) has a well-defined solution provided that

$$\lambda^{(3)} = (x^{(0)}, [A_1 - \lambda^{(1)}]x^{(2)}) - \lambda^{(2)}(x^{(0)}, x^{(1)})\,.$$

This is just like our equations for $\lambda^{(1)}$ and $\lambda^{(2)}$, but here it turns out that we can do better. Using the Hermiticity of A_1, we can write

$$(x^{(0)}, [A_1 - \lambda^{(1)}]x^{(2)}) = ([A_1 - \lambda^{(1)}]x^{(0)}, x^{(2)})\,.$$

Using Eq. (4.55a), we get

$$\begin{aligned}(x^{(0)}, [A_1 - \lambda^{(1)}]x^{(2)}) &= -([A_0 - \lambda^{(0)}]x^{(1)}, x^{(2)}) \\ &= -(x^{(1)}, [A_0 - \lambda^{(0)}]x^{(2)})\,,\end{aligned}$$

again using the Hermiticity of A_0. Finally, Eq. (4.55b) yields

$$(x^{(0)}, [A_1 - \lambda^{(1)}]x^{(2)}) = (x^{(1)}, [A_1 - \lambda^{(1)}]x^{(1)}) - \lambda^{(2)}(x^{(1)}, x^{(0)}) .$$

Putting this into the equation for $\lambda^{(3)}$, we obtain

$$\lambda^{(3)} = (x^{(1)}, [A_1 - \lambda^{(1)}]x^{(1)}) - \lambda^{(2)} [(x^{(0)}, x^{(1)}) + (x^{(1)}, x^{(0)})] . \tag{4.56c}$$

Thus $\lambda^{(3)}$ is determined only by $x^{(0)}$ and $x^{(1)}$, but not by $x^{(2)}$. This is a general feature of perturbation theory. In fact, each $x^{(i)}$ determines two terms in the expansion of λ. We have already seen that $\lambda^{(0)}$ and $\lambda^{(1)}$ are found by just knowing $x^{(0)}$, and $\lambda^{(2)}$ and $\lambda^{(3)}$ are determined by $x^{(0)}$ and $x^{(1)}$. It is a fairly simple exercise to show that $\lambda^{(4)}$ and $\lambda^{(5)}$ can be obtained from $x^{(0)}$, $x^{(1)}$, and $x^{(2)}$, and similarly for terms of higher order.

We may simplify Eqs. (4.56) slightly by noticing that we may choose $(x^{(0)}, x^{(i)}) = 0$, for $i > 0$. This is true because if any one of Eqs. (4.55) for $x^{(i)}$ has some solution $y^{(i)}$, then $y^{(i)} + ax^{(0)}$ is also a solution, because $(A_0 - \lambda^{(0)})x^{(0)} = 0$. Therefore we can always arrange that

$$(x^{(0)}, x^{(i)}) = (x^{(0)}, y^{(i)} + ax^{(0)}) = (x^{(0)}, y^{(i)}) + a = 0$$

by taking as the solutions the vectors $y^{(i)} + ax^{(0)}$, where $a = -(x^0, y^{(i)})$. Then Eqs. (4.56) become

$$\lambda^{(1)} = (x^{(0)}, A_1 x^{(0)}) , \tag{4.57a}$$

$$\lambda^{(2)} = (x^{(0)}, A_1 x^{(1)}) , \tag{4.57b}$$

$$\lambda^{(3)} = (x^{(1)}, [A_1 - \lambda^{(1)}]x^{(1)}) . \tag{4.57c}$$

We should emphasize that these values of $\lambda^{(1)}$, $\lambda^{(2)}$, and $\lambda^{(3)}$ are obtained by requiring that well-defined solutions exist for Eqs. (4.55). With the λ's now given by Eqs. (4.57), we are guaranteed that such solutions do indeed exist. The only task remaining is to see about solving Eqs. (4.55), for which there are several methods available.

The most obvious method would be to solve the matrix equations directly. However, since the main application of perturbation theory is to operators on *infinite*-dimensional spaces, where there is no simple matrix representation, we shall say no more about this direct method other than that it exists and is in principle easy to use on a finite-dimensional space.

If we were fortunate enough to know the complete set of eigenfunctions and eigenvalues of the Hermitian operator A_0, there would be another possibility: we could expand the $x_n^{(i)}$ $(i = 1, 2, \cdots)$ in terms of the eigenfunctions of A_0, say $x_m^{(0)}$. We have now restored the subscript n which indexes *different* eigenvalues and eigenvectors. As an example, we compute $x_n^{(1)}$ in terms of the $x_m^{(0)}$; it is a simple matter to extend this technique to obtain $x_n^{(i)}$, for $i \geq 2$.

First, we assume that $m \neq n$, and take the inner product of both sides of Eq. (4.55a) on the left with $x_m^{(0)}$. Thus we get

$$(x_m^0, [A_0 - \lambda_n^{(0)}]x_n^{(1)}) = -(x_m^{(0)}, [A_1 - \lambda_n^{(1)}]x_n^{(0)}) . \tag{4.58}$$

Since the eigenfunctions of A_0 form an orthonormal set, the right-hand side of Eq. (4.58) is just

$$-(x_m^{(0)}, A_1 x_n^{(0)}) .$$

On the left-hand side of Eq. (4.58), we use the Hermiticity of A_0 and the fact that $x_m^{(0)}$ is an eigenfunction of A_0 with eigenvalue $\lambda_m^{(0)}$ to write it as

$$(\lambda_m^{(0)} - \lambda_n^{(0)})(x_m^{(0)}, x_n^{(1)}) .$$

Then Eq. (4.58) becomes

$$(\lambda_n^{(0)} - \lambda_m^{(0)})(x_m^{(0)}, x_n^{(1)}) = (x_m^{(0)}, A_1 x_n^{(0)}) , \qquad m \neq n .$$

Since there is no degeneracy, $\lambda_n^{(0)} \neq \lambda_m^{(0)}$ so long as $n \neq m$. Therefore

$$(x_m^{(0)}, x_n^{(1)}) = \frac{(x_m^{(0)}, A_1 x_n^{(0)})}{\lambda_n^{(0)} - \lambda_m^{(0)}} \tag{4.59}$$

for $m \neq n$. For $m = n$, we have already decided to take $(x_n^{(0)}, x_n^{(1)}) = 0$. We now use the completeness of the eigenfunctions of A_0 to write

$$x_n^{(1)} = \sum_m (x_m^{(0)}, x_n^{(1)}) x_m^{(0)} .$$

For our particular case, this is just

$$x_n^{(1)} = \sum_{m \neq n} \frac{(x_m^{(0)}, A_1 x_n^{(0)})}{\lambda_n^{(0)} - \lambda_m^{(0)}} x_m^{(0)} . \tag{4.60}$$

This gives us $x_n^{(1)}$; the same method applied to Eqs. (4.55b) and (4.55c) would yield $x_n^{(2)}$ and $x_n^{(3)}$. Note that according to Eqs. (4.57b) and (4.57c), if we know $x_n^{(0)}$ and $x_n^{(1)}$, we can find $\lambda_n^{(2)}$ and $\lambda_n^{(3)}$.

We can illustrate all these operations with the following simple example. We showed in Example 1 of Section 3.10 that the Hermitian Pauli matrix

$$\sigma_x = \begin{bmatrix} 0 & 1 \\ 1 & 0 \end{bmatrix}$$

has eigenvalues $\lambda_1^{(0)} = +1$, $\lambda_2^{(0)} = -1$ and normalized eigenvectors

$$x_1^{(0)} = \frac{1}{\sqrt{2}}\begin{bmatrix} 1 \\ 1 \end{bmatrix}, \qquad x_2^{(0)} = \frac{1}{\sqrt{2}}\begin{bmatrix} 1 \\ -1 \end{bmatrix}.$$

Using nondegenerate perturbation theory, we can compute the eigenvectors of the matrix

$$\sigma = \begin{bmatrix} \epsilon & 1-\epsilon \\ 1-\epsilon & 3\epsilon \end{bmatrix} = \sigma_x + \epsilon \begin{bmatrix} 1 & -1 \\ -1 & 3 \end{bmatrix}$$

through first order and thus obtain the eigenvalues through third order in ϵ.

Since

$$A_1 = \begin{bmatrix} 1 & -1 \\ -1 & 3 \end{bmatrix},$$

$\lambda^{(1)}$ is given by

$$\lambda_1^{(1)} = (x_1^{(0)}, A_1 x_1^{(0)}) = \frac{1}{\sqrt{2}}[1 \quad 1]\begin{bmatrix} 1 & -1 \\ -1 & 3 \end{bmatrix}\begin{bmatrix} 1 \\ 1 \end{bmatrix}\frac{1}{\sqrt{2}}$$

$$= \frac{1}{2}[1 \quad 1]\begin{bmatrix} 0 \\ 2 \end{bmatrix} = 1 ,$$

$$\lambda_2^{(1)} = (x_2^{(0)}, A_1 x_2^{(0)}) = \frac{1}{\sqrt{2}}[1 \quad -1]\begin{bmatrix} 1 & -1 \\ -1 & 3 \end{bmatrix}\begin{bmatrix} 1 \\ -1 \end{bmatrix}\frac{1}{\sqrt{2}}$$

$$= \frac{1}{2}[1 \quad -1]\begin{bmatrix} 2 \\ -4 \end{bmatrix} = 3 .$$

Therefore, through first order,

$$\lambda_1 = 1 + \epsilon , \qquad \lambda_2 = -1 + 3\epsilon .$$

Now, to get the correction to the eigenvector through first order, we use Eq. (4.60):

$$x_1^{(1)} = \sum_{m \neq 1} \frac{(x_m^{(0)}, A_1 x_1^{(0)})}{\lambda_1^{(0)} - \lambda_m^{(0)}} x_m^{(0)} = \frac{(x_2^{(0)}, A_1 x_1^{(0)})}{\lambda_1^{(0)} - \lambda_2^{(0)}} x_2^{(0)}$$

$$= \frac{1}{1-(-1)}\frac{1}{\sqrt{2}}[1 \quad -1]\begin{bmatrix} 1 & -1 \\ -1 & 3 \end{bmatrix}\begin{bmatrix} 1 \\ 1 \end{bmatrix}\frac{1}{\sqrt{2}} x_2^{(0)}$$

$$= -\frac{1}{2} x_2^{(0)} = -\frac{1}{2\sqrt{2}}\begin{bmatrix} 1 \\ -1 \end{bmatrix} .$$

Similarly,

$$x_2^{(1)} = +\frac{1}{2} x_1^{(0)} = \frac{1}{2\sqrt{2}}\begin{bmatrix} 1 \\ 1 \end{bmatrix} .$$

Then, through first order in ϵ,

$$x_1 = \frac{1}{\sqrt{2}}\begin{bmatrix} 1 - \epsilon/2 \\ 1 + \epsilon/2 \end{bmatrix} , \qquad x_2 = \frac{1}{\sqrt{2}}\begin{bmatrix} 1 + \epsilon/2 \\ -1 + \epsilon/2 \end{bmatrix} . \tag{4.61}$$

With $x_1^{(1)}$ and $x_2^{(1)}$, we can now find the second and third order corrections to λ_1 and λ_2. Using Eq. (4.57b), we get

$$\lambda_1^{(2)} = (x_1^{(0)}, A_1 x_1^{(1)}) = \frac{1}{\sqrt{2}}[1 \quad 1]\begin{bmatrix} 1 & -1 \\ -1 & 3 \end{bmatrix}\begin{bmatrix} 1 \\ -1 \end{bmatrix}\frac{(-1)}{2\sqrt{2}}$$

$$= -\frac{1}{4}[1 \quad 1]\begin{bmatrix} 2 \\ -4 \end{bmatrix} = +\frac{1}{2} ,$$

$$\lambda_2^{(2)} = (x_2^{(0)}, A_1 x_2^{(1)}) = \frac{1}{\sqrt{2}}[1 \quad -1]\begin{bmatrix} 1 & -1 \\ -1 & 3 \end{bmatrix}\begin{bmatrix} 1 \\ 1 \end{bmatrix}\frac{1}{2\sqrt{2}}$$

$$= \frac{1}{4}[1 \quad -1]\begin{bmatrix} 0 \\ 2 \end{bmatrix} = -\frac{1}{2} .$$

Also, using Eqs. (4.57c), we obtain

$$\begin{aligned}\lambda_1^{(3)} &= (x_1^{(1)}, A_1 x_1^{(1)}) - \lambda_1^{(1)}(x_1^{(1)}, x_1^{(1)}) \\ &= \frac{-1}{2\sqrt{2}}[1 \quad -1]\begin{bmatrix} 1 & -1 \\ -1 & 3\end{bmatrix}\begin{bmatrix} 1 \\ -1\end{bmatrix}\frac{(-1)}{2\sqrt{2}} \\ &\quad - 1\frac{(-1)}{2\sqrt{2}}(1 \quad -1)\begin{bmatrix} 1 \\ -1\end{bmatrix}\frac{(-1)}{2\sqrt{2}} \\ &= \frac{3}{4} - \frac{1}{4} = \frac{1}{2}.\end{aligned}$$

Similarly, $\lambda_2^{(3)} = -\frac{1}{2}$. Combining all these results, we find, through third order in ϵ, that

$$\begin{aligned}\lambda_1 &= 1 + \epsilon + \tfrac{1}{2}\epsilon^2 + \tfrac{1}{2}\epsilon^3 + \cdots, \\ \lambda_2 &= -1 + 3\epsilon - \tfrac{1}{2}\epsilon^2 - \tfrac{1}{2}\epsilon^3 + \cdots\end{aligned} \tag{4.62}$$

Our final results, Eqs. (4.61) and (4.62), can easily be checked by diagonalizing A directly, that is, by solving

$$\begin{bmatrix} \epsilon & 1-\epsilon \\ 1-\epsilon & 3\epsilon\end{bmatrix} x_n = \lambda_n x_n .$$

Solving the characteristic equation gives

$$\lambda_1 = 2\epsilon + \sqrt{1 - 2(\epsilon - \epsilon^2)}\,, \qquad \lambda_2 = 2\epsilon - \sqrt{1 - 2(\epsilon - \epsilon^2)}\,,$$

and if we expand these in ϵ by using the binomial theorem, we reproduce Eq. (4.62) exactly through third order. We leave it as an exercise to show that the eigenvectors of A, which can now be calculated exactly, since λ_1 and λ_2 are known exactly, agree through first order with Eq. (4.61).

We see that in this case the radius of convergence of perturbation theory is given by the radius of convergence of the binomial expansion of $\sqrt{1 - 2(\epsilon - \epsilon^2)}$. In general, it is very difficult to give a rigorous discussion of this type of perturbation theory, which is often called Rayleigh-Schrodinger perturbation theory, after its principal developers.

4.12 PERTURBATION THEORY—DEGENERATE CASE

In the last section we found that if the zero-order operator, A_0, has a nondegenerate spectrum, the basic results of perturbation theory follow straightforwardly from the completeness of the eigenfunctions of A_0 and Theorem 3.17. We shall now show that the same situation holds when A_0 possesses degenerate eigenvalues. The only real difficulty in this case will be a notational one.

It is tempting to think that we should immediately replace Eq. (4.50) by

$$[A_0 - \lambda_n^{(0)}]\, x_{n,i}^{(0)} = 0\,,$$

where $i = 1, 2, \cdots, \mu_n$, μ_n being the multiplicity of the nth eigenvalue, and then proceed to rewrite Eq. (4.55a) as

$$[A_0 - \lambda_n^{(0)}]\, x_{n,i}^{(1)} = -[A_1 - \lambda_{n,i}^{(1)}]\, x_{n,i}^{(0)} .$$

But at this point we run into the difficulty that, for a solution to exist, Theorem 3.17 requires not only that

$$(x_{n,i}^{(0)}, [A_1 - \lambda_{n,i}^{(1)}]\, x_{n,i}^{(0)}) = 0 ,$$

but also, since $\lambda_n^{(0)}$ has multiplicity μ_n, that

$$(x_{n,j}^{(0)}, [A_1 - \lambda_{n,i}^{(1)}]\, x_{n,i}^{(0)}) = 0 \qquad \text{for all } j .$$

If $j \neq i$, this last equation becomes

$$(x_{n,i}^{(0)}, A_1 x_{n,j}^{(0)}) = 0 ,$$

which looks like a set of restrictions on our perturbation operator, A_1. This is clearly an inadmissible situation. At this point, however, the degeneracy difficulty actually comes to our rescue. What we have forgotten is that the statement "$x_{n,i}^{(0)}$ $(i = 1, 2, \cdots, \mu_n)$ are *the* eigenvectors of A_0 corresponding to $\lambda_n^{(0)}$" is not quite precise, because, in fact, *any* linear combination of the $x_{n,i}^{(0)}$ $(i = 1, 2, \cdots \mu_n)$ is also a perfectly good zero-order solution.

Therefore let us take as zero-order solutions the set $y_{n,i}^{(0)}$ $(i = 1, 2, \cdots, \mu_n)$, where

$$y_{n,i}^{(0)} = \sum_{j=1}^{\mu_n} a_j^{(i)} x_{n,j}^{(0)} . \tag{4.63}$$

We shall see that the $a_i^{(j)}$ can be determined in such a way that we shall not run into difficulty in calculating corrections to $\lambda_n^{(0)}$.

If we require that $(y_{n,j}^{(0)}, y_{n,k}^{(0)}) = \delta_{jk}$, then we must have

$$\sum_i a_i^{(j)*} a_i^{(k)} = \delta_{jk} . \tag{4.64}$$

Thus, instead of Eq. (4.50), we write

$$[A_0 - \lambda_n^{(0)}] y_{n,i}^{(0)} = 0 \tag{4.50'}$$

and replace Eq. (4.52) by

$$\begin{aligned} \lambda_{n,i} &= \lambda_n^{(0)} + \epsilon \lambda_{n,i}^{(1)} + \epsilon^2 \lambda_{n,i}^{(2)} + \cdots, \\ y_{n,i} &= y_{n,i}^{(0)} + \epsilon y_{n,i}^{(1)} + \epsilon^2 y_{n,i}^{(2)} + \cdots, \end{aligned} \tag{4.52'}$$

where $y_{n,i}^{(0)}$ is given by Eq. (4.63). What we are saying is that in solving Eq. (4.50), we may well have found a set of eigenfunctions $x_{n,i}^{(0)}$ which are acceptable as solutions of Eq. (4.50), but, are nevertheless not ideal for handling higher order corrections. We want to find those linear combinations of the $x_{n,i}^{(0)}$ which will do the job. Also, in labeling our eigenvalues, we have included a degeneracy index i on all terms except $\lambda_n^{(0)}$, since we expect that an arbitrary perturbation will split the degeneracy to some extent. Using Eq. (4.52′), we find, just as in the last section, the analogs of Eqs. (4.55):

$$[A_0 - \lambda_n^{(0)}]y_{n,i}^{(1)} = -[A_1 - \lambda_{n,i}^{(1)}]y_{n,i}^{(0)}, \tag{4.55a′}$$

$$[A_0 - \lambda_n^{(0)}]y_{n,i}^{(2)} = -[A_1 - \lambda_{n,i}^{(1)}]y_{n,i}^{(1)} + \lambda_{n,i}^{(2)}y_{n,i}^{(0)}, \tag{4.55b′}$$

$$[A_0 - \lambda_n^{(0)}]y_{n,i}^{(3)} = -[A_1 - \lambda_{n,i}^{(1)}]y_{n,i}^{(2)} + \lambda_{n,i}^{(2)}y_{n,i}^{(1)} + \lambda_{n,i}^{(3)}y_{n,i}^{(0)}. \tag{4.55c′}$$

We now reap the rewards of our reformulation of the problem. For a solution to Eq. (4.55a′) to exist, we require that the vector on the right-hand side of Eq. (4.55a′) have no component in the subspace spanned by the set $\{x_{n,i}^{(0)}\}$ (which is, of course, the same as the subspace spanned by the $\{y_{n,i}^{(0)}\}$). In other words, we require that

$$(x_{n,j}^{(0)}, [A_1 - \lambda_{n,i}^{(1)}]y_{n,i}^{(0)}) = 0 \qquad \text{for all } i, j.$$

But by Eq. (4.63), this is just

$$\sum_k (x_{n,j}^{(0)}, [A_1 - \lambda_{n,i}^{(1)}]x_{n,k}^{(0)})a_k^{(i)} = 0$$

or

$$\sum_k [(x_{n,j}^{(0)}, A_1 x_{n,k}^{(0)}) - \lambda_{n,i}^{(1)}\delta_{jk}]a_k^{(i)} = 0, \tag{4.65a}$$

which is a simple eigenvalue problem of the type we have been discussing in this chapter. This is particularly clear if we define a matrix M whose jk element is given by

$$M_{jk} = (x_{n,j}^{(0)}, A_1 x_{n,k}^{(0)})$$

and a vector $a^{(i)}$ whose kth element is $a_k^{(i)}$. Then Eq. (4.65a) can be written as

$$(M - \lambda_{n,i}^{(1)})a^{(i)} = 0. \tag{4.65b}$$

Since A_1 is Hermitian, so is M, and Eq. (4.65b) is therefore a Hermitian eigenvalue problem. Thus $\lambda_{n,i}^{(i)}$ will be real, and the $a^{(i)}$ can be chosen so that

$$(a^{(i)}, a^{(j)}) = \delta_{ij},$$

in keeping with the requirement of Eq. (4.64). Thus, all at once, we have: (1) found the first corrections to $\lambda_n^{(0)}$, (2) determined the correct linear combinations of the $x_{n,i}^{(0)}$, and (3) guaranteed that Eq. (4.55a′) will have a solution.

In summary, our procedure is the following: We calculate the matrix of A_1 in the subspace spanned by the set $\{x_{n,i}^{(0)}, i = 1, 2, \cdots, \mu_n\}$. We then find the eigenvalues and orthonormal eigenvectors of this $\mu_n \times \mu_n$ matrix. The eigenvalues are the first-order corrections to $\lambda_n^{(0)}$, and the eigenvectors provide the correct linear combinations of the $y_{n,i}^{(0)}$ from Eq. (4.63).

At this point Eq. (4.55a′) can be solved for $y_{n,i}^{(1)}$. For example, if we know *all* the eigenvectors and eigenvalues of A_0, then by the expansion method discussed in Section 4.11, we get, for $m \neq n$,

$$(x_{m,i}^{(0)}, y_{n,j}^{(1)}) = \frac{(x_{m,i}^{(0)}, A_1 y_{n,j}^{(0)})}{\lambda_n^{(0)} - \lambda_m^{(0)}}. \tag{4.66}$$

Note, however, that we do not yet have the μ_n components of $y_{n,j}^{(1)}$ along $x_{n,i}^{(0)}$ $(i = 1, 2, \cdots, \mu_n)$, that is, we do not know $(x_{n,i}^{(0)}, y_{n,j}^{(1)})$. Therefore we cannot yet use the expansion theorem.

In the nondegenerate case, where only one component of the first-order eigenvector was missing, we found that we could, with no risk, set the missing component equal to zero. This is reasonable since we have one loose degree of freedom associated with the normalization of the exact eigenvector (which in physical problems is usually chosen to be normalized to unity). But in the present degenerate case, we will have μ_n components missing, and we clearly cannot get away with setting them all equal to zero.

However, it turns out that the same situation which enabled us to resolve the ambiguity in the choice of zero-order wavefunction will rescue us here. When we go to Eq. (4.55b′) to determine $\lambda_{n,i}^{(2)}$, we shall find in the process the missing components of $y_{n,i}^{(1)}$.

For Eq. (4.55b′) to have a solution, the vectors on the right-hand side must have no component lying in the subspace spanned by the $\{x_{n,j}^{(0)}, j = 1, 2, \cdots, \mu_n\}$, that is,

$$(x_{n,j}^{(0)}, [A_1 - \lambda_{n,i}^{(1)}]y_{n,i}^{(1)}) - \lambda_{n,i}^{(2)}(x_{n,j}^{(0)}, y_{n,i}^{(0)}) = 0$$

for all i and j. Using Eq. (4.63), we obtain

$$a_j^{(i)}\lambda_{n,i}^{(2)} = (x_{n,j}^{(0)}, [A_1 - \lambda_{n,i}^{(1)}]y_{n,i}^{(1)}) \ . \tag{4.67}$$

Now we split $y_{n,i}^{(1)}$ into two parts,

$$y_{n,i}^{(1)} = f_{n,i} + g_{n,i} \ ,$$

when $f_{n,i}$ is the unknown part of $y_{n,i}^{(1)}$ lying in the subspace spanned by the set $\{x_{n,j}^{(0)}, j = 1, 2, \cdots \mu_n\}$ and $g_{n,i}$ is the rest, which according to Eq. (4.66) we already know. In this way, Eq. (4.67) can be rewritten as

$$(x_{n,j}^{(0)}, [A_1 - \lambda_{n,i}^{(1)}]f_{n,i}) = -(x_{n,j}^{(0)}, A_1 g_{n,i}) + a_j^{(i)}\lambda_{n,i}^{(2)} \ . \tag{4.68}$$

Finally, we write $f_{n,i}$ as

$$f_{n,i} = \sum_k b_k^{(i)} x_{n,k}^{(0)} \ , \tag{4.69}$$

and substituting into Eq. (4.68), we get

$$\sum_k (x_{n,j}^{(0)}, [A_1 - \lambda_{n,i}^{(1)}]\, x_{n,k}^{(0)}) b_k^{(i)} = \lambda_{n,i}^{(2)} a_j^{(i)} - (x_{n,j}^{(0)}, A_1 g_{n,i}) \qquad \text{for all } i, j. \tag{4.70a}$$

If the reader will look carefully at Eq. (4.70a), he will see, through the forest of subscripts and superscripts, that it is a simple matrix equation. In fact, if we define a vector $b^{(i)}$ whose components are the μ_n scalars $b_k^{(i)}$ and a vector $a^{(i)}$ whose components are the $a_k^{(i)}$, then Eq. (4.70a) can be written as

$$(M - \lambda_{n,i}^{(1)})b^{(i)} = \lambda_{n,i}^{(2)}a^{(i)} - c^{(i)} \tag{4.70b}$$

for all i, where M is the matrix, defined earlier, whose jk element is $(x_{n,j}^{(0)}, A_1 x_{n,k}^{(0)})$ and $c^{(i)}$ is a vector whose jth component is $(x_{n,j}^{(0)}, A_1 g_{n,i})$. Thus the vectors $b^{(i)}$ are determined by solving a simple matrix equation of the type $Hx = h$. Again according to Eq. (4.70b), H is singular, since it was precisely by diagonalizing the matrix on the left-hand side of Eq. (4.70b) that we determined the $\lambda_{n,i}^{(1)}$ in Eq. (4.65b). Hence our very useful Theorem 3.17 tells us that for a solution to exist, the right-hand side of Eq. (4.70b) must be orthogonal to all the eigenvectors of Eq. (4.65b) belonging to eigenvalue $\lambda_{n,i}^{(1)}$. Let us suppose that perturbation completely removed the degeneracy in first order so none of the $\lambda_{n,i}^{(1)}$ $(i = 1, 2, \cdots, \mu_n)$ are equal. Then the only eigenvector belonging to $\lambda_{n,i}^{(1)}$ is $a^{(i)}$; by the above remarks, a solution to Eq. (4-70b) exists if

$$\lambda_{n,i}^{(2)}(a^{(i)}, a^{(i)}) - (a^{(i)}, c^{(i)}) = 0$$

or in component form,

$$\sum_j a_j^{(i)*}[a_j^{(i)}\lambda_{n,i}^{(2)} - (x_{n,j}^{(0)}, A_1 g_{n,i})] = 0 .$$

Making use of Eq. (4.63) and Eq. (4.64), this reduces to

$$\lambda_{n,i}^{(2)} = (y_{n,i}^{(0)}, A_1 g_{n,i}) . \tag{4.71}$$

If we know all the eigenvectors of A_0 and the corresponding eigenvalues, then by Eq. (4.66),

$$g_{n,j} = \sum_{m \neq n} \sum_{i=1}^{\mu_m} \frac{(x_{m,i}^{(0)}, A_1 y_{n,j}^{(0)})}{\lambda_n^{(0)} - \lambda_m^{(0)}} x_{m,i}^{(0)} , \tag{4.66$'$}$$

so $\lambda_{n,i}^{(2)}$ is determined. We should emphasize that Eq. (4.71) bears a very close resemblance to Eq. (4.57b) of the last section.

Here again, the equation which determines $\lambda_{n,i}^{(2)}$ guarantees the existence of solutions to Eq. (4.70b) for $b^{(i)}$ $(i = 1, 2, \cdots, \mu_n)$, so the missing components of $y_{n,i}^{(1)}$ are found, although the component of $b^{(i)}$ in the direction $a^{(i)}$ is, by Theorem 3.17, undetermined. (It is simple to show that this corresponds to the fact that $(y_{n,i}^{(0)}, y_{n,i}^{(1)})$ is indeterminate.) It is often convenient to pick this inner product equal to zero, which is equivalent to taking $b^{(i)}$ orthogonal to $a^{(i)}$. Finally, with $\lambda_{n,i}^{(2)}$ and $b^{(i)}$ determined in this manner for all i, Theorem 3.17 guarantees that a solution exists to Eq. (4.55b$'$), so $y_{n,i}^{(2)}$ is now known, again with indeterminacy along the directions defined by $\{x_{n,i}^{(0)}, i = 1, 2, \cdots, \mu_n\}$. It is possible, but tedious, to carry on to Eq. (4.51c$'$), remove the indeterminacy in $y_{n,i}^{(2)}$, and at the same time find $\lambda_{n,i}^{(3)}$.

In going through second order, we have assumed that the degeneracy was completely removed in first order. This assumption is *unnecessary*, and we leave it as an exercise in notation for the interested reader to show that if there is some degeneracy left after first order, we must solve an eigenvalue problem for those $\lambda^{(2)}$'s which correspond to degenerate $\lambda^{(1)}$'s, instead of a simple equation for $\lambda_{n,i}^{(2)}$. Again we emphasize that this is a purely notational problem, identical to that encountered in going from nondegenerate to degenerate perturbation theory. In principle, the entire Rayleigh-Schrodinger perturbation theory stems from the existence of an expansion of the type given in Eq. (4.52), followed by repeated application of Theorem 3.17.

Example. As an application of degenerate perturbation theory, consider the linear operator

$$A_0 = \frac{p_x^2}{2m} + \frac{1}{2} m\omega_0^2 x^2 + \frac{p_y^2}{2m} + \frac{1}{2} m\omega_0^2 y^2 . \tag{4.72}$$

This is a sum of two independent linear operators of the type discussed in Section 3.10. We impose the commutation relations

$$[x, p_x] = i\hbar I , \qquad [y, p_y] = i\hbar I ,$$

and all other commutators (e.g., $[x, p_y]$) vanish. As in Section 3.10, we define

$$\begin{aligned} a_x &= (xm\omega_0 + ip_x)(2m\omega_0\hbar)^{-1/2} , & a_y &= (ym\omega_0 + ip_y)(2m\omega_0\hbar)^{-1/2} , \\ a_x^\dagger &= (xm\omega_0 - ip_x)(2m\omega_0\hbar)^{-1/2} , & a_y^\dagger &= (ym\omega_0 - ip_y)(2m\omega_0\hbar)^{-1/2} . \end{aligned} \tag{4.73}$$

An easy computation, like that leading to Eq. (3.47) shows that

$$A_0 = \hbar\omega_0(a_x^\dagger a_x + a_y^\dagger a_y + I) \tag{4.74}$$

and

$$[a_x, a_x^\dagger] = I = [a_y, a_y^\dagger] ,$$

$$[a_x, a_y] = [a_x^\dagger, a_y^\dagger] = [a_x, a_y^\dagger] = [a_x^\dagger, a_y] = 0 .$$

It follows that

$$[a_i^\dagger, A_0] = -\hbar\omega_0 a_i^\dagger , \qquad [a_i, A_0] = \hbar\omega_0 a_i ,$$

for $i = x$ and y. By methods completely identical to those of Section 3.10, it then follows that the eigenfunctions of A_0 are given by

$$\phi_{n,m} = (a_x^\dagger)^n (a_y^\dagger)^m \phi_{0,0} / \sqrt{n!m!} , \tag{4.75a}$$

with corresponding eigenvalues

$$\lambda_N^{(0)} = \hbar\omega_0(n + m + 1) = \hbar\omega_0(N + 1) . \tag{4.75b}$$

If we require that $(\phi_{0,0}, \phi_{0,0}) = 1$, then

$$(\phi_{n,m}, \phi_{\nu,\mu}) = \delta_{n\nu}\delta_{m\mu} .$$

Also, with these definitions, we have

$$\begin{aligned} a_x^\dagger \phi_{n,m} &= (n + 1)^{1/2} \phi_{n+1,m} , & a_y^\dagger \phi_{n,m} &= (m + 1)^{1/2} \phi_{n,m+1} , \\ a_x \phi_{n,m} &= (n)^{1/2} \phi_{n-1,m} , & a_y \phi_{n,m} &= (m)^{1/2} \phi_{n,m-1} . \end{aligned} \tag{4.76}$$

Now suppose that A_0 is perturbed by a term of the form

$$\epsilon A_1 = \epsilon m\omega_0^2 xy .$$

In terms of the creation and annihilation operators, this is

$$\epsilon A_1 = \tfrac{1}{2} \epsilon\hbar\omega_0(a_x^\dagger a_y + a_x a_y^\dagger + a_x^\dagger a_y^\dagger + a_x a_y) . \tag{4.77}$$

The lowest eigenvalue of A_0 corresponds to $n = m = 0$: $\lambda_{0,0} = \hbar\omega_0$; it is nondegenerate. Thus the theory of Section 4.11 is applicable, and we leave it for the reader to show that calculating all terms through ϵ^2, we get

$$\lambda_{0,0} = \lambda_{0,0}^{(0)} + \epsilon\lambda_{0,0}^{(1)} + \epsilon^2\lambda_{0,0}^{(2)} + \cdots$$
$$\lambda_{0,0} = \hbar\omega_0(1 - \epsilon^2/8 + \cdots)\,. \tag{4.78}$$

The next eigenvalue is doubly degenerate, since according to Eqs. (4.75), there are two eigenvectors corresponding to the eigenvalue $2\hbar\omega_0$, namely,

$$\phi_{1,0} = a_x^\dagger\phi_{0,0}\,, \qquad \phi_{0,1} = a_y^\dagger\phi_{0,0}\,.$$

Therefore, to obtain corrections to this eigenvalue, we need degenerate perturbation theory. In keeping with the notation established earlier in this section, we write

$$x_{1,1}^{(0)} = \phi_{1,0}\,, \qquad x_{1,2}^{(0)} = \phi_{0,1}^{(0)}\,, \tag{4.79}$$

According to Eq. (4.65), we must find the matrix of A_1 in the subspace spanned by $x_{1,1}^{(0)}$ and $x_{1,2}^{(0)}$. Using Eqs. (4.76), (4.77), and (4.79), we find for this matrix

$$\tfrac{1}{2}\,\hbar\omega_0\begin{bmatrix}0 & 1\\ 1 & 0\end{bmatrix}.$$

Following Eq. (4.65b), we diagonalize it to get $\lambda_{1,1}^{(1)} = \frac{1}{2}\,\hbar\omega_0$, $\lambda_{1,2}^{(1)} = -\frac{1}{2}\,\hbar\omega_0$, with corresponding eigenvectors,

$$a^{(1)} = \frac{1}{\sqrt{2}}\begin{bmatrix}1\\ 1\end{bmatrix}, \qquad a^{(2)} = \frac{1}{\sqrt{2}}\begin{bmatrix}1\\ -1\end{bmatrix}.$$

Equation (4.63) then gives us the correct zero-order solutions:

$$y_{1,1}^{(0)} = \frac{1}{\sqrt{2}}\,(x_{1,1}^{(0)} + x_{1,2}^{(0)}) = \frac{1}{\sqrt{2}}\,(\phi_{1,0} + \phi_{0,1})\,,$$

$$y_{1,2}^{(0)} = \frac{1}{\sqrt{2}}\,(x_{1,1}^{(0)} - x_{1,2}^{(0)}) = \frac{1}{\sqrt{2}}\,(\phi_{1,0} - \phi_{0,1})\,.$$

These give us through first order in ϵ:

$$\lambda_{1,1} = 2\hbar\omega_0 + \tfrac{1}{2}\,\epsilon\hbar\omega_0\,, \qquad \lambda_{1,2} = 2\hbar\omega_0 - \tfrac{1}{2}\,\epsilon\hbar\omega_0\,.$$

With these results established, we now can solve Eq. (4.55a′) for $y_{1,1}^{(1)}$ and $y_{1,2}^{(1)}$. Since we know all the eigenvectors and eigenvalues of A_0, we use Eq. (4.66) to determine $g_{1,1}$ and $g_{1,2}$, the components of $y_{1,1}^{(1)}$ and $y_{1,2}^{(1)}$ which do not lie in the subspace spanned by $y_{1,1}^{(0)}$ and $y_{1,2}^{(0)}$. According to Eq. (4.71), the g's will determine $\lambda_{1,1}^{(2)}$ and $\lambda_{1,2}^{(2)}$. Using Eqs. (4.66′), (4.75), (4.76), (4.77), and (4.79), we get

$$g_{1,1} = \tfrac{1}{4}\,(\phi_{2,1} + \phi_{1,2})\,, \qquad g_{1,2} = \tfrac{1}{4}\,(\phi_{2,1} - \phi_{1,2})\,.$$

Thus, according to Eq. (4.71),

$$\begin{aligned}\lambda_{1,1}^{(2)} &= (y_{1,1}^{(0)}, A_1 g_{1,1})\\ &= \frac{1}{8\sqrt{2}}\,\hbar\omega_0([\phi_{1,0} + \phi_{0,1}], [a_x^\dagger a_y + a_x a_y^\dagger - a_x^\dagger a_y^\dagger - a_x a_y][\phi_{2,1} + \phi_{1,2}])\\ &= -\tfrac{1}{4}\,\hbar\omega_0 = \lambda_{1,2}^{(2)}\,.\end{aligned}$$

Through second order in ϵ, we have

$$\begin{aligned} \lambda_{1,1} &= \hbar\omega_0(2 + \tfrac{1}{2}\epsilon - \tfrac{1}{4}\epsilon^2 + \cdots) , \\ \lambda_{1,2} &= \hbar\omega_0(2 - \tfrac{1}{2}\epsilon - \tfrac{1}{4}\epsilon^2 + \cdots) , \end{aligned} \tag{4.80}$$

Because of the special form of the interaction, it is possible to solve this problem exactly by using methods very similar to those employed in the solution of the normal modes problem. One can easily show that the *exact* eigenvalues of $A_0 + \epsilon A_1$ are given by

$$\lambda_{\nu,\mu} = \hbar\omega_+(\nu + \tfrac{1}{2}) + \hbar\omega_-(\mu + \tfrac{1}{2}) ,$$

where $\omega_\pm = \omega_0\sqrt{1 \pm \epsilon}$. In particular, we get

$$\lambda_{0,0} = \frac{\hbar}{2}(\omega_+ + \omega_-) = \hbar\omega_0(1 - \epsilon^2/8 + \cdots) ,$$

$$\begin{aligned} \lambda_{1,0} &= \tfrac{3}{2}\hbar\omega_+ + \tfrac{1}{2}\hbar\omega_- = \hbar\omega_0(2 + \tfrac{1}{2}\epsilon - \tfrac{1}{4}\epsilon^2 + \cdots) , \\ \lambda_{0,1} &= \tfrac{1}{2}\hbar\omega_+ + \tfrac{3}{2}\hbar\omega_- = \hbar\omega_0(2 - \tfrac{1}{2}\epsilon - \tfrac{1}{4}\epsilon^2 + \cdots) , \end{aligned}$$

which correspond exactly to our Eqs. (4.78) and (4.80) with a slight change in notation.

Similar calculations apply to the higher levels. The reader may want to examine the triply degenerate states $\phi_{2,0}$, $\phi_{1,1}$, and $\phi_{0,2}$ which all have $\lambda^{(0)} = 3\hbar\omega_0$ (Problem 23). Solutions of the type given above are readily obtained for this case also.

PROBLEMS

1. Prove that if A is self-adjoint, then $\det A$ is real. (This is all one can say about the determinant of a self-adjoint matrix, whereas if B is unitary, then $|\det B| = 1$. Prove this also.)
2. Prove that if the matrices A and B are related by a similarity transformation, $A = P^{-1}BP$, and P is unitary, then the similarity transformation preserves Hermiticity.
3. Prove that the $n \times n$ matrix $(I - iA)$ has an inverse if A is Hermitian.
4. Let B be an $n \times n$ matrix. Define the $n \times n$ matrix $A = e^{iB}$ in terms of the power-series expansion of the exponential. That is,

   $$A = \sum_{n=0}^{\infty} \frac{(iB)^n}{n!} = I + iB + \frac{(iB)^2}{2!} + \cdots$$

   Assuming that the series converges, prove that A is isometric if B is self-adjoint. Also show that any isometry U can be written as $U = \exp(iH)$, where H is Hermitian. Obtain an expression for H in terms of quantities related to U. [*Hint*: the properties of the eigenvectors and eigenvalues of the operators will be helpful.]
5. Prove that the eigenvalues of any anti-self-adjoint matrix are imaginary.

6. Either prove or find a counterexample to the following statements:
   a) If A and B are $n \times n$ matrices, then $AB = 0$ implies that either $A = 0$ or $B = 0$.
   b) An $n \times n$ Hermitian matrix which is completely degenerate (all eigenvalues equal) is necessarily diagonal.
   c) If an $n \times n$ matrix is Hermitian, then all its powers are Hermitian.
   d) $\det(A + B) = \det A + \det B$.
   e) If A and B are $n \times n$ Hermitian matrices, then AB is Hermitian.
   f) $\operatorname{tr} AB = \operatorname{tr} A \operatorname{tr} B$ in a finite-dimensional space.
7. a) Prove that two vectors x and y in a complex inner-product space are orthogonal if and only if $\|\alpha x + \beta y\|^2 = \|\alpha x\|^2 + \|\beta y\|^2$ for all α and β. Does this result hold if we have just $\|x + y\|^2 = \|x\|^2 + \|y\|^2$?
   b) Prove that in any inner-product space,
   $$\|x - y\| \geq |\,\|x\| - \|y\|\,| .$$
   c) Prove the triangle inequality
   $$\|x + y\| \leq \|x\| + \|y\| .$$
8. In classical mechanics, the virial theorem states that the average value of the potential energy of a simple harmonic oscillator is equal to the average value of its kinetic energy. If $(\phi, A\phi)$ denotes the average value in the state ϕ of an observable A (assuming that ϕ is normalized to unity), show that this result is also true for the quantum-mechanical harmonic oscillator. Show that the polarizability of an electron bound by an harmonic force is the same as is found classically, namely e^2/k, where e is the electric charge and k is the spring constant. This means that the dielectric constant for a collection of such objects (density $= n$ per unit volume) at zero frequency is $4\pi ne^2/k$.
9. Define $\exp(A)$, where A is any matrix, by the power series expansion of $\exp(x) = 1 + x + \cdots$. Show that if there exists a matrix P which reduces A to diagonal form by a similarity transform, then one has
   $$\det \exp(A) = \exp \operatorname{tr}(A) .$$
10. Under what circumstances can you prove that AB and BA have the same eigenvalues? If they do have the same eigenvalues, what is the relation between the corresponding eigenvectors?
11. The set of linear transformations on some vector space can itself be considered as a vector space. Show that an inner product for this space is given by $(A, B) = \operatorname{tr}(A^\dagger B)$, where A and B are two linear operators on the space in question, that is, show that this satisfies all the axioms of inner-product spaces.
12. *The Lorentz group.*
    a) Evaluate the determinant of the Lorentz matrix. Comments?
    b) Prove that the set of Lorentz transformations in one velocity direction form a group.
13. *The relativistic law for the sum of two velocities.* Let frame K' move with uniform velocity v parallel to K and let K'' move with uniform velocity v' relative to K'. What is the velocity v'' of K'' relative to K? Prove that $v'' \leqq c$ if v and v' are $\leqq c$. Thus c is an upper limit for the relative velocity of any two frames.

    Suggested method: In proving closure in Problem 12, one finds that the product of one Lorentz transformation (complex orthogonal transformation through "angle" $\theta = \tanh^{-1}\beta$) and another Lorentz transformation (complex rotation through $\theta' = \tanh^{-1}\beta'$) is again a Lorentz transformation (through $\theta + \theta' = \theta'' = \tanh^{-1}\beta''$). Now solve for β'' (that is v'') in terms of v and v'.

14. Derive the matrix A of the general Lorentz transformation corresponding to relative velocity $v/c = \tanh\alpha$ in a direction given by the spherical angles ϕ, θ by a similarity transformation, $R^{-1}LR$, where R is the extension of the rotation matrix, $R(\phi, \theta)$, of Eq. (1.23) to a 4×4 matrix that acts only on the three space coordinates. Note that $R^{-1} = \tilde{R}$. Assume that the axes of K and K' remain parallel during their uniform relative motion. Check your answer to see whether it reduces properly in the 3 special cases: $\phi = \phi$, $\theta = 0$; $\phi = 0$, $\theta = \pi/2$; $\alpha = 0$.

15. Show that the general Lorentz transformation worked out in Problem 14 may be expressed in vector notation as follows:

$$\mathbf{r}' = \mathbf{r} + \frac{(\mathbf{v}\cdot\mathbf{r})\mathbf{v}}{v^2}(\gamma - 1) - \mathbf{v}t\gamma\,, \qquad t' = \left(t - \frac{\mathbf{v}\cdot\mathbf{r}}{c^2}\right)\gamma\,,$$

where $\gamma = 1/\sqrt{1-\beta^2}$, $\beta = v/c$, and $\mathbf{v}$ is the relative velocity of the two frames. These formulas are a compact, convenient way to write the Lorentz transformation between systems with parallel axes moving with arbitrarily directed uniform relative velocity $\mathbf{v}$.

16. a) Find the eigenvalues and eigenvectors of the Lorentz transformation. Verify that the eigenvectors are null vectors.
    b) Evaluate the determinant of the pure Lorentz matrix.
    c) Any other matrix representation of the Lorentz transformation can be found by performing a similarity transformation on the matrix A (Problem 14). What is the value of the determinant of *any* matrix representation of the Lorentz transformation?

17. The *projection* operator P_i is defined as follows:

$$x' = P_i x \equiv x_i(x_i, x)\,,$$

where x_i is a unit vector. This operator is called a projection operator because all x' are in the direction of x_i and the length of x' equals the component of x in the x_i direction, namely (x_i, x). In the following, let the x_i be a set of orthonormal vectors which span the n-dimensional vector space V. *Prove*:
    a) P_i is idempotent.
    b) $P_iP_j = 0$ for $i \neq j$. Interpret geometrically.
    c) P_i has no inverse.
    d) $\sum_{i=1}^{n} P_i = I$.
    e) P_i is Hermitian.
    f) Let A be a self-adjoint linear operator defined on an n-dimensional vector space V with n-eigenvalues, ϵ_i, and n-eigenvectors, x_i. Prove that A may be written as

$$A = \sum_{i=1}^{n} \epsilon_i P_i\,,$$

where P_i is as defined above. This is known as the spectral theorem for self-adjoint linear operators. (The analogous theorem for normal operators can also be easily proved.) Note the consistency of this result with parts (d) and (e) of this problem.

18. Given any linear transformation A on a finite-dimensional vector space, prove that in a given coordinate system, the adjoint matrix always exists.
[*Hint*: Modify the reasoning of Theorem 4.7.]

19. Obtain an upper bound for the lowest eigenvalue of the quantum harmonic oscillator by taking as a trial vector (function)

$$\phi = c \exp(-\alpha x^2/2) .$$

Note that this is exactly the result derived in Section 3.10.

20. In the text we obtained the first-order correction to the eigenfunctions of a linear operator with a nondegenerate spectrum. Using these methods, obtain the second-order correction. You may assume that the second-order correction is orthogonal to the zeroth order state from which you start.

21. Suppose that the harmonic oscillator Hamiltonian,

$$A_0 = \frac{p^2}{2m} + \frac{1}{2}m\omega^2 x^2 , \qquad [x, p] = i\hbar ,$$

is perturbed by a small term $A_1 = \epsilon x$. Find the first and second-order corrections to the eigenvalues of A_0 produced by A_1. Since this problem can be solved exactly, you can check your answers.

22. Find the first few corrections in perturbation theory to the lowest eigenvalue and corresponding eigenvector for the coupled harmonic-oscillator problem discussed in Section 4.12.

23. Using degenerate perturbation theory, find the corrections to the eigenvalues and eigenvectors of the coupled harmonic oscillator problem which correspond to the unperturbed eigenvalues $\lambda^{(0)} = 3\hbar\omega_0$.

24. Define an operator P_n by

$$P_n x = \sum_{i=1}^{\mu_n} x_{n,i}^{(0)}(x_{n,i}^{(0)}, x) ,$$

where $i = 1, \cdots, \mu_n$ is the degree of degeneracy of the nth eigenvalue of some operator A_0, and $x_{n,i}^{(0)}$ are the corresponding eigenvectors. Let ϵA_1 be the perturbation. Show that if we define

$$A_0' \equiv A_0 + \epsilon \sum_n P_n^\dagger A_1 P_n , \qquad A_1' \equiv A_1 - \sum_n P_n^\dagger A_1 P_n ,$$

then it is possible to solve for the eigenfunctions and eigenvalues of A_0' in terms of the eigenfunctions and eigenvalues of A_0. If the degeneracy of A_0 is completely lifted in first order by A_1, then show that we can do ordinary nondegenerate perturbation theory on A_1'. Show that the first-order corrections to the eigenvectors of A_0'. are all zero. Make any comments that seem relevant about the relative sizes of terms in the first few orders of perturbation theory.

25. Consider a four-dimensional real vector space with basis vectors $\{e_0, e_1, e_2, e_3\}$ and define a multiplication of basis vectors as follows:

$$e_0 e_i = e_i e_0 = e_i \qquad (i = 1, 2, 3)$$

$$e_0 e_0 = e_0, \qquad e_i e_i = -e_0 \qquad (i = 1, 2, 3)$$

$$e_i e_j = \sum_{k=1}^{3} \epsilon_{ijk} e_k \qquad (i \neq j;\ i, j = 1, 2, 3)$$

Because of this last equation it is customary to write $e_1 = \mathbf{i}$, $e_2 = \mathbf{j}$ and $e_3 = \mathbf{k}$ and hence to write a general vector in this space as

$$\begin{aligned} q &= q_0e_0 + q_1e_1 + q_2e_2 + q_3e_3 \\ &= q_0e_0 + q_1\mathbf{i} + q_2\mathbf{j} + q_3\mathbf{k} \\ &= q_0e_0 + \mathbf{q} \,. \end{aligned}$$

(a) Extend the definition of vector multiplication to all vectors in the space by linearity. (Following Sir W. R. Hamilton we call the elements of the space *quaternions*.) Let $q = q_0e_0 + \mathbf{q}$ and $p = p_0e_0 + \mathbf{p}$ be any two quaternions. Show that $pq = (p_0q_0 - \mathbf{p}\cdot\mathbf{q})e_0 + (p_0\mathbf{q} + q_0\mathbf{p} + \mathbf{p} \times \mathbf{q})$.

(b) When does $pq = qp$?

(c) Show that the unit quaternion is equal to e_0, that is, $e_0q = q = qe_0$.

(d) Prove that if $pq = 0$, then either $p = 0$ or $q = 0$.

(e) Show that $(pq)r = p(qr)$. Hence the set of quaternions forms what is called an *algebra*.

(f) Let $q = q_0e_0 + \mathbf{q}$ be a quaternion; define its quaternion *conjugate* to be $q^* = q_0e_0 - \mathbf{q}$. Show, using this definition, that if $q \neq 0$, there exists a unique inverse quaternion, q^{-1}, for which $qq^{-1} = e_0 = q^{-1}q$. This shows that the quaternions form a non-commutative field.

(g) Show that a realization of the quaternion field is obtained by writing

$$e_0 = \begin{pmatrix} 1 & 0 \\ 0 & 1 \end{pmatrix}, \quad e_1 = -i\begin{pmatrix} 0 & 1 \\ 1 & 0 \end{pmatrix}, \quad e_2 = -i\begin{pmatrix} 0 & -i \\ i & 0 \end{pmatrix}, \quad e_3 = -i\begin{pmatrix} 1 & 0 \\ 0 & -1 \end{pmatrix}$$

where $i = \sqrt{-1}$.

26. The quantum mechanical Hamiltonian for a charged particle in a magnetic field is $H = (\mathbf{p} - (e/c)\mathbf{A})^2/2m$, where $\mathbf{A}$ is the vector potential. We assume that the magnetic field is constant in magnitude and direction.

(a) Show that a vector potential for such a field is $\mathbf{A} = \mathbf{B} \times \mathbf{r}/2$, where $\mathbf{B}$ is the constant field, i.e., show that $\nabla \times \mathbf{A} = \mathbf{B}$. Show also that $\nabla \cdot \mathbf{A} = 0$.

(b) Assuming for convenience that the magnetic field is in the z-direction and that the particle's momentum in the z-direction is zero, that is, $p_z = 0$, show that if we take $p_x = (\hbar/i)d/dx$ and $p_y = (\hbar/i)d/dy$ then the Hamiltonian can be written as

$$H = \frac{p_x^2}{2m} + \frac{p_y^2}{2m} + \frac{1}{2}m\omega^2x^2 + \frac{1}{2}m\omega^2y^2 - \omega L_z \,,$$

where $\mathbf{L} = \mathbf{r} \times \mathbf{p}$ and $\omega = eB/2mc$.

(c) Using the fact that p_x, p_y, x and y are self-adjoint operators, define annihilation operators a_x and a_y by

$$a_x = (2m\hbar\omega)^{-1/2}(xm\omega + ip_x) \text{ and } a_y = (2m\hbar\omega)^{-1/2}(ym\omega + ip_y)$$

so that we have $[a_x, a_x^\dagger] = [a_y, a_y^\dagger] = I$ and $[a_x, a_y] = [a_x, a_y^\dagger] = 0$. Show that H takes the form

$$H = \hbar\omega[a_x^\dagger a_x + a_y^\dagger a_y + 1 + i(a_x^\dagger a_y - a_y^\dagger a_x)]$$

(d) Define $b = (a_x + ia_y)/\sqrt{2}$. Show that

(i) $H = 2\hbar\omega(b^\dagger b + 1/2)$

(ii) $[b, b^\dagger] = I$

(iii) $[b, H] = 2\hbar\omega b$, $[b^\dagger, H] = -2\hbar\omega b^\dagger$.

(e) Now solve the eigenvalue problem $H\phi = E\phi$. Assuming that there is a lowest eigenvalue show that it equals $\hbar\omega$ and that the corresponding eigenvector(s) must satisfy $b\phi_0 = 0$. Show that the remaining (normalized) eigenvectors and eigenvalues are given by

$$\phi_n = (b^\dagger)^n \phi_0 / \sqrt{n!}$$
$$E_n = 2\hbar\omega(n + 1/2)\ ,$$

or, in a more conventional notation, $E_n = \hbar\omega_L(n + 1/2)$, where $\omega_L = 2\omega = eB/mc$ is the Larmor frequency.

(f) Let u_0 be the unique solution to $a_x u_0 = 0$ and let v_0 be the unique solution to $a_y v_0 = 0$. (See Section 3.10; u_0 and v_0 are simply normalized Gaussians in the x and y variables, respectively.) Then clearly the function $\phi_0^{(0)} = u_0 v_0$ satisfies $b\phi_0^{(0)} = 0$, so we have easily found one solution for the ground state. However, this is not the *only* solution. Define an operator, d, by $d = (a_x - ia_y)/\sqrt{2}$ and show that $[b, d] = 0$ and $[b, d^\dagger] = 0$. Hence show that if $\phi_0^{(0)}$ is defined as above, then

$$\phi_0^{(m)} = (d^\dagger)^m \phi_0^{(0)} / \sqrt{m!}$$

also satisfies $b\phi_0^{(m)} = 0$ for *any* m, and $(\phi_0^{(m)}, \phi_0^{(m')}) = \delta_{mm'}$. Thus, the ground state is *infinite-fold degenerate*, and hence so is each excited state. In fact, we have

$$\phi_n^{(m)} = (b^\dagger)^n \phi_0^{(m)} / \sqrt{n!} = (d^\dagger)^m \phi_n^{(0)} / \sqrt{m!}\ .$$

27. For a vector space, V, with a pseudo inner product defined by Eqs. (4.23), (4.24) and (4.25) we make the following definitions: For any operator A, we *define* the transpose operator $\tilde{A}$ by $[x, Ay] = [\tilde{A}x, y]$ for all x and y. S is said to be symmetric if $\tilde{S} = S$. O is said to be orthogonal if $\tilde{O}O = I$. Two vectors, $x \neq y$, are said to be *pseudo-orthogonal* if $[x, y] = 0$, and finally a set of *non-null* vectors, x_i, is said to be a complete, pseudo-orthonormal set if the x_i span the space and if $[x_i, x_j] = \delta_{ij}$.

(a) Prove that if $X = \{x_i\}$ is a complete, pseudo-orthonormal set, then for any x in V

(i) $x = \sum_{i=1}^{n} [x_i, x] x_i$

(ii) $[x, y] = \sum_{i=1}^{n} [x, x_i][x_i, y]\ .$

(b) Show that the matrix of a symmetric operator with respect to a complete, pseudo-orthonormal basis satisfies $S_{ij} = S_{ji}$. Show that the matrix elements of an orthogonal operator O satisfy $\sum_{k=1}^{n} O_{ik} O_{kj} = \delta_{ij}$.

(c) Show that if S is a symmetric operator and $Sx_1 = \lambda_1 x_1$ and $Sx_2 = \lambda_2 x_2$, then $[x_1, x_2] = 0$ for $\lambda_1 \neq \lambda_2$. Show that if all the eigenvalues of S are *distinct*, then the eigenvectors of S form a complete, pseudo-orthonormal set. [Hint: you must exclude the possibility of a null-eigenvector.]

(d) Give an example of a symmetric 2×2 matrix whose eigenvalues are *not* real, but whose eigenvectors are nevertheless pseudo-orthogonal.

(e) Show that every eigenvalue of an orthogonal operator belonging to a *non-null* eigenvector is either 1 or -1.

(f) Prove that if O is an orthogonal transformation and $Ox_1 = \lambda_1 x_1$ and $Ox_2 = \lambda_2 x_2$, then $[x_1, x_2] = 0$ if $\lambda_1 \neq \lambda_2^{-1}$. Show further that if either x_1 or x_2 is non-null, then $[x_1, x_2] = 0$ if $\lambda_1 \neq \lambda_2$.

FURTHER READINGS

HALMOS, P. R., *Finite-Dimensional Vector Space*, Second Edition. Princeton, N. J.: D. Van Nostrand, 1958.

HERSTEIN, I. N., *Topics in Algebra*, Waltham, Mass.: Blaisdell, 1964.

HOFFMAN, K., and R. KUNZE, *Linear Algebra*, Englewood Cliffs, N. J.: Prentice Hall, 1961.

LANG, S., *Linear Algebra*, Reading, Mass.: Addison-Wesley, 1965.

CHAPTER 5

HILBERT SPACE—COMPLETE ORTHONORMAL SETS OF FUNCTIONS

INTRODUCTION

The principal strands of mathematics—algebra, geometry, and analysis—are joined when functions are viewed as vectors in a vector space. This unification is the subject of this chapter; none of the results or methods of mathematics has more relevance to modern physics.

Throughout this chapter we shall be using concepts, terminology, and theorems from Chapters 3 and 4. The present chapter represents the infinite—dimensional generalization of certain of the results obtained there. However, the fact that the vectors are now functions means that there are a number of additional considerations and possibilities in the development of the theory. These result primarily from the attempt to represent a function as a linear combination of some given set of functions—i. e., the problem of series expansions. All the characteristic questions of analysis, such as those of convergence, therefore become relevant.

In particular, the so-called "special functions" of mathematical physics—spherical harmonics, Legendre, Hermite, and Laguerre polynomials, to name a few—are all conveniently treated within the framework of Hilbert space. As we deal with the various special functions, we shall discuss the differential equations of which they are solutions. But the treatment will not be based on the differential equations. The framework of Hilbert space is a much more comprehensive one, and unifies what is otherwise a bewildering maze of special cases and properties.

There will be three stages in the discussion of these functions. In the first stage, where we introduce the basic notions of Hilbert space, the treatment will be quite abstract; the special functions will be mentioned only in passing. Then we shall consider the special functions one by one, somewhat inductively, de-emphasizing their common origins and properties. Finally, we shall return to a more abstract level and systematize the properties of these functions in a single comprehensive framework.

Hilbert space is the mathematical setting for quantum physics. Physical observables are represented by operators in Hilbert space, and physical states are vectors (functions) in Hilbert space. There is no more important property of the functions that describe the possible states of a physical system than that they form a *complete* set. The whole of quantum theory is based on this fact. Thus most of this chapter will in one way or another have to do with the completeness of sets of functions.

At the end of the chapter we shall bring together the key results of the theory of Hilbert space that have been developed throughout the chapter, and emphasize their role in the formulation of quantum mechanics.

5.1 FUNCTION SPACE AND HILBERT SPACE

In discussing finite-dimensional vector spaces, we referred several times to P_n, the vector space of polynomials of degree $\leqq n$. The vectors in that vector space are a class of simple functions, the polynomials.

We are now going to define another vector space whose elements are functions. The elements of this space are the complex-valued functions of a real variable x, defined on the closed interval $[a, b]$,* which are *square integrable*.† We shall show that the set of square integrable functions forms a vector space. This space is called L_2 by mathematicians: we shall call it *function space*. It will be found to be an infinite-dimensional space.

Intuitively, it would seem that function space is far "larger" than the finite-dimensional space P_n, and in a sense this is so. However, suppose that the basis functions for P_n—the set $\{x^m, m = 0, 1, \cdots, n\}$—were extended to contain all possible powers of x by letting $n \to \infty$. Weierstrass's theorem, the central result of this chapter, shows that this infinite set of functions is not as poverty-stricken as it might appear. We shall shortly give a precise account of its latent powers. First we return to the definition of function space.

Addition of the two vectors, f_1 and f_2, in function space is defined according to the natural rule:

$$(f_1 + f_2)(x) \equiv f_1(x) + f_2(x) ,$$

and multiplication by a complex scalar α is defined as

$$(\alpha f)(x) \equiv \alpha f(x) .$$

The only possible difficulty in showing that these operations satisfy the various axioms that define a vector space is establishing closure. That is, are the sums and scalar multiples of square-integrable functions also square-integrable? The

* A *closed interval*, written $[a, b]$, is the set of all points $\{x\}$ such that $a \leq x \leq b$. An *open interval*, which we write as (a, b), is the set of points $\{x\}$ such that $a < x < b$. A closed interval is always finite; if either a or b is infinite, the interval is open at that end. Later in the chapter, when we deal with infinite intervals, we shall alert the reader by explicit mention of the shift from finite to infinite intervals.

† A function is *square integrable* on $[a, b]$ if $\int_a^b |f(x)|^2\,dx$ exists and is finite.

answer is yes, and so the space is in fact a vector space. We prove closure of the sum.

$$\begin{aligned} |f_1 + f_2|^2 &= |f_1|^2 + |f_2|^2 + f_1^* f_2 + f_1 f_2^* \\ &= |f_1|^2 + |f_2|^2 + 2\text{Re}\,(f_1^* f_2) \\ &\leqq |f_1|^2 + |f_2|^2 + 2|f_1^* f_2| \\ &\leqq |f_1|^2 + |f_2|^2 + 2|f_1||f_2|\,. \end{aligned} \tag{5.1}$$

Also,

$$0 \leqq (|f_1| - |f_2|)^2 = |f_1|^2 + |f_2|^2 - 2|f_1||f_2|\,,$$

so

$$|f_1|^2 + |f_2|^2 \geqq 2|f_1||f_2|\,.$$

We use this last inequality to replace $2|f_1||f_2|$ in Eq. (5.1) with something larger, thereby preserving the inequality. Thus the inequality

$$|f_1 + f_2|^2 \leqq 2|f_1|^2 + 2|f_2|^2\,, \tag{5.2}$$

holds at every point in $[a, b]$. Integrating over both sides, we see that square integrability of f_1 and f_2 ensures square integrability of their sum.

We now add an inner product to function space.

Definition 5.1. The inner product of two functions, f_1 and f_2, belonging to function space, is defined by

$$(f_1, f_2) \equiv \int_a^b f_1^*(x) f_2(x)\, dx\,. \tag{5.3}$$

Note that square integrability implies that $(f, f) \equiv ||f||^2 = \int_a^b |f|^2\, dx < \infty$. The quantity $||f||$ is called the norm of f. The inner product for any pair of square-integrable functions also exists. Since

$$|f_1^* f_2| = |f_1||f_2| \leqq \tfrac{1}{2}(|f_1|^2 + |f_2|^2)\,,$$

it follows that

$$\int_a^b |f_1^* f_2|\, dx \leqq \tfrac{1}{2}(||f_1||^2 + ||f_2||^2) < \infty\,;$$

but any function whose absolute value is integrable is itself integrable, so (f_1, f_2) exists.

Verification that (f_1, f_2) as defined in Eq. (5.3) is an inner product proceeds very smoothly until we come to the question of positive-definiteness. Clearly,

$$(f, f) \equiv ||f||^2 = \int_a^b |f|^2\, dx \geqq 0\,.$$

But does $(f, f) = 0$ imply that $f(x) = 0$ for *all* x in $[a, b]$? Not quite, for the function $f(x)$ can be nonzero at any finite number of points, and the integral will not "notice" this. That is, there will be no contribution to the integral even though the integrand is not identically zero throughout $[a, b]$.

The discussion of situations like these is facilitated if a slightly more general notion of integration is used. The Riemann integral suffers from several difficulties; for example, let us consider an extreme case by looking at the bizarre function $f(x)$ defined to be 1 for every rational number in $[0, 1]$ and 0 for every non-rational number. Now, since there are "very few" rationals, only a countable number in fact, we strongly suspect that the integral of this function is zero. However, if we form the upper and lower Riemann integrals by partitioning $[0, 1]$ into small segments Δx_i, and write

$$\overline{\int} f(x)\, dx = \sum_i \Delta x_i \max\,[f(x)], \qquad x_i \leq x \leq x_i + \Delta x_i$$

$$\underline{\int} f(x)\, dx = \sum_i \Delta x_i \min\,[f(x)], \qquad x_i \leq x \leq x_i + \Delta x_i$$

in the usual way, we see that no matter how small the subinterval, Δx_i, the maximum of $f(x)$ on this interval is always 1 and the minimum is always zero. Thus

$$\overline{\int} f(x)\, dx = 1 \qquad \text{and} \qquad \underline{\int} f(x)\, dx = 0\,,$$

so the Riemann integral does not exist.

In the theory of Lebesgue integration, this integral does exist and it equals zero. We say that $f(x) = 0$ except on a set of points of *measure zero*, or $f(x) = 0$ *almost everywhere*. The intuitive content of this phrase is simple. If we have a countable number of points on the real line and are given a small strip of paper of length ϵ, then we can paste a small piece of the strip over each element of the set by dividing it into a countable number of pieces of width $\epsilon/2^n$. Since $\sum_{n=1}^{\infty} \epsilon/2^n = \epsilon$, we use up only our given strip in the process. But since the original strip can be arbitrarily small, the set of points on which $f(x)$ is nonzero is negligible with respect to the set on which it is zero, despite the fact that every real number is arbitrarily close to some rational. Thus, the rationals are a set of measure zero on the real line.

We shall go no further with the notion of the Lebesgue integral, since all the functions which we will actually want to integrate in this book will be Riemann integrable. For example, the Riemann integral always exists for the class of piecewise continuous functions defined on closed intervals. In any case, the Riemann integral, when it exists, is equal to the Lebesgue integral.

To summarize: If f is a function in function space and $(f, f) = 0$, then $f(x)$ need not be identically zero for all x, but it can differ from zero only on a set of measure zero. We say that $(f, f) = 0$ implies $f(x) = 0$ almost everywhere. Any function which equals zero almost everywhere is called the zero function. With this broadened concept of the zero function, Axiom 3 for an inner product is satisfied, so Definition 5.1 does define an inner product.

Thus we have a complex inner product defined in function space, which is itself a complex vector space of complex square-integrable functions of a *real* variable ranging through the closed interval $[a, b]$.

To be useful to the physicist, the inner-product space must also be *complete.* A complete space is one in which there exists no Cauchy sequence of elements of the space which tend toward limits outside the space.* An elementary example of an incomplete space is the set of rational numbers. The sequence of partial sums, $S_N = \sum_{n=0}^{N} 1/n!$ is a sequence of rationals, but it converges to the number e, which is irrational. However, a basic theorem of real analysis states that the set of all real numbers is complete; one cannot get out of the set by taking the limits of Cauchy sequences of real numbers.

Similarly, we want to discover a class of *functions* having the property that there will be no Cauchy sequence of functions in this class whose limits do not belong to the class. Such a class of functions is *complete.* This absolutely fundamental problem in analysis is solved by the Riesz-Fischer theorem, which states that the space of square-integrable functions, i. e., functions with a finite norm, is complete. The proof may be found in any book on functional analysis (e. g., Riesz and Nagy, Rudin). The theorem may be stated as follows.

Riesz-Fischer Theorem. Let the functions $f_1(x), f_2(x), \cdots$ be elements in function space. If

$$\lim_{n,\, m\to\infty} ||f_n - f_m||^2 \equiv \lim_{n,\, m\to\infty} \int_a^b |f_n(x) - f_m(x)|^2\, dx = 0\,,$$

then there exists a square (Lebesgue) integrable function $f(x)$ to which the sequence $f_n(x)$ converges "in the mean"; i. e., there exists an f such that

$$\lim_{n\to\infty} \int_a^b |f(x) - f_n(x)|^2\, dx = 0\,.$$

We shall discuss convergence in the mean in greater detail in the following section. It should be emphasized that this theorem is not true unless the integral used is the Lebesgue integral.

Thus function space, as we have defined it, is in fact complete. Henceforth we shall call this complete inner-product space by its usual name, *Hilbert space,* although function space is not the only Hilbert space.

The notions of orthogonality, normalization, and orthonormal sets of functions are defined exactly as for vectors. Thus the set of functions $\{f_i\}$ is said to be *orthonormal* if

$$(f_i, f_j) \equiv \int_a^b f_i^*(x) f_j(x)\, dx = \delta_{ij}\,.$$

* A *Cauchy sequence* is a sequence $\{s_n\}$ which has the property that given an arbitrary number $\epsilon > 0$, there is an index $N(\epsilon)$ such that if m and n are larger than $N(\epsilon)$,

$$||s_n - s_m|| < \epsilon.$$

If the sequence in question is a sequence of numbers, then the norm is interpreted as the absolute value. If, as in our case, the sequence is a sequence of functions, then the norm is given by

$$||f|| \equiv \left[\int_a^b |f|^2\, dx\right]^{1/2}.$$

Example. *The Fourier functions.* Let $f_n(x) = e^{inx}/\sqrt{2\pi}$, where $n = 0, \pm 1, \pm 2, \cdots$ The set $\{f_n\}$ is orthonormal over the interval $[-\pi, \pi]$.

Proof.

$$(f_n, f_m) = \int_{-\pi}^{+\pi} f_n^* f_m \, dx = \frac{1}{2\pi} \int_{-\pi}^{\pi} e^{i(m-n)x} \, dx$$

$$= \frac{1}{2\pi(m-n)i} e^{i(m-n)x} \Big]_{-\pi}^{\pi} = 0. \qquad \textit{if } m \neq n \,;$$

$$(f_n, f_n) = \frac{1}{2\pi} \int_{-\pi}^{\pi} 1 \, dx = 1, \qquad \textit{if } m = n \,.$$

Therefore

$$(f_n, f_m) = \delta_{nm} \,.$$

An often used generalization of orthonormality is the following:

Definition 5.2. A set of functions $\{f_n\}$ is orthonormal with respect to a nonnegative weight function $w(x)$ on $[a, b]$ if

$$(f_n, f_m) \equiv \int_a^b f_n^*(x) f_m(x) w(x) \, dx = \delta_{nm} \,.$$

We shall work almost exclusively with orthonormal sets of functions, because, as in finite-dimensional vector spaces, they greatly simplify computations.

5.2 COMPLETE ORTHONORMAL SETS OF FUNCTIONS

In the theory of finite-dimensional vector spaces, we found a number of equivalent alternative characterizations of a complete set of basis vectors (Theorem 4.3). The corresponding problem in Hilbert space is that of representing a function as a linear combination of some given set of functions, or, in other words, the problem of series expansions of functions in terms of a given set. The prototype of all such series expansions is the Fourier series. Our treatment, however, will be a general one, embracing many of the functions of mathematical physics; Fourier series will be treated within this framework as a special case.

The first question we must face is that of defining the *completeness of an orthonormal set of functions* in Hilbert space. (Completeness of a *set* of functions is not the same as completeness of a *space*, mentioned briefly in the last section; however, they are intimately connected as we shall see.) We could say that an orthonormal set of functions $\{f_i(x)\}$ is complete if any function $f(x)$ in Hilbert space is expressible as a linear combination of the $f_i(x)$:

$$f(x) = \sum_{i=1}^{\infty} c_i f_i(x) \,,$$

the series converging to f at every point x. This would provide a close analog of the idea of completeness of a set of basis vectors in finite-dimensional spaces. But this criterion of convergence is, for many purposes, unnecessarily severe and exclusive. As we shall see, by this criterion there would exist no complete orthonormal set of functions in Hilbert space. So, instead of demanding strict pointwise convergence, we shall weaken the convergence criterion, and in this way permit the existence of complete sets of functions.

The appropriate weakening of the convergence criterion is suggested by the slight difficulty we encountered with the positive-definiteness of the inner product. There we found that

$$(f, f) = \int_a^b |f|^2\,dx = 0$$

implied that $f(x)$ vanished not at *every* point x in $[a, b]$, but rather, almost everywhere in $[a, b]$; that is, at all but a set of points of measure zero. Similarly, we shall say that $\sum_i c_i f_i(x)$ converges "in the mean" to $f(x)$ if

$$\lim_{n\to\infty} \int_a^b |f(x) - \sum_i^n c_i f_i(x)|^2\,dx = 0\,.$$

This allows the series to differ from $f(x)$ on a set of measure zero.

In order to develop the notion of completeness of an orthonormal set of functions further, we need to define the different kinds of convergence we shall be using:

1. Pointwise convergence;
2. Uniform convergence;
3. Convergence in the mean.

Definition 5.3. A sequence of functions, $h_n(x)$, converges pointwise to $h(x)$ on $[a, b]$ if for every x in $[a, b]$ and every $\epsilon > 0$ there exists an integer $N(x, \epsilon)$ such that for $n > N$,

$$|h(x) - h_n(x)| < \epsilon\,.$$

The $h_n(x)$ may themselves be the partial sums of another sequence; that is,

$$h_n(x) = \sum_{i=1}^n k_i(x)\,.$$

The definition of pointwise convergence, and the other types of convergence to follow, may all be stated equivalently in terms of the infinite series $\sum_{i=1}^\infty k_i(x)$, which is the limit of the sequence $h_n(x)$ of partial sums. We say that this series *converges pointwise* to $h(x)$ on $[a, b]$ if for every x in $[a, b]$ and every $\epsilon > 0$, there exists an integer $N(x, \epsilon)$ such that for $n > N$,

$$|h(x) - h_n(x)| = |h(x) - \sum_{i=1}^n k_i(x)| < \epsilon\,.$$

If there is a single N that works in the above definition for all x in $[a, b]$, the convergence is said to be *uniform*. Formally, we have:

Definition 5.4. A sequence of functions, $h_n(x)$, converges *uniformly* to $h(x)$ on $[a, b]$ if for every $\epsilon > 0$ there exists an integer $N(\epsilon)$, *independent of* x, such that for $n > N$, $|h(x) - h_n(x)| < \epsilon$, for all x in $[a, b]$. Clearly, uniform convergence implies pointwise convergence.

Note that this definition makes explicit mention of the limit $h(x)$ of the sequence of functions, $h_n(x)$. The *Cauchy criterion* for uniform convergence supplies us with a useful alternative to this definition, in which knowledge of the limit function is *not* assumed. Cauchy's criterion is proved in analysis courses; we state it here.

Theorem 5.1. The sequence of functions $h_n(x)$ converges uniformly on $[a, b]$ if for every $\epsilon > 0$ there exists an integer $N(\epsilon)$ such that for all $r > N$, $s > N$, and x in $[a, b]$, $|h_r(x) - h_s(x)| < \epsilon$.
In terms of the partial sums, $h_n(x) \equiv \sum_{i=1}^{n} k_i(x)$, this becomes

$$|h_r(x) - h_s(x)| = |\sum_{i=1}^{r} k_i - \sum_{i=1}^{s} k_i| = |\sum_{i=r+1}^{s} k_i(x)| < \epsilon .$$

If we have uniform or pointwise convergence, we may write

$$h(x) = \lim_{n\to\infty} h_n(x) = \sum_{i=1}^{\infty} k_i(x) .$$

The weaker convergence we discussed earlier can now be formalized precisely.

Definition 5.5. A sequence of functions $h_n(x)$, converges *in the mean* to $h(x)$ on $[a, b]$ if

$$\lim_{n\to\infty} \int_a^b |h(x) - h_n(x)|^2 \, dx = 0,$$

that is, if for every ϵ there exists an $N(\epsilon)$ such that for $n > N$,

$$\int_a^b |h(x) - h_n(x)|^2 \, dx < \epsilon .$$

The series $\sum_{i=1}^{\infty} k_i(x)$ converges in the mean to $h(x)$ if

$$\lim_{n\to\infty} \int_a^b |h(x) - \sum_{i=1}^{n} k_i(x)|^2 \, dx = 0 .$$

It is easy to see that uniform convergence implies mean convergence. For if convergence is uniform, then for any ϵ there exists an N such that for $n > N$, $|h - h_n| < \epsilon$ for all x in $[a, b]$. Therefore

$$\int_a^b |h - h_n|^2 \, dx < \int_a^b \epsilon^2 \, dx = \epsilon^2(b - a) ,$$

so the integral can be made arbitrarily small by choice of ϵ. This is just the statement that

$$\lim_{n\to\infty} \int_a^b |h - h_n|^2 \, dx = 0,$$

and so we have mean convergence. Note, however, that pointwise convergence does not imply mean convergence (Problem 19).

It is in terms of *mean* convergence that the completeness of an orthonormal set of functions is defined.

Definition 5.6. Let $g(x)$ be any function in Hilbert space (i. e., any square-integrable function) and let $\{f_i(x)\}$ be an orthonormal set of functions in Hilbert space. If there exist constant coefficients $\{a_i\}$ such that the sequence of partial sums $g_n(x) \equiv \sum_{i=1}^{n} a_i f_i(x)$ converges in the mean to $g(x)$, then the *set of functions* $\{f_i\}$ *is a complete orthonormal set.* Equivalently, if the mean square error can be made arbitrarily small,

$$\lim_{n\to\infty} \int_a^b |g - g_n|^2 \, dx = \lim_{n\to\infty} \int_a^b |g - \sum_{i=1}^{n} a_i f_i|^2 \, dx = 0 ,$$

then the set $\{f_i\}$ is a *complete orthonormal set of functions.* It should be noted that the coefficients $\{a_i\}$ are independent of n. Thus as n increases and we include more terms in the partial sum approximating g, the earlier coefficients do not change. We may say that as we extend the sum to infinity, the infinite *series*

$$\sum_{i=1}^{\infty} a_i f_i$$

approximates the arbitrary function g in the mean. We shall write this as

$$g(x) \doteq \sum_{i=1}^{\infty} a_i f_i(x) .$$

The dot over the equal sign serves to distinguish mean convergence from pointwise convergence.

Since mean convergence does not necessarily imply pointwise or uniform convergence, it should be clear that the completeness of an orthonormal set of functions $\{f_i\}$, expressed by the relation

$$\lim_{n\to\infty} \int_a^b |f - \sum_{i=1}^{n} a_i f_i|^2 \, dx = 0 ,$$

or symbolically,

$$f(x) \doteq \sum_{i=1}^{\infty} a_i f_i(x) ,$$

does not imply that

$$f(x) = \sum_{i=1}^{\infty} a_i f_i(x) . \tag{5.4}$$

We may only *equate* $f(x)$ and the expansion series if the series converges pointwise or uniformly to $f(x)$.

Let $f(x)$ be an arbitrary function in Hilbert space, and assume for the moment that we have an orthonormal set of functions $\{f_i(x)\}$ such that the series

$\sum_{i=1}^{\infty} c_i f_i(x)$ converges uniformly to $f(x)$:

$$f(x) = \sum_{i=1}^{\infty} c_i f_i(x) .$$

The coefficients c_i are called the *generalized Fourier coefficients*, or *expansion coefficients*. The formula for them is especially simple because $\{f_i\}$ is an orthonormal set:

$$(f_n, f) = \sum_{i=1}^{\infty} c_i (f_n, f_i) = \sum_{i=1}^{\infty} c_i \delta_{ni} = c_n . \tag{5.5}$$

Since the convergence is uniform, mean convergence is also implied. Therefore

$$\lim_{n \to \infty} \int_a^b |f - \sum_{i=1}^{n} c_i f_i|^2 \, dx = 0 ,$$

[where the c_i are given by Eq. (5.5)], and, consequently, the set of functions $\{f_i\}$ is complete.

Now consider the nonnegative quantity

$$M_n = \int_a^b |f(x) - \sum_{i=1}^{n} a_i f_i(x)|^2 \, dx \geq 0 , \tag{5.6}$$

where $\{f_i\}$ is an orthonormal set, and f is any function in Hilbert space. We want to know: What values of the coefficients a_i will minimize M_n? Or, to put it in the language of the physicist: What values of the a_i will give the best least-squares fit to the arbitrary function $f(x)$?

To answer this question, we expand Eq. (5.6) to obtain

$$\begin{aligned} M_n &= \int_a^b \left(f^*f - \sum_{i=1}^{n} a_i f^* f_i - \sum_{i=1}^{n} a_i^* f f_i^* + \sum_{i,j=1}^{n} a_i^* a_j f_i^* f_j\right) dx \\ &= (f, f) - \sum_{i=1}^{n} a_i c_i^* - \sum_{i=1}^{n} a_i^* c_i + \sum_{i,j=1}^{n} a_i^* a_j \delta_{ij} , \end{aligned}$$

where $c_i \equiv (f_i, f)$. Adding and subtracting the quantity $\sum_{i=1}^{n} |c_i|^2$, we have

$$M_n = (f, f) + \sum_{i=1}^{n} |a_i - c_i|^2 - \sum_{i=1}^{n} |c_i|^2 \geq 0 .$$

It is clear that M_n is minimized by choosing $a_i = c_i$. We then have

$$\begin{aligned} M_n &= \int_a^b |f(x) - \sum_{i=1}^{n} c_i f_i(x)|^2 \, dx \\ &= (f, f) - \sum_{i=1}^{n} |c_i|^2 \geq 0 , \end{aligned} \tag{5.7}$$

which may be rewritten as

$$(f, f) \geq \sum_{i=1}^{n} |c_i|^2 = \sum_{i=1}^{n} |(f_i, f)|^2 . \tag{5.8}$$

Writing $s_n = \sum_{i=1}^{n} |c_i|^2$, we see that the sequence $s_1, s_2, \cdots, s_n, \cdots$ is monotonically increasing and bounded above by (f, f), (Eq. 5.8). Therefore s_n tends toward a limit as n becomes arbitrarily large, and we can write

$$(f, f) \geq \sum_{i=1}^{\infty} |c_i|^2 . \tag{5.9}$$

Hence the infinite series $\sum_{i=1}^{\infty} |c_i|^2$ converges. Equation (5.9) is just Bessel's inequality (compare Theorem 4.2) in an infinite-dimensional space.

The orthonormal set $\{f_i\}$ is complete if and only if there exists a set $\{a_i\}$ such that $\lim_{n\to\infty} M_n = 0$. If the set $\{f_i\}$ is complete, then $a_i = c_i$ and the equal sign holds in Bessel's inequality:

$$(f, f) = \sum_{i=1}^{\infty} |c_i|^2 = \sum_{i=1}^{\infty} |(f_i, f)|^2 , \tag{5.10}$$

for every f. This was also true in the finite-dimensional case (see Theorem 4.3). Equation (5.10) is called the *completeness relation*. As in the finite-dimensional case, it can also be stated in the form of Parseval's equation:

$$(f, g) = \sum_{i=1}^{\infty} (f, f_i)(f_i, g) . \tag{5.11}$$

We leave the proof to the reader (Problem 14).

Another characterization of a complete set of n vectors is that there exists no nonzero vector orthogonal to all n vectors in the set. The corresponding statement holds for a complete orthonormal set of functions. We shall present it as a therorem. It will prove very useful in what follows. To facilitate the statement of the theorem, we first make a definition.

Definition 5.7. A set of orthonormal functions is said to be *closed* if no nonzero function is orthogonal to every function in the set.

Theorem 5.2. A set of orthonormal functions in Hilbert space is complete if and only if it is closed.

Proof. We first prove that completeness of the set implies that the set is closed. Assume that there is a nonzero function $f(x)$ (and let it be normalized), such that

$$(f_i, f) \equiv c_i = \int_a^b f_i^*(x) f(x)\, dx = 0$$

for all i. Then

$$\lim_{n\to\infty} \int_a^b |f - \sum_{i=1}^{n} c_i f_i|^2\, dx = \int_a^b |f|^2\, dx = 1 \neq 0 ,$$

(since f is normalized), so the set $\{f_i\}$ is not complete. Thus completeness of an orthonormal set of functions implies that there are no functions that are orthogonal to every member of the set.

We now prove the converse: if the orthonormal set is closed, it is complete. If it is not complete, then the completeness relation, Eq. (5.10), is not satisfied. Thus there exists some function $f(x)$ such that

$$||f||^2 > \sum_{n=1}^{\infty} |c_n|^2 ,$$

where $c_n = (f_n, f)$. But since the above infinite series is convergent, the sequence $\{g_m(x)\}$, where

$$g_m(x) = \sum_{n=1}^{m} c_n f_n(x)$$

is a Cauchy sequence in Hilbert space, and therefore, because of the *completeness of the space*, the $g_m(x)$ must converge in the mean to a limit *in the space*, call it $g(x)$, such that $c_n = (f_n, g)$. Therefore $(f_n, g) = (f_n, f)$ so $(f_n, f\text{-}g) = 0$. Thus $f(x)\text{-}g(x)$ is orthogonal to $f_n(x)$ for all n. We now show that the norm of $f(x)\text{-}g(x)$ is not equal to zero, so $\{f_n(x)\}$ is not closed, contrary to our assumption. It will then follow by contradiction that the set $\{f_n(x)\}$ is complete and the proof will be finished.

Using the inequality

$$||x - y|| \geq |\, ||x|| - ||y||\, |$$

(see Problem 4.7b), we have

$$||f - g|| = ||f - g_m - (g - g_m)|| \geq \Big|\, ||f - g_m|| - ||g - g_m|| \,\Big|$$

for all m. Now as $m \to \infty$, we know that $||g - g_m|| \to 0$, whereas

$$||f - g_m||^2 = ||f - \sum_{n=1}^{m} c_n f_n||^2 = ||f||^2 - \sum_{n=1}^{m} |c_n|^2 > 0$$

for all m by assumption. Thus $||f - g|| > 0$ and the proof is complete.

This theorem plays a key role in establishing the completeness of the various orthonormal sets of functions which we shall treat in this chapter. Note the crucial use that is made of the completeness of the Hilbert space in proving it.

We conclude this section with some observations on the uniqueness of the representations of functions in series expansions. We first prove that a function in Hilbert space is uniquely determined almost everywhere by its expansion coefficients with respect to a given complete orthonormal set of functions $\{f_i\}$. Suppose that two functions, f and g, have the same expansion coefficients, that is,

$$c_i = (f_i, f) = (f_i, g) .$$

Thus, $(f_i, f - g) = 0$, so by Theorem 5.2, $f - g = 0$, and hence $f = g$.

Now consider the converse problem. Does a given function have a unique set of expansion coefficients? Assume that

$$\lim_{n\to\infty} ||f - \sum_{i=1}^{n} c_i f_i|| = \lim_{n\to\infty} ||f - \sum_{i=1}^{n} d_i f_i|| = 0 ;$$

that is, assume that there are two partial sums with different expansion coefficients that converge in the mean to the same function f. If the expansion coefficients are unique, then $c_i = d_i$. To prove this, we observe that

$$\begin{aligned} \|\sum_{i=1}^{n} c_i f_i - \sum_{i=1}^{n} d_i f_i\| &= \|\sum_i c_i f_i - f + f - \sum_i d_i f_i\| \\ &\leq \|f - \sum_i c_i f_i\| + \|f - \sum_i d_i f_i\| , \end{aligned}$$

where we have used the triangle inequality (see Problem 4.7c). Now given any ϵ, we can choose by assumption an n large enough so that both the last two norms are less than $\epsilon/2$. Therefore, for such an n,

$$\|\sum_i c_i f_i - \sum_i d_i f_i\| = \|\sum_i (c_i - d_i) f_i\| = \left[\sum_i (c_i - d_i)^2\right]^{1/2} < \epsilon .$$

But this can only be true if $c_i = d_i$. Therefore the expansion coefficients of a given function are unique. Since the set $\{f_i\}$ is a complete orthonormal set of functions, it follows from our earlier remarks that $c_i = d_i = (f_i, f)$, the Fourier coefficients.

5.3 THE DIRAC δ-FUNCTION

In this section we introduce Dirac's δ-function in an informal way. The δ-function will play an important role in this book as it usually does in physics.

The first thing to understand about the δ-function is that it is not a function at all. A function is a rule that assigns another number to each number in a set of numbers. The δ-function, as used in physics, is instead a shorthand notation for a rather complicated limiting process whose use greatly simplifies calculations. It takes on a meaning only when it appears under an integral sign, in which case it has the following effect:

$$\int_{-\infty}^{\infty} f(x)\delta(x)\, dx = f(0) . \tag{5.12}$$

A special case of this is $f(x) = 1$. Then

$$\int_{-\infty}^{\infty} \delta(x)\, dx = 1 . \tag{5.13}$$

If the singular point is located at an arbitrary point x, then

$$\int_{-\infty}^{\infty} f(x')\delta(x' - x)\, dx' = f(x) . \tag{5.14}$$

Except at the singular point $x = 0$,

$$\delta(x) = 0 . \tag{5.15}$$

Thus $\delta(x)$ behaves as an ordinary function almost everywhere. It vanishes at all points where its argument is not zero, but at that one point it is undefined. Nevertheless its behavior near this point is all that matters.

Now the integral of any real function which vanishes everywhere except at one point must be zero, regardless of the value of the function at the singular point. Thus no *function* which satisfies Eq. (5.15) could possibly satisfy Eqs. (5.12) or (5.13). These equations must be interpreted as a symbolic notation for a process of the following type.

Let $\delta_\alpha(x)$ be a set of functions parametrized by the index α, which have the properties

$$\lim_{\alpha\to 0} \delta_\alpha(x) = 0 \qquad \text{for all } x \neq 0 ,$$
$$\lim_{\alpha\to 0} \int_{-\infty}^{+\infty} f(x)\delta_\alpha(x)\, dx = f(0) . \tag{5.16}$$

Our earlier equations result if we denote "$\lim_{\alpha\to 0} \delta_\alpha(x)$" by $\delta(x)$, and interchange the order of the limiting process and the integration, a procedure which is not, in general, valid. The original equations defining the δ-function must be interpreted as standing for the limiting processes of Eqs. (5.16).

Let us look at several sets of functions that have the properties described in Eqs. (5.16).

1. The simplest possible set of functions which has the proper limiting behavior is depicted in Fig. 5.1(a). The function $\delta_c(x)$ is defined (for $c>0$) by

$$\delta_c(x) \equiv \begin{cases} \dfrac{1}{c} & \text{for } |x| \leq \dfrac{c}{2} , \\ 0 & \text{for } |x| > \dfrac{c}{2} . \end{cases} \tag{5.17}$$

Clearly, $\lim_{c\to 0} \delta_c(x) = 0$, at all $x \neq 0$. Also, $\int_{-\infty}^{+\infty} \delta_c(x)\, dx = 1$, independent of c. The function $\delta_c(x)$ (and it *is* a function) is defined for all $c \neq 0$, and the limit

$$\lim_{c\to 0} \int_{-\infty}^{\infty} \delta_c(x)\, dx \tag{5.18}$$

is defined and equals 1. Also,

$$\lim_{c\to 0} \int_{-\infty}^{\infty} f(x)\delta_c(x)\, dx = f(0) , \tag{5.19}$$

which may be shown formally for continuous functions $f(x)$ as follows:

$$\lim_{c\to 0} \int_{-\infty}^{\infty} f(x)\delta_c(x)\, dx = \lim_{c\to 0} \int_{-c/2}^{c/2} f(x)\delta_c(x)\, dx = \lim_{c\to 0} \frac{1}{c} \int_{-c/2}^{c/2} f(x)\, dx ;$$

now, by the mean value theorem of integral calculus,

$$\int_{-c/2}^{c/2} f(x) dx = f(\xi c) \int_{-c/2}^{c/2} dx = c f(\xi c) ,$$

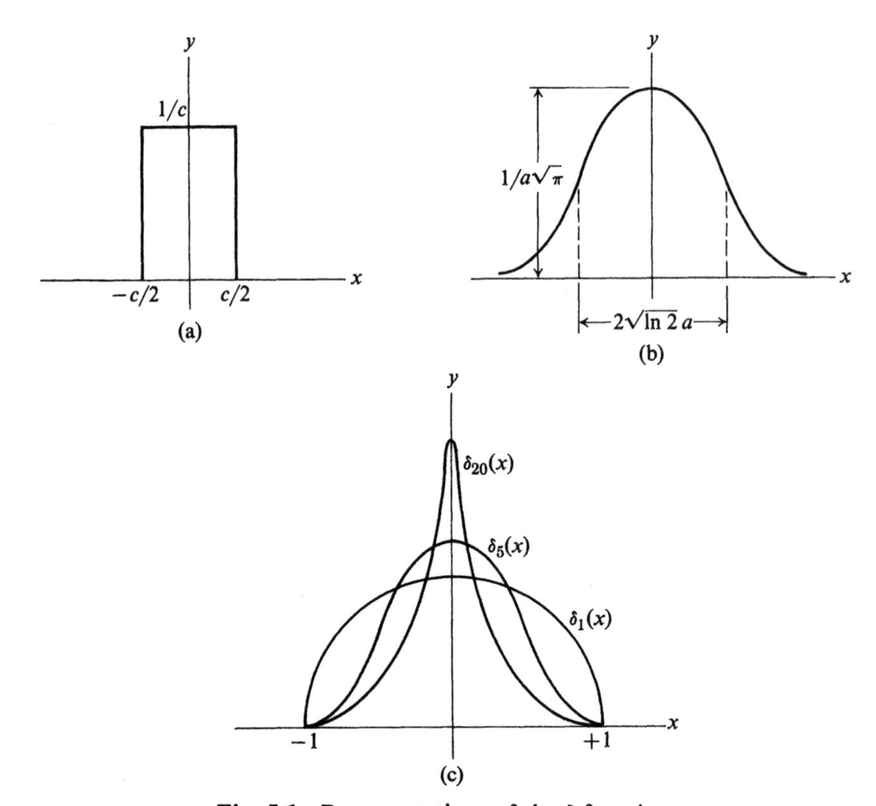

Fig. 5.1. Representations of the δ-function.

where $-1/2 < \xi < 1/2$. Letting $c \to 0$, we obtain

$$\lim_{c \to 0} \int_{-\infty}^{\infty} f(x)\delta_c(x)\, dx = f(0)\ .$$

2. The sequence of Gaussian distribution functions

$$\delta_a(x) \equiv \frac{1}{a\sqrt{\pi}} e^{-x^2/a^2}$$

provides another representation of the δ-function (see Fig. 5.1b). Note that

$$\lim_{a \to 0} \delta_a(x) = 0 \qquad \text{for all } x \neq 0\ ;$$

$$\int_{-\infty}^{\infty} \delta_a(x)\, dx = 1, \qquad \text{independent of } a\ ;$$

$$\lim_{a \to 0} \int_{-\infty}^{\infty} f(x)\delta_a(x)\, dx = f(0)\ .$$

The entire contribution to the integral, as $a \to 0$, comes from the neighborhood of $x = 0$. Therefore we may write symbolically,

$$\delta(x) = \lim_{a \to 0} \delta_a(x) = \lim_{a \to 0} \frac{1}{a\sqrt{\pi}} e^{-x^2/a^2}. \tag{5.20}$$

3. Another useful representation for the δ-function is

$$\delta(x) = \lim_{\epsilon\to 0} \delta_\epsilon(x) \equiv \lim_{\epsilon\to 0} \frac{1}{\pi} \frac{\epsilon}{x^2+\epsilon^2}, \tag{5.21}$$

which the reader can establish as in the above example.

4. The final representation of the δ-function is slightly different from the preceding three. It will play a central role in the proof of Weierstrass's theorem, the basic result of this chapter. It is defined as

$$\delta_n(x) = \begin{cases} c_n(1-x^2)^n & \text{for } 0 \le |x| \le 1, \quad n = 1, 2, 3, \cdots, \\ 0 & \text{for } |x| > 1, \end{cases} \tag{5.22}$$

where the constant c_n must be determined so that

$$\int_{-1}^{1} \delta_n(x)\, dx = 1 . \tag{5.23}$$

The functions $\delta_n(x)$ form a sequence whose limit is a δ-function (see Fig. 5.1c). We shall show, first informally, then rigorously in the proof of Weierstrass's theorem, that

$$\lim_{n\to\infty} \int_{-1}^{1} f(x)\delta_n(x)\, dx = f(0) . \tag{5.24}$$

Thus

$$\lim_{n\to\infty} \delta_n(x) = \delta(x) . \tag{5.25}$$

This representation of the δ-function differs from the others in that the defining parameter n takes on integral values that increase to infinity, instead of decreasing continuously to zero.

First we determine the normalization constant c_n. We have

$$1/c_n = \int_{-1}^{1} (1-x^2)^n\, dx = 2\int_0^1 (1-x^2)^n\, dx . \tag{5.26}$$

Making the change of variable $x = \sin\theta$, we obtain

$$\frac{1}{c_n} = 2\int_0^{\pi/2} \cos^{2n+1}\theta\, d\theta = \frac{2^{n+1}n!}{1\cdot 3\cdot 5\cdots(2n+1)} . \tag{5.27}$$

From Eq. (5.27), it follows that

$$c_n = (2n+1)!/2^{2n+1}(n!)^2 . \tag{5.28}$$

It is not clear from this expression for c_n how it behaves as $n\to\infty$. We can estimate its behavior from Eq. (5.26). Picking up again from there, we have

$$1/c_n = 2\int_0^1 (1-x^2)^n\, dx \ge 2\int_0^{1/\sqrt{n}} (1-x^2)^n\, dx , \tag{5.29}$$

since $1/\sqrt{n} \le 1$ for all $n = 1, 2, \cdots$, and since the integrand is positive throughout [0, 1]. Now we shall show that for all n, the integrand $(1-x^2)^n \ge 1 - nx^2$

for all x in [0, 1]. Consider the function $g(x) \equiv (1 - x^2)^n - (1 - nx^2)$. Since $g(0) = 0$, and

$$g'(x) = 2nx[1 - (1 - x^2)^{n-1}] > 0 \qquad \text{for all } 0 < x \leq 1,$$

$g(x)$ must be monotonically increasing in the interval [0, 1]. Therefore $g(x) \geq 0$, or $(1 - x^2)^n \geq (1 - nx^2)$, for all x in [0, 1]. Using this inequality in Eq. (5.29), we have

$$1/c_n \geq 2\int_0^{1/\sqrt{n}} (1 - nx^2)\, dx = 4/3n^{1/2} > 1/n^{1/2}.$$

Therefore

$$c_n < n^{1/2}. \tag{5.30}$$

This result may also be obtained from Stirling's formula.

As $n \to \infty$, the contribution to the integral $\int_{-1}^{1} \delta_n(x)\, dx$ comes increasingly from the neighborhood surrounding the origin. To see this, note that for $0 < \delta < 1$,

$$\int_{-1}^{-\delta} \delta_n(x)\, dx = \int_{\delta}^{1} \delta_n(x)\, dx, \tag{5.31}$$

since $\delta_n(x)$ is an even function of x. Now

$$\int_{\delta}^{1} \delta_n(x)\, dx < n^{1/2}(1 - \delta^2)^n(1 - \delta) < n^{1/2}(1 - \delta^2)^n, \tag{5.32}$$

since $c_n < n^{1/2}$ [Eq. (5.30)], and $(1 - x^2)^n$ takes on its maximum value at $x = \delta$, and therefore $\int_\delta^1 (1 - x^2)^n\, dx$ is bounded by the area of a rectangle of height $(1 - \delta^2)^n$ and base $(1 - \delta)$, since $0 < \delta < 1$. It is well known that the nth power of any positive number less than 1 will decrease more rapidly with n than any power of n will increase. In particular, the behavior of the term $(1 - \delta^2)^n$ will dominate the term $n^{1/2}$ as $n \to \infty$, and therefore

$$\lim_{n\to\infty} \int_{\delta}^{1} \delta_n(x)\, dx = 0.$$

Since $\delta_n(x)$ is continuous and never negative, it follows that $\lim_{n\to\infty} \delta_n(x) = 0$ for $0 < x \leqq 1$. Since we have already arranged that $\int_{-1}^{1} \delta_n(x)\, dx = 1$ by our choice of c_n (Eq. 5.28), we obtain the result

$$\lim_{n\to\infty} \int_{-1}^{1} f(x)\delta_n(x)\, dx = f(0). \tag{5.33}$$

5.4 WEIERSTRASS'S THEOREM: APPROXIMATION BY POLYNOMIALS

Weierstrass's famous approximation theorem proves that one can construct from the set of powers of x a sequence of polynomials which converges *uniformly* to any function that is continuous on the finite closed interval $[a, b]$. From this result, we can prove that there exists a complete orthonormal set of polynomials on any interval $[a, b]$. Weierstrass's theorem is the starting point for proving the completeness properties of many of the functions of mathematical physics, such as the Legendre polynomials, the trigonometric functions (Fourier series), and the spherical harmonics.

Weierstrass's Theorem. If $f(x)$ is continuous on the closed interval $[a, b]$, there exists a sequence of polynomials $P_n(x)$ such that

$$\lim_{n\to\infty} P_n(x) = f(x)$$

uniformly on $[a, b]$.

Proof. We may assume without loss of generality that $f(x)$ is defined on $[0, 1]$. Suppose that $f(x)$ were defined on $[a, b]$. Then consider the function h defined by

$$h\left(\frac{x-a}{b-a}\right) \equiv f(x) .$$

Clearly, $f(a) = h(0)$, $f(b) = h(1)$, and any x in the interval $[a, b]$ corresponds to a z in $[0, 1]$. Thus if $h(z)$ can be approximated by polynomials in z, then since any polynomial in $z = (x-a)/(b-a)$ is also a polynomial in x, we can go from the polynomial approximating h to a polynomial approximating f. Furthermore, we can assume that $h(z)$ vanishes at $z = 0$ and $z = 1$, for if it does not, we define

$$g(z) = h(z) - h(0) - z[h(1) - h(0)]$$

for z in $[0, 1]$. Clearly, $g(0) = 0$ and $g(1) = 0$. Since $g(z)$ and $h(z)$ differ only by a polynomial, if we can approximate $g(z)$ by a polynomial, we can approximate $h(z)$ by that same polynomial plus the polynomial $h(0) + [h(1) - h(0)]z$.

Therefore we assume our original function $f(x)$ to be defined on $[0, 1]$ and to vanish at $x = 0$ and $x = 1$. We may define $f(x)$ as we choose outside $[0, 1]$, and we define it to be identically zero there.

Now we set

$$P_n(x) = \int_{-1}^{1} f(x+t)\delta_n(t)\,dt, \qquad 0 \leq x \leq 1 , \tag{5.34}$$

where $\delta_n(t)$ is the sequence of functions, defined in Eq. (5.22), that we claimed represented a δ-function:

$$\delta_n(t) = \begin{cases} c_n(1-t^2)^n & \text{for } 0 \leq |t| \leq 1 , \\ 0 & \text{for } |t| > 1 . \end{cases}$$

If that claim were true, then

$$\lim_{n\to\infty} P_n(x) = f(x) ,$$

and the proof would be complete. We mention this just to indicate the motivation behind the rigorous proof we now present. This proof will also demonstrate that the sequence $\delta_n(t)$ as $n \to \infty$ really does have the properties that characterize a δ-function.

We are assuming that $f(x)$ vanishes outside $[0, 1]$. We may therefore write Eq. (5.34) as

$$P_n(x) = \int_{-x}^{1-x} f(x+t)\delta_n(t)\,dt\,,$$

where $f(x+t) \equiv 0$ whenever $t \leqq -x$ or $t \geqq 1-x$. By a simple change of variable $(t \to t - x)$, we obtain

$$P_n(x) = \int_0^1 f(t)\delta_n(t-x)\,dt = \int_0^1 f(t)c_n[1-(t-x)^2]^n\,dt\,.$$

This last integral shows clearly that $P_n(x)$ *is a polynomial* (of degree $2n$) in x. The coefficients of the powers of x are definite integrals over t. Thus $P_n(x)$ is a sequence of polynomials. We shall prove that this sequence converges uniformly to $f(x)$.

In analysis it is shown that a function which is continuous on a finite *closed* interval is uniformly continuous there. For those unfamiliar with this result, its meaning may be abstracted from the following example. The function $q(x) = x^{-1}$ is continuous on the *open* interval $(0, 1)$, but is not uniformly continuous there because

$$|q(x+\delta) - q(x)| = \frac{\delta}{x(x+\delta)}\,,$$

and this difference *cannot* be made arbitrarily small *for all* x by a *single* choice of δ. In fact, by choosing x sufficiently close to zero—yet still in $(0, 1)$—the difference can be made arbitrarily large for any given $\delta > 0$. Thus there does not exist a $\delta > 0$ *independent of* x such that $|q(x+\delta) - q(x)|$ is arbitrarily small, and the continuity is not uniform.

Since $f(x)$ is continuous on the closed interval $[0, 1]$, it is uniformly continuous there. As will be seen, *uniform* continuity is essential to this proof. Therefore we know that given any $\epsilon > 0$, there exists a δ such that

$$|f(x+\delta) - f(x)| < \epsilon$$

for all x in $[0, 1]$.

Now, using Eq. (5.34) for $P_n(x)$, we form the quantity

$$\begin{aligned} |P_n(x) - f(x)| &= \left|\int_{-1}^{1} [f(x+t) - f(x)]\delta_n(t)\,dt\right| \\ &\leq \int_{-1}^{1} |f(x+t) - f(x)|\delta_n(t)\,dt\,, \end{aligned}$$

since $\delta_n(t) \geq 0$ for all t in $[-1, 1]$. We now break up the range of integration into three parts:

$$\int_{-1}^{1} |f(x+t) - f(x)|\delta_n(t)\,dt = \int_{-1}^{-\delta} + \int_{-\delta}^{\delta} + \int_{\delta}^{1}.$$

Since $f(x)$ is continuous on a closed interval, it is bounded there. Let the maximum value of $|f(x)| = M$. Then both $\int_{-1}^{-\delta}$ and $\int_{\delta}^{1}$ are bounded by the quantity $2Mn^{1/2}(1-\delta^2)^n$. We may see this as follows:

$$\int_{\delta}^{1} |f(x+t) - f(x)|\delta_n(t)dt \leq \int_{\delta}^{1} |f(x+t)|\delta_n(t)\, dt + \int_{\delta}^{1} |f(x)|\delta_n(t)\, dt$$

$$\leq 2M\int_{\delta}^{1} \delta_n(t)\, dt < 2Mn^{1/2}(1-\delta^2)^n .$$

Here we have used the fact that $|f| < M$ and Eqs. (5.31) and (5.32).

We may estimate the remaining integral, $\int_{-\delta}^{\delta}$, by using the uniform continuity of $f(x)$, and interpreting the limits of integration as a δ which guarantees

$$|f(x+t) - f(x)| < \epsilon/2, \qquad \text{for } |t| < \delta.$$

We find

$$\int_{-\delta}^{\delta} |f(x+t) - f(x)|\delta_n(t)\, dt < \epsilon/2 \int_{-\delta}^{\delta} \delta_n(t)\, dt < \epsilon/2 ,$$

since $\int_{-\delta}^{\delta} \delta_n(t)\, dt < 1$.

Collecting these results, we have

$$|P_n(x) - f(x)| < 4Mn^{1/2}(1-\delta^2)^n + \epsilon/2 .$$

The value of $n^{1/2}(1-\delta^2)^n$ for $0 < \delta < 1$ can be made arbitrarily small for large enough n and, in particular, smaller than $\epsilon/2$. Therefore there exists an N such that for $n > N$,

$$|P_n(x) - f(x)| < \epsilon ,$$

for any arbitrarily small preassigned ϵ, that is,

$$\lim_{n\to\infty} |P_n(x) - f(x)| = 0.$$

This means that the sequence of polynomials $P_n(x)$ converges uniformly to the continuous function $f(x)$ on $[0, 1]$, and so the proof is complete. In fact, this holds for an arbitrary continuous function on an arbitrary finite closed interval $[a, b]$, as was demonstrated at the outset. QED

Thus Weierstrass's theorem tells us that there exists a set of coefficients

$$a_{nm}; \qquad n = 0, \cdots, \infty, \quad m = 0, \cdots, \mu = 2n ,$$

such that $\sum_{m=0}^{m=\mu} a_{nm} x^m$ tends uniformly to $f(x)$ as $n \to \infty$. It most emphatically does not guarantee a uniformly convergent power series, which would consist of a set of coefficients a_m $(m = 0, \cdots, \infty)$ such that $\sum_{m=0}^{m=n} a_m x^m$ tends uniformly to $f(x)$ as $n \to \infty$. The point is that the Weierstrass coefficients a_{nm} are not independent of n for fixed m. As the approximation improves, by going to polynomials of higher degree, the earlier coefficients change. The theorem would be false if it claimed to produce a uniformly convergent power series. For example, there is no *power* series that converges uniformly to the continuous function $\sqrt{x}$ in the interval [0, 1].

Weierstrass's theorem for approximation by a sequence of polynomials is in one sense, much stronger than Taylor's theorem for expansion in power series. To expand a function in a Taylor series, its derivatives of all orders must exist: it must be "analytic." In Weierstrass's theorem, only continuity is needed. Furthermore, Weierstrass's theorem demonstrates the existence of polynomial approximations outside the radius of convergence of a Taylor series. But there is, in general, no possibility of rearranging the uniformly convergent sequence of *polynomials* that approximate any continuous function so as to produce a convergent *power* series (Taylor expansion).

In Taylor expansions we need to know the function and its derivatives *locally*, at a point; the radius of convergence of the expansion may be finite or infinite. Weierstrass's theorem applies only to finite intervals, and we need to know the function, but not its derivatives, *globally*, over the entire interval.

As a by-product of this theorem, we have established rigorously that the representation of Dirac's delta function given in Eq. (5.22) has the properties claimed for it [Eq. (5.24)]. Thus Weierstrass, who proved this theorem in 1885 by the above method, anticipated Dirac considerably. In fact, he was not the only one. Heaviside used a closely related symbolic function before Dirac. Yet it was with Dirac's introduction of symbolic functions in his classic book, *The Principles of Quantum Mechanics*, which appeared in 1930, that their use became widespread. Only relatively recently was the theory of δ-functions, and other related symbolic functions, established rigorously by the mathematician Laurent Schwartz in his theory of distributions.

Weierstrass's theorem may also be extended to functions of several variables. By a straightforward generalization of the proof, it can be shown that if a function $f(x_1, x_2, \cdots, x_m)$ is continuous in each variable x_i for x_i in $[a_i, b_i]$ $(i = 1, 2, \cdots, m)$, f may be approximated uniformly by the polynomials

$$P_n(x_1, x_2, \cdots, x_m) = \int_{a_1}^{b_1}\int_{a_2}^{b_2}\cdots\int_{a_m}^{b_m} f(t_1, t_2, \cdots, t_m)\delta_n(t_1 - x_1)$$
$$\delta_n(t_2 - x_2)\cdots\delta_n(t_m - x_m)\, dt_1\, dt_2\cdots dt_m\,. \tag{5.35}$$

The completeness of the trigonometric functions follows from the special case $m = 2$, as we shall see in Section 5.6.

We close this section with the statement of an important consequence of our basic result. From Section 5.2 it follows that if the set P_n defined in the proof of Weierstrass's theorem approximate any continuous function f uniformly, they also approximate any continuous function *in the mean*; that is, given any ϵ, there exists an n such that $\|f - P_n\| < \epsilon$. Now it can be shown that any function in the Hilbert space of square-integrable functions can be approximated arbitrarily closely in the mean by a continuous function. This is discussed further in Section 8.7. It follows that *any* function in this Hilbert space can be approximated *in the mean* by some P_n. This can be seen as follows.

Let ϕ be a function in Hilbert space. Then the difference $\phi - P_n$ can be written as

$$\phi - P_n = (\phi - f) + (f - P_n) ,$$

where f is a continuous function. By the triangle inequality, we have

$$||\phi - P_n|| \leq ||\phi - f|| + ||f - P_n|| .$$

Now f can be chosen so that $||\phi - f||$ is as small as we please; and by Weierstrass's theorem, n can be chosen so that $||f - P_n||$ is as small as we please. Thus P_n approximates ϕ, an arbitrary function in Hilbert space, arbitrarily closely in the mean. This fact will be needed in what follows. A complete proof can be found in most analysis texts, e.g., Riesz and Nagy (Section 46) or Rudin (Theorem 10.38).

5.5 LEGENDRE POLYNOMIALS

We are now in a position to demonstrate that there exists a *complete* orthonormal set of polynomials on the finite closed interval $[a, b]$. The proof of completeness follows from Weierstrass's theorem. After establishing the general result, we shall examine in detail one very important special case—the complete orthonormal set of polynomials on the interval $[-1, 1]$, which, apart from constant normalization factors, is the set of Legendre polynomials. We shall construct the first three functions in this set explicitly, and then provide a formula for calculating all the others.

Weierstrass's theorem tells us that any continuous function f on $[a, b]$ can be approximated uniformly by a sequence of polynomials:

$$P_n(x) = \sum_{m=0}^{2n} a_{nm} x^m .$$

Since the set $\{x^n, n = 0, 1, \cdots\}$ is linearly independent, we may apply the Gram-Schmidt orthogonalization process to construct from it an orthonormal basis $\{Q_n(x)\}$, where $Q_n(x)$ is a polynomial of degree n. We may now express the original functions as finite linear combinations of the orthonormal set:

$$x^m = \sum_{i=0}^{m} c_{mi} Q_i(x) .$$

It follows that the polynomials P_n, which approximate the function f uniformly, may be expressed as

$$P_n(x) = \sum_{m=0}^{2n} a_{nm} \sum_{i=0}^{m} c_{mi} Q_i(x) .$$

We now prove that the orthonormal set $\{Q_n\}$ is complete by showing that it is closed (Theorem 5.2). The orthonormal set $\{Q_n\}$ will be closed if the only function orthogonal to all the Q_n is the zero function. Assume that we have a function in Hilbert space such that $(f, Q_n) = 0$ for all n. It follows immediately from the above equation that $(f, P_n) = 0$ for all n, since the P_n are linear combinations of the Q_n. However, we know from Weierstrass's theorem that f may be approximated in the mean by the P_n. Thus for any ϵ there exists an n such that

$||f - P_n|| < \epsilon$. But since $(f, P_n) = 0$,

$$||f - P_n|| \equiv (f - P_n, f - P_n)^{1/2} = [||f||^2 + ||P_n||^2]^{1/2}.$$

Therefore

$$||f||^2 + ||P_n||^2 < \epsilon^2.$$

But this implies that $||f||$ is arbitrarily small, and hence f is equal to zero (almost everywhere). Therefore the orthonormal set $\{Q_n\}$ is closed and hence, by Theorem 5.2, it is a complete orthonormal set of polynomials on $[a, b]$.

The completeness of the set Q_i means (according to Definition 5.6) that there exists a set of of constants $\{a_i\}$ such that any function g in the Hilbert space can be approximated in the mean by the sequence of partial sums

$$g_n = \sum_{i=0}^{n} a_i Q_i(x) .$$

But the a_i are independent of n. Thus as we extend the sum to infinity, the approximation improves *without the earlier a_i changing*. Therefore we may say that there exists an infinite *series*

$$\sum_{i=0}^{\infty} a_i Q_i$$

which approximates g in the mean. Symbolically, we write

$$g(x) \doteq \sum_{i=0}^{\infty} a_i Q_i(x) .$$

We have shown that if we have a complete orthonormal set of functions, the coefficients which provide the best approximation in the mean are the Fourier coefficients. Thus the expansion coefficients in the infinite series are given by

$$a_i = (Q_i, g) .$$

Let us now apply these considerations to a concrete case. We shall use the Gram-Schmidt orthogonalization process applied to the basis $\{1, x, x^2, \cdots\}$ to construct the set of orthonormal polynomials on the interval $[-1, 1]$. By the previous result we know that this set will be complete.

The first member of the orthonormal set is $\bar{P}_0(x) = (\frac{1}{2})^{1/2}$, which is evidently normalized for x in $[-1, 1]$. Using the prescription of Section 4.3, we have

$$\bar{P}_0(x) = (\tfrac{1}{2})^{1/2},$$

$$\bar{P}_1 = \frac{x - 1/2^{1/2}\int_{-1}^{1} 1/2^{1/2}x\,dx}{\left\|x - 1/2^{1/2}\int_{-1}^{1} 1/2^{1/2}x\,dx\right\|} = \left(\frac{3}{2}\right)^{1/2} x ,$$

$$\bar{P}_2 = \frac{x^2 - 1/2^{1/2}\int_{-1}^{1} 1/2^{1/2}x^2\,dx - (\frac{3}{2})^{1/2}x\int_{-1}^{1} (\frac{3}{2})^{1/2}x^3\,dx}{\left\|x^2 - 1/2^{1/2}\int_{-1}^{1} 1/2^{1/2}x^2\,dx - (\frac{3}{2})^{1/2}x\int_{-1}^{1} (\frac{3}{2})^{1/2}x^3\,dx\right\|} = \left(\frac{5}{2}\right)^{1/2}\left(\frac{3}{2}x^2 - \frac{1}{2}\right). \tag{5.36}$$

This is a tedious process; we have carried it out for these three cases only to show that one can construct orthonormal functions directly. One would hope that there is a better way to compute the orthonormal polynomials of higher order, and, indeed, there is.

We claim that a general formula for the complete orthonormal set of polynomials on $[-1, 1]$ is

$$\bar{P}_n(x) = \left(\frac{2n+1}{2}\right)^{1/2} \frac{1}{2^n n!} \frac{d^n}{dx^n}(x^2-1)^n . \tag{5.37}$$

The orthogonal, but unnormalized, polynomials

$$P_n(x) = \frac{1}{2^n n!} \frac{d^n}{dx^n}(x^2-1)^n \tag{5.38}$$

are known as the *Legendre polynomials*. Equation (5.38) is called *Rodrigues' formula*.

Thus the Legendre polynomials differ from the orthonormal set $\{\bar{P}_n(x)\}$ by constant multiplicative factors. The problem now is to show that the Rodrigues formula gives the *same* set of orthonormal polynomials as the Gram-Schmidt orthogonalization process. Direct computation verifies that they agree through the polynomials of second degree. To show that the Rodrigues formula holds for all orders, we must show that:

1. On any fixed interval there can be only one complete orthonormal set of polynomials in which the nth polynomial is of degree n.
2. The $\bar{P}_n(x)$ given by the Rodrigues formula do in fact form a complete orthonormal set on $[-1, 1]$.

The uniqueness property of the complete orthonormal set of polynomials is embodied in the Gram-Schmidt process, which can generate one and only one complete orthonormal set of polynomials (up to a phase factor) on $[-1, 1]$, or, in fact, on any given interval. Hence we have, in fact, already proved condition (1).

To demonstrate condition (2), we now show that the set of polynomials generated by the Rodrigues formula for the interval $[-1, 1]$ is orthonormal, and that the function $P_n(x)$ given by Rodrigues' formula is a polynomial of degree n. Then completeness follows by our earlier proof. It is immediately clear that $P_n(x)$ is a polynomial of degree n. We prove orthonormality in the following theorem.

Theorem 5.3.

$$(\bar{P}_n, \bar{P}_m) = \int_{-1}^{1} \bar{P}_n(x)\bar{P}_m(x)\, dx = \delta_{nm} . \tag{5.39}$$

Proof. We first prove orthogonality. We denote d^n/dx^n by d^n, and suppose that $n > m$. Dropping constant factors, where we have integrated by parts, we have

$$\int_{-1}^{1} \bar{P}_n \bar{P}_m \, dx = \int_{-1}^{1} [d^n(x^2 - 1)^n] [d^m(x^2 - 1)^m] \, dx$$

$$= [d^{n-1}(x^2 - 1)^n] [d^m(x^2 - 1)^m] \Big]_{-1}^{1}$$

$$- \int_{-1}^{1} [d^{n-1}(x^2 - 1)^n] [d^{m+1}(x^2 - 1)^m \, dx] ,$$

Since

$$d^{n-1}(x^2 - 1)^n = (\text{a polynomial}) \cdot (x^2 - 1),$$

the first term vanishes upon putting in the limits ± 1, leaving the second term. Therefore, after n partial integrations, we have

$$\int_{-1}^{1} (-1)^n (x^2 - 1)^n d^{m+n} (x^2 - 1)^m \, dx ,$$

since the term which is evaluated at ± 1 always vanishes because it is proportional to some power of $(x^2 - 1)$. Now, since $n > m$, $n + m > 2m$ and so $d^{n+m}(x^2 - 1)^m = 0$. Therefore

$$\int_{-1}^{1} \bar{P}_n \bar{P}_m \, dx = 0 \qquad \text{for } m \neq n .$$

If $m = n$, then as before (but putting in constant factors),

$$\int_{-1}^{1} \bar{P}_n \bar{P}_n \, dx = \frac{(2n + 1)(-1)^n}{2^{2n+1}(n\,!)^2} \int_{-1}^{1} (x^2 - 1)^n d^{2n} (x^2 - 1)^n \, dx .$$

But $(x^2 - 1)^n$ is a polynomial of degree $2n$, so the $(2n)^{\text{th}}$ derivative is just $(2n)!$. The integral becomes

$$\frac{(-1)^n (2n + 1)!}{2^{2n+1}(n!)^2} \int_{-1}^{1} (x^2 - 1)^n \, dx = 1 ,$$

where we have used Eqs. (5.26) and (5.28).

The functions $\bar{P}_n(x) = [(2n + 1)^{1/2}/2] \, P_n(x)$ are therefore a complete orthonormal set on the interval $[-1, 1]$, and they are the only such set.

We shall now work out several other results pertaining to the Legendre polynomials $P_n(x)$. Our approach will be inductive and straightforward, a continuation of the techniques used to prove orthonormality. There are more elegant ways to obtain most of these results which will be used later to deal with both the Legendre polynomials and other complete sets of functions. But for the moment we proceed in a more pedestrian manner.

First we shall derive from the Rodrigues formula an explicit expression for $P_n(x)$ as a polynomial. We apply the binomial theorem to the factor $(x^2 - 1)^n$ to obtain

$$(x^2 - 1)^n = \sum_{m=0}^{n} \binom{n}{m} x^{2m} (-1)^{n-m} ,$$

where $\binom{n}{m}$ is the binomial coefficient:

$$\binom{n}{m} \equiv \frac{n!}{m!(n-m)!} . \tag{5.40}$$

Then, taking the nth derivative, we obtain

$$P_n(x) = \frac{1}{2^n n!} d^n (x^2 - 1)^n = \frac{1}{2^n n!} \sum_{m \geq p}^{n} (-1)^{n-m} \binom{n}{m} \frac{(2m)!}{(2m-n)!} x^{2m-n} , \tag{5.41}$$

where $p = n/2$ if n is even and $p = (n+1)/2$ if n is odd.

We next prove that $P_n(1) = 1$ for all n. In itself this is not of staggering importance, although it is interesting. We prove it now to illustrate an important technique. (We shall have an easier way later.) We write

$$P_n(x) = \frac{1}{2^n n!} d^n[(x-1)^n(x+1)^n] ,$$

and evaluate the nth derivative of the product by Leibnitz's rule (the "binomial expansion" for the nth derivative of a product):

$$d^n(uv) = \sum_{m=0}^{n} \binom{n}{m} d^m u d^{n-m} v = u d^n v + nd\, u d^{n-1} v + \frac{n(n-1)}{2!} d^2 u d^{n-2} v + \cdots + \frac{n!}{k!(n-k)!} d^k u d^{n-k} v + \cdots + v d^n u. \tag{5.42}$$

We now take $u = (x-1)^n$ and $v = (x+1)^n$. It is clear that the only nonzero term in the Leibnitz expansion at $x = 1$ is the term $m = n$, in which the factor $u = (x-1)^n$ is "differentiated down" to the constant $n!$ and hence does not vanish at $x = 1$. Therefore

$$d^n[(x-1)^n(x+1)^n]\Big|_{x=1} = d^n(x-1)^n d^0(x+1)^n\Big|_{x=1} = n!2^n ,$$

and consequently,

$$P_n(1) = 1 \qquad \text{for all } n . \tag{5.43}$$

The Legendre polynomials arose originally as the solution to a differential equation, now called Legendre's equation:

$$(1 - x^2)P_n''(x) - 2xP_n'(x) + n(n+1)P_n(x) = 0 . \tag{5.44}$$

To prove that the elements of the complete orthonormal set of polynomials on $[-1, 1]$ satisfy this equation, we begin with the identity

$$(x^2 - 1)\frac{d}{dx}(x^2 - 1)^n = 2nx(x^2 - 1)^n .$$

Now differentiate both sides of this identity $n + 1$ times. With the help of

Leibnitz's formula (Eq. 5.42), we obtain for the left-hand side

$$d^{n+1}[(x^2-1)d(x^2-1)^n] = (x^2-1)d^{n+2}(x^2-1)^n + 2(n+1)xd^{n+1}(x^2-1)^n + n(n+1)d^n(x^2-1)^n\,,$$

and for the right-hand side, we have

$$d^{n+1}[2nx(x^2-1)^n] = 2nx\,d^{n+1}(x^2-1)^n + 2n(n+1)d^n(x^2-1)^n\,.$$

The difference of these two equations must vanish. Therefore

$$(x^2-1)d^{n+2}(x^2-1)^n + 2x\,d^{n+1}(x^2-1)^n - n(n+1)d^n(x^2-1)^n = 0\,.$$

Now, using the Rodrigues formula for $P_n(x)$, we obtain

$$(x^2-1)P_n'' + 2x\,P_n' - n(n+1)P_n = 0\,, \tag{5.45}$$

which is equivalent to Legendre's equation (5.44). It may also be written in the form

$$[(1-x^2)P_n'(x)]' + n(n+1)P_n(x) = 0\,. \tag{5.46}$$

Thus the $P_n(x)$ are solutions of Legendre's equations; but they are not the only solutions. There exists another linearly independent solution called, collectively, the Legendre *functions* of the second kind, denoted by $Q_n(x)$. However, these functions are not finite at $x = \pm 1$ as are the $P_n(x)$, and so are excluded as possible solutions in many physical applications; they are not polynomials.

There is an alternative proof of the orthogonality of the Legendre polynomials which follows directly from the differential equation (5.46) for $P_n(x)$. We know that P_n and P_m $(n \neq m)$ satisfy the equations

$$\frac{d}{dx}[(1-x^2)P_n'] + n(n+1)P_n = 0\,,$$

$$\frac{d}{dx}[(1-x^2)P_m'] + m(m+1)P_m = 0\,.$$

Multiplying these equations by P_m and P_n respectively, subtracting them, and integrating from -1 to $+1$, we obtain

$$\begin{aligned}(n-m)(n+m+1)&\int_{-1}^{1} P_nP_m\,dx \\ &= \int_{-1}^{1}\left[P_n\frac{d}{dx}((1-x^2)P_m') - P_m\frac{d}{dx}((1-x^2)P_n')\right]dx \\ &= \left\{\left[P_n(1-x^2)P_m'\right]_{-1}^{1} - \int_{-1}^{1} dx\,(1-x^2)P_m'P_n'\right. \\ &\quad \left. - \left[P_m(1-x^2)P_n'\right]_{-1}^{1} + \int_{-1}^{1} dx\,(1-x^2)P_n'P_m'\right\} = 0\,;\end{aligned}$$

here the two integrals cancel and the two other terms vanish. Since $n \neq m$, it follows that

$$(P_n, P_m) \equiv \int_{-1}^{1} P_n P_m \, dx = 0 \, .$$

This is actually a special case of a proof that holds for a wide variety of functions arising as solutions to the Sturm-Liouville problem, which we shall discuss in Section 5.10. The proof depends on the fact that the differential equations obeyed by orthogonal functions have a certain form.

We have sampled only a few of the properties of the complete orthonormal set of polynomials on $[-1, 1]$. We shall encounter them many times in the course of the book (we met them once before in the multipole expansion (Eqs. 1.100, 1.101, 1.102), and shall discuss their properties in more detail. But our main concern for the moment is the *completeness* of this set of polynomials, and other sets of functions—not all the special properties these sets of functions happen to have.

We now turn to the single most important complete set of functions, the ancestor of them all: the trigonometric functions and Fourier series.

5.6 FOURIER SERIES

The completeness properties of the set of trigonometric functions $\{\sin n\theta, \cos n\theta, n = 0, 1, 2, \cdots, \infty\}$ can be deduced from Weierstrass's theorem for two variables. Equation 5.35 tells us that any function $g(x, y)$ which is continuous in both variables on specified finite closed intervals may be approximated uniformly by the the sequence of functions

$$g_N(x, y) = \sum_{n,m=0}^{N} A_{nm}^{(N)} x^n y^m \; ;$$

that is, $\lim_{N\to\infty} g_N(x, y) = g(x, y)$, uniformly in x and y. The coefficients $A_{nm}^{(N)}$ are not independent of N for fixed n and m, so this is *not* a power-series expansion.

If we change to polar coordinates and restrict the domain of definition to the unit circle, then $x = \cos\theta$ and $y = \sin\theta$ so that

$$g(\cos\theta, \sin\theta) = f(\theta) = \lim_{N\to\infty} \sum_{n,m=0}^{N} A_{nm}^{(N)} \cos^n\theta \sin^m\theta \, . \tag{5.47}$$

Clearly, the only functions $f(\theta)$ which can satisfy this last equation are periodic functions with periodicity 2π; this is a consequence of restricting x and y to the unit circle. We shall generalize later to include functions having periodicities different from 2π.

Using Euler's formula,

$$e^{i\theta} = \cos\theta + i \sin\theta \, ,$$

we obtain expressions for the nth powers of $\sin\theta$ and $\cos\theta$:

$$\cos^n\theta = \left[\frac{1}{2}\left(e^{i\theta} + e^{-i\theta}\right)\right]^n , \qquad \sin^n\theta = \left[\frac{1}{2i}\left(e^{i\theta} - e^{-i\theta}\right)\right]^n .$$

Then Eq. (5.47) can be rewritten in the form

$$f(x) = \lim_{M\to\infty} f_M(x) \equiv \lim_{M\to\infty} \sum_{n=-M}^{M} \frac{c_n^{(M)}}{(2\pi)^{1/2}} e^{inx}, \tag{5.48a}$$

where we have inserted the factor $(2\pi)^{1/2}$ for later convenience, and have replaced the variable θ by x to emphasize the general nature of the result. Clearly, we may write equivalently

$$f(x) = \lim_{M\to\infty} f_M(x) = \lim_{M\to\infty} \left[\frac{a_0^{(M)}}{2} + \sum_{n=1}^{M} (a_n^{(M)} \cos nx + b_n^{(M)} \sin nx)\right]. \tag{5.48b}$$

The complex expansion coefficients in the exponential form of the series, and the real coefficients in the trigonometric form, are related as follows:

$$\begin{aligned} a_0^{(M)} &= \left(\frac{2}{\pi}\right)^{1/2} c_0^{(M)}, & & & c_0^{(M)} &= \left(\frac{\pi}{2}\right)^{1/2} a_0^{(M)}, \\ a_n^{(M)} &= \frac{1}{(2\pi)^{1/2}} (c_n^{(M)} + c_{-n}^{(M)}), & &\text{or} & c_n^{(M)} &= \left(\frac{\pi}{2}\right)^{1/2} (a_n^{(M)} - ib_n^{(M)}), \\ b_n^{(M)} &= \frac{i}{(2\pi)^{1/2}} (c_n^{(M)} - c_{-n}^{(M)}), & & & c_{-n}^{(M)} &= \left(\frac{\pi}{2}\right)^{1/2} (a_n^{(M)} + ib_n^{(M)}). \end{aligned} \tag{5.49}$$

The superscript M reminds us that the coefficients in the sequence are *not* independent of M; the coefficients of earlier terms in the sequence may change as M increases.

Thus any continuous function $f(x)$, for which $f(x) = f(x + 2\pi)$, can be approximated uniformly by a sequence of trigonometric polynomials:

$$f_M(x) = \sum_{n=-M}^{M} \frac{c_n^{(M)}}{(2\pi)^{1/2}} e^{inx} = \frac{a_0^{(M)}}{2} + \sum_{n=1}^{M} (a_n^{(M)} \cos nx + b_n^{(M)} \sin nx).$$

Now suppose that $f(x)$ is any continuous function on the closed interval $[a, a + 2\pi]$, and assume further that $f(x)$ does not satisfy the periodicity condition $f(a) = f(a + 2\pi)$ at the endpoints of the interval. It is clear that for any such function f, we can find a continuous function $g(x)$ that both satisfies the periodicity requirement and is such that the quantity

$$\int_a^{a+2\pi} |f(x) - g(x)|^2 \, dx$$

can be made arbitrarily small. (See Courant-Hilbert for proof.) Therefore the trigonometric sequence that converges uniformly to $g(x)$ will converge *in the mean* to $f(x)$ on the interval $[a, a + 2\pi]$. In fact, since any function in Hilbert space can be approximated in the mean to arbitrary precision by some continuous function (see Section 5.4), this same result holds for any function in Hilbert space.

We saw in Section 5.1 that the set of functions

$$\left\{\frac{1}{(2\pi)^{1/2}} e^{inx}, n = 0, \pm 1, \cdots\right\}$$

is orthonormal on the interval $[-\pi, \pi]$. We shall denote this orthonormal set of functions by $\{T_n\}$. We shall now show that this set is a *complete* orthonormal set. We do this, exactly as we did for the orthonormal set of polynomials $\{Q_n\}$ in the last section, by proving that the orthonormal set is closed. Thus, assuming that $(f, T_n) = 0$ for all n, we want to prove $f = 0$. It follows from Eq. (5.48a) that $(f, f_M) = 0$ for all M, because f_M is just a linear combination of the orthonormal funtions T_n. However, we know that any function f may be approximated in the mean by the sequence f_M; that is,

$$||f - f_M|| = [||f||^2 + ||f_M||^2]^{1/2} < \epsilon .$$

Thus $f = 0$ (almost everywhere). Therefore the orthonormal set $\{T_n\}$ is closed, and hence by Theorem 5.2, it is a complete orthonormal set on $[-\pi, \pi]$. Clearly, the set of trigonometric functions which is orthonormal on $[-\pi, \pi]$,

$$\left\{\frac{1}{(2\pi)^{1/2}}, \quad \frac{1}{\pi^{1/2}} \sin nx, \quad \frac{1}{\pi^{1/2}} \cos nx, \quad n = 1, 2, \cdots\right\},$$

is also complete.

It follows from the completeness of the orthonormal set $\{T_n\}$ that we may approximate an arbitrary function f in the mean by an infinite series of the T_n. We write symbolically,

$$f(x) \doteq \sum_{n=-\infty}^{\infty} c_n T_n(x) = \sum_{n=-\infty}^{\infty} \frac{c_n}{(2\pi)^{1/2}} e^{inx} = \frac{a_0}{2} + \sum_{n=1}^{\infty} (a_n \cos nx + b_n \sin nx). \quad (5.50)$$

The expansion coefficients which here are constants, independent of M (compare Eq. 5.48a), are given by

$$c_n = (T_n, f) = \frac{1}{(2\pi)^{1/2}} \int_{-\pi}^{\pi} f(x) e^{-inx}\, dx . \quad (5.51)$$

The coefficients a_n and b_n of the trigonometric series may be computed using Eqs. (5.49):

$$\begin{aligned} a_n &= \frac{1}{(2\pi)^{1/2}} (c_n + c_{-n}) = \left(\frac{1}{(2\pi)^{1/2}}\right)^2 \int_{-\pi}^{\pi} f(x)\,(e^{-inx} + e^{inx})\, dx \\ &= \frac{1}{\pi} \int_{-\pi}^{\pi} f(x) \cos nx\, dx \qquad \text{for } n = 0, 1, \cdots \\ b_n &= \frac{1}{\pi} \int_{-\pi}^{\pi} f(x) \sin nx\, dx \qquad \text{for } n = 1, 2, \cdots \end{aligned} \quad (5.52)$$

The series (5.50) with the coefficients (5.52) is known as the *Fourier series*.

Since the functions

$$T_n(x) = \frac{1}{(2\pi)^{1/2}} e^{inx}, \qquad n = 0, \pm 1 \cdots$$

form a complete orthonormal set, the completeness relation (Eq. 5.10) must hold for it. This gives

$$(f, f) = \int_{-\pi}^{\pi} |f|^2 \, dx = \sum_{n=-\infty}^{\infty} |c_n|^2 = \pi\left[\frac{a_0^2}{2} + \sum_{n=1}^{\infty} (a_n^2 + b_n^2)\right], \tag{5.53}$$

where we have used Eqs. (5.49).

Let us now summarize the situation. We have shown that the Fourier series converges in the mean to any function $f(x)$ in Hilbert space on $[-\pi, \pi]$. We also know that there exists *some trigonometric sequence* that converges *uniformly* to any *continuous* function $f(x)$ on $[-\pi, \pi]$ for which $f(-\pi) = f(\pi)$. It is plausible therefore that the result for Fourier series, i. e., the particular trigonometric series with the Fourier coefficients [Eqs. (5.48)], can be strengthened. However, it most definitely is not guaranteed that the particular choice of coefficients which guarantees convergence in the mean, also yields uniform convergence. The converse is easily proved, however; namely, if a trigonometric series converges uniformly, then its coefficients are the Fourier coefficients. This may be demonstrated by multiplying [Eq. (5.50)] by cos mx or sin mx and integrating the uniformly convergent series term by term. But establishing the conditions for uniform convergence is the real problem. We begin our investigation of this question, which is the central question in the theory of Fourier series, by proving a theorem.

Theorem 5.4. The convergence of the Fourier series to $f(x)$ is uniform in the closed interval $[-\pi, \pi]$ if $f(x)$ is continuous and its derivative is piecewise continuous in this interval, and $f(-\pi) = f(\pi)$. If, in addition, $f(x + 2\pi) = f(x)$, the convergence will be uniform everywhere.

An example of a continuous function with a piecewise continuous derivative is given in Fig. 5.2. Such functions are often referred to as *piecewise smooth* functions.

Proof. We shall denote the Fourier coefficients of $f'(x)$ by a_n' and b_n'. Integrating by parts, we have

$$a_n' \equiv \frac{1}{\pi}\int_{-\pi}^{\pi} f'(x) \cos nx \, dx = \frac{n}{\pi}\int_{-\pi}^{\pi} f(x) \sin nx \, dx = nb_n \,,$$

$$b_n' \equiv \frac{1}{\pi}\int_{-\pi}^{\pi} f'(x) \sin nx \, dx = -\frac{n}{\pi}\int_{-\pi}^{\pi} f(x) \cos nx \, dx = -n\, a_n \,, \tag{5.54}$$

$$a_0' = \int_{-\pi}^{\pi} f'(x) \, dx = f(\pi) - f(-\pi) = 0 \,.$$

We have used the fact that $f(\pi) = f(-\pi)$ in several places; the Fourier coefficients of $f(x)$ are denoted by a_n and b_n as before.

Since $f'(x)$ is piecewise continuous on $[-\pi, \pi]$, and hence square integrable, it must satisfy Bessel's inequality, Eqs. (5.9) and (5.53):

$$(f', f') \geq \pi\left[\frac{a_0'^2}{2} + \sum_{n=1}^{\infty} (a_n'^2 + b_n'^2)\right] = \pi \sum_{n=1}^{\infty} n^2(b_n^2 + a_n^2) \,, \tag{5.55}$$

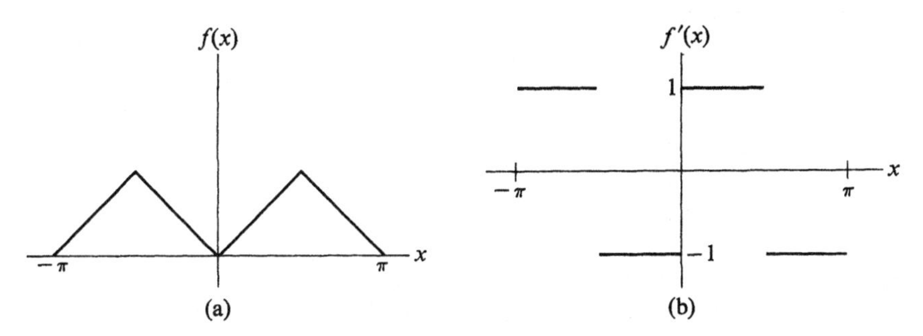

Fig. 5.2. Zig-zag function and its derivative.

where we have used Eqs. (5.54). [In fact, we know that the equality sign holds by the completeness relation, Eq. (5.10)].

The preceding results are all preliminary. We now use the Cauchy criterion to test for the uniform convergence of the Fourier series. Let

$$S_n = \frac{a_0}{2} + \sum_{p=1}^{n} a_p \cos px + \sum_{p=1}^{n} b_p \sin px .$$

By Theorem 5.1, if we can show that $|S_n - S_m| < \epsilon$ for all x in$[-\pi, \pi]$, and for all n, m larger than some $N(\epsilon)$, then we will have established uniform convergence. We have

$$\begin{aligned} |S_n - S_m| &= \left| \sum_{p=m+1}^{n} (a_p \cos px + b_p \sin px) \right| \\ &= \left| \sum_{p=m+1}^{n} \frac{1}{p} (pa_p \cos px + pb_p \sin px) \right| \\ &\leq \sqrt{\sum_{p=m+1}^{n} \left|\frac{1}{p}\right|^2 \sum_{p=m+1}^{n} \left| pa_p \cos px + pb_p \sin px \right|^2} , \end{aligned}$$

where we have used Schwarz's inequality [Eq. (4.8)]. This last expression may be written as

$$\begin{aligned} |S_n - S_m| &\leq \sqrt{\sum_{p=m+1}^{n} \frac{1}{p^2}} \cdot \sqrt{\sum_{p=m+1}^{n} p^2 |a_p \cos px + b_p \sin px|^2} \\ &\leq \sqrt{\sum_{p=m+1}^{n} \frac{1}{p^2}} \cdot \sqrt{\sum_{p=m+1}^{n} p^2 (a_p^2 + b_p^2)} , \end{aligned}$$

since $|a_p \cos px + b_p \sin px|$ can be written as

$$|\sqrt{a_p^2 + b_p^2} \cos (px - \theta)| \leq |\sqrt{a_p^2 + b_p^2}| ,$$

where $\theta = \tan^{-1} b_p/a_p$. Therefore

$$\left| S_n - S_m \right| \leq \sqrt{\sum_{p=m+1}^{n} \frac{1}{p^2}} \cdot \sqrt{\frac{1}{\pi} \int_{-\pi}^{\pi} |f'(x)|^2 \, dx} ,$$

where the term $\sqrt{\sum_{p=m+1}^{n} p^2(a_p^2 + b_p^2)}$ has been replaced by the equal or larger term

$$\sqrt{\sum_{p=1}^{\infty} p^2(a_p^2 + b_p^2)} \leq \sqrt{(1/\pi)\int_{-\pi}^{\pi} |f'|^2\, dx}\,, \tag{5.56}$$

from Eq. (5.55). Now the series $\sum_{p=1}^{\infty}(1/p^2)$ converges (it is, in fact, the Riemann zeta function evaluated at 2 and converges to $\pi^2/6$ as we shall show in Section 6.9). If we let

$$M = \sqrt{(1/\pi)\int_{-\pi}^{\pi} |f'(x)|^2\, dx}\,,$$

then we know that M is finite, because $f'(x)$ is piecewise continuous and always finite.

The series $\sum_{p=1}^{\infty}(1/p^2)$, converges, so it must satisfy the Cauchy criterion. Then for any positive number, which we take to be ϵ^2/M^2, there exists an integer N such that $\sum_{p=m+1}^{n}(1/p^2) < \epsilon^2/M^2$ when $n, m > N$. Taking the n and m that appear in $|S_n - S_m|$ larger than this N, we then have

$$|S_n - S_m| < (\epsilon^2/M^2)^{1/2} M = \epsilon\,,$$

and the proof of uniform convergence is complete. We know that the series converges *to* the function $f(x)$ because its convergence in the mean to $f(x)$ has already been established.

Note that the completeness relation could have been used earlier in this proof; there would then be an equals sign where we have an inequality in Eq. (5.56). It *is* valid to use the completeness relation because completeness has been proved already. But in order to demonstrate that prior knowledge of completeness is not required to prove uniform convergence, we have used Bessel's inequality instead. The piecewise continuity of $f'(x)$ implies that M is finite, and this is all that is needed to prove uniform convergence. However, to show that the uniform convergence is *to* $f(x)$ (since Cauchy's criterion makes no mention of the limit function) the convergence in the mean to $f(x)$ (completeness) was used at the very end of the proof. It is possible to prove independently that the uniform convergence is to $f(x)$, and then this result leads to an independent proof of completeness.

This theorem may be extended to deal with the piecewise continuous functions, such as step functions, which may have a finite number of finite discontinuities in the interval $[-\pi, \pi]$.

Theorem 5.5. If $f(x)$ is piecewise continuous in $[-\pi, \pi]$, and has a piecewise continuous derivative there, then its Fourier series converges uniformly to $f(x)$ in every closed subinterval of $[-\pi, \pi]$ in which $f(x)$ is continuous. At points of discontinuity of $f(x)$, its Fourier series converges to the arithmetic mean of the left- and right-hand limits of the function. If $f(x + 2\pi) = f(x)$, then these statements are true everywhere on the real line.

It is interesting that although we know that there exists some *trigonometric* sequence that converges uniformly to every continuous function on a closed interval, continuity of $f(x)$ alone is not sufficient to prove the covergence of its *Fourier* series. There exist continuous functions which are not differentiable, and their *Fourier* series need not converge. The additional condition of continuity of $f'(x)$, however, ensures convergence. Also, the assumption that the intervals be closed is vital to proving *uniform* convergence. The convergence of the Fourier series for piecewise continuous functions—such as step functions—will certainly not be uniform in open intervals (a, b), where one of the end-points is a point of discontinuity. In fact, it may be shown that just before the Fourier series passes over the discontinuity, it differs from the function by a finite amount. This overshooting effect is called the Gibbs phenomenon (see Problem 4c).

It is primarily the smoothness of the function that determines the size of the Fourier coefficients—the smoother the function, the more rapidly these coefficients decrease and the more rapid is the convergence. The theory of Fourier series is largely the study of the interplay between assumptions about smoothness and conclusions about convergence.

Fejer has constructed a theory of Fourier series based on a special type of series summation (called Cesaro summation) in which one considers the sequence of *arithmetic means* of the partial sums. If a series is convergent in the usual sense, then it is Cesaro summable to the same value. But many divergent series are Cesaro summable; thus Cesaro summability is a natural generalization since it preserves the usual summation as a special case. It was shown by Fejer that the Fourier series of any continuous function $f(x)$ is uniformly Cesaro summable to $f(x)$ (see Rudin or Apostol). No assumptions on smoothness are required. This result is to be contrasted to the fact that there are continuous functions whose Fourier series are divergent at a point.

So far we have restricted our attention to functions defined on the interval $[-\pi, \pi]$. If these functions are periodic of period 2π, the expansions are as good everywhere as within the interval $[-\pi, \pi]$. It is a simple matter to generalize these results to the interval $[-l, l]$, and functions of period $2l$. But the requirement of periodicity cannot be removed if the series is to converge outside the basic interval.

Clearly the set of functions

$$\left\{\frac{e^{in\pi x/l}}{(2l)^{1/2}}\right\}, \qquad n = 0, \pm 1, \pm 2, \cdots$$

is a complete orthonormal set on the interval $[-l, l]$. All our results go through for this set of functions exactly as before. For the complex exponential series, we obtain

$$\begin{aligned} f(x) &= \sum_{n=-\infty}^{\infty} c_n \frac{e^{in\pi x/l}}{(2l)^{1/2}}, \\ c_n &= \frac{1}{(2l)^{1/2}} \int_{-l}^{l} f(x) e^{-in\pi x/l}\, dx . \end{aligned} \tag{5.57}$$

Similarly, for the trigonometric series,

$$
\begin{aligned}
f(x) &= \frac{a_0}{2} + \sum_{n=1}^{\infty} a_n \cos\frac{n\pi}{l}x + \sum_{n=1}^{\infty} b_n \sin\frac{n\pi}{l}x\,, \\
a_n &= \frac{1}{l}\int_{-l}^{l} f(x)\cos\frac{n\pi}{l}x\,dx \qquad \text{for } n = 0, 1, \cdots, \\
b_n &= \frac{1}{l}\int_{-l}^{l} f(x)\sin\frac{n\pi}{l}x\,dx \qquad \text{for } n = 1, 2, \cdots
\end{aligned}
\tag{5.58}
$$

These formulas hold in the interval $[-l, l]$ under the conditions of the theorems, and will hold outside this interval only if $f(x + 2l) = f(x)$.

For any l (including $l = \pi$), if $f(x)$ is an even function, then $b_n = 0$ for all n and we have a Fourier cosine series; and if $f(x)$ is an odd function, $a_n = 0$ for each n and we have a Fourier sine series.

We close this section with a very simple, but very important observation. It will be recalled that the Legendre polynomials, which we have shown to be a complete orthonormal set, satisfy a second-order differential equation (Eq. 5.44). In this section we have been discussing the more familiar trigonometric functions, and have found that they also are a complete orthonormal set of functions. As we study other such sets of functions we shall always make note of the fact that these functions are the solutions of certain differential equations. Our treatment will not be derived from these differential equations, but rather will culminate in them, deriving them as a by-product of other considerations. Finally, by way of a summary, we shall, in Section 5.11, view all the special polynomials in their capacity as the solutions of second-order differential equations.

The trigonometric functions will, of course, not be included in this summary, since they are not polynomials. For completeness sake, and to help motivate the next section on Fourier integrals, we state the well-known differential equation satisfied by the sine and cosine functions:

$$\frac{d^2}{dx^2}u + \omega^2 u = 0\,. \tag{5.59}$$

This second-order differential equation has the two linearly independent solutions

$$u_1 = \sin\omega x, \qquad u_2 = \cos\omega x\,. \tag{5.60}$$

If $\omega^2 = n^2$, where n is an integer, these solutions give the Fourier functions as n runs from 0 to ∞. It is instructive to prove the orthogonality of these functions from the form of the differential equation as was done for the Legendre polynomials.

5.7. FOURIER INTEGRALS

The restriction of the validity of Fourier series expansions to the basic interval $[-l, l]$ unless the expanded function is periodic is annoying because many functions are not periodic. However, a way around this difficulty is suggested by the generalization we made in the last section: If we want to expand a non-periodic function over any specified finite range, all we have to do is expand the function in an interval $[-l, l]$ large enough to contain the specified range. But what if we want an expansion for a nonperiodic function that is valid everywhere, over the whole real line? Throwing all considerations of rigor aside, we shall here derive heuristically a plausible answer to this question. (When a physicist speaks of a "heuristic" derivation, he is usually apologizing in advance for a sloppy, nonrigorous one; actually, the word means "serving to help toward discovery.") We shall take up these matters again, in a more rigorous fashion, in Chapter 9 (corollary to Theorem 9.11).

It is reasonable to try to get around the periodicity requirement by letting $l \to \infty$ in Eqs. (5.57). To carry this out, set $(\pi/l)^{1/2}x = y$ and $n(\pi/l)^{1/2} = k_n$, so $(n\pi/l)\,x = k_n y$ and $\Delta k_n \equiv k_{n+1} - k_n = (\pi/l)^{1/2}$, and $1/(2l)^{1/2} = \Delta k_n/(2\pi)^{1/2}$. Then, replacing c_n by g_{k_n}, Eqs. (5.57) become

$$f(y) = \frac{1}{(2\pi)^{1/2}} \sum_{k_n=-\infty}^{\infty} g_{k_n} e^{ik_n y}\, \Delta k_n ,$$

where

$$g_{k_n} = \frac{1}{(2\pi)^{1/2}} \int_{-\sqrt{\pi l}}^{\sqrt{\pi l}} f(y) e^{-ik_n y}\, dy ,$$

and k_n changes by steps of $\Delta k_n = (\pi/l)^{1/2}$, corresponding to steps of $\Delta n = 1$ in the original sum. Now let $l \to \infty$. As $l \to \infty$, $\Delta k_n \to 0$ and k_n takes on all real values. Therefore the sum over k_n becomes an integral over a continuous variable (which we denote by k), and we have, replacing y by x,

$$\begin{aligned} f(x) &= \frac{1}{(2\pi)^{1/2}} \int_{-\infty}^{\infty} g(k) e^{ikx}\, dk , \\ g(k) &= \frac{1}{(2\pi)^{1/2}} \int_{-\infty}^{\infty} f(x) e^{-ikx}\, dx . \end{aligned} \tag{5.61}$$

The function $g(k)$ is called the *Fourier transform* of $f(x)$, and vice versa. Two functions which satisfy Eqs. (5.61) are called a Fourier transform pair. This pair may be regarded as the statement and solution of an integral equation of a particular type.

By an easy extension of these results, we get the three-dimensional Fourier integral:

$$\begin{aligned} F(\mathbf{r}) &= \frac{1}{(2\pi)^{3/2}} \int_{-\infty}^{\infty} G(\mathbf{k}) e^{i\mathbf{k}\cdot\mathbf{r}}\, d\mathbf{k} , \\ G(\mathbf{k}) &= \frac{1}{(2\pi)^{3/2}} \int_{-\infty}^{\infty} F(\mathbf{r}) e^{-i\mathbf{k}\cdot\mathbf{r}}\, d\mathbf{r} . \end{aligned} \tag{5.62}$$

Here $\mathbf{r}$ stands for x, y, z, and $d\mathbf{r} = dx\,dy\,dz$; similarly, $\mathbf{k}$ stands for k_x, k_y, k_z, and $d\mathbf{k} = dk_x\,dk_y\,dk_z$. Thus $\mathbf{k}\cdot\mathbf{r} = k_x x + k_y y + k_z z$. The integrals are three-fold. We shall use this three-dimensional Fourier transform pair to solve some of the partial differential equations of physics in Chapter 7.

Equation (5.61) may be proved rigorously if one assumes that $f(x)$ is piecewise smooth and that $\int_{-\infty}^{\infty} |f(x)|\, dx$ exists (see Courant-Hilbert), but we shall not go into these issues here (see Section 9.6). Instead, we shall focus on the uses of the Fourier transform in physics.

Example. An interesting result may be obtained if we combine Eqs. (5.61) into the single equation

$$f(x) = \frac{1}{2\pi}\int_{-\infty}^{\infty}\left[\int_{-\infty}^{\infty} f(x')e^{-ikx'}\, dx'\right]e^{ikx}\, dk\,,$$

and then interchange the orders of integration to get

$$f(x) = \int_{-\infty}^{\infty} f(x')\left[\frac{1}{2\pi}\int_{-\infty}^{\infty} e^{ik(x-x')}\, dk\right]dx'\,.$$

The expression in brackets within the integration over x' must be a δ-function if this equation is to hold. Therefore we have an integral representation of the δ-function:

$$\delta(x - x') = \frac{1}{2\pi}\int_{-\infty}^{\infty} e^{ik(x-x')}\, dk\,. \tag{5.63}$$

This equation is to be interpreted as meaning that the expression on either side of it will have the same effect under an integral sign. This "derivation" shows clearly that the δ-functions (as naively conceived and used) originate as a consequence of the illegitimate business of interchanging orders of integration; the original crime must be paid for by the introduction of the extraordinary new mathematical objects called δ-functions. Their symbolic use is so simple, however, that one is apt to get the impression that the crime paid. In a sense it did: violations of the existing rules in mathematics and physics are always required to reach the new, which then turns out to hide the old as a special case.

The δ-function in three dimensions is

$$\delta^3(\mathbf{r} - \mathbf{r}_0) \equiv \delta(x - x_0)\delta(y - y_0)\delta(z - z_0) = \frac{1}{(2\pi)^3}\int_{-\infty}^{\infty} e^{i\mathbf{k}\cdot(\mathbf{r}-\mathbf{r}_0)}\, d\mathbf{k}\,. \tag{5.64}$$

The integral representation of the δ-function can be used to prove the completeness relation (sometimes called Parseval's theorem) for the Fourier integral. Thus

$$\begin{aligned}
\int_{-\infty}^{\infty} |f(x)|^2\, dx &= \int_{-\infty}^{\infty} dx\, \frac{1}{(2\pi)^{1/2}}\int_{-\infty}^{\infty} g^*(k)e^{-ikx}\, dk\, \frac{1}{(2\pi)^{1/2}}\int_{-\infty}^{\infty} g(k')e^{ik'x}\, dk' \\
&= \int_{-\infty}^{\infty} dk\, g^*(k)\int_{-\infty}^{\infty} dk'\, g(k')\left[\frac{1}{2\pi}\int_{-\infty}^{\infty} e^{i(k'-k)x}\, dx\right] \\
&= \int_{-\infty}^{\infty} dk\, g^*(k)\int_{-\infty}^{\infty} dk'\, g(k')\delta(k' - k) \\
&= \int_{-\infty}^{\infty} g^*(k)\, g(k)\, dk = \int_{-\infty}^{\infty} |g(k)|^2\, dk\,.
\end{aligned}$$

If x is the time t, and k is the frequency υ, then Parseval's theorem has a simple physical interpretation. It reads

$$\int_{-\infty}^{\infty} |f(t)|^2 \, dt = \int_{-\infty}^{\infty} |g(\upsilon)|^2 \, d\upsilon \, .$$

If, for example, $f(t)$ is a radiated electric field, then $|f(t)|^2$ is proportional to the total radiated power, and the integral is a measure of the total radiated energy. On the right-hand side is the integral over all frequencies of the spectrum amplitude squared, $|g(\upsilon)|^2$, which is proportional to the energy radiated per unit frequency interval. Thus Parseval's theorem expresses the conservation of energy.

Before turning to some applications, we state and prove the convolution theorem for Fourier transforms. We shall use it in Chapter 7. Let $f_1(x)$ and $f_2(x)$ be two functions whose Fourier transforms, $g_1(k)$ and $g_2(k)$, are given by Eq. (5.61).

The quantity

$$F(x) \equiv \frac{1}{(2\pi)^{1/2}} \int_{-\infty}^{\infty} f_1(t) f_2(x-t) \, dt$$

is called the *convolution* of the functions f_1 and f_2. The convolution theorem states that

$$F(x) = \frac{1}{(2\pi)^{1/2}} \int_{-\infty}^{\infty} g_1(k) g_2(k) e^{ikx} \, dk \, . \tag{5.65}$$

By taking the Fourier transform of this last equation, and thus solving for the product $g_1(k)\, g_2(k)$, we see that the convolution theorem may also be expressed as follows: The Fourier transforms of the convolution of two functions is the product of the Fourier transforms of these two functions. Denoting the Fourier transform of $F(x)$ by $G(k)$ this may be written

$$G(k) = g_1(k)\, g_2(k) \, .$$

The proof follows immediately from the properties of the Fourier-transform pair. Thus

$$\begin{aligned} F(x) &= \frac{1}{(2\pi)^{1/2}} \int_{-\infty}^{\infty} f_1(t) f_2(x-t) \, dt \\ &= \frac{1}{(2\pi)^{1/2}} \int_{-\infty}^{\infty} f_1(t) \left[\frac{1}{(2\pi)^{1/2}} \int_{-\infty}^{\infty} g_2(k) e^{ik(x-t)} \, dk \right] dt \\ &= \frac{1}{(2\pi)^{1/2}} \int_{-\infty}^{\infty} \left[\frac{1}{(2\pi)^{1/2}} \int_{-\infty}^{\infty} f_1(t) e^{-ikt} \, dt \right] g_2(k) e^{ikx} \, dk \\ &= \frac{1}{(2\pi)^{1/2}} \int_{-\infty}^{\infty} g_1(k) g_2(k) e^{ikx} \, dk \, , \end{aligned}$$

which establishes the convolution theorem.

We shall now examine, in some detail, an example of the use of Fourier transforms, and incidentally, the calculation of some specific transforms.

We consider an undamped, one-dimensional harmonic oscillator acted upon by a time-varying but spatially uniform (no x-dependence) external force $F(t)$. This might be, for example, a spring hanging from a board which is jerked upward. We want to compute the energy transferred to the oscillator by such a force. Throughout this example, we shall limit ourselves to force functions $F(t)$ for which the Fourier transform exists.

The equation of motion is

$$\ddot{x} + \omega^2 x = (1/m)F(t) , \tag{5.66}$$

where m is the mass and ω is the natural frequency of the oscillator. We may rewrite this as

$$\dot{z} - i\omega z = (1/m)F(t) , \tag{5.67}$$

where $z = \dot{x} + i\omega x$. The energy of the oscillator at any time t is given by

$$E(t) = m\dot{x}^2/2 + m\omega^2 x^2/2 = (m/2)|z(t)|^2 .$$

Let us now assume that the oscillator is originally at rest ($x = \dot{x} = z = 0$ for $t \leq T_1$). Since the oscillator initially has zero energy, the energy ΔE transferred to the oscillator by $F(t)$ is $E(\infty)$. We now compute this quantity; we shall find that it depends in a simple way on the Fourier transform of $F(t)$.

Multiplying both sides of Eq. (5.67) by $\exp(-i\omega t)$ and integrating from T_1 to T_2 (arbitrary times), we obtain

$$\int_{T_1}^{T_2} (\dot{z} - i\omega z)\, e^{-i\omega t}\, dt = (1/m) \int_{T_1}^{T_2} F(t) e^{-i\omega t}\, dt .$$

Integration by parts of the term on the left gives

$$z(T_2)e^{-i\omega T_2} = (1/m) \int_{T_1}^{T_2} F(t)e^{-i\omega t}\, dt ,$$

where we have used the boundary condition that z vanishes for $t \leq T_1$. Since $F(t)$ vanishes for $t \leq T_1$ we may extend the lower limit of the integration to $-\infty$. The energy transfer is then given by

$$\begin{aligned} \Delta E &= \lim_{T_2 \to \infty} \frac{m}{2} |z(T_2)|^2 \\ &= \lim_{T_2 \to \infty} \frac{m}{2} \left| e^{i\omega T_2} \frac{1}{m} \int_{-\infty}^{T_2} F(t)e^{-i\omega t}\, dt \right|^2 \\ &= \frac{1}{2m} \left| \int_{-\infty}^{\infty} F(t)e^{-i\omega t}\, dt \right|^2 \\ &\equiv \frac{1}{2m} |f(\omega)|^2 , \end{aligned} \tag{5.68}$$

where $f(\omega)$ is the Fourier transform of $F(t)$ evaluated at the natural frequency of the oscillator:

$$f(\omega) \equiv \int_{-\infty}^{\infty} F(t)e^{-i\omega t}\, dt .$$

Thus $F(t)$ is given in terms of $f(\omega)$ by

$$F(t) = \frac{1}{2\pi}\int_{-\infty}^{\infty} f(\omega)e^{i\omega t}\,d\omega\,.$$

We have consolidated the factors of $(2\pi)^{-1/2}$ in one member of this transform pair, which therefore differs slightly in appearance from the symmetric form of Eqs. (5.61). Thus the energy transfer is essentially the absolute value squared of the Fourier component of the force function whose frequency is the natural frequency of the oscillator. The energy transfer can only occur at the resonant frequency ω, because only the ω-component of the Fourier transform of $F(t)$ enters into the formula for ΔE.

To solve the energy transfer problem for a specific force function, we must find its Fourier transform. We develop here some formulas which facilitate the determination of Fourier transforms. Let

$$F_n(t) \equiv \frac{d^{n-1}F(t)}{dt^{n-1}}\,, \qquad F_1(t) \equiv F(t)\,,$$

$$f_n(\omega) = \int_{-\infty}^{\infty} F_n(t)e^{-i\omega t}\,dt\,.$$

An integration by parts gives

$$f_n(\omega) = \left(\frac{1}{i\omega}\right)f_{n+1}(\omega) - \frac{1}{i\omega}\,F_n(t)\,e^{-i\omega t}\Big]_{-\infty}^{\infty}\,.$$

This formula will be useless unless $F_n(-\infty) = F_n(+\infty) = 0$. We therefore consider only such cases. Then, by repeated integration by parts (i.e., iteration of the above formula), we can express $f_1(\omega)$, the Fourier transform needed for this problem, in terms of the Fourier transform of some higher derivative of $F(t)$. This may facilitate the determination of $f_1(\omega)$ if the Fourier transform of a higher derivative is known. We obtain for $f_1(\omega) \equiv f(\omega)$,

$$f(\omega) = \left(\frac{1}{i\omega}\right)^{n-1} f_n(\omega)\,. \tag{5.69}$$

The index n is arbitrary, but in using the formula one naturally chooses the smallest n for which $f_n(\omega)$ is known.

We shall now compute the Fourier transforms of several force functions, and the energy transferred to an oscillator subjected to these forces.

Example 1. *Impulsive force*: $F(t) = P_0\delta(t)$. We obtain

$$\Delta E = \frac{1}{2m}\left|\int_{-\infty}^{\infty} P_0\delta(t)e^{-i\omega t}\,dt\right|^2 = P_0^2/2m\,.$$

This result may also be obtained by simply dropping the term $\omega^2 x$ in the equation of motion (5.66), the so-called impulse approximation.

Example 2. *Gaussian force*: $F(t) = (P_0/\sqrt{\pi}\tau)e^{-t^2/\tau^2}$ (Fig. 5.3). The width of the Gaussian pulse at half-maximum is $2\sqrt{\ln 2}\,\tau$. The Fourier transform may

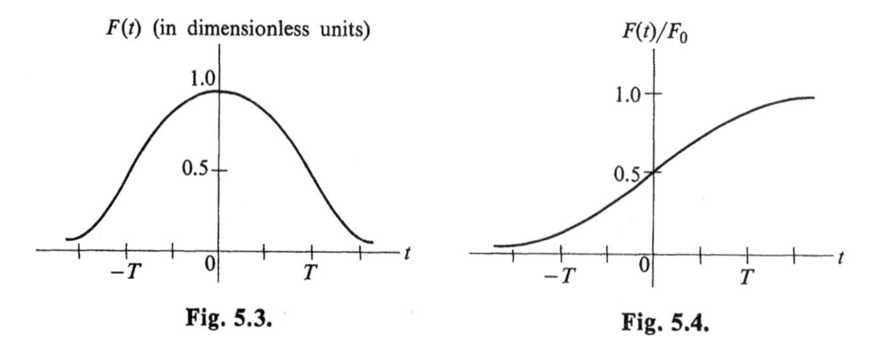

Fig. 5.3. **Fig. 5.4.**

be computed directly by completing the square in the exponent:

$$f(\omega) = \frac{P_0}{\sqrt{\pi}\tau}\int_{-\infty}^{\infty} e^{-t^2/\tau^2}e^{-i\omega t}\,dt$$

$$= \frac{P_0}{\sqrt{\pi}\tau}\,e^{-\omega^2\tau^2/4}\int_{-\infty}^{\infty} e^{-(t+i\omega\tau^2/2)^2/\tau^2}\,dt$$

$$= \frac{P_0}{\pi^{1/2}}\,e^{-\omega^2\tau^2/4}\int_{-\infty}^{\infty} e^{-x^2}\,dx = P_0 e^{-\omega^2\tau^2/4}\ .$$

By using techniques that will be developed in Section 6.9, we can show that this formal procedure for evaluating the integral is in fact valid. Note that the Fourier transform of a Gaussian is a Gaussian. For the energy transferred to the oscillator, we find that

$$\Delta E = (P_0^2/2m)e^{-\omega^2\tau^2/2}\ .$$

In the "sudden" limit ($\tau << \omega^{-1}$), $\Delta E = (P_0^2/2m)$, as in Example 1. This is to be expected because in the limit $\tau \to 0$, the Gaussian force function becomes a δ-function (see Example 2, Section 5.3). In the adiabatic limit ($\tau >> \omega^{-1}$), there is no energy transferred to the oscillator which also makes sense physically.

Example 3. *Probability integral:*

$$F(t) = (F_0/\sqrt{\pi}\tau)\int_{-\infty}^{t} e^{-t'^2/\tau^2}\,dt'$$

(Fig. 5.4). In this case the Fourier integral does not really exist because the function assumes a nonzero value (F_0) asymptotically as $t \to \infty$. We may still use Eqs. (5.68) and (5.69) to calculate the energy transfer, imagining that the constant force is turned off adiabatically at very large times. This will not affect the net energy transfer, and will bring about convergence of the Fourier integral. We have already determined the Fourier transform of

$$dF(t)/dt \equiv F_2(t) = (F_0/\sqrt{\pi}\tau)e^{-t^2/\tau^2}$$

in Example 2. Using Eq. (5.69) with $n = 2$ and Eq. (5.68), we obtain for the energy transfer

$$\Delta E = \frac{1}{2m}\left| f(\omega) \right|^2 = \frac{1}{2m}\left|\left(\frac{1}{i\omega}\right)f_2(\omega)\right|^2 = \frac{1}{2m}(F_0/\omega)^2 e^{-\omega^2\tau^2/2}.$$

Again, in the adiabatic limit, $\Delta E \to 0$; in the sudden limit, $\Delta E \to 1/2m\,(F_0/\omega)^2$. In the latter case, the mass, because of its inertia, suddenly finds itself displaced a distance $l_0 = (F_0/m\omega^2)$ from equilibrium and hence acquires the energy $(\frac{1}{2}m\omega^2 l_0^2)$ corresponding to the full displacement. It has no time to follow the shifting equilibrium position in the sudden limit. This result for the sudden limit should agree with the limiting behavior of the energy transfer for the force functions of Problems 7 and 8.

5.8 SPHERICAL HARMONICS AND ASSOCIATED LEGENDRE FUNCTIONS

We have derived the completeness of the Legendre polynomials from Weierstrass's theorem in one variable. From the two-variable generalization of Weierstrass's theorem, we proved the completeness of the trigonometric functions. We now derive the completeness of the spherical harmonics from Weierstrass's for three variables. It tells us that a function F of x, y, z (that is, $\mathbf{r}$) can be approximated uniformly by a sequence of partial sums as follows:

$$F(\mathbf{r}) = \lim_{M\to\infty} F_M(\mathbf{r}) = \lim_{M\to\infty} \sum_{m,n,p=0}^{M} c_{mnp}^{(M)} x_1^m x_2^n x_3^p. \tag{5.70}$$

We may also express $F_M(\mathbf{r})$ in terms of the three variables:

$$z_1 \equiv x_1 + ix_2 = r\sin\theta e^{i\phi},$$

$$z_2 \equiv x_1 - ix_2 = r\sin\theta e^{-i\phi},$$

$$z_3 \equiv x_3 = r\cos\theta,$$

which are linear combinations of x_1, x_2, x_3. Thus

$$F_M(\mathbf{r}) = \sum_{\alpha,\beta,\gamma=0}^{M} A_{\alpha\beta\gamma}^{(M)} z_1^\alpha z_2^\beta z_3^\gamma$$

$$= \sum_{l=0}^{3M} r^l \sum_{\substack{\alpha,\beta,\gamma=0 \\ (\alpha+\beta+\gamma=l)}}^{M} A_{\alpha\beta\gamma}^{(M)} e^{i(\alpha-\beta)\phi} \sin^{(\alpha+\beta)}\theta \cos^\gamma\theta. \tag{5.71}$$

In the last expression we have converted the unrestricted sum over all values of α, β, γ to a sum over just those combinations of α, β, γ that sum to the integer l. But then we sum over all possible restrictions (all l), which in effect removes the restriction on α, β, γ and gives the same results as the original unrestricted sum.

We now restrict $F(\mathbf{r})$ to the unit sphere by requiring that $r = 1$; also, we set $(\alpha - \beta) = m$. We shall need the following facts: Since α, β, and γ are all positive or zero, we have

$$\alpha + \beta = m + 2\beta \geq 0 .$$

Also,

$$\alpha + \beta \geq |\alpha - \beta| = |m| \Longrightarrow \alpha + \beta - |m| = m - |m| + 2\beta \geq 0 .$$

Furthermore, $(\alpha + \beta - |m|)$ is always even because 2β is even and $m - |m| = 0$ if $m \geq 0$ and $m - |m| = -2m$ if $m < 0$. We now rewrite Eq. (5.71) in the form

$$F_M(\theta, \phi) = \sum_{l=0}^{3M} \sum_{\substack{\alpha, \beta, \gamma=0 \\ (\alpha+\beta+\gamma=l)}}^{M} A_{\alpha\beta\gamma}^{(M)} e^{im\phi} \sin^{(\alpha+\beta-|m|)} \theta \cos^\gamma \theta \sin^{|m|} \theta .$$

By a trigonometric identity,

$$\sin^{(\alpha+\beta-|m|)}\theta \cos^\gamma \theta = (1 - \cos^2 \theta)^{(\alpha+\beta-|m|)/2} \cos^\gamma \theta ,$$

which is a polynomial in $\cos\theta$ of maximum degree $\alpha + \beta + \gamma - |m| = l - |m|$, since $(\alpha + \beta - |m|)$ is even. Denoting this polynomial by $f_{lm}(\cos\theta)$, we get

$$F_M(\theta, \phi) = \sum_{l=0}^{3M} \sum_{m} B_{lm}^{(M)} e^{im\phi} \sin^{|m|} \theta f_{lm}(\cos\theta) .$$

The range of the summation over m is still to be specified. Since $m = \alpha - \beta$, $m \leq l$; also, since we sum only over nonnegative α, β, γ, $\alpha + \beta + \gamma - |m| = l - |m| \geq 0$. Therefore, in changing to a sum over m, we must only sum over those m's such that $|m| \leq l$. Thus the last equation becomes

$$F_M(\theta, \phi) = \sum_{l=0}^{3M} \sum_{m=-l}^{m=l} B_{lm}^{(M)} e^{im\phi} \sin^{|m|} \theta f_{lm}(\cos\theta) . \tag{5.72}$$

Therefore the sequence of functions

$$\bar{Y}_{lm}(\theta, \phi) \equiv e^{im\phi} \sin^{|m|} \theta f_{lm}(\cos\theta) ,$$

where $f_{lm}(\cos\theta)$ is a polynomial in $\cos\theta$ of degree $(l - |m|)$, provides a uniform approximation to any continuous function defined on the unit sphere—that is, over the range $0 \leq \theta \leq \pi$ and $0 \leq \phi \leq 2\pi$. If we can now construct from this set of functions an *orthonormal* set of functions, then by reasoning exactly analogous to that applied to the Legendre polynomials and the trigonometric functions, we will know that this set of functions is complete. This is just another instance of the use of our basic result on completeness: whenever we can uniformly approximate an arbitrary function in Hilbert space by a sequence of partial sums which are themselves linear combinations of a set of orthonormal functions, then this set is a complete orthonormal set in Hilbert space.

Thus let us look for functions Y_{lm} satisfying

$$\int_\Omega Y_{l'm'}^* Y_{lm} \, d\Omega \equiv \int_0^{2\pi} d\phi \int_0^\pi \sin\theta \, d\theta \, Y_{l'm'}^* Y_{lm} = \delta_{ll'} \delta_{mm'} . \tag{5.73}$$

This orthonormality condition uniquely determines the functions Y_{lm} up to a phase factor, exactly as in the case of the normalized Legendre polynomials, $\bar{P}_l$. Thus there can only be one complete set of orthonormal functions defined on the unit sphere. We shall calculate a few of the Y_{lm} explicitly.

If $l = m = 0$, we obtain $Y_{00} = (1/4\pi)^{1/2}$. If $l = 1$, then m may equal -1, 0, or $+1$. Recalling that $f_{lm}(\cos\theta)$ is a polynomial in $\cos\theta$ of degree $l - |m|$, we obtain

$$\begin{aligned} Y_{10} &= a_1 \cos\theta + a_2 , \\ Y_{11} &= a_3 e^{i\phi} \sin\theta , \\ Y_{1,-1} &= a_4 e^{-i\phi} \sin\theta . \end{aligned}$$

The constants a_1, a_2, a_3, a_4 are determined by imposing orthonormality. Thus

$$\begin{aligned} 0 = \delta_{01}\delta_{00} &= \int Y_{00}^* Y_{10}\, d\Omega \\ &= \frac{1}{(4\pi)^{1/2}} \int_0^{2\pi} d\phi \int_0^{\pi} d\theta \sin\theta (a_1 \cos\theta + a_2) \Longrightarrow a_2 = 0 , \\ 1 = \delta_{10}\delta_{10} &= \int |Y_{10}|^2\, d\Omega \Longrightarrow a_1 = \left(\frac{3}{4\pi}\right)^{1/2} . \end{aligned}$$

Similarly, $a_3 = -a_4 = -(3/8\pi)^{1/2}$. We choose the minus sign to be consistent with the convention to be adopted later, in Eq. (5.75). Therefore the first few members of the complete orthonormal set of functions on the unit sphere are

$$\begin{aligned} Y_{00} &= \left(\frac{1}{4\pi}\right)^{1/2} , & Y_{11} &= -\left(\frac{3}{8\pi}\right)^{1/2} e^{i\phi} \sin\theta , \\ Y_{10} &= \left(\frac{3}{4\pi}\right)^{1/2} \cos\theta , & Y_{1,-1} &= \left(\frac{3}{8\pi}\right)^{1/2} e^{-i\phi} \sin\theta . \end{aligned} \tag{5.74}$$

Note that for a given l there are $2l + 1$ functions Y_{lm}. The functions Y_{lm} are called *spherical harmonics*. They are a sort of two-dimensional generalization and combination of the Fourier functions and the Legendre polynomials, since they do for the spherical *surface* what the Fourier functions and the Legendre polynomials do for their respective linear intervals.

We give next a general formula for the Y_{lm}. Since we know that the Y_{lm} are unique, the formula is true if we establish orthonormality—just as in the case of the Legendre polynomials. The formula is

$$Y_{lm}(\theta, \phi) = (-1)^m \left[\frac{2l+1}{4\pi} \frac{(l-m)!}{(l+m)!}\right]^{1/2} P_l^m(\cos\theta) e^{im\phi} , \qquad m \geq 0 , \tag{5.75}$$

$$Y_{l,-m} = (-1)^m Y_{lm}^* , \qquad m \geq 0 ,$$

where

$$P_l^m(x) = (1 - x^2)^{m/2} \frac{d^m}{dx^m} P_l(x) , \qquad m \geq 0 , \tag{5.76}$$

are called the *associated Legendre functions*. Thus the functions

$$Y_{l0}(\theta, \phi) = \left(\frac{1}{2\pi}\right)^{1/2}\left(\frac{2l+1}{2}\right)^{1/2} P_l\,(\cos\theta)$$

are an orthonormal set of functions on the unit sphere.

We shall first show that if u is a solution of Legendre's differential equation, that is,

$$(1 - x^2)u'' - 2xu' + l(l+1)u = 0\,, \tag{5.77}$$

then $v \equiv (1 - x^2)^{m/2}(d^m/dx^m)u$, for integral $m \geq 0$, is a solution of the equation

$$(1 - x^2)v'' - 2xv' + [l(l+1) - m^2/(1 - x^2)]v = 0\,, \tag{5.78}$$

known as Legendre's associated equation. Let $d^m u/dx^m \equiv d^m u \equiv w$. Then

$$v = (1 - x^2)^{m/2}w\,, \qquad v' = (1 - x^2)^{m/2}w' - mx(1 - x^2)^{(m/2)-1}w\,,$$

$$v'' = (1 - x^2)^{m/2}w'' - 2mx(1 - x^2)^{(m/2)-1}w' - m(1 - x^2)^{(m/2)-1}w$$
$$+ mx^2(m-2)(1 - x^2)^{(m/2)-2}w\,.$$

Substituting into (5.78) and replacing w by $d^m u$, the left-hand side of Eq. (5.78) becomes

$$(1 - x^2)d^{m+2}u - 2(m+1)x\,d^{m+1}u + [l(l+1) - m(m+1)]d^m u\,. \tag{5.79}$$

We shall now show that this expression can be written in the form

$$d^m[(1 - x^2)u'' - 2xu' + l(l+1)u]\,, \tag{5.80}$$

which vanishes, by comparison with Eq. (5.77). This completes the proof that $v = (1 - x^2)^{m/2}(d^m u/dx^m)$ is a solution of Legendre's associated equation. Using Leibnitz's formula [Eq. (5.42)], we obtain

$$d^m[(1 - x^2)u''] = \sum_{n=0}^{m}\binom{m}{n} d^n(1 - x^2)d^{m-n+2}u$$
$$= (1 - x^2)d^{m+2}u - 2xm\,d^{m+1}u - m(m-1)d^m u\,,$$

and also

$$\frac{d^m}{dx^m}[-2xu'] = -2xd^{m+1}u - 2md^m u\,.$$

Combining the last two lines and the term $l(l+1)d^m u$ from Eq. (5.80) establishes the equivalence of Eqs. (5.79) and (5.80) and completes the proof that the functions

$$P_l^m(x) = (1 - x^2)^{m/2}\frac{d^m}{dx^m}P_l(x) = \frac{1}{2^l l!}(1 - x^2)^{m/2}\frac{d^{l+m}}{dx^{l+m}}(x^2 - 1)^l \tag{5.81}$$

are solutions of Legendre's associated equation (5.78). We note that $P_l^0(x) = P_l(x)$.

Using the same procedure that led to the orthonormality relations for the Legendre polynomials, we can show that

$$\int_{-1}^{1} P_l^m(x) P_{l'}^m \, dx = \frac{(l+m)!}{(l-m)!} \frac{2}{2l+1} \delta_{ll'} . \tag{5.82}$$

Once this result has been established, the normalization constant of the Y_{lm} (Eq. 5.75) follows immediately, as the integration over ϕ merely produces a factor of 2π. There is, of course, a free choice of phase factor; ours is a common choice in the physics literature. However, one must be careful, because different authors choose different phase factors for the spherical harmonics. The $P_l^m(x)$ are *not* another orthonormal set of polynomials on $[-1, 1]$, challenging the uniqueness of the Legendre polynomials, for the simple reason that they are not polynomials at all! Equation (5.81) shows this most clearly.

The $P_l^m(x)$ are not the only solutions to Legendre's associated equation. Just as there is a second set of solutions to Legendre's equation (5.77), so too there is a second set of solutions to Eq. (5.78). However, they also are not finite at $x = \pm 1$.

At this point we state an important result; the addition theorem for spherical harmonics. If two vectors $\mathbf{x}(r, \theta, \phi)$ and $\mathbf{x}'(r', \theta', \phi')$ have an angle γ between them, then

$$P_l(\cos \gamma) = \frac{4\pi}{2l+1} \sum_{m=-l}^{l} Y_{lm}^*(\theta', \phi') Y_{lm}(\theta, \phi) , \tag{5.83}$$

where

$$\cos \gamma = \frac{\mathbf{x} \cdot \mathbf{x}'}{|\mathbf{x}| \, |\mathbf{x}'|} = \cos \theta \cos \theta' + \sin \theta \sin \theta' \cos (\phi - \phi') . \tag{5.84}$$

Thus the addition theorem expresses a Legendre polynomial of order l in the angle γ in terms of a sum over products of spherical harmonics of the angles θ, ϕ and θ', ϕ'. We omit the proof; note, however, that for the case $l = 1$, Eq. (5.83) is just a statement of the familiar result of Eq. (5.84).

There is a raft of other properties of the spherical harmonics and associated Legendre functions that are straightforward, but messy, generalizations of the identities proved for Legendre polynomials. We shall not go into these further.

We devote the remainder of this section to a discussion of the spherical harmonics in their capacity as solutions to the angular part of Laplace's equation in spherical coordinates. Since the angular part of the Laplacian is essentially the quantum-mechanical operator that represents the total angular momentum squared, the spherical harmonics play a prominent role in quantum theory. The operator that represents the square of the total angular momentum is

$$L^2 = -\hbar^2 \left(\frac{1}{\sin \theta} \frac{\partial}{\partial \theta} \left(\sin \theta \frac{\partial}{\partial \theta} \right) + \frac{1}{\sin^2 \theta} \frac{\partial^2}{\partial \phi^2} \right) . \tag{5.85}$$

Let us first see why this is the angular-momentum operator. Classically, $\mathbf{L} = \mathbf{r} \times \mathbf{p}$. If we make the usual Schrödinger substitution for $\mathbf{p}$: $\mathbf{p} \to i\hbar\nabla$, we find that

$$L_x = -i\hbar\left(y\frac{\partial}{\partial z} - z\frac{\partial}{\partial y}\right),$$
$$L_y = -i\hbar\left(z\frac{\partial}{\partial x} - x\frac{\partial}{\partial z}\right), \tag{5.86}$$
$$L_z = -i\hbar\left(x\frac{\partial}{\partial y} - y\frac{\partial}{\partial x}\right).$$

Now we transform to spherical coordinates. The equations of transformation of coordinates are:

$$x = r\sin\theta\cos\phi, \qquad r = (x^2+y^2+z^2)^{1/2},$$
$$y = r\sin\theta\sin\phi, \qquad \phi = \tan^{-1} y/x,$$
$$z = r\cos\theta, \qquad \theta = \tan^{-1}\frac{(x^2+y^2)^{1/2}}{z}.$$

Therefore

$$\frac{\partial}{\partial x} = \frac{\partial}{\partial r}\frac{\partial r}{\partial x} + \frac{\partial}{\partial \theta}\frac{\partial \theta}{\partial x} + \frac{\partial}{\partial \phi}\frac{\partial \phi}{\partial x}$$
$$= \sin\theta\cos\phi\frac{\partial}{\partial r} + \frac{\cos\theta\cos\phi}{r}\frac{\partial}{\partial\theta} - \frac{1}{r}\frac{\sin\phi}{\sin\theta}\frac{\partial}{\partial\phi}.$$

Similarly,

$$\frac{\partial}{\partial y} = \sin\theta\sin\phi\frac{\partial}{\partial r} + \frac{1}{r}\cos\theta\sin\phi\frac{\partial}{\partial\theta} + \frac{1}{r}\frac{\cos\phi}{\sin\theta}\frac{\partial}{\partial\phi},$$
$$\frac{\partial}{\partial z} = \cos\theta\frac{\partial}{\partial r} - \frac{\sin\theta}{r}\frac{\partial}{\partial\theta}.$$

Using these relations in Eqs. (5.86), we obtain after some manipulation,

$$L_x = +i\hbar\left(\sin\phi\frac{\partial}{\partial\theta} + \cos\phi\cot\theta\frac{\partial}{\partial\phi}\right),$$
$$L_y = -i\hbar\left(\cos\phi\frac{\partial}{\partial\theta} - \sin\phi\cot\theta\frac{\partial}{\partial\phi}\right), \tag{5.87}$$
$$L_z = -i\hbar\frac{\partial}{\partial\phi}.$$

Now we form L^2:

$$L^2 = L_x^2 + L_y^2 + L_z^2 = -\hbar^2\left[\frac{\partial^2}{\partial\theta^2} + \cot\theta\frac{\partial}{\partial\theta} + (1+\cot^2\theta)\frac{\partial^2}{\partial\phi^2}\right]$$
$$= -\hbar^2\left[\frac{1}{\sin\theta}\frac{\partial}{\partial\theta}\left(\sin\theta\frac{\partial}{\partial\theta}\right) + \frac{1}{\sin^2\theta}\frac{\partial^2}{\partial\phi^2}\right]. \tag{5.88}$$

We observe [see Eq. (1.68)] that the L^2 operator is essentially the angular part of the Laplacian in spherical coordinates:

$$\nabla^2 = \frac{1}{r^2}\frac{\partial}{\partial r}\left(r^2\frac{\partial}{\partial r}\right) - \frac{L^2}{\hbar^2 r^2} . \tag{5.89}$$

Thus we may write Laplace's equation in the form

$$\nabla^2\psi = \frac{1}{r^2}\frac{\partial}{\partial r}\left(r^2\frac{\partial\psi}{\partial r}\right) - \frac{L^2\psi}{\hbar^2 r^2} = 0 ,$$

or (5.90)

$$\frac{\partial}{\partial r}\left(r^2\frac{\partial\psi}{\partial r}\right) = \frac{1}{\hbar^2}L^2\psi .$$

In order to solve this equation, let us assume that $\psi(r, \theta, \phi)$ is a product of two functions, one depending on r alone, and the other on θ and ϕ:

$$\psi(r, \theta, \phi) = R(r)\,Y(\theta, \phi) .$$

If we substitute a solution of this form into Eq. (5.90), and divide the equation by $\psi = RY$, we obtain

$$\frac{1}{R(r)}\frac{\partial}{\partial r}\left(r^2\frac{\partial R(r)}{\partial r}\right) = \frac{1}{\hbar^2 Y(\theta, \phi)}L^2 Y(\theta, \phi) .$$

The left-hand side is a function of r alone, and the right-hand side is a function only of θ and ϕ. This can only hold if both sides equal the same constant, for otherwise we could vary r and yet the right-hand side, depending only on θ and ϕ, would not change. Let us call the constant λ. Then we have the two equations

$$\frac{d}{dr}\left(r^2\frac{dR(r)}{dr}\right) = \lambda R(r) , \tag{5.91}$$

$$L^2 Y(\theta, \phi) = \hbar^2\lambda Y(\theta, \phi) . \tag{5.92}$$

It will be shown below that a solution for the angular equation (5.92) is just the spherical harmonic, Y_{lm} with $\lambda = l(l+1)$. In other words, the spherical harmonics are eigenvectors of the operator L^2 corresponding to eigenvalues $\hbar^2 l(l+1)$. For $\lambda = l(l+1)$, the radial equation (5.91) may be solved by letting $R(r) = U(r)/r$. It then becomes

$$\frac{d^2}{dr^2}U(r) - l(l+1)\frac{U(r)}{r^2} = 0 , \tag{5.93}$$

which has the solution

$$U(r) = rR(r) = Ar^{l+1} + Br^{-l} , \tag{5.94}$$

where A and B are constants. Together with the spherical harmonics, these radial functions give a solution to Laplace's equation in spherical coordinates. A function which satisfies Laplace's equation is called a harmonic function. This explains the origin of the name "spherical harmonics": these functions are *harmonic*, since they satisfy the angular part of Laplace's equation, and they are defined on the surface of the unit *sphere*.

The most general solution is a linear combination of solutions for given values of l and m. Thus

$$\psi(r, \theta, \phi) = \sum_{l=0}^{\infty} \sum_{m=-l}^{l} [A_{lm}r^l + B_{lm}r^{-l-1}]Y_{lm}(\theta, \phi) . \tag{5.95}$$

It is of interest to separate the variables again, in Eq. (5.92). That is, let $Y_{lm} = \Phi(\phi)P(\theta)$, substitute into the equation, and divide through by $Y = \Phi P$. The result may be written

$$\frac{-\sin\theta}{P}\frac{d}{d\theta}\left(\sin\theta\frac{dP}{d\theta}\right) - l(l+1)\sin^2\theta = \frac{1}{\Phi}\frac{d^2\Phi}{d\phi^2} .$$

The θ- and ϕ-dependences have thus been isolated and so the expressions on both sides of the equation must equal the same constant, call it $-m^2$. We then have the two equations

$$\frac{d^2\Phi}{d\phi^2} = -m^2\Phi , \tag{5.96}$$

$$\frac{1}{\sin\theta}\frac{d}{d\theta}\left(\sin\theta\frac{dP}{d\theta}\right) + \left[l(l+1) - \frac{m^2}{\sin^2\theta}\right]P = 0 . \tag{5.97}$$

The azimuthal equation (in ϕ), Eq. (5.96), is immediately solved:

$$\Phi(\phi) = Ae^{\pm im\phi} . \tag{5.98}$$

For $\Phi(\phi)$ to be single-valued over the range $[0, 2\pi]$, m must be an integer. This result was anticipated in choosing the form of the separation constant. The equation in θ (Eq. 5.97) will assume a recognizable form if we simply change variables. Letting $\cos\theta = x$, so $dx = d(\cos\theta)$, and $\sin\theta = (1 - x^2)^{1/2}$, it becomes

$$(1 - x^2)P'' - 2xP' + \left[l(l+1) - \frac{m^2}{1 - x^2}\right]P = 0 , \tag{5.99}$$

which, due to our earlier choice of $\lambda = l(l+1)$, is just Legendre's associated equation (5.78). The solutions are the associated Legendre functions $P(x) = P_l^m(x)$ given in Eq. (5.81). This proves that the spherical harmonics are eigenvectors of L^2 with eigenvalues $\hbar^2 l(l+1)$, as stated above.

It should be noted that we really found *two* solutions to the second-order differential equation (5.91), $R(r) = r^l$ and $R(r) = r^{-l-1}$. This was also true for the differential equation (5.96), where we found the two solutions given by Eq. (5.98). However, for Eq. (5.99), which is also a second order differential equation, we have stated only one solution, $P_l^m(x)$. There is, of course, a second solution, just as there was to the other two equations. The second solution is rarely used in physical problems, however, because it is singular at $x = \pm 1$ $(\theta = 0, \pi)$ and must usually be rejected in order to satisfy boundary conditions.

Usually one first encounters all these functions in their capacity as solutions to differential equations that arise in physics. Our approach has been different.

We have emphasized the *completeness* properties of these sets of functions, touching only incidentally on them as solutions to differential equations.

5.9 HERMITE POLYNOMIALS

In the next section we shall develop the properties of the various sets of orthogonal polynomials as special cases of a single comprehensive framework. But in order for us to be able to see the trees, and not just the forest, we shall first work out some of the properties of the Hermite polynomials in an inductive way. We first encountered these polynomials in the discussion of the eigenvalue problem for the quantum-mechanical harmonic oscillator (see Section 3.10).

The Hermite polynomials, $H_n(x)$, are the orthogonal polynomials on the infinite interval $[-\infty, \infty]$ with respect to the nonnegative weight function e^{-x^2}. That is, for $m \neq n$,

$$\int_{-\infty}^{\infty} H_n(x) H_m(x) e^{-x^2}\, dx = 0 . \tag{5.100}$$

Thus they differ in two important ways from the orthonormal sets of functions dealt with so far. First, they are defined from $-\infty$ to $+\infty$. Therefore Weierstrass' theorem, which deals only with finite intervals, does not apply. Secondly, they are orthogonal with respect to a weight function (see Definition 5.2). The Hermite polynomials are most conveniently *defined* in terms of a generating function $\psi(x, t)$:

$$\psi(x, t) \equiv e^{-t^2+2tx} = e^{x^2} e^{-(t-x)^2} \equiv \sum_{n=0}^{\infty} \frac{H_n(x)}{n!} t^n , \tag{5.101}$$

From this implicit definition of $H_n(x)$, we can get an explicit formula. It is clear that

$$\begin{aligned} H_n(x) &= \left[\frac{\partial^n \psi(x, t)}{\partial t^n}\right]_{t=0} = \left[e^{x^2} \frac{\partial^n}{\partial t^n} e^{-(t-x)^2}\right]_{t=0} \\ &= \left[e^{x^2} \frac{d^n}{dy^n} e^{-y^2}\right]_{y=-x} = (-1)^n e^{x^2} \frac{d^n}{dx^n} e^{-x^2} . \end{aligned} \tag{5.102}$$

This is the Rodrigues formula for the Hermite polynomials. It shows that they are indeed polynomials of degree n.

We next derive some recursion relations for the Hermite polynomials. Since it is true of the generating function that

$$\frac{\partial \psi(x, t)}{\partial x} = 2t\, \psi(x, t) ,$$

it follows that

$$\sum_n \frac{H_n'(x) t^n}{n!} = \sum_n \frac{2H_n(x) t^{n+1}}{n!} .$$

Equating the coefficients of equal powers of t, we get

$$H_n'(x) = 2nH_{n-1}(x), \qquad n \geq 1 . \tag{5.103}$$

Likewise, from

$$\frac{\partial\phi(x, t)}{\partial t} + 2(t - x)\,\phi(x, t) = 0\,,$$

it follows that

$$H_{n+1}(x) - 2xH_n(x) + 2nH_{n-1}(x) = 0, \qquad n \geq 1\,. \tag{5.104}$$

The reader may show that, by differentiating these recursion relations and combining them properly, the following differential equation for the $H_n(x)$ can be derived:

$$H_n''(x) - 2x\,H_n'(x) + 2nH_n(x) = 0, \qquad n \geq 0\,. \tag{5.105}$$

There are many ways to prove orthogonality of the $H_n(x)$ with respect to the weight function e^{-x^2}. The two methods we have used for the Legendre polynomials—repeated integration by parts, and the proof based directly on the differential equation—also work for the Hermite polynomials. However, we shall use a different method, based on the generating function, in order to illustrate still another technique. Consider the integral

$$I = \int_{-\infty}^{\infty} e^{-t^2+2tx}e^{-s^2+2sx}e^{-x^2}\,dx = \sum_{n,m=0}^{\infty} \frac{t^n s^m}{n!m!} \int_{-\infty}^{\infty} H_n H_m e^{-x^2}\,dx\,.$$

We want to prove that the integral on the right is zero for $m \neq n$. The integral on the left is

$$I = e^{-(t^2+s^2)} \int_{-\infty}^{\infty} e^{-x^2+2(s+t)x}\,dx = e^{-(t^2+s^2)}\pi^{1/2}e^{(s+t)^2} = \pi^{1/2}e^{2st}\,;$$

the integral is done by completing the square. Expanding this expression for I in a Taylor series, we have

$$I = \pi^{1/2} \sum_{n=0}^{\infty} \frac{(2st)^n}{n!} = \sum_{n,m=0}^{\infty} \frac{t^n s^m}{n!\,m!} \int_{-\infty}^{\infty} H_n H_m\, e^{-x^2}\,dx\,.$$

If equal powers of s and t are equated in this identity, we find that

$$\int_{-\infty}^{\infty} H_n H_m e^{-x^2}\,dx = \pi^{1/2}2^n n!\,\delta_{nm}\,. \tag{5.106}$$

Thus the functions $u_n(x) = N_n H_n(x)e^{-x^2/2}$ with $N_n = (1/\pi^{1/2}2^n n!)^{1/2}$ are orthonormal on the interval $(-\infty, \infty)$. The functions $u_n(x)$ are the eigenfunctions of the quantum oscillator. We may say equivalently that the set of Hermite polynomials $H_n(x)$ are orthonormal on the interval $(-\infty, \infty)$ with respect to the weight function $N_n^2 e^{-x^2}$.

Another important set of polynomials are the Laguerre polynomials, $L_n(x)$. They are orthogonal on the interval $[0, \infty]$ with respect to the weight function e^{-x}, that is

$$\int_0^{\infty} L_n(x)L_m(x)e^{-x}\,dx = 0 \qquad \text{for } n \neq m\,. \tag{5.107}$$

Everything that we have done for Hermite polynomials can be done for Laguerre polynomials (Problem 6.36).

5.10 STURM-LIOUVILLE SYSTEMS—ORTHOGONAL POLYNOMIALS

All the special functions we have studied so far are solutions to differential equations. These equations and their solutions look rather different at first, but, in fact, they have a great deal in common. In this section we shall view them as special cases in a single general framework.

All the equations are second-order, linear differential equations of the form

$$Lu = \lambda u \,, \tag{5.108}$$

where

$$L = \alpha(x) \frac{d^2}{dx^2} + \beta(x) \frac{d}{dx} + \gamma(x), \tag{5.109}$$

and λ is a constant; α, β, and γ are real functions of x. The operators L have an important property in common that until now we have not mentioned. Namely, they are *Hermitian* if (1) we use the inner product

$$(f, g) \equiv \int_{-\infty}^{\infty} f^*(x) g(x) w(x) \, dx \,, \tag{5.110}$$

[where the weight function $w(x) \geq 0$], and, (2) the functions u satisfy appropriate boundary conditions. We have not yet committed ourselves to a choice of $w(x)$, and we will find that this freedom is crucial in attaining the generality which we want.

In quantum mechanics, all the operators which correspond to physical observables are Hermitian. It will be recalled that a Hermitian operator H satisfies the equation

$$(Hx, y) = (x, Hy) \,,$$

where x and y are vectors.†

Now suppose that f and g are two vectors (functions) in Hilbert space. For the inner product we defined above, the condition that L be Hermitian is

$$(Lf, g) \equiv \int_{-\infty}^{\infty} (Lf(x))^* g(x) w(x) \, dx = \int_{-\infty}^{\infty} (f(x))^* (Lg(x)) w(x) \, dx \equiv (f, Lg) \,. \tag{5.111}$$

Henceforth, we restrict attention to such operators.

To investigate the hermiticity of L, we compute (f, Lg) and (Lf, g) and then form their difference; L is Hermitian if and only if this difference is zero.

† Operators, H, satisfying $(Hx, y) = (x, Hy)$ on an infinite dimensional vector space are called *symmetric* operators by mathematicians. We will refer to them as Hermitian operators. They clearly have real eigenvalves and orthonormal eigenfunctions according to our work in Chapter 4.

We have

$$(f, Lg) = \int_{-\infty}^{\infty} wf^*\alpha g''\,dx + \int_{-\infty}^{\infty} wf^*\beta g'\,dx + \int_{-\infty}^{\infty} wf^*\gamma g\,dx\,.$$

Integrating the first two integrals by parts once, we get

$$\begin{aligned}(f, Lg) = [w\alpha f^*g']_{-\infty}^{\infty} &- \int_{-\infty}^{\infty} [(w\alpha)'f^*g' + (w\alpha)f^{*\prime}g']\,dx \\ &+ [w\beta f^*g]_{-\infty}^{\infty} - \int_{-\infty}^{\infty} [(w\beta)'f^*g + (w\beta)f^{*\prime}g]\,dx + \int_{-\infty}^{\infty} wf^*\gamma g\,dx\,.\end{aligned}$$

We now subtract the corresponding expression for (Lf, g), which may be obtained from the above by simply interchanging f^* and g, and obtain

$$\begin{aligned}(f, Lg) - (Lf, g) &= [w\alpha(f^*g' - f^{*\prime}g)]_{-\infty}^{\infty} \\ &\quad - \int_{-\infty}^{\infty} [(w\alpha)' - (w\beta)](f^*g' - gf^{*\prime})\,dx\,.\end{aligned}$$

For L to be Hermitian, this last expression must be equal to zero for *all* f and g. Therefore the necessary and sufficient conditions for L to be Hermitian are

$$1. \qquad [w\alpha(f^*g' - gf^{*\prime})]_{-\infty}^{\infty} = 0\,, \tag{5.112}$$

$$2. \qquad (w\alpha)' = w\beta\,. \tag{5.113}$$

Clearly, the weight function $w(x)$ plays an important role in the Hermiticity requirement. In fact, the second equation above actually determines $w(x)$ up to a multiplicative constant, since $\alpha(x)$ and $\beta(x)$ are given functions. The first equation illustrates the importance of the endpoint conditions in determining Hermiticity. For example, if the functions on which L operates vanish sufficiently rapidly at the boundaries, then the first condition above is satisfied. The reader can easily show that there are several less stringent conditions which will fulfill condition one. On the other hand, if $w\alpha$ vanishes sufficiently rapidly at both endpoints, then the functions upon which L operates need not satisfy any special boundary conditions and condition (1) will still be satisfied. The requirement that $w\alpha$ vanish sufficiently rapidly at the endpoints can be met either by having $w\alpha$ identically zero outside some infinite interval $[a, b]$ or having $w\alpha$ fall off sufficiently rapidly at arbitrarily large distances. In what follows, we shall have occasion to utilize both possibilities.

Since $w(x)$ has not been specified, we may consider condition (2) as an equation to be solved for $w(x)$. We rewrite it as

$$(w\alpha)' = \frac{\beta}{\alpha}\,w\alpha\,. \tag{5.114}$$

A simple integration gives

$$w\alpha = C\cdot\exp\left[\int \frac{\beta}{\alpha}\,dx\right] \tag{5.115}$$

in any region in which $(w\alpha)'$ is continuous. The integral is an indefinite integral; C is an arbitrary constant. Note in particular that $C = 0$ ($w\alpha \equiv 0$) is a solution. For future use, we should point out that it is possible to join a solution $w\alpha \not\equiv 0$ to a solution $w\alpha \equiv 0$ at some point $x = a$, if $w\alpha$ is continuous at this point, that is, $w\alpha = 0$ just to the right *and* the left of $x = a$ on the real line.

In the region where $w\alpha \not\equiv 0$, $\alpha(x)$ must either be nonnegative or nonpositive since we require $w(x) > 0$. In the former case, we must choose C positive, in the latter case, C must be negative. Without loss of generality, we can take $\alpha(x) \geq 0$ and $C > 0$. In summary, for any given $\alpha(x)$ and $\beta(x)$ there exists a weight function $w(x)$ given in any region where $w\alpha \not\equiv 0$ by Eq. (5.115). This guarantees that condition (2) is satisfied.

Any equation of the form we are studying,

$$Lu = \alpha(x)u'' + \beta(x)u' + \gamma(x)u = \lambda u\,, \tag{5.116}$$

can be written in the equivalent form

$$\frac{d}{dx}\left[w\alpha\,\frac{du}{dx}\right] + (\gamma - \lambda)wu = 0 \tag{5.117}$$

by choosing the weight function $w(x)$ according to Eq. (5.113). An equation of the form (5.117), together with the boundary condition stated in Eq. (5.112), is called a *Sturm-Liouville system*. In Eq. (5.46), Legendre's differential equation appears in Sturm-Liouville form. We used this form of Legendre's equation to prove the orthogonality of eigenfunctions belonging to distinct eigenvalues. We shall now give a general proof of this result, valid for any Sturm-Liouville system.

Let λ_m and λ_n ($\lambda_m \neq \lambda_n$) be two eigenvalues of Eq. (5.117). Then

$$\frac{d}{dx}\left[(w\alpha)\,\frac{du_m}{dx}\right] + (\gamma - \lambda_m)wu_m = 0\,,$$

and

$$\frac{d}{dx}\left[(w\alpha)\,\frac{du_n^*}{dx}\right] + (\gamma - \lambda_n)wu_n^* = 0\,.$$

Multiplying the first of these equations by u_n^*, the second by u_m, and forming their difference, we obtain

$$u_n^*\,\frac{d}{dx}\,[(w\alpha)u_m'] - u_m\,\frac{d}{dx}\,[(w\alpha)u_n^{*\prime}] = -(\lambda_n - \lambda_m)wu_n^*u_m\,.$$

The left-hand side of this expression is a total derivative. Rewriting this, and integrating over both sides, we find that

$$[w\alpha(u_n^*u_m' - u_mu_n^{*\prime})]_{-\infty}^{\infty} = -(\lambda_n - \lambda_m)\int_{-\infty}^{\infty} wu_n^*u_m\,dx\,.$$

The left-hand side vanishes identically by assumption [Eq. (5.112)]; and since

$\lambda_n \neq \lambda_m$, it follows that

$$(u_m, u_n) = \int_{-\infty}^{\infty} u_m^*(x) u_n(x) w(x)\, dx = 0\,.$$

Therefore eigenfunctions belonging to distinct eigenvalues are orthogonal with respect to the weight function $w(x)$. This result merely translates into function language the result on the orthogonality of the eigenvectors of Hermitian operators belonging to distinct eigenvalues (Theorem 4.18).

Let us now restrict our attention to *polynomial* Sturm-Liouville systems. That is, we shall assume that the functions $u_n(x)$ are polynomials of degree n. We denote them by $Q_n(x)$. Then

$$LQ_n = \lambda_n Q_n\,, \tag{5.118}$$

where

$$L = \alpha(x)\frac{d^2}{dx^2} + \beta(x)\frac{d}{dx} + \gamma(x)\,, \tag{5.119}$$

as before, and

$$\int_{-\infty}^{\infty} Q_n^*(x) Q_m(x) w(x)\, dx = 0 \tag{5.120}$$

if $m \neq n$, in keeping with the above discussion of orthogonality.

For this case, where Q_n is a polynomial of degree n, the real functions $\alpha(x)$, $\beta(x)$, and $\gamma(x)$ must have the form

$$\alpha(x) = \alpha_0 x^2 + \alpha_1 x + \alpha_2\,, \tag{5.121}$$

$$\beta(x) = \beta_0 x + \beta_1\,, \tag{5.122}$$

$$\gamma(x) = \gamma_0\,. \tag{5.123}$$

That this is the most general form possible is a consequence of the fact that the differential equation is of second order and that we are considering only *polynomial* solutions. We prove this by letting $n = 0$, 1, and 2 in Eq. (5.118). Then

$$\gamma Q_0 = \lambda_0 Q_0 \Longrightarrow \gamma = \lambda_0 = \gamma_0\,,$$

$$\beta Q_1' + \gamma Q_1 = \lambda_1 Q_1 \Longrightarrow \beta = \frac{\lambda_1 - \lambda_0}{Q_1'} Q_1 = \beta_0 x + \beta_1\,,$$

$$\alpha Q_2'' + \beta Q_2' + \gamma Q_2 = \lambda_2 Q_2 \Longrightarrow \alpha = \frac{\lambda_2 - \lambda_0}{Q_2''} Q_2 - \frac{\lambda_1 - \lambda_0}{Q_1' Q_2''} Q_1 Q_2'$$
$$= \alpha_0 x^2 + \alpha_1 x + \alpha_2\,.$$

We note that β can never be identically zero because if it were, then it follows from Eq. (5.113) that $w\alpha = \text{const}$. But this makes it impossible to satisfy the other requirement for Hermiticity (Eq. 5.112), since polynomials do not vanish at ∞.

Thus there are no polynomial solutions to the Sturm-Liouville problem (Hermitian L) if $\beta(x) \equiv 0$. If we do not insist on L being Hermitian, there are polynomial solutions of all degrees with eigenvalues equal to $n(n-1)\alpha_0$. The eigenvectors will not be orthogonal. The general expression for $\beta(x)$ above shows that $\lambda_1 = \lambda_0$ if $\beta(x) \equiv 0$. Conversely, if $\beta(x) \not\equiv 0$, then $\lambda_1 \neq \lambda_0$ and $\beta_0 \neq 0$ because Q_1 has a leading term proportional to x.

The restriction to polynomial solutions yields only a few of the possible Sturm-Liouville systems. For example, the well-known Bessel functions do not fall into this category. To ensure the Hermiticity of L, we shall satisfy Eq. (5.112) by choosing $w\alpha$ to vanish as $|x| \to \infty$. In fact, since in Eq. (5.112) $w\alpha$ is multiplied by f^*g' and $gf^{*\prime}$ which are now by assumption polynomials in x, we see that $w\alpha$ must actually tend to zero faster than any inverse power of x as $|x|$ tends to infinity. This also follows from the requirement that $(f, f) < \infty$ for all f in our space.

With these assumptions, all the properties of a polynomial Sturm-Liouville system now are determined by the six parameters α_0, α_1, α_2, β_0, β_1, and γ_0. However, there are four degrees of freedom for inessential changes in L:

1. L can be multiplied by a constant C_1. This leaves Q_n unchanged and multiplies the eigenvalues by C_1.
2. The independent variable x may be shifted by some constant C_2. Thus $Q_n(x)$ is replaced by $Q_n(x + C_2)$, and the eigenvalues are unchanged.
3. The independent variable x may be scaled by a constant C_3: $x \to C_3 x$. Thus $Q_n(x)$ is replaced by $Q_n(C_3 x)$, and the eigenvalues are left unchanged.
4. A constant may be added to L, that is, to γ_0. This leaves Q_n unchanged, and λ_n is replaced by $\lambda_n + C_4$. Throughout this section we shall, for convenience, choose C_4 so that $\gamma_0 \equiv 0$.

With γ_0 fixed (equal to zero), we have only five parameters left which characterize the Sturm-Liouville system (α_0, α_1, α_2, β_0, and β_1); but there are three inessential degrees of freedom remaining. Thus, in fact, the polynomial Sturm-Liouville system is completely characterized by $5 - 3 = 2$ parameters. We now consider the following three cases:

1. $\alpha(x)$ is quadratic;
2. $\alpha(x)$ is linear ($\alpha_0 = 0$), and
3. $\alpha(x)$ is a constant ($\alpha_0 = 0 = \alpha_1$).

Our discussion will completely exhaust the possible *polynomial* solutions to the second-order Sturm-Liouville differential equation. The reader will never be able to discover a new system of Sturm-Liouville polynomials and have them named after himself!

CASE 1. $\alpha(x)$ is quadratic. In this case, Eq. (5.115) is

$$w\alpha = C \cdot \exp\left[\int \frac{\beta_0 x + \beta_1}{\alpha_0 x^2 + \alpha_1 x + \alpha_2}\, dx\right], \tag{5.124}$$

where $\alpha_0 \neq 0$. There are two subcases to distinguish, depending on whether $\alpha(x)$ has real or complex roots. Let us consider the case of complex roots first, and for convenience pick C_1 so that $\alpha_0 = 1$. Then $\alpha(x) = (x - \kappa)(x - \kappa^*)$, where κ and κ^* are the two complex roots of $\alpha(x)$ (the asterisk, as usual, denotes complex conjugation). The integral in Eq. (5.124) is

$$w\alpha = C[\alpha(x)]^{\beta_0/2} \exp\left[\frac{\beta_1 + \beta_0 \operatorname{Re}\kappa}{|\operatorname{Im}\kappa|} \tan^{-1}\left(\frac{x - \operatorname{Re}\kappa}{|\operatorname{Im}\kappa|}\right)\right],$$

where $\operatorname{Re}\kappa$ and $\operatorname{Im}\kappa$ denote, respectively, the real and imaginary parts of κ. Since $\alpha(x)$ is a polynomial in x, $w\alpha$ cannot vanish faster than some inverse power of x as x tends to infinity (the precise power being determined by β_0), so it fails to satisfy our previously derived endpoint requirements, Eq. (5.112). Also, $w\alpha$ is never zero, so we cannot join it onto a solution $w\alpha \equiv 0$ outside some finite interval. Thus our only hope is to consider the case where $\alpha(x)$ has two real roots. For convenience in this case, let us choose the three inessential parameters so that $\alpha_0 = -1$ and the roots of $\alpha(x)$ are at $x = 1$ and $x = -1$. Thus $\alpha(x) = 1 - x^2$. Furthermore, let us adopt as a notational convenience the relations $\beta_1 = q - p$, $\beta_0 = -(p + q + 2)$; this particular convention is made so that $w(x)$ will work out in a nice form; p and q are real but need not be integers. Thus

$$\frac{\beta}{\alpha} = \frac{-(p + q + 2)x + q - p}{1 - x^2} = \frac{q + 1}{1 + x} - \frac{p + 1}{1 - x},$$

and by doing the simple integral in Eq. (5.115) we get

$$w\alpha = C(1 + x)^{q+1}(1 - x)^{p+1} \qquad \text{or} \qquad w(x) = C(1 + x)^q(1 - x)^p .$$

Once again it turns out that, at large distances, $w\alpha$ can tend to zero only as fast as some fixed inverse power of x (depending on the specific values of p and q), so in this form our endpoint requirements cannot be satisfied. However, if $q > -1$ and $p > -1$, $w\alpha$ vanishes at $x = 1$ and $x = -1$, so we can join the above expression for $w(x)$ on $[-1, 1]$ to the solution $w(x)$ identically equal to zero outside $[-1, 1]$. (Note that α_0 and β_0 always have the same sign, a fact we shall use subsequently.) Therefore all the objections we raised in the case of complex roots are overcome, and we have an acceptable weight function for the interval $[-1, 1]$:

$$w(x) = (1 + x)^q(1 - x)^p ,$$

where we have set $C = 1$.

The $Q_n(x)$ that correspond to these choices of the essential parameters for the case of $\alpha(x)$ quadratic are called *Jacobi polynomials* of index p, q, and they shall be designated $J_n^{(p,q)}(x)$. (The upper indices distinguish them from Bessel functions.) By making a linear change of the independent variable, they may be recast in another form. Letting $x \to 1 - 2x$ in the differential equation (5.119), we obtain

$$\alpha(x) = x(1 - x) \geq 0 \text{ in } [0, 1], \qquad \beta(x) = -(p + q + 2)x + (p + 1) .$$

Thus the left-hand endpoint is now $x = 0$; $\beta_1 = \beta(0) = p + 1$; and

$$\beta_0 = \beta(1) - \beta(0) = -(p + q + 2)$$

is unchanged. It follows that $\alpha w = x^{p+1}(1 - x)^{q+1}$, so $p, q > -1$, as before. The weight function is, therefore, now

$$w(x) = x^p(1 - x)^q .$$

The set of polynomials which are orthogonal to the interval $[-1, 1]$ with respect to the weight function $w(x) = (1 - x)^p(1 + x)^q$, and the polynomials orthogonal on $[0, 1]$ with respect to $w(x) = x^p(1 - x)^q$, differ only by a linear change of variable. Both are referred to as the Jacobi polynomials of index p, q.

We now deal with a number of special cases of the Jacobi polynomials as defined on $[-1, 1]$. For $p = q = m (m > -1)$, $w(x) = (1 - x^2)^m$. The corresponding polynomials are called *ultraspherical* or *Gegenbauer polynomials* of integral index m. We designate them by

$$G_n^m(x) = J_n^{(m, m)}(x) .$$

Note that $p = q$ implies $\beta_1 = 0$, which means that the operator L is unaltered by the change of variable $x \to -x$. If we call P the operator which changes x into $-x$, then the invariance of L under $x \to -x$ may be expressed in operator notation as $PLf(x) = LPf(x)$ for all $f(x)$, so

$$PL = LP ,$$

which according to Theorem (4.22) tells us that any eigenfunction of L can be chosen to be simultaneously an eigenfunction of P. Since $P^2f(x) = f(x)$ for all $f(x)$, we see that $P^2 = I$, the identity operator. This places severe limitations on the eigenvalues of P. Consider $Pf(x) = \lambda f(x)$. This implies

$$P^2f(x) = \lambda Pf(x) = \lambda^2 f(x) = f(x) ,$$

since $P^2 = I$. Thus $\lambda^2 = 1$, so $\lambda = \pm 1$. This means that any Gegenbauer polynomial can be characterized by a subscript $+$ or $-$, depending on whether it is an eigenfunction of P with eigenvalue $+1$ or -1 (see Problem 5.15).

P is called the *parity operator*. It plays a fundamental role in many areas of physics. According to the above discussion, one speaks of eigenvectors of positive or negative parity. Since the leading term in $G_n^m(x)$ is proportional to x^n and since $Px^n = (-1)^n x^n$, we see that the Gegenbauer polynomials have positive or negative parity depending on whether n is even or odd. If n is even, $G_n^m(x)$ contains *only* even powers of x, whereas if n is odd, $G_n^m(x)$ contains *only* odd powers of x. The Gegenbauer polynomials occur in the solution of certain relativistically invariant equations in the quantum theory of elementary particles.

If we consider the Jacobi polynomials with $p = q = -\frac{1}{2}$, then $w(x) = (1 - x^2)^{-1/2}$. We designate these polynomials

$$T_n(x) \equiv J_n^{(-1/2, -1/2)}(x) .$$

The orthogonality relation is

$$(T_{n_1}, T_{n_2}) = \int_{-1}^{1} T_{n_1}^*(x) T_{n_2}(x) (1 - x^2)^{-1/2}\, dx = 0\,, \qquad (n_1 \neq n_2).$$

If we set $x = \cos\theta$, we get the alternate expression

$$(T_{n_1}, T_{n_2}) = \int_{-\pi}^{\pi} T_{n_1}^* (\cos\theta) T_{n_2}^* (\cos\theta)\, d\theta = 0\,.$$

The only set of polynomials in $\cos\theta$ which obeys this relationship on $[-\pi, \pi]$ is

$$T_n (\cos\theta) \propto \cos n\theta\,.$$

Therefore in the x-variable on $[-1, 1]$,

$$T_n(x) \propto \cos (n \cos^{-1} x)\,,$$

so this set of polynomials is determined explicitly up to a constant factor. These polynomials are known as the *Tschebycheff polynomials*. It is probably correct to say that, until now, no two authors have ever spelled their name the same way.

The most familiar of all the polynomials, the Legendre polynomials, are sure to be found somewhere in this cascade of special cases. We see that they result if we take $p = q = m = 0$. Then $w(x) = 1$; thus $J_n^{(0,0)}(x) = P_n(x)$ up to constant factor.

There is a theorem (Problem 20) which tells us that if $\{Q_n\}$ is an orthonormal system of Sturm-Liouville polynomials, then the set $\{Q'_n, n > 1\}$ is a system of orthogonal polynomials with weight function $w\alpha$. In the case of the Legendre polynomials, where $w = 1$ in $[-1, 1]$ and $\alpha = (1 - x^2)$, this result, applied μ times, tells us immediately that the set of polynomials $(d^\mu/dx^\mu)P_n$ is orthogonal on $[-1, 1]$ with respect to the weight function $\alpha^\mu = (1 - x^2)^\mu$. The uniqueness of sets of orthonormal polynomials on given intervals means that these polynomials are none other than the Gegenbauer polynomials:

$$G_{n-\mu}^\mu = \frac{d^\mu}{dx^\mu} P_n\,,$$

which is the only orthonormal set of polynomials with weight function $w = (1 - x^2)^\mu$ on $[-1, 1]$.

It also follows that the set of *functions*

$$P_n^\mu(x) = (1 - x^2)^{\mu/2} \frac{d^\mu}{dx^\mu} P_n(x)$$

is orthogonal with weight function 1 on $[-1, 1]$. These functions are the associated Legendre functions.

CASE 2. $\alpha(x)$ is a linear function. Under these circumstances, $\alpha(x) = x + \alpha_2$, where we have chosen C_1 so that $\alpha_1 = 1$. Also, $\alpha(x)$ has one real zero at $x = -\alpha_2$; adjusting our constant C_2, let us shift the zero to the origin, so finally

$\alpha(x) = x$. Thus

$$w\alpha = C \cdot \exp\left[\int \frac{\beta_0 x + \beta_1}{x}\, dx\right]$$
$$= C \cdot \exp\left(\beta_0 x + \beta_1 \ln x\right) = C x^{\beta_1} e^{\beta_0 x} .$$

If we choose $\beta_0 < 0$, then $w\alpha \to 0$ as $x \to +\infty$, but $w\alpha$ grows exponentially as $x \to -\infty$. Thus to satisfy the boundary conditions, $w(x)$ must be chosen to vanish identically for all $x < a$, if possible. a is still to be determined. If we choose $\beta_1 > 0$, then $w\alpha = 0$ at $x = 0$ and the integral of $w\beta$ over an interval of length ϵ around $x = 0$ goes to zero as ϵ tends to zero, so we may correctly join our identically zero solution (for $x < 0$) to the above nontrivial solution at this point. By adjusting C_3 we may set $\beta_0 = -1$ for convenience. (Note that if we choose $\beta_0 > 0$, this would just reverse the situation, and $w\alpha$ would have to vanish on the positive x-axis. The choice $\beta_0 < 0$ is conventional.) Calling $\beta_1 = s + 1$, $s > -1$, we have

$$w(x) = x^s e^{-x}$$

for positive x and $w(x) = 0$ for negative x. The Q_n that correspond to these conditions are the *associated Laguerre polynomials* of order s. They are designated by $L_n^s(x)$. If $s = 0$, the functions $L_n^0(x) \equiv L(x)$ are called the *Laguerre polynomials*. The presence of only one free parameter (s) reflects the fact that $\alpha(x)$ is *linear* in x: $\alpha_0 = 0$, and since three of the four remaining coefficients in $\alpha(x)$ and $\beta(x)$ may be chosen arbitrarily, there is just a single free parameter left.

CASE 3. $\alpha(x)$ is a constant. Here $\alpha_0 = 0 = \alpha_1$, and by a judicious choice of C_1 we can arrange that $\alpha_2 = 1$. Thus

$$w\alpha = C \cdot \exp\left[\int (\beta_0 x + \beta_1)\, dx\right] = C \cdot \exp\left(\frac{\beta_0}{2} x^2 + \beta_1 x\right)$$
$$= C' \cdot \exp\left[\frac{\beta_0}{2}\left(x + \frac{\beta_1}{\beta_0}\right)^2\right] .$$

If we choose $\beta_0 < 0$, then $w\alpha = w$ falls off exponentially as $|x| \to \infty$, so this expression for $w(x)$ is acceptable as it stands; choosing C_2 and C_3 appropriately, we can make $\beta_0 = -2$ and $\beta_1 = 0$, so setting $C' = 1$, we get

$$w(x) = e^{-x^2} .$$

The corresponding $Q_n(x)$ are the Hermite polynomials. We have now run through *all* the possibilities: there are no more Sturm-Liouville polynomials that are solutions of second-order differential equations. Table 5.1 summarizes the results obtained thus far.

We now work out the eigenvalues belonging to each of the three basic cases. We need only collect the terms in x^n in the differential equation. This gives

$$\lambda_n = \alpha_0 n(n-1) + \beta_0 n = n(\alpha_0 n + \beta_0 - \alpha_0) . \tag{5.125}$$

It follows immediately that the eigenvalues are spaced linearly for the Hermite and Laguerre equations (for which $\alpha_0 = 0$), and quadratically for Jacobi's equation, and all the special cases of it.

In particular, for Hermite's equation we have $\lambda_n = -2n$, and for Laguerre's equation we have $\lambda_n = -n$. For Jacobi's equation the eigenvalues are $\lambda_n = -n(n+p+q+1)$ regardless of whether the equation is the one whose solutions are orthogonal on $[0, 1]$ or the equation whose solutions are orthogonal on $[-1, 1]$. The Gegenbauer special case ($p = q = m$) has eigenvalues $\lambda_n = -n(n+2m+1)$. The two further specialized cases, Tschebycheff's ($m = -\frac{1}{2}$) and Legendre's ($m = 0$) equations, have eigenvalues $\lambda_n = -n^2$ and $\lambda_n = -n(n+1)$, respectively. It should be clear that whether the eigenvalues are positive or negative is merely a matter of the specific sign conventions adopted elsewhere in the treatment.

We conclude this discussion of polynomial Sturm-Liouville systems by proving a general Rodrigues formula which provides explicit expressions for all the polynomials up to a constant factor. The general formula is

$$Q_n(x) = K_n \frac{1}{w} \frac{d^n}{dx^n} (\alpha^n w) , \tag{5.126}$$

where the choice of the constant K_n depends on the physical application. The proof we shall give is a general version of the one in Section 5.5 for the Legendre polynomials. Any set of polynomials is uniquely determined by the requirement that $Q_n(x)$ is a polynomial of nth degree that satisfies the orthogonality relation

$$\int_a^b Q_m^* Q_n w \, dx = 0 , \qquad \text{for } n \neq m . \tag{5.127}$$

Thus we only have to show that the $Q_n(x)$, as given by the generalized Rodrigues formula [Eq. (5.126)], are polynomials of degree n, and satisfy Eq. (5.127) and we shall have proved Eq. (5.126).

We first prove a preliminary result: If $f(x) = \alpha^k w r(x)$, where $r(x)$ is a polynomial of degree l, then $f'(x) = \alpha^{k-1} w s_{l+1}(x)$, where $s_{l+1}(x)$ is a polynomial of degree $l+1$. To prove this we simply differentiate $f(x)$:

$$\begin{aligned} f'(x) &= (\alpha^{k-1})'(\alpha w) r + \alpha^{k-1}(\alpha w)' r + \alpha^{k-1}(\alpha w) r' \\ &= (k-1)\alpha^{k-2}\alpha'(\alpha w) r + \alpha^{k-1}(\beta w) r + \alpha^{k-1}(\alpha w) r' \\ &= \alpha^{k-1} w[\beta r + (k-1)\alpha' r + \alpha r'] \\ &= \alpha^{k-1} w[r_0(\beta_0 + (2k+l-2)\alpha_0) x^{l+1} + \cdots] \\ &= \alpha^{k-1} w s_{l+1}(x) . \end{aligned}$$

We have written $r(x) = r_0 x^l + \cdots$. Since $\beta_0 \neq 0$, and since α_0 and β_0 have the same sign if $\alpha_0 \neq 0$, then the expression in brackets is a polynomial of degree $l+1$ if $k \geq 1$ which is the only situation of interest. From this result applied n times to $\alpha^n w$, ($r = 1$), we infer that

$$\frac{d^n}{dx^n} (\alpha^n w) = \alpha^0 w t(x) = w t_n(x) ,$$

Table 5.1

THE STURM-LIOUVILLE SYSTEMS OF ORTHOGONAL POLYNOMIALS

Name of polynomials	Differential equation	α_0 α_1 α_2	β_0	β_1	$w\alpha$	w	$[a, b]$
(I) α *quadratic*							
Jacobi	$(1-x^2)u'' + [-(p+q+2)x + (q-p)]u' + n(n+p+q+1)u = 0$	-1 0 1	$-(p+q+2)$	$q-p$	$(1-x)^{p+1}(1+x)^{q+1}$	$(1-x)^p(1+x)^q$	$[-1, 1]$
Jacobi	$(x-x)u'' + [-(p+q+2)x + (p+1)]u' + n(n+p+q+1)u = 0$	-1 1 0	$-(p+q+2)$	$p+1$	$x^{p+1}(1-x)^{q+1}$	$x^p(1-x)^q$	$[0, 1]$
Gegenbauer	$(1-x^2)u'' - 2(m+1)xu' + n(n+2m+1)u = 0$	-1 0 1	$-2(m+1)$	0	$(1-x^2)^{m+1}$	$(1-x^2)^m$	$[-1, 1]$
Tschebycheff	$(1-x^2)u'' - xu' + n^2u = 0$	-1 0 1	-1	0	$(1-x^2)^{1/2}$	$(1-x^2)^{-1/2}$	$[-1, 1]$
Legendre	$(1-x^2)u'' - 2xu' + n(n+1)u = 0$	-1 0 1	-2	0	$1-x^2$	1	$[-1, 1]$
(II) α *linear*							
Laguerre	$xu'' - [x-(s+1)]u' + nu = 0$	0 1 0	-1	$s+1$	$x^{s+1}e^{-x}$	x^se^{-x}	$[0, \infty]$
(III) α constant							
Hermite	$u'' - 2xu' + 2nu = 0$	0 0 1	-2	0	e^{-x^2}	e^{-x^2}	$[-\infty, \infty]$

where $t(x)$ is a polynomial of degree n. It follows from Eq. (5.126) that $Q_n(x)$ is indeed a polynomial of degree n.

We can now prove the orthogonality of the $Q_n(x)$ with respect to $w(x)$. We have

$$\int_{-\infty}^{\infty} Q_m^*(x)Q_n(x)w(x)\,dx = \int_{-\infty}^{\infty} Q_m^*(x)\,\frac{d^n}{dx^n}\,(\alpha^n w)\,dx$$
$$= \sum_{k=0}^{n-1}\left[(-1)^k\left(\frac{d^k}{dx^k}\,Q_m^*\right)\frac{d^{n-1-k}}{dx^{n-1-k}}\,(\alpha^n w)\right]_{-\infty}^{\infty}$$
$$+\,(-1)^n\int_{-\infty}^{\infty}\left(\frac{d^n}{dx^n}\,Q_m^*\right)\alpha^n w\,dx\,. \tag{5.128}$$

Here we have integrated by parts n times. If $n > m$, the integral vanishes, so orthogonality will be satisfied if we can show that the terms $d^l/dx^l(\alpha^n w)$, $0 \leq l \leq n-1$, all vanish at the endpoints faster than any inverse power of x. But this follows from the preliminary result which tells us that

$$\frac{d^l}{dx^l}\,(\alpha^n w) = \alpha^{n-l} w v_l(x)\,,$$

where $v_l(x)$ is a polynomial of degree l. Since by assumption αw vanishes at the endpoints faster than any inverse power of x, so will $\alpha^{n-l} w v_l(x)$ for all $0 \leq l \leq n-1$, and the proof of Eq. (5.126) is complete.

In carrying out this proof we have shown that, in general,

$$\int_{-\infty}^{\infty} R(x)Q_n(x)w(x)\,dx = (-1)^n\int_{-\infty}^{\infty}\frac{d^n R}{dx^n}\,\alpha^n w\,dx\,. \tag{5.129}$$

This relation can be useful in the evaluation of integrals. In particular, if we take $R(x) \equiv Q_n(x)$, the normalization integral is

$$\int_{-\infty}^{\infty} Q_n^2(x)w(x)\,dx = (-1)^n n!\,q_n\int_{-\infty}^{\infty}\alpha^n w\,dx\,,$$

where q_n is the leading coefficient of $Q_n(x)$, as given in Eq. (5.126).

We can now use this generalized Rodrigues formula to obtain expressions for the polynomials we have considered. The Rodrigues formula will only provide us with expressions up to constant factors, which may then be selected in several ways. In some books they are selected in such a way as to make the coefficient of the highest power in $Q_n(x)$ equal to one for all n. Sometimes they are chosen so as to normalize the polynomials with respect to the corresponding weight function; and sometimes in still other ways. For example, the Legendre polynomials defined in Eq. (5.38) are not normalized on $[-1, 1]$. The normalized Legendre polynomials are given by $(n + \frac{1}{2})^{1/2}\,P_n(x)$. One must beware of the multiple inconsistent conventions that exist in the literature in defining the various sets of orthogonal polynomials. We shall not put in explicit expressions for the constant factors in the list that follows, except for the Legendre and Hermite polynomials, which we have worked out earlier.

Now, using the generalized Rodrigues formula (Eq. 5.126), we obtain for the Jacobi polynomials,

$$J_n^{(p,q)}(x) = A_n^{(p,q)}(1-x)^{-p}(1+x)^{-q}\frac{d^n}{dx^n}\Big[(1-x^2)^n(1-x)^p(1+x)^q\Big], \tag{5.130}$$

where $A_n^{(p,q)}$ is a constant factor depending on p, q, and n.

For the Gegenbauer polynomials, $p = q = m$, so this becomes

$$G_n^m(x) = B_n^m(1-x^2)^{-m}\frac{d^n}{dx^n}(1-x^2)^{n+m}, \tag{5.131}$$

where B_n^m is a constant factor depending on m and n.

For the Tschebycheff ($m = -\frac{1}{2}$) and Legendre ($m = 0$) polynomials, we have

$$T_n(x) = C_n(1-x^2)^{1/2}\frac{d^n}{dx^n}(1-x^2)^{n-1/2}, \tag{5.132}$$

where C_n is a constant factor depending on n;

$$P_n(x) = \frac{(-1)^n}{2^n n!}\frac{d^n}{dx^n}(1-x^2)^n. \tag{5.133}$$

For the associated Laguerre polynomials, we obtain

$$L_n^s(x) = D_n^s x^{-s} e^x \frac{d^n}{dx^n}(x^{n+s}e^{-x}), \tag{5.134}$$

where D_n^s is a constant factor which may depend on s and n; conventions vary widely in its choice. The Laguerre polynomials are the case $s = 0$:

$$L_n(x) = D_n e^x \frac{d^n}{dx^n}(n^x e^{-x}); \tag{5.135}$$

here D_n depends only on n.

Finally, the Hermite polynomials are given by

$$H_n(x) = (-1)^n e^{x^2}\frac{d^n}{dx^n}e^{-x^2}. \tag{5.136}$$

We close this section by showing that all the Sturm-Liouville polynomials form complete sets in Hilbert space when the inner product in Hilbert space is given by

$$(f, g) \equiv \int_{-\infty}^{\infty} f^*(x)g(x)w(x)\,dx,$$

with $w(x)$ given for the various polynomials in Table 5.1. In this discussion we shall take into account the Hermite and Laguerre polynomials which are defined on infinite intervals and therefore do not fall within the scope of Weierstrass's Theorem. The proof of completeness is based on Theorem 5.2 which says that if a set of orthonormal functions is closed, then it is complete.

Completeness Theorem. The orthonormal set of Sturm-Liouville polynomials $\{Q_n(x)\}$ is complete in Hilbert space.

Proof. We shall show that the set $\{Q_n(x)\}$ is closed. Suppose there is some function $f(x)$ orthogonal to all the $Q_n(x)$, so that $(f, Q_n) = 0$ for all n. Since x^m can be written as a finite linear combination of the $Q_n(x)$, the first $m+1$ to be exact, we see immediately that $(x^m, f) = 0$ for all m. Now consider

$$g(k) \equiv (e^{ikx}, f) = \int_{-\infty}^{\infty} f(x) w(x) e^{-ikx}\, dx\,,$$

for all real k. Because of the presence of $w(x)$ in the inner product, e^{-ikx} will belong to Hilbert space for any $w(x)$ appropriate to a Sturm-Liouville polynomial; so will x^m. Now, if we expand e^{-ikx} as

$$e^{-ikx} = \sum_{m=0}^{\infty} \frac{(-ik)^m}{m!} x^m$$

and use $(x^m, f) = 0$, we see that $g(k) = 0$ for all k. But $g(k)$ is precisely the Fourier transform of $f(x)\, w(x)$. Hence by our discussion in Section 5.7, $f(x)\, w(x) = 0$ almost everywhere. Thus we conclude that on any interval where $w(x) \neq 0$, $f(x) = 0$ almost everywhere. We are only interested in $f(x)$ on intervals where $w(x) \neq 0$, since we can define $f(x)$ to be anything we like where $w(x) = 0$. Another way to put this is to say that the Hilbert space is really defined only on the interval where $w(x) \neq 0$, since when $w(x) = 0$ on some finite interval, the inner product gives no structure there at all.

In summary, we conclude that $f(x) = 0$ almost everywhere, so the set $\{Q_n(x)\}$ is closed, and therefore it is a complete orthonormal set. QED

These sets of orthogonal polynomials have many additional polynomials. For example, using the generalized Rodrigues formula, one can derive recursion relations for the various sets of polynomials. Also, the zeros of the polynomials can be characterized in some detail. The interested reader is referred to Lyusternik or Wilf. In Section 6.10 we shall go on to establish the equivalence of the explicit Rodrigues representation of some of these polynomials and their implicit representation in terms of a generating function.

All these sets of functions are solutions to differential equations, as are the common trigonometric functions. All of them are related by various identities, just as the sine and cosine functions are. It is useful to think about them in this light. We study sines and cosines first, but these particular functions are not essentially different from Legendre or Hermite polynomials, or any of the other special functions of physics. They are more familiar to us only because we can apply them to the measurement of distances and because they turn up in macroscopic simple harmonic motion. If we imagine very small intelligent beings who do not measure large distances and instead first investigate quantum effects, it is reasonable to suppose that they discover Hermite polynomials at an early date by investigating the quantum oscillator. After centuries of work, they spend half their Gross National Product to build a (to us) macroscopic oscillator (to them it will be absolutely gigantic), and, in the course of this project, they discover the trigonometric functions. To them, sines and cosines will seem complicated and troublesome compared to the old familiar Hermite polynomials.

5.11 A MATHEMATICAL FORMULATION OF QUANTUM MECHANICS

In this and in the preceding chapters we have developed many of the mathematical ideas used in the formulation of quantum mechanics, namely, the concepts of vector spaces and, in particular, Hilbert space. We shall now write and interpret the axioms of quantum mechanics in terms of these mathematical tools. What we give here is not a rigorous axiomatization, but rather an introduction to a profound and beautiful subject. For more complete treatments, the reader should see J. M. Jauch's book, *Foundations of Quantum Mechanics*, or J. Von Neumann's classic work, *Mathematical Foundations of Quantum Mechanics*.

The mathematical setting for quantum mechanics is Hilbert space, that is, a complete complex inner-product space. As we have seen, the elements of this space are functions. To give a mathematical formulation of quantum mechanics, we must also make use of *operators* on Hilbert space. In our discussion of the properties of these operators, we will rely heavily on the results of Chapters 3 and 4. We have already noted that the validity of many of the theorems in Chapters 3 and 4 is not restricted to finite-dimensional spaces, and we shall make use of these theorems where they are applicable. We shall also use them to arrive intuitively at certain important generalizations of finite-dimensional results which will be central to our treatment.

We begin by discussing the notion of a *projection* operator (Problem 4.17), which will prove very useful in the mathematical formulation of quantum mechanics. We do this in the context of some familiar finite-dimensional results. Consider an Hermitian operator, A, whose eigenvalues and eigenvectors are $\{\lambda_n\}$ and $\{\phi_n\}$ respectively ($n = 1, 2, \cdots, N$). We assume, for notational convenience, that λ_n is nondegenerate. We now define the operator I by

$$I\psi \equiv \sum_{n=1}^{N} \phi_n(\phi_n, \psi)$$

for any ψ in the finite-dimensional space in question. According to the completeness theorem (Theorem 4.20), this sum is equal to ψ, so that the operator defined in this way is in fact the identity operator, as we had anticipated by our choice of notation. If we define P_n by

$$P_n\psi \equiv \phi_n(\phi_n, \psi) \ , \tag{5.137}$$

then I can be written as

$$I = \sum_{n=1}^{N} P_n \ .$$

The reader can easily verify that $P_n^2 = P_n$, $P_n^\dagger = P_n$, and $P_n P_m = 0$ if $m \neq n$. P_n is a *projection operator*, so called because it projects any vector into the one-dimensional subspace of V spanned by ϕ_n.

What can we say about the operator, B, defined by

$$B \equiv \sum_{n=1}^{N} \lambda_n P_n \,?$$

Since

$$B\psi = \sum_{n=1}^{N} \lambda_n \phi_n (\phi_n, \psi) \;,$$

it is immediately clear that B has eigenvalues $\{\lambda_n\}$ with corresponding eigenvectors $\{\phi_n\}$. It must therefore be equal to A (the reader will find it instructive to prove this). In summary, we have

$$I = \sum_{n=1}^{N} P_n \;, \qquad A = \sum_{n=1}^{N} \lambda_n P_n \;. \tag{5.138}$$

Thus we may restate the completeness theorem as follows: For every self-adjoint operator, A, there exists a set of orthogonal projections, $\{P_n\}$, such that Eqs. (5.138) are satisfied. By *orthogonal* projections, we mean projections which satisfy $P_n P_m = 0$ for $m \neq n$. It is in this language that one can generalize the completeness theorem to infinite-dimensional spaces. However, it turns out that only for a special type of infinite-dimensional operator (the completely continuous type, which we will discuss in Chapters 8 and 9) do Eqs. (5.138) generalize directly.

In an infinite-dimensional space, there arises the possibility of "eigenvalues" which are infinitely closely spaced, so that the above formulation needs some alteration. To get the flavor of what these alterations involve, we must say a word about an important generalization of the Riemann integral—the Stieltjes integral. We define

$$\int_a^b f(x)\, dg(x) \equiv \lim \sum_{i=1}^{N} f(\bar{x}_i)[g(x_{i+1}) - g(x_i)] \;, \tag{5.139}$$

where the points $x_1, x_2, \cdots, x_{n+1}$ represent a partition of the interval $[a, b]$, and $\bar{x}_i$ is any point in the ith interval. "Lim" denotes the passage to very small partition intervals. Obviously, if $g(x)$ is a nice differentiable function, we may write

$$\int_a^b f(x)\, dg(x) = \int_a^b f(x) \frac{dg}{dx}\, dx \;,$$

and we get back to the familiar Riemann integral. However, $g(x)$ need not be differentiable or even continuous. Consider the function

$$g(x) = \begin{cases} 1 & \text{if} \quad x \geq 0 \\ 0 & \text{if} \quad x < 0 \;. \end{cases}$$

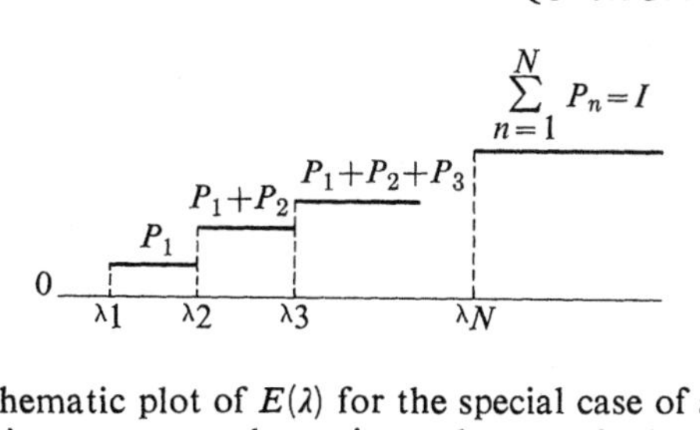

Fig. 5.5. A schematic plot of $E(\lambda)$ for the special case of a finite-dimensional self-adjoint operator whose eigenvalues are $\lambda_1, \lambda_2, \cdots, \lambda_N$.

According to the definition of Eq. (5.139), we have simply

$$\int_{-\infty}^{\infty} f(x)\, dg(x) = f(0) ,$$

since only an arbitrarily small interval around $x = 0$ can contribute to the sum in Eq. (5.139). This result is a δ-function type equation, which is not surprising since in the imprecise δ-function language,

$$\frac{d}{dx} g(x) = \delta(x) .$$

However, in the Stieltjes treatment, we never go beyond the domain of very traditional mathematics.

Now let us define an operator-valued function $E(\lambda)$ as follows:

$$E(\lambda) = \begin{cases} 0 & \text{for } \lambda < \lambda_1 , \\ \sum_{n=1}^{\nu} P_n & \text{for } \lambda_\nu \leqslant \lambda < \lambda_{\nu+1} , \qquad \nu = 1, \cdots, N-1 , \\ \sum_{n=1}^{N} P_n & \text{for } \lambda \geq \lambda_N ; \end{cases}$$

the P_n and λ_n are defined above. This operator-valued function is "plotted" schematically in Fig. 5.5. Note that

$$E(-\infty) = 0 \qquad \text{and} \qquad E(+\infty) = I.$$

The reader can easily verify that

$$E(\lambda_1) E(\lambda_2) = E(\bar{\lambda}),$$

where $\bar{\lambda}$ is the smaller of λ_1 and λ_2 (this follows from the fact that the P_n are *orthogonal* projections), and that

$$[E(\lambda)]^2 = E(\lambda).$$

Using the definition of the Stieltjes integral (Eq. 5.139), we see that

$$\int_{-\infty}^{\lambda} dE(\lambda) = \sum_{n=1}^{\nu} P_n$$

if λ lies between λ_ν and $\lambda_{\nu+1}$. The only contributions to this integral come from the points at which $E(\lambda)$ changes discontinuously; at each such point, λ_i (which must be an eigenvalue of A by construction), we pick up the projection operator onto the one-dimensional subspace of V which is spanned by ϕ_i. In this new way of writing things, the first of Eqs. (5.138) becomes

$$I = \int_{-\infty}^{\infty} dE(\lambda) ,$$

and the second of Eqs. (5.138) can be written as

$$A = \int_{-\infty}^{\infty} \lambda dE(\lambda) .$$

Thus we have succeeded in making the simple finite-dimensional completeness theorem look extremely complicated.

Now let us look at the virtues of this complexity. Consider the familiar problem

$$\frac{d^2\phi_n(x)}{dx^2} = \lambda_n \phi_n(x) .$$

We know that with periodic boundary conditions on $[-L, L]$

$$\phi_n = \frac{1}{\sqrt{2L}} e^{i(n\pi/L)x} , \qquad \lambda_n = \frac{n^2\pi^2}{L^2} , \qquad n = 0, \pm 1, \pm 2, \cdots$$

As we let L become very large, the eigenvalues cluster very close together, so that in Fig. 5.5 the intervals of constancy of $E(\lambda)$ become increasingly small. We may imagine that in this limit $E(\lambda)$ tends to a *continuous* operator-valued function. It is this phenomenon which is given the name of "continuous spectrum," and it is very familiar in quantum mechanics—indeed, it may be said to lie at the heart of quantum mechanics.

Having introduced the function $E(\lambda)$ on a finite-dimensional space, we will generalize, without proof, to an infinite-dimensional Hilbert space. The following theorem is one of the most powerful results obtained in mathematics and is the key to a rigorous formulation of quantum mechanics.

Spectral Theorem. To every self-adjoint operator, A, on a Hilbert space, H, there corresponds an operator-valued function (unique), $E(\lambda)$, such that

1. $E(\lambda_1)E(\lambda_2) = E(\bar{\lambda}) \; ; \quad \bar{\lambda} = \min(\lambda_1, \lambda_2) ,$
2. $\lim_{\lambda \to -\infty} E(\lambda) = 0 , \quad \lim_{\lambda \to \infty} E(\lambda) = I ,$
3. $I = \int_{-\infty}^{\infty} dE(\lambda) ,$
4. $A = \int_{-\infty}^{\infty} \lambda dE(\lambda) .$

$E(\lambda)$ is called a *resolution of the identity* belonging to A. The set of points at which $E(\lambda)$ is nonconstant is referred to as the *spectrum* of A. For all λ, $E(\lambda)$ commutes with A and with any transformation which commutes with A.

Obviously, the spectral theorem reduces to the familiar completeness theorem in a finite-dimensional space. There are several remarks which should be made at this point. First, note that we have not been precise in saying what we mean by the limit of a sequence of operators (see Part 2 of the above theorem). We shall say more on this point in Chapters 8 and 9, but it is worth remarking that in nonrelativistic quantum mechanics, the spectrum is always bounded below (i.e., we cannot have bound states with arbitrarily large negative energy) so that the limit $\lambda \to -\infty$ is unambiguous. Second, we remark that if A is an *unbounded* operator (i.e., if there are elements of H on which A is *not defined*) then self-adjoint does not mean merely $(Af, g) = (f, Ag)$ whenever Af and Ag are defined. We will not dwell on this point; suffice it to say that all interesting operators which occur in nonrelativistic quantum mechanics are self-adjoint.

With these mathematical preliminaries out of the way, we now proceed to discuss the axioms of quantum mechanics.

Axiom I. Any physical system is completely described by a normalized vector ψ (the *state vector* or *wave function*) in Hilbert space. All possible information about the system can be derived from this state vector by rules which will be given in the following axioms.

Axiom II. To every physical observable there corresponds a self-adjoint operator on Hilbert space.

Examples of observables are position, momentum, energy, angular momentum, spin, etc. This axiom brings with it the powerful machinery associated with the spectral theorem. Notice that Axiom II does not specify how one is to discover the self-adjoint operator which corresponds to a given observable. It merely says that one exists. In practice, the use of classical expressions such as $\mathbf{L} = \mathbf{r} \times \mathbf{p}$, $E = p^2/2m + V(\mathbf{r})$, etc., along with the operator substitutions $\mathbf{p} \to -i\hbar\nabla$ and $\mathbf{r} \to \mathbf{r}$ has been very successful. However, in the case of electron spin, one is forced beyond classical analogy to find the required operator.

Axiom III. The only allowed physical results of measurements of the observable A are the elements of the spectrum of the operator which corresponds to A.

We have already discussed some examples in which the spectrum is discrete. For instance, the operator L^2 of Section 5.8, which corresponds to the square of the total angular momentum of a particle, satisfies

$$L^2 Y_{lm} = \hbar^2 l(l+1)\, Y_{lm} \; .$$

Here there are $2l + 1$ eigenvectors corresponding to each eigenvalue, $\hbar^2 l(l+1)$. Another system whose spectrum is particularly simple is the quantum harmonic oscillator. The classical Hamiltonian of this system is

$$\mathscr{H} = p^2/2m + \tfrac{1}{2}kx^2,$$

and by making the operator substitutions suggested above, we find

$$H = -\frac{\hbar^2}{2m}\frac{d^2}{dx^2} + \frac{1}{2}kx^2$$

for the *operator* H corresponding to the energy. We saw in Sections 3.10 and 4.4 that this operator has eigenvalues $\lambda_n = \hbar\omega_0(n + \frac{1}{2})$, where $\omega_0 = (k/m)^{1/2}$, and normalized eigenvectors

$$\phi_n(x) = (\sqrt{\pi}\, 2^n n!)^{-1/2} H_n(\alpha x) e^{-\alpha^2 x^2/2} \; ,$$

where

$$\alpha = (m\omega_0/\hbar)^{1/2} \; ,$$

and H_n is the nth Hermite polynomial. We have already seen in this chapter that this set of ϕ_n is a complete orthonormal set in Hilbert space.

Now let us look at the situation for a very simple quantum-mechanical system whose spectrum is strictly continuous. The operator corresponding to the momentum of a particle is $p_x = -i\hbar\, d/dx$ (we consider just the x-component). In the space of square-integrable functions (such functions must tend to zero at large distances), p_x is self-adjoint, so according to the spectral theorem, we expect that we should be able to find a resolution of the identity belonging to p_x. If we were very free and easy about this problem, we would simply say that p_x had "eigenfunctions" $\phi_k = (2\pi)^{-1/2} e^{ikx}$ and "eigenvalues" $\hbar k$ for all k. The factor $(2\pi)^{-1/2}$ in ϕ_k is chosen to "normalize e^{ikx} to a δ-function," that is,

$$\int_{-\infty}^{\infty} \phi_k^*(x)\phi_{k'}(x)\, dx = \frac{1}{2\pi}\int_{-\infty}^{\infty} e^{i(k'-k)x}\, dx = \delta(k - k') \; .$$

Also,

$$\int_{-\infty}^{\infty} \phi_k^*(x)\phi_k(x')\, dk = \frac{1}{2\pi}\int_{-\infty}^{\infty} e^{ik(x'-x)} = \delta(x - x') \; ,$$

which one often thinks of as being analogous to

$$\sum_{n=1}^{N} P_n = I$$

in finite-dimensional spaces. That is, we imagine a P_k defined by

$$P_k f = \phi_k(\phi_k, f) \; ,$$

where k is a continuous variable $(-\infty < k < \infty)$. Note, however, that P_k is not a conventional operator, since it pushes vectors *out of Hilbert space*. This is due to the fact that ϕ_k is not an element of Hilbert space, and hence is not an eigenvector of p_x in the customary sense. However, this approach does suggest what a correct solution of the problem would be. On the basis of our earlier work, we expect that if we want to obtain a resolution of the identity, $E(k)$, then this resolution should be the sum of all the projection operators P_k, with $k' < k$. Thus we might try

$$E(k)f = \int_{-\infty}^{k} \phi_{k'}(\phi_{k'}, f)\, dk' .$$

Written out in detail, this is

$$E(k)f(x) = \frac{1}{\sqrt{2\pi}} \int_{-\infty}^{k} e^{ik'x} \left[\frac{1}{\sqrt{2\pi}} \int_{-\infty}^{\infty} e^{-ik'y} f(y)\, dy \right] dk' ,$$

or

$$E(k)f(x) = \frac{1}{\sqrt{2\pi}} \int_{-\infty}^{k} e^{ik'x} \hat{f}(k')\, dk' , \tag{5.140}$$

where $\hat{f}(k')$ is the Fourier transform of $f(x)$. We have already indicated that the Fourier transform of every element in the space of square-integrable functions exists and is itself square-integrable (we will prove this rigorously in Chapter 9). Since $\hat{f}(k')$ is square-integrable, the integral in Eq. (5.140) clearly exists. It is a simple exercise to show that the vector $E(k)f(x)$ lies in Hilbert space.

We see by inspection that $E(\infty) = I$, by our main result on the inversion of Fourier transforms, and since $\hat{f}(k')$ is square-integrable, $E(-\infty) = 0$. Now we must show that $E(k_1)E(k_2) = E(k_1)$ (where we assume without loss of generality that $k_1 < k_2$). We reason as follows:

$$\begin{aligned} E(k_1)E(k_2)f &= E(k_1) \frac{1}{\sqrt{2\pi}} \int_{-\infty}^{k_2} e^{ik'x} \hat{f}(k')\, dk' \\ &= \frac{1}{2\pi} \int_{-\infty}^{k_1} e^{ikx} F(k)\, dk , \end{aligned}$$

where $F(k)$ denotes the Fourier transform of

$$\int_{-\infty}^{k_2} e^{ik'x} \hat{f}(k')\, dk'$$

considered as a function of x. This transform is readily computed:

$$F(k) = \frac{1}{\sqrt{2\pi}} \int_{-\infty}^{\infty} e^{-ikx} \int_{-\infty}^{k_2} e^{ik'x} \hat{f}(k') dk'\, dx = \begin{cases} \sqrt{2\pi}\, \hat{f}(k) & \text{if } k_2 > k , \\ 0 & \text{if } k_2 < k . \end{cases}$$

This result can be most simply obtained by the use of Eq. (5.63). Thus, since $k_1 < k_2$,

$$E(k_1)E(k_2)f = \frac{1}{\sqrt{2\pi}} \int_{-\infty}^{k_1} e^{ikx} \hat{f}(k)\, dk = E(k_1)f ,$$

as required. Finally, the reader can verify for himself that

$$\int_{-\infty}^{\infty} dE\,(k) = I\,,$$

so we have indeed found a resolution of the identity.

To complete our task, we need only show that the above resolution belongs to p_x, that is, that

$$p_x = \hbar \int_{-\infty}^{\infty} k\, dE\,(k)\ .$$

This is easily done. We write

$$\begin{aligned}
\left[\hbar \int_{-\infty}^{\infty} k\, dE\,(k)\right] f &= \frac{\hbar}{\sqrt{2\pi}} \int_{-\infty}^{\infty} k\, d\left[\int_{-\infty}^{k} e^{ik'x} \hat{f}(k')\, dk'\right] \\
&= \frac{\hbar}{\sqrt{2\pi}} \int_{-\infty}^{\infty} k \frac{d}{dk}\left[\int_{-\infty}^{k} e^{ik'x} \hat{f}(k')\, dk'\right] dk \\
&= \frac{\hbar}{\sqrt{2\pi}} \int_{-\infty}^{\infty} k\, e^{ikx} \hat{f}(k)\, dk \\
&= \frac{-i\hbar}{\sqrt{2\pi}} \int_{-\infty}^{\infty} \frac{d}{dx} e^{ikx} \hat{f}(k)\, dk \\
&= \frac{-i\hbar}{\sqrt{2\pi}} \frac{d}{dx} \int_{-\infty}^{\infty} e^{ikx} \hat{f}(k)\, dk = -i\hbar \frac{df}{dx} = p_x f
\end{aligned}$$

for any f for which df/dx lies in our Hilbert space (under these conditions the above interchange of integration and differentiation is permissible). Thus we have found the resolution of the identity belonging to p_x. Of course, the spectral theorem tells us that such a resolution must exist; we have now explicitly constructed it. As we expect, the spectrum of p_x is purely continuous, consisting of all real numbers. Using the informal ideas above, the reader should be able to show that the resolution of the identity to the position operator X, defined by $Xf(x) \equiv xf(x)$, is the operator-valued function $E(\xi)$, specified by

$$E(\xi)f(x) = \begin{cases} f(x) & \text{if } x \leqslant \xi\,, \\ 0 & \text{if } x > \xi\,. \end{cases}$$

For more complicated self-adjoint operators, it is more difficult to find an explicit resolution of I, but it is always possible in principle.

We are now in a position to state the central axiom of quantum mechanics.

Axiom IV. If, on a system in the state ψ, we make a measurement of the observable A, then the probability that the value λ obtained by this measurement will lie between λ_1 and λ_2 ($\lambda_2 > \lambda_1$) is given by

$$P(\lambda_1, \lambda_2) = ||[E(\lambda_2) - E(\lambda_1)]\psi||^2\,,$$

when $E(\lambda)$ is the resolution of the identity belonging to A (by Axiom II plus the Spectral theorem).

Since $||\psi|| = 1$, we see that the probability of finding λ between $-\infty$ and $+\infty$ is 1, as it must be if the theory is to make sense physically. If $E(\lambda)$ is constant between λ_1 and λ_2 (i.e., if there are no points of the spectrum of A between λ_1 and λ_2), then $E(\lambda_2) - E(\lambda_1) = 0$, so a value can never be found in this range. This is in accord with Axiom III; in fact, it is clear that Axiom III is contained in Axiom IV. We have put in Axiom III only because its statement is such a familiar part of elementary quantum mechanics; if we were trying for minimal redundancy, we would omit Axiom III.

A particularly important case is that in which $E(\lambda)$ is constant between λ_1 and λ_2 *except* for a discontinuous jump at $\bar{\lambda}$. We will look at this case in some detail both for its importance to physics and because it enables us to show how a discrete spectrum can emerge from the projection-operator formalism. Intuitively, we should expect from our original motivations that whenever we have a discontinuity in $E(\lambda)$ at, say, $\bar{\lambda}$, on each side of which are intervals of constancy of $E(\lambda)$, we should find an eigenfunction of A. This is easily shown to be the case. Let λ_+ be any point in the interval of constancy to the right of $\bar{\lambda}$, and let λ_- be any point in the interval of constancy to the left of $\bar{\lambda}$. By analogy with our starting point in finite-dimensional spaces, we expect that if ψ is any element of H such that $\phi_{\bar{\lambda}} \equiv [E(\lambda_+) - E(\lambda_-)]\psi \neq 0$, then $\phi_{\bar{\lambda}}$ is an eigenfunction of A. We argue as follows:

$$\begin{aligned} A[E(\lambda_+) - E(\lambda_-)]\psi &= \int_{-\infty}^{\infty} \lambda\, d\{E(\lambda)[E(\lambda_+) - E(\lambda_-)]\psi\} \\ &= \int_{-\infty}^{\infty} \lambda\, d[E(\lambda)E(\lambda_+)\psi] - \int_{-\infty}^{\infty} \lambda\, d[E(\lambda)E(\lambda_-)\psi] \\ &= \int_{-\infty}^{\lambda_+} \lambda\, d[E(\lambda)\psi] - \int_{-\infty}^{\lambda_-} \lambda\, d[E(\lambda)\psi] \\ &= \int_{\lambda_-}^{\lambda_+} \lambda\, d[E(\lambda)\psi]\,. \end{aligned}$$

But by our assumption of intervals of constancy on either side of $\bar{\lambda}$, we have

$$\int_{\lambda_-}^{\lambda_+} \lambda\, d[E(\lambda)\psi] = \bar{\lambda}[E(\lambda_+) - E(\lambda_-)]\psi = \bar{\lambda}\phi_{\bar{\lambda}}\,.$$

Hence

$$A\phi_{\bar{\lambda}} = \bar{\lambda}\phi_{\bar{\lambda}}\,,$$

so $\bar{\lambda}$ is an eigenvalue of A, and $\phi_{\bar{\lambda}}$ is an eigenvector belonging to that eigenvalue.

Of course, $\phi_{\bar{\lambda}}$ may not be the *only* eigenvector belonging to $\bar{\lambda}$. If one picked a different ψ to start with (note that it is always possible to find a ψ such that $[E(\lambda_+) - E(\lambda_-)]\psi \neq 0$), one might find a vector $\tilde{\phi}_{\bar{\lambda}}$ which belongs to $\bar{\lambda}$ but is not just a multiple of $\phi_{\bar{\lambda}}$. In this case, we say that $\bar{\lambda}$ has multiplicity $\mu > 1$. We leave it as an exercise for the reader to show that if $\Phi_{\bar{\lambda}}$ is the unique normalized eigenvector belonging to $\bar{\lambda}$ (that is, $\bar{\lambda}$ has multiplicity 1), then for any $\psi \in H$,

$$[E(\lambda_+) - E(\lambda_-)]\psi = \Phi_{\bar{\lambda}}(\Phi_{\bar{\lambda}}, \psi)\,, \tag{5.141}$$

where λ_+ and λ_- are as chosen above. Thus, just as we would expect from our original finite-dimensional analogy, the discontinuity in $E(\lambda)$ at an eigenvalue $\bar{\lambda}$ is just the projection operator onto the one-dimensional subspace of H spanned by $\Phi_{\bar{\lambda}}$. If $\bar{\lambda}$ has multiplicity μ greater than one, then the discontinuity in $E(\lambda)$ at $\bar{\lambda}$ is just the projection operator onto the μ-dimensional subspace of H spanned by the μ eigenvectors belonging to $\bar{\lambda}$. Thus, in general, we may write

$$[E(\lambda_+) - E(\lambda_-)]\psi = \sum_{\nu=1}^{\mu} \Phi_{\bar{\lambda}}^{(\nu)} (\Phi_{\bar{\lambda}}^{(\nu)}, \psi) \,, \tag{5.142}$$

where μ is the multiplicity of $\bar{\lambda}$ (see Problem 5.25).

Returning now to our interpretation of Axiom IV, we see that in the case of a discontinuous jump in $E(\lambda)$ between λ_1 and λ_2, the probability of obtaining a measured value of A between λ_1 and λ_2 is, according to Eq. (5.141), just

$$P(\lambda_1, \lambda_2) = |(\Phi_{\bar{\lambda}}, \psi)|^2 \,,$$

where $\bar{\lambda}$ is the point of discontinuity; we have assumed that $\bar{\lambda}$ has multiplicity 1. Thus $P(\lambda_1, \lambda_2)$ is just the modulus squared of the component of ψ along the "direction" in Hilbert space defined by the eigenvector $\Phi_{\bar{\lambda}}$ belonging to $\bar{\lambda}$. More generally, if $\bar{\lambda}$ has multiplicity μ,

$$P(\lambda_1, \lambda_2) = \sum_{\nu=1}^{\mu} |(\Phi_{\bar{\lambda}}^{(\nu)}, \psi)|^2 \,. \tag{5.143}$$

Hence Eqs. (5.141) and (5.142) lead to very simple results for the probability of finding an observable A in the discrete spectrum.

If we choose an observable with a continuous spectrum, the situation is similar. Consider the x-component of the momentum, p_x. By our previous work, the probability of finding p_x between $\hbar k_1$ and $\hbar k_2$ is just

$$\begin{aligned} P(k_1, k_2) &= \|[E(k_2) - E(k_1)]\psi\| \\ &= \left\| \frac{1}{\sqrt{2\pi}} \int_{k_1}^{k_2} e^{ikx} \hat{\psi}(k)\, dk \right\| \\ &= \int_{k_1}^{k_2} |\hat{\psi}(k)|^2\, dk \,, \end{aligned}$$

where we have again used Eq. (5.63). In a more conventional language, $P(k_1, k_2)$ is just the continuous superposition of "plane-wave" probability amplitudes.

We now define the *expectation value* of A.

Definition. $\langle A \rangle \equiv \lim \sum_i \bar{\lambda}_i \|[E(\lambda_i + \Delta) - E(\lambda_i)]\psi\|^2$ is the *expectation value* of the observable A for the system described by the state vector ψ. Here we have divided the spectrum into intervals of length Δ, and $\bar{\lambda}_i$ is any point in the ith interval. By lim we mean the limit as the interval size tends to zero.

The physical meaning of the expectation value is very simple. It is just the value that would be found by taking the average of many measurements of

A on a large collection of systems all in the state ψ. Each possible result, $\bar{\lambda}_i$, of a measurement is weighted by the probability of finding a result in the vicinity of $\bar{\lambda}_i$. Note that $\langle A \rangle$ is a number, not an operator.

Theorem. $\langle A \rangle = (\psi, A\psi)$.

Proof. According to the definition (Eq. 5.139) of the Stieltjes integral,

$$\begin{aligned}(\psi, A\psi) &= \int_{-\infty}^{\infty} \lambda \, d[(\psi, E(\lambda)\psi)] \\ &= \lim \sum_i \bar{\lambda}_i (\psi, [E(\lambda_i + \Delta) - E(\lambda_i)]\psi) \, .\end{aligned}$$

But $[E(\lambda_i + \Delta) - E(\lambda_i)]^2 = [E(\lambda_i + \Delta) - E(\lambda_i)]$, so using the self-adjointness of E, we have

$$\begin{aligned}(\psi, A\psi) &= \lim \sum_i \bar{\lambda}_i ([E(\lambda_i + \Delta) - E(\lambda_i)]\psi, [E(\lambda_i + \Delta) - E(\lambda_i)]\psi) \\ &= \lim \sum_i \bar{\lambda}_i || (E(\lambda_i + \Delta) - E(\lambda_i)]\psi||^2 = \langle A \rangle\end{aligned}$$

by definition.

In terms of the expectation value of A, we define the mean-square deviation $(\Delta A)^2$, which measures the dispersion around the mean value $\langle A \rangle$.

Definition. $(\Delta A)^2$ is the expectation value of $[A - \langle A \rangle]^2$ in the state ψ in which $\langle A \rangle$ is computed.

Theorem. $(\Delta A)^2 = \langle A^2 \rangle - \langle A \rangle^2$.

Proof.

$$\begin{aligned}(\Delta A)^2 &= \langle [A - \langle A \rangle]^2 \rangle = (\psi, [A - \langle A \rangle]^2 \psi) \\ &= (\psi, [A^2 - 2A\langle A \rangle + \langle A \rangle^2]\psi) \\ &= (\psi, A^2\psi) - 2(\psi, A\psi)\langle A \rangle + \langle A \rangle^2 \\ &= \langle A^2 \rangle - \langle A \rangle^2 \, ,\end{aligned}$$

as required. We have used the previous theorem several times in obtaining this result.

Note that according to our definition,

$$(\Delta A)^2 = (\psi, [A - \langle A \rangle]^2 \psi) = ([A - \langle A \rangle]\psi, [A - \langle A \rangle]\psi) \, ,$$

because A is self-adjoint and $\langle A \rangle$ is real. Thus

$$(\Delta A)^2 = ||[A - \langle A \rangle]\psi||^2 \, ,$$

so if $\Delta A = 0$ for some ψ, we must have $[A - \langle A \rangle]\psi = 0$; that is, ψ is an eigenvector of A with eigenvalue $\langle A \rangle$. This has the important implication that unless ψ is an eigenstate of A, one cannot make a perfectly precise measurement of A in that state. Thus only values of A lying in the discrete spectrum of A can be determined with perfect accuracy. For other values, one must always have some dispersion, ΔA, associated with measurements.

It is instructive to check that if $\psi = \Phi_n$, a normalized eigenstate of A with eigenvalue λ_n, then indeed $\Delta A = 0$. We have

$$\langle A \rangle = (\Phi_n, A\Phi_n) = \lambda_n ,$$
$$\langle A^2 \rangle = (\Phi_n, A^2\Phi_n) = (A\Phi_n, A\Phi_n) = \lambda_n^2 ,$$

so

$$\langle \Delta A \rangle^2 = \langle A^2 \rangle - \langle A \rangle^2 = \lambda_n^2 - \lambda_n^2 = 0 .$$

Furthermore, if $\psi = \Phi_n$, the probability of obtaining a value λ between λ_1 and λ_2 when A is measured is zero if $[\lambda_1, \lambda_2]$ does not contain λ_n, and this probability is 1 if $[\lambda_1, \lambda_2]$ contains λ_n. This follows simply by showing that the probability of finding λ in a small interval around λ_n is 1. According to Axiom IV,

$$\begin{aligned} P(\lambda_n - \epsilon, \lambda_n + \epsilon) &= ||[E(\lambda_n + \epsilon) - E(\lambda_n - \epsilon)]\Phi_n||^2 \\ &= |(\Phi_n, \Phi_n)|^2 = 1 , \end{aligned}$$

where we have made use of Eq. (5.142). Thus the probability of the measured value of A being λ_n is 1. In other words, if a system is in an eigenstate, so that its state vector happens to be an eigenvector of A and not a superposition of many eigenvectors, then the result of a measurement of A is certain to be the eigenvalue corresponding to that eigenvector.

Having considered in detail the case of a single measurement, let us mention briefly an important generalization of Axiom IV which admits the possibility of simultaneous measurement of several observables.

Axiom IV′. Let A, B, and C be observables whose corresponding linear operators commute, that is, $[A, B] = [A, C] = [B, C] = 0$. Then the probability that a simultaneous measurement of A, B, and C in a system whose state vector is ψ will yield a value of A between a_1 and a_2, B between b^1 and b_2, and C between c_1 and c_2 is

$$\begin{aligned} P(a_1, a_2; b_1, b_2; c_1, c_2) = ||[E_A(a_2) - E_A(a_1)][E_B(b_2) - E_B(b_1)] \\ \times [E_C(c_2) - E_C(c_1)]\psi||^2 , \end{aligned}$$

where $E_A(a)$, $E_B(b)$, and $E_C(c)$ are the resolutions of the identity belonging to A, B, and C, respectively.

In the case of a single measurement, Axiom IV′ is clearly equivalent to Axiom IV.

Note that since A, B, and C commute, so do $E_A(a)$, $E_B(b)$, and $E_C(c)$, and therefore the ordering of the projection operators on the right-hand side of the above equation is immaterial. Clearly, if their order were significant, the concept of simultaneous measurement would be untenable. The possibility of simultaneous measurement of *noncommuting* observables is denied in quantum mechanics; thus p_x and x, which satisfy $[x, p_x] = i\hbar$, cannot be measured simultaneously. The importance of simultaneous measurement of observables will become apparent when we discuss Axiom VI below.

Everything said so far has been concerned with the properties of a system at a fixed time. We now must say what happens to the system as the time changes.

Axiom V. For every system, there exists an Hermitian operator, H, the Hamiltonian or energy operator, which determines the time development of the state vector, Ψ, of the system through the time-dependent Schrödinger equation:

$$H\Psi(x, t) = i\hbar \frac{\partial \Psi(x, t)}{\partial t}, \tag{5.144}$$

provided that the system is not disturbed (by a measurement, for example). Here $\hbar$ is a constant equal to 1.054×10^{-27} erg-sec.

Axiom I says that the state vector Ψ is normalized. To be certain that our axioms are consistent, we must therefore show that this normalization is preserved in time by the above equation which governs the time dependence of the state vector. Thus we prove the following result:

Theorem.

$$\frac{\partial}{\partial t}(\Psi, \Psi) = 0 .$$

Proof.

$$\frac{\partial}{\partial t}(\Psi, \Psi) = \left(\frac{\partial \Psi}{\partial t}, \Psi\right) + \left(\Psi, \frac{\partial \Psi}{\partial t}\right) .$$

But

$$\frac{\partial \Psi}{\partial t} = \frac{1}{i\hbar} H\Psi ,$$

so,

$$\begin{aligned}
\frac{\partial}{\partial t}(\Psi, \Psi) &= \left(\frac{1}{i\hbar} H\Psi, \Psi\right) + \left(\Psi, \frac{1}{i\hbar} H\Psi\right) \\
&= -\frac{1}{i\hbar}(H\Psi, \Psi) + (\Psi, H\Psi)\frac{1}{i\hbar} \\
&= -\frac{i}{i\hbar}(\Psi, H\Psi) + (\Psi, H\Psi)\frac{1}{i\hbar} = 0 ,
\end{aligned}$$

since H is self-adjoint. Thus Axiom I is consistent with Axiom V.

In general, the solution of Schrödinger's time-dependent equation for a Hamiltonian with complicated time dependence is a difficult job. However, if the Hamiltonian is time-independent, then matters simplify considerably. In this case, it is elementary to verify that the solution to Eq. (5.144), which reduces to $\Psi(x, 0)$ at $t = 0$, is simply

$$\Psi(x, t) = e^{-iHt/\hbar}\Psi(x, 0) . \tag{5.145}$$

Since H is self-adjoint, we have a resolution of the identity belonging to H, and therefore Eq. (5.145) can be written as

$$\Psi(x, t) = \int_{-\infty}^{\infty} e^{-i\epsilon t/\hbar} d[E(\epsilon)\Psi(x, 0)] . \tag{5.146}$$

Again, it is elementary to verify directly that this $\Psi(x, t)$ satisfies Eq. (5.144). Note that if the spectrum of H is purely discrete, then according to Eq. (5.143), Eq. (5.146) takes on the simple form

$$\Psi(x, t) = \sum_n e^{-i\epsilon_n t/\hbar}(\phi_n, \Psi(t = 0))\phi_n(x) , \tag{5.147}$$

where the ϕ_n are the eigenvectors of H and the ϵ_n are its eigenvalues. According to Axiom IV, the probability of finding a value ϵ_n for the energy of the system whose state vector is $\Psi(x, 0)$ is just

$$P(\epsilon_n; t = 0) = |(\phi_n, \Psi(t = 0))|^2 .$$

Using Eq. (5.147), we see that at time t, the probability of measuring H and finding ϵ_n is

$$P(\epsilon_n; t) = |e^{-i\epsilon_n t/\hbar}(\phi_n, \Psi(t = 0)|^2 = |(\phi_n, \Psi(t = 0))|^2 = P(\epsilon_n; t = 0) ;$$

that is, the probability of finding a particular value of the energy does not change with time.

Let us now see what happens for more general observables. Suppose that A is any observable, with associated resolution of the identity, $E_A(\lambda)$. According to Axiom IV, the probability of finding a value of A between λ_1 and λ_2 is just

$$P(\lambda_1, \lambda_2; t) = ||[E_A(\lambda_2) - E_A(\lambda_1)]e^{-iHt/\hbar}\Psi(x, 0)||^2 ,$$

where we have made use of Eq. (5.145) and are continuing to assume that H is independent of time. It is clear that, in general, $P(\lambda_1, \lambda_2; t)$ will depend on time, but there are two important situations in which $P(\lambda_1, \lambda_2; t)$ will be time-independent.

1. $\Psi(x, 0) = \phi_n(x)$, where $\phi_n(x)$ is an eigenvector of H.

   In this case,

$$P(\lambda_1, \lambda_2; t) = ||E_A(\lambda_2) - E_A(\lambda_1)]e^{-i\epsilon_n t/\hbar}\phi_n(x)||^2 ,$$

but since $e^{-i\epsilon_n t/\hbar}$ is just a complex number with modulus 1, it cannot affect the norm of a vector, so

$$P(\lambda_1, \lambda_2; t) = ||[E_A(\lambda_2) - E_A(\lambda_1)]\phi_n||^2 = P(\lambda_1, \lambda_2; t = 0) .$$

2. H commutes with A.

   In this case, H also commutes with $E_A(\lambda)$, so

$$P(\lambda_1, \lambda_2; t) = ||e^{-iHt/\hbar}[E_A(\lambda_2) - E_A(\lambda_1)]\Psi(x, 0)||^2 .$$

But according to Theorem 4.13, this is simply

$$P(\lambda_1, \lambda_2;\ t) = ||[E_A(\lambda_2) - E_A(\lambda_1)]\Psi(x,0)||^2 = P(\lambda_1, \lambda_2;\ t = 0)\ ,$$

since $\exp[-iHt/\hbar]$ is a unitary operator. This tells us that the results of measuring an observable which commutes with the Hamiltonian are independent of the time at which they are measured.

By way of illustration, let us now investigate the situation for a particular observable, the position of a particle. We have already introduced the operator X defined by $Xf(x) \equiv xf(x)$ and have stated (see Problem 5.26) that its resolution of the identity is given by

$$E_X(\xi)f(x) = \begin{cases} f(x) & \text{if} \quad x \leq \xi\ , \\ 0 & \text{if} \quad x > \xi\ . \end{cases}$$

We now ask, what is the probability of obtaining a value of the position between ξ_1 and ξ_2 for a particle in a state Ψ? According to Axiom IV,

$$P(\xi_1, \xi_2;\ t) = ||[E_X(\xi_2) - E_X(\xi_1)]\,\Psi(x,\ t)||^2\ .$$

From the definition of $E_X(\xi)$, we see that the vector whose norm is to be evaluated is just the function which is zero outside $[\xi_1, \xi_2]$ and equal to $\Psi(x, t)$ inside $[\xi_1, \xi_2]$. Thus

$$P(\xi_1, \xi_2;\ t) = \int_{\xi_1}^{\xi_2} |\Psi(x,\ t)|^2\,dx\ .$$

Note that the probability of finding the particle *someplace* is 1, since Ψ is normalized. If we set $\xi_1 = \xi - \Delta/2$ and $\xi_2 = \xi + \Delta/2$, then for Δ small, we see that the probability of finding the particle in a small interval of length Δ about ξ is just $P = |\Psi(\xi, t)|^2\Delta$; thus $|\Psi(x, t)|^2$ is called the *probability density*, $\rho(x, t)$. This result is often stated at the beginning of quantum mechanics courses. It arose first, historically, in the pioneering work of Born on the probabilistic interpretation of quantum mechanics. Here the result is presented as a special case of Axiom IV. Note that, in general, the probability of finding the particle in a particular region varies with time; however, if $\Psi(x, t)$ is an eigenstate of the Hamiltonian, then $\Psi(x, t)$ is related to $\Psi(x, 0)$ just by a multiplicative phase factor, so $|\Psi(x, t)|^2 = |\Psi(x, 0)|^2$, and the probability of finding the particle in any region is constant in time. This is why eigenstates of the Hamiltonian are often called *stationary states*.

Finally, let us use Axiom V to obtain $P(\xi_1, \xi_2;\ t)$ in terms of a particular initial state, $\Psi(x, t = 0)$. For simplicity, we assume that $\Psi(x, 0)$ is given by

$$\Psi(x,\ 0) = (1/\sqrt[4]{\pi a^2})e^{-x^2/2a^2}\ .$$

Thus the initial probability density is equal to $(1/\pi^{1/2}a)e^{-x^2/a^2}$, so that the probability of finding the particle at a distance greater than a from the origin is very small compared with the probability of finding the particle within a distance a on either side of the origin. Let us assume that we are dealing with a free particle (and confine our attention to the x-dependence of the problem). The

Hamiltonian is just $H = p_x^2/2m$, where m is the particle's mass, so according to Axiom V and Eq. (5.145),

$$\Psi(x, t) = e^{-ip_x^2 t/2m\hbar} \frac{1}{\sqrt[4]{\pi a^2}} e^{-x^2/2a^2} .$$

Using Eq. (5.146) and the spectral resolution which we have already determined for p_x we have immediately

$$\Psi(x, t) = \frac{1}{\sqrt[4]{\pi a^2}} \int_{-\infty}^{\infty} e^{i\hbar k^2 t/2m} \, d[E(k) \, e^{-x^2/2a^2}] .$$

But according to our previous work,

$$E(k) f(x) = \frac{1}{\sqrt{2\pi}} \int_{-\infty}^{k} e^{ik'x} \hat{f}(k') \, dk' ,$$

where $\hat{f}$ denotes the Fourier transform of f. The Fourier transform of $\Psi(x, 0)$ is easily computed; we find

$$\hat{\Psi}(k, 0) = \sqrt[4]{\frac{a^2}{\pi}} \, e^{-a^2k^2/2} .$$

Thus

$$\begin{aligned} \Psi(x, t) &= \sqrt[4]{\frac{a^2}{4\pi^3}} \int_{-\infty}^{\infty} e^{-i\hbar k^2 t/2m} \, d\left[\int_{-\infty}^{k} e^{ik'x} e^{-a^2k'^2/2} \, dk'\right] \\ &= \sqrt[4]{\frac{a^2}{4\pi^3}} \int_{-\infty}^{\infty} e^{-i\hbar k^2 t/2m} \, e^{-a^2k^2/2} \, e^{ikx} \, dk . \end{aligned}$$

This last integral is readily evaluated. We find that

$$\Psi(x, t) = \frac{1}{\pi^{1/4}} \frac{1}{\sqrt{a + i\dfrac{\hbar t}{ma}}} \exp\left[-\frac{x^2}{2a(a + i\hbar t/ma)}\right] ,$$

and therefore the probability density is given by

$$\rho(x, t) = |\Psi(x, t)|^2 = (1/\sqrt{\pi\Delta^2}) \, e^{-x^2/\Delta^2} ,$$

where $\Delta^2 = a^2 + (\hbar t/ma)^2$. Thus the width of the probability distribution increases in time. This phenomenon is known as "the spreading of the wave packet." The initial width was equal to $2a$; the width at time t is $2\sqrt{a^2 + (\hbar t/ma)^2}$. To get a feeling for the magnitudes involved, suppose that we are dealing with electrons and that $a \equiv 1$ mm. This might describe a case in which one creates a beam of electrons collimated by a slit of width approximately equal to one millimeter. Then if we use $m = 9 \times 10^{-28}$ gm and $\hbar = 10^{-27}$ erg-sec, we find that the width has doubled in about 15 msec. This is a very short time on a macroscopic scale, but on a scale appropriate to electrons, it is a very large time. For example, a velocity of 10^8 cm/sec is not unusual for an electron; in 15 msec, such an electron would travel 1.5×10^6 cm, or 15 km! Thus a beam of electrons initially having a width of 1 mm and moving with a velocity of 10^8 cm/sec would travel 15 km before doubling in width. This assumes, of course, that the electrons are truly free; i.e., we neglect collisions, electromagnetic forces, gravitational forces, etc.

Before we can claim to have a complete system of axioms, we must give some method by which one can gain information about the state vector Ψ that figures so prominently in the previous axioms. The key to this problem, which is clearly of paramount practical importance, is the following axiom.

Axiom VI. If at time $t = 0$ one measures simultaneously the commuting observables A, B, and C, and finds with certainty that the values of these observables lie between a_1 and a_2, b_1 and b_2, and c_1 and c_2 respectively, then immediately after the measurement, the state vector of the system satisfies the relation

$$[E_A(a_2) - E_A(a_1)][E_B(b_2) - E_B(b_1)][E_C(c_2) - E_C(c_1)]\Psi = \Psi \,. \quad (5.148)$$

We may say that the measurements project the original wave function into the subspace of Hilbert space associated with the projection operator

$$[E_A(a_2) - E_A(a_1)][E_B(b_2) - E_B(b_1)][E_C(c_2) - E_C(c_1)] \,.$$

Note that because of Axiom IV, Eq. (5.148) implies that if we remeasure A (or B or C) immediately after our first measurement, we will find a value between a_1 and a_2 (or b_1 and b_2, or c_1 and c_2) with certainty.

Let us first consider the measurement of a single observable, say A. If there is only one point, a, in the spectrum of A between a_1 and a_2, then according to Eq. (5.141),

$$\Psi = e^{i\alpha}\Phi_a \,,$$

if a is nondegenerate, when Φ_a is the normalized eigenvector of A belonging to a and $e^{i\alpha}$ is a phase factor which is indeterminate. Clearly, according to our previous axioms, this phase factor can have no observable consequences. In this case, the measurement of A has, so to speak, "forced" the system into an eigenstate of A. Note that this does *not* tell us that the system was in the state Φ_a *before* the measurement, but only that it is in the state Φ_a *after* the measurement. We have "prepared" the system in the state Φ_a. If the eigenvalue a has multiplicity $\mu_a > 1$, then according to Eq. (5.142), after the measurement of A, we can only say that the state vector lies in the subspace of Hilbert space spanned by the μ_a orthonormal eigenvector of A belonging to a, that is,

$$\Psi = \sum_{\nu=1}^{\mu_a} c_\nu \Phi_a^{(\nu)} \,,$$

where

$$\sum_{\nu=1}^{\mu_a} |c_\nu|^2 = 1 \,.$$

Thus in this case we have not completely specified the state vector by measuring A.

However, according to Axiom VI, we have additional resources at our disposal; we can choose *another* observable, B, whose self-adjoint operator commutes with A, and measure A and B simultaneously. Suppose we discover that the value of A lies with certainty between a_1 and a_2 and the value of B lies with certainty between b_1 and b_2. Again, we imagine that a is the only point in the spectrum of A between a_1 and a_2 and that a has multiplicity μ_a. Hence

$$[E_A(a_2) - E_A(a_1)]\Psi = \sum_{\nu=1}^{\mu_a} c_\nu \Phi_a^{(\nu)} . \tag{5.149}$$

We assume also that there is only one point, b, between b_1 and b_2 in the spectrum of B, and that b has multiplicity μ_b. Using Eqs. (5.142) and (5.149), we find that

$$[E_B(b_2) - E_B(b_1)][E_A(a_2) - E_A(a_1)]\Psi = \sum_{\nu=1}^{\mu_a} \sum_{\nu=1}^{\mu_b} c_\nu (\Phi_b^{(\lambda)}, \Phi_a^{(\nu)})\, \Phi_b^{(\lambda)} .$$

We know from Theorem 4.22 that every eigenvector of A must be an eigenvector of B *and vice versa*, so $(\Phi_b^{(\lambda)}, \Phi_a^{(\nu)}) = 1$ if $\Phi_b^{(\lambda)} = \Phi_a^{(\nu)}$, and vanishes otherwise. It may happen that there is only one pair, $(\bar{\lambda}, \bar{\nu})$, for which $(\Phi_b^{(\lambda)}, \Phi_a^{(\nu)}) = 1$. In this case, Axiom VI tells us that

$$\Psi = e^{i\beta}\Phi_b^{(\bar{\lambda})} = e^{i\beta}\Phi_a^{(\bar{\nu})} ,$$

so that we have uniquely specified the state vector of the system.

If there is more than one pair for which $(\Phi_b^{(\lambda)}, \Phi_a^{(\nu)})$ fails to vanish, then calling these pairs (λ_i, ν_i), we have

$$\Psi = \sum_i c_{\nu_i} \Phi_b^{(\lambda_i)} = \sum_i c_{\nu_i} \Phi_a^{(\nu_i)} ;$$

that is, Ψ lies in the subspace of Hilbert space spanned by the mutual eigenvectors of A and B belonging to a and b. In this last case, the state vector is not uniquely determined, so we must find a new observable, C, which commutes with A and B, and repeat the process. In this manner we eventually will specify the state uniquely.

A collection of commuting observables which has the property that simultaneous measurement of the observables specifies uniquely the state vector of a system is called a *complete set* of commuting observables. For example, if we consider an electron in a coulomb field and neglect electron spin, then H, L^2, and L_z (where $H = p^2/2m - e^2/r$ and $\mathbf{L}$ is the orbital angular momentum of the electron) constitute such a set; H and L^2 alone do not constitute a complete set of observables, for after a measurement of the energy and the total angular momentum, there is still a degeneracy corresponding to the possible values of the z-component of the angular momentum. If we include the spin of the electron, then H, L^2, S^2, J^2, and J_z constitute a complete set, where $\mathbf{S}$ is the electron's spin angular momentum, and $\mathbf{J} = \mathbf{L} + \mathbf{S}$ is the total angular momentum of the electron. This set is not unique; H, L^2, L_z, S^2 and S_z would do just as well. Which set is more useful depends on the practical situation. If we wish to take into account relativistic effects by including magnetic interactions via a term in the Hamiltonian of the form $A\mathbf{L}\cdot\mathbf{S}/r^3$, then the second set mentioned above is no longer a complete commuting set, whereas the first is (because

$$\mathbf{L}\cdot\mathbf{S} = (J^2 - L^2 - S^2)/2) .$$

Thus Axiom VI enables us to perform the all-important task of controlling the initial conditions of an experiment by the process of preparing quantum states. Without this axiom, Axiom IV would have no content, since if one never knew what the state vector of a system was, one could never hope to compute the probability of obtaining a particular result in a measurement.

PROBLEMS

1. Prove that $\int_{-1}^{1} P_n(x)dx = 0$ for $n \neq 0$ in two ways.

2. A metal spherical shell of radius R with its center at $x = y = z = 0$ is cut in half along its intersection with the plane $z = 0$. The two halves are separated by an infinitesimal gap and the upper and lower hemispheres are brought to voltages $+V$ and $-V$ respectively. Show that the potential inside the sphere is

$$\psi(r, \theta) = V \sum_{l=0}^{\infty} (-1)^l \left(\frac{r}{R}\right)^{2l+1} \frac{(2l)!}{(2^l l!)^2} \frac{4l+3}{2l+2} P_{2l+1}(\cos\theta) ,$$

where θ is the polar angle measured relative to the positive z-axis.
Hint: $\nabla^2\psi = 0$ inside and on a spherical shell. ψ is clearly independent of ϕ. Following the usual procedure of separation of variables, we find that the solution is of the form

$$\psi(r, \theta) = \sum_{l=0}^{\infty} [A_l r^l + B_l r^{-l-1}][C_l P_l(\cos\theta) + D_l Q_l(\cos\theta)] .$$

Immediately set $B_l = 0$ (all l) so $\psi(0, \theta)$ is finite and set $D_l = 0$ (all l) so $\psi(r, 0)$ and $\psi(r, \pi)$ are finite. The problem is now one of mathematics—determining the constants $A_l C_l \equiv \alpha_l$ such that the boundary conditions are satisfied:

$$\psi(R, \theta) = \begin{cases} +V & \text{for} \quad 0 \le \theta < \pi/2 , \\ -V & \text{for} \quad \pi/2 < \theta \le \pi . \end{cases}$$

3. A conducting sphere of radius d sits in a charge-free region in an electric field. Its surface is kept at a fixed distribution of electric potential $V = F(\theta)$, where (r, ϕ, θ) are spherical coordinates with the origin at the center of the sphere. Determine the potential at all points in the space inside and outside the sphere; i.e., solve Laplace's equation $\nabla^2 V = 0$, assuming the following boundary conditions:

   i. $\lim_{r\to d} V(r, \theta) = F(\theta), \qquad 0 < \theta < \pi ,$

   ii. $\lim_{r\to\infty} V(r, \theta) = 0$

   (potential vanishes at points infinitely far away from sphere).

Hint: Use the method of separation of variables. Solve the r equation by making the substitution $r = e^t$. When it is solved, set the exponent $-\frac{1}{2} + (\lambda + \frac{1}{4})^{1/2} = n$ so that $\lambda = n(n+1)$ (λ = separation constant). Solve the θ equation by an appropriate change of variable.

Answer:

Inside sphere: $$V(r,\theta) = \sum_{n=0}^{\infty} \frac{2n+1}{2} \frac{r^n}{d^n} P_n(\cos\theta) \int_{-1}^{1} F(x) P_n(x)\, dx$$

$$r \leq d,\ F(\theta) = f(\cos\theta)\,.$$

Outside sphere: $$V(r,\theta) = \sum_{n=0}^{\infty} \frac{2n+1}{2} \frac{d^{n+1}}{r^{n+1}} P_n(\cos\theta) \int_{-1}^{1} F(x) P_n(x)\, dx\,.$$

4. a) Show that the 2π periodic Fourier series representation of the square wave (step function)

$$f(x) = \begin{cases} -1, & -\pi < x < 0\,, \\ +1, & 0 < x < \pi\,, \end{cases} \quad \text{is} \quad f(x) = \frac{4}{\pi} \sum_{n=1}^{\infty} \frac{1}{2n-1} \sin(2n-1)x\,.$$

Compare this result with the expansion of the same function in Legendre polynomials—see Jackson, *Classical Electrodynamics*, pp. 58–59.

b) Derive from the Fourier series obtained in (a) a simple infinite series expression for π; preferably one with rapid convergence. It is instructive to plot several of the approximations to $f(x)$; that is, plot

$$S_m(x) = \frac{4}{\pi} \sum_{n=1}^{m} \frac{1}{2n-1} \sin(2n-1)x$$

for several values of m to see how $S_m(x) \to f(x)$ with increasing m.

c) *Gibbs' phenomenon*: Show that the Fourier series overshoots the function just to the right and left of the origin by about 18%. This is not easy. Guidance may be found in Sommerfeld's *Partial Differential Equations in Physics*, pp. 7–12, or in Morse and Feshbach, p. 747.

5. Prove that if $f(x)$ is real,

$$f(x) = \frac{1}{\pi} \int_0^{\infty} du \int_{-\infty}^{\infty} f(t) \cos u(t-x)\, dt\,.$$

6. Solve the integral equation

$$\int_0^{\infty} g(t) \cos xt\, dt = \begin{cases} e^{-\alpha x} & \text{for } x > 0 \\ e^{\alpha x} & \text{for } x < 0 \end{cases} \qquad (\alpha > 0)\,.$$

Answer:

$$g(t) = \frac{2}{\pi} \frac{\alpha}{\alpha^2 + t^2}\,.$$

7. a) Show that the energy transferred to a spring by a time-dependent force $F(t)$ defined by (Fig. 5.6)

$$F(t) = \begin{cases} 0, & t < -\tau \\ F_0(t/2\tau + 1/2), & |t| \leq \tau \\ F_0, & t > \tau \end{cases} \quad \text{is} \quad \Delta E = \frac{1}{2m} (F_0/\omega)^2 \left(\frac{\sin\omega\tau}{\omega\tau}\right)^2.$$

Hint: Take derivatives of $F(t)$ until you reach a function whose Fourier transform is known.

b) Show that the energy transfers in the adiabatic and sudden limits agree with these limits as determined for the probability integral force function (Example 3, Section 5.7).

c) For certain finite values of $\omega\tau$ there will be no energy transfer to the spring. Determine these values and explain their origin physically.

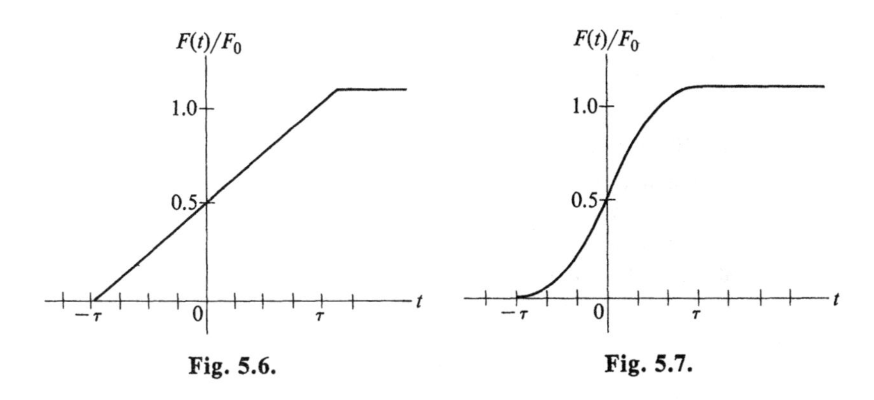

Fig. 5.6. **Fig. 5.7.**

8. Work Problem 7 for the force function (Fig. 5.7)

$$F(t) = \begin{cases} 0\,, & t < -\tau\,, \\ F_0\left[\frac{1}{2} + \frac{1}{4}\left(\left(\frac{3t}{\tau}\right) - \left(\frac{t}{\tau}\right)^3\right)\right], & |t| \leqslant \tau\,. \\ F_0\,, & t > \tau\,. \end{cases}$$

Answer:

$$\Delta E = \frac{1}{2m}\,(F_0/\omega)^2\,\frac{9}{(\omega_0\tau)^6}\,(\sin\omega_0\tau - \omega_0\tau\cos\omega_0\tau)^2\,.$$

The adiabatic and sudden limits must and do (prove this) agree with those of Problem 7. For part (c), settle for a determination of the position of the first node of ΔE. You should get

$$\omega\tau \cong 4.49\,.$$

9. Compute the three-dimensional Fourier transforms of the following functions:

a) $H(\mathbf{r}) = \dfrac{e^{-\lambda|\mathbf{r}|}}{|\mathbf{r}|}$

b) $F(\mathbf{r}) = \dfrac{1}{|\mathbf{r}|}$ [To save work consider $F(\mathbf{r})$ as a special case of $H(\mathbf{r})$.]

c) $Y(\mathbf{r}) = e^{-\lambda|\mathbf{r}|}$ [The Fourier transform of this function can be obtained from that of $H(\mathbf{r})$ also.]

10. Maxwell's equations in free space in Gaussian units are:

$$\text{curl}\,\mathbf{E} = -\frac{1}{c}\frac{\partial\mathbf{H}}{\partial t}, \qquad \text{div}\,\mathbf{H} = 0\,, \qquad \text{curl}\,\mathbf{H} = \frac{1}{c}\frac{\partial\mathbf{E}}{\partial t}, \qquad \text{div}\,\mathbf{E} = 0\,.$$

At any time t the electric (and magnetic) fields may be Fourier analyzed:

$$\mathbf{E}(\mathbf{r}, t) = \frac{1}{(2\pi)^{3/2}} \int_{-\infty}^{\infty} \mathbf{E}(k, t) e^{i\mathbf{k}\cdot\mathbf{r}}\, d\mathbf{k}, \quad \text{etc.}$$

[Here the Fourier components of $\mathbf{E}(\mathbf{r}, t)$ have been written $\mathbf{E}(\mathbf{k}, t)$; the argument identifies $\mathbf{E}(\mathbf{k}, t)$ as the Fourier component of the field, distinguishing it from the field itself, which has argument $\mathbf{r}$. If you don't like this notation, adopt another.]

a) Prove that the Fourier components of the electric and magnetic fields satisfy the following equations:

i) $\dot{\mathbf{H}}(\mathbf{k}, t) = -ic(\mathbf{k} \times \mathbf{E}(\mathbf{k}, t))$
ii) $\mathbf{k}\cdot\mathbf{E}(\mathbf{k}, t) = 0$
iii) $\dot{\mathbf{E}}(\mathbf{k}, t) = ic(\mathbf{k} \times \mathbf{H}(\mathbf{k}, t))$
iv) $\mathbf{k}\cdot\mathbf{H}(\mathbf{k}, t) = 0$.

These four equations are equivalent to Maxwell's equations. This reformulation is well-suited as a starting point for quantum mechanical considerations of the electromagnetic field.

b) Using the results of part(a), prove that the Fourier components of the electric field satisfy the homogeneous wave equation. That is, prove

$$\left[\frac{\partial^2}{\partial t^2} + c^2 k^2\right] \mathbf{E}(\mathbf{k}, t) = 0\,.$$

11. Show informally that

$$\lim_{g\to\infty} \frac{\sin gx}{\pi x}, \qquad g > 0\,,$$

is a representation of the Dirac δ-function.

12. Prove that

$$\delta(x) = \begin{cases} \lim\limits_{n\to\infty} \left(\dfrac{1}{|a|} c_n \left[1 - \left(\dfrac{x}{a}\right)^2\right]^n\right) & \text{if} \quad 0 \le |x| \le |a|\,, \\ 0 & \text{otherwise} \end{cases}$$

is also a representation of the δ-function, where c_n is given in Eq. (5.28). Show, therefore, that $\delta(ax) = (1/|a|)\,\delta(x)$.

13. Show that if $r = (x^2 + y^2 + z^2)^{1/2}$, then

$$\nabla^2\left(\frac{1}{r}\right) = -\,4\pi\delta^3(\mathbf{r})$$

[*Hint*: Use Green's Theorem.]

14. Prove Parseval's equation (5.11) in Hilbert space by applying Bessel's completeness relation (5.10) to the function $f + g$ and then subtracting the corresponding equations for f and g.

15. The parity operator P plays a fundamental role in many areas of physics. Show that the Gegenbauer polynomials are eigenfunctions of P; that is, if the leading power of x in $G_n^m(x)$ is even, then $G_n^m(x)$ contains *only* even powers of x. One way to do this is to assume the contrary. Write $G_n^m(x) = g_n^m(x) + u_n^m(x)$, where $g_n^m(x)$ contains only even powers of x and $u_n^m(x)$ contains only odd powers, and conclude that $(L - \lambda_n)g_n^m = 0$ and $(L - \lambda_n)u_n^m = 0$, where L is the Sturm-Liouville operator which has the Gegenbauer polynomials as eigenfunctions. Thus $g_n^m(x)$ and $u_n^m(x)$

are linearly independent (because they belong to different eigenvalues of P) solutions of the Gegenbauer equation, and they are both polynomial solutions. From this conclude that $u_n^m(x) = 0$. [*Hint*: we know that for the Gegenbauer polynomials $\lambda_n = -n(n+2m+1)$.] Thus $G_n^m(x)$ contains only even powers of x and has parity eigenvalue $+1$.

16. Prove that the complete set of orthonormal functions on the unit sphere, Y_{lm}, is unique up to a phase factor. Use induction on l for $m = m'$ fixed.
17. If $[x_i, p_j] = i\hbar\delta_{ij}$, where the indices run from one to three, and if we define $\mathbf{L} = \mathbf{r} \times \mathbf{p}$, then show that
$$[L_i, L_j] = i\hbar \sum_k \epsilon_{ijk} L_k .$$
Defining $L_+ = L_x + iL_y$ and $L_- = L_x - iL_y$, show that $[L_+, L_-] = 2\hbar L_z$. Show that if we define
$$\phi_m = L_+^{l+m}\phi_{-l}, \qquad L_-\phi_{-l} = 0,$$
where l is an integer, then
$$L_-\phi_m = \hbar(l+m)(l-m+1)\phi_{m-1} .$$
18. $H_n(x)$ are the Hermite polynomials. Prove for arbitrary n:

a) $H_{2n}(0) = (-1)^n \dfrac{(2n)!}{n!}$, b) $H'_{2n}(0) = 0$,

c) $H_{2n+1}(0) = 0$, d) $H'_{2n+1}(0) = (-1)^n 2\dfrac{(2n+1)!}{n!}$.

19. It should be clear from the definitions that pointwise convergence does not imply uniform convergence. The question comes up, Does pointwise convergence imply convergence in the mean? Show that the sequence of functions
$$f_n(x) = \frac{2n^{1/2}}{(\pi/2)^{1/4}} nxe^{-(nx)^2}$$
converges pointwise for all x to the function zero, that is,
$$\lim_{n\to\infty} f_n(x) = 0,$$
but does not converge in the mean to the function zero; i. e.,
$$\lim_{n\to\infty}\int_{-\infty}^{\infty} |0 - f_n|^2\, dx = 1 \neq 0 .$$
20. Let $\{Q_n\}$ be an orthonormal system of Sturm-Liouville polynomials. Prove that the functions $Q_1', Q_2', \cdots$ form an orthogonal system of polynomials with weight function $w\alpha$, where the notation is that of Section 5.10.
21. Show that the set of functions $[x^m, x^{m+1}, \cdots]$ is a complete set in the mean on $[-1, +1]$ for $m \geqslant 1$. (Of course, by Weierstrass's theorem, the above is true for $m = 0$.)
22. Obtain the first four orthonormal polynomials for the complete set $[x, x^2, x^3, \cdots]$ on the interval $[-1, +1]$, and write the first few terms of the expansion of $f(x) = 1$ in terms of this set. What do the first two approximations look like graphically?
23. Consider Bessel's equation $Bf_n = n^2 f_n$ on the interval $[0, \infty]$, where
$$B = x^2\frac{d^2}{dx^2} + x\frac{d}{dx} + x^2 .$$

a) Using the methods discussed in Section 5.10, find the weight function appropriate to Bessel's equation.

b) Show that if $n \neq m$ and $w(x)$ is the weight function of Part (a), then

$$\int_0^\infty J_{2n}(x)J_{2m}(x)w(x)\,dx = \int_0^\infty J_{2n+1}(x)J_{2m+1}(x)w(x)\,dx = 0\,.$$

You may make use of the fact that for small x,

$$J_n(x) = \left(\frac{x}{2}\right)^n \left[1 - \frac{x^2}{2(2n+2)} + \cdots\right],$$

while for large x,

$$J_n(x) = \sqrt{\frac{2}{\pi x}}\left[\cos\left(x - \frac{n\pi}{2} - \frac{\pi}{4}\right)(1 + O(x^2)) - \frac{4n^2-1}{8}\frac{1}{x}\sin\left(x - \frac{n\pi}{2} - \frac{\pi}{4}\right)(1 + O(x^2))\right].$$

Both of these results can be readily derived from complex variable techniques, which will be discussed in Chapter 6.

c) Using the asymptotic formulas given in Part (b), evaluate

$$\int_0^\infty J_{2n+1}(x)J_{2m}(x)w(x)\,dx\,.$$

d) Is the operator, B, in Bessel's equation an Hermitian operator with respect to the space of functions which satisfy

$$\int_0^\infty |f(x)|^2 w(x)\,dx < \infty\,,$$

where $w(x)$ is given by Part (a)? [It is understood, of course, that B can act only on functions f having the property that Bf lies in the Hilbert space if f does.]

24. We have shown that the Legendre polynomials, $P_l(x)$, are solutions to

$$(1 - x^2)f''(x) - 2xf'(x) + l(l+1)f(x) = 0\,, \qquad l = 0, 1, 2, \cdots,$$

which are orthogonal on $[-1, 1]$. However, these are not the only solutions; they are merely the only polynomial solutions.

a) To reassure yourself on this point, show that the function

$$\tanh^{-1} x = \frac{1}{2}\log\left(\frac{1+x}{1-x}\right)$$

satisfies Legendre's equation with $l = 0$. Note that this function is singular at $x = \pm 1$.

b) Show that if we look for a solution of Legendre's equation of the form

$$Q_l(x) = P_l(x)\tanh^{-1} x + \Pi_l(x)\,,$$

then a solution of this form always exists where $\Pi_l(x)$ is a polynomial of degree $l - 1$. Find the inhomogeneous differential equation that $\Pi_l(x)$ must satisfy, and show that this equation always has a solution. What type of boundary condition has been applied in taking $\Pi_l(x)$ to be a polynomial of order $l - 1$?

c) Since $\Pi_l(x)$ is a polynomial of degree $l - 1$ on $[-1, 1]$, it can be written as a linear combination of Legendre polynomials. Find an expression for the coefficients in this linear combination. To be precise, show that

$$\Pi_l(x) = 2 \sum_{n=0,1}^{l-1}{}' \frac{2n+1}{(n-l)(n+l+1)} P_n(x) ,$$

where the sum begins at $n = 0$ if l is odd and at $n = 1$ if l is even. The prime on the summation indicates that the sum runs in steps of two.

d) Show that $(Q_l, P_l) = 0$.
[*Hint*: Don't do any integrals.]

e) From their form it is clear that the $Q_l(x)$ lie in the Hilbert space of square-integrable functions on $[-1, 1]$. Do they form an orthogonal set of functions on $[-1, 1]$? Explain.

25. Let us define $E(\lambda)$, an operator-valued function, as in Section 5.11. Denote by λ_n and ϕ_n ($n = 1, 2, \cdots, N$) the nondegenerate eigenvalues and corresponding eigenvectors of an Hermitian operator, A, and call P_n the projection operator associated with ϕ_n. Then define

$$E(\lambda) = \begin{cases} 0 \text{ for } \lambda < \lambda_1 , \\ \sum_{n=1}^{\nu} P_n \text{ for } \lambda_\nu \leq \lambda < \lambda_{\nu+1} , \qquad \nu = 1, 2, \cdots, N-1 , \\ \sum_{n=1}^{N} P_n = I \text{ for } \lambda \geq \lambda_N . \end{cases}$$

Show that

i) $E(\lambda_1)E(\lambda_2) = E(\lambda_{\min})$, where $\lambda_{\min} = \min(\lambda_1, \lambda_2)$, ii) $E(\lambda)^2 = E(\lambda)$.

26. Let X be the self-adjoint operator on the Hilbert space of square-integrable functions defined by $Xf(x) \equiv xf(x)$.

i) Is X well defined on all elements of the space?
If not, characterize the elements for which it is defined.

ii) Prove that $E_X(\xi)$, defined by

$$E_X(\xi)f(x) = \begin{Bmatrix} f(x) & \text{if } x \leq \xi \\ 0 & \text{if } x > \xi \end{Bmatrix} ,$$

is the resolution of the identity belonging to X.

27. Show that if $E_A(\lambda)$ suffers a discontinuous jump at an eigenvalue $\bar{\lambda}$ of A, then

$$[E_A(\lambda_+) - E_A(\lambda_-)]\psi = \sum_{\nu=0}^{\mu} \Phi_{\bar{\lambda}}^{(\nu)} (\Phi_{\bar{\lambda}}^{(\nu)}, \psi) ,$$

where μ is the multiplicity of $\bar{\lambda}$ and $\{\Phi_{\bar{\lambda}}^{(\nu)}: \nu = 1, 2, \cdots, \mu\}$ is a set of μ orthonormal eigenvectors of A belonging to $\bar{\lambda}$, and λ_+ and λ_- are points on either side of $\bar{\lambda}$ in the intervals of constancy of $E(\lambda)$; that is, $E(\lambda)$ is constant in the intervals $[\bar{\lambda}, \lambda_+]$ and $[\lambda_-, \bar{\lambda}]$. A is any self-adjoint operator.

28. Given that $E_A(\lambda)$ is the resolution of the identity belonging to A, show that

$$[E_A(\lambda_2) - E_A(\lambda_1)]^2 = [E_A(\lambda_2) - E_A(\lambda_1)]$$

if $\lambda_2 > \lambda_1$.

29. i) What is the expectation value of the operator X of Problem 5.26 in the state $\Psi(x) = (1/\sqrt[4]{\pi a^2})e^{-x^2/2a^2}$?
ii) What is the dispersion, ΔX, about this average value?
iii) What is the expectation value of $p_x = -i\hbar\, d/dx$ in the above state?
iv) What is the dispersion, Δp_x, about this average value?

Note that as the dispersion in X gets smaller (i.e., as a gets smaller, which means that the wave function Ψ becomes more sharply peaked about $x = 0$), the dispersion in p_x gets larger. The product of the two dispersions, $\Delta X \Delta p_x$, is a constant, equal to $\hbar/2$. One can show that for any state Ψ we must always have $\Delta X \Delta p_x \geqslant \hbar/2$.

30. Verify that for all z not in the spectrum of A,

$$(A - zI)^{-1} = \int_{-\infty}^{\infty} \frac{1}{\lambda - z}\, dE_A(\lambda) ,$$

where A is a self-adjoint operator whose associated spectral resolution is $E_A(\lambda)$.

31. Suppose that we wish to find an eigenvector and an eigenvalue of a self-adjoint operator $A = A_0 + \epsilon A_1$. Imagine that we know the resolution of the identity, $E_{A_0}(\lambda)$, belonging to A_0, and assume that for ϵ sufficiently small any eigenvalue of A can be written as

$$\lambda_n = \lambda_n^{(0)} + \epsilon\lambda_n^{(1)} + \epsilon^2\lambda_n^{(2)} + \cdots$$

and that the corresponding eigenvector can be written as

$$\phi_n = \phi_n^{(0)} + \epsilon\phi_n^{(1)} + \epsilon^2\phi_n^{(2)} + \cdots$$

where $A_0\phi_n^{(0)} = \lambda_n^{(0)}\phi_n^{(0)}$. Show in the manner of Section 4.11 that we can find an inhomogeneous equation for $\phi_n^{(1)}$, and that the solution to this equation which satisfies $(\phi_n^{(0)}, \phi_n^{(1)}) = 0$ is

$$\phi_n^{(1)} = -\int_{S_0} \frac{1}{\lambda - \lambda_n^{(0)}}\, d[E_{A_0}(\lambda)(A_1 - \lambda_n^{(1)})\phi_n^{(0)}] , \qquad (1)$$

where $\lambda_n^{(1)} = (\phi_n^{(0)}, A_1\phi_n^{(0)})$ and S_0 denotes the entire spectrum of A_0 except for an arbitrarily small closed interval about the point $\lambda_n^{(0)}$. We assume that the spectra of A and A_0 are nondegenerate. Show that in the case of a finite-dimensional vector space, the results of Section 4.11 are obtained from Eq. (1).

32. i) Let A be a self-adjoint operator on a Hilbert space H. Assume that the spectrum of A consists of a finite set of points, $\{\lambda_i < 0\colon\ i = 1, 2, \cdots, N\}$, plus the positive real line. Show that if the lowest (i.e., most negative) eigenvalue of A, λ_1, is nondegenerate, then $\lambda_1 = \min_{\psi \in H} (\psi, A\psi)$,
where ψ is constrained by $||\psi|| = 1$.

[*Hint*: Note that any vector in H can be written as

$$\psi = c_0\phi_0 + \sum_{i=1}^{N} c_i\phi_i ,$$

where ϕ_i are the eigenvectors of A, and ϕ_0 is a vector orthogonal to all the ϕ_i $(i = 1, 2, \cdots)$. Thus show that

$$(\psi, A\psi) = \sum_{i=0}^{N} |c_i|^2\lambda_i ,$$

where $\{\lambda_i: i = 1, 2, \cdots\}$ is the set of eigenvalues of A and λ_0 is a *positive* number. Now proceed as in Section 4.9.]

ii) Write down and prove a result analogous to Eq. (1) for λ_2. How would the result generalize for λ_n?

33. Consider the quantum-mechanical observables L_x, L_y, and L_z, which are represented as matrices as follows:

$$L_x = \begin{bmatrix} 0 & 1 & 0 \\ 1 & 0 & 1 \\ 0 & 1 & 0 \end{bmatrix}, \quad L_y = \begin{bmatrix} 0 & -i & 0 \\ i & 0 & -i \\ 0 & i & 0 \end{bmatrix}, \quad L_z = \begin{bmatrix} 1 & 0 & 0 \\ 0 & 0 & 0 \\ 0 & 0 & -1 \end{bmatrix}.$$

The Hamiltonian for this system has the form $H = H_0 + \alpha L_z$, where H_0 commutes with L_x, L_y, and L_z.

a) Suppose that at $t = 0$ we have prepared the system in an eigenstate of L_z, namely, the state belonging to the eigenvalue $m = +1$ of L_z. If we measure L_z at the later time $t = T$, what is the probability that we will find the value $+1$? the value 0? the value -1?

b) Suppose that instead of measuring L_z at time $t = T$, we measure L_x. What are the possible values of L_x which we can find? What is the probability of finding each of these values?

c) Suppose that at $t = T$ we measure L_x and find with certainty the value -1. After a time τ has elapsed (i.e., at $t = T + \tau$), we remeasure L_x. What is the probability of finding the various allowed values of L_x?

FURTHER READINGS

Churchill, R. V., *Fourier Series and Boundary Value Problems.* New York: McGraw-Hill, 1944. An exceptionally readable book. Good for self-study.

Courant, R., and D. Hilbert, *Methods of Mathematical Physics*, Vol. I, New York: Wiley-Interscience, 1961.

Davis, P. J., *Interpolation and Approximation.* Waltham, Mass.: Blaisdell 1963. An advanced account of the various types of convergence and their roles in the approximation of functions.

Dirac, P. A. M. *The Principles of Quantum Mechanics*, Fourth edition. New York: Oxford University Press, 1958.

Dunford, N., and J. T. Schwartz, *Linear Operators*, Vol. II: Spectral Theory. New York: Wiley-Interscience, 1967. A definitive account of the theory of self-adjoint operators in Hilbert space.

Hochstadt, H., *Special Functions of Mathematical Physics.* New York: Holt, Rinehart and Winston, 1961. A brief, very readable treatment of orthogonal polynomials, Legendre, Bessel, and Mathieu functions. On p. 20 there is a highly intuitive probabilistic proof of Weierstrass's theorem in terms of the Bernstein polynomials.

Jackson, J. D. *Mathematics for Quantum Mechanics.* New York: W. A. Benjamin, 1962.

Jauch, J.M., *Foundations of Quantum Mechanics.* Reading, Mass: Addison-Wesley, 1968.

VOLUME TWO

CHAPTER 6

ELEMENTS AND APPLICATIONS OF THE THEORY OF ANALYTIC FUNCTIONS

INTRODUCTION

The role played by the theory of analytic functions in physics has changed considerably over the past few decades. It no longer suffices to be able to work out residue integrals; a deeper understanding of the mathematical ideas has become essential if one wants to follow current applications to physical theory. Therefore the emphasis here will be on introducing the mathematical concepts and the logical structure of the theory of analytic functions. Assuming only that the reader is familiar with the properties of complex numbers, we aim to present a self-contained account of this theory in a way that prepares one to cope with modern applications of the theory as well as those of the past.

"Imaginary" numbers were discovered in the Middle Ages in the search for a general solution of quadratic equations. It is clear from the name given them that they were regarded with suspicion. Gauss, in his doctoral thesis of 1799, gave the now familiar geometrical representation of complex numbers, and thus helped to dispel some of the mystery about them. In this century, the trend has been toward defining complex numbers as abstract symbols subject to certain formal rules of manipulation. Thus complex numbers never have taken on the "earthy" qualities of real numbers. In fact, more nearly the opposite has occurred: we have come to view real numbers abstractly as symbols obeying their own set of axioms, just like complex numbers. We now speak of *number fields*: the real field and the complex field. The axioms which define a field were stated in Chapter 3 on vector spaces.

The theory of complex numbers can be developed by viewing them as ordered pairs of real numbers, written (x, y). Let (a, b) and (c, d) be two different complex numbers, and let K be a real number. Then we define addition, multiplication of a real and a complex number, and multiplication of two complex numbers by the following rules:

1. $(a, b) + (c, d) = (a + b, c + d)$,
2. $K \cdot (a, b) = (Ka, Kb)$,
3. $(a, b) \cdot (c, d) = (ac - bd, bc + ad)$.

From these definitions, we see that the set of all complex numbers—the complex plane—has the same mathematical structure as the set of all vectors in a plane.

This approach is followed in Landau's *Foundations of Analysis*, in which the various number systems are built up logically from Peano's five axioms; the imaginary number i is never mentioned. However, if we write the ordered pair (a, b) as $a + ib$, where $i^2 = -1$, then the above rule of complex-number multiplication is obeyed if we simply multiply out the product $(a + ib)(c + id)$ according to the usual rules of multiplication of reals. The introduction of the symbol i subsumes the ordering aspect of the ordered pair of real numbers, while extending the formal rules of arithmetic from real to complex numbers.

From the complex numbers constructed as ordered pairs of reals, where $(a, b) = a + ib$, it is possible to generalize to hypercomplex numbers of three or more components, for example $(a, b, c) = a + ib + kc$. The four-component quaternions, a type of hypercomplex number which satisfies all the rules of arithmetic except the commutative law of multiplication, are useful in dealing with rotations of a rigid body. The four 4 × 4 Dirac matrices, $\gamma_i (i = 1, 2, 3, 4)$, form a set of hypercomplex numbers which satisfy the anticommutative relations

$$\gamma_i \gamma_j + \gamma_j \gamma_i = 2\delta_{ij} .$$

It can be shown that no matter how we define addition and multiplication for these hypercomplex numbers, it is impossible to retain all the usual rules of arithmetic. As Weyl points out, the complex numbers form a natural boundary for the extension of the number concept in this respect.

6.1 ANALYTIC FUNCTIONS—THE CAUCHY-RIEMANN CONDITIONS

If to each complex number z in a certain domain there corresponds another complex number w, then w is a function of the complex variable z: $w = f(z)$. If the correspondence is one to one, we can view this as a mapping from one plane (or part of it), the z-plane, to another, the w-plane. The complex functions thus defined are equivalent to ordered pairs of real functions of two variables, because w is a complex number depending on $z = x + iy$ and therefore can be written in the form

$$w(z) = u(x, y) + iv(x, y) .$$

However, this class of functions is too general for our purposes. We are interested only in functions which are differentiable with respect to the complex variable z—a restriction which is much stronger than the condition that u and v be differentiable with respect to x and y. Therefore one of our first tasks in the study of complex function theory will be to determine the necessary and sufficient conditions for a complex function to have a derivative with respect to the complex variable z. Single-valued functions of a complex variable which have derivatives throughout a region of the complex plane are called *analytic* functions. We shall restrict our attention to this special class of complex functions.

Two examples of complex functions (both written in the form $w = u + iv$) are

1. $w = z^* = x - iy$,
2. $w = z^2 = (x + iy)^2 = x^2 - y^2 + i2xy$.

Presently, we shall show that (1) is not an analytic function, but that (2) is analytic everywhere in the complex plane; i.e., its derivative exists at all points.

Before stating exactly what is meant by the derivative of a function of a complex variable, we must have a notion of *continuity* for these functions.

In the definition that follows, mention is made of the ***absolute value*** of a complex number, denoted by $|z|$. The reader will recall that $|z| \equiv (zz^*)^{1/2} = (x^2 + y^2)^{1/2}$. The absolute value is sometimes called the *modulus*.

Definition. A complex function $w = f(z)$ is continuous at the point z_0 if, given any $\epsilon > 0$, there exists a δ such that $|f(z) - f(z_0)| < \epsilon$, when $|z - z_0| < \delta$, or $f(z)$ is continuous at z_0 if

$$\lim_{z \to z_0} f(z) = f(z_0) .$$

This definition is formally exactly like the definition of continuity for real functions of a real variable. However, here the absolute value signs mean that whenever z lies within a *circle* of radius δ centered at z_0 in the complex z-plane, then $f(z)$ lies within a *circle* of radius ϵ centered at $f(z_0)$ in the complex w-plane. If $f(z) = u(x, y) + iv(x, y)$, then $f(z)$ is continuous at $z_0 = x_0 + iy_0$ if u and v are continuous at (x_0, y_0).

From the class of single-valued, continuous complex functions, we now want to select those that can be differentiated. Patterning the definition of a derivative after that of real analysis, we have

Definition. $f(z)$ is differentiable at the point z_0 if the limit

$$\lim_{z \to z_0} \frac{f(z) - f(z_0)}{z - z_0} = \lim_{\Delta z \to 0} \frac{\Delta f}{\Delta z}$$

exists. We shall denote this limit, the derivative of $f(z)$ at z_0, by $f'(z_0)$.

A very important feature of the limits that occur in the definitions of continuity and the derivative is that z may approach z_0 from *any direction* on the plane. When we say the limit exists, we therefore mean that the same number must result from the limiting process regardless of how the limit is taken. This is also true in real analysis, but in that case there are only two possible directions of approach in taking the limit: from the left or the right on the real line. In real analysis, the limiting process is one-dimensional; in complex analysis, it is two-dimensional.

The equation that defines the derivative means that given any $\epsilon > 0$, there exists a δ such that

$$\left| f'(z) - \frac{f(z) - f(z_0)}{z - z_0} \right| < \epsilon$$

provided $|z - z_0| < \delta$. The requirement that the ratio $[f(z) - f(z_0)]/(z - z_0)$ always tends to the same limiting value, no matter along what path z approaches z_0, is an extremely exacting condition. The theory of analytic functions contains a number of amazing theorems, and they all result from this stringent initial requirement that the functions possess "isotropic" derivatives.

A single-valued function of z is said to be *analytic* (or *regular*) at a point z_0 if it has a derivative at z_0 and at all points in some neighborhood z_0. Thus a slight distinction is drawn between differentiability and analyticity. It pays to do this, because although there exist functions which have derivatives at certain points, or even along certain curves, no interesting results can be obtained unless functions are differentiable throughout a region, i.e., unless they are analytic. Thus if we say a function is analytic on a curve, we mean that it has a derivative at all points in a two-dimensional strip containing the curve. If a function is not analytic at a point or on a curve, we say it is *singular* there.

We shall now examine the two complex functions mentioned earlier for differentiability and analyticity. We write the derivative at z_0 in the form

$$f'(z_0) = \lim_{\Delta z \to 0} \frac{f(z_0 + \Delta z) - f(z_0)}{\Delta z},$$

by letting $z = z_0 + \Delta z$ in the original definition. For $f(z) = z^2$, we have

$$f'(z_0) = \lim_{\Delta z \to 0} \frac{(z_0 + \Delta z)^2 - z_0^2}{\Delta z} = \lim_{\Delta z \to 0} (2z_0 + \Delta z) = 2z_0,$$

a result which is clearly independent of the path along which $\Delta z \to 0$, so $f(z) = z^2$ is differentiable and analytic everywhere. The result parallels exactly the result for the derivative of the real function $f(x) = x^2$.

On the other hand, if $f(z) = z^*$, we have

$$f'(z_0) = \lim_{\Delta z \to 0} \frac{z_0^* + \Delta z^* - z_0^*}{\Delta z} = \lim_{\Delta z \to 0} \frac{\Delta z^*}{\Delta z}.$$

Now if $\Delta z \to 0$ along the real x-axis, then $\Delta z = \Delta x$ and $\Delta z^* = \Delta x^* = \Delta x$, so $f'(z_0) = +1$. However, if Δz approaches zero along the imaginary y-axis, then $\Delta z = i\Delta y$ so $\Delta z^* = -i\Delta y = -\Delta z$, so $f'(z_0) = -1$. Since at any point z_0 the limit as $z \to z_0$ depends on the direction of approach, the function is not differentiable or analytic anywhere. [As a general rule, $\Delta z^*/\Delta z = e^{-2i\theta}$, where $\theta = \tan^{-1}(\Delta y/\Delta x)$, which manifestly involves the direction of approach (θ) in taking the limit.]

Many of the theorems on differentiability in real analysis have analogs in complex analysis. For example:

1. A constant function is analytic.
2. $f(z) = z^n$ $(n = 1, 2, \cdots)$ is analytic.
3. The sum, product, or quotient of two analytic functions is analytic, provided, in the case of the quotient, that the denominator does not vanish anywhere in the region under consideration.
4. An analytic function of an analytic function is analytic.

The proofs go through exactly as in the real case.

We now determine the necessary and sufficient conditions for a function $w(z) = u(x, y) + iv(x, y)$ to be differentiable at a point. First, we assume that

$w(z)$ is in fact differentiable for some $z = z_0$. Then

$$w'(z_0) = \lim_{\Delta z \to 0} \frac{\Delta w}{\Delta z} = \lim_{\Delta z \to 0} \left(\frac{\Delta u}{\Delta z} + i \frac{\Delta v}{\Delta z} \right).$$

Since $w'(z_0)$ exists, it is independent of how $\Delta z \to 0$; that is, it is independent of the ratio $\Delta y / \Delta x$. If the limit is taken along the real axis, $\Delta y = 0$, and $\Delta z = \Delta x$. Then

$$w'(z_0) = \lim_{\Delta x \to 0} \left(\frac{\Delta u}{\Delta x} + i \frac{\Delta v}{\Delta x} \right) = \frac{\partial u}{\partial x} + i \frac{\partial v}{\partial x}.$$

On the other hand, if we approach the origin along the imaginary axis, $\Delta x = 0$ and $\Delta z = i \Delta y$. Now

$$w'(z_0) = \lim_{\Delta y \to 0} \left(\frac{\Delta v}{\Delta y} - i \frac{\Delta u}{\Delta y} \right) = \frac{\partial v}{\partial y} - i \frac{\partial u}{\partial y}.$$

But by the assumption of differentiability, these two limits must be equal. Therefore, equating real and imaginary parts, we have

$$\frac{\partial u}{\partial x} = \frac{\partial v}{\partial y} \quad \text{and} \quad \frac{\partial v}{\partial x} = -\frac{\partial u}{\partial y}. \tag{6.1}$$

Equations (6.1) are known as the Cauchy-Riemann equations. They give a *necessary* condition for differentiability. We have determined this condition from special cases of the requirement of differentiability; therefore it is not surprising that these conditions alone are not sufficient.

The sufficient conditions for the differentiability of $w(z)$ at z_0 are, first, that the Cauchy-Riemann equations hold there, and second, that the first partial derivatives of $u(x, y)$ and $v(x, y)$ exist and be continuous at z_0.

The proof is straightforward. To begin, u is continuous at (x_0, y_0) because it is differentiable there; the partial derivatives of u are continuous by hypothesis. Under these assumptions, it follows from the calculus of functions of several variables* that

$$\begin{aligned} \Delta u &= u(x_0 + \Delta x, y_0 + \Delta y) - u(x_0, y_0) \\ &= \frac{\partial u}{\partial x} \Delta x + \frac{\partial u}{\partial y} \Delta y + \epsilon_1 \Delta x + \epsilon_2 \Delta y, \end{aligned}$$

where $\partial u / \partial x$ and $\partial u / \partial y$ are the partial derivatives evaluated at the point (x_0, y_0) and where ϵ_1 and ϵ_2 go to zero as both Δx and Δy go to zero. Using a similar formula for $v(x, y)$, we have

$$\begin{aligned} \Delta w &= w(z_0 + \Delta z) - w(z_0) = \Delta u + i \Delta v \\ &= \frac{\partial u}{\partial x} \Delta x + \frac{\partial u}{\partial y} \Delta y + \epsilon_1 \Delta x + \epsilon_2 \Delta y + i \left(\frac{\partial v}{\partial x} \Delta x + \frac{\partial v}{\partial y} \Delta y + \epsilon_3 \Delta x + \epsilon_4 \Delta y \right). \end{aligned}$$

* See, for example, G. B. Thomas, Jr. *Calculus and Analytic Geometry*, 4th Ed., Addison-Wesley Publishing Co., 1968, Section 15–4, p. 503 Eq. 4, or W. Kaplan, *Advanced Calculus*, Addison-Wesley Publishing Co., 1953, Section 2–6, p. 84.

Now using the Cauchy-Riemann equations, which by assumption hold at the point (x_0, y_0), we have

$$\Delta w = \frac{\partial u}{\partial x}(\Delta x + i\Delta y) + i\frac{\partial v}{\partial x}(\Delta x + i\Delta y) + \Delta x(\epsilon_1 + i\epsilon_3) + \Delta y(\epsilon_2 + i\epsilon_4) .$$

Therefore

$$\frac{\Delta w}{\Delta z} = \frac{\partial u}{\partial x} + i\frac{\partial v}{\partial x} + (\epsilon_1 + i\epsilon_3)\frac{\Delta x}{\Delta z} + (\epsilon_2 + i\epsilon_4)\frac{\Delta y}{\Delta z} .$$

Since $|\Delta z| = [(\Delta x)^2 + (\Delta y)^2]^{1/2}$, $|\Delta x| \leq |\Delta z|$ and $|\Delta y| \leq |\Delta z|$, and so $|\Delta x/\Delta z| \leq 1$ and $|\Delta y/\Delta z| \leq 1$. Since these factors are bounded, the last two terms in the above equation tend to zero with Δz because ϵ_1, ϵ_2, ϵ_3, and ϵ_4 go to zero as Δz goes to zero. Therefore at z_0

$$w'(z_0) = \frac{\partial u}{\partial x} + i\frac{\partial v}{\partial x} ; \tag{6.2}$$

the limit is independent of the path followed, so the derivative exists. Using the Cauchy-Riemann conditions, we also have

$$w'(z_0) = \frac{\partial v}{\partial y} - i\frac{\partial u}{\partial y} . \tag{6.3}$$

Example. Consider the function z^3. We have

$$z^3 = (x^3 - 3xy^2) + i(3x^2y - y^3) = u + iv .$$

Thus

$$\frac{\partial u}{\partial x} = 3x^2 - 3y^2 = \frac{\partial v}{\partial y}, \qquad \text{and} \qquad \frac{\partial v}{\partial x} = 6xy = -\frac{\partial u}{\partial y} .$$

Thus the Cauchy-Riemann equations hold everywhere. Since the partial derivatives are continuous, the function z^3 is, in fact, analytic everywhere. A function which is analytic in the entire complex plane is said to be an *entire* function. The derivative of z^3 may be found using Eq. (6.2) or (6.3). We obtain

$$\frac{\partial z^3}{\partial z} = \frac{\partial u}{\partial x} + i\frac{\partial v}{\partial x} = 3[(x^2 - y^2) + 2ixy] = 3z^2 ,$$

a satisfying result. As a second example, we leave it to the reader to show that the function $|z|^2 \equiv zz^*$ is differentiable only at the origin, and therefore is analytic nowhere.

One remarkable result which points to connections with physics follows immediately from the Cauchy-Riemann equations. Assuming that they hold in a region, we have

$$\frac{\partial^2 u}{\partial x^2} = \frac{\partial^2 v}{\partial x\,\partial y} = \frac{\partial^2 v}{\partial y\,\partial x} = -\frac{\partial^2 u}{\partial y^2} \Longrightarrow \frac{\partial^2 u}{\partial x^2} + \frac{\partial^2 u}{\partial y^2} = \nabla^2 u = 0 \tag{6.4}$$

if the second partial derivatives are continuous, so we can interchange the orders of differentiation in the mixed partial derivative. It follows in the same way that the function v also satisfies the two-dimensional Laplace equation. Thus both the real and imaginary parts of an analytic function with continuous second partial derivatives satisfy the two-dimensional Laplace equation. We shall later prove, using integration theory, that the second partial derivatives of an analytic function are necessarily continuous, so this qualification can be dropped. (It is interesting that these theorems about derivatives can be proved only by integration.) Any function ϕ satisfying $\nabla^2\phi = 0$ is called a *harmonic function*. If $f = u + iv$ is an analytic function, then $\nabla^2 u = \nabla^2 v = 0$, and u and v are called *conjugate* harmonic functions.

Given one of two conjugate harmonic functions, the Cauchy-Riemann equations can be used to find the other, up to a constant. For example, the function $u(x, y) = 2x - x^3 + 3xy^2$ is easily seen to be harmonic. To find its harmonic conjugate, we proceed as follows:

$$\frac{\partial u}{\partial x} = \frac{\partial v}{\partial y} = 2 - 3x^2 + 3y^2 \Longrightarrow v = 2y - 3x^2 y + y^3 + \phi(x) ,$$

where $\phi(x)$ is some function of x. Now, using the other Cauchy-Riemann equation, we obtain

$$\frac{\partial v}{\partial x} = -\frac{\partial u}{\partial y} \Longrightarrow -6xy + \phi'(x) = -6xy \Longrightarrow \phi' = 0 .$$

Thus $\phi(x)$ must be a constant, and the harmonic conjugate of u is

$$v = 2y - 3x^2 y + y^3 + \text{const.}$$

Note that the function $w = u + iv = 2z - z^3 + C$ is an analytic function, as we know it must be.

Before leaving the Cauchy-Riemann conditions, let us take advantage of being physicists to present another, shorter derivation of these conditions, based on the use of infinitesimals. Let $w = u + iv$ and $w' = p + iq$. Then $\delta w = w'\delta z$, or, taking real and imaginary parts,

$$\delta u = p\delta x - q\delta y , \qquad \delta v = p\delta y + q\delta x .$$

It follows immediately that

$$\frac{\partial u}{\partial x} = \frac{\partial v}{\partial y} = p , \qquad \frac{\partial v}{\partial x} = -\frac{\partial u}{\partial y} = q .$$

These equations are identical to the Cauchy-Riemann equations (6.1).

Continuing in this informal spirit, we may derive another closely related result which provides some insight into the meaning of analyticity. Again, let $w(z) = w(x, y) = u(x, y) + iv(x, y)$. We now show that $\partial w/\partial z^* = 0$ if and only if the Cauchy-Riemann equations hold. We shall not worry about the meaning of this derivative with respect to z^*, but just differentiate formally, treating the derivative as symbolic. Using the expressions

$$x = (z + z^*)/2 \qquad \text{and} \qquad y = (z - z^*)/2i$$

we have

$$\begin{aligned}
\frac{\partial w}{\partial z^*} &= \frac{\partial w}{\partial x}\frac{\partial x}{\partial z^*} + \frac{\partial w}{\partial y}\frac{\partial y}{\partial z^*} \\
&= \left(\frac{\partial u}{\partial x} + i\frac{\partial v}{\partial x}\right)\tfrac{1}{2} + \left(\frac{\partial u}{\partial y} + i\frac{\partial v}{\partial y}\right)\left(-\frac{1}{2i}\right) \\
&= \tfrac{1}{2}\left(\frac{\partial u}{\partial x} - \frac{\partial v}{\partial y}\right) + \frac{i}{2}\left(\frac{\partial v}{\partial x} + \frac{\partial u}{\partial y}\right).
\end{aligned}$$

If the Cauchy-Riemann equations hold, this last expression vanishes. If, on the other hand, $\partial w/\partial z^* = 0$, then both the real and imaginary parts of the last expression must vanish, so the Cauchy-Riemann equations hold.

This purely formal result, which can be made rigorous, is trying to tell us that analytic functions are independent of z^*: they are functions of z *alone*. Thus analytic functions are true functions of a *complex* variable, not just complex functions of two real variables (see, for example, Problem 1), which will in general depend on z^* as well as z according to

$$f(x, y) = f\left(\frac{z + z^*}{2}, \frac{z - z^*}{2i}\right).$$

6.2 SOME BASIC ANALYTIC FUNCTIONS

One of the most useful functions in the complex domain is the exponential function which we define for $z = x + iy$ by

$$e^z \equiv e^x (\cos y + i \sin y) . \tag{6.5}$$

It follows easily from this definition and our earlier work that e^z is an entire function and that

$$\frac{d}{dz} e^z = e^z .$$

The other familiar properties of exponentials, in particular, $e^{z_1+z_2} = e^{z_1} e^{z_2}$, follow readily from Eq. (6.5). We note that e^z is a periodic function of period $2\pi i$:

$$e^{z+2\pi i} = e^z e^{2\pi i} = e^z (\cos 2\pi + i \sin 2\pi) = e^z .$$

From Eq. (6.5) we see that

$$e^{iy} = \cos y + i \sin y ,$$

so it follows that

$$\cos y = \frac{e^{iy} + e^{-iy}}{2}, \qquad \sin y = \frac{e^{iy} - e^{-iy}}{2i}.$$

These relations suggest that for an arbitrary complex z we define

$$\cos z \equiv \frac{e^{iz} + e^{-iz}}{2}, \tag{6.6}$$

$$\sin z \equiv \frac{e^{iz} - e^{-iz}}{2i}. \tag{6.7}$$

Since

$$\frac{d}{dz} e^z = e^z,$$

it is a simple matter to calculate the derivatives of $\cos z$ and $\sin z$. We find that

$$\frac{d}{dz} \cos z = \frac{ie^{iz} - ie^{-iz}}{2} = -\sin z,$$

$$\frac{d}{dz} \sin z = \frac{ie^{iz} + ie^{-iz}}{2i} = \cos z,$$

as we might expect from experience with the real variable case. Using Eqs. (6.6) and (6.7), it is a simple matter to verify that all the familiar trigonometric identities, such as

$$\cos(z_1 + z_2) = \cos z_1 \cos z_2 - \sin z_1 \sin z_2,$$

continue to be valid for complex variables.

The complex functions sine and cosine may, of course, be put in the form $u(x, y) + iv(x, y)$. For example,

$$\begin{aligned} \sin z &= \frac{1}{2i}\left[e^{i(x+iy)} - e^{-i(x+iy)}\right] \\ &= \frac{1}{2i} e^{-y}(\cos x + i \sin x) - \frac{1}{2i} e^{y}(\cos x - i \sin x) \\ &= \sin x(e^y + e^{-y})/2 + i \cos x(e^y - e^{-y})/2. \end{aligned}$$

Therefore

$$\sin z = \cosh y \sin x + i \sinh y \cos x. \tag{6.8}$$

Similarly,

$$\cos z = \cosh y \cos x - i \sinh y \sin x. \tag{6.9}$$

Setting $x = 0$, we obtain the useful relations $\sin(iy) = i \sinh y$ and $\cos(iy) = \cosh y$. We also see that the Cauchy-Riemann conditions are satisfied everywhere, as we know they must be. Other properties which follow directly from Eqs. (6.8) and (6.9) are

$$\begin{aligned} (\sin z)^* &= \sin(z^*), \\ \sin(-z) &= -\sin(z), \\ \sin(z + 2\pi) &= \sin(z). \end{aligned}$$

Using the sine and cosine functions, we can define the other familiar trigonometric functions. For example,

$$\tan z = \sin z/\cos z \ ;$$

similar extensions of the real case are defined for the cotangent, secant, and cosecant. These functions differ from the sine and cosine in that they are *not* analytic everywhere. The tangent, being the ratio of two analytic functions, will be analytic everywhere *except* at points where $\cos z = 0$. Using the real and imaginary parts of the cosine, we can rewrite this condition as

$$\cosh y \cos x = 0 \ , \qquad \sinh y \sin x = 0 \ .$$

Now $\cosh y \geq 1$ for all real y, so the first equation has a solution whenever $\cos x = 0$, or $x = (2n + 1)\pi/2$, $n = 0, \pm 1, \pm 2, \cdots$. At these points, $\sin x = \pm 1$, so the second equation requires that $\sinh y = 0$, that is, $y = 0$. Thus the tangent function is singular at the points $(2n + 1)\pi/2$, $(n = 0, \pm 1, \cdots)$ on the real axis, and only at these points. Therefore $\tan z$ becomes infinite at precisely those points where $\tan x$ (real x) becomes infinite and *only* at those points.

On the basis of the above discussion, one might be tempted to think that the complex trigonometric functions are "just the same thing" as their real counterparts. However, the reader can easily show that

$$|\sin z|^2 = \sin^2 x + \sinh^2 y \ ,$$

and this expression increases *without limit* as y tends to infinity. This is in marked contrast with the real case, where $|\sin x| \leq 1$ for all real x.

The functions which we have discussed thus far all have the property that if we pick any point z_0 in the complex plane and follow any path from z_0 through the plane back to z_0, then the value of the function changes continuously along the path, returning to its original value at z_0. For example, suppose that we consider the function $w(z) = e^z$ and start at the point $z_0 = 1$, encircling the origin in the z-plane counterclockwise along the unit circle. Figure 6.1(a) shows the circular path in the z-plane, and Fig. 6.1(b) shows the corresponding path in the w-plane. [The use of two complex planes to "graph" the function $w(z)$ is often employed in complex variable theory.] We note that both paths are closed, which is just the geometrical statement of the fact that if we start at a point z_0, where the function has the value $w(z_0)$, then when we move along a closed curve back to z_0, the functional values also follow a smooth path back to $w(z_0)$.

Now for e^z this result is hardly surprising since we have *defined* e^z in such a way as to ensure this behavior, letting ourselves be guided by the properties of the real exponential function. Now if we look at another simple function, namely, the square root, we see that things do not always go so smoothly. Let us write formally

$$w(z) \equiv \sqrt{z} \equiv \sqrt{x + iy} \ .$$

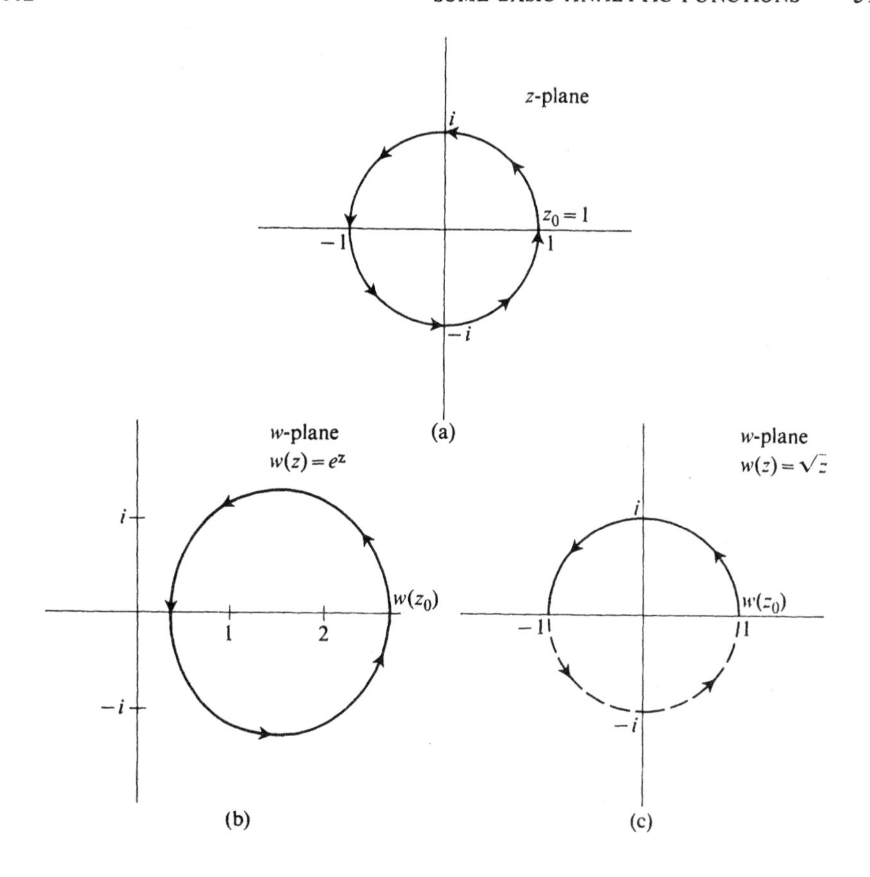

Fig. 6.1(a) A circular contour in the z-plane about the origin. **6.1(b)** The mapping of the contour of Figure 6.1(a) by the function e^z. **Fig. 1(c)** The mapping of the contour of Figure 6.1(a) by the function $\sqrt{z}$.

We observe that this definition is empty, since there is no set of operations presently at our disposal which will enable us to find $w(z)$ for some given x and y (unless $y = 0$). This is in contrast with the situation in Eq. (6.5) where all problems of evaluation can be handled by familiar real-variable operations.

Fortunately, in the case of the square root there is another possibility, namely, we can write z in polar form† as $z = re^{i\theta}$. In this form, a logical extension of the square root to the complex domain is contained in the definition

$$w(z) \equiv \sqrt{z} \equiv \sqrt{r}\, e^{i\theta/2} = \sqrt{r}\,[\cos(\theta/2) + i\sin(\theta/2)]\,.$$

† If we write $z = x + iy$ and make the familiar change to polar coordinates ($x = r\cos\theta$, $y = r\sin\theta$) we obtain $z = r(\cos\theta + i\sin\theta) = re^{i\theta}$, where $r = (x^2 + y^2)^{1/2}$ and $\theta = \tan^{-1}(y/x)$.

Clearly, this function satisfies $w^2 = z$, which is certainly a minimum requirement for any sensible square root. Using this definition, let us vary z along the same path chosen in Fig. 6.1 (a), starting at $r = 1$, $\theta = 0$. Figure 6.1 (c) shows the corresponding path in the w-plane. Note that it is *not* a closed path; after making a complete circle around the origin in the z-plane, we arrive at the point $w = -1$ in the w-plane, not at $w = +1$. In order to get back to $w = +1$, we must let θ go from 2π to 4π; that is, make the circular trip in the z-plane one more time [see the dotted curve in Fig. 6.1 (c)]. Actually this is not quite the best way to describe the situation; we do not want to think of tracing the circular path in the original z-plane a second time, but rather of tracing an identical circular path in a *different* z-plane. This corresponds to the fact that in the first circuit, θ went from 0 to 2π, whereas in the second circuit, it went from 2π to 4π.

This is not so different from the case of functions like e^z as one might imagine at first glance. We can write in polar variables

$$e^z = e^{r \cos \theta} [\cos (r \sin \theta) + i \sin (r \sin \theta)] ,$$

and then trace out a circular path as many times as we please ($\theta = 0 \to \theta = 2\pi$, $\theta = 2\pi \to \theta = 4\pi$, etc.). In this case we get the same values of e^z for each circuit. Therefore no information about e^z is lost if we identify the z-planes corresponding to $\theta = 0 \to \theta = 2\pi$, $\theta = 2\pi \to \theta = 4\pi$, etc., with each other. However, in the case of $w(z) = \sqrt{z}$ we need two planes, usually referred to as *Riemann sheets,* to characterize the values of $w(z)$ in a single-valued manner. Two planes are clearly sufficient: when we let θ range from 4π to 6π we obtain the same values as we did when we let θ range from 0 to 2π.

It is important to remember that the path of Fig. 6.1 (a) encloses the origin. If we choose a closed path which neither encloses the origin nor intersects the positive real axis, then we also obtain a closed path in the w-plane. Fig. 6.2(a) and (b) illustrates the situation for $w(z) = \sqrt{z}$, starting from the $z_0 = \sqrt{5}\, e^{i\phi_0}$, where $\phi_0 = \tan^{-1} 2$ (we adopt the usual trigonometric convention that $\tan^{-1} x$ takes on values between 0 and $\pi/2$). In Fig. 6.2(a) we may say that we start at $z_0 = 1 + 2i$ on the *first Riemann sheet* and return to that point without encircling the origin. If we do the same thing for $z_0 = 1 + 2i$ on the *second Riemann sheet* (that is, $\phi_0 = \tan^{-1} 2 + 2\pi$), we obtain the corresponding closed curve traced out in the w-plane (Fig. 6.2(c)).

It is readily seen that the difficulties described above for $w(z) = \sqrt{z}$ will persist for any path beginning on the positive real axis and returning to the original point along a path enclosing the origin. Thus if we wish to consider $w(z) = \sqrt{z}$ in the simple fashion that we used for e^z, then we conclude that $\sqrt{z}$ is not continuous along the positive real axis and is not analytic there. However, to avoid this dilemma, we can say that when we come back to the real axis after a circuit of 2π radians, we transfer continuously onto the *second* Riemann sheet. If we go around $z = 0$ once more on the second sheet, when we return toward the positive real axis, we transfer continuously back to the *first* Riemann sheet. Thus the two sheets can be imagined to be *cut* along the positive real axis and

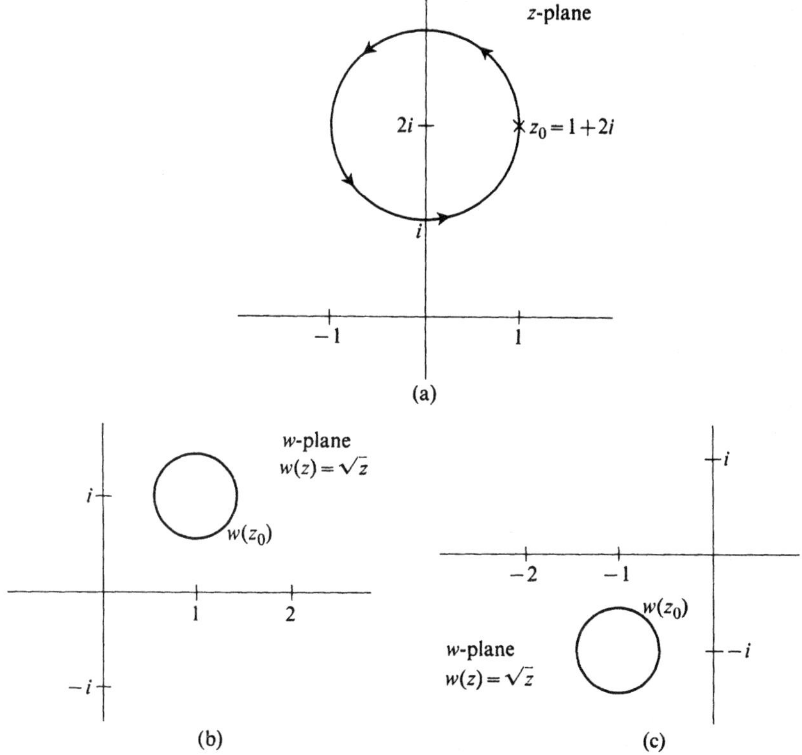

Fig. 6.2(a) A closed contour in the z-plane which does not enclose the origin. **6.2(b)** The mapping of the contour of Figure 6.2(a) by the function $\sqrt{z}$, if the contour of Figure 6.2(a) is imagined to lie on the *first* Riemann sheet. **6.2(c)** The mapping of the contour of Figure 6.2(a) by the function $\sqrt{z}$, if the contour of Figure 6.2(a) is imagined to lie on the *second* Riemann sheet.

joined in the manner illustrated in Fig. 6.3. With this construction, the function $w(z) = \sqrt{z}$ is seen to be single valued everywhere [on both sheets we set $w(0) = 0$] and analytic everywhere except at the origin, where $\sqrt{z}$ suffers from the same difficulty as does $\sqrt{x}$ in the real variable case. Thus the origin is a singular point for $w(z) = \sqrt{z}$.

In general, suppose that we have a singular point, z_0, of some function $w(z)$, and a path starting at z_1, which encircles z_0. If we must sweep through an angle

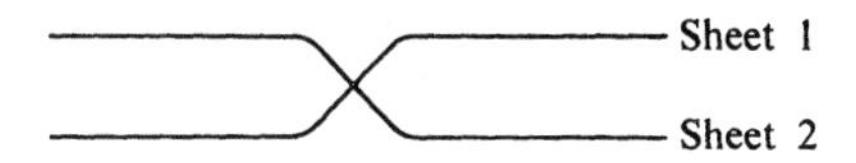

Fig. 6.3 A side view of the two Riemann sheets, looking down the real axis towards the origin, for the function $w(z) = \sqrt{z}$.

greater than 2π in order to return to the original value at z_1, then z_0 is called a *branch point* of $w(z)$. The cut which emanates from this point is called a *branch cut*. In the simple case discussed above $[w(z) = \sqrt{z}]$, the value of $w(z_1)$ with z_1 on the *second* sheet is just the negative of $w(z_1)$ with z_1 on the *first* sheet. Note that it is possible for a point z_1 on the *first* sheet to be a point at which some function $w(z)$ is analytic, whereas the point z_1 on the *second* sheet is a singular point. The function

$$w(z) = \frac{1}{i + \sqrt{z}}$$

is an example of such a function; $w(z)$ is a single-valued function over a two-sheeted Riemann surface cut along the positive real axis and joined as in Fig. 6.3. It is analytic at $z = -1\,(\phi = \pi)$ on the first Riemann sheet and is singular at $z = -1\,(\phi = 3\pi)$ on the second Riemann sheet.

In the above discussions we could, of course, have insisted that θ, the argument of z, range only through 2π radians. Then we could say that

$$w_1(z) = \sqrt{r}\, e^{i\theta/2}\,, \qquad 0 \leq \theta < 2\pi\,,$$

defines a single-valued function, analytic everywhere in the complex z-plane *except along the positive real axis* (including $z = 0$), and that

$$w_2(z) = \sqrt{r}\, e^{i(\theta+2\pi)/2} = -\sqrt{r}\, e^{i\theta/2}\,, \qquad 0 \leq \theta < 2\pi\,,$$

also defines such a function. Both $w_1(z)$ and $w_2(z)$ satisfy $w_1^2 = w_2^2 = z$ and are referred to as *single-valued branches* of $\sqrt{z}$. Clearly, if we defined

$$w_3(z) = \sqrt{r}\, e^{i(\theta+4\pi)/2}\,, \qquad 0 \leq \theta < 2\pi\,,$$

we would find $w_3(z) = w_1(z)$, so we do not obtain a new branch in this manner. We leave it as an exercise to show that one can define *three* single-valued branches of $w(z) = \sqrt[3]{z}$ and that, in this case, if one wants to define $w(z)$ as a single-valued function, analytic everywhere except at $z = 0$, a three-sheeted Riemann surface is necessary.

It should be noted that the choice of the real axis as the branch cut for $w(z) = \sqrt{z}$ was entirely arbitrary. Any other ray, say $\theta = \theta_0$, will serve equally well. The only thing which is *not* arbitrary is the choice of $z = 0$ as a branch point; $z = 0$ is a bona fide singular point for $w(z) = \sqrt{z}$, and this cannot be changed. However, in the functions $w_1(z)$ and $w_2(z)$ defined above, the singular line $\theta = 0$ is, apart from $z = 0$, a line of "man-made" singularities; we could equally well choose the line $\theta = \theta_0$ to be the singular line. For example, we could define

$$w_1(z) = \sqrt{r}\, e^{i\theta/2}\,, \qquad \theta_0 \leq \theta < \theta_0 + 2\pi\,,$$

and, similarly,

$$w_2(z) = \sqrt{r}\, e^{i(\theta+2\pi)/2} = -\sqrt{r}\, e^{i\theta/2}, \qquad \theta_0 \leq \theta < \theta_0 + 2\pi\,.$$

If we want a single-valued function which is analytic everywhere except at $z = 0$, then we can construct a two-sheeted Riemann surface, cut and joined along the line $\theta = \theta_0$. On the counterclockwise edge of the cut, $\sqrt{z} = \sqrt{r}\, e^{i\theta_0/2}$. After going through a circuit of 2π radians, we do not return to this value, but to $-\sqrt{r}\, e^{i\theta_0/2}$, and we pass onto the second sheet. Note that, in this case, some of the values of $w(z) = \sqrt{z}$ which were on the *first* Riemann sheet when the cut was made along $\theta = 0$ now find themselves on the *second* Riemann sheet and vice versa. This brings home the fact that the Riemann construction is merely a way to write a collection of values of a function in a single-valued manner. Distinctions between the first sheet and the second sheet are purely matters of convention. In fact, it should not be difficult for the reader to imagine that any reasonable curve from the origin to infinity could serve as an acceptable cut along which the two Riemann sheets of $w(z) = \sqrt{z}$ can be joined.

Before leaving this example, let us first propose an argument which might appear at first sight to contradict what we have been saying. Consider any point $x_0 \neq 0$ on the positive real axis. We should imagine that there exists a neighborhood of x_0 in which we can write $z = x_0 + \rho$, where ρ is some complex number, and then define $\sqrt{z}$ by the power series

$$\sqrt{z} \equiv \sqrt{x_0}\left(1 + \tfrac{1}{2}\frac{\rho}{x_0} - \tfrac{1}{8}\frac{\rho^2}{x_0^2} + \cdots\right),$$

whenever the series converges (we will see later that the series converges whenever $|\rho| < x_0$). This is a single-valued function which defines $\sqrt{z}$ continuously across a part of the positive real axis. However, this definition does not apply to the whole complex plane since the series does not converge everywhere. It turns out that it is impossible to extend (or "continue") this function to all points of the z-plane in such a way that $\sqrt{z}$ is single-valued and analytic. We will return to this point when we discuss the principle of analytic continuation later in this chapter. The continuity of the series definition across the positive real axis does not contradict our original positioning of a cut along the positive real axis, because, as we have seen above, the cut could be positioned anywhere so long as it begins at $z = 0$. In particular, the cut can be chosen so that it lies completely outside the domain of convergence of the series used to define $\sqrt{z}$ (for example, the cut could be chosen to lie along the negative real axis). We may remark in passing that we *could* successfully define e^z in this manner. The function e^x possesses an everywhere convergent power series for real x, and it is not hard to believe that a complex power series with the same coefficients $(1/n!)$ will converge *everywhere* in the complex plane.

As another example of a multivalued function, we consider the logarithm. Again using $z = re^{i\theta}$, we define

$$\log z \equiv \ln r + i\theta\,,$$

where ln denotes the usual natural logarithm of a positive real number. Note that

$$e^{\log z} = e^{\ln r + i\theta} = e^{\ln r} e^{i\theta} = re^{i\theta} = z$$

and also that

$$\log(z_1 z_2) = \ln(r_1 r_2) + i(\theta_1 + \theta_2) = \ln r_1 + i\theta_1 + \ln r_2 + i\theta_2 = \log z_1 + \log z_2 ,$$

so the logarithm has the main properties that one would expect by analogy with the real variable case. With the logarithm, the multivaluedness difficulties described above are more striking, since no matter how many times one encircles the origin, starting, say, at some point on the positive real axis, one *never* returns to the original value of the logarithm. The logarithm increases by $2\pi i$ on each circuit (or decreases by $2\pi i$ if one moves in the direction of decreasing θ). Thus an infinite number of Riemann sheets, each one joined to the one below it via a cut along the positive real axis, is necessary to turn $\log z$ into a single-valued function. When this is done, $\log z$ is analytic everywhere except at $z = 0$, where we assign the value $\log(z = 0) = -\infty$ on all sheets. We can also form an infinite number of single-valued branches of the logarithm:

$$w_n(z) = \ln r + i\theta + 2\pi n i , \qquad 0 \leq \theta < 2\pi ,$$

where $n = 0, \pm 1, \pm 2, \cdots$, and $w_n(z)$ is a single-valued function, analytic everywhere except at $z = 0$ and along the positive real axis. Just as before,

$$e^{w_n(z)} = z ,$$

but

$$w_n(z_1 z_2) = w_n(z_1) + w_n(z_2) - 2\pi n i .$$

The branch

$$w_0(z) = \ln r + i\theta , \qquad 0 \leq \theta < 2\pi ,$$

is called the principal value or principal branch of the logarithm and is usually denoted by $\operatorname{Log} z$. We have

$$e^{\operatorname{Log} z} = z ,$$

$$\operatorname{Log}(z_1 z_2) = \operatorname{Log} z_1 + \operatorname{Log} z_2 .$$

As before, the choice of the ray $\theta = 0$ as the line of singularities is entirely arbitrary.

From the preceding examples, it is a simple matter to build up to more complicated cases. For example, the function

$$w(z) = \sqrt{z - a}$$

is a single-valued function on the two-sheeted Riemann surface cut from a to infinity. The point $z = a$ is a branch point singularity; if we choose the branch cut to lie parallel to the real axis, we obtain the picture of Fig. 6.4. We thus may define, using the notation of Fig. 6.4,

$$w(z) \equiv |z - a|^{1/2} e^{i\theta/2} .$$

A more challenging problem is provided by the function

$$w(z) = \sqrt{(z - a)(z - b)} .$$

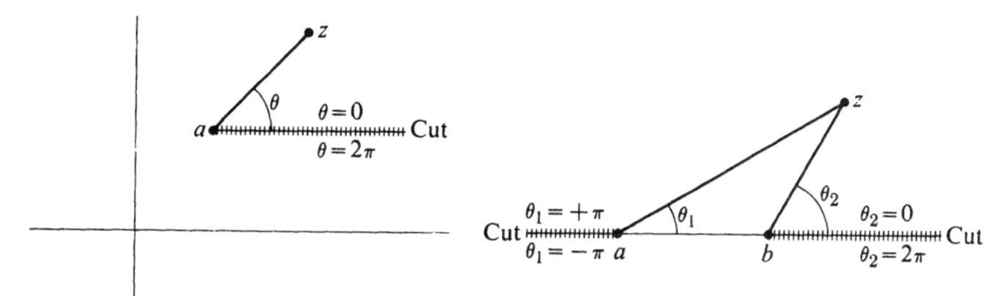

Fig. 6.4 The appropriate cut for the function $w(z) = \sqrt{z-a}$.

Fig. 6.5 The appropriate cuts for the function $w(z) = \sqrt{(z-a)(z-b)}$.

To simplify the geometry, we will consider the special case when a and b are real. The most obvious way to proceed is simply to put in two cuts from the branch points $z = a$ and $z = b$ as shown in Fig. 6.5. We have chosen these two simple directions for the cuts because of pictorial convenience. The left-hand cut corresponds to what we would have for the function $\sqrt{z-a}$, the right-hand cut to what we would have for $\sqrt{z-b}$. Here we take θ_1 to begin on the "bottom" of the left-hand cut at $-\pi$ and go to $+\pi$ on the "top" of the cut. At π we transfer to the second sheet and θ_1 continues to 3π, where we return to the first sheet. As we look down the negative real axis toward the origin, the two sheets are joined as shown in Fig. 6.3. Similarly, for the right-hand cut, we start at the "top" of the cut at $\theta_2 = 0$ and move counterclockwise, passing to the second sheet at $\theta_2 = 2\pi$, and finally return to the first sheet when $\theta_2 = 4\pi$. Along the positive real axis, the two sheets are joined as shown in Fig. 6.3. Using the above conventions, we define

$$w(z) \equiv |z-a|^{1/2}|z-b|^{1/2}e^{i\theta_1/2}e^{i\theta_2/2}.$$

As shown in Fig. 6.6, any point z_0 on the second sheet can be reached from the point z_0 on the first sheet either by going via the left-hand cut or the right-hand cut. These two options differ only in the sense that in the first case θ_1 increases by 2π, whereas in the second case θ_2 increases by 2π. In both cases, the value

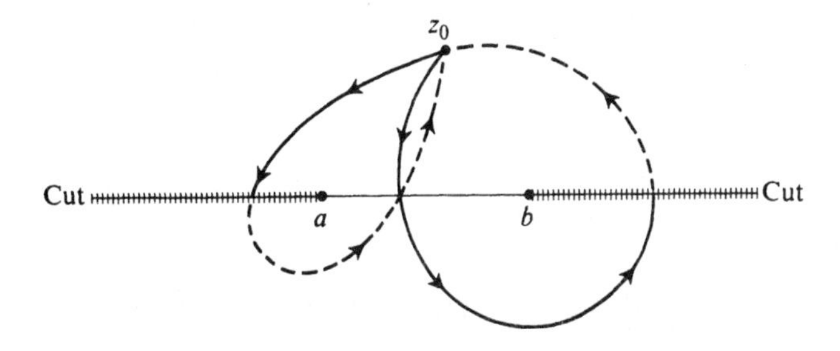

Fig. 6.6 Two different paths by which one can go from $z = z_0$ on the first sheet to $z = z_0$ on the second sheet. The function in question is $w(z) = \sqrt{(z-a)(z-b)}$.

of $w(z)$ is the same (the negative of what it would be for the corresponding point on the first sheet), as it must be if we are to have a single-valued function. With this construction, $w(z) = \sqrt{(z-a)(z-b)}$ becomes a single-valued function, analytic everywhere except at $z = a$ and $z = b$.

Note that in the above example, if we go through 2π radians in both θ_1 and θ_2, starting for example on the "top" of the right-hand cut, then we go down to the second sheet via the left-hand cut and return to our starting point on the first sheet via the right-hand cut. This suggests that it should equally well be possible to cut and join the two sheets along the real axis from a to b. The reader may find it interesting to show that this is indeed the case.

Using the above ideas, we can also obtain sensible expressions for the inverse trigonometric functions. For example, consider

$$w = \tan^{-1} z \,.$$

Writing this as

$$z = \tan w = \frac{1}{i}\frac{e^{iw} - e^{-iw}}{e^{iw} + e^{-iw}},$$

we obtain readily

$$(1 - iz)e^{iw} = (1 + iz)e^{-iw},$$

and hence

$$e^{2iw} = \frac{1 + iz}{1 - iz}.$$

Taking the logarithms of both sides, we find that

$$w = \frac{1}{2i}\log\left(\frac{1+iz}{1-iz}\right) = \frac{1}{2i}\log\left(\frac{i-z}{i+z}\right).$$

Thus

$$\tan^{-1} z = \frac{i}{2}\log\left(\frac{i+z}{i-z}\right).$$

Just as we did in discussing the logarithm, we can speak of single-valued branches of $\tan^{-1} z$ corresponding to single-valued branches of $\log z$. The principal branch of the inverse tangent is defined in the obvious manner:

$$\mathrm{Tan}^{-1} z = \frac{i}{2}\left[\mathrm{Log}\,(i+z) - \mathrm{Log}\,(i-z)\right].$$

Similar considerations apply to the functions $\sin^{-1} z$ and $\cos^{-1} z$.

6.3 COMPLEX INTEGRATION—THE CAUCHY-GOURSAT THEOREM

We now come to integration of complex functions, the part of the theory that makes the subject really interesting to both mathematicians and physicists. Because of the correspondence between complex numbers and two-dimensional vectors, we might expect to be able to define the line integral of a complex function along a curve in the z-plane.

Let t be a real parameter ranging from t_A to t_B, and let $z = z(t)$ be a curve, or *contour* C, in the complex plane, with endpoints $A = z(t_A)$ and $B = z(t_B)$. (See Fig. 6.7.) Now we mark off a number of points t_i between t_A and t_B, and approximate the curve by a series of straight lines drawn from each $z(t_i)$ to $z(t_{i+1})$.

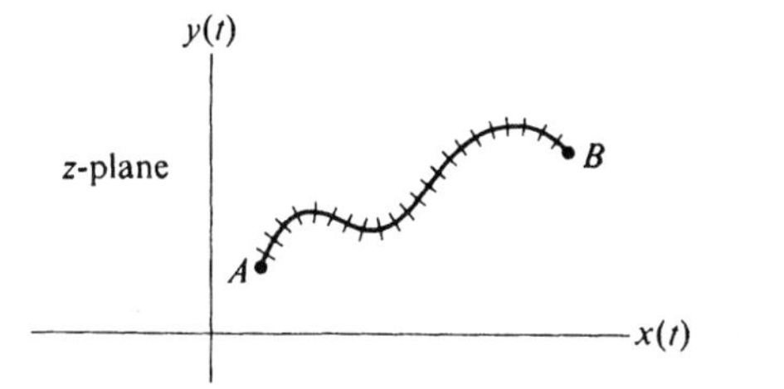

Fig. 6.7

To define the integral of a function w of a complex variable, we form the quantity

$$\lim_{|\Delta z_i| \to 0} \sum_{i=0}^{n} w(z_i) \Delta z_i \equiv \int_C w(z) dz ,$$

where $\Delta z_i = z(t_{i+1}) - z(t_i)$, and $w(z_i)$ is the function evaluated at a point z_i on C between $z(t_{i+1})$ and $z(t_i)$. The sum is evaluated in the limit of an arbitrarily fine partition of the range through which the real parameter t moves as it generates the contour from A to B; that is, as $n \to \infty$, or, what is the same thing, in the limit of arbitrarily small $|\Delta z_i|$ for all i.

Writing $w(z) = u(x, y) + iv(x, y)$ and $dz = dx + i\, dy$, we have

$$\int_C w(z)\, dz = \int_C (u\, dx - v\, dy) + i \int_C (u\, dy + v\, dx) . \tag{6.10}$$

We can also write this in parametric form. Then

$$dx = \frac{dx}{dt} dt , \qquad dy = \frac{dy}{dt} dt$$

and so

$$\int_C w(z)\, dz = \int_{t_A}^{t_B} \left(u \frac{dx}{dt} - v \frac{dy}{dt} \right) dt + i \int_{t_A}^{t_B} \left(u \frac{dy}{dt} + v \frac{dx}{dt} \right) dt .$$

For a given contour C running from A to B we define the opposite contour, written as $-C$, to be the same curve but traversed from B to A. The integral of $w(z)$ along $-C$ is clearly given by the above equation but with t_A and t_B interchanged. Thus,

$$\int_C = - \int_{-C} . \tag{6.11}$$

Also it follows that

$$\int_{C_1} + \int_{C_2} = \int_{C_1+C_2} . \tag{6.12}$$

If C is a closed curve that does not intersect itself, we shall always interpret $\oint_C$ to mean the integral taken counterclockwise along the closed contour C.

Another property of the integral that we shall need very often is

$$\left|\int_C f(z)\, dz\right| \leq \int_C |f(z)|\, |dz| \leq ML , \tag{6.13}$$

where M is the maximum value of $|f(z)|$ on C, and the length of C is L. The first inequality in Eq. (6.13) is a generalization of $|z_1 + z_2| \leq |z_1| + |z_2|$, the triangle inequality; both inequalities are derived exactly as in the real case.

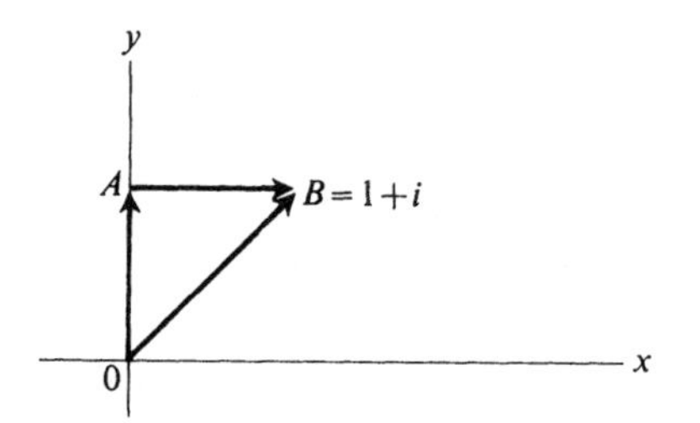

Fig. 6.8

Before presenting the general theorems on the integration of functions of a complex variable, we work out two examples of such integrals using only the results obtained so far. First, we evaluate the integral

$$I = \int_C \sin z\, dz$$

over the two paths shown in Fig. 6.8: (1) $C_1 = OB$, (2) $C_2 = OA + AB$. Since

$$\sin z = \cosh y \sin x + i \sinh y \cos x ,$$

we have, using Eq. (6.10),

$$I \equiv \int_C \sin z\, dz = \int_C [\cosh y \sin x\, dx - \sinh y \cos x\, dy]$$
$$+ i \int_C [\cosh y \sin x\, dy + \sinh y \cos x\, dx] .$$

Along the curve C_1, $x = y$. Therefore

$$I_1 = (1 + i) \int_0^1 \cosh x \sin x\, dx - (1 - i) \int_0^1 \sinh x \cos x\, dx$$
$$= (1 - \cosh 1 \cos 1) + i(\sinh 1 \sin 1) .$$

Now we compute I along C_2. Along the path from O to A, $x = 0$ and $dx = 0$, and along the path from A to B, $y = 1$ and $dy = 0$. Therefore

$$I_2 = \int_{C_2} \sin z\, dz = -\int_0^1 \sinh y\, dy + \int_0^1 \cosh 1 \sin x\, dx + i\int_0^1 \sinh 1 \cos x\, dx$$
$$= 1 - \cosh 1 \cos 1 + i \sinh 1 \sin 1 = I_1 .$$

The integral from O to B is the same for both paths. In fact, we shall prove later that it is the same for any path whatsoever—it depends only on the two endpoints. Also, the definite integral around the closed contour consisting of C_1 and $-C_2$ (that is, C_2 traveled backward) is zero. We shall show that this result holds for any function which is analytic on and inside the closed contour.

Note that if we evaluate formally, according to the rule of real calculus,

$$I_1 = I_2 = \int_0^{1+i} \sin z\, dz = -\cos z\Big]_0^{1+i}$$
$$= 1 - \cosh 1 \cos 1 + i \sinh 1 \sin 1 ,$$

where we have used Eq. (6.9). We shall also prove this "fundamental theorem of complex calculus," which holds in any region in which the integrand is analytic.

As a second example, let us integrate the function $f(z) = z^*$ counterclockwise around the unit circle centered at the origin. The values of z on this curve are given by $z = e^{i\theta}$, $\theta = 0$ to 2π. Therefore

$$I = \oint_C z^*\, dz = \int_0^{2\pi} e^{-i\theta} i e^{i\theta}\, d\theta = 2\pi i .$$

Since $zz^* = 1$ on C, we also obtain the result

$$\oint_C \frac{1}{z}\, dz = 2\pi i .$$

Neither integral around the closed contour is zero. The reason, as we shall see, is that z^* is not analytic anywhere, and therefore not within and on C, and z^{-1} is not analytic at $z = 0$, which is within C.

Both these examples are explained by

Cauchy's Theorem. If a function $f(z)$ is analytic within and on a closed contour C, *and $f'(z)$ is continuous throughout this region,* then

$$\oint_C f(z)\, dz = 0 .$$

We shall give two proofs of this theorem.

Proof 1.

$$\oint_C f(z)\, dz = \oint_C (u\, dx - v\, dy) + i\oint_C (u\, dy + v\, dx) .$$

To evaluate the two line integrals on the right, we use Green's theorem for line integrals. It states that if the derivatives of P and Q are continuous functions within and on a closed contour C, then

$$\oint_C (P\,dx + Q\,dy) = \int_R\int \left(\frac{\partial Q}{\partial x} - \frac{\partial P}{\partial y}\right) dx\,dy,$$

where R is the surface bounded by C. By hypothesis, $f'(z)$ is continuous, so the first partial derivatives of u and v are also continuous; then Green's theorem yields

$$\oint_C (u\,dx - v\,dy) + i\oint_C (u\,dy + v\,dx)$$
$$= \int_R\int \left(\frac{\partial u}{\partial y} + \frac{\partial v}{\partial x}\right) dx\,dy + i\int_R\int \left(\frac{\partial u}{\partial x} - \frac{\partial v}{\partial y}\right) dx\,dy\,.$$

But since the Cauchy-Riemann equations hold, the integrands above all vanish. Therefore

$$\oint_C f(z)\,dz = 0\,. \qquad \text{QED}$$

Proof 2. Cauchy's theorem may also be proved if we use Stokes's theorem, which is closely related to Green's theorem and is perhaps more familiar. We write

$$\oint_C f(z)\,dz = \oint_C \mathbf{F}\cdot d\mathbf{l} + i\oint_C \mathbf{G}\cdot d\mathbf{l}\,,$$

where

$$\mathbf{F} = u\mathbf{i} - v\mathbf{j}\,, \qquad \mathbf{G} = v\mathbf{i} + u\mathbf{j}\,, \qquad \text{and} \qquad d\mathbf{l} = dx\mathbf{i} + dy\mathbf{j}\,.$$

Let S be the region interior to and including C. Since the Cauchy-Riemann conditions hold throughout S, it follows that $(\nabla\times\mathbf{F})_z = 0$ and $(\nabla\times\mathbf{G})_z = 0$ throughout S, where the subscript z identifies the $\mathbf{k}$ component of the curl; for example,

$$(\nabla\times\mathbf{F})_z = \frac{\partial(-v)}{\partial x} - \frac{\partial u}{\partial y} = 0\,,$$

by virtue of the Cauchy-Riemann equations. Now, using Stokes's theorem, the validity of which depends on the continuity of the four first partials of u and v,

$$\oint_C f(z)\,dz = \oint_C \mathbf{F}\cdot d\mathbf{l} + i\oint_C \mathbf{G}\cdot d\mathbf{l} = \int_S (\nabla\times\mathbf{F})\cdot d\mathbf{S} + i\int_S (\nabla\times\mathbf{G})\cdot d\mathbf{S}$$
$$= \int_S (\nabla\times\mathbf{F})_z\,dS + i\int_S (\nabla\times\mathbf{G})_z\,dS = 0\,,$$

where $d\mathbf{S} = \mathbf{k}\,dS$. QED

It is possible to prove Cauchy's theorem without assuming the continuity of $f'(z)$. This is because any function which is analytic in a region *necessarily* has a continuous derivative. In fact, we shall prove that an analytic function has derivatives of all orders, and therefore all its derivatives are continuous, the continuity of the nth derivative being a consequence of the existence of the derivative of order $n + 1$. But we shall only be able to establish this result on higher derivatives *after* we have shown that the continuity of $f'(z)$ is not needed in the proof of Cauchy's theorem.

This relaxation or weakening of the hypotheses under which

$$\oint_C f(z)\,dz = 0$$

is therefore of the utmost importance—it is, in fact, the centerpiece of the theory of analytic functions. Some authors (never mathematicians) define an analytic function as a differentiable function *with a continuous derivative*. Then the central result of the theory follows trivially, as we have seen in the previous theorem. But this is a mathematical fraud of cosmic proportions. It was Goursat who first proved that the condition that $f'(z)$ be continuous is superfluous. It is Goursat's result that really distinguishes the theory of integration of a function of a complex variable from the theory of line integrals in the real plane. Although the theorem is often simply called Cauchy's theorem, it is the "-Goursat" half that gives it real mathematical power. In our proof we follow the presentations of Franklin and of Knopp.

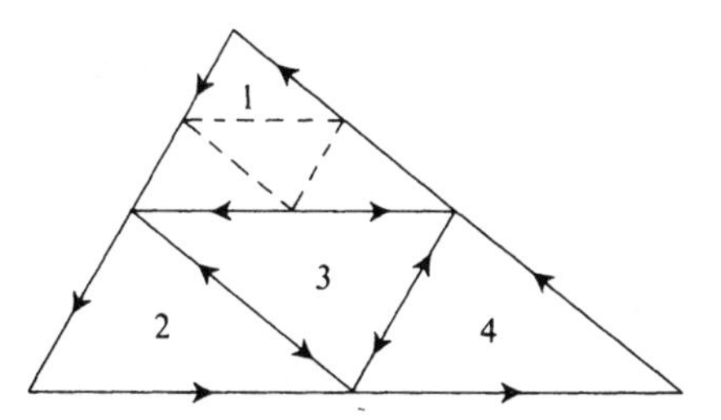

Fig. 6.9

Cauchy-Goursat Theorem. If a function $f(z)$ is analytic within and on a closed contour C, then $\oint_C f(z)\,dz = 0$.

Proof. We shall first prove the theorem for a triangular region; then it is very easily extended to an arbitrary region. Let R denote the closed region consisting of the points interior to and on the triangle bounded by the closed contour T of total length L. Since $f(z)$ is analytic in R, $f'(z)$ exists throughout R, and therefore, $f(z)$ is continuous in R. We now begin subdividing R into smaller triangles as shown in Fig. 6.9. Each subtriangle is similar to the original triangle but its sides (and perimeter) are only one-half as long. The boundaries of the subtriangles are denoted by T_i ($i = 1, 2, 3, 4$). Clearly

$$\oint_T f(z)\,dz = \sum_{i=1}^{4} \oint_{T_i} f(z)\,dz\,,$$

where all contours are traversed in a counterclockwise direction. All three of the boundaries of triangle 3 in Fig. 6.9 are traversed in both directions and therefore cancel out. Applying the triangle inequality to this equation, we obtain

$$\left|\oint_T f(z)\,dz\right| \leq \sum_{i=1}^{4} \left|\oint_{T_i} f(z)\,dz\right| .$$

The object of the proof is to show that the quantity on the left is arbitrarily small.

Now let C_1 denote the triangle which contributes the largest term to the above sum. Then we have

$$\left|\oint_T f(z)\,dz\right| \leq 4 \left|\oint_{C_1} f(z)\,dz\right| ,$$

where the length of $C_1 \equiv L_1 = L/2$. We now repeat this process on the subtriangle bounded by C_1. That is, we find a contour C_2, bounding a "sub-subtriangle," such that

$$\left|\oint_{C_1} f(z)\,dz\right| \leq 4 \left|\oint_{C_2} f(z)\,dz\right| ,$$

where the length of $C_2 \equiv L_2 = L_1/2 = L/2^2$. If the subdivision is repeated n times, we obtain a nested sequence of triangular contours C_n, such that

$$\left|\oint_T f(z)\,dz\right| \leq 4^n \left|\oint_{C_n} f(z)\,dz\right| , \tag{6.14}$$

where the length of $C_n \equiv L_n = L/2^n$. In order to finish the proof, we have to show that

$$\left|\oint_{C_n} f(z)\,dz\right|$$

is decreasing with n more rapidly than 4^n is increasing.

We let R_n denote the closed region consisting of C_n and the interior points of the (sub)$^{n\text{th}}$-triangle bounded by C_n. Clearly, each point of the region R_{n+1} is a point of R_n, and as n goes to infinity this nested sequence of closed sets closes down on a single point z_0 which is in each R_n, and R itself. (If R is the continent of Africa and we select R_n as that subregion which contains the biggest lion in Africa, we have an algorithm for capturing a big lion—we simply build a cage about the point z_0.)

Since $f'(z)$ exists, it follows by definition that for any $\epsilon > 0$, there exists a δ, such that when $0 < |z - z_0| < \delta$,

$$\left|\frac{f(z) - f(z_0)}{z - z_0} - f'(z_0)\right| \leq \epsilon .$$

Now consider the function $g(z)$ defined by

$$g(z) = \begin{cases} \dfrac{f(z) - f(z_0)}{z - z_0} - f'(z_0) & \text{for } z \neq z_0 , \\ 0 & \text{for } z = z_0 . \end{cases}$$

Note that $|g(z)| \leq \epsilon$ if $|z - z_0| < \delta$; $g(z)$ is therefore continuous at $z = z_0$.

Now $f(z)$ is given for all z in R by

$$f(z) = f(z_0) + (z - z_0)f'(z_0) + (z - z_0)g(z);$$

we use this relation to evaluate $\oint_{C_n} f(z)\,dz$. The first two terms are entire functions of z (z_0 is constant) with *derivatives* that are *continuous* everywhere. Therefore we may apply the earlier version of Cauchy's theorem to deduce that

$$\oint_{C_i} [f(z_0) + (z - z_0)f'(z_0)]\,dz = 0\,, \qquad \text{for all } i\,.$$

Therefore

$$\oint_{C_n} f(z)\,dz = \oint_{C_n} (z - z_0)g(z)\,dz\,.$$

We now subdivide enough times (i.e., we choose n large enough) so that $2^n > L/\delta$. Then $L_n = L/2^n < \delta$. Furthermore, for any point z on C_n, $|z - z_0| < L_n < \delta$, since z_0 is inside R_n, and the distance from any interior point to any point on the boundary of a triangle is clearly less than the perimeter of the triangle. Therefore, since $|z - z_0| < \delta$, $|g(z)| < \epsilon$, and $|(z - z_0)g(z)| \leq L_n\epsilon$. Consequently,

$$\left|\oint_{C_n} f(z)\,dz\right| = \left|\oint_{C_n} (z - z_0)g(z)\,dz\right| \leq L_n^2\epsilon = \epsilon(L/2^n)^2 = \epsilon L^2/4^n\,,$$

where we have used Eq. (6.13). It now follows from Eq. (6.14) that

$$\left|\oint_T f(z)\,dz\right| \leq \epsilon L^2\,.$$

Since L is the fixed finite perimeter of the triangular region R, and ϵ is arbitrary, we can make the quantity ϵL^2 smaller than any preassigned number ϵ'. Thus

$$\left|\oint_T f(z)\,dz\right| = 0\,, \qquad \text{and hence} \qquad \oint_T f(z)\,dz = 0\,.$$

This proves the Cauchy-Goursat theorem for triangular contours.

We shall not give a formal proof of the extension of this result to arbitrary regions, because the method and result are simple and clear. Given an arbitrary C, we inscribe a polygon in C. Any polygon may be decomposed into a sum of triangles, so we know the theorem holds for polygons of any number of sides. It is clear that the difference

$$\left|\oint_C f(z)\,dz - \oint_P f(z)\,dz\right|\,,$$

where P is the perimeter of the polygon inscribed in C, can be made arbitrarily small by simply choosing a polygon with a sufficiently large number of sides. This establishes the Cauchy-Goursat theorem for a region of arbitrary shape.

Throughout the proof we have tacitly assumed that the region R is a *simply-connected* region. This means that any closed contour in R encloses only points belonging to R. Suppose, however, that R were a region with one or more subregions "punched out." Then it would be possible to construct curves around these holes in such a way that the curves would lie entirely in R, but enclose points *not* belonging to R. Such regions are called *multiply-connected*. Cauchy's theorem does not hold for arbitrary contours in multiply-connected regions.

6.4 CONSEQUENCES OF CAUCHY'S THEOREM

The hardest work is behind us; we turn now to an examination of some of the main consequences of Cauchy's theorem.

Path Independence

We first prove that if $f(z)$ is analytic in the region R and C_1 and C_2 lie in R and have the same endpoints, then

$$\int_{C_1} f\,dz = \int_{C_2} f\,dz\,.$$

The proof follows immediately by applying Cauchy's theorem to the closed contour consisting of C_2 and $-C_1$ as shown in Fig. 6.10;

$$\int_{C_2} + \int_{-C_1} = 0 \Longrightarrow \int_{C_2} = -\int_{-C_1} = \int_{C_1}$$

by Eq. (6.11).

Fundamental Theorem of Calculus

From our discussion of path independence it follows that the equation

$$F(z) \equiv \int_{z_0}^{z} f(z')\,dz'$$

defines a unique function of z if $f(z')$ is analytic throughout the region containing the path between z_0 and z.

Theorem. $F(z)$ is analytic and $F'(z) = f(z)$.

Proof.

$$F(z + \Delta z) - F(z) = \int_{z}^{z+\Delta z} f(z')\,dz'\,,$$

where the path from z to $(z + \Delta z)$ may be taken to be a straight line. We can write

$$f(z) = \frac{f(z)}{\Delta z}\int_{z}^{z+\Delta z} dz' = \frac{1}{\Delta z}\int_{z}^{z+\Delta z} f(z)\,dz'\,,$$

and it follows that

$$\frac{F(z + \Delta z) - F(z)}{\Delta z} - f(z) = \frac{1}{\Delta z}\int_{z}^{z+\Delta z} [f(z') - f(z)]\,dz'\,.$$

Now $f(z)$ is continuous because it is analytic; therefore, for all $\epsilon > 0$, there exists a $\delta > 0$ such that if $|z' - z| < \delta$, then $|f(z') - f(z)| < \epsilon$. Now take $0 < |\Delta z| < \delta$. Then

$$\left|\frac{F(z + \Delta z) - F(z)}{\Delta z} - f(z)\right| < \epsilon\frac{1}{|\Delta z|}\int_{z}^{z+\Delta z} |dz'| = \epsilon\,.$$

That is,

$$F'(z) \equiv \lim_{\Delta z \to 0} \frac{F(z + \Delta z) - F(z)}{\Delta z} = f(z) ,$$

so $F(z)$ is analytic and its derivative is $f(z)$.

Thus the integral $F(z)$ of an analytic function $f(z)$ is an analytic function of its upper limit, provided the path of integration is confined to a region R within which the integrand is analytic. The fundamental theorem of calculus follows immediately from this result.

$$\int_a^b f(z)\, dz = \int_{z_0}^b f(z)\, dz - \int_{z_0}^a f(z)\, dz = F(b) - F(a) ,$$

where a and b are points in R, and $F'(z) = f(z)$, that is, $F(z)$ is an antiderivative of $f(z)$. We have already noticed that this method of evaluating integrals worked in a special case: the integral of $\sin z$ from $a = 0$ to $b = 1 + i$ (Section 6.3).

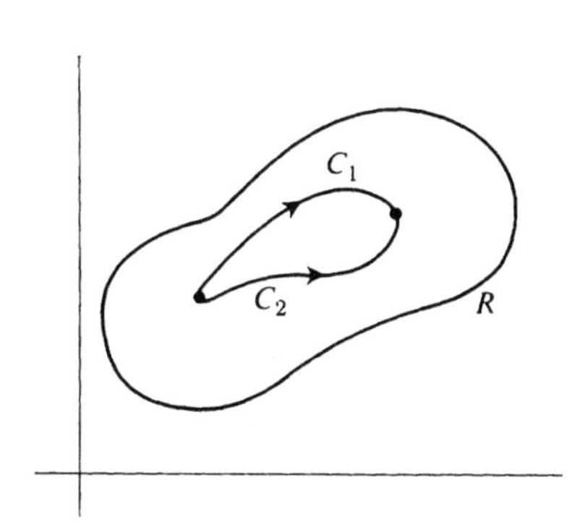

Fig. 6.10

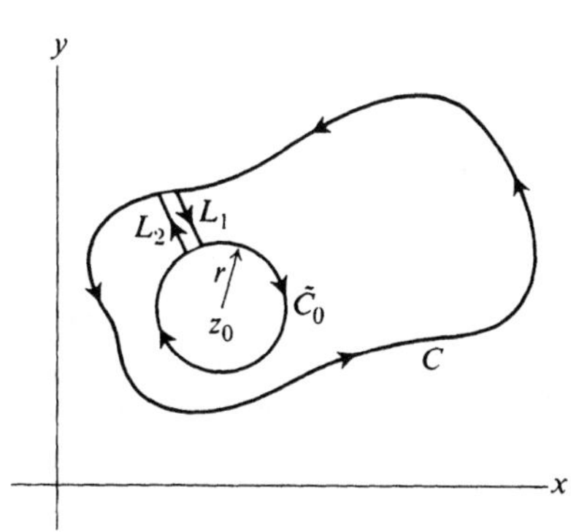

Fig. 6.11

Cauchy's Integral Formula

We now prove one of the most useful results in all mathematical physics.

> **Theorem.** If $f(z)$ is analytic within and on a closed contour C, then for *any* point z_0, interior to C,
>
> $$f(z_0) = \frac{1}{2\pi i} \oint_C \frac{f(z)}{z - z_0}\, dz . \tag{6.15}$$

Proof. Inside the contour C, draw a circle $\tilde{C}_0$ of radius r about z_0, and consider the contour shown in Fig. 6.11. It consists of the circle $\tilde{C}_0$ and the contour C joined by two straight line segments, L_1 and L_2, which lie arbitrarily close to each other. Let us call this entire contour C'. Now consider

$$\oint_{C'} \frac{f(z)}{z - z_0}\, dz = \oint_C \frac{f(z)}{z - z_0}\, dz + \int_{L_1} \frac{f(z)}{z - z_0}\, dz + \oint_{\tilde{C}_0} \frac{f(z)}{z - z_0}\, dz + \int_{L_2} \frac{f(z)}{z - z_0}\, dz .$$

Inside C', $f(z)/(z - z_0)$ is analytic, so by the Cauchy-Goursat theorem,

$$\oint_{C'} \frac{f(z)}{z - z_0}\, dz = 0 .$$

Now, as we bring the line segments L_1 and L_2 arbitrarily close together,

$$\int_{L_1} \frac{f(z)}{z - z_0}\, dz \to - \int_{L_2} \frac{f(z)}{z - z_0}\, dz ,$$

since the lines are traversed in opposite directions. Thus in this limit we have

$$\oint_{C'} \frac{f(z)}{z - z_0}\, dz = 0 = \oint_{C} \frac{f(z)}{z - z_0}\, dz + \oint_{\tilde{C}_0} \frac{f(z)}{z - z_0}\, dz ,$$

so that

$$\oint_{C} \frac{f(z)}{z - z_0}\, dz = - \oint_{\tilde{C}_0} \frac{f(z)}{z - z_0}\, dz .$$

At this point we note that considered as a contour in its own right, i.e., not just as a part of C', $\tilde{C}_0$ is traversed in a clockwise direction. Let us therefore define $C_0 = -\tilde{C}_0$ so that C_0 is a counterclockwise contour (as is C). Then we may write

$$\oint_{C} \frac{f(z)}{z - z_0}\, dz = \oint_{C_0} \frac{f(z)}{z - z_0}\, dz .$$

(Note that for the purposes of what we have just done, C_0 need not be a circle; it could equally well be *any* closed contour lying completely inside C and oriented in the same sense as C.) We may rewrite the last equation as

$$\oint_{C} \frac{f(z)}{z - z_0}\, dz = f(z_0) \oint_{C_0} \frac{dz}{z - z_0} + \oint_{C_0} \frac{f(z) - f(z_0)}{z - z_0}\, dz .$$

We now use the fact that C_0 is a circle to write $z - z_0 = re^{i\theta}$ on C_0. Thus the first integral on the right becomes

$$\oint_{C_0} \frac{dz}{z - z_0} = \int_0^{2\pi} \frac{ire^{i\theta} d\theta}{re^{i\theta}} = 2\pi i , \qquad \text{for all } r > 0 \text{ within } C .$$

Cauchy's formula will therefore be established if we can show that the second integral vanishes for some choice of the contour C_0. The continuity of $f(z)$ at z_0 tells us that for all $\epsilon > 0$ there exists a δ such that if $|z - z_0| \leq \delta$, then $|f(z) - f(z_0)| < \epsilon$. So by taking $r = \delta$, we satisfy the equation $|z - z_0| = \delta$, which in turn implies that

$$\left| \oint_{C_0} \frac{f(z) - f(z_0)}{z - z_0}\, dz \right| \leq \oint_{C_0} \frac{|f(z) - f(z_0)|}{|z - z_0|}\, |dz| < \frac{\epsilon}{\delta}\, (2\pi\delta) = 2\pi\epsilon .$$

Thus by taking r small enough, but still greater than zero, the absolute value of the integral can be made smaller than any preassigned number. Thus

$$\oint_{C} \frac{f(z)}{z - z_0}\, dz = 2\pi i\, f(z_0) .$$

This result gives us another hint of the amazingly strong inner structure of analytic functions. It means that if a function is analytic within and on a contour C, its value at every point inside C is determined by its values on the bounding curve C. There is a familiar equivalent result from electrostatics: If a real-valued function $u(x, y)$ is fixed on some boundary, and if $\nabla^2 u = 0$, then u is determined everywhere inside the boundary. An analytic function is built out of a pair of such harmonic functions. We could, if we liked, study the "theory of harmonic functions" instead of "analytic function theory".

Derivatives of Analytic Functions

Using Cauchy's integral formula, we can prove that all the derivatives of an analytic function are analytic. The corresponding result for real variables fails: a function which is once differentiable in some region is not necessarily infinitely differentiable in that region. The function $f(x) = x|x|$, for example, has as its derivative $f'(x) = 2|x|$, which is continuous everywhere; $f'(x)$ is not differentiable at the origin, however.

If we differentiate both sides of Cauchy's integral formula, interchanging the orders of integration and differentiation, we get

$$f'(z_0) = \frac{1}{2\pi i} \oint_C \frac{f(z)}{(z - z_0)^2} \, dz \,. \tag{6.16}$$

Since z_0 is any point inside C, we may take it as a variable. To establish this formula in a rigorous manner, note that by using Cauchy's integral formula, we may write

$$\begin{aligned} f'(z_0) = \lim_{z_1 \to z_0} \frac{f(z_1) - f(z_0)}{z_1 - z_0} &= \frac{1}{2\pi i} \lim_{z_1 \to z_0} \oint_C \left[\frac{f(z)}{z - z_1} - \frac{f(z)}{z - z_0} \right] \frac{dz}{z_1 - z_0} \\ &= \frac{1}{2\pi i} \lim_{z_1 \to z_0} \oint_C \frac{f(z)}{(z - z_1)(z - z_0)} \, dz \,. \end{aligned}$$

Hence

$$\begin{aligned} f'(z_0) - \frac{1}{2\pi i} \oint_C \frac{f(z)}{(z - z_0)^2} \, dz &= \frac{1}{2\pi i} \lim_{z_1 \to z_0} \oint_C f(z) \left[\frac{1}{(z - z_1)(z - z_0)} - \frac{1}{(z - z_0)^2} \right] dz \\ &= \frac{1}{2\pi i} \lim_{z_1 \to z_0} (z_1 - z_0) \oint_C \frac{f(z)}{(z - z_1)(z - z_0)^2} \, dz \,. \end{aligned}$$

Calling $z_1 - z_0 = \epsilon e^{i\theta}$, we have

$$\left| f'(z_0) - \frac{1}{2\pi i} \oint_C \frac{f(z)}{(z - z_0)^2} \, dz \right| \leq \frac{1}{2\pi} \lim_{\epsilon \to 0} \epsilon \oint_C \frac{|f(z)| \, |dz|}{|(z - z_0) - \epsilon e^{i\theta}| \, |z - z_0|^2} \,.$$

Replacing $|z - z_0|$ by its minimum value, say μ, and $|f(z)|$ by its maximum value M, we obtain

$$\left| f'(z_0) - \frac{1}{2\pi i} \oint_C \frac{f(z)}{(z - z_0)^2} \, dz \right| \leq \frac{1}{2\pi} \frac{ML}{\mu^2} \lim_{\epsilon \to 0} \frac{\epsilon}{\mu - \epsilon} = 0 \,,$$

where L is the length of the contour. Thus we have proved Eq. (6.16). Repeating the process, we obtain

$$f''(z_0) = \frac{2!}{2\pi i}\oint_C \frac{f(z)}{(z - z_0)^3}\, dz\,,$$

and, in general, for the nth derivative,

$$f^{(n)}(z_0) = \frac{n!}{2\pi i}\oint_C \frac{f(z)}{(z - z_0)^{n+1}}\, dz\,. \tag{6.17}$$

This result is readily established by induction in the same manner we used to prove Eq. (6.16). Thus $f(z)$ has derivatives of all orders within C. The kth derivative of $f(z)$ is continuous within C because the $(k + 1)$ derivative exists. Thus if we write $f(z) = u(x, y) + iv(x, y)$, the partial derivatives of u and v of all orders are continuous whenever f is analytic. We can therefore drop in our derivation of Eq. (6.4) the *requirement* that the second partial derivatives be continuous: they are guaranteed to be continuous because f is analytic.

Liouville's Theorem

Theorem. If $f(z)$ is entire and $|f(z)|$ is bounded for all values of z, then $f(z)$ is a constant.

Proof. From Cauchy's integral formula, we have found that

$$f'(z_0) = \frac{1}{2\pi i}\oint_C \frac{f(z)}{(z - z_0)^2}\, dz\,.$$

If we take C to be the circle $|z - z_0| = r_0$, then

$$|f'(z_0)| \leq \left|\frac{1}{2\pi i}\right| \oint_{C_0} \frac{|f(z)|}{|(z - z_0)^2|}\, |dz|$$

$$< \frac{1}{2\pi r_0^2}\, M 2\pi r_0 = \frac{M}{r_0}\,,$$

where $|f(z)| < M$ within and on C_0. Therefore $|f'(z_0)| < M/r_0$, and we may take r_0 as large as we like because $f(z)$ is entire. So taking r_0 large enough, we can make $|f'(z_0)| < \epsilon$, for any preassigned ϵ. That is, $|f'(z_0)| = 0$, which implies that $f'(z_0) = 0$ for all z_0, so $f(z_0) =$ constant. QED

Example. The entire functions $\sin z$ and $\cos z$ must not be bounded. It is clear from Eqs. (6.8) and (6.9) that they are not.

Fundamental Theorem of Algebra

This theorem, which is difficult to prove algebraically, follows easily from Liouville's theorem, and provides a remarkable tie-up between analysis and algebra. We include the proof because of its great simplicity and beauty.

Theorem. If $P(z) = a_0 + a_1 z + \cdots + a_m z^m$ is a polynomial in z of degree one or greater, then the equation $P(z) = 0$ has at least one root.

Proof. Assume the contrary, namely that $P(z) \neq 0$ for any z. Then the function $1/P(z)$ is entire. Furthermore $|1/P(z)| \to 0$ as $|z| \to \infty$ so $|1/P(z)|$ is bounded for all z. Therefore, by Liouville's theorem, $1/P(z) = \text{const}$, a contradiction, since $P(z)$ is of degree one or greater. Hence $P(z) = 0$, for at least one value of z. QED

6.5 HILBERT TRANSFORMS AND THE CAUCHY PRINCIPAL VALUE

It is often the case that in the study of some physical system one has to deal with complex-valued functions—indices of refraction, susceptibilities, scattering amplitudes, impedances, etc.—which have a physical meaning only when the argument of the function (which might, for example, be a frequency or an energy) takes on *real* values. In many cases it is possible to obtain, from the laws governing the system, information about the general properties of such functions when the argument is *complex*; for example, it may be that the function is analytic in some region of the complex plane. Since experimental data can only be obtained for real values of the argument, it is of interest to see whether we can use general properties such as analyticity to deduce relations between real quantities of direct physical significance. The key to such a program can be found in the study of Hilbert transform pairs, which we shall investigate in this section.

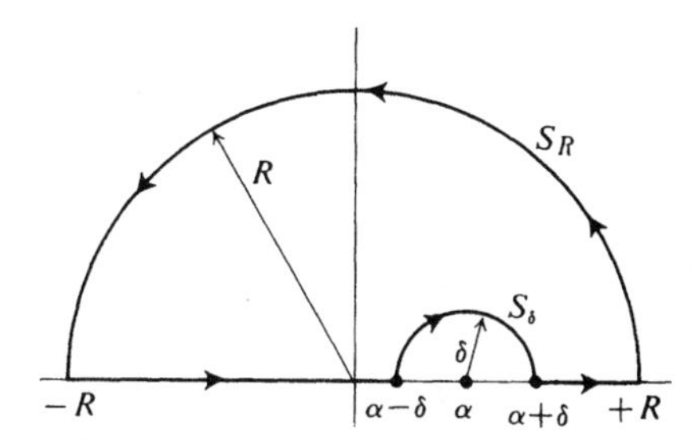

Fig. 6.12 The contour, C, used to obtain Eq. (6.18). The radius, R, of the semi-circle, S_r, may be made as large as necessary, and the radius, S, of the semi-circle, S_δ, may be made as small as we please.

Let us begin by considering a function $f(z)$, which is analytic in the upper half of the complex plane, and which is such that $|f(z)| \to 0$ as $|z| \to \infty$ in the upper half-plane. (Note that the only function which can satisfy these conditions in the *entire* plane is $f \equiv 0$). Now consider the contour integral

$$\oint_C \frac{f(z)}{z - \alpha}\, dz\,,$$

where C is the contour shown in Fig. 6.12 and α is real. By assumption, $f(z)$ is analytic within and on C; so is $1/(z - \alpha)$. Thus

$$\oint_C \frac{f(z)}{z - \alpha}\, dz = 0\,.$$

Let us break this up as follows:

$$\oint_C \frac{f(z)}{z-\alpha}\,dz \equiv \int_{-R}^{\alpha-\delta} \frac{f(x)}{x-\alpha}\,dx + \int_{S_\delta} \frac{f(z)}{z-\alpha}\,dz + \int_{\alpha+\delta}^{R} \frac{f(x)}{x-\alpha}\,dx + \int_{S_R} \frac{f(z)}{z-\alpha}\,dz = 0\,.$$

Here δ is the radius of the small semicircle S_δ, centered at $x = \alpha$, and R is the radius of the large semicircle S_R, centered at the origin, as shown in Fig. 6.12. The radius δ can be chosen as small as we please, and R can be chosen as large as we please. In the limit of arbitrarily small δ, the quantity

$$\int_{-R}^{\alpha-\delta} \frac{f(x)}{x-\alpha}\,dx + \int_{\alpha+\delta}^{R} \frac{f(x)}{x-\alpha}\,dx$$

is called the *principal-value integral* of $f(x)/(x-\alpha)$ and is denoted by

$$P\int_{-R}^{R} \frac{f(x)}{x-\alpha}\,dx\,.$$

We will say more about this integral below. Now, along the large semicircle S_R, we set $z = Re^{i\theta}$, so that

$$\int_{S_R} \frac{f(z)}{z-\alpha}\,dz = i\int_0^{\pi} \frac{f(Re^{i\theta})}{Re^{i\theta}-\alpha}\,Re^{i\theta}\,d\theta\,,$$

and hence

$$\left|\int_{S_R} \frac{f(z)}{z-\alpha}\,dz\right| \le \frac{R}{|R-\alpha|}\int_0^{\pi} |f(Re^{i\theta})|\,d\theta\,,$$

since $|Re^{i\theta}-\alpha| = [R^2+\alpha^2-2R\alpha\cos\theta]^{1/2} \ge [R^2+\alpha^2-2R\alpha]^{1/2} = |R-\alpha|$. But as $R\to\infty$, $|f(z)|\to 0$ and $R/|R-\alpha|\to 1$. Therefore the integral over the semicircle of radius R can be made arbitrarily small by choosing R sufficiently large. Thus we may write

$$\lim_{R\to\infty} P\int_{-R}^{R} \frac{f(x)}{x-\alpha}\,dx = -\int_{S_\delta} \frac{f(z)}{z-\alpha}\,dz = -f(\alpha)\int_{S_\delta} \frac{dz}{z-\alpha} - \int_{S_\delta} \frac{f(z)-f(\alpha)}{z-\alpha}\,dz\,,$$

where we have added and subtracted the term $\int_{S_\delta}[f(\alpha)/(z-\alpha)]\,dz$. Setting $z-\alpha = \delta e^{i\theta}$ in the first integral on the right-hand side of this equation, we find that

$$-f(\alpha)\int_{S_\delta} \frac{dz}{z-\alpha} = -if(\alpha)\int_{\pi}^{0} d\theta = i\pi f(\alpha)\,.$$

Thus

$$\lim_{R\to\infty} P\int_{-R}^{R} \frac{f(x)}{x-\alpha}\,dx = i\pi f(\alpha) - \int_{S_\delta} \frac{f(z)-f(\alpha)}{z-\alpha}\,dz\,.$$

Since $f(z)$ is continuous at $z = \alpha$, the argument used in deriving Cauchy's integral formula tells us that this last integral over S_δ vanishes. Hence

$$\lim_{R\to\infty} P\int_{-R}^{R} \frac{f(x)}{x-\alpha}\,dx = i\pi f(\alpha)\ .$$

For the sake of brevity, we write this simply as

$$P\int_{-\infty}^{\infty} \frac{f(x)}{x-\alpha}\,dx = i\pi f(\alpha)\ , \tag{6.18}$$

where $f(x)$ is a complex-valued function of a real variable. We may write it as

$$f(x) \equiv f_R(x) + if_I(x)\ .$$

Equating real and imaginary parts in Eq. (6.18), we get

$$f_R(\alpha) = \frac{1}{\pi} P\int_{-\infty}^{\infty} \frac{f_I(x)}{x-\alpha}\,dx\ , \tag{6.19a}$$

$$f_I(\alpha) = -\frac{1}{\pi} P\int_{-\infty}^{\infty} \frac{f_R(x)}{x-\alpha}\,dx\ . \tag{6.19b}$$

Any pair of functions which satisfy Eqs. (6.19a) and (6.19b) is called a *Hilbert transform pair*. Note that these equations tell us that if $f_I(x) \equiv 0$, then $f_R(x) \equiv 0$.

The principal-value integral is seen to be a way of avoiding singularities on a path of integration: one integrates to within δ of the singularity in question, skips over the singularity, and begins integrating again a distance δ beyond the singularity. This prescription enables one to make sense out of integrals like

$$\int_{-R}^{R} \frac{dx}{x}\ .$$

One would like this integral to be zero, since we are integrating an odd function over a symmetric domain. However, unless we insert a P in front of this integral, the singularity at the origin makes the integral meaningless. Following the prescription for principal-value integrals, we can easily evaluate the above integral. We have

$$P\int_{-R}^{R} \frac{dx}{x} = \lim_{\delta\to 0}\left[\int_{-R}^{-\delta} \frac{dx}{x} + \int_{\delta}^{R} \frac{dx}{x}\right].$$

In the first integral on the right-hand side, set $x = -y$. Then

$$P\int_{-R}^{R} \frac{dx}{x} = \lim_{\delta\to 0}\left[\int_{R}^{\delta} \frac{dy}{y} + \int_{\delta}^{R} \frac{dx}{x}\right].$$

The sum of the two integrals inside the brackets is zero, since

$$\int_{R}^{\delta} = -\int_{\delta}^{R}.$$

Thus

$$P\int_{-R}^{R} \frac{dx}{x} = 0\ .$$

In a similar manner, we may evaluate

$$P\int_{-R}^{R}\frac{dx}{x-a}$$

when $-R < a < R$. As above, we write

$$P\int_{-R}^{R}\frac{dx}{x-a}=\lim_{\delta\to 0}\left[\int_{-R}^{a-\delta}\frac{dx}{x-a}+\int_{a+\delta}^{R}\frac{dx}{x-a}\right].$$

Again setting $x = -y$ in the first integral on the right-hand side, we find that

$$\begin{aligned}P\int_{-R}^{R}\frac{dx}{x-a}&=\lim_{\delta\to 0}\left[\int_{R}^{\delta-a}\frac{dy}{y+a}+\ln(R-a)-\ln\delta\right]\\&=\lim_{\delta\to 0}\left[\ln\delta-\ln(R+a)+\ln(R-a)-\ln\delta\right].\end{aligned}$$

Thus

$$P\int_{-R}^{R}\frac{dx}{x-a}=\ln\left(\frac{R-a}{R+a}\right),\qquad -R<a<R. \tag{6.20}$$

For the case

$$P\int_{-R}^{R}\frac{f(x)}{x-a},$$

the result of Eq. (6.20) leads to

$$P\int_{-R}^{R}\frac{f(x)}{x-a}\,dx=P\int_{-R}^{R}\frac{f(a)}{x-a}\,dx+P\int_{-R}^{R}\frac{f(x)-f(a)}{x-a}\,dx,$$

or

$$P\int_{-R}^{R}\frac{f(x)}{x-a}\,dx=f(a)\ln\left(\frac{R-a}{R+a}\right)+P\int_{-R}^{R}\frac{f(x)-f(a)}{x-a}\,dx. \tag{6.21}$$

It will often happen that the second integral on the right-hand side of Eq. (6.21) will not be singular at $x = a$ [for example, this will be the case if $f(x)$ is differentiable at $x = a$] so the P symbol there can be dropped. We leave it to the reader to obtain the closely related result:

$$P\int_{0}^{R}\frac{f(x)}{x^2-a^2}\,dx=f(a)\frac{1}{2a}\ln\left(\frac{R-a}{R+a}\right)+P\int_{0}^{R}\frac{f(x)-f(a)}{x^2-a^2}\,dx. \tag{6.22}$$

To illustrate the use of the principal-value method and also the use of Eqs. (6.19a) and (6.19b), we consider the function $f(z) = 1/(z + i)$. This function satisfies all the hypotheses made in deriving Eqs. (6.19a) and (6.19b). We see that

$$f_R(x)=\frac{x}{x^2+1},\qquad f_I(x)=-\frac{1}{x^2+1}.$$

Let us examine Eq. (6.19b). We write it as

$$\lim_{R\to\infty}P\int_{-R}^{R}\frac{f_R(x)}{x-\alpha}\,dx=-\pi f_I(\alpha).$$

We want to see if $f(z) = 1/(z+i)$ satisfies this relation. Using Eq. (6.21), we have

$$\lim_{R\to\infty} P\int_{-R}^{R} \frac{f_R(x)}{x-\alpha}\,dx = \lim_{R\to\infty}\left[f_R(\alpha)\ln\left(\frac{R-\alpha}{R+\alpha}\right) + P\int_{-R}^{R}\frac{f_R(x)-f_R(\alpha)}{x-\alpha}\,dx\right]. \tag{6.23}$$

Now, since $f_R(x) = x/(x^2+1)$, we find that

$$\frac{f_R(x)-f_R(\alpha)}{x-\alpha} = \frac{1-\alpha x}{(\alpha^2+1)(x^2+1)},$$

so we may drop the principal-value sign on the right-hand side of Eq. (6.23). Also,

$$\lim_{R\to\infty}\ln\left(\frac{R-\alpha}{R+\alpha}\right) = 0,$$

so Eq. (6.23) becomes

$$\lim_{R\to\infty} P\int_{-R}^{R}\frac{f_R(x)}{x-\alpha}\,dx = \frac{1}{\alpha^2+1}\lim_{R\to\infty}\int_{-R}^{R}\frac{1-\alpha x}{x^2+1}\,dx = \frac{1}{\alpha^2+1}\lim_{R\to\infty}\int_{-R}^{R}\frac{dx}{x^2+1},$$

since $x/(x^2+1)$ is an odd function. Thus

$$\lim_{R\to\infty} P\int_{-R}^{R}\frac{f_R(x)}{x-\alpha}\,dx = \frac{2}{\alpha^2+1}\lim_{R\to\infty}\tan^{-1}R = \frac{\pi}{\alpha^2+1} = -\pi f_I(\alpha),$$

so Eq. (6.19b) is indeed satisfied. We leave it to the reader to show that Eq. (6.19a) is also satisfied. Clearly Eq. (6.21) is very useful in conjunction with Eqs. (6.19a) and (6.19b) because under the assumptions made in deriving these equations, $f(z)$ is differentiable at $z = \alpha$. Thus

$$f_R(\alpha) = \frac{1}{\pi}\int_{-\infty}^{\infty}\frac{f_I(x)-f_I(\alpha)}{x-\alpha}\,dx, \tag{6.24a}$$

$$f_I(\alpha) = -\frac{1}{\pi}\int_{-\infty}^{\infty}\frac{f_R(x)-f_R(\alpha)}{x-\alpha}\,dx, \tag{6.24b}$$

where, in case of any ambiguity,

$$\int_{-\infty}^{\infty} \equiv \lim_{R\to\infty}\int_{-R}^{R}.$$

These methods can sometimes be used to evaluate real definite integrals. For example, consider the function $f(z) = e^{iz}$. This function is analytic everywhere, and if we write $z = Re^{i\theta}$, then $|f(z)| \to 0$ as $R\to\infty$ for all θ such that $0<\theta<\pi$. This is not quite what we used above to show that the integral around a large semicircle of $f(z)/(z-\alpha)$ vanishes (although it would be the same if we had $0\le\theta\le\pi$). However, the reader can show for $f(z) = e^{iz}$ that the contribution from the large semicircle vanishes and Eqs. (6.19a) and (6.19b) are satisfied. In this case, $f_R(x) = \cos x$ and $f_I(x) = \sin x$, so using Eq.

(6.24a), we obtain

$$\cos\alpha = \frac{1}{\pi}\int_{-\infty}^{\infty} \frac{\sin x - \sin\alpha}{x - \alpha}\, dx\,.$$

Since $\sin x - \sin\alpha = 2\sin\frac{1}{2}(x-\alpha)\cos\frac{1}{2}(x+\alpha)$, we see that indeed there is no singularity of the integrand at $x = \alpha$. For the special case $\alpha = 0$, we find that

$$1 = \frac{1}{\pi}\int_{-\infty}^{\infty} \frac{\sin x}{x}\, dx\,,$$

that is,

$$\int_{-\infty}^{\infty} \frac{\sin x}{x}\, dx = \pi\,.$$

From this result, we also obtain by symmetry

$$\int_{0}^{\infty} \frac{\sin x}{x}\, dx = \frac{\pi}{2}\,.$$

Here we see that the oscillations of $\sin x$, when x is large, make the integral converge, even though

$$\int_{1}^{\infty} \frac{1}{x}\, dx$$

diverges. This is analogous to the fact that the *alternating* series

$$\sum_{n=1}^{\infty} (-1)^{n+1}\frac{1}{n} \;(= \ln 2)$$

converges, whereas $\sum_{n=1}^{\infty}(1/n)$ diverges.

6.6 AN INTRODUCTION TO DISPERSION RELATIONS

As mathematical results, the equations derived in the previous section are interesting in their own right. However, a scientist naturally wants to know if there are any physical systems to which these results can be applied. What we shall now show is that under fairly broad physically motivated assumptions, one can find physical quantities which possess the analytic properties necessary for them to satisfy a Hilbert transform relation. In our detailed applications, we will focus our attention on electromagnetic theory, but many of our results will be more general than this.

We begin by considering a physical system for which an input, $I(t)$, is related to a response, $R(t)$, in the following *linear* manner:

$$R(t) = \frac{1}{\sqrt{2\pi}}\int_{-\infty}^{\infty} G(t-t')I(t')\, dt'\,. \tag{6.25}$$

For example, $I(t')$ might be the electric field at a time t', and $R(t)$ might be the resulting polarization field at time t. We have assumed that G depends only on

$t - t'$ because we want the system to respond to a sharp input at t_0, $I(t') = I_0\delta(t' - t_0)$, in the same way it would respond to a sharp input at $t_0 + \tau$, that is, at a time τ later. For the first case, we have

$$R_1(t) = \frac{1}{\sqrt{2\pi}} \int_{-\infty}^{\infty} G(t - t') I_0 \delta(t' - t_0)\, dt' = \frac{1}{\sqrt{2\pi}} I_0 G(t - t_0)\ .$$

For the second case, we have

$$R_2(t) = \frac{1}{\sqrt{2\pi}} \int_{-\infty}^{\infty} G(t - t') I_0 \delta(t' - t_0 - \tau)\, dt' = \frac{1}{\sqrt{2\pi}} I_0 G(t - t_0 - \tau)\ ,$$

or, in other words,

$$R_2(t + \tau) = \frac{1}{\sqrt{2\pi}} I_0 G(t - t_0) = R_1(t)\ .$$

Thus if we shift the input by τ, then we shift the response by τ.

Now, what can we say about $G(\tau)$ on general physical grounds? First we see that an input at t should not give rise to a response at times prior to t, that is, $G(\tau) = 0$ for $\tau < 0$, so

$$R(t) = \int_{-\infty}^{t} G(t - t') I(t')\, dt'\ ,$$

which shows that the response at t is the weighted linear superposition of all inputs *prior* to t. This is the *causality requirement*. The possibility that $G(\tau)$ is singular for any finite τ is excluded since the response from a sharp input, $I(t') = I_0\delta(t' - t_0)$, is

$$R(t) = \frac{1}{\sqrt{2\pi}} I_0 G(t - t_0)\ , \qquad t > t_0\ ,$$

and since on physical grounds we require that this response always be finite, $G(\tau)$ is finite for all τ. Furthermore, we make the physically reasonable assumption that the effect of an input in the *remote* past does not appreciably influence the present. This may be stated as the requirement that $G(\tau) \to 0$ as $\tau \to \infty$ since, from the previous equation, it amounts to the assumption that the response to any impulse dies down after a sufficiently long time (i.e., any system has some dissipative mechanism).

Now consider the Fourier transform of Eq. (6.25). Using the convolution theorem (see Section 5.7), we find that

$$r(\omega) = g(\omega) i(\omega)\ ,$$

where

$$r(\omega) = \frac{1}{\sqrt{2\pi}} \int_{-\infty}^{\infty} R(t) e^{i\omega t}\, dt\ , \qquad g(\omega) = \frac{1}{\sqrt{2\pi}} \int_{-\infty}^{\infty} G(t) e^{i\omega t}\, dt\ ,$$

$$i(\omega) = \frac{1}{\sqrt{2\pi}} \int_{-\infty}^{\infty} I(t) e^{i\omega t}\, dt\ .$$

In electromagnetic theory, where I is the applied electric field $\mathbf{E}$, and R is the polarization field $\mathbf{P}$, it is customary to denote $g(\omega)$ by $\chi(\omega)$, which is referred to as the electric susceptibility. Thus

$$P(\omega) = \chi(\omega)E(\omega)\ .$$

By assuming that $G(\tau)$ satisfies

$$\int_0^\infty |G(\tau)|\, d\tau < \infty\ ,$$

we can guarantee the existence of a bounded $g(\omega)$ for all ω. We may now summarize our physically motivated assumptions on $G(\tau)$:

a) $G(\tau)$ is bounded for all τ;
b) $|G(\tau)|$ is integrable, so $G(\tau) \to 0$ faster than $1/\tau$ as $\tau \to \infty$;
c) $G(\tau) = 0$ for $\tau < 0$.

We may remark that (a) and (b) taken together imply that $G(\tau)$ is square integrable and hence (see Section 9.6) $g(\omega)$ is square integrable.

We now want to show that we can extend $g(\omega)$ into the complex z-plane in such a way that $g(z)$ satisfies the conditions under which we derived the Hilbert transform pair [Eqs. (6.19a) and (6.19b)] of the previous section. First, since $G(\tau) = 0$ for $\tau < 0$, we write

$$g(\omega) = \frac{1}{\sqrt{2\pi}} \int_0^\infty G(t) e^{i\omega t}\, dt\ .$$

We extend this relation into the complex plane by using the definition

$$\begin{aligned} g(z) &\equiv \frac{1}{\sqrt{2\pi}} \int_0^\infty G(t) e^{izt}\, dt \\ &= \frac{1}{\sqrt{2\pi}} \int_0^\infty G(t) e^{i\omega t} e^{-\omega' t}\, dt\ , \end{aligned}$$

where we have written $z = \omega + i\omega'$. We now restrict our attention to the upper half-plane ($\omega' > 0$) where, because of the causality requirement given in assumption (c) above, ($t > 0$ in the above integral), the term $e^{-\omega' t}$ is a *decaying* exponential. For $0 < \theta < \pi$, we have

$$|g(z)| \le \frac{1}{\sqrt{2\pi}} M_G \int_0^\infty e^{-[|z| \sin\theta]t}\, dt\ ,$$

where we have replaced $G(t)$ by its maximum value, M_G [assumption (a) above]. Thus

$$|g(z)| \le \frac{M_G}{\sqrt{2\pi}\, |z| \sin\theta}\ ,$$

which tends to zero when $|z| \to \infty$. For $\theta = 0$ or π, we have

$$g(\omega, \omega' = 0) = \frac{1}{\sqrt{2\pi}} \int_0^\infty G(t) e^{i\omega t}\, dt\ .$$

Since $G(t)$ is square integrable, so is $g(\omega, \omega' = 0)$ as a function of ω (see Section 9.6), and hence $|g(\omega, \omega' = 0)|$ tends to zero as $\omega \to \infty$. Thus in any direction in the upper half-plane, $|g(z)| \to 0$ as $|z| \to \infty$.

Now we want to show that $g(z)$ is analytic in the upper half-plane. Using

$$g(z) = \frac{1}{\sqrt{2\pi}} \int_0^\infty G(t) e^{izt}\, dt = \frac{1}{\sqrt{2\pi}} \int_0^\infty G(t) e^{i\omega t} e^{-\omega' t}\, dt\,, \tag{6.26}$$

we see that for $\omega' > 0$,

$$\frac{d^n g}{dz^n} = \frac{1}{\sqrt{2\pi}} \int_0^\infty G(t) \frac{d^n}{dz^n} e^{izt}\, dt = \frac{i^n}{\sqrt{2\pi}} \int_0^\infty t^n G(t) e^{i\omega t} e^{-\omega' t}\, dt\,, \tag{6.27}$$

since in this case the integrals in both (6.26) and (6.27) are uniformly convergent because of the term $e^{-\omega' t}(\omega' > 0, t > 0)$. Thus $g(z)$ is analytic in the upper half-plane ($\omega' > 0$). However, it is clear that our assumptions on $G(t)$ do not enable us to extend the domain of analyticity to $\omega' \geq 0$. Nevertheless, we can say that $g(z)$ is bounded on the real axis, so the only singularities when $\omega' = 0$ will be of the branch point variety, and even then such branch singularities as $1/\sqrt{z}$ or $\log z$ are excluded by the boundedness requirement. The reader can see by looking back at the derivation of the original pair of Hilbert transform equations [Eqs. (6.19a) and (6.19b)] that it can be modified to include bounded branch point singularities on the real axis by taking a small semicircular detour around any such point. Thus Eqs. (6.19a) and (6.19b) remain unaltered by the presence of such singularities (the branches can always be chosen to avoid the upper half-plane). To eliminate the possibility of these branch singularities, one would have to assume an exponential type falloff of $G(\tau)$ as $\tau \to \infty$. Hence for any $g(z)$ arising from a $G(t)$ which satisfies assumptions (a), (b), and (c), we may write

$$g_R(\omega) = \frac{1}{\pi} P \int_{-\infty}^{\infty} \frac{g_I(\bar{\omega})}{\bar{\omega} - \omega}\, d\bar{\omega}\,, \tag{6.28a}$$

$$g_I(\omega) = -\frac{1}{\pi} P \int_{-\infty}^{\infty} \frac{g_R(\bar{\omega})}{\bar{\omega} - \omega}\, d\bar{\omega}\,. \tag{6.28b}$$

Thus by making a few very reasonable assumptions about the system in question, we can show that the real and imaginary parts of the physical quantity $g(\omega)$ are intimately related to each other for *real* values of the argument by what is essentially a dispersion relation. The key assumption is the *causality requirement*; we may say that causality implies the existence of dispersion relations in the case we have considered. In actual practice, one often restricts the term "dispersion relation" to mean an integral relation between two observable quantities which involves only an integration over values of the argument which are physically meaningful. Thus in Eqs. (6.28a) and (6.28b) only positive frequencies are accessible to experiment, so they are not directly useful as they stand. However, $G(t)$ is real, so we may proceed as follows:

$$g(z) = \frac{1}{\sqrt{2\pi}} \int_0^\infty G(t) e^{izt}\, dt\,,$$

$$g^*(z) = \frac{1}{\sqrt{2\pi}} \int_0^\infty G^*(t) e^{-iz^* t}\, dt = \frac{1}{\sqrt{2\pi}} \int_0^\infty G(t) e^{-iz^* t}\, dt = g(-z^*)\,.$$

Thus we have $g^*(z) = g(-z^*)$, which is often referred to as the *reality condition*. If z is real ($z = \omega$), we find that

$$g_R(\omega) - ig_I(\omega) = g_R(-\omega) + ig_I(-\omega) ,$$

or

$$g_R(\omega) = g_R(-\omega) , \tag{6.29a}$$

$$g_I(\omega) = -g_I(-\omega) , \tag{6.29b}$$

that is, g_R is an *even* function of ω and g_I is an *odd* function of ω. Note that if Eqs. (6.29a) and (6.29b) are satisfied, then the function

$$G(t) = \frac{1}{\sqrt{2\pi}} \int_{-\infty}^{\infty} g(\omega) e^{-i\omega t} \, d\omega$$

is a real function.

Now in Eq. (6.28a), let us write

$$g_R(\omega) = \frac{1}{\pi} P \int_{-\infty}^{0} \frac{g_I(\bar{\omega})}{\bar{\omega} - \omega} \, d\bar{\omega} + \frac{1}{\pi} P \int_{0}^{\infty} \frac{g_I(\bar{\omega})}{\bar{\omega} - \omega} \, d\bar{\omega} .$$

In the first integral, we let $\bar{\omega} \to -\bar{\omega}$. Thus

$$g_R(\omega) = \frac{1}{\pi} P \int_{\infty}^{0} \frac{g_I(-\bar{\omega})}{\bar{\omega} + \omega} \, d\bar{\omega} + \frac{1}{\pi} P \int_{0}^{\infty} \frac{g_I(\bar{\omega})}{\bar{\omega} - \omega} \, d\bar{\omega} .$$

Using Eq. (6.29b), we finally obtain

$$g_R(\omega) = \frac{2}{\pi} P \int_{0}^{\infty} \frac{\bar{\omega} g_I(\bar{\omega})}{\bar{\omega}^2 - \omega^2} \, d\bar{\omega} , \tag{6.30a}$$

and in an identical manner,

$$g_I(\omega) = -\frac{2\omega}{\pi} P \int_{0}^{\infty} \frac{g_R(\bar{\omega})}{\bar{\omega}^2 - \omega^2} \, d\bar{\omega} . \tag{6.30b}$$

These expressions involve only positive, experimentally accessible frequencies. For the electric susceptibility, for example, we have

$$\chi_R(\omega) = \frac{2}{\pi} P \int_{0}^{\infty} \frac{\bar{\omega} \chi_I(\bar{\omega})}{\bar{\omega}^2 - \omega^2} \, d\bar{\omega} , \tag{6.31a}$$

$$\chi_I(\omega) = -\frac{2\omega}{\pi} P \int_{0}^{\infty} \frac{\chi_R(\bar{\omega})}{\bar{\omega}^2 - \omega^2} \, d\bar{\omega} . \tag{6.31b}$$

Equations (6.31a) and (6.31b) were first derived by H. A. Kramers and R. de L. Kronig and are referred to as the Kramers-Kronig dispersion relations.

Now, according to electromagnetic theory, we may write

$$n^2(\omega) = 1 + 4\pi\chi(\omega) ,$$

where $n(\omega)$ is the (complex) index of refraction. Since $\chi(z)$ is analytic in the upper half-z-plane, so is $n^2(z)$, and the function

$$n(z) = \sqrt{1 + 4\pi\chi(z)} \tag{6.32}$$

is also analytic in this region *if* $1 + 4\pi\chi(z)$ has no zeroes in the upper half-plane. If $1 + 4\pi\chi(z)$ vanishes for some $z = z_0$ in the upper half-plane, then according to the reality condition

$$\chi^*(z_0) = \chi(-z_0^*) ,$$

we find that $1 + 4\pi\chi(z)$ also vanishes at $z = -z_0^*$. Since $-z_0^*$ lies in the upper half-plane with the imaginary part equal to Im z_0, there will be a cut in the upper half-plane running from $-z_0^*$ to z_0. In this case, the dispersion relations of Eqs. (6.31a) and (6.31b) would have to be modified. We will assume that $n(z)$ has no zeroes in the upper half-plane and is therefore analytic in this region. However, because of Eq. (6.32), $|n(z)|$ does not tend to zero as $|z| \to \infty$. In fact,

$$n(z) \to 1 \qquad \text{as} \qquad |z| \to \infty$$

in the upper half-plane. This necessitates a modification in the treatment of the term coming from the large semicircle of Fig. 6.12 in obtaining Eq. (6.18). We have, in the case of $n(z)$,

$$\int_{S_R} \frac{n(z)}{z - \alpha}\, dz \xrightarrow{|z| \to \infty} \int_0^\pi \frac{dz}{z} = i\pi$$

in the notation of the previous section (see Fig. 6.12). Thus, for the case of $n(z)$, Eq. (6.18) must be replaced by

$$i\pi n(\omega) = i\pi + P\int_{-\infty}^{\infty} \frac{n(\bar{\omega})}{\bar{\omega} - \omega}\, d\bar{\omega} .$$

Separating real and imaginary parts, we obtain

$$n_R(\omega) = 1 + \frac{1}{\pi} P\int_{-\infty}^{\infty} \frac{n_I(\bar{\omega})}{\bar{\omega} - \omega}\, d\bar{\omega} , \tag{6.33a}$$

$$n_I(\omega) = -\frac{1}{\pi} P\int_{-\infty}^{\infty} \frac{n_R(\bar{\omega})}{\bar{\omega} - \omega}\, d\bar{\omega} . \tag{6.33b}$$

Making use of the reality condition, we can write these equations as

$$n_R(\omega) = 1 + \frac{2}{\pi} P\int_0^{\infty} \frac{\bar{\omega} n_I(\bar{\omega})}{\bar{\omega}^2 - \omega^2}\, d\bar{\omega} , \tag{6.34a}$$

$$n_I(\omega) = -\frac{2\omega}{\pi} P\int_0^{\infty} \frac{n_R(\bar{\omega})}{\bar{\omega}^2 - \omega^2}\, d\bar{\omega} . \tag{6.34b}$$

The quantity $\mu(\omega) \equiv (2\omega/c) n_I(\omega)$, when c is the speed of light in vacuum, is called the *absorption coefficient* and is the inverse of the distance a wave

$$\psi(x, t) = A \exp\left\{ i\omega\left[\frac{n(\omega)}{c} x - t \right] \right\}$$

travels before its intensity drops to $1/e$ of its value at $x = 0$. In terms of μ, we have

$$n_R(\omega) = 1 + \frac{c}{\pi} P\int_0^{\infty} \frac{\mu(\bar{\omega})}{\bar{\omega}^2 - \omega^2}\, d\bar{\omega} . \tag{6.35}$$

Thus the real part of the index of refraction is completely specified by knowing the absorption coefficient at all frequencies! It is from Eq. (6.35) that dispersion relations derive their name. Equation (6.35) relates a substance's absorption to its dispersive effects, i.e., to the way the real index of refraction varies with frequency. It is this variation with frequency which produces the well-known separation (dispersion) of different wavelengths of light by a prism.

In more recent times, the term dispersion relation has continued to be used to denote any relationship between real and imaginary parts of a physical quantity (a scattering amplitude in quantum mechanics, for example) which has the general appearance of a Hilbert transform. Note that since $\mu(\omega)$ must be positive for all frequencies on physical grounds (i.e., we do not expect to find waves which grow in time as they pass through a substance), Eq. (6.35) specifies that

$$n_R(0) = 1 + \frac{c}{\pi} P \int_0^\infty \frac{\mu(\bar{\omega})}{\bar{\omega}^2} d\bar{\omega} , \tag{6.36}$$

so we see that $n_R(0) > 1$. Thus the familiar static dielectric constant ϵ, given by $\epsilon = n^2(0)$ is always greater than unity. Since $\chi_I(0) = 0$, we see that the integral in Eq. (6.36) converges without our using the principal-value technique. Thus using $n(0) = \sqrt{\epsilon}$, we may write Eq. (6.36) as

$$\sqrt{\epsilon} - 1 = \frac{c}{\pi} \int_0^\infty \frac{\mu(\bar{\omega})}{\bar{\omega}^2} d\bar{\omega} .$$

This relation is known as a "sum rule" for the absorption coefficient; it relates the weighted integral over all values of the absorption coefficient to a simple, experimentally accessible constant.

The derivation of the dispersion relation for the index of refraction exhibits certain features which are often encountered in deriving dispersion relations. Namely, it often happens that the quantity in question does not tend toward zero as $|z|$ tends toward infinity, and, furthermore, one is not usually fortunate enough to know the precise behavior of the quantity as $|z|$ tends to infinity, except, for example, to say that it is *bounded* for large values of $|z|$. In this case, we can proceed as follows. Let α_0 be some point on the real axis at which $f(z)$ is analytic. Then the function

$$\frac{f(z) - f(\alpha_0)}{z - \alpha_0} \equiv \phi(z)$$

is not singular at $z = \alpha_0$, and $|\phi(z)| \to 0$ as $|z| \to \infty$. Also, if $f(z)$ is analytic in the upper half-plane, so is $\phi(z)$, and we can write a dispersion relation for $\phi(z)$:

$$i\pi \left[\frac{f(\alpha) - f(\alpha_0)}{\alpha - \alpha_0} \right] = P \int_{-\infty}^{\infty} \frac{f(x) - f(\alpha_0)}{(x - \alpha)(x - \alpha_0)} dx .$$

But

$$\frac{1}{(x - \alpha)(x - \alpha_0)} = \frac{1}{\alpha - \alpha_0} \left[\frac{1}{x - \alpha} - \frac{1}{x - \alpha_0} \right] .$$

Therefore, our dispersion relation takes the form

$$i\pi f(\alpha) = i\pi f(\alpha_0) + (\alpha - \alpha_0) P \int_{-\infty}^{\infty} \frac{f(x)\, dx}{(x-\alpha)(x-\alpha_0)} - f(\alpha_0) P \int_{-\infty}^{\infty} \frac{dx}{x-\alpha}$$
$$+ f(\alpha_0) P \int_{-\infty}^{\infty} \frac{dx}{x-\alpha_0} .$$

According to the work of the previous section, these last two principal-value integrals vanish, so we have just

$$i\pi f(\alpha) = i\pi f(\alpha_0) + (\alpha - \alpha_0) P \int_{-\infty}^{\infty} \frac{f(x)\, dx}{(x-\alpha)(x-\alpha_0)} .$$

Separating the real and imaginary parts, we finally obtain

$$f_R(\alpha) = f_R(\alpha_0) + \frac{1}{\pi} (\alpha - \alpha_0) P \int_{-\infty}^{\infty} \frac{f_I(x)}{(x-\alpha)(x-\alpha_0)}\, dx , \qquad (6.37a)$$

$$f_I(\alpha) = f_I(\alpha_0) - \frac{1}{\pi} (\alpha - \alpha_0) P \int_{-\infty}^{\infty} \frac{f_R(x)}{(x-\alpha)(x-\alpha_0)}\, dx . \qquad (6.37b)$$

Relations of the type of Eqs. (6.37a) and (6.37b) are referred to as *once-subtracted dispersion relations*. For them to be of use in a particular physical problem, one must have a means of determining, say, $f_R(\alpha_0)$ for *some* α_0 in addition to possessing the usual information required by an ordinary dispersion relation of the type of Eq. (6.19a). If the properties of $f(z)$ for large $|z|$ are even "worse" than assumed above [for example, suppose that $|f(z)/z|$ tended toward a nonzero constant as $|z|$ tended toward infinity], then one could introduce more subtraction points, α_1, α_2, etc, in a similar manner.

We have already seen that causality implies certain analyticity properties; we conclude this section by showing that the converse is also true, namely, analyticity implies causality! We will do this by using the analytic properties of $n(z)$ to show that electromagnetic signals will not propagate in any medium faster than the speed of light. Consider a wave front traveling in the x-direction in a dielectric medium with a complex index of refraction, $n(z)$. Assume that at $x = 0$ there is no disturbance before $t = 0$, that is, $\psi(0, t) = 0$ for $t < 0$. We can write a general wave, as $\psi(x, t)$, a superposition of plane waves of all frequencies:

$$\psi(x, t) = \int_{-\infty}^{\infty} \psi(\omega) \exp\left\{ i\omega\left[\frac{n(\omega)}{c} x - t \right]\right\} d\omega .$$

Note that

$$\psi(0, t) = \int_{-\infty}^{\infty} \psi(\omega) e^{-i\omega t}\, d\omega ,$$

so

$$\psi(\omega) = \frac{1}{2\pi} \int_{-\infty}^{\infty} \psi(0, t) e^{i\omega t}\, dt . \qquad (6.38)$$

We may now use Eq. (6.38) to define $\psi(z)$ for z complex:

$$\psi(z) \equiv \frac{1}{2\pi} \int_{0}^{\infty} \psi(0, t) e^{i\omega t} e^{-\omega' t}\, dt ,$$

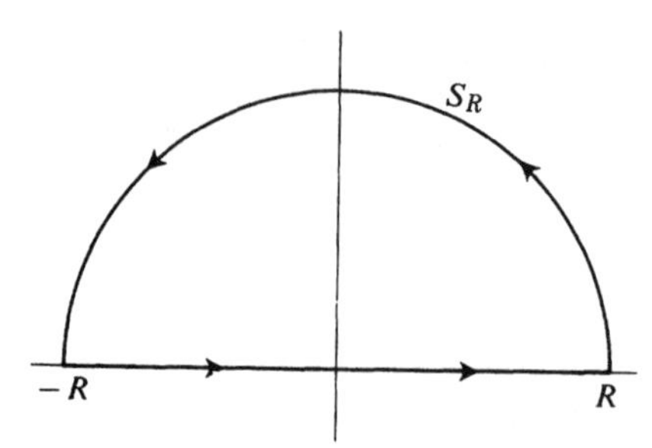

Fig. 6.13 The contour, C, used in evaluating Eq. (6.39). R may become as large as we please.

where we have made use of the fact that $\psi(0, t) = 0$ for $t < 0$ and have written, as usual, $z = \omega + i\omega'$. By our previous arguments, $\psi(z)$ is analytic in the upper half-plane and tends in norm toward zero as $|z|$ tends toward infinity in this region. Now let us consider

$$\oint_C \psi(z) \exp\left[iz\left(\frac{n(z)}{c}x - t\right)\right] dz$$

around the contour shown in Fig. 6.13. Since $\psi(z)$ and $n(z)$ are analytic in the upper half-plane,

$$\oint_C \psi(z) \exp\left[iz\left(\frac{n(z)}{c}x - t\right)\right] dz = 0 .$$

Thus

$$\int_{-R}^{R} \psi(\omega) \exp\left[i\omega\left(\frac{n(\omega)}{c}x - t\right)\right] d\omega + \int_{S_R} \psi(z) \exp\left[iz\left(\frac{n(z)}{c}x - t\right)\right] dz = 0 . \tag{6.39}$$

Calling the second integral in Eq. (6.39) I_R, we have

$$|I_R| \leq \int_0^{\pi} |\psi(z)| \exp\left[-R \sin\theta\left(\frac{n_R}{c}x - t\right) - R\cos\theta \frac{n_I}{c} x\right] R\, d\theta ,$$

where R is some arbitrarily large number. Now as $R \to \infty$, $n_I \to 0$ and $n_R \to 1$, as we have seen above. Thus, under these circumstances,

$$|I_R| \leq 2R \int_0^{\pi/2} |\psi(z)| \exp\left[-R\left(\frac{x}{c} - t\right)\sin\theta\right] d\theta .$$

But $\sin\theta \geq 2\theta/\pi$ for $0 \leq \theta \leq \pi/2$. (To see this, note that $g(\theta) = \sin\theta - 2\theta/\pi$ vanishes at $\theta = 0$ and $\theta = \pi/2$. It is positive at $\theta = \pi/4$, so if it is to become zero or negative inside $[0, \pi/2]$, it must take on a minimum in this region. However, $g''(\theta)$ is always *negative* in $[0, \pi/2]$ so the function cannot possibly take on a minimum value. This can also be seen by drawing a graph of $\sin\theta$ and $2\theta/\pi$.) Therefore, if $(x/c - t)$ is positive,

$$|I_R| \leq 2R \int_0^{\pi/2} |\psi(z)| \exp\left[-\frac{2R}{\pi}\left(\frac{x}{c} - t\right)\theta\right] d\theta .$$

Now $|\phi(z)| \to 0$ as $R \to \infty$, so let us write for large R, $|\phi(z)| \leq aR^{-\lambda} (\lambda > 0)$. Then

$$|I_R| \leq 2aR^{1-\lambda} \int_0^{\pi/2} \exp\left[-\frac{2}{\pi} R \left(\frac{x}{c} - t\right) \theta\right] d\theta ,$$

or

$$|I_R| \leq \frac{\pi a}{R^\lambda[(x/c) - t]} \left[1 - \exp\left\{-R\left(\frac{x}{c} - t\right)\right\}\right].$$

If we assume that $x/c > t$, then as $R \to \infty$, we see that $|I_R| \to 0$. Note that if $x/c < t$, we cannot draw this conclusion because of the *growing* exponential in the above inequality. Thus as $R \to \infty$, Eq. (6.39) becomes

$$\phi(x, t) = \int_{-\infty}^{\infty} \phi(\omega) \exp\left\{i\omega\left[\frac{n(\omega)}{c} x - t\right]\right\} d\omega = 0$$

for $x/c > t$. Thus we reach the satisfying conclusion that if no signal is present at $x = 0$ when $t = 0$, then there will be no signal at $x = x_0 > 0$ before $t = x_0/c$, that is, a signal can propagate with at most the speed of light, c, even though $c/n(\omega)$ may be greater than c, since $n(\omega)$ is known experimentally to become less than 1 at high frequencies [it clearly *cannot* do so at very low frequencies, since we have already shown that $n(0) \geq 1$].

6.7 THE EXPANSION OF AN ANALYTIC FUNCTION IN A POWER SERIES

We now come to one of the most important applications of the Cauchy-Goursat theorem, namely, the possibility of expanding an analytic function in a power series. The main result may be stated as follows:

Laurent's theorem. Let $f(z)$ be analytic throughout the closed annular region between the two circles C_1 and C_2 with common center z_0. Then at each point in this annulus

$$f(z) = \sum_{n=-\infty}^{\infty} A_n (z - z_0)^n , \tag{6.40}$$

with the series converging uniformly in any closed region, R, lying wholly within the annulus. Here

$$A_n = \frac{1}{2\pi i} \oint_C \frac{f(z')}{(z' - z_0)^{n+1}} dz' , \tag{6.41}$$

for $n = 0, \pm 1, \pm 2, \cdots$, and C is any closed contour in the annulus which encircles z_0.

Proof. Consider the contour K enclosing the region R as shown in Fig. (6.14); it may be written symbolically as $K = C_1 + L_1 - C_2 + L_2$. Here we adhere to the convention that simple circular contours are always traversed in a counterclockwise direction. Since the inner circle is traversed in a clockwise direction

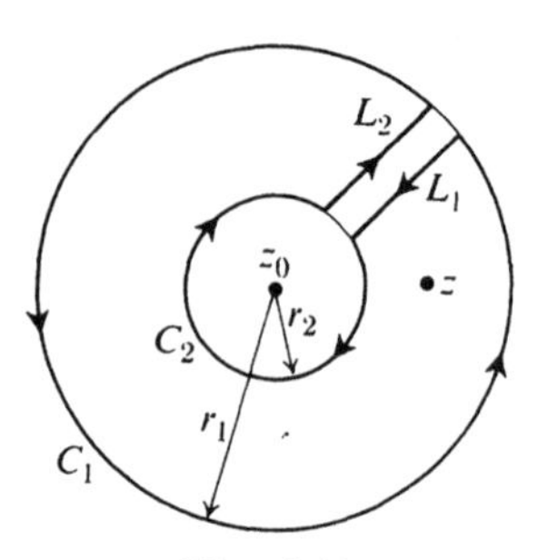

Fig. 6.14

when considered as a part of K, we write it is $-C_2$, where C_2 is a conventionally oriented curve. Since $f(z)$ is analytic within and on K,

$$2\pi i f(z) = \oint_K \frac{f(z')}{z' - z} dz' = \int_{C_1} \frac{f(z')}{z' - z} dz'$$
$$+ \int_{L_1} \frac{f(z')}{z' - z} dz + \int_{-C_2} \frac{f(z')}{z' - z} dz' + \int_{L_2} \frac{f(z')}{z' - z} dz',$$

where $z \in R$. By the same argument used in deriving Cauchy's integral formula we see that if L_1 and L_2 are taken to be arbitrarily close together the integrals along L_1 and L_2 cancel, and

$$f(z) = \frac{1}{2\pi i} \oint_{C_1} \frac{f(z')}{z' - z} dz' - \frac{1}{2\pi i} \oint_{C_2} \frac{f(z')}{z' - z} dz', \tag{6.42}$$

where we have used Eq. (6.11). Equation (6.42) is the starting point for the proof of Laurent's theorem. To proceed further we make use of the identity (for $\alpha \neq 0$)

$$\frac{1}{\alpha - \beta} = \frac{1}{\alpha} + \frac{\beta}{\alpha(\alpha - \beta)}.$$

The second term is just $(\beta/\alpha)[1/(\alpha - \beta)]$, that is, β/α times the originally expanded quantity, so upon iterating N times we get

$$\frac{1}{\alpha - \beta} = \sum_{n=0}^{N} \frac{\beta^n}{\alpha^{n+1}} + \left(\frac{\beta}{\alpha}\right)^{N+1} \frac{1}{\alpha - \beta}. \tag{6.43}$$

Keeping this last equation in mind, let us write Eq. (6.42) in the form

$$f(z) = \frac{1}{2\pi i} \oint_{C_1} \frac{f(z')}{[(z' - z_0) - (z - z_0)]} dz' + \frac{1}{2\pi i} \oint_{C_2} \frac{f(z')}{[(z - z_0) - (z' - z_0)]} dz'.$$

Using Eq. (6.43) we can rewrite this as

$$f(z) = \frac{1}{2\pi i}\sum_{n=0}^{N}\oint_{C_1}\frac{(z-z_0)^n}{(z'-z_0)^{n+1}}f(z')\,dz' + \frac{1}{2\pi i}\oint_{C_1}\left[\frac{z-z_0}{z'-z_0}\right]^{N+1}\frac{f(z')}{z'-z}\,dz'$$
$$+\frac{1}{2\pi i}\sum_{m=0}^{N}\oint_{C_2}\frac{(z'-z_0)^m}{(z-z_0)^{m+1}}f(z')\,dz' + \frac{1}{2\pi i}\oint_{C_2}\left[\frac{z'-z_0}{z-z_0}\right]^{N+1}\frac{f(z')}{z-z'}\,dz'. \tag{6.44}$$

According to by now familiar arguments the contours C_1 and C_2 in the first and third terms of the above equation may be replaced by any contour C about z_0 which is contained in the annulus bounded by C_1 and C_2. Making the change of index $m = n + 1$ in the third term, we obtain

$$f(z) = \sum_{n=-N-1}^{N} A_n(z-z_0)^n + R_N(z)\,,$$

where A_n is given by Eq. (6.41) and

$$R_N(z) = \frac{1}{2\pi i}\oint_{C_1}\left[\frac{z-z_0}{z'-z_0}\right]^{N+1}\frac{f(z')}{z'-z}\,dz' + \frac{1}{2\pi i}\oint_{C_2}\left[\frac{z'-z_0}{z-z_0}\right]^{N+1}\frac{f(z')}{z-z'}\,dz'.$$

If we can show that the magnitude of $R_N(z)$ can be made less than any pre-assigned ϵ for N sufficiently large, where N is independent of z, then the proof will be complete. Letting r_1 be the radius of C_1 and r_2 be the radius of C_2, we have

$$|R_N(z)| \le \frac{1}{2\pi}\left[\int_0^{2\pi}\frac{|z-z_0|^{N+1}}{r_1^{N+1}}\frac{|f(z')|}{|z'-z|}r_1\,d\theta + \int_0^{2\pi}\frac{r_2^{N+1}}{|z-z_0|^{N+1}}\frac{|f(z')|}{|z-z'|}r_2\,d\theta\right].$$

Now we define

$$M_1 \equiv \operatorname*{Max}_{z\in C_1}|f(z)|\,, \qquad M_2 \equiv \operatorname*{Max}_{z\in C_2}|f(z)|\,,$$
$$l_1 \equiv \operatorname*{Max}_{z\in R}|z-z_0|\,, \qquad l_2 \equiv \operatorname*{Min}_{z\in R}|z-z_0|\,,$$
$$d_1 \equiv \operatorname*{Min}_{\substack{z\in R\\ z'\in C_1}}|z'-z|\,, \qquad d_2 \equiv \operatorname*{Min}_{\substack{z\in R\\ z'\in C_2}}|z'-z|\,.$$

Since R is a domain within the annulus bounded by C_1 and C_2,

$$l_1 < r_1\,, \qquad l_2 > r_2\,, \tag{6.45}$$

Thus,

$$|R_N(z)| \le \frac{M_1r_1}{d_1}\left(\frac{l_1}{r_1}\right)^{N+1} + \frac{M_2r_2}{d_2}\left(\frac{r_2}{l_2}\right)^{N+1}. \tag{6.46}$$

But according to Eq. (6.45), $(l_1/r_1) < 1$ and $(r_2/l_2) < 1$, independent of z, so the magnitude of $R_N(z)$ can be made as small as we please for N sufficiently large. Since the bound of Eq. (6.46) is independent of z we have completed the proof of the uniform convergence of the series in Eq. (6.40). This series is known as the *Laurent expansion* of $f(z)$ about the point z_0.

Example 1. Consider the function

$$f(z) = \frac{z^3 + 2z^2 + 4}{(z-1)^3} .$$

We shall obtain the Laurent expansion of $f(z)$ about the singular point $z = 1$. Writing $z \equiv (z-1) + 1$, we have

$$\begin{aligned} f(z) &= \frac{[(z-1)+1]^3 + 2[(z-1)+1]^2 + 4}{(z-1)^3} \\ &= 1 + \frac{5}{z-1} + \frac{7}{(z-1)^2} + \frac{7}{(z-1)^3} . \end{aligned}$$

Clearly this is the Laurent expansion of $f(z)$ about $z = 1$, and the reader is urged to verify this by evaluating the coefficients A_n of Eq. (6.41) which are non-vanishing for this particular $f(z)$.

Example 2. We now consider the Laurent series for a less trivial case, namely the function $\cosh(z + 1/z)$. The hyperbolic cosine is an entire function. Its argument, $z + 1/z$, is analytic everywhere except at the origin, and therefore $\cosh(z + 1/z)$ is analytic everywhere except at the origin. Thus we can pick C_2 to be an arbitrarily small circle about the origin and C_1 to be an arbitrarily large circle about the origin. Then

$$\cosh\left(z + \frac{1}{z}\right) = \sum_{-\infty}^{\infty} A_n z^n , \quad \text{where} \quad A_n = \frac{1}{2\pi i} \oint_C \frac{\cosh(z' + 1/z')}{(z')^{n+1}} dz' ,$$

and C is any closed contour about the origin. Let C be the unit circle. Then $z' = e^{i\theta}$ on C, so the integral becomes

$$\begin{aligned} A_n &= \frac{1}{2\pi} \int_{-\pi}^{\pi} \cosh(2\cos\theta) e^{-in\theta}\, d\theta \\ &= \frac{1}{\pi} \int_0^{\pi} \cosh(2\cos\theta) \cos n\theta\, d\theta . \end{aligned}$$

This integral may be evaluated by using the integral representation of the Bessel function, which will be discussed in Section 6.9. We will obtain A_n by still another method later in this section but will give the result here for the sake of completeness:

$$\begin{aligned} A_{2n} &= \sum_{m=0}^{\infty} \frac{1}{m!(m + 2|n|)!} , \\ A_{2n+1} &= 0 \quad \text{for} \quad n = 0, \pm 1, \pm 2, \cdots \end{aligned} \tag{6.47}$$

Example 3. To illustrate another kind of problem involving Laurent expansions, consider the function $f(z) = (z^2 - 1)^{-1/2}$. According to the discussion in

Section 6.2, this function has branch points at $z = -1$ and $z = +1$. We can choose the cut to run between these two points along the real axis, defining a single-valued branch by

$$f(z) = (\rho_1\rho_2)^{-1/2} e^{-i\phi_1/2} e^{-i\phi_2/2},$$

where $0 \le \phi_1 < 2\pi$, $0 \le \phi_2 < 2\pi$, $\rho_1 = |z - 1|$, $\rho_2 = |z + 1|$ and $(\rho_1 \rho_2)^{-1/2}$ denotes the positive square root of $1/\rho_1\rho_2$ (see Fig. (6.15)). With these definitions $f(z)$ is analytic everywhere except on the real axis between $z = -1$ and $z = +1$. Thus, if we take C_2 to be a circle centered at $z = 0$ with radius infinitesimally larger than 1 and C_1 to be a circle of arbitrarily large radius, we can obtain a Laurent expansion of $f(z)$ in the annulus defined by C_1 and C_2. As usual

$$(z^2 - 1)^{-1/2} = \sum_{n=-\infty}^{\infty} A_n z^n,$$

where

$$A_n = \frac{1}{2\pi i}\oint_C \frac{(z'^2 - 1)^{-1/2}}{(z')^{n+1}} dz'.$$

For $n \ge 0$ we choose C to be a circle of arbitrarily large radius; it is clear that for this contour the relevant integral vanishes, so $A_n = 0$ for $n \ge 0$. To deal with the case of negative n, we choose as contour any circle with radius greater than 1 (see Fig. 6.15). According to our results on path independence, this contour can be deformed into a "dogbone" contour as shown in Fig. 6.15. The

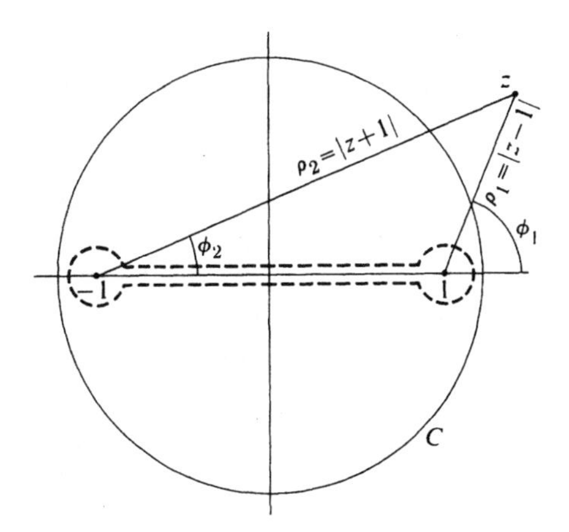

Fig. 6.15 The circular contour C is the starting point for the evaluation of the Laurent series coefficients, A_n of $(z^2 - 1)^{-1/2}$ when n is negative. ρ_1, ρ_2, ϕ_1, and ϕ_2 define the single-valued branch of this function, and the dashed "dogbone" contour is the deformation of C which enables us to evaluate the coefficients.

contribution to A_n from the infinitesimally small circles at the ends of the bone is vanishingly small, so we have simply

$$A_n = \frac{1}{2\pi i}\left[\int_1^{-1} \frac{1}{\sqrt{1-x^2}} e^{-i\pi/2} x^m\, dx + \int_{-1}^{1} \frac{1}{\sqrt{1-x^2}} e^{-i(\pi+2\pi)/2} x^m\, dx\right]$$

$$= \frac{1}{\pi}\int_{-1}^{1} \frac{x^m}{\sqrt{1-x^2}}\, dx\,,$$

where $m = -n - 1$ (note that for the range of n in question m is never negative). Clearly, for m odd (n even) A_n vanishes. For m even (n odd) the integral is elementary; we have

$$A_n = \frac{2}{\pi}\int_0^1 \frac{x^m}{\sqrt{1-x^2}}\, dx = \frac{2}{\pi}\int_0^{\pi/2} \sin^m \theta\, d\theta = \frac{m!}{2^m[(m/2)!]^2}\,.$$

Thus,

$$(z^2 - 1)^{-1/2} = \sum_{m=0}^{\infty}{}' \frac{m!}{2^m[(m/2)!]^2} \frac{1}{z^{m+1}}\,,$$

where the prime indicates that only even values of m are included in the summation. Setting $m = 2\nu$, we have finally

$$(z^2 - 1)^{-1/2} = \sum_{\nu=0}^{\infty} \frac{(2\nu)!}{4^\nu(\nu!)^2} \frac{1}{z^{2\nu+1}} = \frac{1}{z} + \frac{1}{2}\frac{1}{z^3} + \frac{3}{8}\frac{1}{z^5} + \frac{5}{16}\frac{1}{z^7} + \cdots\,.$$

This result is not surprising: it is just what we would have obtained by expanding $(z^2 - 1)^{-1/2}$ using real variable techniques. Note that in keeping with our definition of the single-valued branch of $(z^2 - 1)^{-1/2}$ the Laurent series gives a value of the square root which is positive on the positive real axis and negative on the negative real axis.

Until now we have spoken of singularities as points where the function in question is not analytic, but have made no attempt to classify singular points. There are differences, however, as we might expect if we look at the singular point $z = 0$ of the three functions $1/z$, $1/z^2$, and $\cosh(z + 1/z)$. We feel that the singularity of $1/z^2$ is a little worse than that of $1/z$ and that the singularity of $\cosh(z + 1/z)$ is terrible. A way of classifying the singularities is provided by the Laurent series of each function.

Suppose that $f(z)$ is analytic in a domain R except at $z = z_0$. Expand $f(z)$ in a Laurent series about z_0 inside R:

$$f(z) = \sum_{n=0}^{\infty} a_n(z - z_0)^n + \sum_{n=1}^{\infty} b_n \frac{1}{(z - z_0)^n}\,.$$

If $b_n = 0$ for $n = N + 1, N + 2, \cdots, \infty$, that is, if the series of negative powers of $(z - z_0)$ terminates with the Nth power, then $f(z)$ is said to have a *pole of order* N at z_0. The functions $1/z$ and $1/z^2$ have poles of order 1 and 2 respectively. But the Laurent expansion of $\cosh(z + 1/z)$ has an *infinite* number of negative powers of z; so does the Laurent expansion of $e^{1/z}$ as we shall soon see. The singularities of these functions at the origin are called *essential* singularities. They cannot be removed by multiplying the function by some finite power of z, as is the case with the poles of $1/z$ and $1/z^2$. Notice, however, that the presence of an infinite number of negative powers of z in a Laurent series does not guarantee the presence of an essential singularity. Example 3 above shows that a branch cut singularity can also give rise to an infinite number of negative powers of z in a Laurent series. Only if we know that the singularity in question is confined within an arbitrarily small region can we conclude that an infinite number of negative powers of z implies the existence of an essential singularity. A function with no essential singularities in a region—although it may have poles there—is said to be a *meromorphic* function.

A particularily important consequence of Laurent's theorem arises when there are *no singularities* contained within the inner circle C_2, that is, if $f(z)$ is analytic everywhere inside C_1. Using Laurent's theorem, we can write

$$f(z) = \sum_{n=-\infty}^{\infty} A_n (z - z_0)^n ,$$

where

$$A_n = \frac{1}{2\pi i} \oint_C \frac{f(z')}{(z' - z_0)^{n+1}} \, dz' . \tag{6.48}$$

But if $f(z')$ is analytic inside C_1, then for $n = -1, -2, \cdots$, the integrand in Eq. (6.48) is analytic within and on C (since C lies within C_1). Therefore, according to the Cauchy-Goursat theorem, $A_n = 0$ for $n \leqslant -1$. Furthermore, for $n \geqslant 0$, we have, using Eq. (6.17),

$$\frac{1}{2\pi i} \oint_C \frac{f(z')}{(z' - z_0)^{n+1}} \, dz' = \frac{1}{n!} f^{(n)}(z_0) ,$$

where $f^{(n)}(z_0)$ denotes the nth derivative of $f(z)$ at $z = z_0$ and $f^{(0)}(z) \equiv f(z_0)$. Thus,

$$f(z) = \sum_{n=0}^{\infty} \frac{1}{n!} f^{(n)}(z_0)\,(z - z_0)^n . \tag{6.49}$$

This result is known as Taylor's theorem, and the series in Eq. (6.49) is called Taylor's series for $f(z)$ about z_0. This very important result may be stated formally as follows:

Taylor's Theorem. If $f(z)$ is analytic at all points interior to a circle C centered about z_0 then in any closed region contained wholly inside C

$$f(z) = \sum_{n=0}^{\infty} \frac{1}{n!} f^{(n)}(z_0)\,(z - z_0)^n , \tag{6.50}$$

and the series converges uniformly.

We can now express the analytic function $f(z)$ as a uniformly convergent series of analytic functions (it is a simple matter to show that *any* uniformly convergent series of analytic functions is analytic). Because of the uniformity of the convergence, it can readily be shown that the integral of $f(z)$ along any path in the region of convergence of the power series in Eq. (6.49) can be evaluated by integrating the series term-by-term (see Problem 6.28). Using Eq. (6.17), we see that this also means that the derivatives of $f(z)$ can be evaluated similarly by term-by-term differentiation. Thus Eq. (6.49) provides $f(z)$ and all its derivatives throughout the region of convergence.

Examples

1. $e^z = \sum_{n=0}^{\infty} \frac{z^n}{n!}, \qquad \text{for} \quad |z| < \infty .$

2. $\sin z = \sum_{n=0}^{\infty} (-1)^n \frac{z^{2n+1}}{(2n+1)!}, \qquad \text{for} \quad |z| < \infty .$

3. $(1-z)^{-1} = \sum_{n=0}^{\infty} z^n, \qquad \text{for} \quad |z| < 1 .$

These series all reduce to familiar results for real values of z.

It is often possible to obtain the *Laurent* expansion of some function by using a related Taylor series. Thus, Example 1, above, gives us

$$e^{1/z} = \sum_{n=0}^{\infty} \frac{1}{n!\, z^n} \qquad (|z| > 0) .$$

This show clearly that $e^{1/z}$ has an essential singularity at $z = 0$. A slightly more substantial example is provided by the Taylor expansion of cosh z. It is a simple matter to show that

$$\cosh z = \sum_{n=0}^{\infty} \frac{z^{2n}}{(2n)!}$$

For $|z| < \infty$. Thus for $|z| > 0$ we have

$$\cosh (z + 1/z) = \sum_{n=0}^{\infty} \frac{1}{(2n)!} (z + 1/z)^{2n} .$$

Using the binomial theorem, we find

$$\cosh (z + 1/z) = \sum_{n=0}^{\infty} \sum_{m=0}^{2n} \frac{z^{2(n-m)}}{(2n-m)!\, m!} = \sum_{n=0}^{\infty} \sum_{\mu=-n}^{n} \frac{z^{2\mu}}{(n+\mu)!\,(n-\mu)!} .$$

If in Eq. (6.50) we interchange the order of summation and then make the change of variable $\nu = n - \mu$, we obtain

$$\cosh (z + 1/z) = \sum_{\mu=-\infty}^{\infty} z^{2\mu} \left[\sum_{\nu=0}^{\infty} \frac{1}{(\nu + 2|\mu|)!\, \nu!} \right],$$

in agreement with the result of Eqs. (6.47).

The region in which a Taylor series expansion of a function about a point z_0 is valid is limited by the presence of a singularity of $f(z)$. We have seen that if this singularity is at the point α_0, the Taylor's expansion holds within a circular region of radius $|\alpha_0 - z_0|$ about z_0. Now the Taylor expansion gives the values of the analytic function and all its derivatives at every point in the region of analyticity. In particular, these quantities are known at a point z_1 (see Fig. 6.16) near the region's border. We may now use the point z_1 as a point about which to expand the function in another Taylor series. We can do this because $f^{(n)}(z_1)$ is known for all n from the first Taylor expansion about z_0. The radius of convergence of this second series expansion about z_1 is determined by the distance from z_1 to the nearest singularity. Continuing in this way (Fig. 6.16), we can determine the function throughout the entire plane except the points at which it is singular. To get started, we need only know the values of the analytic function in some *region*, however small. This process is called *analytic continuation*. It is as though a paleontologist could reconstruct a whole dinosaur from the fossil remains of a single toenail.

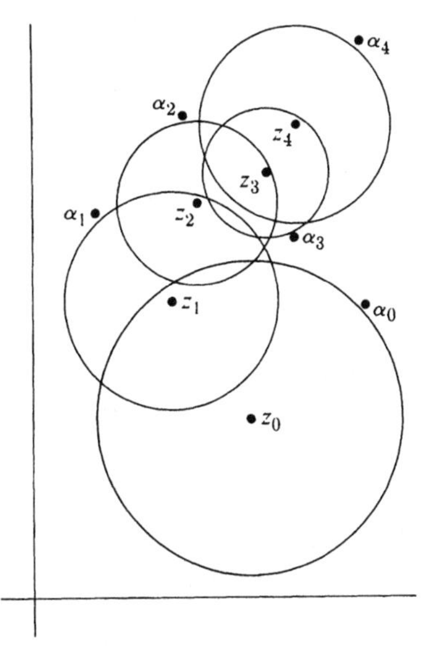

Fig. 6.16 A sequence of Taylor expansions which analytically continue a function originally known in some region around z_0. The first expansion about z_0 is limited in its radius of expansion by the singularity at α_0. The next Taylor expansion about z_1 (inside the first expansion's radius of convergence) is limited by a second singularity at α_1, and so on.

The process of analytic continuation is the best demonstration of the rigid inner structure of *analytic* functions. Again we see how interdependent are the values of an analytic function: its values in *any* region on the plane determine its values *everywhere* that it is analytic. This blueprint for the construction of all the values of an analytic function also demonstrates the necessity of the property that all derivatives of analytic functions are analytic. For if this were not the case, we could not continue a function analytically by a chain of Taylor expansions. To do this requires that we be able to approximate the function arbitrarily well, and this in turn means that all its derivatives must exist so the Taylor series can be extended to achieve arbitrary accuracy.

An immediate consequence of the fact that an analytic function may be continued is that a function which is zero along any curve in a region R, throughout which it is analytic, must be zero everywhere in R. This means that if two functions $f(z)$ and $g(z)$ are equal on any curve inside a simply-connected region R in which both are single-valued and analytic, then they are equal everywhere in R, since $f(z) - g(z)$ is zero on the curve and therefore throughout R. It will be recalled that the analytic functions e^z, $\sin z$, $\sinh z$, etc., were defined for complex arguments in a way that reduced to the usual definitions on the real axis. The principle discussed above shows that these functions could not have been defined otherwise away from the real axis and still be analytic. Furthermore, this explains why all the familiar identities satisfied by the functions for real values continue to hold throughout the complex plane.

It should also be noted that the process of analytic continuation is closely related to the problem of multivalued functions and Riemann surfaces discussed in Section 6.2. Suppose that we are given an analytic power series for a function in some region, for example,

$$w(z) = \sqrt{z} = \sqrt{1 - (1 - z)} = 1 - \tfrac{1}{2}(1 - z) - \tfrac{1}{8}(1 - z)^2 + \cdots$$

in the region consisting of the interior of a circle of unit radius about the point $z = 1$. If we try to continue such a function along certain paths in the z-plane (in the case above, any path enclosing the origin), it may happen that upon returning to the original region of definition, we do *not* return to the original values of the function. This leads in a natural way to the construction of Riemann surfaces of the type discussed in Section 6.2.

6.8 RESIDUE THEORY—EVALUATION OF REAL DEFINITE INTEGRALS AND SUMMATION OF SERIES

There is really nothing fundamentally new in this section. All the theorems have been proved; here we just apply them in certain ways to determine the values of some real definite integrals.

> **The Residue Theorem.** The integral of $f(z)$ around a closed contour C containing a finite number n of singular points of $f(z)$ equals the sum of n integrals of $f(z)$ about n circles, each enclosing one (and only one) of the n singular points.

Proof. If we apply Cauchy's theorem to the region shown in Fig. 6.17, we obtain

$$\int_C f(z)\,dz + \int_{C_1} f(z)\,dz + \int_{C_2} f(z)\,dz + \cdots + \int_{C_n} f(z)\,dz = 0.$$

Note that the contours C_j are traversed clockwise in Fig. (6.17). It follows that

$$\oint_C f(z)\,dz = 2\pi i \sum_{j=1}^{n} R_j\,,$$

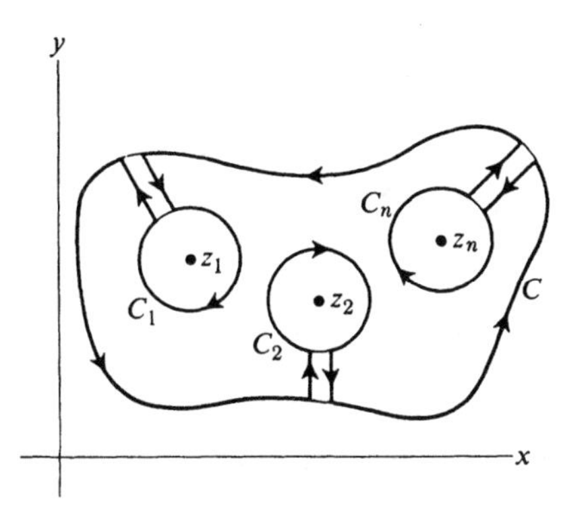

Fig. 6.17

where $R_j = (1/2\pi i)\oint_{C_j} f(z)\,dz$ is called the *residue* at the point z_j. In this equation for R_j the integral sign has its conventional meaning, that is, the contours C_j are traversed counterclockwise.

There is nothing new in this theorem except the name "residue." To compute the residues we shall often use Cauchy's integral formula or the formula for the derivative of an analytic function derived from Cauchy's formula by differentiation.

There is another way to compute residues, however, which is sometimes useful. We expand $f(z)$ in a Laurent series about the singular point z_0:

$$f(z) = \sum_{n=0}^{\infty} a_n(z - z_0)^n + \sum_{n=1}^{\infty} b_n(z - z_0)^{-n} .$$

Now $b_1 = (1/2\pi i)\oint_{C_0} f(z)\,dz = R_0$. It is instructive to check this by integrating both sides of the Laurent series expansion about a circle which includes z_0 but no other singularities. Interchanging summation and integration, which is permissible because these series converge uniformly, we get

$$\oint_{C_0} f(z)\,dz = \sum_{n=0}^{\infty} a_n \oint_{C_0} (z - z_0)^n\,dz + \sum_{n=1}^{\infty} b_n \oint_{C_0} \frac{dz}{(z - z_0)^n} .$$

The first integral vanishes for all n because $(z - z_0)^n$ is analytic. For $n = 1$, the second integral gives $2\pi i b_1$ by Cauchy's integral formula. For $n > 1$, we write $b_n \equiv g(z)$; then by Eq. (6.17)

$$\oint_{C_0} \frac{b_n\,dz}{(z - z_0)^n} = \oint_{C_0} \frac{g(z)\,dz}{(z - z_0)^n} = \frac{2\pi i}{(n-1)!} g^{(n-1)}(z_0) = 0 ,$$

since $g(z) = b_n = \text{const}$. Thus $(1/2\pi i)\oint_{C_0} f(z)\,dz = b_1 =$ the residue at z_0, so the residue at a point can be found by expanding the function about the point in a Laurent series and picking out the coefficient of the term in $(z - z_0)^{-1}$. There is nothing mysterious about this; it is a simple consequence of Cauchy's integral formula.

Examples.

1. $I = \oint_C e^{1/z}\, dz$, C = unit circle about the origin.

$$e^{1/z} = \sum_{n=0}^{\infty} \frac{1}{n!\, z^n}, \qquad \text{so } b_1 = 1 .$$

Therefore $I = 2\pi i$.

2. $$I = \oint_C \frac{3z^2 + 2}{z(z + 1)}\, dz ,$$

where C is the circle: $|z| = 3$. The quickest way to do this integral is to write it as

$$\oint_C = \oint_{C_0} + \oint_{C_1} = \oint_{C_0} \frac{(3z^2 + 2)/(z + 1)}{z} dz + \oint_{C_1} \frac{(3z^2 + 2)/z}{(z + 1)} dz ,$$

where C_0 is a little circle inside C enclosing $z = 0$, but not $z = -1$, and C_1 is a little circle inside C enclosing $z = -1$, but not $z = 0$. Then using Cauchy's integral formula to evaluate both integrals, we obtain $I = 2\pi i(2 - 5) = -6\pi i$.

3. $$I = \oint_C \frac{3z + 2}{z(z + 1)^3}\, dz ,$$

where C is the circle $|z| = 3$. We break up the integral as before:

$$\begin{aligned} I &= \oint_{C_0} \frac{(3z + 2)/(z + 1)^3}{z} dz + \oint_{C_1} \frac{(3z + 2)/z}{(z + 1)^3} dz \\ &= 2\pi i[2] + 2\pi i \left[\frac{f''(z)}{2!}\right]_{z=-1} \end{aligned}$$

where $f(z) = (3z + 2)/z$. Therefore $f''(-1) = -4$, so the residue at $z = -1$ is -2, thus $I = 2\pi i(2 - 2) = 0$.

In this last example we evaluated the residue of a function of the form $f(z)/(z - z_0)^n$ at the singular point z_0 by using Cauchy's formula for the derivative of an analytic function. In general, if $f(z)$ is analytic within and on some contour C surrounding z_0, then the residue of the function $f(z)/(z - z_0)^n$ at z_0 is

$$\frac{1}{2\pi i} \oint_C \frac{f(z)}{(z - z_0)^n} dz = \frac{f^{n-1}(z_0)}{(n - 1)!} .$$

An instructive way to view this formula is to expand $f(z)$ in a Taylor series about z_0. This is possible because $f(z)$ is analytic in some region about z_0, since the singularities are isolated. Then

$$f(z) = f(z_0) + f'(z_0)(z - z_0) + \cdots + \frac{f^{n-1}(z_0)(z - z_0)^{n-1}}{(n - 1)!} + \cdots .$$

If we now form the quantity $f(z)/(z - z_0)^n$ by dividing both sides by $(z - z_0)^n$, the coefficient of the term in $1/(z - z_0)$, which is just the residue of $f(z)/(z - z_0)^n$, is $f^{(n-1)}(z_0)/(n - 1)!$, in agreement with the above.

We now use these techniques to evaluate various definite integrals.

a) $\int_0^{2\pi} R(\cos\theta, \sin\theta)\, d\theta$, where R is a rational function, that is,

$$R = \frac{a_1\cos\theta + a_2\sin\theta + a_3\cos^2\theta + \cdots}{b_1\cos\theta + b_2\sin\theta + b_3\cos^2\theta + b_4\sin^2\theta + \cdots}.$$

To evaluate this integral, let $z = e^{i\theta}$, so that

$$\cos\theta = \frac{1}{2}\left(z + \frac{1}{z}\right), \qquad \sin\theta = \frac{1}{2i}\left(z - \frac{1}{z}\right),$$

and $d\theta = -i(dz/z)$. The integral becomes

$$-i\oint_C R\left[\frac{1}{2}\left(z + \frac{1}{z}\right), \frac{1}{2i}\left(z - \frac{1}{z}\right)\right]\frac{dz}{z},$$

where C is the unit circle.

Example.

$$I = \int_0^{2\pi} \frac{d\theta}{a + \cos\theta}, \qquad a > 1.$$

$$I = -i\oint_{|z|=1} \frac{1}{\left(a + \frac{z}{2} + \frac{1}{2z}\right)}\frac{dz}{z} = -2i\oint_{|z|=1} \frac{dz}{z^2 + 2az + 1}.$$

The denominator can be factored into $(z - \alpha)(z - \beta)$, where

$$\alpha = -a + (a^2 - 1)^{1/2}, \qquad \beta = -a - (a^2 - 1)^{1/2}.$$

Since $a > 1$, it follows readily that $|\alpha| < 1$ and $|\beta| > 1$. Thus the integrand has one singularity, at $z = \alpha$, within the unit circle, and

$$I = -2i(2\pi i)\frac{1}{\alpha - \beta} = \frac{2\pi}{(a^2 - 1)^{1/2}}.$$

Next we consider integrals of the form:

b) $\int_{-\infty}^{\infty} R(x)\, dx$, where $R(x)$ is a rational function (i.e., a ratio of two polynomials), without poles on the real axis.

If there are no poles on the real axis, then this integral exists if the degree of the denominator of $R(x)$ is at least two units higher than the degree of the numerator. This means that $|R(z)| \to 1/|z^2|$ as $|z| \to \infty$. Now

$$\oint_C R(z)\, dz = \int_{-\rho}^{\rho} R(x)\, dx + \int_{\text{semicircle}} R(z)\, dz$$

(see Fig. 6.18). As $\rho \to \infty$, the closed contour C encloses all the singularities of $R(z)$ in the upper half-plane, so

$$\oint_C R(z)\, dz = 2\pi i \sum_{y>0} \text{Res}\, R(z),$$

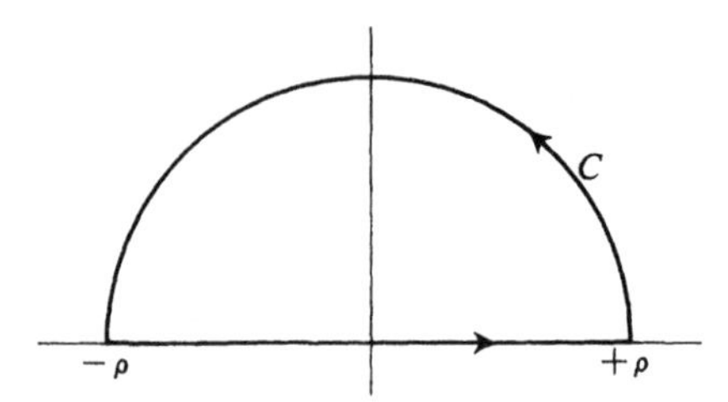

Fig. 6.18

where this notation means to form the sum of all the residues of $R(z)$ in the upper half-plane. Now in the limit of large $|z|$,

$$\left|\int_{\text{semicircle}} R(z)\, dz\right| \leq \int \frac{|\text{const}|\, \rho\, d\theta}{\rho^2} = \frac{|\text{const}|}{\rho},$$

which goes to zero as $\rho \to \infty$. Thus

$$\oint_C R(z)\, dz = \int_{-\infty}^{\infty} R(x)\, dx = 2\pi i \sum_{y>0} \operatorname{Res} R(z).$$

When we say that an integral over a "contour at infinity" vanishes, we really mean, of course, that given any $\epsilon > 0$ there exists a *finite* contour C such that $\left|\oint_C f(z)\, dz\right| < \epsilon$. If this is so, we say that the integral over the contour at ∞ is zero; all contours are really finite.

Example.

$$I = \int_{-\infty}^{\infty} \frac{dx}{(x^2 + 1)^2}.$$

To find the zeros of the denominator in the upper half-plane, we must solve $z^2 = -1$. We get $z = \pm i$. Only the root $+i$ is in the upper half-plane. Therefore

$$I = \oint \frac{dz}{(z + i)^2 (z - i)^2} = \frac{2\pi i}{1!} \frac{d}{dz}\left(\frac{1}{(z + i)^2}\right)\Bigg]_{z=i}$$

$$= (2\pi i) \cdot (-2) \cdot \frac{1}{(z + i)^3}\Bigg]_{z=i} = \frac{\pi}{2}.$$

c) Another very important class of integrals is $\int_{-\infty}^{\infty} R(x) e^{ix}\, dx$. This is the Fourier integral of the rational function $R(x)$. Its real and imaginary parts determine the integrals:

$$\int_{-\infty}^{\infty} R(x) \cos x\, dx \qquad \text{and} \qquad \int_{-\infty}^{\infty} R(x) \sin x\, dx.$$

We retain the assumption that there are no poles on the real axis. Here, too, we consider the integral $\oint_C R(z) e^{iz}\, dz$ over a semicircle. Since $|e^{iz}| = e^{-y} \leq 1$

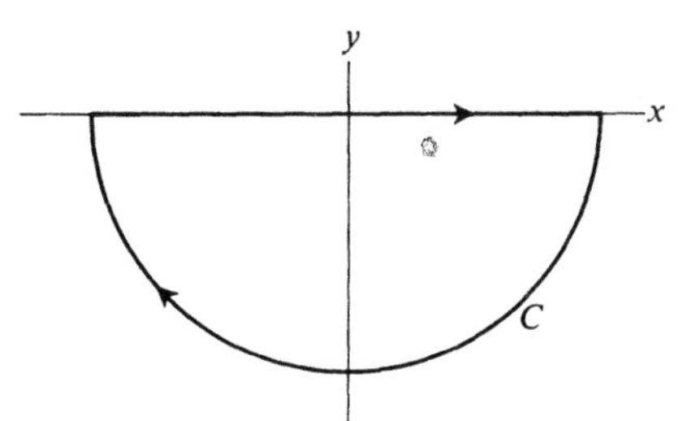

Fig. 6.19

for any point in the upper half-plane, this integral will exist if the rational function $R(z)$ has a zero of at least order two at infinity. Then, just as before, the integral over the infinite semicircle vanishes, leaving

$$\oint_C R(z)e^{iz}\,dz = \int_{-\infty}^{\infty} R(x)e^{ix}\,dx = 2\pi i \sum_{y>0} \text{Res}\,[R(z)e^{iz}]\,.$$

Example.

$$I = \int_{-\infty}^{\infty} \frac{e^{ikr}}{k^2+\mu^2}\,dk\,.$$

If $r > 0$, then letting $x = rk$, we find that the integral becomes

$$I = r\int_{-\infty}^{\infty} \frac{e^{ix}}{x^2+(\mu r)^2}\,dx = r\oint \frac{e^{iz}}{z^2+(\mu r)^2}\,dz = 2\pi i r \sum_{y>0} \text{Res}\left(\frac{e^{iz}}{z^2+(\mu r)^2}\right)$$

$$= 2\pi i r \left.\frac{e^{iz}}{z+i\mu r}\right]_{z=i\mu r} = \pi\frac{e^{-\mu r}}{\mu}\,.$$

Note that

$$I = \int_{-\infty}^{\infty} \frac{\cos kr}{k^2+\mu^2}\,dk$$

and hence the result above is actually independent of the sign of r. Therefore

$$\int_{-\infty}^{\infty} \frac{e^{ikr}}{k^2+\mu^2}\,dk = \pi\frac{e^{-\mu|r|}}{\mu}\,.$$

Another way to see this is to compute the integral assuming $r < 0$. Then we must take the contour in the *lower* half-plane, since $|e^{ikr}| = e^{-(\text{Im}\,k)r}$, which is bounded only for Im $(k) < 0$, if $r < 0$. As before, for a large enough semicircle (Fig. 6.19),

$$\oint_C = \int_{-\infty}^{\infty} = -2\pi i \sum_{y<0} \text{Res}\,.$$

The minus sign arises because the contour C is traversed clockwise. Thus

$$I = -2\pi i r \left.\frac{e^{iz}}{z-i\mu r}\right]_{z=-i\mu r} = \frac{\pi e^{\mu r}}{\mu}\,,$$

which for $r < 0 = \pi e^{-\mu|r|}/\mu$. Here the residue was computed at the only pole in the lower half-plane, $z = -i\mu r$.

d) There arise integrals of the form $\int_{-\infty}^{\infty} R(x)e^{i\alpha x}\,dx$, where the rational function $R(x)$ has a zero of order one at infinity. The integrals

$$\int_{-\infty}^{\infty} R(x)\cos\alpha x\,dx \qquad \text{and} \qquad \int_{-\infty}^{\infty} R(x)\sin\alpha x\,dx$$

are the real and imaginary parts, respectively, of this integral.

From the preceding discussion, it is not clear that these integrals exist. Jordan's lemma, which we now prove, tells us that they do. Again we assume that there are no poles on the real axis.

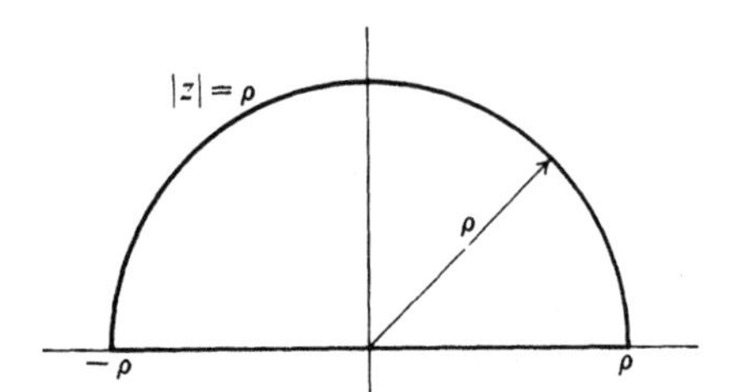

Fig. 6.20

Jordan's Lemma. If as $\rho \to \infty$ (in Fig. 6.20), $|R(z)| \to 0$ uniformly in θ for $0 < \theta < \pi$, then $\lim_{\rho\to\infty} \int_{|z|=\rho} R(z)e^{i\alpha z}\,dz = 0$, for $\alpha > 0$.

Proof. We have almost proved this result in the section on dispersion relations. There, however, the analyticity of $R(z)$ in the upper half-plane was used. Here we use the hypothesis on the uniformity of the approach of $|R(z)|$ to zero. Let $M(\rho)$ be the maximum of $|R(z)|$ on the semicircle $|z| = \rho$. Then the statement that $|R(z)| \to 0$ uniformly means that $|R(z)| \leq M(\rho)$, where $\lim_{\rho\to\infty} M(\rho) = 0$, *independent of θ.*

Let

$$I = \int_{|z|=\rho} R(z)e^{i\alpha z}\,dz\,.$$

Then

$$|I| \leq M(\rho)\int_0^{\pi} |e^{i\alpha(\rho\cos\theta + i\rho\sin\theta)}|\,|\rho i e^{i\theta}|\,d\theta$$
$$= \rho M(\rho)\int_0^{\pi} e^{-\alpha\rho\sin\theta}\,d\theta\,.$$

Treating this integral exactly as in Section 6.6, we obtain

$$|I| < \pi M(\rho)\,\frac{(1 - e^{-\alpha\rho})}{\alpha} \to 0 \qquad \text{as } \rho \to \infty\,,$$

since $\alpha > 0$, and $M(\rho) \to 0$ as $\rho \to \infty$. Thus we arrive at the formula

$$\int_{-\infty}^{\infty} R(x)e^{i\alpha x}\,dx = 2\pi i \sum_{y>0} \text{Res}\,[R(z)e^{i\alpha z}]$$

for $\alpha > 0$. For $\alpha < 0$, the derivation works only in the lower half-plane $(y < 0)$, and changes in the above formula must be made accordingly.

Example.

$$\begin{aligned} I &= \int \frac{e^{i\mathbf{k}\cdot\mathbf{r}}}{k^2 + \mu^2}\, d\mathbf{k} \\ &= \int_0^{2\pi} k\, d\phi \int_0^{\pi} k \sin\theta\, d\theta \int_0^{\infty} dk\, \frac{e^{ik|r|\cos\theta}}{k^2 + \mu^2} \\ &= 2\pi \int_{-1}^{1} dt \int_0^{\infty} dk\, \frac{k^2 e^{ik|r|t}}{k^2 + \mu^2} \\ &= \frac{2\pi}{i|r|} \int_0^{\infty} dk\, \frac{k(e^{ik|r|} - e^{-ik|r|})}{k^2 + \mu^2} \\ &= \frac{2\pi}{i|r|} \int_{-\infty}^{\infty} dk\, \frac{ke^{ik|r|}}{k^2 + \mu^2}. \end{aligned}$$

Jordan's lemma tells us that this integral for complex k vanishes over the infinite semicircle. Therefore

$$I = \frac{2\pi}{i|r|}\, 2\pi i \sum_{y>0} \operatorname{Res} \frac{ke^{ik|r|}}{k^2 + \mu^2} = \frac{4\pi^2}{|r|} \frac{i\mu e^{i\cdot i\mu|r|}}{2i\mu} = \frac{2\pi^2}{|r|} e^{-\mu|r|}, \tag{6.51}$$

since there is only one pole in the upper half-plane $(k = +i\mu)$.

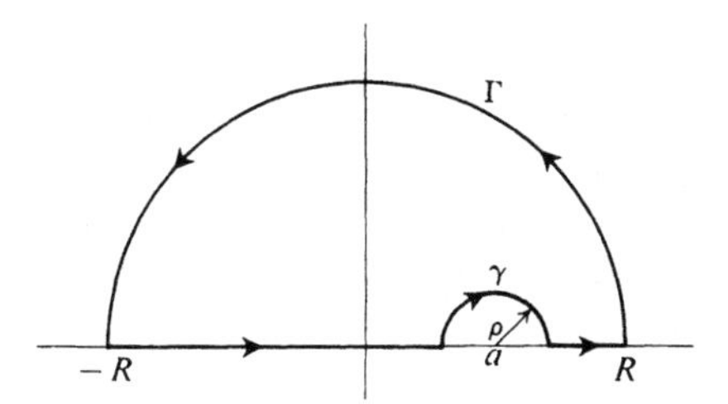

Fig. 6.21

e) We have not as yet allowed the integrand to have poles on the real axis. If $Q(z)$ is a meromorphic function in the upper half-plane, if it only has poles of order one (i.e., simple poles) on the real axis, and if $Q(z)$ behaves at infinity like any of the integrands of class b, c, or d, then the techniques of these cases can be extended to evaluate integrals of the form

$$P\int_{-\infty}^{\infty} Q(x)\, dx,$$

where the P stands for Cauchy's principal value, which arises because of the presence of the poles on the real axis.

Let $Q(z)$ have a single simple pole on the real axis at $z = a$. Consider the indented contour C (Fig. 6.21) consisting of the small semicircle γ about point a, the large semicircle Γ about the origin, and the two straight line segments on the real axis from $-R$ to $a - \rho$ and from $a + \rho$ to R. We take γ small enough so it encloses only the pole at a; Γ is taken large enough to enclose all the poles in the upper half-plane and large enough so the integral over Γ approaches 0 as $R \to \infty$. We have

$$\left(\int_\Gamma + \int_{-R}^{a-\rho} + \int_\gamma + \int_{a+\rho}^{R}\right) Q(z)\, dz = 2\pi i \sum_{y>0} \operatorname{Res} Q(z)\,.$$

Taking the limit as $R \to \infty$, we get

$$\left[\lim_{R\to\infty}\left(\int_{-R}^{a-\rho} + \int_{a+\rho}^{R}\right) + \int_\gamma\right] Q(z)\, dz = P\int_{-\infty}^{\infty} Q(x)\, dx + \int_\gamma Q(z)\, dz$$
$$= 2\pi i \sum_{y>0} \operatorname{Res} Q(z)\,.$$

We now consider $\int_\gamma Q(z)\, dz$. On γ, $z = a + \rho e^{i\theta}$, so that

$$\int_\gamma Q(z)\, dz = \int_\pi^0 Q(a + \rho e^{i\theta}) \rho e^{i\theta} i\, d\theta\,.$$

Since $Q(z)$ has a simple pole at $z = a$, it contains the factor $(z - a)^{-1}$. We may therefore write $Q(z) = \phi(z)/(z - a) + \psi(z)$, where $\phi(z)$ and $\psi(z)$ are analytic at and near $z = a$. Clearly $\psi(z)$ does not contribute to the integral over γ as $\rho \to 0$, so since $z - a = \rho e^{i\theta}$, we have

$$\int_\gamma Q(z)\, dz = \int_\pi^0 \phi(a + \rho e^{i\theta}) i\, d\theta\,.$$

We now expand ϕ in a Taylor series about a (it is analytic there):

$$\phi(a + \rho e^{i\theta}) = \phi(a) + \text{terms in } \rho\,.$$

Therefore, letting $\rho \to 0$, we get

$$\int_\gamma Q(z)\, dz = \int_\pi^0 \phi(a) i\, d\theta = -i\pi\phi(a)\,.$$

Now $\phi(a)$ is the residue of $Q(z) = \phi(z)/(z - a)$ at $z = a$, so the final answer may be written as

$$P\int_{-\infty}^{\infty} Q(x)\, dx = 2\pi i \sum_{y>0} \operatorname{Res} Q(z) + \pi i \sum_{y=0} \operatorname{Res} Q(z)\,,$$

where $\sum_{y=0} \operatorname{Res} Q(z)$ denotes the sum of the residues of $Q(z)$ at each of its simple poles on the real axis (generalizing from one to several simple poles).

Example.

$$I = \int_{-\infty}^{\infty} \frac{\sin x}{x}\, dx\,.$$

The value of this integral may be determined from the following simple result:

$$P\int_{-\infty}^{\infty} \frac{e^{ix}}{x} dx = \pi i \sum_{y=0} \operatorname{Res}\left(\frac{e^{iz}}{z}\right) = \pi i\,.$$

Here the only pole is on the real axis at $z = x = 0$.

Equating real and imaginary parts, we have

$$P\int_{-\infty}^{\infty} \frac{\cos x}{x}\,dx = 0\,,$$

which is trivial because the integrand is an odd function, and

$$P\int_{-\infty}^{\infty} \frac{\sin x}{x}\,dx = \pi\,,$$

a result derived previously from Hilbert transform theory. In this formula the P is superfluous because the integrand has no pole. Many integrals involving $\sin x$ or $\cos x$ may be evaluated, as we have done here, by replacing the trigonometric function with e^{iz}, and later taking real and imaginary parts.

f) Next we consider integrals of the form $\int_0^\infty x^{\lambda-1} R(x)\,dx$, where $R(z)$ is rational, analytic at $z = 0$, and has no poles on the positive real axis, and where $|z^\lambda R(z)| \to 0$ uniformly as $|z| \to 0$ and as $|z| \to \infty$. Since the case of integral λ can be handled by the methods described earlier, we will assume that λ is not equal to an integer.

This problem involves branch points and branch cuts because $z^{\lambda-1}$ is not in general, a single-valued function. The power function $z^{\lambda-1}$ has a branch point at the origin. Let $z^{\lambda-1}$ denote the following branch of the power function:

$$z^{\lambda-1} = \exp\left[\,(\lambda - 1)\log z\right] = \exp\left[\,(\lambda - 1)(\log r + i\theta)\right],$$

where $0 < \theta < 2\pi$, $r > 0$. Then

$$z^{\lambda-1} = r^{\lambda-1}e^{i(\lambda-1)\theta}\,, \qquad 0 < \theta < 2\pi,\ r > 0\,.$$

The branch cut has been chosen to be the positive real axis. For $\theta = 0$, the power function has the value

$$z^{\lambda-1} = r^{\lambda-1} = x^{\lambda-1}\,.$$

Now consider the contour integral $\oint_{C'} z^{\lambda-1}R(z)\,dz$, where the closed contour C' is shown in Fig. 6.22. Here C' consists of a small circle about $z = 0$, whose radius will later be shrunk to zero, a large circle whose radius will later be expanded to ∞, and two integrals along the positive real axis in opposite directions and on opposite sides of the branch cut. Since the integrand is discontinuous along the branch cut, these two integrals will not cancel. To do the integral, the phase of $z^{\lambda-1}$ must be prescribed everywhere. We have done this above by defining it to be 0 on the positive real axis so that $z^{\lambda-1} = x^{\lambda-1}$ there, which is the usual convention. Therefore, on the line just below the real axis, $z^{\lambda-1} = x^{\lambda-1}e^{i2\pi(\lambda-1)}$. The integrals over the little and big circles vanish, because $|z^\lambda R(z)| \to 0$ as $|z| \to 0$, and as $|z| \to \infty$. Therefore

$$\oint_{C'} z^{\lambda-1}R(z)\,dz = -\int_0^\infty e^{2\pi i(\lambda-1)}x^{\lambda-1}R(x)\,dx + \int_0^\infty x^{\lambda-1}R(x)\,dx\,,$$

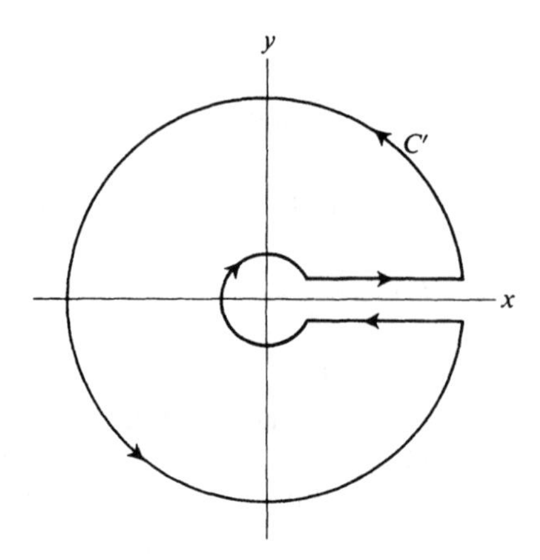

Fig. 6.22

where the first integral on the right is the contribution from the line below the positive x-axis and the second is the contribution from the line above the x-axis. Together these integrals give

$$[1 - e^{2\pi i(\lambda-1)}]\int_0^\infty x^{\lambda-1}R(x)\,dx = \left(\frac{-2i\sin\pi\lambda}{e^{-\pi i\lambda}}\right)\int_0^\infty x^{\lambda-1}R(x)\,dx\,.$$

Since

$$\oint_{C'} z^{\lambda-1}R(z)\,dz = 2\pi i \sum_{\text{inside } C'} \text{Res}\,[z^{\lambda-1}R(z)\,]\,,$$

we obtain

$$\int_0^\infty x^{\lambda-1}R(x)\,dx = \frac{-\pi e^{-\pi i\lambda}}{\sin\pi\lambda}\sum_{\text{inside } C'} \text{Res}\,[z^{\lambda-1}R(z)\,] \tag{6.52}$$

$$= \frac{\pi(-1)^{\lambda-1}}{\sin\pi\lambda}\sum_{\text{inside } C'} \text{Res}\,[z^{\lambda-1}R(z)\,]\,,$$

since $e^{-i\pi\lambda} = (-1)^\lambda$. This is as far as we can go without choosing a specific function $R(z)$. A very simple example of this type is the following integral:

$$I = \int_0^\infty \frac{x^{\lambda-1}}{1+x}\,dx\,, \qquad 0 < \lambda < 1\,.$$

After evaluating it, we shall use it to prove some results involving the beta and gamma functions.

First since $0<\lambda<1$, $|z^\lambda/(1+z)| \to 0$ as $|z| \to \infty$ and as $z \to 0$. Therefore

$$\int_0^\infty \frac{x^{\lambda-1}}{1+x}\,dx = \frac{\pi(-1)^{\lambda-1}}{\sin\pi\lambda}\,\text{Res}\left(\frac{z^{\lambda-1}}{1+z}\right)_{z=-1} = \frac{\pi}{\sin\pi\lambda}\,.$$

The beta function $\beta(x, y)$ is defined as

$$\beta(x, y) \equiv \int_0^1 t^{x-1}(1-t)^{y-1}\,dt = \frac{\Gamma(x)\Gamma(y)}{\Gamma(x+y)}\,, \qquad x, y > 0\,, \tag{6.53}$$

where $\Gamma(x)$ is the gamma function:

$$\Gamma(x) = \int_0^\infty t^{x-1}e^{-t}\,dt\,. \tag{6.54}$$

The proof of the identity in Eq. (6.53) may be found in many books on analysis. We note that

$$\beta(x, 1-x) = \Gamma(x)\Gamma(1-x) = \int_0^1 t^{x-1}(1-t)^{-x}\,dt\,, \qquad \text{if } 0 < x < 1.$$

If in the last integral we make the change of variable

$$t = \frac{u}{1+u} = 1 - \frac{1}{1+u}\,,$$

it becomes

$$\int_0^\infty \frac{u^{x-1}}{1+u}\,du\,,$$

which is precisely the integral we have just evaluated. Thus

$$\beta(x, 1-x) = \Gamma(x)\Gamma(1-x) = \frac{\pi}{\sin \pi x}\,, \qquad \text{for } 0 < x < 1\,. \tag{6.55}$$

This formula may be extended by analytic continuation to all z in the complex plane. In particular, for $x = \frac{1}{2}$, we have $[\Gamma(\frac{1}{2})]^2 = \pi$. Therefore

$$\pi^{1/2} = \Gamma(\tfrac{1}{2}) \equiv \int_0^\infty t^{-1/2}e^{-t}\,dt = 2\int_0^\infty e^{-u^2}du = \int_{-\infty}^\infty e^{-u^2}\,du\,, \tag{6.56}$$

where we have made the substitution $t = u^2$.

We close this section on residue theory with an illustration of how series can be summed by contour integration. The result we shall obtain depends on the fact that $\pi \cot \pi z$ has poles of order one at the zeros of $\sin \pi z$, (that is, at $z = n$, $n = 0, \pm 1, \pm 2, \cdots$), and the fact that the residue at each of these poles is 1.

Theorem. Let $f(z)$ be a meromorphic function and let C be a contour which encloses the zeros of $\sin \pi z$, located at $z = \rho, \rho + 1, \cdots, n$. If we assume that the poles of $f(z)$ and $\sin \pi z$ are distinct, then

$$\sum_{m=\rho}^{n} f(m) = \frac{1}{2\pi i}\oint_C \pi \cot \pi z\, f(z)\,dz - \sum_{\substack{\text{poles of } f(z) \\ \text{inside } C}} \text{Res}\,[\pi \cot(\pi z)\, f(z)\,]\,. \tag{6.57}$$

Proof.

$$\oint_C \pi \cot \pi z f(z)\,dz = 2\pi i \sum \text{(Residues at all the poles of the integrand)}$$

$$= 2\pi i\left[\sum_{m=\rho}^{n} f(m) + \sum_{\substack{\text{poles of } f(z) \\ \text{inside } C}} \text{Res}\,[\pi \cot \pi z\, f(z)\,]\right]. \quad \text{QED}$$

Example. We shall use this theorem to establish the equivalence of Langevin's function ($\coth x - 1/x$) and the sum

$$\sum_{n=1}^{\infty} \frac{2x}{x^2 + n^2\pi^2}.$$

We shall then use this result to derive an infinite *product* expression for $\sin\theta$, which is of considerable interest in its own right.

Letting $f(z) = 2x/(x^2 + z^2\pi^2)$, and using the formula above, we obtain

$$\sum_{m=-N}^{N} \frac{2x}{x^2 + m^2\pi^2} = \frac{1}{2\pi i}\oint_C \pi \cot \pi z\, f(z)\, dz - \sum_{\substack{\text{poles of } f(z)\\ \text{inside } C}} \operatorname{Res}\left[\pi \cot \pi z\, f(z)\right],$$

where C is a closed contour, say, a rectangle, enclosing the points $z = -N, -N+1, \cdots, 0, 1, \cdots, N-1, N$. Now let the length and width of the rectangle C approach ∞. As this happens,

$$\left|\frac{1}{2\pi i}\oint_C \pi \cot \pi z\, f(z)\, dz\right| \leq \tfrac{1}{2}\oint_C |\cot \pi z| \left|\frac{2x}{x^2 + z^2\pi^2}\right| |dz| \to 0.$$

To see that this integral vanishes as $z \to \infty$, we observe that

$$|\cot \pi z| = \frac{|\cos \pi z|}{|\sin \pi z|} = \left(\frac{\cos^2 \pi x + \sinh^2 \pi y}{\sin^2 \pi x + \sinh^2 \pi y}\right)^{1/2}.$$

Now, arbitrarily high accuracy can be achieved in summing the series by choosing the rectangle so that its vertical sides cross the x-axis at a large enough half-integer, for example $(10^{97} + \frac{1}{2})$, where $\cos \pi(10^{97} + \frac{1}{2}) = 0$ and $\sin \pi(10^{97} + \frac{1}{2}) = 1$. Then, over these sides of the rectangle

$$|\cot \pi z| = \left|\left(\frac{\sinh^2 \pi y}{1 + \sinh^2 \pi y}\right)^{1/2}\right| = |\tanh \pi y| \leq 1.$$

Over the horizontal sides of the rectangle, $\lim_{z\to\infty} |\cot \pi z| = 1$. Thus the integrand goes as $|1/z^2|$ as $|z| \to \infty$, and the integral vanishes.

If we take an infinite rectangular contour, then

$$\sum_{m=-\infty}^{\infty} \frac{2x}{x^2 + m^2\pi^2} = -\operatorname{Res}\left[\frac{(\pi \cot \pi z)2x}{x^2 + z^2\pi^2}\right]_{z=\pm ix/\pi}$$

$$= -\frac{2x}{\pi}\cdot\left[\frac{\cot \pi(ix/\pi)}{2ix/\pi} + \frac{\cot \pi(-ix/\pi)}{-2ix/\pi}\right] = 2i \cot ix = 2 \coth x.$$

Therefore

$$2\sum_{m=1}^{\infty} \frac{2x}{x^2 + m^2\pi^2} + \frac{2}{x} = 2 \coth x$$

or

$$\coth x - \frac{1}{x} = \sum_{m=1}^{\infty} \frac{2x}{x^2 + m^2\pi^2}, \tag{6.58}$$

which establishes the result stated at the outset.

Aside from illustrating this summation technique, this particular result is not of much interest. However, if we integrate both sides from 0 to x, we get

$$\sum_{m=1}^{\infty} \ln\left(1 + \frac{x^2}{m^2\pi^2}\right) = \ln\left(\frac{\sinh x}{x}\right).$$

Hence

$$\ln\left(\frac{\sinh x}{x}\right) = \sum_{m=1}^{\infty} \ln\left(1 + \frac{x^2}{m^2\pi^2}\right) = \ln \prod_{m=1}^{\infty}\left(1 + \frac{x^2}{m^2\pi^2}\right),$$

so

$$\frac{\sinh x}{x} = \prod_{m=1}^{\infty}\left(1 + \frac{x^2}{m^2\pi^2}\right). \tag{6.59}$$

We may extend this result to all z in the complex plane by analytic continuation. Then setting $x = i\theta$ (θ real), we obtain

$$\sin\theta = \theta \prod_{n=1}^{\infty}\left(1 - \frac{\theta^2}{n^2\pi^2}\right). \tag{6.60}$$

This infinite product formula displays *explicitly* all the zeros of $\sin\theta$. It represents the complete factorization of the Taylor series. It can, in fact, be taken as the definition of the sine function.

By equating coefficients of the θ^3 term of both sides of the above equation, we obtain a useful sum:

$$\sum_{n=1}^{\infty} \frac{1}{n^2} = \frac{\pi^2}{6}. \tag{6.61}$$

This is a special value of the Riemann zeta function,

$$\zeta(z) = \sum_{n=1}^{\infty} \frac{1}{n^z}. \tag{6.62}$$

There are many other special tricks for evaluating integrals and summing series. We have surveyed only the principal techniques, but the reader will have no trouble understanding any particular evaluation procedure he may encounter if he understands this section. Problem 31 provides applications of all the basic techniques of residue calculus developed above.

6.9 APPLICATIONS TO SPECIAL FUNCTIONS AND INTEGRAL REPRESENTATIONS

In this section we use the formula for the derivative of an analytic function to find generating functions for certain special functions from their Rodrigues formulas. Also, we shall derive integral representations for Bessel functions and Legendre polynomials.

Bessel Functions

The function $e^{(1/2)z(w-1/w)}$ is analytic everywhere in the w-plane except at $w = 0$, so it can be expanded in a Laurent series in any annulus $R_2 < |w| < R_1$, no matter how small $R_2(> 0)$ or how large R_1. Denoting the expansion coefficients by $J_n(z)$, we have

$$e^{(1/2)z(w-1/w)} = \sum_{n=-\infty}^{\infty} J_n(z) w^n . \tag{6.63}$$

The expansion coefficients are functions of the complex variable z. The notation chosen for them anticipates the fact that they will turn out to be Bessel functions of integral order. In other words, $e^{(1/2)z(w-1/w)}$ is a generating function for Bessel functions.

We now prove that the functions $J_n(z)$ do satisfy Bessel's differential equation:

$$z^2 \frac{d^2 J_n}{dz^2} + z \frac{dJ_n}{dz} + (z^2 - n^2) J_n = 0 . \tag{6.64}$$

Along the way to this result we shall find an integral representation for Bessel functions and also an explicit formula for them.

By Laurent's theorem,

$$J_n(z) = \frac{1}{2\pi i} \oint_C \frac{f(w)}{(w - w_0)^{n+1}} dw , \tag{6.65}$$

where C is any closed contour in the annulus. We take C to be the unit circle: $w = e^{i\theta}$. Then in Eq. (6.65), $w_0 = 0$ and $f(w) = e^{(1/2)z(w-1/w)}$. On the unit circle $w - 1/w = 2i \sin \theta$. Therefore

$$\begin{aligned} J_n(z) &= \frac{1}{2\pi} \int_{-\pi}^{\pi} e^{iz \sin \theta} e^{-in\theta} d\theta = \frac{1}{2\pi} \int_{-\pi}^{\pi} e^{-i(n\theta - z \sin \theta)} d\theta \\ &= \frac{1}{2\pi} \int_{-\pi}^{\pi} \cos (n\theta - z \sin \theta) d\theta - \frac{i}{2\pi} \int_{-\pi}^{\pi} \sin (n\theta - z \sin \theta) d\theta . \end{aligned}$$

But the second integral is zero because the integrand is odd in θ. Since the first integrand is even in θ,

$$J_n(z) = \frac{1}{\pi} \int_0^{\pi} \cos (n\theta - z \sin \theta) d\theta . \tag{6.66}$$

This is an integral representation for the function $J_n(z)$, which we now prove is a solution of Bessel's equation by substituting it into Eq. (6.64). The verification is trickier than one might expect. We therefore resist the temptation to "leave it to the reader." He is, of course, welcome to have a try at it before reading further. We have

$$J_n'(z) = \frac{1}{\pi} \int_0^{\pi} \sin \theta \sin (n\theta - z \sin \theta) d\theta \tag{6.67a}$$

$$\begin{aligned} &= -\frac{1}{\pi} \cos \theta \sin (n\theta - z \sin \theta) \Big]_0^{\pi} \\ &\quad + \frac{1}{\pi} \int_0^{\pi} \cos \theta \cos (n\theta - z \sin \theta)(n - z \cos \theta) d\theta , \end{aligned} \tag{6.67b}$$

where we have integrated by parts. The first term is zero. Differentiating Eq. (6.67a) again, we have

$$J_n''(z) = -\frac{1}{\pi}\int_0^\pi \sin^2\theta \cos(n\theta - z\sin\theta)d\theta .$$

We now form the quantity $z^2(J_n'' + J_n) + zJ_n' - n^2J_n$, using Eq. (6.67b) for $J_n'(x)$. It must be identically zero if the $J_n(z)$ are Bessel functions. We get

$$\frac{1}{\pi}\int_0^\pi \cos(n\theta - z\sin\theta)[z^2 - z^2\sin^2\theta + z\cos\theta(n - z\cos\theta) - n^2]\,d\theta$$

$$= \frac{-n}{\pi}\int_0^\pi [\cos(n\theta - z\sin\theta)(n - z\cos\theta)]\,d\theta .$$

The function $\sin(n\theta - z\sin\theta)$ is an antiderivative of the integrand. It vanishes at both 0 and π, so the proof is complete: the $J_n(z)$ are solutions of Bessel's equation. It is easy to show that

$$\begin{aligned} J_0(0) &= 1, \qquad J_n(0) = 0 \quad \text{for } n \neq 0, \\ J_1'(0) &= \tfrac{1}{2}, \qquad J_n'(0) = 0 \quad \text{for } n \neq 1, \end{aligned} \tag{6.68}$$

so these functions are indeed that solution of Bessel's equation, known as Bessel functions of nth (integral) order of the *first* kind (which are analytic at the origin).

Usually, the solution of Bessel's equation is expressed as an infinite series, not as an integral. We now derive the infinite series solution from the generating function:

$$\begin{aligned} e^{(1/2)z(w-1/w)} &= e^{(1/2)zw}e^{-z/2w} \\ &= \sum_{r=0}^\infty \frac{(z/2)^r w^r}{r!}\sum_{m=0}^\infty \frac{(-z/2)^m w^{-m}}{m!} = \sum_{n=-\infty}^\infty w^n J_n(z) . \end{aligned}$$

To obtain $J_n(z)$, the coefficient of the term in w^n $(n \geq 0)$, we multiply each term in w^{-m} in the second series by the term w^{n+m} in the first series, and then sum over all m:

$$\begin{aligned} J_n(z)w^n &= \sum_{m=0}^\infty \frac{(z/2)^{n+m}w^{n+m}(-z/2)^m w^{-m}}{(n+m)!\,m!} \\ &= \sum_{m=0}^\infty \left[\frac{(-1)^m(z/2)^{2m+n}}{(n+m)!\,m!}\right] w^n . \end{aligned}$$

Thus

$$J_n(z) = \sum_{m=0}^\infty \frac{(-1)^m(z/2)^{2m+n}}{(n+m)!\,m!} . \tag{6.69}$$

This is the infinite series solution for integral $n \geq 0$. The reader may show that, for integral n,

$$J_{-n}(z) = (-1)^n J_n(z) . \tag{6.70}$$

Thus these solutions are linearly dependent, and there must exist another set of linearly independent solutions. They are called the Neumann functions, and are denoted by $N_n(z)$. A whole family of equations and solution functions may be constructed from Bessel's equations and Bessel functions by permitting the index n to be nonintegral and half-integral. Also, various linear combinations of Bessel and Neumann functions of real and imaginary argument give rise to the modified Bessel functions and the Hankel functions. The spherical Bessel functions (which we discuss in Chapter 7) are still another set of functions that are defined in terms of the Bessel functions. Each of these sets of functions is a Sturm-Liouville system. It is not our purpose to provide an exhaustive (and exhausting) account of them, but merely to derive and discuss the fundamental results so the reader can dig the particular facts he needs for his work out of a treatise on the subject.

Legendre polynomials

Previously we have shown that the Legendre polynomials, defined by the Rodrigues formula

$$P_n(x) = \frac{1}{2^n n!}\frac{d^n}{dx^n}(x^2-1)^n ,$$

are an orthogonal set on the interval $[-1, 1]$ and satisfy Legendre's equation:

$$(1-x^2)P_n'' - 2xP_n' + n(n+1)P_n = 0 .$$

It was mentioned in Section 1.8, but not proved, (except for P_0, P_1, and P_2), that a generating function for the Legendre polynomials is

$$\frac{1}{(1-2xt+t^2)^{1/2}} = \sum_{n=0}^{\infty} P_n(x)t^n , \qquad 0 < t < 1 , \qquad |x| \leq 1 . \tag{6.71}$$

In the comparison of this equation with Eq. (1.101), the reader will note that we have set $t \equiv r'/r$ and $x \equiv \cos\theta$. We now prove this relation for all n. We begin by expressing the Rodrigues formula as a contour integral by making use of the formula for the nth derivative of an analytic function:

$$P_n(x) = \frac{(-1)^n}{2^n n!}\frac{d^n}{dx^n}(1-x^2)^n = \frac{(-1)^n}{2^n}\frac{1}{2\pi i}\oint_C \frac{(1-z^2)^n}{(z-x)^{n+1}}\,dz . \tag{6.72}$$

Here we have used the fact that the function $(1-z^2)^n$ is entire (C is a closed contour that encloses the point x). This integral representation of the Legendre polynomials is known as Schläfli's integral formula. If we now form the series $S = \sum_{n=0}^{\infty} P_n(x)t^n$, using this integral representation for $P_n(x)$, and interchange the order of summation and integration, we get

$$S = \frac{1}{2\pi i}\oint_C \frac{dz}{z-x}\sum_{n=0}^{\infty}\frac{(-1)^n}{2^n}t^n\left(\frac{1-z^2}{z-x}\right)^n .$$

This is a simple geometric series, which converges and is easily summed if

$$\left|\frac{t}{2}\frac{1-z^2}{z-x}\right| < 1 .$$

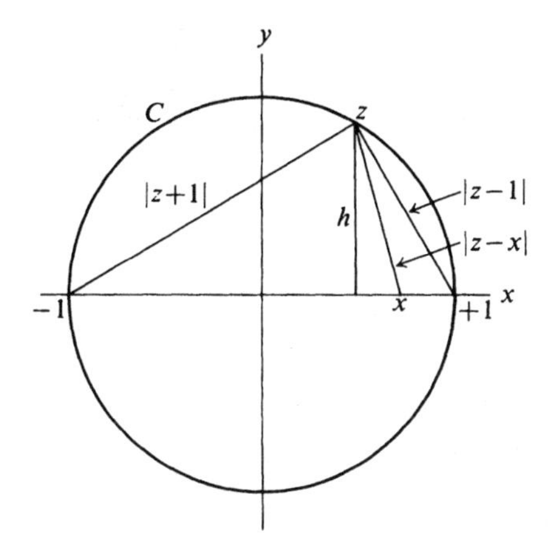

Fig. 6.23

We must show, however, that a contour C, enclosing the point x, can always be chosen so that this is true for all x and t in their prescribed ranges. Let C be the unit circle; z is a point anywhere on C, and x is on the real axis inside or on C as shown in Fig. 6.23. The area A of the right triangle with vertices at -1, z, and $+1$ is given by

$$A = \tfrac{1}{2}\,|z+1|\,|z-1| = \tfrac{1}{2}\,|1-z^2|\,.$$

It is also given by $\frac{1}{2}(2h) = h$. But $|z-x| \geq h$ for all z and x, so

$$|z-x| \geq h = A = \tfrac{1}{2}\,|1-z^2| \Longrightarrow \left|\tfrac{1}{2}\cdot\frac{1-z^2}{z-x}\right| \leq 1\,.$$

Consequently,

$$\left|\frac{t}{2}\,\frac{1-z^2}{z-x}\right| < 1$$

since $t < 1$. A purely algebraic proof is also possible, but this geometric one is simpler. We now sum the geometric series to obtain

$$S = \frac{1}{2\pi i}\oint_C \frac{dz}{z-x}\,\frac{1}{1-\left[\dfrac{-t}{2}\dfrac{(1-z^2)}{(z-x)}\right]} = -\frac{1}{2\pi i}\oint_C \frac{2/t\,dz}{z^2-\dfrac{2}{t}z-\left(1-\dfrac{2}{t}x\right)}\,.$$

The denominator has two roots;

$$z_\pm = \frac{1}{t} \pm \left(\frac{1}{t^2}-\frac{2x}{t}+1\right)^{1/2} = \frac{1}{t}\,[1 \pm (1-2xt+t^2)^{1/2}]\,.$$

Remembering that $0 < t < 1$ and $-1 \leq x \leq 1$, it is easy for us to show that $z_+ > 1$ and $-1 < z_- < 1$, so z_- is the root enclosed by the contour C. Evaluating the residue at z_-, we have

$$S = \sum_{n=0}^{\infty} P_n(x)t^n = \frac{-1}{2\pi i}\,\frac{2}{t}\,2\pi i\,\frac{1}{z_- - z_+} = \frac{1}{(1-2xt+t^2)^{1/2}}\,. \qquad \text{QED}$$

Table 6.1

PROPERTIES OF ORTHOGONAL FUNCTIONS ARISING FROM STURM-LIOUVILLE SYSTEMS

Name and physical application	Rodrigues formula	Generating function
Legendre polynomials: 1) Multipole expansion 2) ∇^2 in spherical coordinates	$P_n(x) = \frac{1}{2^n n!}\frac{d^n}{dx^n}(x^2-1)^n$	$\sum_{n=0}^{\infty} P_n(x)t^n = \frac{1}{\sqrt{1-2xt+t^2}}$ $(0 < t < 1)$
Hermite polynomials: Quantum oscillator	$H_n(x) = (-1)^n e^{x^2}\frac{d^n e^{-x^2}}{dx^n}$	$\sum_{n=0}^{\infty} \frac{H_n(x)}{n!} t^n = e^{-t^2+2tx}$ $(t > 0)$
Laguerre polynomials: Hydrogen atom (radial equation)	$L_n(x) = e^x \frac{d^n}{dx^n} x^n e^{-x}$	$\sum_{n=0}^{\infty} \frac{L_n(x)}{n!} t^n = \frac{e^{-xt/(1-t)}}{1-t}$ $(0 < t < 1)$
	Series representation	
Bessel's function (of integral order) ∇^2 in cylindrical coordinates	$J_n(x) = \sum_{m=0}^{\infty} \frac{(-1)^m (x/2)^{2m+n}}{(n+m)!\, m!}$	$\sum_{n=-\infty}^{\infty} J_n(x)t^n = e^{(1/2)x(t-1/t)}$ $(t > 0)$
Trigonometric functions: Classical oscillator	$f_n = \sin nx = \sum_{m=0}^{\infty} (-1)^m \frac{(nx)^{2m+1}}{(2m+1)!}$ $g_n = \cos nx = \sum_{m=0}^{\infty} (-1^m) \frac{(nx)^{2m}}{(2m)!}$	

Differential equation	Differential equation in Sturm-Liouville form	Orthonormality
$(1 - x^2)P_n'' - 2xP_n' + n(n+1)P_n = 0$	$\frac{d}{dx}((1 - x^2)P_n') + n(n+1)P_n = 0$	$\int_{-1}^{1} P_n P_m \, dx = \delta_{nm} \frac{2}{2n+1}$
$H_n'' - 2xH' + 2nH_n = 0$	$\frac{d}{dx}(e^{-x^2}H_n') + 2ne^{-x^2}H_n = 0$	$\int_{-\infty}^{\infty} H_n H_m e^{-x^2} \, dx = \delta_{nm}\sqrt{\pi}2^n n!$
$xL_n'' + (1 - x)L_n' + nL_n = 0$	$\frac{d}{dx}(xe^{-x}L_n') + ne^{-x}L_n = 0$	$\int_0^{\infty} L_n L_m e^{-x} \, dx = \delta_{nm}(n!)^2$
$x^2J_n'' + xJ_n' + (x^2 - n^2)J_n = 0$	$\frac{d}{dx}(xJ_n') + \left(x - \frac{n^2}{x}\right)J_n = 0$	See Problem 5.23
$f_n'' + n^2f_n = 0$ $g_n'' + n^2g_n = 0$	$\frac{d}{dx}(f_n') + n^2f_n = 0$ $\frac{d}{dx}(g_n') + n^2g_n = 0$	$\int_{-\pi}^{\pi} f_n f_m \, dx = \delta_{nm}\pi$ $\int_{-\pi}^{\pi} g_n g_m \, dx = \delta_{nm}\pi$ $\int_{-\pi}^{\pi} f_n g_m \, dx = 0$

In exactly the same way, the generating functions for the Hermite polynomials and the Laguerre polynomials can be derived from their Rodrigues formulas (see Problems 36 and 37).

Let us return to Schläfli's integral representation of the Legendre polynomials:

$$P_n(x) = \frac{1}{2\pi i}\oint_C \frac{(z^2-1)^n}{2^n(z-x)^{n+1}}\,dz\,. \tag{6.73}$$

It can be proved directly from this integral representation that the $P_n(x)$ satisfy Legendre's differential equation. But we have already proved that the $P_n(x)$ as given by Rodrigues' formula satisfy Legendre's equation, so we omit this proof. Here we derive another integral representation: a real integral.

For the contour C, we take a circle about x of radius $|(x^2-1)^{1/2}|$. Then for any point z on C, we may put $z = x + (x^2-1)^{1/2}e^{i\theta}$, where θ increases from $-\pi$ to π. Making this substitution in Schläfli's formula, we obtain

$$P_n(x) = \frac{1}{2\pi i}\int_{-\pi}^{\pi}\left[\frac{(x-1+(x^2-1)^{1/2}e^{i\theta})(x+1+(x^2-1)^{1/2}e^{i\theta})}{2(x^2-1)^{1/2}e^{i\theta}}\right]^n i\,d\theta\,,$$

where we have written $(z^2-1)^n$ in the form $[(z-1)(z+1)]^n$ before substituting $z = x + (x^2-1)^{1/2}e^{i\theta}$. Now everything in square brackets above simplifies after a little algebra to $x + (x^2-1)^{1/2}\cos\theta$, an even function of θ. Thus we have a real integral representation of the Legendre polynomials due to Laplace:

$$P_n(x) = \frac{1}{\pi}\int_0^{\pi}[x+(x^2-1)^{1/2}\cos\theta]^n\,d\theta\,, \qquad \text{for } |x| \le 1\,. \tag{6.74}$$

Note how obvious it is from this representation that $P_n(1) = 1$, and $P_n(-1) = (-1)^n$. It is instructive to compute P_0, P_1, and P_2 from this formula. Although $(x^2-1)^{1/2}$ is pure imaginary for $|x| < 1$, all the terms in $(x^2-1)^{1/2}$ vanish when the integral is performed because $\int_0^{\pi}\cos^m\theta\,d\theta = 0$ for odd m.

We conclude this section with Table 6.1. It summarizes some of the more useful information concerning the special functions of mathematical physics.

PROBLEMS

1. Let

$$f(z) = \frac{xy^2(x+iy)}{x^2+y^2} \qquad \text{for } z \neq 0\,, \quad f(0) = 0\,.$$

Determine where, if anywhere, this function is a) differentiable, b) analytic.

2. Show that the complex numbers z_1, z_2, z_3 lie on a straight line if and only if $(z_1 - z_3)/(z_2 - z_3)$ is a real number.

[*Hint*: This problem can be done very simply if one thinks of the complex numbers geometrically (as vectors). It can be done algebraically also, but the geometric proof is certainly easier.]

3. Prove that $u = \sin x \cosh y + 2 \cos x \sinh y + x^2 - y^2 + 4xy$ is a harmonic function and find a conjugate harmonic function v. Find a complex function of the complex variable z such that $f(z) = u + iv$.

4. a) Determine all the values of the constants a, b, c, d for which the polynomial $u(x, y) = ax^3 + bx^2y + cxy^2 + dy^3$ is harmonic, i.e., satisfies the two-dimensional Laplace equation in the entire plane.

   b) Find a harmonic conjugate, $v(x, y)$, of $u(x, y)$.

   c) Find an analytic function $f(z) = u(x, y) + iv(x, y)$, where $z = x + iy$.

5. A complex function $E^*(z)$ is defined by $E^*(z) = E_x(x, y) - iE_y(x, y)$, where E_x and E_y are the components of the electric field in two dimensions. Show that the static Maxwell equations in free space imply that $E^*(z)$ is an analytic function of z.

6. Show that if we write $z = re^{i\theta}$, the Cauchy-Riemann equations become, in terms of r and θ,

$$\frac{\partial u}{\partial r} = \frac{1}{r}\frac{\partial v}{\partial \theta}, \qquad \frac{\partial v}{\partial r} = -\frac{1}{r}\frac{\partial u}{\partial \theta}.$$

7. Analyze the function $z^{1/3}$ in terms of the Riemann sheet concepts discussed in Section 6.2.

8. Using the results of Problem 6.6, show that $\log z$, $z^{1/2}$, and $z^{1/3}$ are analytic everywhere on their appropriate Riemann sheets except at $z = 0$.

9. Show that the two-sheeted Riemann surface for $w(z) = \sqrt{(z - a)(z - b)}$ can be cut and joined along the line segment $[a, b]$ of the real axis (assume for convenience that a and b are real). Make an appropriate definition of $w(z)$ in terms of the polar representation of complex numbers, including a definition of the range of the phases.

10. Find the Riemann surface on which $\sqrt{z - a} + \sqrt{z - b}$ (a, b real and positive) is a single-valued function, analytic except at $z = a$ and $z = b$.

11. Find the Riemann surface on which $\sqrt{(z - 1)(z - 2)(z - 3)}$ is a single-valued function, analytic except at $z = 1, 2, 3$.

12. Show that $w(z) = z^\alpha$ ($0 < \alpha < 1$), where α is irrational, can be made single-valued only on an infinite-sheeted Riemann surface. Discuss how the sheets should be connected.

13. Show that, according to the definitions of Section 6.2,

$$\tan(\tan^{-1} z) = z, \qquad \log(e^z) = z + 2\pi ni.$$

14. *The mean-value theorem.* Prove that for charge-free two-dimensional space the value of the electrostatic potential at any point is equal to the average of the potential over the surface of *any* circle centered on that point. Do this by considering the electrostatic potential as the real part of an analytic function.

15. Find all the singularities of $\tanh z$.

16. Explain why one of the following definite integrals is meaningful and the other meaningless. Evaluate the one that makes sense, writing the answer in the form $a + bi$.

$$\text{a)} \int_{-1}^{1} z^*\,dz, \qquad \text{b)} \int_0^i \sin 2z\,dz.$$

17. Compute, using Cauchy's integral formula,

 a) $\oint_C \frac{e^z\,dz}{z - \pi i/2}$, C = boundary of a square with sides $x = \pm 2$, $y = \pm 2$.

 b) $\oint_C \frac{dz}{z^2 + 2}$, C = circle of radius 1, center i.

 (Both contours are oriented counterclockwise.)

18. Show that

 a) $\oint_C \frac{z^4 + 2z + 1}{(z - z_0)^4}\,dz = 8\pi i z_0$, where C is any closed contour containing z_0.

 b) $\oint_C \frac{\cosh z}{z^{n+1}}\,dz = \frac{2\pi i}{n!}\frac{[1 + (-1)^n]}{2}$, where C is any closed contour containing the origin.

19. *Cauchy's integral formula for a polynomial.* Let $P(z)$ be a polynomial. Using Cauchy's Theorem, prove that

$$\frac{1}{2\pi i}\oint_C \frac{P(z)}{z - a}\,dz = P(a)\,,$$

where C encloses the point a. [*Hint*: Consider

$$Q(z) = \frac{P(z) - P(a)}{z - a},$$

which is also a polynomial (why ?), and evaluate the integral you want from this expression.]

20. Prove that if $f(z)$ is analytic within and on a closed contour C, and α and β are two distinct points within C, then

$$\frac{1}{2\pi i}\oint_C \frac{f(z)}{(z - \alpha)(z - \beta)}\,dz = \frac{f(\alpha)}{\alpha - \beta} + \frac{f(\beta)}{\beta - \alpha}.$$

From this result, deduce Liouville's theorem, which states that a function must be a constant. [*Hint*: If for arbitrary α and β, $f(\alpha) = f(\beta)$, then $f(z) = \text{const.}$]

21. *Morera's theorem.* Morera's theorem is a kind of converse to the Cauchy-Goursat theorem. It states that if $f(z)$ is continuous in a region R and if $\oint f(z)\,dz = 0$ for any C inside R, then $f(z)$ is analytic inside R. Prove this theorem. Morera's theorem gives us another way of establishing the analyticity of a function. We can use it instead of checking the Cauchy-Riemann conditions *and* the sometimes troublesome additional requirement for analyticity, the continuity of the derivative of $f(z) = u + iv$, that is, continuity of the four first partial derivatives of u and v.

22. *Conformal mappings.* Consider the analytic function $w = f(z)$ as transforming the complex z-plane into a complex w-plane. Suppose that two curves in the z-plane, F_z^1 and F_z^2, meet at an angle α at the point z_0 in the z-plane. Prove that the transformed curves in the w-plane, F_w^1 and F_w^2, meet at the same angle in the w-plane if $f'(z_0) \neq 0$. The requirement that $f'(z_0) \neq 0$ is essential for conformality at z_0. Consider the analytic function $w = z^2$, for which $w'(0) = 0$. The coordinate axes themselves, which pass through the origin and are separated by an angle of 90°, are mapped into lines that are separated by 180°. The angle between any two straight lines through the origin in the z-plane will be doubled in the w-plane. Thus the mapping is not conformal at the origin. However, it is conformal everywhere else.

23. *Poisson's formula*

a) Let $f(z)$ be an analytic function within and on the circle C of radius a. Prove that

$$f(z') = \frac{1}{2\pi i}\oint_C \left[\frac{f(z)}{z - z'} - \frac{f(z)}{z - (a^2/z'^*)}\right] dz\,,$$

if $z' = re^{i\phi}, r < a$, in polar coordinates located at the center of C.

b) From the above formula, derive Poisson's formula:

$$f(re^{i\phi}) = \frac{1}{2\pi}\int_0^{2\pi} \frac{a^2 - r^2}{a^2 + r^2 - 2ar\cos(\theta - \phi)} f(ae^{i\theta})\, d\theta\,.$$

c) From Poisson's formula, prove that if a function is analytic throughout and on a circle, then the value of the function at the center of the circle, $f(0)$, is the average of its boundary values on the circle (mean value theorem).

d) Now, starting with the equation

$$f(z') = \frac{1}{2\pi i}\oint_C \left[\frac{f(z)}{z - z'} + \frac{f(z)}{z - (a^2/z'^*)}\right] dz\,,$$

[which differs from the equation of part (a) by the + sign, but holds for the same reason], derive [using part (c) eventually]

$$f(z') = f(0) - \frac{iar}{\pi}\int_0^{2\pi} \frac{\sin(\theta - \phi) f(ae^{i\theta})}{a^2 + r^2 - 2ar\cos(\theta - \phi)}\, d\theta\,.$$

e) Letting $f(z') = f(re^{i\phi}) = u(r, \phi) + iv(r, \phi)$, etc., deduce from (d) the formulas

$$u(a, \phi) = u(0) + \frac{1}{2\pi} P\int_0^{2\pi} v(a, \theta) \cot\left(\frac{\theta - \phi}{2}\right) d\theta\,,$$

$$v(a, \phi) = v(0) - \frac{1}{2\pi} P\int_0^{2\pi} u(a, \theta) \cot\left(\frac{\theta - \phi}{2}\right) d\theta\,.$$

These formulas express the real part of an analytic function on a circle in terms of its imaginary part on the circle, and vice versa. Many further applications of these formulas are possible.

24. *An alternative approach to Hilbert transforms*

a) Show that if $f(z)$ is analytic in the upper half z-plane, then if $a = \alpha + i\beta$ $(\beta > 0)$,

$$\frac{1}{2\pi i}\oint_C \left[\frac{f(z)}{z - a} + \frac{f(z)}{z - a^*}\right] dz = f(a)\,,$$

where C is a semicircle of arbitrarily large radius with the real axis for its base.

b) If $|f(z)| \to 0$ as $|z| \to \infty$ in the upper half-plane, then show that

$$\frac{1}{2\pi i}\int_{-\infty}^{\infty} \left[\frac{f(x)}{x - a} + \frac{f(x)}{x - a^*}\right] dx = f(a)\,.$$

Combine fractions to obtain

$$\frac{1}{i\pi}\int_{-\infty}^{\infty} \frac{(x - \alpha) f(x)}{(x - \alpha)^2 + \beta^2}\, dx = f(a)\,.$$

c) If $f(z) = u(x, y) + iv(x, y)$, show that (b) reduces to

$$u(\alpha, \beta) = \frac{1}{\pi}\int_{-\infty}^{\infty} \frac{(x - \alpha) v(x, 0)}{(x - \alpha)^2 + \beta^2}\, dx\,, \quad v(\alpha, \beta) = -\frac{1}{\pi}\int_{-\infty}^{\infty} \frac{(x - \alpha) u(x, 0)}{(x - \alpha)^2 + \beta^2}\, dx\,.$$

d) Hence show that if we let $\beta = 0$,

$$u(\alpha, 0) = \frac{1}{\pi} P\int_{-\infty}^{\infty} \frac{v(x, 0)}{x - \alpha} dx, \qquad v(\alpha, 0) = -\frac{1}{\pi} P\int_{-\infty}^{\infty} \frac{u(x, 0)}{x - \alpha} dx,$$

which is, apart from notation, the Hilbert transform pair derived in Section 6.5 [Eqs. (6.19a) and (6.19b)].

25. Consider the integral

$$I_{\pm} = \lim_{\epsilon \to 0} \int_{-A}^{+B} \frac{f(x)}{x \pm i\epsilon} dx,$$

where A and B are positive real numbers and $\epsilon > 0$.

a) Show that $I_{\pm}$ can be written as

$$I_{\pm} = \lim_{\epsilon \to 0} \int_{-A}^{-\delta} \frac{xf(x)}{x^2 + \epsilon^2} dx + \lim_{\epsilon \to 0} \int_{+\delta}^{+B} \frac{xf(x)}{x^2 + \epsilon^2} dx + \lim_{\epsilon \to 0} \int_{-\delta}^{+\delta} \frac{xf(x)}{x^2 + \epsilon^2} dx \mp i \lim_{\epsilon \to 0} \epsilon \int_{-A}^{B} \frac{f(x)}{x^2 + \epsilon^2} dx,$$

where δ is a small positive number.

b) Use plausible arguments (of the type found, for example, in Section 5.3) to show that $I_{\pm}$ can be written as

$$I_{\pm} = \int_{-A}^{-\delta} \frac{f(x)}{x} dx + \int_{\delta}^{+B} \frac{f(x)}{x} dx + f(0) \lim_{\epsilon \to 0} \int_{-\delta}^{\delta} \frac{x\, dx}{x^2 + \epsilon^2} \mp i f(0) \lim_{\epsilon \to 0} \epsilon \int_{-A}^{B} \frac{dx}{x^2 + \epsilon^2}.$$

c) Show that the third term of the equation in (b) vanishes by symmetry. Doing the fourth integral explicitly, show that

$$I_{\pm} = P\int_{-A}^{B} \frac{f(x)}{x} dx \mp i f(0) \lim_{\epsilon \to 0} \left[\tan^{-1}\left(\frac{B}{\epsilon}\right) + \tan^{-1}\left(\frac{A}{\epsilon}\right) \right].$$

d) Let $\epsilon \to 0$ and use the definition of the δ-function to write, finally,

$$\lim_{\epsilon \to 0} \int_{-A}^{+B} \frac{f(x)}{x \pm i\epsilon} dx = P\int_{-A}^{+B} \frac{f(x)}{x} dx \mp i\pi f(0) = P\int_{-A}^{B} \frac{f(x)}{x} dx \mp i\pi \int_{-A}^{B} \delta(x) f(x)\, dx.$$

This is often written, rather cryptically, as the symbolic formula

$$\lim_{\epsilon \to 0} \frac{1}{x \pm i\epsilon} = P\frac{1}{x} \mp i\pi\delta(x).$$

26. Solve the integral equation

$$\frac{1}{\pi} P\int_{-\infty}^{\infty} \frac{u(x)}{x - t} dx = \frac{1}{1 + t^2} \equiv -v(t)$$

by finding the Hilbert transform of $v(t)$. Find a complex function $f(z) = u(x, y) + iv(x, y)$ such that $u(x, 0) \equiv u(x) = Re(f(z))$ and $v(x, 0) \equiv v(x) = Im(f(z))$ and verify that this complex function has the two properties (what are they?) that ensure that $u(x)$ and $v(x)$ are a Hilbert transform pair.

27. Derive a dispersion relation for $f(z)$ when $|f(z)/z|$ tends to a constant as $|z| \to \infty$.

28. Prove that a uniformly convergent series of analytic functions can be integrated term by term. Hence show that a uniformly convergent series of analytic functions is an analytic function.

[*Hint*: Use Morera's theorem of Problem 21.]

29. Find the order of the pole at the origin, the residue there, and the integral around a (small) path C enclosing the origin, but no other singularities, for each of the following functions:

a) $\cot z$, b) $\csc^2 z \log(1-z)$, c) $\dfrac{z}{\sin z - \tan z}$.

30. What are the positions and natures of the singularities and the residues at these singularities of the following functions in the z-plane, excluding the point at infinity?

a) $f_1(z) = \dfrac{\cot \pi z}{(z-1)^2}$, b) $f_2(z) = \dfrac{1}{z(e^z - 1)}$.

[*Note*: Both these functions have nonisolated and hence essential singularities at infinity because the origin for $f_1(1/z)$ and $f_2(1/z)$ is a limit point of poles.]

31. Problems on the evaluation of real definite integrals by contour integration:

a) $\displaystyle\int_0^{2\pi} \frac{d\theta}{a + b\sin\theta} = \frac{2\pi}{(a^2 - b^2)^{1/2}}, \qquad a > b > 0.$

b) $\displaystyle\int_0^{2\pi} \frac{\sin^2\theta\, d\theta}{a + b\cos\theta} = \frac{2\pi}{b^2}[a - (a^2 - b^2)^{1/2}], \qquad a > b > 0.$

c) $\displaystyle\int_0^{2\pi} \frac{d\theta}{(a + b\cos\theta)^2} = \frac{2\pi a}{(a^2 - b^2)^{3/2}}, \qquad a > b > 0.$

d) $\displaystyle\int_0^{\infty} \frac{dx}{1 + x^4} = \frac{\pi}{2\sqrt{2}}.$ e) $\displaystyle\int_0^{\infty} \frac{x^2\, dx}{(a^2 + x^2)^3} = \frac{\pi}{16a^3}.$

f) $\displaystyle\int_0^{\infty} \frac{\sin x\, dx}{x(a^2 + x^2)} = \frac{\pi}{2a^2}(1 - e^{-a}).$ g) $\displaystyle\int_{-\infty}^{\infty} \frac{\sin^2 x}{x^2}\, dx = \pi.$

h) $\displaystyle\int_0^{\infty} \frac{x^{2a-1}}{b^2 + x^2}\, dx = \frac{\pi b^{2(a-1)}}{2} \csc \pi a, \qquad 0 < a < 1.$

i) $\displaystyle\int_0^{\infty} \frac{\log x}{b^2 + x^2}\, dx = \frac{\pi \log b}{2b}.$

j) $\displaystyle P\int_{-\infty}^{\infty} \frac{1}{(\omega' - \omega_0)^2 + a^2}\,\frac{1}{\omega' - \omega}\, d\omega' = \frac{\pi}{a}\,\frac{\omega_0 - \omega}{(\omega_0 - \omega)^2 + a^2}.$

k) $\displaystyle\int_0^{\infty} \frac{dx}{x^3 + a^3} = \frac{2\pi}{3a^2\sqrt{3}}.$

l) Consider the real integral

$$G(x, x', \tau) = \frac{1}{2\pi}\int_{-\infty}^{\infty} e^{ik(x-x')} e^{-k^2\tau}\, dk,$$

where τ is real. In this case it is easy to show that

$$G(x, x', \tau) = \frac{1}{(4\pi\tau)^{1/2}} e^{-(x-x')^2/4\tau}.$$

Prove that this result still holds when τ is pure imaginary. [*Hint*: The value of the integral

$$\int_0^{\infty} e^{-iu^2}\, du$$

may be found by considering the integral of e^{-z^2} around the boundary of the circular sector $0 \le \theta \le \pi/4$, $0 \le r \le R$. In the limit $R \to \infty$, ***prove*** that the contribution to the integral over the circular arc goes to zero. The contributions over the remaining straight segments of the contour provide the value of the integral needed in the problem.]

Note: This contour integral provides the values of the two real definite integrals:

$$\int_{-\infty}^{\infty} \sin(x^2)\,dx = \int_{-\infty}^{\infty} \cos(x^2)\,dx = (\pi/2)^{1/2},$$

by taking real and imaginary parts. By making the change of variable $x^2 = t$, this result can be transformed into

$$\int_0^{\infty} \frac{\sin t}{\sqrt{t}}\,dt = \int_0^{\infty} \frac{\cos t}{\sqrt{t}}\,dt = (\pi/2)^{1/2}.$$

32. We have seen how the function $\pi \cot \pi z$ may be used to sum certain series. But it is useless if the series is an alternating one. Show how alternating series may be summed by using the function $\pi \csc(\pi z)$ in place of $\pi \cot(\pi z)$. Use the general result to prove that

$$\sum_{n=1}^{\infty} (-1)^{n+1} \frac{1}{n^2} = \pi^2/12.$$

33. Consider the problem of evaluating the integral of $z^{1/2}$ from A to D along the circular path C, shown in Fig. 6.24. We take the cut to lie along the ray $\theta = \alpha$ and define $z^{1/2} \equiv r^{1/2}e^{i\theta/2}$ for $\alpha \leq \theta < \alpha + 2\pi$. Compare this with the value obtained by integrating $z^{1/2}$ from A to D along the "keyhole" path $ABCD$. Do the results agree? Should they agree? Why?

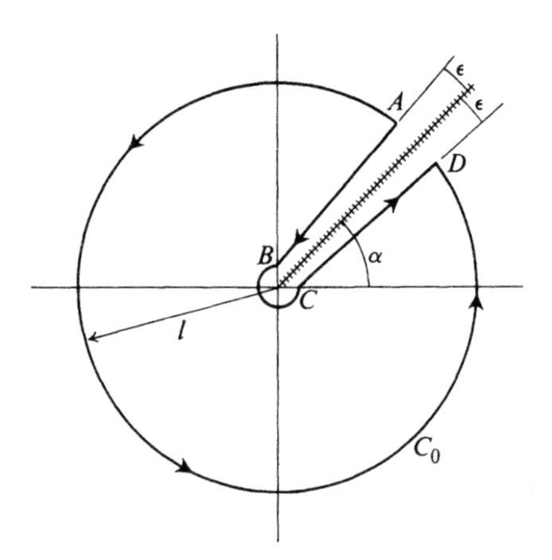

Fig. 6.24 The contours for Problem 6-34. The angles, ϵ, for the "keyhole" path are arbitrarily small. The radius of the circle, C_0, is l, and the cut is along the ray $\theta = \alpha$.

34. Consider the following response function, $G(t)$, which vanishes for negative values of the argument and which, for positive values, is given by

$$G(t) = G_0 \frac{\sin^2 \mu t}{t^{3/2}}.$$

Note that this function satisfies conditions (a), (b), (c) at the beginning of Section

6.6. Show that the Fourier transform, $g(\omega)$, is given by

$$g(\omega) = \frac{i-1}{2}G_0[\sqrt{\omega} - \tfrac{1}{2}\sqrt{\omega+2\mu} - \tfrac{1}{2}\sqrt{\omega-2\mu}], \qquad \omega \geq 2\mu ;$$

$$= \frac{i-1}{2}G_0[\sqrt{\omega} - \tfrac{1}{2}\sqrt{\omega+2\mu}] + \frac{i+1}{2}G_0[\tfrac{1}{2}\sqrt{2\mu-\omega}], \qquad 0 \leq \omega \leq 2\mu ,$$

$$g(-\omega) = g^*(\omega) .$$

Show that the complex-valued function of z which reduces to the above result on the real axis is

$$g(z) = \frac{i-1}{2}G_0[\sqrt{z} - \tfrac{1}{2}\sqrt{z+2\mu} - \tfrac{1}{2}\sqrt{z-2\mu}] .$$

Thus we see that $g(z)$ has branch points on the real axis,a possibility which is *not* excluded by conditions (a), (b), and (c). Note that we can choose the cuts to lie in the lower half-plane, so $g(z)$ is analytic in the upper half-plane.
[*Hint*: the integrals mentioned in the note following Problem 6.31 will be of use.]

35. The θ-function (or step function) is defined as

$$\theta(x) = \begin{cases} 0 & \text{for } x < 0 , \\ \frac{1}{2} & \text{for } x = 0 , \\ 1 & \text{for } x > 0 . \end{cases}$$

Prove

a) $\theta(x) = \dfrac{1}{2} + \dfrac{1}{2\pi i}P\displaystyle\int_{-\infty}^{\infty} \frac{e^{i\omega x}}{\omega}\, d\omega ,$ b) $\dfrac{d\theta(x)}{dx} = \delta(x) .$

[*Hint*: See Eq. (5.63)]

36. The Laguerre polynomials are generated by the generating function

$$\psi(x, t) = \frac{e^{-xt/(1-t)}}{1-t} = \sum_{n=0}^{\infty} \frac{L_n(x)}{n!}t^n , \qquad 0 < t < 1 .$$

They are given directly by the formula

$$L_n(x) = e^x \frac{d^n}{dx^n}(x^n e^{-x}) , \qquad 0 \leqq x \leqq \infty ,$$

analogous to Rodrigues' formula for Legendre polynomials.

a) By differentiating the generating function with respect to x [getting $(1-t)\psi' = -t\psi$], derive the recursion relation

$$L_n' - nL_{n-1}' = -nL_{n-1} .$$

By differentiating $\psi(x, t)$ with respect to t, another recursion relation can be derived. Show that

$$L_{n+1} - (2n+1-x)L_n + n^2 L_{n-1} = 0 .$$

b) From these two recursion relations, derive Laguerre's differential equation

$$xL_n'' + (1-x)L_n' + nL_n = 0 .$$

c) Show that $L_n(0) = n!$

d) Derive the generating function from Rodrigues' formula by contour integration.

37. In Section 5.10, we defined the Hermite polynomials in terms of the generating function

$$e^{-t^2+2tx} = \sum_{n=0}^{\infty} \frac{H_n(x)}{n!} t^n , \qquad t > 0 ,$$

and deduced from this the Rodrigues formula:

$$H_n(x) = (-1)^n e^{x^2} \frac{d^n}{dx^n} e^{-x^2} .$$

Now prove the converse: that Rodrigues' formula implies the generating function. Enroute, establish the integral representation

$$H_n(x) = \frac{(-1)^n n!}{2\pi i} e^{x^2} \oint_C \frac{e^{-z^2}}{(z - x)^{n+1}} dz ,$$

where C encloses the point x.

38. Prove the integral representation for the Hermite polynomials:

$$H_n(x) = \frac{2^n}{\pi^{1/2}} \int_{-\infty}^{\infty} (x + it)^n e^{-t^2} dt .$$

Use this result to derive an explicit series for the Hermite polynomials:

$$H_n(x) = \sum_{\substack{r=0 \\ (2r \leq n)}}^{n} \frac{(-1)^r n!}{r!(n - 2r)!} (2x)^{n-2r} .$$

39. Show that
    a) $\Gamma(x + 1) = x\Gamma(x)$ for $x > 0$
    b) $\Gamma(n + 1) = n!$, where n is a positive integer.

FURTHER READINGS

CHURCHILL, R. V., *Complex Variables and Applications*. Second edition. New York: McGraw-Hill, 1960. An excellent text of great clarity, it is ideal for self-study. Churchill's books have influenced a generation of physicists (including the authors).

FRANKLIN, P., *Functions of Complex Variables*. Englewood Cliffs, N. J.: Prentice Hall, 1958.

GOURSAT, E., *Functions of a Complex Variable*. New York: Dover Publications, 1966. Very readable.

HILLE, E., *Analytic Function Theory*, Volumes I and II. Boston: Ginn, 1959. An exhaustive account.

KNOPP, K., *Theory of Functions*, Parts I and II. New York: Dover, 1945. There are two accompanying problem books.

LANDAU, E., *Foundations of Analysis*. New York: Chelsea, 1960. An axiomatic treatment of the various number systems, including complex numbers.

SESHU, S., and N. BALABARIAN, *Linear Network Analysis*. New York: Wiley, 1959. This book treats engineering applications of dispersion relations.

SCHWEBER, S., *An Introduction to Relativistic Quantum Field Theory.* Evanston, Ill.: Row Peterson, 1961. See p. 411 for a discussion of causality and dispersion relations.

TITCHMARSH, E. C., *Introduction to the Theory of Fourier Integrals.* New York: Oxford University Press, 1937. Chapter 5 provides a rigorous treatment of the Hilbert transform pair.

WEYL, H., *Philosophy of Mathematics and Natural Science.* Princeton, N. J.: Princeton University Press, 1949. Chapter II is devoted to the various number systems.

CHAPTER 7

GREEN'S FUNCTIONS

INTRODUCTION

In this chapter, we turn our attention to integral operators. It turns out that there is a particularly close connection between differential and integral operators; and the boundary conditions with which one is familiar from the theory of differential equations play a crucial role in linking the two. The connection between them is essentially the theory of Green's functions, an area in theoretical physics which has become particularly important in connection with modern developments in quantum mechanics. We cannot hope to give a comprehensive treatment of this enormous subject in just one chapter, but we shall try to illustrate the main ideas. Green's functions can be useful in classical physics, but our emphasis here will be primarily on their role in quantum mechanics, for it has been largely through the formulation in terms of Green's functions that the mathematical structure of quantum mechanics, both in the relativistic and nonrelativistic regimes, has been clarified.

7.1 A NEW WAY TO SOLVE DIFFERENTIAL EQUATIONS

The simplest of all differential equations is undoubtedly

$$-i\frac{dy}{dx} = f(x) . \tag{7.1}$$

(We have put the i in front of the derivative for future convenience; note the analogy with the quantum-mechanical operator $p_x = -i\hbar d/dx$.) If we choose the initial condition to be $y(a) = y_0$, the solution to Eq. (7.1) may be written immediately:

$$y(x) = y_0 + i\int_a^x f(x')\,dx' . \tag{7.2}$$

Thus we have already solved a simple differential equation by integration.

But this is really a trivial example, because in these terms every integral one writes may be dignified by the title "A Solution to a Differential Equation." However, let us push the problem a bit further. Imagine that x lies within some closed interval $[a, b]$. Then we may rewrite Eq. (7.2) as

$$y(x) = y_0 + i\int_a^b \theta(x - x')f(x')\,dx' , \tag{7.3}$$

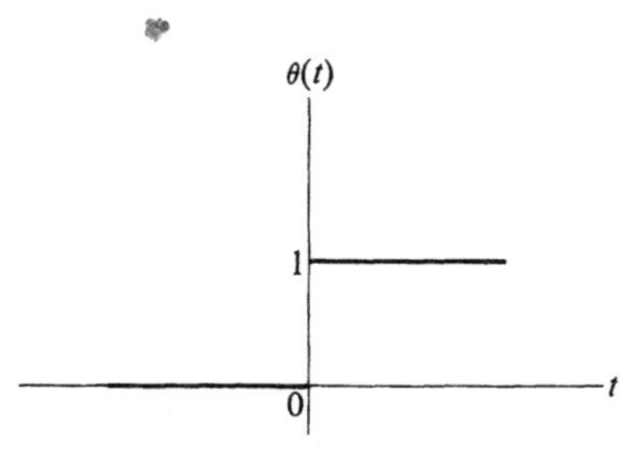

Fig. 7.1 The Heaviside function.

where $\theta(t)$ is the discontinuous step function defined by

$$\theta(t) = \begin{cases} 1, & t > 0, \\ 0, & t \leq 0, \end{cases} \tag{7.4}$$

$\theta(t)$ is sometimes called the Heaviside function (Fig. 7.1). Its presence in the integrand in Eq. (7.3) serves to cut off the integration over x' at $x = x'$; thus Eq. (7.3) is equivalent to Eq. (7.2). We note that $y(x)$ has the form

$$y(x) = y_0 + Kf(x), \tag{7.5}$$

where K is an integral operator defined by

$$Kf(x) \equiv i\int_a^b \theta(x - x')f(x')\,dx'. \tag{7.6}$$

Here $i\theta(x - x')$ is the *kernel* of the integral operator K, and when the kernel comes from the solution of an equation involving a differential operator, it is often referred to as the *Green's function* of that differential operator *for the relevant boundary conditions.* Thus

$$G_1(x, x') = i\theta(x - x') \tag{7.3'}$$

is the Green's function belonging to id/dx for a system subject to the boundary condition $y(a) = y_0$.

A more interesting example is the equation

$$\frac{d^2y}{dx^2} = f(x), \tag{7.7}$$

where $y(x)$ is subject to the one-point boundary conditions, $y(a) = y_0$, $y'(a) = \bar{y}_0$. Integrating Eq. (7.7) once, we obtain

$$\frac{dy}{dx} = \bar{y}_0 + \int_a^x f(x')\,dx';$$

a second integration gives

$$y(x) = y_0 + \bar{y}_0(x - a) + \int_a^x dx'' \int_a^{x''} dx'\,f(x').$$

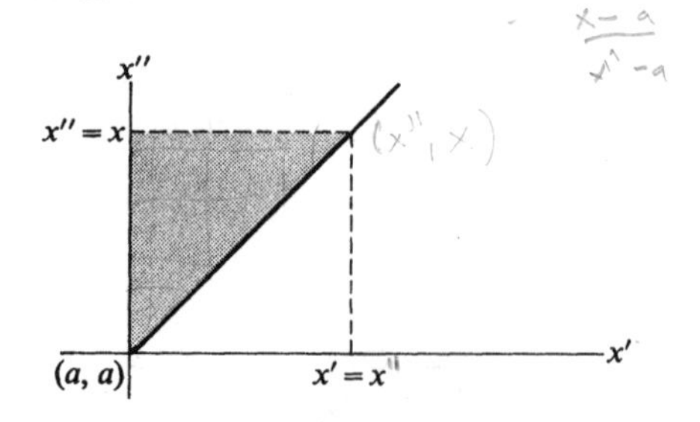

Fig. 7.2

In this expression we are integrating over a triangular region in the $x'x''$ plane (the shaded region in Fig. 7.2) by first integrating x' from $x' = a$ to $x' = x''$, and then integrating x'' from a to x. However, we can equally well integrate x'' from $x'' = x'$ to $x'' = x$ and then integrate x' from a to x. Since the function being integrated depends only on x', it is advantageous to change the order and limits of integration in this manner. We get

$$y(x) = y_0 + \dot{y}_0(x - a) + \int_a^x dx' \, f(x') \int_{x'}^x dx''$$

so that

$$y(x) = y_0 + \dot{y}_0(x - a) + \int_a^x (x - x')f(x') \, dx' \,. \tag{7.8}$$

The reader may check this by differentiating $y(x)$ twice and comparing the result with Eq. (7.3). Again, if we restrict x to the interval $[a, b]$, we can write

$$y(x) = y_0 + \dot{y}_0(x - a) + \int_a^b (x - x')\theta(x - x')f(x') \, dx' \,. \tag{7.9}$$

This solution for $y(x)$ has exactly the same form as the solution to Eq. (7.1). Using the terminology introduced earlier, we may say that the function

$$G_2(x, x') = (x - x')\theta(x - x') \tag{7.9'}$$

is the Green's function corresponding to the operator d^2/dx^2 for the boundary conditions $y(a) = y_0$, $y'(a) = \dot{y}_0$. Note that this Green's function is continuous while the Green's function of the operator d/dx is discontinuous (at $x = x'$).

In both Eqs. (7.3) and (7.9), the term on the right-hand side which is not under the integral is a solution of the homogeneous equation for the given boundary condition. We have

$$\frac{dy}{dx} = 0$$

in the case of Eq. (7.3), and

$$\frac{d^2y}{dx^2} = 0$$

in the case of Eq. (7.9). Therefore, the general solution to Eq. (7.7) is

$$y(x) = \alpha + \beta x + \int_a^b (x - x')\theta(x - x')f(x')\,dx' , \tag{7.10}$$

where α and β are to be determined by the boundary conditions.

For example, if we have a two-point boundary condition

$$y(0) = y_0,\ y(1) = y_1 ,$$

the solution interior to the interval [0, 1] will be

$$y(x) = \alpha + \beta x + \int_0^1 (x - x')\theta(x - x')f(x')\,dx' ,$$

where α and β are determined via the equations

$$y_0 = \alpha - \int_0^1 x'\theta(-x')f(x')\,dx' = \alpha ,$$

$$y_1 = \alpha + \beta + \int_0^1 (1 - x')\theta(1 - x')f(x')\,dx' .$$

Thus

$$\alpha = y_0 , \qquad \beta = y_1 - y_0 - \int_0^1 (1 - x')f(x')dx' .$$

Hence the solution looks like

$$y(x) = y_0 + (y_1 - y_0)x - \int_0^1 x(1 - x')f(x')\,dx' + \int_0^1 (x - x')\theta(x - x')f(x')\,dx' .$$

Calling

$$G_3(x, x') = -x(1 - x') + (x - x')\theta(x - x') ,$$

we have

$$y(x) = y_0 + (y_1 - y_0)x + \int_0^1 G_3(x, x')f(x')\,dx' . \tag{7.11}$$

Let us look at $G_3(x, x')$ more closely. When $0 \le x' \le x$,

$$G_3(x, x') = -x(1 - x') + (x - x') = -x'(1 - x) ;$$

when $x \le x' \le 1$,

$$G_3(x, x') = -x(1 - x') .$$

To summarize, the solution to Eq. (7.7) is given by Eq. (7.11), where $G_3(x, x')$ is

$$G_3(x, x') = \begin{cases} -x'(1 - x) , & 0 \le x' \le x , \\ -x(1 - x') , & x \le x' \le 1 . \end{cases} \tag{7.11'}$$

Using the function $\theta(x)$, $G_3(x, x')$ can also be written conveniently as

$$G_3(x, x') = xx' - x'\theta(x - x') - x\theta(x' - x) . \qquad (7.11'')$$

Note that $G_3(x, x')$ vanishes at $x = 0$ and $x = 1$, and also at $x' = 0$ and $x' = 1$.

It is interesting to observe the dramatic effect which the boundary conditions have on the form of the Green's function. For one thing, the functional form changes. Moreover, the Green's function of Eq. (7.11″) is a symmetric function of x and x', since

$$G_3(x, x') = G_3(x', x);$$

but the Green's function of Eq. (7.9′) is not symmetric. However, both $G_2(x, x')$ and $G_3(x, x')$ are Green's functions of the operator d^2/dx^2; the only difference lies in the boundary conditions.

Let us now examine the effect of changing boundary conditions in a first-order equation. We return to Eq. (7.1) and this time try the boundary condition $y(1) = Cy(0)$. The solution will have the general form

$$y(x) = A + i \int_0^1 \theta(x - x')f(x')\, dx' .$$

Now we require that

$$y(1) = A + i \int_0^1 f(x')\, dx' = Cy(0) = CA;$$

here we have substituted both $x=1$ and $x=0$ into the above equation. From this we find immediately that

$$A = -\frac{i}{1 - C} \int_0^1 f(x')\, dx' .$$

Thus

$$y(x) = -\frac{i}{1 - C} \int_0^1 f(x')\, dx' + i \int_0^1 \theta(x - x')f(x')\, dx' ,$$

or

$$y(x) = \int_0^1 G_4(x, x')f(x')\, dx' , \qquad (7.12)$$

where

$$G_4(x, x') = -i\left[\frac{1}{1 - C} - \theta(x-x')\right].$$

Thus

$$G_4(x, x') = \begin{cases} -\dfrac{iC}{1 - C}, & x > x' , \\ -\dfrac{i}{1 - C}, & x \leq x' . \end{cases}$$

If we demand that $|C| = 1$, then

$$G_4(x, x') = \begin{cases} -\dfrac{i}{C^* - 1}, & x > x', \\ +\dfrac{i}{C - 1}, & x \leq x', \end{cases} \tag{7.12'}$$

from which we see that $G_4(x, x')$ is symmetric in the sense that

$$G(x, x') = G(x', x)^* .$$

This definition of symmetry is motivated by the definition of a Hermitian matrix as being one whose elements satisfy

$$a_{ij} = a_{ji}^* .$$

As we shall see, the relationship between symmetric functions like $G_4(x, x')$ and Hermitian matrices is a very close and fruitful one. For the present, we may conclude that even in the simple case of a single derivative, the choice of boundary condition can mean the difference between a symmetric and a non-symmetric Green's function [since $G_1(x, x')$ of Eq. (7.3′) is not symmetric].

The fact that certain boundary conditions lead to a Green's function which is symmetric while others do not is readily made plausible. The reader can easily show that on the space of square-integrable functions which satisfy $f(1) = Cf(0)$ with $C = e^{i\phi}$ the differential operator $p = -id/dx$ is symmetric in the sense that $(f, pg) = (pf, g)$. Here $(\ ,\ )$ denotes the inner product of Eq. (5.3) with $a = 0$ and $b = 1$. However, p will not be symmetric if we require only that the functions in the space take on a fixed value at $x = 0$. Similar comments apply to the operator d^2/dx^2, if one remembers that in finding a Green's function for d^2/dx^2 acting on the space of square-integrable functions which take on fixed values at $x = 0$ and $x = 1$ one can always choose the endpoint values to be zero. This follows simply from the fact that if we solve

$$\frac{d^2\eta}{dx^2} = f(x)$$

subject to $\eta(0) = 0$ and $\eta(1) = 0$, then the solution to

$$\frac{d^2y}{dx^2} = f(x)$$

with $y(0) = y_0$ and $y(1) = y_1$ is just $y(x) = y_0 + (y_1 - y_0)x + \eta(x)$. Because the Green's functions $G_3(x, x')$ and $G_4(x, x')$ are symmetric functions of x and x', it is a simple matter to see that the integral operators K_3 and K_4 which have $G_3(x, x')$ and $G_4(x, x')$, respectively, as kernels are symmetric operators in the same sense that p was symmetric, that is, $(f, K_3g) = (K_3f, g)$ and $(f, K_4g) = (K_4f, g)$. Thus, it is not surprising that there appears to be a close relationship between symmetric differential operators and symmetric Green's functions.

The reader may well wonder why we have bothered developing such an elaborate way of integrating simple differential equations. One very important reason is the following. Suppose that, instead of Eq. (7.1), we had to solve

$$-i\frac{dy}{dx} = f(x, y)\ . \tag{7.13}$$

This is a true differential equation in the sense that it is not immediately integrable. By our previous work we have

$$y(x) = \alpha + \int_0^1 G(x, x')f[x', y(x')]\, dx'\ . \tag{7.14}$$

If the boundary condition is $y(0) = y_0$, then $\alpha = y_0$, and $G(x, x') = -i\theta(x - x')$; if the boundary condition is $y(1) = Cy(0)$, where $|C| = 1$, then $\alpha = 0$, and $G(x, x')$ is given by the symmetric function $G_4(x, x')$ of Eq. (7.12). Since the unknown function $y(x)$ now appears under the integral sign, we have transformed the *differential* equation into an *integral* equation. Similarly, if we have

$$\frac{d^2y}{dx^2} = f(x, y)$$

with $y(0) = y_0$, $y(1) = y_1$, we get as the corresponding integral equation

$$y(x) = y_0 + (y_1 - y_0)x + \int_0^1 G_3(x, x')f[x', y(x')]\, dx'\ ,$$

where $G_3(x, x')$ is the symmetric function given by Eq. (7.11′). For example, the well-known second-order linear differential equation

$$\frac{d^2y}{dx^2} + \lambda y = 0\ , \qquad y(0) = 0 = y(1)$$

is transformed into the integral equation

$$y(x) = -\lambda \int_0^1 G_3(x, x')y(x')\, dx'\ .$$

There is one very interesting and tricky aspect of these equations which should be mentioned here. Crudely speaking, the differential operator id/dx is *unbounded* in the sense that it can act on functions of unit norm in Hilbert space and transform them into functions with arbitrarily large norm. For example, the norm of e^{inx} in the Hilbert space of square-integrable functions on $[0, 1]$ is 1, but the norm of

$$i\frac{d}{dx}[e^{inx}] = -ne^{inx}$$

is n, which can be as large as we like. As we shall see later, the theory of such unbounded operators is rather subtle.

However, the integral operator whose kernel is $G_4(x, x')$ is *bounded*, in the sense that it transforms any square-integrable function of unit norm into a function whose norm is less than some given number. As a general rule, let

$$f(x) = \int_0^1 K(x, y)\, g(y)\, dy\ .$$

If we assume that $g(y)$ is square-integrable and that $K(x, y)$ is square-integrable in y, for all x, then by Schwartz's inequality (Eq. 4.10),

$$|f(x)| \leq \left[\int_0^1 |K(x, y)|^2 \, dy\right]^{1/2} \left[\int_0^1 |g(y)|^2 \, dy\right]^{1/2} .$$

Thus, if g has unit norm, we get

$$|f(x)| \leq \left[\int_0^1 |K(x, y)|^2 \, dy\right]^{1/2} .$$

But this means that

$$\int_0^1 |f(x)|^2 \, dx \leq \int_0^1 \int_0^1 |K(x, y)|^2 \, dx \, dy .$$

If $K(x, y)$ is square integrable in the xy-plane, which is the case for all the Green's functions discussed above, then calling

$$\int_0^1 \int_0^1 |K(x, y)|^2 \, dx \, dy = C ,$$

we have

$$\int_0^1 |f(x)|^2 \, dx \equiv ||f||^2 \leq C$$

for *any* function $g(x)$ of unit norm. Kernels of the square-integrable type actually have many other nice properties. For instance, the kernel $G_4(x, x')$ transforms the sequence of functions of unit norm $\{e^{inx}\}$ into a sequence of functions whose norms tend toward zero. In fact, for kernels of this type, *any* infinite bounded sequence of functions (i.e., functions $f_n(x)$ satisfying $||f_n|| \leq C$ for all n) will be transformed into a sequence containing a subsequence which is convergent in the mean. Operators of this type are said to be *completely continuous*, and we shall have a great deal more to say about them in Chapter 9. The analysis of these operators is much more straightforward than that of unbounded operators, and, in fact, bears a very close resemblance to ordinary matrix theory. Thus by turning a differential equation into an integral equation, or, in other words, by turning a differential operator into an integral operator, it is often possible to effect significant mathematical simplifications in a given problem.

7.2 GREEN'S FUNCTIONS AND DELTA FUNCTIONS

With the general idea of a Green's function in mind, let us now approach the problem of solving a differential equation in a slightly different manner. Suppose that we have the equation

$$Ly = f , \tag{7.15}$$

where L is some linear ordinary differential operator and f is a given function. Throwing aside questions of rigor, suppose now that L possesses a complete, orthonormal set of eigenfunctions $\{\phi_n(x)\}$ so that

$$L\phi_n(x) = \lambda_n\phi_n(x) .$$

Under these circumstances we could write (omitting the dots over the equal signs which denote mean convergence)

$$y(x) = \sum_{n=1}^{\infty} \alpha_n\phi_n(x) , \tag{7.16a}$$

$$f(x) = \sum_{n=1}^{\infty} \beta_n\phi_n(x) . \tag{7.16b}$$

The sum runs to infinity, which reflects the fact that in an infinite-dimensional Hilbert space, a complete orthonormal set contains an infinite number of elements.

If we now substitute Eqs. (7.16a) and (7.16b) into Eq. (7.15), we get

$$Ly = L\sum_{n=1}^{\infty} \alpha_n\phi_n = \sum_{n=1}^{\infty} L\alpha_n\phi_n = \sum_{n=1}^{\infty} \alpha_n\lambda_n\phi_n = f = \sum_{n=1}^{\infty} \beta_n\phi_n .$$

Thus

$$\sum_{n=1}^{\infty} (\alpha_n\lambda_n - \beta_n)\phi_n = 0 .$$

Since the ϕ_n are linearly independent, we conclude that

$$\alpha_n = \beta_n/\lambda_n ;$$

thus

$$y(x) = \sum_{n=1}^{\infty} \frac{1}{\lambda_n} \beta_n\phi_n(x) , \tag{7.17}$$

where $\beta_n = (\phi_n, f)$. If some $\lambda_n = 0$, then a solution exists only if the corresponding $\beta_n = 0$. In this case, the solution will not be unique; an arbitrary multiple of the ϕ_n corresponding to $\lambda_n = 0$ can be added to any solution (see Theorem 3.17). For the moment, then, we assume that $\lambda_n \neq 0$ for all n.

Let us write Eq. (7.17) in more detail:

$$\begin{aligned} y(x) &= \sum_{n=1}^{\infty} \frac{1}{\lambda_n} \phi_n(x)(\phi_n, f) \\ &= \sum_{n=1}^{\infty} \frac{1}{\lambda_n} \phi_n(x) \int \phi_n^*(x') f(x')\, dx' \\ &= \int \sum_{n=1}^{\infty} \frac{\phi_n(x)\phi_n^*(x')}{\lambda_n} f(x')\, dx' . \end{aligned}$$

Thus

$$y(x) = \int G(x, x') f(x')\, dx' , \tag{7.18}$$

where

$$G(x, x') = \sum_{n=1}^{\infty} \frac{\phi_n(x)\phi_n^*(x')}{\lambda_n} . \tag{7.18'}$$

(Here we have interchanged sums and integrals in a rather cavalier manner, i.e., without establishing the uniform convergence of the sum.) Thus we obtain a representation of the Green's function in terms of an infinite eigenfunction expansion. If the λ_n are real, then $G(x, x')$ is symmetric:

$$G(x, x') = G(x', x)^* .$$

As an example, consider the operator

$$L = \frac{d^2}{dx^2}$$

for x in $[0, 1]$. The normalized eigenfunctions of L on $[0, 1]$, which vanish at the endpoints, are $\phi_n(x) = \sqrt{2} \sin(n\pi x)$, $n = 1, 2, \cdots$, with eigenvalues $\lambda_n = -n^2\pi^2$. Thus the Green's function for this case is

$$G(x, x') = -\frac{2}{\pi^2} \sum_{n=1}^{\infty} \frac{\sin(n\pi x) \sin(n\pi x')}{n^2} . \tag{7.19}$$

Since the series is dominated term by term by $1/n^2$, we have uniform convergence to a continuous function. In fact, the function to which the series of Eq. (7.19) converges must be precisely the function $G_3(x, x')$ given by Eq. (7.11'), since we have already determined by another (more rigorous) method that this is the Green's function of d^2/dx^2 with the two-endpoint boundary condition. It is interesting to calculate $G_3(x, x')$ at a few points to see whether Eqs. (7.11') and (7.19) agree. For example, at $x = x' = \frac{1}{2}$, $G_3(\frac{1}{2}, \frac{1}{2}) = -\frac{1}{4}$, while Eq. (7.19) gives

$$G\left(\frac{1}{2}, \frac{1}{2}\right) = -\frac{2}{\pi^2} \sum_{n=1}^{\infty}{}' \frac{1}{n^2} ,$$

where the prime on the summation sign means that the sum runs only over odd values of n. Using the results of Eq. (6.61) and Problem (6.32) one readily obtains the result

$$\sum_{n=1}^{\infty}{}' \frac{1}{n^2} = \frac{\pi^2}{8} ,$$

so that $G(\frac{1}{2}, \frac{1}{2}) = -\frac{1}{4}$, in agreement with the previously obtained result for $G_3(\frac{1}{2}, \frac{1}{2})$.

It is also instructive to look at the operator $p = -id/dx$ acting on the space of square-integrable functions on $[0, 1]$ which satisfy $f(1) = Cf(0)$, where $C = e^{i\phi}$ $(0 \leq \phi < 2\pi)$. We have considered this operator from a different point of view in the previous section. The normalized eigenfunctions of p are readily seen to be $\phi_n = e^{i(2\pi n + \phi)x}$ $(n = 0, \pm 1, \pm 2, \cdots)$; the corresponding eigenvalues are $\lambda_n = 2\pi n + \psi$. Thus, the Green's function for this operator is

$$G(x, x') = \sum_{n=-\infty}^{\infty} \frac{e^{i(2\pi n+\psi)(x-x')}}{2\pi n + \psi} .$$

Note that if $\psi = 0$, that is, if $C = 1$, this expression for $G(x, x')$ is ill-defined because of a zero denominator when $n = 0$. This illustrates the type of difficulty that arises when an operator has an eigenvector with a zero eigenvalue. In the above case, $\lambda_0 = 0$, so the solution of Eq. (7.17) breaks down, and the problem posed in Eq. (7.15) does not have a solution unless $(\phi_0, f) = 0$. Even if $(\phi_0, f) = 0$, the solution to Eq. (7.15) is not unique. However, if $\psi \neq 0$, the problem is well-posed, and it is, in fact, a simple matter to determine $G(x, x')$. We shall see in the next section that the Green's function for p must be constant except when $x = x'$. To evaluate $G(x, x')$ when $x > x'$ we may take $x - x' = 1/2$ without loss of generality. Thus,

$$G(x, x') = \frac{1}{2\pi} e^{i\psi/2} \sum_{n=-\infty}^{\infty} \frac{(-1)^n}{n + \psi/2\pi} \qquad (x > x') ,$$

and similarly, taking $x - x' = -\frac{1}{2}$,

$$G(x, x') = \frac{1}{2\pi} e^{-i\psi/2} \sum_{n=-\infty}^{\infty} \frac{(-1)^n}{n + \psi/2\pi} \qquad (x < x') .$$

From the result of Problem 6.32 (see also Eq. (6.57)) the sum in question can be readily evaluated; the reader will find that the Green's function obtained in this way agrees precisely with the Green's function $G_4(x, x')$ of Eq. (7.12′) which was obtained in a more elementary fashion for the operator p.

An interesting way of viewing the Green's function is to operate on $G(x, x')$ with L:

$$LG(x, x') = L \sum_{n=1}^{\infty} \frac{\phi_n(x)\phi_n^*(x')}{\lambda_n} .$$

If we interchange summation and the operation denoted by L, we have, using $L\phi_n(x) = \lambda_n \phi_n(x)$,

$$LG(x, x') = \sum_{n=1}^{\infty} \frac{L\phi_n(x)\phi_n^*(x')}{\lambda_n} = \sum_{n=1}^{\infty} \phi_n(x)\phi_n^*(x') \equiv I(x, x') .$$

For any function $f(x)$,

$$\begin{aligned} \int I(x, x')f(x')\, dx' &= \sum_{n=1}^{\infty} \phi_n(x) \int \phi_n^*(x')f(x')\, dx' \\ &= \sum_{n=1}^{\infty} \phi_n(x)(\phi_n, f) . \end{aligned}$$

But since the ϕ_n are assumed to be a complete orthonormal set, this equation may be written

$$\int I(x, x')f(x')\, dx = f(x)\,.$$

We have already seen in Chapter 5 that the "function" $I(x, x')$ which has this property is just the Dirac δ-function, $\delta(x - x')$. Thus we may write

$$LG(x, x') = \delta(x - x')\,, \tag{7.20}$$

an equation which is perhaps better written in the form

$$LK = I\,, \tag{7.21}$$

where K is the integral operator defined by

$$Kf = \int G(x, x')f(x')\, dx'\,.$$

Thus, in terms of Eq. (7.21), finding the Green's function of an operator L is tantamount to *inverting* L, so that equations like $Ly = f$ can be solved according to

$$y = L^{-1}f = Kf = \int G(x, x')f(x')\, dx'\,.$$

This analysis tells us that if we go back to our simplest of Green's functions, $G_1(x, x')$ of Eq. (7.3′), we should find that

$$-i\frac{d}{dx}G_1(x, x') = \delta(x - x')\,.$$

Using Eq. (7.3′), we find that this becomes

$$\frac{d}{dx}\theta(x - x') = \delta(x - x')$$

or

$$\frac{d}{dt}\theta(t) = \delta(t)\,. \tag{7.22}$$

Let us see whether $d\theta/dt$ has the properties of a δ-function discussed in Section 5.3. Using the definition of $\theta(t)$, we see that

$$\int_{-\epsilon}^{\epsilon}\frac{d\theta(t)}{dt}\, dt = \theta(\epsilon) - \theta(-\epsilon) = 1\,,$$

so one basic property of the δ-function is trivially satisfied. Also, since $\theta(t)$ is a constant everywhere except at $t = 0$, $d\theta/dt$ vanishes everywhere except at $t = 0$. Now we investigate the effect of $d\theta(t)/dt$ as a factor in an integrand:

$$\int_{-\epsilon}^{\epsilon}\frac{d\theta(t)}{dt}f(t)\, dt = [\theta(t)f(t)\,]_{-\epsilon}^{\epsilon} - \int_{-\epsilon}^{\epsilon}\theta(t)f'(t)\, dt\,,$$

where we have integrated once by parts. Using the definition of $\theta(t)$, we get

$$\int_{-\epsilon}^{\epsilon} \frac{d\theta}{dt} f(t)\, dt = f(\epsilon) - \int_0^{\epsilon} f'(t)\, dt$$
$$= f(\epsilon) - [f(\epsilon) - f(0)] = f(0) .$$

Hence $d\theta/dt$ has the most important of the δ-function properties, namely,

$$\int_{-\epsilon}^{\epsilon} \delta(t) f(t)\, dt = f(0) .$$

An intuitive way of seeing that $d\theta/dt = \delta(t)$ is to write an approximate $\theta(t)$ as

$$\theta_\epsilon(t) = \begin{cases} 0, & t < -\epsilon, \\ \dfrac{1}{2} + \dfrac{1}{2\epsilon} t, & -\epsilon \leq t \leq \epsilon, \\ 1, & t > \epsilon. \end{cases}$$

Clearly, $\theta_\epsilon(t) \to \theta(t)$ as $\epsilon \to 0$. Now we compute $d\theta_\epsilon(t)/dt$ and find that

$$\theta'_\epsilon(t) = \begin{cases} 0, & t < -\epsilon, \\ 1/2\epsilon, & -\epsilon \leq t \leq \epsilon, \\ 0, & t > \epsilon. \end{cases}$$

The sequence of functions $\{\theta'_\epsilon(t)\}$ is one of those we showed in Section 5.3, which approximated the δ-function. Figure 7.3 shows the function $\theta_\epsilon(t)$ and $\theta'_\epsilon(t)$. With these properties of the δ-function in mind, the reader should verify that all the Green's functions derived in the last section satisfy Eq. (7.20).

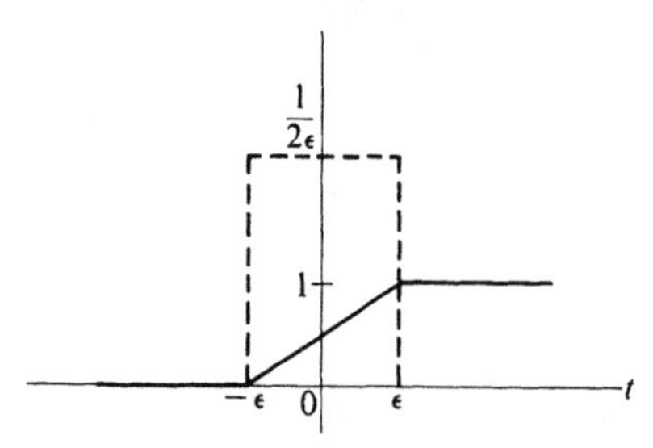

Fig. 7.3 The solid line represents the function $\theta_\epsilon(t)$; the dashed line represents the function $\theta'_\epsilon(t)$.

The differential representation of the δ-function in Eq. (7.22) is very useful, and for future reference we note a few of its simple consequences. First, since

$$\theta(at) = \theta(t) \quad \text{if} \quad a > 0,$$
$$\theta(at) = 1 - \theta(t) \quad \text{if} \quad a < 0,$$

we obtain by differentiation

$$a\frac{d\theta(at)}{d(at)} = \frac{d\theta}{dt}, \qquad a > 0,$$

$$a\frac{d\theta(at)}{d(at)} = -\frac{d\theta}{dt}, \qquad a < 0.$$

Thus

$$a\delta(at) = \delta(t), \qquad a > 0,$$

$$a\delta(at) = -\delta(t), \qquad a < 0,$$

or

$$\delta(at) = \frac{1}{|a|}\delta(t). \tag{7.23}$$

A second useful result concerns the function signum(x), which is usually written sgn (x). This is a function which equals $+1$ when x is positive and -1 when x is negative. Clearly,

$$\text{sgn}\,(x) = \theta(x) - \theta(-x),$$

so

$$\frac{d}{dx}[\text{sgn}\,(x)] = \frac{d\theta(x)}{dx} + \frac{d\theta(-x)}{d(-x)} = \delta(x) + \delta(-x) = 2\delta(x),$$

that is,

$$\frac{d}{dx}\,\text{sgn}\,(x) = 2\delta(x). \tag{7.24}$$

As an immediate consequence, we see that

$$\frac{d^2}{dx^2}|x| = \frac{d^2}{dx^2}[x\,\text{sgn}\,(x)] = \frac{d}{dx}[\text{sgn}\,(x) + 2x\delta(x)] = \frac{d}{dx}[\text{sgn}\,(x)] = 2\delta(x).$$

Here we have used the fact that $x\delta(x) = 0$ and Eq. (7.24).

With these rather informal interpretations of Green's functions and the δ-function in mind, we turn to some applications.

7.3 GREEN'S FUNCTIONS IN ONE DIMENSION

Having studied a few simple operators by way of introduction, we shall now try our hand at operators of a more general type and at the same time introduce another technique for the study of Green's functions. We shall begin with an example from classical mechanics, but before we are finished, it will be apparent that the applications arising from this special case are considerable indeed.

We consider the equation of the forced, damped harmonic oscillator,

$$\ddot{x}(t) + 2\gamma\dot{x}(t) + \omega_0^2 x(t) = F(t), \tag{7.25}$$

where we indicate differentiation with respect to time by a dot. We have chosen units of mass such that $m = 1$; γ is essentially the coefficient of frictional drag, ω_0 is the square root of the (positive) spring constant, and $F(t)$ is the external driving force. We assume that $F(t)$ has a Fourier transform,

$$\hat{F}(k) = \frac{1}{\sqrt{2\pi}} \int_{-\infty}^{\infty} e^{ikt} F(t) dt \, .$$

Since we have assumed that $F(t)$ has a Fourier transform, $F(t)$ must go to zero for large $|t|$. This, coupled with the presence of damping, makes it reasonable to assume that $x(t)$ and its first derivative go to zero for large $|t|$, and therefore their Fourier transforms exist. Taking the Fourier transform of both sides of Eq. (7.25), we get

$$\frac{1}{\sqrt{2\pi}} \int_{-\infty}^{\infty} (\ddot{x} + 2\gamma \dot{x} + \omega_0^2 x) e^{ikt} \, dt = \hat{F}(k) \, .$$

Let

$$\hat{x}(k) = \frac{1}{\sqrt{2\pi}} \int_{-\infty}^{\infty} e^{ikt} x(t) \, dt;$$

then

$$\frac{1}{\sqrt{2\pi}} \int_{-\infty}^{\infty} \frac{d^2x}{dt^2} e^{ikt} \, dt + 2\gamma \frac{1}{\sqrt{2\pi}} \int_{-\infty}^{\infty} \frac{dx}{dt} e^{ikt} \, dt + \omega_0^2 \hat{x}(k) = \hat{F}(k) \, . \tag{7.26}$$

In the first term we integrate by parts once to get

$$\begin{aligned} \int_{-\infty}^{\infty} \frac{d^2x}{dt^2} e^{ikt} dt &= \left[\frac{dx}{dt} e^{ikt} \right]_{-\infty}^{\infty} - \int_{-\infty}^{\infty} \left(\frac{dx}{dt} \right) \frac{d}{dt} (e^{ikt}) \, dt \\ &= -ik \int_{-\infty}^{\infty} \left(\frac{dx}{dt} \right) e^{ikt} \, dt \, . \end{aligned}$$

Here the contribution from the endpoints vanishes since we assume that $x(t)$ and $\dot{x}(t)$ may be Fourier transformed and therefore vanish at large distances. Thus, Eq. (7.26) implies that

$$\frac{(2\gamma - ik)}{\sqrt{2\pi}} \int_{-\infty}^{\infty} \left(\frac{dx}{dt} \right) e^{ikt} \, dt + \omega_0^2 \hat{x}(k) = \hat{F}(k) \, . \tag{7.27}$$

Finally, we integrate by parts once more and use the boundary conditions at infinity to find

$$\frac{1}{\sqrt{2\pi}} \int_{-\infty}^{\infty} \left(\frac{dx}{dt} \right) e^{ikt} = -ik \frac{1}{\sqrt{2\pi}} \int_{-\infty}^{\infty} x(t) e^{ikt} \, dt = -ik\hat{x}(k) \, .$$

Hence Eq. (7.27) becomes

$$(-k^2 - 2i\gamma k + \omega_0^2) \hat{x}(k) = \hat{F}(k) \, . \tag{7.28}$$

Since $(\omega_0^2 - 2i\gamma k - k^2)$ has no zeroes on the real line,

$$\hat{x}(k) = \frac{\hat{F}(k)}{-k^2 - 2i\gamma k + \omega_0^2},$$

and by the basic result on Fourier transforms,

$$x(t) = \frac{1}{\sqrt{2\pi}} \int_{-\infty}^{\infty} \frac{\hat{F}(k)}{-k^2 - 2i\gamma k + \omega_0^2} e^{-ikt} \, dk \, .$$

Using the definition of $\hat{F}(k)$, we get

$$x(t) = \frac{1}{2\pi} \int_{-\infty}^{\infty} dt' \int_{-\infty}^{\infty} dk \, \frac{F(t') e^{ikt'}}{-k^2 - 2i\gamma k + \omega_0^2} e^{-ikt} \, .$$

Finally, we can add an arbitrary linear combination of solutions to the homogeneous equation to get the general solution

$$x(t) = Ax_1(t) + Bx_2(t) + \int_{-\infty}^{\infty} G(t, t') F(t') \, dt' \, , \tag{7.29a}$$

where

$$G(t, t') = \frac{1}{2\pi} \int_{-\infty}^{\infty} \frac{e^{-ik(t-t')}}{-k^2 - 2i\gamma k + \omega_0^2} dk \, , \tag{7.29b}$$

and A and B must be chosen to satisfy the boundary conditions. For the two linearly independent solutions of the homogeneous equation, we may take

$$\begin{aligned} x_1(t) &= e^{-\gamma t} \sin [\sqrt{\omega_0^2 - \gamma^2}\, t] \, , \\ x_2(t) &= e^{-\gamma t} \cos [\sqrt{\omega_0^2 - \gamma^2}\, t] \, . \end{aligned} \tag{7.30}$$

Let us now see if we can evaluate $G(t, t')$. First we write

$$G(t, t') = -\frac{1}{2\pi} \int_{-\infty}^{\infty} \frac{e^{-ik(t-t')} \, dk}{(k - k_1)(k - k_2)} \, , \tag{7.31}$$

where $k_1 = (\omega_0^2 - \gamma^2)^{1/2} - i\gamma$ and $k_2 = -(\omega_0^2 - \gamma^2)^{1/2} - i\gamma$. We will assume for definiteness that $\omega_0 > \gamma$, but this is in no way essential in what follows. The important fact about k_1 and k_2 is that their imaginary parts are always *negative*. Our work on complex variable theory puts us in a position to do this integral easily: we consider the integral in Eq. (7.31) as part of a contour integral in the complex k-plane. The relevant contour—which will be labeled C—runs along the real axis from $-K$ to $+K$ and then closes in the lower half-plane by

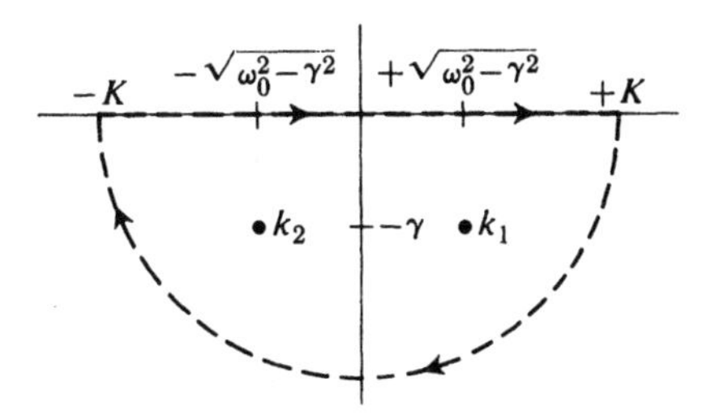

Fig. 7.4 The complex k-plane showing the contour used in evaluating the Green's function for the damped harmonic oscillator.

means of a semicircle of radius K about the origin (Fig. 7.4). We shall see that when $t - t' > 0$, this is the appropriate contour, whereas when $t - t' < 0$, the appropriate contour is one that closes in the upper half-plane:

$$I \equiv -\frac{1}{2\pi}\oint_C \frac{e^{-ik(t-t')}\,dk}{(k-k_1)(k-k_2)} = -\frac{1}{2\pi}\int_{-K}^{K} \frac{e^{-ik(t-t')}}{(k-k_1)(k-k_2)}\,dk$$
$$-\frac{1}{2\pi}\int_0^{-\pi} \frac{e^{-iK(t-t')e^{i\phi}}}{(Ke^{i\phi}-k_1)(Ke^{i\phi}-k_2)}\,iKe^{i\phi}\,d\phi\,. \tag{7.32}$$

In the first term on the right of Eq. (7.32), the integral is taken along the real axis; in the second term we integrate along the semicircle of radius K, on which

$$k = Ke^{i\phi}\,, \qquad dk = iKe^{i\phi}\,d\phi\,,$$

and ϕ ranges from $\phi = 0$ to $\phi = -\pi$. Now we let K, the radius of the contour, become very large. Then since the imaginary parts of k_1 and k_2 are both in the lower half-plane, the contour C eventually encloses both of them, and we can evaluate the integral by the residue theorem. We get

$$-2\pi I = \lim_{K\to\infty}\oint_C \frac{e^{-ik(t-t')}}{(k-k_1)(k-k_2)}\,dk = (-1)(2\pi i)\left[\frac{e^{-ik_1(t-t')}}{k_1-k_2} + \frac{e^{-ik_2(t-t')}}{k_2-k_1}\right].$$

The factor (-1) enters because the contour is closed in the *lower* half-plane, and so is traversed *clockwise*. Inserting the values of k_1 and k_2, we obtain

$$I = \frac{e^{-\gamma(t-t')}\sin\left[\sqrt{\omega_0^2-\gamma^2}(t-t')\right]}{\sqrt{\omega_0^2-\gamma^2}}\,.$$

As $K \to \infty$, the first term on the right-hand side of Eq. (7.32) becomes just $G(t, t')$, Eq. (7.31), so

$$\frac{e^{-\gamma(t-t')}\sin\left[\sqrt{\omega_0^2-\gamma^2}(t-t')\right]}{\sqrt{\omega_0^2-\gamma^2}} = G(t, t') + \lim_{K\to\infty} R\,, \tag{7.33}$$

where

$$R = -\frac{1}{2\pi}\int_0^{-\pi} \frac{e^{-iKe^{i\phi}(t-t')}}{(Ke^{i\phi}-k_1)(Ke^{i\phi}-k_2)}\,iKe^{i\phi}\,d\phi\,.$$

But this integral vanishes as $K \to \infty$ by Jordan's lemma (Section 6.8). Therefore

$$G(t, t') = \frac{e^{-\gamma(t-t')}\sin\left[\sqrt{\omega_0^2-\gamma^2}(t-t')\right]}{\sqrt{\omega_0^2-\gamma^2}} \qquad t > t'\,. \tag{7.34}$$

Now if $t < t'$, the above argument fails because the integrals involved in estimating R grow exponentially. In this case, however, we choose a contour that closes in the upper half-plane. But then the integrand in I of Eq. (7.32) has no singularities inside the contour, so

$$G(t, t') = 0\,, \qquad t < t'\,.$$

Thus we may summarize by saying

$$G(t, t') = \theta(t - t') \frac{e^{-\gamma(t-t')} \sin [\sqrt{\omega_0^2 - \gamma^2} (t - t')]}{\sqrt{\omega_0^2 - \gamma^2}}, \tag{7.35a}$$

and writing the general solution as

$$x(t) = Ax_1(t) + Bx_2(t) + \int_{t_0}^{t} \frac{e^{-\gamma(t-t')} \sin [\sqrt{\omega_0^2 - \gamma^2}(t - t')] F(t')\, dt'}{\sqrt{\omega_0^2 - \gamma^2}}, \tag{7.35b}$$

where t_0 is the time at which the initial conditions are applied (i. e., $F(t')$ may be taken to be zero at times prior to t_0). Note that the term involving the integral vanishes at $t = t_0$, as does its first derivative with respect to t. Thus A and B are determined from the initial conditions, $x(t_0) = x_0$, $\dot{x}(t_0) = \dot{x}_0$, just as they would be if there were no inhomogeneous term present. We anticipated this result in referring to the kernel of the integral as $G(t, t')$ at the beginning of this section.

Finally, we leave it to the reader to show that

$$\left[\frac{d^2}{dt^2} + 2\gamma \frac{d}{dt} + \omega_0^2\right] G(t, t') = \delta(t - t'), \tag{7.36}$$

just as we expect.

As an example, suppose that the forcing term is

$$F(t) = F_0 e^{-\alpha t}$$

starting at $t = 0$. The system is assumed to be at rest in its equilibrium position at $t = 0$, so that $A = 0 = B$ in Eq, (7.35b). Thus

$$x(t) = F_0 e^{-\gamma t} \int_0^t \frac{1}{\sqrt{\omega_0^2 - \gamma^2}} \sin [\sqrt{\omega_0^2 - \gamma^2}(t - t')] e^{(\gamma - \alpha)t'}\, dt'.$$

All the integrals are straightforward, and we obtain

$$x(t) = \frac{F_0}{\sqrt{\omega_0^2 - \gamma^2}} \frac{\sin [\sqrt{\omega_0^2 - \gamma^2}\, t - \delta]}{\sqrt{\omega_0^2 + \alpha^2 - 2\alpha\gamma}} e^{-\gamma t} + \frac{F_0}{\omega_0^2 + \alpha^2 - 2\alpha\gamma} e^{-\alpha t},$$

where $\tan \delta = \sqrt{\omega_0^2 - \gamma^2}/(\alpha - \gamma)$ with δ in the first or second quadrant. Note that the boundary conditions $x(0) = 0$, $\dot{x}(0) = 0$ are satisfied. It is also easy to check that the $x(t)$ found in this way actually does satisfy Eq. (7.25).

If we let the damping go to zero, we find that

$$x(t) = \frac{F_0}{\omega_0} \frac{\sin (\omega_0 t - \delta)}{\sqrt{\omega_0^2 + \alpha^2}} + \frac{F_0}{\omega_0^2 + \alpha^2} e^{-\alpha t}.$$

After a long time has elapsed and the force $F(t)$ has become very small, we obtain pure oscillatory motion. Thus, in the limit of no damping and of long time, we get simply

$$x(t) = \frac{F_0}{\omega_0} \frac{\sin (\omega_0 t - \delta)}{\sqrt{\omega_0^2 + \alpha^2}}.$$

From this result we can compute the final energy of the system. The system had initially zero energy, so the final energy is equal to the energy transferred to the system. Now

$$E = \tfrac{1}{2}\dot{x}^2 + \tfrac{1}{2}\omega_0^2 x^2 ,$$

for a particle of unit mass. Thus

$$E = \frac{F_0^2}{2(\omega_0^2 + \alpha^2)} .$$

According to Section 5.7,

$$E = \frac{1}{2}\left|\int_{-\infty}^{\infty} e^{-i\omega_0 t} F(t)\right|^2 = \frac{1}{2} F_0^2 \left|\int_0^{\infty} e^{-i\omega_0 t} e^{-\alpha t}\, dt\right|^2 = \frac{F_0^2}{2(\omega_0^2 + \alpha^2)} ,$$

which agrees with our result.

In looking at Eq. (7.35a), one is struck by the fact that $G(t, t')$ bears a close resemblance to the solutions of the homogeneous equation

$$\frac{d^2x}{dt^2} + 2\gamma \frac{dx}{dt} + \omega_0^2 x = 0 . \tag{7.37}$$

In fact, considered as a function of t, $G(t, t')$ for $t > t'$ is a linear combination of the two linearly independent solutions of Eq. (7.37) given in Eq. (7.30). (The coefficients of this linear combination depend, of course, on t'.) For this case,

$$G(t, t') = A(t')e^{-\gamma t} \sin [\sqrt{\omega_0^2 - \gamma^2}\, t] + B(t')e^{-\gamma t} \cos [\sqrt{\omega_0^2 - \gamma^2}\, t] ,$$

where

$$A(t') = \frac{e^{\gamma t'} \cos [\sqrt{\omega_0^2 - \gamma^2}\, t']}{\sqrt{\omega_0^2 - \gamma^2}} , \qquad B(t') = -\frac{e^{\gamma t'} \sin [\sqrt{\omega_0^2 - \gamma^2}\, t']}{\sqrt{\omega_0^2 - \gamma^2}} .$$

The reason for this is easily seen by looking at Eq. (7.36).

Everywhere, except at $t = t'$, $G(t, t')$ must satisfy Eq. (7.37), since the δ-function vanishes except at $t = t'$. Therefore, for $t > t'$, $G(t, t')$ is some linear combination of the two solutions to Eq. (7.37) and similarly for $t < t'$.

To put the matter in a less specialized setting, let us consider the most general second-order linear differential operator

$$L \equiv f_0(t) \frac{d^2}{dt^2} + f_1(t) \frac{d}{dt} + f_2(t) .$$

Let $x_1(t)$ and $x_2(t)$ be two linearly independent solutions to

$$Lx(t) = 0 .$$

We wish to solve

$$LG(t, t') = f_0(t) \frac{d^2G(t, t')}{dt^2} + f_1(t) \frac{dG(t, t')}{dt} + f_2(t)G(t, t') = \delta(t - t') . \tag{7.38}$$

For $t > t'$, we write

$$G(t, t') = a_1x_1(t) + a_2x_2(t) , \tag{7.39a}$$

and for $t < t'$,

$$G(t, t') = b_1x_1(t) + b_2x_2(t) . \tag{7.39b}$$

Now let us try to determine a_1, a_2, b_1, and b_2. At $t = t'$, $G(t, t')$ must be continuous; if it were not, dG/dt would contain a δ-function and therefore d^2G/dt^2 would contain the *derivative* of a δ-function. However, on the right-hand side of Eq. (7.38) there is just a δ-function, so we conclude that $G(t, t')$ is continuous at $t = t'$. However, dG/dt will not be continuous at $t = t'$, and from Eq.(7.38) we can determine the value of

$$\left[\frac{dG(t, t')}{dt}\right]_{t=t'+\epsilon} - \left[\frac{dG(t, t')}{dt}\right]_{t=t'-\epsilon}$$

by integrating both sides of Eq, (7.38) from $t = t' - \epsilon$ to $t = t' + \epsilon$. On the right-hand side, the δ-function yields 1, so

$$\int_{t'-\epsilon}^{t'+\epsilon} f_0(t) \frac{d^2G(t, t')}{dt^2} dt + \int_{t'-\epsilon}^{t'+\epsilon} f_1(t) \frac{dG(t, t')}{dt} dt + \int_{t'-\epsilon}^{t'+\epsilon} f_2(t)G(t, t')\, dt = 1 .$$

Let us assume that f_0, f_1, and f_2 are continuous throughout the region of definition of the Green's function. Then since ϵ is very small, these functions will vary negligibly in the region of integration, and we can replace them by their values at $t = t'$:

$$f_0(t') \int_{t'-\epsilon}^{t'+\epsilon} \frac{d^2G(t, t')}{dt^2} dt + f_1(t') \int_{t'-\epsilon}^{t'+\epsilon} \frac{dG(t, t')}{dt} dt + f_2(t') \int_{t'-\epsilon}^{t'+\epsilon} G(t, t')\, dt = 1.$$

Since G is continuous at $t = t'$, and since the domain of integration can be made arbitrarily small, the last term on the left of this equation vanishes. Doing the remaining two integrals, we get

$$f_0(t') \left\{\left[\frac{dG(t, t')}{dt}\right]_{t=t'+\epsilon} - \left[\frac{dG(t, t')}{dt}\right]_{t=t'-\epsilon}\right\}$$
$$+ f_1(t') [G(t' + \epsilon, t') - G(t' - \epsilon, t')] = 1.$$

Since $G(t, t')$ is continuous at $t = t'$, the second term on the left-hand side vanishes and we are left with

$$\left[\frac{dG(t, t')}{dt}\right]_{t=t'+\epsilon} - \left[\frac{dG(t, t')}{dt}\right]_{t=t'-\epsilon} = \frac{1}{f_0(t')} . \tag{7.40}$$

Using this relation and the fact that $G(t, t')$ is continuous at $t = t'$, we can obtain useful information about the coefficients a_1, a_2, b_1, and b_2 in Eqs. (7.39a) and (7.39b).

From continuity at $t = t'$, we get

$$a_1x_1(t') + a_2x_2(t') = b_1x_1(t') + b_2x_2(t') ,$$

while from Eq. (7.40), we obtain

$$a_1\dot{x}_1(t') + a_2\dot{x}_2(t') - [b_1\dot{x}_1(t') + b_2\dot{x}_2(t')] = 1/f_0(t') .$$

Combining terms, we find the following set of equations for the differences $(a_1 - b_1)$ and $(a_2 - b_2)$:

$$(a_1 - b_1)x_1(t') + (a_2 - b_2)x_2(t') = 0 ,$$
$$(a_1 - b_1)\dot{x}_1(t') + (a_2 - b_2)\dot{x}_2(t') = 1/f_0(t') .$$

The solutions are

$$a_1 - b_1 = -\frac{x_2(t')}{f_0(t')\,W(t')}, \qquad a_2 - b_2 = \frac{x_1(t')}{f_0(t')\,W(t')},$$

where $W(t')$ is the *Wronskian* of x_1 and x_2, defined by

$$W(t') = x_1(t')\dot{x}_2(t') - \dot{x}_1(t')x_2(t') = \begin{vmatrix} x_1(t') & x_2(t') \\ \dot{x}_1(t') & \dot{x}_2(t') \end{vmatrix} .$$

In this determinant form, the Wronskian generalizes to differential equations of any order. Putting the above result into Eqs. (7.39a, b) we see that $G(t, t')$ may be written as

$$\begin{aligned} G(t, t') &= b_1x_1(t) + b_2x_2(t) - \frac{x_1(t)x_2(t') - x_2(t)x_1(t')}{f_0(t')\,W(t')}, \qquad t > t' \\ G(t, t') &= b_1x_1(t) + b_2x_2(t) , \qquad t < t' . \end{aligned} \tag{7.41a}$$

It is left as an exercise (Problem 6) to show that $f_0(t')\,W(t')$ never vanishes. The two remaining constants, b_1 and b_2, can now be chosen to satisfy the appropriate boundary conditions. The form which the Green's function finally takes is strongly dependent on the type of boundary conditions involved. If the boundary conditions are of the single point variety typical of classical mechanics, then we shall require that $b_1 = 0 = b_2$ in Eq. (7.41a). If we do this, the Green's function is given by

$$G(t, t') = -\theta(t - t')\,\frac{x_1(t)x_2(t') - x_2(t)x_1(t')}{f_0(t')\,W(t')}, \tag{7.41b}$$

and the solution in terms of the Green's function is

$$x(t) = Ax_1(t) + Bx_2(t) + \int_{t_0}^{t} G(t, t')F(t')\,dt' . \tag{7.41c}$$

The term involving the integral vanishes at $t = t_0$ and so does its derivative with respect to t. Thus, when we apply the one-point boundary conditions, $x(t_0) = x_0$, $\dot{x}(t_0) = \dot{x}_0$, the constants A and B in Eq. (7.41c) are determined just as though the integral term were not present.

However, if the boundary conditions are of the two-point type, say

$$x(t_0) = x_0, \quad x(t_1) = x_1 ,$$

then we write

$$x(t) = Ax_1(t) + Bx_2(t) + \int_{t_0}^{t_1} G(t, t')F(t')\, dt' \,. \qquad (7.42)$$

Then it would seem reasonable to choose b_1 and b_2 in Eq. (7.41a) so that $G(t_0, t') = 0 = G(t_1, t')$ and therefore A and B of Eq. (7.42) are determined by the boundary condition, just as though the integral term were not present.

To illustrate this case and also to demonstrate a simple calculation of a Green's function by our new method, let us consider

$$L = \frac{d^2}{dt^2} + \omega_0^2 \,.$$

We already know that for one-point (classical mechanical) boundary conditions, the Green's function is

$$G(t, t') = \theta(t - t')\, \frac{\sin [\omega_0(t - t')]}{\omega_0} \,,$$

which is readily obtained by letting $\gamma = 0$ in Eq. (7.35a). Now let us use another method and apply a two-point boundary condition. We may choose

$$x_1(t) = \sin \omega_0 t \,, \qquad x_2(t) = \cos \omega_0 t \,,$$

so that

$$W(t') = x_1(t)\dot{x}_2(t) - \dot{x}_1(t)x_2(t) = -\omega_0 \,.$$

Since $f_0 = 1$, we obtain from Eq. (7.41a)

$$G(t, t') = b_1 \sin \omega_0 t + b_2 \cos \omega_0 t + \frac{\sin \omega_0(t - t')}{\omega_0} \,, \qquad t > t' \,, \qquad (7.43a)$$

$$G(t, t') = b_1 \sin \omega_0 t + b_2 \cos \omega_0 t \,, \qquad t < t' \,. \qquad (7.43b)$$

For convenience, let us choose the endpoints to be $t_0 = 0$ and $t_1 = 1$. Then applying the condition $G(t_0, t') \equiv G(0, t') = 0$ to Eq. (7.43b), we get $b_2 = 0$. Similarly, applying the condition $G(t_1, t') \equiv G(1, t') = 0$ to Eq. (7.43a), we find that

$$b_1 \sin \omega_0 + \frac{1}{\omega_0} \sin \omega_0(1 - t') = 0 \,,$$

so that

$$b_1 = \frac{1}{\omega_0} \cot \omega_0 \sin \omega_0 t' - \frac{1}{\omega_0} \cos \omega_0 t' \,.$$

With b_1 and b_2 evaluated, Eqs. (7.43a) and (7.43b) become

$$G(t, t') = \frac{1}{\omega_0} [\cot \omega_0 \sin \omega_0 t - \cos \omega_0 t] \sin \omega_0 t' \qquad t > t' \,, \qquad (7.44a)$$

$$G(t, t') = \frac{1}{\omega_0} [\cot \omega_0 \sin \omega_0 t' - \cos \omega_0 t'] \sin \omega_0 t \qquad t < t' \,. \qquad (7.44b)$$

Note how different this result is from that found for the case of one-point boundary conditions. It is interesting to note that if we let $\omega_0 \to 0$, we recover the function $G_3(t, t')$ of Eq. (7.11′) for the Green's function of d^2/dt^2 with two-point boundary conditions.

Note that when $\omega_0 = n\pi$ $(n = 1, 2, \cdots)$ in Eqs. (7.44a) and (7.44b), the function $G(t, t')$ becomes infinite and our method fails! This is because, for $\omega_0 = n\pi$,

$$L = \frac{d^2}{dt^2} + (n\pi)^2$$

has an eigenfunction with zero eigenvalue, namely, $\sin(n\pi t)$, which satisfies the boundary condition that it vanish at both $t = 0$ and $t = 1$. In this case the eigenfunction expansion of Eq. (7.18′) does not work because of a zero eigenvalue in the denominator of the expression for $G(t, t')$. And then, of course, there is *no unique* solution to the equation

$$Lx = F;$$

we can add an arbitrary multiple of $\sin(n\pi t)$ to any solution and thus get another perfectly acceptable solution.

An analysis identical to the one given here can be carried out for the general nth order linear differential operator, again with the restriction $f_0(t) \neq 0$ and with the proviso ruling out zero eigenvalues.

As was pointed out at the end of Section 7.1, the use of Green's functions is by no means restricted to solving differential equations with inhomogeneous terms which are independent of x. One of the most important applications occurs when the inhomogeneous term itself depends on the unknown function $x(t)$. For example, the differential equation

$$Lx = f(t, x)$$

can be rewritten as an integral equation of the form

$$x(t) = \xi(t) + \int G(t, t') f[t', x(t')]\, dt',$$

where $\xi(t)$ is a linear combination of the solutions to the homogeneous equation,

$$Lx = 0,$$

which satisfies the boundary conditions, and $G(t, t')$ is the Green's function belonging to L which satisfies its own appropriate boundary conditions. For example, for a simple harmonic oscillator, perturbed by some nonharmonic force of the form $\phi(t)x(t)$ or $[x(t)]^2$, we have

$$x(t) = x_0 \cos\omega_0 t + \frac{\dot{x}_0}{\omega_0}\sin\omega_0 t + \int_0^t \frac{1}{\omega_0}\sin\omega_0(t - t') f[t', x(t')]\, dt',$$

for the one-point boundary conditions $x(0) = x_0$, $\dot{x}(0) = \dot{x}_0$. A perturbation of the form $\phi(t)x(t)$ indicates that the "spring constant" of the harmonic oscillator is not constant in time, but instead has the form $k = k_0 + \phi(t)$.

Similarly, if the force were not perfectly harmonic, a term of the form $[x(t)]^2$ might be necessary to achieve an accurate representation of the force; this is often the case for the forces which bind solids, where small terms proportional to the square of the displacement play an important role. In the next chapter, we shall discuss methods by which such integral equations can be solved.

7.4 GREEN'S FUNCTIONS IN THREE DIMENSIONS

The method described at the end of the last chapter for obtaining Green's functions in one dimension is not generally suited for three dimensional problems. Since three dimensions corresponds to the real world, it is not surprising that the problems are harder in this case. However, there is a particularly important operator in three dimensions which yields readily to the Fourier transform method described at the beginning of the last section. The operator in question is

$$H_0 = \nabla^2 + \lambda ,$$

where ∇^2 is the familiar Laplacian operator discussed in Chapter 1. We shall employ the Fourier transform method to solve

$$H_0\psi(\mathbf{r}) = F(\mathbf{r}) , \tag{7.45}$$

assuming that $\psi(\mathbf{r})$ and $F(\mathbf{r})$ may be Fourier transformed. Let

$$\hat{F}(\mathbf{k}) = \frac{1}{(2\pi)^{3/2}} \int e^{-i\mathbf{k}\cdot\mathbf{r}} F(\mathbf{r}) d^3\mathbf{r} , \tag{7.46a}$$

$$\hat{\psi}(\mathbf{k}) = \frac{1}{(2\pi)^{3/2}} \int e^{-i\mathbf{k}\cdot\mathbf{r}} \psi(\mathbf{r}) d^3\mathbf{r} , \tag{7.46b}$$

then from Eq. (7.45),

$$\frac{1}{(2\pi)^{3/2}} \int e^{-i\mathbf{k}\cdot\mathbf{r}} \nabla^2\psi(\mathbf{r}) d^3\mathbf{r} + \lambda\hat{\psi}(\mathbf{k}) = \hat{F}(\mathbf{k}) . \tag{7.47}$$

Now we use Green's theorem, which states that

$$\int_V (F\nabla^2 G - G\nabla^2 F) d^3\mathbf{r} = \int_S (F\nabla G - G\nabla F)\cdot\mathbf{n}\, dS ,$$

where S is the surface enclosing the volume V, and $\mathbf{n}$ is the unit vector normal to S, pointing outward. Thus we conclude that

$$\begin{aligned} \frac{1}{(2\pi)^{3/2}} \int_V e^{-i\mathbf{k}\cdot\mathbf{r}} \nabla^2\psi(\mathbf{r}) d^3\mathbf{r} &= \frac{1}{(2\pi)^{3/2}} \int_V \psi(\mathbf{r}) \nabla^2 e^{-i\mathbf{k}\cdot\mathbf{r}} d^3\mathbf{r} \\ &+ \frac{1}{(2\pi)^{3/2}} \int_S [e^{-i\mathbf{k}\cdot\mathbf{r}} \nabla\psi(\mathbf{r}) - \psi(\mathbf{r}) \nabla e^{-i\mathbf{k}\cdot\mathbf{r}}]\cdot\mathbf{n}\, dS . \end{aligned} \tag{7.48}$$

The region of integration is the entire three-dimensional space, so in evaluating the surface integral, let us take the surface to be a sphere of radius R and take the limit as $R \to \infty$. In this case, $\mathbf{n}$ is just a unit vector in the radial direction, so the surface term is

$$\frac{1}{(2\pi)^{3/2}}\int_S [e^{-i\mathbf{k}\cdot\mathbf{r}}\nabla\psi(\mathbf{r}) - \psi(\mathbf{r})\nabla e^{-i\mathbf{k}\cdot\mathbf{r}}]\cdot\mathbf{n}\,dS$$

$$= \frac{1}{(2\pi)^{3/2}}\lim_{R\to\infty} R^2\left\{\int\left[e^{-i\mathbf{k}\cdot\mathbf{r}}\frac{d\psi}{dr} - \psi\frac{d}{dr}e^{-i\mathbf{k}\cdot\mathbf{r}}\right]d\Omega\right\}_{r=R}, \tag{7.48'}$$

where $d\Omega = \sin\theta\,d\theta\,d\phi$. If $\psi(\mathbf{r})$ tends to zero sufficiently rapidly as $|\mathbf{r}|\to\infty$ (roughly, if ψ vanishes faster than $1/r$), then the surface term will vanish, so Eq. (7.48) reduces to

$$\frac{1}{(2\pi)^{3/2}}\int_V e^{-i\mathbf{k}\cdot\mathbf{r}}\nabla^2\psi(\mathbf{r})d^3\mathbf{r} = -k^2\hat{\psi}(\mathbf{k}),$$

and Eq. (7.47) becomes

$$(-k^2 + \lambda)\hat{\psi}(\mathbf{k}) = \hat{F}(\mathbf{k}). \tag{7.49}$$

There are now two cases to be distinguished: $\lambda \geq 0$ and $\lambda < 0$. Let us consider the latter case first, and write $\lambda = -\kappa^2$. Then $(-k^2 - \kappa^2)$ never vanishes, so

$$\hat{\psi}(\mathbf{k}) = -\frac{\hat{F}(\mathbf{k})}{k^2 + \kappa^2}, \tag{7.50}$$

and by the Fourier integral theorem,

$$\psi(\mathbf{r}) = \xi(\mathbf{r}) - \frac{1}{(2\pi)^{3/2}}\int\frac{\hat{F}(\mathbf{k})}{k^2+\kappa^2}e^{i\mathbf{k}\cdot\mathbf{r}}d^3\mathbf{k},$$

where $\xi(\mathbf{r})$ represents an arbitrary linear combination of solutions of the homogeneous equation

$$H_0\phi(\mathbf{r}) = (\nabla^2 - \kappa^2)\phi(\mathbf{r}) = 0.$$

Using Eq. (7.45a), we get

$$\psi(\mathbf{r}) = \xi(\mathbf{r}) + \int G(\mathbf{r},\mathbf{r}')F(\mathbf{r}')d^3\mathbf{r}' \tag{7.51a}$$

with

$$G(\mathbf{r},\mathbf{r}') = -\frac{1}{(2\pi)^3}\int\frac{e^{i\mathbf{k}\cdot(\mathbf{r}-\mathbf{r}')}}{k^2+\kappa^2}d^3\mathbf{k}. \tag{7.51b}$$

Using the result of Eq. (6.51) we obtain

$$G(\mathbf{r},\mathbf{r}') = -\frac{1}{4\pi}\frac{e^{-\kappa|\mathbf{r}-\mathbf{r}'|}}{|\mathbf{r}-\mathbf{r}'|}. \tag{7.52}$$

To complete the solution, we need only determine $\xi(\mathbf{r})$ in Eq. (7.51a). In most physical problems we want $\psi(\mathbf{r})$ to remain bounded as $|\mathbf{r}|$ becomes very large. Let us see if we can solve

$$H_0\phi(\mathbf{r}) = (\nabla^2 - \kappa^2)\phi(\mathbf{r}) = 0 \tag{7.53}$$

subject to the requirement that $\phi(\mathbf{r})$ be bounded as $|\mathbf{r}| \to \infty$. Clearly, the solutions to Eq. (7.53) are just

$$\phi(\mathbf{r}) = e^{-\kappa_1 x} e^{-\kappa_2 y} e^{-\kappa_3 z}, \qquad \kappa^2 = \kappa_1^2 + \kappa_2^2 + \kappa_3^2. \tag{7.54}$$

But all these solutions "blow up" as $|\mathbf{r}|$ becomes large, except in the directions in which $\kappa_1 x$, $\kappa_2 y$, and $\kappa_3 z$ are positive. Thus there are no acceptable solutions to the homogeneous equation, and we may write the final result as

$$\phi(\mathbf{r}) = -\frac{1}{4\pi} \int \frac{e^{-\kappa|\mathbf{r}-\mathbf{r}'|}}{|\mathbf{r}-\mathbf{r}'|} F(\mathbf{r}') d^3\mathbf{r}'. \tag{7.55}$$

If $F(\mathbf{r}')$ decreases sufficiently rapidly as $|\mathbf{r}'| \to \infty$ (for example, if $F(\mathbf{r}')$ vanishes when $|\mathbf{r}'| > R$ for some R), then we see that as $|\mathbf{r}| \to \infty$,

$$\phi(\mathbf{r}) \to C \frac{e^{-\kappa r}}{r}.$$

In this case our neglect of surface terms in Eq. (7.48) was justified, since $\phi(\mathbf{r})$ drops off exponentially for large values of $|\mathbf{r}|$.

An interesting application of Eq. (7.55) occurs in electromagnetic theory, the electromagnetic potential, $\phi(\mathbf{r})$ is related to the charge distribution, $\rho(\mathbf{r})$, by Poisson's equation:

$$\nabla^2 \phi(\mathbf{r}) = -4\pi\rho(\mathbf{r}).$$

Equation (7.55), in the limit $\kappa = 0$, tells us that

$$\phi(\mathbf{r}) = \int \frac{\rho(\mathbf{r}')}{|\mathbf{r}-\mathbf{r}'|} d^3\mathbf{r}',$$

which leads to the familiar multipole expansion for $\phi(\mathbf{r})$. For $|\mathbf{r}|$ large enough, $\phi(\mathbf{r}) \to Q/r$, where

$$Q = \int \rho(\mathbf{r}') d^3\mathbf{r}'.$$

This result is sensible, since at large enough distances any charge distribution should look like a point charge.

Another important application of Eq. (7.55) is to Schrödinger's equation,

$$\left(\nabla^2 + \frac{2mE}{\hbar^2}\right) \phi(\mathbf{r}) = \frac{2m}{\hbar^2} V(\mathbf{r}) \phi(\mathbf{r}). \tag{7.56}$$

For bound states in an attractive potential $[V(\mathbf{r}) \leq 0]$, E is negative. Hence, writing $(-2mE/\hbar^2) \equiv \kappa^2$, we get from Eq. (7.56), using Eq. (7.55),

$$\phi(\mathbf{r}) = -\frac{m}{2\pi\hbar^2} \int \frac{e^{-\kappa|\mathbf{r}-\mathbf{r}'|}}{|\mathbf{r}-\mathbf{r}'|} V(\mathbf{r}') \phi(\mathbf{r}') d^3\mathbf{r}'. \tag{7.57}$$

Thus the linear, differential eigenvalue problem has been transformed into a linear, integral eigenvalue problem. Because of the complexity of the Green's function, it is often preferable to write equations like this in "momentum space," i.e., to write instead an integral equation for the Fourier transform of

$\psi(\mathbf{r})$. Such an equation is given straightaway by Eq. (7.50). For Schrödinger's equation,

$$F(\mathbf{r}) = \frac{2m}{\hbar^2} V(\mathbf{r})\psi(\mathbf{r}) ,$$

and using Eq. (7.46a) and the convolution theorem [see Eq. (5.65)], we get

$$\hat{F}(\mathbf{k}) = \frac{2m}{(2\pi)^{3/2}\hbar^2} \int \hat{V}(\mathbf{k} - \mathbf{p})\hat{\psi}(\mathbf{p})\, d^3\mathbf{p} ,$$

so Eq. (7.50) becomes

$$\hat{\psi}(\mathbf{k}) = -\frac{2m}{(2\pi)^{3/2}\hbar^2} \int \frac{\hat{V}(\mathbf{k} - \mathbf{p})}{k^2 + \kappa^2} \hat{\psi}(\mathbf{p}) d^3\mathbf{p} , \tag{7.57'}$$

which, of course, could also have been obtained by taking the Fourier transform of both sides of Eq. (7.57) and making use of the convolution theorem [Eq. (5.65)] twice.

As a specific example, suppose that $V(\mathbf{r})$ is given by the Yukawa potential

$$V(\mathbf{r}) = -g^2 \frac{e^{-\mu r}}{r} . \tag{7.58}$$

A straightforward integration shows that

$$\hat{V}(\mathbf{q}) = -\left(\frac{2}{\pi}\right)^{1/2} g^2 \frac{1}{q^2 + \mu^2} , \tag{7.58'}$$

so for this case Schrödinger's integral equation is

$$\hat{\psi}(\mathbf{k}) = \frac{mg^2}{\pi^2\hbar^2} \int \frac{\hat{\psi}(\mathbf{p})}{(k^2 + \kappa^2)[(\mathbf{k} - \mathbf{p})^2 + \mu^2]} d^3\mathbf{p} . \tag{7.59}$$

In this form it is more convenient to regard κ as fixed and g^2 to be determined as an eigenvalue; that is, it is easier to ask what potential strengths can produce a given binding energy, $\hbar^2\kappa^2/2m$, rather than what binding energies will be produced by a given potential strength, g^2. For a central potential, the wave functions have the angular dependence of the spherical harmonics. Therefore

$$\hat{\psi}(\mathbf{k}) = \phi_l(k)\, Y_{lm}(\theta_k, \phi_k) , \tag{7.60}$$

and it is a simple matter to obtain a collection of integral equations, one for each $\phi_l(k)$. For example, when $l = 0$, Y_{lm} is a constant. If we substitute Eq. (7.60) into both sides of Eq. (7.59), the angular integration is straightforward, and we get as our equation for $\phi_0(k)$,

$$\phi_0(k) = \frac{mg^2}{\pi^2\hbar^2} \int_0^\infty \frac{1}{k(k^2 + \kappa^2)} \ln\left[\frac{(k + p)^2 + \mu^2}{(k - p)^2 + \mu^2}\right] \phi_0(p) p\, dp . \tag{7.61}$$

One-dimensional integral equations of this kind can be solved by using techniques that will be developed in the next two chapters. Just to assure himself that this equation really does have something in common with Schrödinger's equation, the reader who enjoys doing arduous integrations can substitute the values

$$\phi_0(p) = \frac{C}{\left(p^2 + \frac{1}{a_0^2}\right)^2}, \qquad E_0 = -\frac{\hbar^2}{2ma_0^2}, \qquad k^2 = -\frac{2mE_0}{\hbar^2} = +\frac{1}{a_0^2} \tag{7.62}$$

(where $a_0 = \hbar^2/mg^2$ is the Bohr radius of the hydrogen atom) into Eq. (7.61). Upon doing the integration and letting $\mu \to 0$ *afterward*, he will find that this $\phi_0(p)$ is in fact a solution to Eq. (7.61). [*Hint*: Integrate by parts once to get rid of the log term and then do a contour integral.] This is as it should be, since $\phi_0(p)$, as given in Eq. (7.62), is proportional to the Fourier transform of the ground state wave function of the hydrogen atom, and E_0 is the corresponding ground state energy. Note in passing that Eq. (7.61) tells us that for a Yukawa potential (and also for a Coulomb potential),

$$\phi_0(k) \sim 1/k^4$$

for k very large, as may be seen by expanding the log term in Eq. (7.61). This is true for any $l = 0$ state.

Before the reader objects that there is nothing new in all this, since everyone knows that there exist exact solutions for all the bound states of hydrogen, let us point out that the Coulomb potential is one of the very few potentials which can be dealt with in closed form. Many potentials of interest in physics do not have closed-form solutions, so other techniques, such as the ones being developed here, must be employed.

Having disposed of the case $\lambda < 0$, let us now look at the more subtle case, $\lambda \geq 0$. Now the quantity $(-k^2 + \lambda)$ in Eq. (7.49) has a zero for $k = \pm\sqrt{\lambda}$, so we cannot invert it as it stands. We shall first give a prescription for avoiding this difficulty and discuss its legitimacy later. The idea is to assume that $\sqrt{\lambda}$ is the sum of a positive real number and an imaginary number, so that

$$\lambda = (q \pm i\epsilon)^2, \qquad \epsilon > 0. \tag{7.63}$$

The use of ϵ expresses the hope that when we are finished we can let ϵ tend to zero and obtain a well-behaved result. Thus, Eq. (7.49) yields

$$\hat{\psi}_\pm(\mathbf{k}) = -\frac{\hat{F}(\mathbf{k})}{k^2 - (q \pm i\epsilon)^2}, \tag{7.64}$$

and proceeding just as before, we arrive at

$$\psi_\pm(\mathbf{r}) = \xi(\mathbf{r}) + \int G_\pm(\mathbf{r}, \mathbf{r}')F(\mathbf{r})d^3\mathbf{r}', \tag{7.65}$$

with

$$G_\pm(\mathbf{r}, \mathbf{r}') = -\frac{1}{(2\pi)^3}\int \frac{e^{i\mathbf{k}\cdot(\mathbf{r}-\mathbf{r}')}}{k^2 - (q \pm i\epsilon)^2} d^3\mathbf{k}.$$

Because of the imaginary part which we have inserted, this integral has no poles on the path of integration, so we may proceed just as in the $\lambda < 0$ case.

Doing the angular integration and extending the k integral to $-\infty$ as we did in obtaining Eq. (7.52), we obtain

$$G_{\pm}(\mathbf{r}, \mathbf{r}') = -\frac{1}{8\pi^2 i|\mathbf{r}-\mathbf{r}'|}\int_{-\infty}^{\infty}\left[\frac{e^{ik|\mathbf{r}-\mathbf{r}'|}}{(k-q\mp i\epsilon)(k+q\pm i\epsilon)} - \frac{e^{-ik|\mathbf{r}-\mathbf{r}'|}}{(k-q\mp i\epsilon)(k+q\pm i\epsilon)}\right]k\,dk\,.$$

There are two integrals in question. We consider first

$$J_{\pm} = \int_{-\infty}^{\infty} k\frac{e^{ik|\mathbf{r}-\mathbf{r}'|}}{(k-q\mp i\epsilon)(k+q\pm i\epsilon)}\,dk\,.$$

By arguments used before in similar circumstances, we can evaluate this integral by contour integration, closing the contour in the upper half-plane. Again, the contribution from the semicircle vanishes just as in the $\lambda < 0$ case, so that, by the residue theorem,

$$\begin{aligned} J_{\pm} &= 2\pi i\frac{(\pm q + i\epsilon)e^{i(\pm q+i\epsilon)|\mathbf{r}-\mathbf{r}'|}}{2(\pm q + i\epsilon)} \\ &= \pi i e^{\pm iq|\mathbf{r}-\mathbf{r}'|}e^{-\epsilon|\mathbf{r}-\mathbf{r}'|}\,. \end{aligned}$$

Likewise,

$$K_{\pm} = \int_{-\infty}^{\infty} k\frac{e^{-ik|\mathbf{r}-\mathbf{r}'|}}{(k-q\pm i\epsilon)(k+q\pm i\epsilon)}\,dk = -\pi i e^{\pm iq|\mathbf{r}-\mathbf{r}'|}e^{-\epsilon|\mathbf{r}-\mathbf{r}'|}\,.$$

Hence

$$G_{\pm}(\mathbf{r}, \mathbf{r}') = -\frac{1}{4\pi}\frac{e^{\pm iq|\mathbf{r}-\mathbf{r}'|}}{|\mathbf{r}-\mathbf{r}'|}e^{-\epsilon|\mathbf{r}-\mathbf{r}'|}\,. \tag{7.66}$$

Returning to Eq. (7.65), we find that

$$\psi_{\pm}(\mathbf{r}) = \xi(\mathbf{r}) - \frac{1}{4\pi}\int\frac{e^{i(\pm q+i\epsilon)|\mathbf{r}-\mathbf{r}'|}}{|\mathbf{r}-\mathbf{r}'|}F(\mathbf{r}')d^3\mathbf{r}'\,.$$

Thus the choice of positive or negative imaginary part in Eq. (7.63) has a profound effect on the solution. In one case, as $|\mathbf{r}| \to \infty$,

$$\psi_{+}(\mathbf{r}) \to \xi(\mathbf{r}) + C\frac{e^{iqr}}{r}\,,$$

whereas, in the other case,

$$\psi_{-}(\mathbf{r}) \to \xi(\mathbf{r}) + C\frac{e^{-iqr}}{r}\,.$$

In both these cases, C may depend on θ and ϕ, but not on r. Thus, in cases where the boundary condition can be put in the form of a restriction on the behavior of the solutions at large distances (such as in quantum mechanical scattering problems), the choice of $\lambda = (q + i\epsilon)^2$ or $\lambda = (q - i\epsilon)^2$ corresponds to the application of a particular boundary condition at all points of a two-dimensional surface.

Now we must examine the possibilities for $\xi(\mathbf{r})$, the appropriate solution of the homogeneous equation

$$(\nabla^2 + q^2)\phi(\mathbf{r}) = 0 .$$

This equation has as its solution $e^{i\mathbf{q}\cdot\mathbf{r}}$, where $\mathbf{q}$ is a vector whose magnitude is q and whose direction is arbitary. Thus the complete solution reads

$$\psi_\pm(\mathbf{r}) = \frac{A}{(2\pi)^{3/2}} e^{i\mathbf{q}\cdot\mathbf{r}} - \frac{1}{4\pi}\int \frac{e^{i(\pm q + i\epsilon)|\mathbf{r}-\mathbf{r}'|}}{|\mathbf{r}-\mathbf{r}'|} F(\mathbf{r}')d^3\mathbf{r}' . \tag{7.67}$$

The constant A and the direction of $\mathbf{q}$ are determined by the initial conditions.

A particularly important example is again provided by Schrödinger's equation, which we write in the form

$$\left(\nabla^2 + \frac{2mE}{\hbar^2}\right)\psi(\mathbf{r}) = \frac{2m}{\hbar^2} V(\mathbf{r})\psi(\mathbf{r}) .$$

If we look at the so-called *scattering case* ($E \geq 0$), then we have precisely the situation envisioned in deriving Eq. (7.67). We find immediately the following *integral equation* for the wave function:

$$\psi_\pm(\mathbf{r}) = \frac{A}{(2\pi)^{3/2}} e^{i\mathbf{q}_i\cdot\mathbf{r}} - \frac{m}{2\pi\hbar^2}\int \frac{e^{\pm iq_i|\mathbf{r}-\mathbf{r}'|}}{|\mathbf{r}-\mathbf{r}'|} V(\mathbf{r}')\psi_\pm(\mathbf{r}')d^3\mathbf{r}' . \tag{7.68}$$

where $q_i = \sqrt{2mE/\hbar^2}$. This equation is called the Lippmann-Schwinger equation; it is the starting point for most modern studies of quantum-mechanical scattering phenomena. We have dropped the ϵ in anticipation of the fact that the limit $\epsilon \to 0$ will pose no difficulties. We shall return to this point later.

The physical picture here is that of a plane wave traveling in the direction $\mathbf{q}_i$, with energy $\hbar^2 q_i^2/2m$, incident on a scattering center represented by $V(\mathbf{r})$. The intensity is determined by A, but is largely irrelevant to the physics of the problem. At large distances, we expect to find not only the incident wave, but also an outgoing spherical wave due to the scattering. In this case the boundary condition demands that we have an asymptotic behavior of the type

$$\psi_+(\mathbf{r}) = \frac{A}{(2\pi)^{3/2}} e^{i\mathbf{q}_i\cdot\mathbf{r}} + C\frac{e^{iq_ir}}{r} .$$

That this is the appropriate solution which contains *outgoing* spherical waves can be seen by remembering that Schrödinger's time-dependent equation is

$$H\Psi = i\hbar\frac{\partial\Psi}{\partial t} .$$

If H is time independent,

$$\Psi(\mathbf{r}, t) = \psi(\mathbf{r})e^{-iEt/\hbar} = \psi(\mathbf{r})e^{-i\omega t} ,$$

where $\omega = E/\hbar$ and $\psi(\mathbf{r})$ satisfies Schrödinger's time-independent equation,

$$H\psi(\mathbf{r}) = E\psi(\mathbf{r})\ .$$

Thus the *asymptotic* solution to the time-dependent equation is

$$\Psi_+(\mathbf{r}, t) = \frac{A}{(2\pi)^{3/2}} e^{i(\mathbf{q}_i\cdot\mathbf{r}-\omega t)} + C\frac{e^{i(q_i r-\omega t)}}{r}, \tag{7.69}$$

where $\omega = \hbar q_i^2/2m$. For the phase to remain constant as t increases, r must increase; thus Eq. (7.69) describes an *outgoing* wave.

Conversely, the asymptotic form

$$\psi_-(\mathbf{r}) = \frac{A}{(2\pi)^{3/2}} e^{i\mathbf{q}_i\cdot\mathbf{r}} + C\frac{e^{-iq_i r}}{r}$$

corresponds to an incoming spherical wave. This boundary condition is also relevant in quantum mechanics when discussing final states of scattering processes. Since it involves no extra difficulty to do so, we shall discuss both conditions together.

It is interesting to look a bit more closely at the asymptotic form of Eq. (7.68). If we write

$$|\mathbf{r} - \mathbf{r}'| = \sqrt{r^2 - 2\mathbf{r}\cdot\mathbf{r}' + r'^2} = r\left[1 - \frac{2\mathbf{r}\cdot\mathbf{r}'}{r^2} + \frac{r'^2}{r^2}\right]^{1/2},$$

and assume that r is very large, then a binomial expansion yields

$$|\mathbf{r} - \mathbf{r}'| = r - \mathbf{n}\cdot\mathbf{r}' + \mathcal{O}(1/r);$$

$\mathcal{O}(1/r)$ represents terms which vanish at least as fast as $1/r$ for large r, and $\mathbf{n}$ is a unit vector in the radial direction. Thus for large r, Eq. (7.68) becomes

$$\psi_\pm(\mathbf{r}) \to \frac{A}{(2\pi)^{3/2}} e^{i\mathbf{q}_i\cdot\mathbf{r}} - \frac{m}{2\pi\hbar^2}\frac{e^{\pm iq_i r}}{r}\int e^{\mp iq_i\mathbf{n}\cdot\mathbf{r}'} V(\mathbf{r}')\psi_\pm(\mathbf{r}')d^3\mathbf{r}'\ .$$

Setting $\mathbf{q}_f = q_i\mathbf{n}$, we have

$$\psi_\pm(\mathbf{r}) = \frac{A}{(2\pi)^{3/2}}\left[e^{i\mathbf{q}_i\cdot\mathbf{r}} + f_\pm^s(\mathbf{q}_f)\frac{e^{\pm iq_i r}}{r}\right], \tag{7.70}$$

where

$$f_\pm^s(\mathbf{q}_f) = -\frac{(2\pi)^{1/2}m}{A\hbar^2}\int e^{\mp i\mathbf{q}_f\cdot\mathbf{r}'} V(\mathbf{r}')\psi_\pm(\mathbf{r}')d^3\mathbf{r}'; \tag{7.71}$$

$f_\pm^s(\mathbf{q}_f)$ is called the scattering amplitude. It measures the ratio of the flux of scattered particles to that of incident particles. It is clearly independent of A, since $\psi(\mathbf{r})$ will be proportional to A. Since $\psi(\mathbf{r})$ and A have the same dimensions, we see from Eq. (7.70) that $f_\pm^s(\mathbf{q}_f)$ must have the dimensions of length. The quantity $f_\pm^s(\mathbf{q}_f)$ is of great importance in quantum mechanics, since its modulus squared is what is measured in scattering experiments. We shall soon obtain an integral equation for this important quantity. For the moment, we merely remark that since $f_\pm^s(\mathbf{q}_f)$ is, apart from constant factors, just the Fourier transform of $V(\mathbf{r})\psi(\mathbf{r})$, we can use the convolution theorem to obtain

$$f_{\pm}^{s}(\mathbf{q}_f) = -\frac{(2\pi)^{1/2}m}{A\hbar^2}\int \hat{V}(\pm\mathbf{q}_f - \mathbf{p})\hat{\psi}(\mathbf{p})d^3\mathbf{p}\,. \tag{7.72}$$

The fact that $f_{\pm}^{s}(\mathbf{q}_f)$ is simply related to the Fourier transform of $\psi(\mathbf{r})$ suggests that it may be useful to obtain an integral equation for $\hat{\psi}(\mathbf{p})$ just as we did in Eq. (7.57′) for the bound state ($\lambda < 0$) case. The only difference is that we now have an inhomogeneous term. By looking at Eq. (7.68), we see that this must be the Fourier transform of $[A/(2\pi)^{3/2}]e^{i\mathbf{q}_i\cdot\mathbf{r}}$, which is

$$\frac{A}{(2\pi)^3}\int e^{i\mathbf{q}_i\cdot\mathbf{r}}e^{-i\mathbf{k}\cdot\mathbf{r}}d^3\mathbf{r} = A\delta(\mathbf{q}_i - \mathbf{k})\,.$$

(This representation of the δ-function was introduced in Chapter 5, Eq. 5.63.) Now proceeding exactly as we did in obtaining Eq. (7.57′), we find that

$$\hat{\psi}_{\pm}(\mathbf{k}) = A\delta(\mathbf{q}_i - \mathbf{k}) - \frac{2m}{(2\pi)^{3/2}\hbar^2}\int\frac{\hat{V}(\mathbf{k}-\mathbf{p})}{k^2 - (q_i \pm i\epsilon)^2}\hat{\psi}_{\pm}(\mathbf{p})d^3\mathbf{q}\,. \tag{7.73}$$

This is the momentum-space version of the Lippmann-Schwinger equation. Equation (7.72) suggests that we define a new function $f_{\pm}(\mathbf{q})$ by

$$f_{\pm}(\mathbf{q}) = -\frac{(2\pi)^{1/2}m}{A\hbar^2}\int \hat{V}(\mathbf{q}-\mathbf{p})\hat{\psi}_{\pm}(\mathbf{p})d^3\mathbf{p}\,. \tag{7.74}$$

Multiplying both sides of Eq. (7.73) by

$$-\frac{(2\pi)^{1/2}m}{A\hbar^2}\hat{V}(\mathbf{q}-\mathbf{k})\,,$$

and integrating over all $\mathbf{k}$-space, we obtain

$$f_{\pm}(\mathbf{q}) = -\frac{(2\pi)^{1/2}m}{\hbar^2}\hat{V}(\mathbf{q}-\mathbf{q}_i) - \frac{2m}{(2\pi)^{3/2}\hbar^2}\int\hat{V}(\mathbf{q}-\mathbf{k})\frac{f_{\pm}(\mathbf{k})}{k^2 - (q_i \pm i\epsilon)^2}d^3\mathbf{k}\,. \tag{7.75}$$

Comparing Eqs. (7.72) and (7.74), we see that the scattering amplitude is given simply by

$$f_{\pm}^{s}(\mathbf{q}_f) = f_{\pm}(\pm\mathbf{q}_f)\,,$$

where

$$\mathbf{q}_f = q_i\mathbf{n} = \sqrt{2mE/\hbar^2}\,\mathbf{n}\,.$$

The function $f_{\pm}$ has physical significance only at a particular value of $\mathbf{q}$; when we evaluate $f_{\pm}$ at this particular value, we often say that we evaluate $f_{\pm}$ "on the energy shell."

It is interesting to see what Eq. (7.75) looks like for the Yukawa potential of Eq. (7.58), whose Fourier transform is given by Eq. (7.58′). Putting this into Eq. (7.75), we get

$$f_{\pm}(\mathbf{q}) = \frac{2mg^2}{\hbar^2}\frac{1}{(\mathbf{q}-\mathbf{q}_i)^2 + \mu^2} + \frac{mg^2}{\pi^2\hbar^2}\int\frac{f_{\pm}(\mathbf{k})d^3\mathbf{k}}{[(\mathbf{q}-\mathbf{k})^2 + \mu^2][k^2 - (q_i \pm i\epsilon)^2]}\,. \tag{7.76}$$

The first term of this equation for $f_\pm$ [generally, the first term of Eq. (7.75)] is called the *Born approximation* to the scattering amplitude when evaluated at $\mathbf{q} = \pm\mathbf{q}_f$. It is not unreasonable to imagine that if the potential is weak (say, g^2 is small), then this might be a good approximation for the scattering amplitude. We shall see later in what sense this is actually the case. If we let $g^2 = e^2$, where e is the electronic charge, and set $\mu = 0$, then the Born approximation to the scattering amplitude for a Coulomb potential is

$$f^s = \frac{2me^2}{\hbar^2}\frac{1}{(\mathbf{q}_f - \mathbf{q}_i)^2} .$$

Since $|\mathbf{q}_i| = |\mathbf{q}_f|$, we may write this as

$$f^s = \frac{me^2}{\hbar^2 q_i^2}\frac{1}{1 - \cos\theta} ,$$

where θ is the *scattering angle*, i.e., the angle between the incident direction of the plane wave, $\mathbf{q}_i$, and the direction of the outgoing particle, $\mathbf{q}_f$. Writing $1 - \cos\theta = 2\sin^2(\theta/2)$, we have

$$f^s = \frac{me^2}{2\hbar^2 q_i^2}\frac{1}{\sin^2(\theta/2)} ,$$

and therefore

$$|f^s|^2 = \left(\frac{e^2}{2mv_i^2}\right)^2 \cdot \frac{1}{\sin^4(\theta/2)} , \tag{7.77}$$

since $\hbar^2 q_i^2/m = mv_i^2$, where v_i is the velocity of the incident particle. Equation (7.77) gives the relative probability of finding a particle scattered through an angle θ. It is the famous Rutherford differential cross section, first derived by Lord Rutherford from classical mechanics.

7.5 RADIAL GREEN'S FUNCTIONS

In the last section we saw how Schrödinger's momentum-space integral equation could be reduced to a one-variable integral equation by factoring out the known angular dependence of the solutions given by the spherical harmonics. This suggests that a similar reduction could be carried out at the very outset starting from Schrödinger's differential equation in position space. This can, in fact, be done, and the manner in which the Green's function is handled in this case demonstrates the relationship between the last two sections.

Writing Schrödinger's equation as

$$-\frac{\hbar^2}{2m}\nabla^2\psi(\mathbf{r}) + V(r)\psi(\mathbf{r}) = E\psi(\mathbf{r}) ,$$

where $V(r)$ depends only on the magnitude of $\mathbf{r}$, we look for solutions of the form

$$\psi(\mathbf{r}) = \phi_l(r)\, Y_{lm}(\theta, \phi) .$$

Now, according to Section 5.8, the spherical harmonics are eigenfunctions of

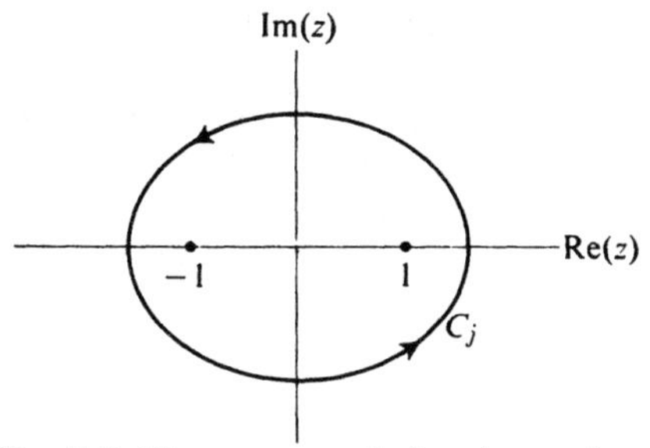

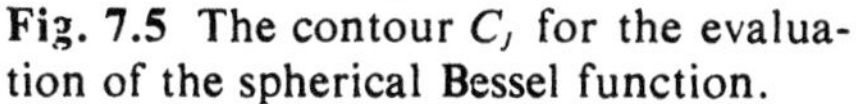
Fig. 7.5 The contour C_j for the evaluation of the spherical Bessel function.

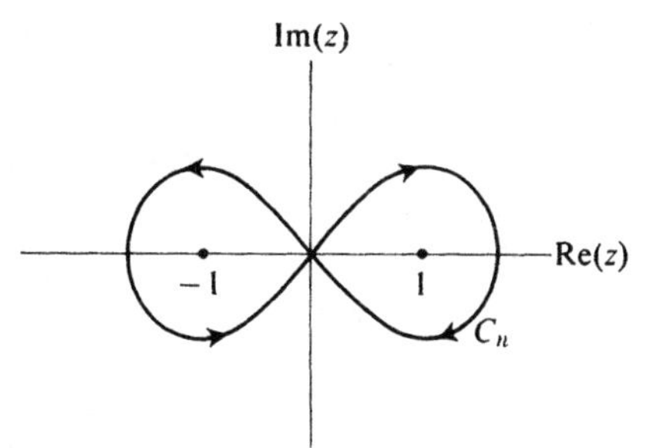

Fig. 7.6 The contour C_n for the evaluation of the spherical Neumann function.

the angular part of ∇^2 with eigenvalues $l(l+1)$. Thus $\phi_l(r)$ satisfies

$$\frac{d^2\phi_l}{dr^2} + \frac{2}{r}\frac{d\phi_l}{dr} + \frac{2mE}{\hbar^2}\phi_l(r) - \frac{l(l+1)}{r^2}\phi_l(r) = \frac{2m}{\hbar^2}V(r)\phi_l(r)\ . \tag{7.78}$$

Let us consider the homogeneous equation

$$\frac{d^2\phi_l}{dr^2} + \frac{2}{r}\frac{d\phi_l}{dr} + \left[\frac{2mE}{\hbar^2} - \frac{l(l+1)}{r^2}\right]\phi_l(r) = 0\ . \tag{7.79}$$

If we make the change of variable

$$x = [2mE/\hbar^2]^{1/2}r\ , \tag{7.80}$$

the equation for ϕ_l becomes

$$\frac{d^2\phi_l}{dx^2} + \frac{2}{x}\frac{d\phi_l}{dx} + \left[1 - \frac{l(l+1)}{x^2}\right]\phi_l(x) = 0\ . \tag{7.81}$$

When x tends to zero, a solution to Eq. (7.81) can tend to either x^l or x^{-l-1}. The reader should find it instructive to verify this. The regular solution which behaves like x^l near $x = 0$ is called the *spherical Bessel function* of order l, and the solution which blows up like x^{-l-1} is called the *spherical Neumann function* of order l.

Power series solutions to equations of the form of Eq. (7.81) are well known. However, we shall not proceed in this manner; instead, we shall write two linearly independent solutions in the form of complex *integral representations*. In particular, we claim that

$$j_l(x) = \frac{1}{2\pi}\frac{(-2)^l l!}{x^{l+1}}\oint_{C_j}\frac{e^{-ixz}}{(z+1)^{l+1}(z-1)^{l+1}}\,dz \tag{7.82}$$

is a solution to Eq. (7.81), when C_j is any contour enclosing the poles at $z = \pm 1$ in the manner shown in Fig. 7.5. Another solution is given by taking a different contour, C_n, shown in Fig. 7.6. We write

$$n_l(x) = \frac{1}{2\pi i}\frac{(-2)^l l!}{x^{l+1}}\oint_{C_n}\frac{e^{-ixz}}{(z+1)^{l+1}(z-1)^{l+1}}\,dz\ . \tag{7.83}$$

We adopt this method in order to demonstrate the power of complex variable techniques in the domain of differential equations. It is an interesting exercise to derive these integral representations starting with power series solutions corresponding to the two different types of behavior near the origin; however, at this point we content ourselves with showing that the integral representations above do indeed satisfy Eq. (7.81).

To show this, let us compute the first two derivatives of

$$f_l(x) = \frac{C_l}{x^{l+1}} \oint \frac{e^{-ixz}}{(z+1)^{l+1}(z-1)^{l+1}}\, dz\,,$$

where the contour is any closed curve containing at least one of the points $z = \pm 1$. Clearly, if it contains neither $z = +1$ nor $z = -1$, then by the Cauchy theorem, $f_l(x) \equiv 0$. We shall now show that such an $f_l(x)$ satisfies Eq. (7.81). For the first derivative we have, omitting the constant C_l,

$$f_l'(x) = -\frac{(l+1)}{x^{l+2}} \oint \frac{e^{-ixz}}{(z^2-1)^{l+1}}\, dz - \frac{i}{x^{l+1}} \oint \frac{ze^{-ixz}}{(z^2-1)^{l+1}}\, dz\,.$$

Similarly,

$$f_l''(x) = \frac{(l+1)(l+2)}{x^{l+3}} \oint \frac{e^{-ixz}\, dz}{(z^2-1)^{l+1}} + \frac{2i(l+1)}{x^{l+2}} \oint \frac{ze^{-ixz}\, dz}{(z^2-1)^{l+1}} - \frac{1}{x^{l+1}} \oint \frac{z^2 e^{-ixz}\, dz}{(z^2-1)^{l+1}}\,.$$

Now we form the quantity

$$q_l(x) = f_l''(x) + \frac{2}{x} f_l'(x) + \left[1 - \frac{l(l+1)}{x^2}\right] f_l(x)\,.$$

To show that $f_l(x)$ satisfies Eq. (7.81), we now need only show that $q_l(x) = 0$. Using the expressions for $f_l'(x)$ and $f_l''(x)$, we get

$$q_l(x) = \frac{2il}{x^{l+2}} \oint \frac{ze^{-ixz}}{(z^2-1)^{l+1}}\, dz - \frac{1}{x^{l+1}} \oint \frac{e^{-ixz}}{(z^2-1)^{l}}\, dz\,.$$

But

$$\frac{2lze^{-ixz}}{(z^2-1)^{l+1}} = -\frac{d}{dz}\left[\frac{e^{-ixz}}{(z^2-1)^l}\right] - ix \frac{e^{-ixz}}{(z^2-1)^l}\,,$$

so

$$q_l(x) = -\frac{i}{x^{l+2}} \oint \frac{d}{dz}\left[\frac{e^{-ixz}}{(z^2-1)^l}\right] dz = 0\,,$$

since the integral of an exact derivative around a closed contour vanishes.

Hence $f_l(x)$ is a solution to Eq. (7.81) for any closed contour containing a singularity. We have taken the two contours of Figs. 7.5 and 7.6 to conform to the conventional definitions. This is also the reason for our particular choice of constants multiplying the integrals in Eqs. (7.82) and (7.83). Naturally,

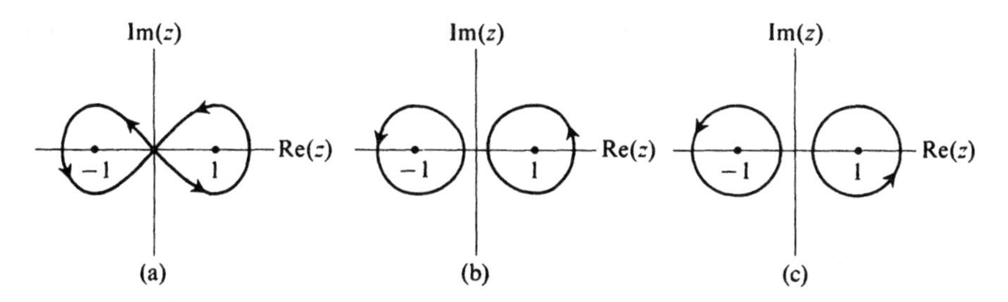

Fig. 7.7 The pinching of the contour of Figure 7–5 to obtain two contours for the evaluation of the spherical Bessel function.

the shape of the contour does not matter; Cauchy's theorem tells us that only the singularities which it encloses are relevant. For example, we have chosen for C_j a circular contour, as shown in Fig. 7.5, which contains both $z = 1$ and $z = -1$. But we can just as well pinch this contour as shown in Fig. 7.7(a), then separate the contour into two parts as shown in Fig. 7.7(b), and finally deform it into the form of *two* circular contours as shown in Fig. 7.7(c), each traced out in a counterclockwise direction, one about $z = -1$, and the other about $z = +1$. Similarly, the figure-eight contour defining $n_l(x)$ can be turned into two circular contours, the one about $z = -1$ being traced out counterclockwise, the one about $z = +1$ being traced out *clockwise*. Thus the function defined by

$$h_l^{(1)}(x) \equiv j_l(x) + in_l(x) , \tag{7.84a}$$

has the representation

$$h_l^{(1)}(x) = \frac{1}{\pi}\frac{(-2)^l l!}{x^{l+1}}\oint_{C_1}\frac{e^{-ixz}}{(z^2-1)^{l+1}}\,dz , \tag{7.84b}$$

where C_1 is a contour encircling only $z = -1$, in the counterclockwise direction. [Remember that we defined $n_l(x)$ with an extra factor of i^{-1} relative to $j_l(x)$.] Similarly,

$$h_l^{(2)}(x) \equiv j_l(x) - in_l(x) \tag{7.85a}$$

has the representation

$$h_l^{(2)}(x) = \frac{1}{\pi}\frac{(-2)^l l!}{x^{l+1}}\oint_{C_2}\frac{e^{-ixz}}{(z^2-1)^{l+1}}\,dz , \tag{7.85b}$$

where C_2 is a contour encircling only $z = 1$ in the counterclockwise sense. $h_l^{(1)}(x)$ and $h_l^{(2)}(x)$ are called the *spherical Hankel functions* of the first and second kind, respectively. Had we so desired, we could equally well have chosen $h_l^{(1)}(x)$ and $h_l^{(2)}(x)$ to be our basic solutions rather than $j_l(x)$ and $n_l(x)$.

We can now obtain the properties of $j_l(x)$ and $n_l(x)$ from the integral representations in a very simple way. These properties will in turn lead us quickly to the desired Green's function for Eq. (7.81). First of all, let us obtain explicit formulas for the two functions. Using Leibnitz's rule for differentiating the product of two functions,

$$\frac{d^n}{dx^n}[f(x)g(x)] = \sum_{m=0}^{n} \frac{n!}{m!\,(n-m)!}\left[\frac{d^{n-m}}{dx^{n-m}} f(x)\right]\frac{d^m}{dx^m} g(x) ,$$

along with the residue theorem for a pole of order $l + 1$, [Eq. (6.17)], we find that

$$j_l(x) = \frac{i(-2)^l l!}{l!\,x^{l+1}} \sum_{m=0}^{l} \frac{l!}{m!(l-m)!}\left\{\left[\frac{d^{l-m}}{dz^{l-m}} e^{-ixz} \frac{d^m}{dz^m}\left(\frac{1}{z+1}\right)^{l+1}\right]_{z=1} + \left[\frac{d^{l-m}}{dz^{l-m}} e^{-ixz} \frac{d^m}{dz^m}\left(\frac{1}{z-1}\right)^{l+1}\right]_{z=-1}\right\} .$$

The expression for $n_l(x)$ differs from the above only by a factor of i^{-1} and the presence of a minus sign in front of the term coming from the pole at $z = +1$, since this pole is encircled in a clockwise direction in the case of $n_l(x)$. The derivatives are elementary, and we find that

$$\begin{aligned} j_l(x) &= \frac{i(-2)^l}{x^{l+1}} \sum_{m=0}^{l} \frac{l!}{m!(l-m)!} (-ix)^{l-m}(-1)^m \frac{(l+m)!}{l!}\left[\frac{e^{-ix}}{2^{l+m+1}} + \frac{e^{ix}}{(-2)^{l+m+1}}\right] \\ &= \sum_{m=0}^{l} \frac{(l+m)!}{m!(l-m)!}\frac{1}{2^m}\frac{1}{x^{m+1}}\, i^{l-m+1}\left[\frac{(-1)^{-(l-m+1)}e^{ix} + e^{-ix}}{2}\right], \end{aligned}$$

where we have several times made use of the fact that $(-1)^m = (-1)^{-m}$. Writing $(-1) = e^{i\pi}$, $i = e^{i\pi/2}$, we get finally

$$j_l(x) = \sum_{m=0}^{l} \frac{(l+m)!}{m!(l-m)!}\frac{1}{2^m}\frac{\cos\left(x - \dfrac{l-m+1}{2}\pi\right)}{x^{m+1}} . \tag{7.86a}$$

Changing the sign of e^{-ix} in the previous expression and dividing by i, we get

$$n_l(x) = \sum_{m=0}^{l} \frac{(l+m)!}{m!(l-m)!}\frac{1}{2^m}\frac{\sin\left(x - \dfrac{l-m+1}{2}\pi\right)}{x^{m+1}} . \tag{7.86b}$$

According to our definitions of the spherical Hankel functions, we have immediately

$$h_l^{(1)}(x) = \sum_{m=0}^{l} \frac{(l+m)!}{m!(l-m)!}\frac{1}{2^m}\frac{\exp\left(i\left[x - \dfrac{l-m+1}{2}\pi\right]\right)}{x^{m+1}} , \tag{7.86c}$$

$$h_l^{(2)}(x) = \sum_{m=0}^{l} \frac{(l+m)!}{m!(l-m)!}\frac{1}{2^m}\frac{\exp\left(-i\left[x - \dfrac{l-m+1}{2}\pi\right]\right)}{x^{m+1}} . \tag{7.86d}$$

Thus, after a few simple manipulations, we have explicit expressions for all the spherical Bessel functions and spherical Hankel functions. Let us now examine their asymptotic behavior. First of all, it is clear that as x tends toward infinity, the dominant term in each sum will be $m = 0$, so

$$j_l(x) \underset{x\to\infty}{\longrightarrow} \frac{\cos\left(x - \frac{l+1}{2}\pi\right)}{x}, \tag{7.87a}$$

$$n_l(x) \underset{x\to\infty}{\longrightarrow} \frac{\sin\left(x - \frac{l+1}{2}\pi\right)}{x}, \tag{7.87b}$$

$$h_l^{(1)}(x) \underset{x\to\infty}{\longrightarrow} \frac{\exp\left(i\left[x - \frac{l+1}{2}\pi\right]\right)}{x}, \tag{7.87c}$$

$$h_l^{(2)}(x) \underset{x\to\infty}{\longrightarrow} \frac{\exp\left(-i\left[x - \frac{l+1}{2}\pi\right]\right)}{x}. \tag{7.87d}$$

As $x \to 0$, the term $m = l$ in the sum for $n_l(x)$ dominates, and we find that

$$n_l(x) \underset{x\to 0}{\longrightarrow} \frac{(2l)!}{2^l l!}\frac{\sin(x - \pi/2)}{x^{l+1}}.$$

Thus

$$n_l(x) \underset{x\to 0}{\longrightarrow} -\frac{(2l)!}{2^l l!}\frac{1}{x^{l+1}}, \tag{7.88a}$$

since $\sin(x - \pi/2) = -\cos x$, which tends toward -1 as x tends toward zero.

Now, at first glance it would appear reasonable to do the same thing for $j_l(x)$. If we proceed in this manner, we are rapidly led to

$$j_l(x) \underset{x\to 0}{\longrightarrow} \frac{(2l)!}{2^l l!}\frac{\cos(x - \pi/2)}{x^{l+1}}.$$

But $\cos(x - \pi/2) = \sin x$, and since $\sin x$ varies like x as x tends to zero, we find that $j_l(x)$ appears to vary like x^{-l}. Before we conclude that this is the limiting behavior, we must check the term $m = l - 1$ in the sum for $j_l(x)$. For x very small this term varies like

$$\frac{(2l-1)!}{(l-1)!\,2^{l-1}}\frac{\cos(x-\pi)}{x^l} = -\frac{(2l-1)!}{(l-1)!\,2^{l-1}}\frac{1}{x^l}.$$

However,

$$\frac{(2l-1)!}{(l-1)!\,2^{l-1}} = \frac{(2l)!}{l!\,2^l},$$

so this term ($m = l - 1$) exactly cancels the leading contribution from the term $m = l$. That this should be the case is really not surprising, since we have seen earlier that the two solutions to Eq. (7.91) vary like either x^{-l-1} or x^l. We have seen that $n_l(x)$ varies like x^{-l-1}, so we suspect that if all is well, $j_l(x)$ should vary like x^l for small x. Hence the expression Eq. (7.86a) for $j_l(x)$ is rather misleading as far as small x behavior is concerned. This is not to say that Eq. (7.86a) is incorrect, but merely that when x becomes small there must be a great deal of cancellation (the cancellation increases rapidly with l) among the various terms contributing to $j_l(x)$, which ultimately produces the desired x^l behavior for small x.

To see that $j_l(x)$ actually does vary like x^l for small x let us return to the integral representation, Eq. (7.82), and make the change of variable $\xi = xz$. Then

$$j_l(x) = \frac{1}{2\pi} \frac{(-2)^l l!}{x^{l+2}} \oint_{C_j} \frac{e^{-i\xi}}{(\xi^2/x^2 - 1)^{l+1}} \, d\xi \,,$$

where C_j is any counterclockwise contour enclosing the singularities at $\xi = \pm x$. We rewrite this as

$$j_l(x) = \frac{1}{2\pi} (-2)^l l! \, x^l \oint_{C_j} \frac{e^{-i\xi}}{\xi^{2l+2}} \left[1 - \frac{x^2}{\xi^2}\right]^{-l-1} d\xi \,.$$

But in any closed region outside the circle of radius $|\xi| = x$, the quantity in square brackets has a uniformly convergent power series:

$$\begin{aligned} \left[1 - \frac{x^2}{\xi^2}\right]^{-l-1} &= 1 + (l+1)\frac{x^2}{\xi^2} + \frac{1}{2}(l+1)(l+2)\frac{x^4}{\xi^4} + \cdots \\ &= \sum_{m=0}^{\infty} \frac{(l+m)!}{m!\,l!} \frac{x^{2m}}{\xi^{2m}} \,. \end{aligned}$$

Since we are free to do so, let us choose C_j to be a circle lying outside the disc $\xi = x$. Because of uniform convergence, we may interchange summation and integration to obtain

$$\begin{aligned} j_l(x) &= \frac{1}{2\pi} (-2)^l l! \, x^l \left[\oint_{C_j} \frac{e^{-i\xi}}{\xi^{2l+2}} \, d\xi + (l+1) x^2 \oint_{C_j} \frac{e^{-i\xi}}{\xi^{2l+4}} \, d\xi + \cdots \right] \\ &= i(-2)^l l! \, x^l \left[\frac{(-i)^{2l+1}}{(2l+1)!} + (l+1) \frac{(-i)^{2l+3}}{(2l+3)!} x^2 + \cdots \right] \end{aligned}$$

by a simple application of the residue calculus. Thus

$$j_l(x) = \frac{2^l l!}{(2l+1)!} \left[x^l - \frac{x^{l+2}}{2(2l+3)} + \cdots \right] . \tag{7.88b}$$

The leading term is proportional to x^l, as we had anticipated. We leave it to the interested reader to show that if one keeps all terms in the expansion of

$[1-x^2/\xi^2]^{-l-1}$, then

$$j_l(x) = 2^l x^l \sum_{m=0}^{\infty} (-1)^m \frac{(l+m)!\,x^{2m}}{m!(2l+2m+1)!}\,.$$

$$= \frac{\sqrt{\pi}}{2} \sum_{m=0}^{\infty} (-1)^m \frac{(x/2)^{2m+l}}{m!\,\Gamma(l+m+3/2)}\,, \tag{7.88b$'$}$$

where we have used Eq. (6.54) as the definition of $\Gamma(x)$. It follows readily from Eq. (6.54) [see, for example, the derivation of Eq. (6.56)] that

$$\Gamma(n+3/2) = \sqrt{\pi}\,\frac{(2n+1)!}{2^{2n+1}\,n!}\,,$$

which result has been used in obtaining Eq. (7.88b$'$).

If the reader will glance back at Eq. (6.69), he will note a rather strong similarity between the above expression for $j_l(x)$ and the expression for $J_n(x)$ given there. For integral values of m and n, we know that $(m+n)! = \Gamma(m+n+1)$, and once this substitution is made, it is apparent that Eq. (6.69) generalizes to non-integral values of n. Then using Eq. (7.88b$'$), it takes only a few steps to show that

$$j_l(x) = \sqrt{\frac{\pi}{2x}}\,J_{l+1/2}(x)\,.$$

This is why the functions $j_l(x)$ are called spherical *Bessel* functions; they are closely related to the Bessel functions of half-integral order.

With exact expressions and all the asymptotic information for the various solutions to Eq. (7.81), we are finally ready to determine the Green's function for Eq. (7.79). Let us take the two linearly independent solutions of Eq. (7.81) to be $j_l(x)$ and $n_l(x)$. The corresponding solutions to Eq. (7.79) are $j_l(kr)$ and $n_l(kr)$, where we assume first that $E \geq 0$ and define, according to Eq. (7.80),

$$k \equiv \sqrt{2mE/\hbar^2}\,.$$

Thus, according to Eq. (7.41), the Green's function for Eq. (7.79) is

$$g_l(r, r') = b_1 j_l(kr) + b_2 n_l(kr) - \frac{1}{W(r')}\,[j_l(kr)n_l(kr') - j_l(kr')n_l(kr)] \tag{7.89a}$$

if $r > r'$, and

$$g(r, r') = b_1 j_l(kr) + b_2 n_l(kr) \tag{7.89b}$$

if $r < r'$. $W(r')$ is the Wronskian, given by

$$W(r') = j_l(kr')\,\frac{d}{dr'}\,n_l(kr') - n_l(kr')\,\frac{d}{dr'}\,j_l(kr')\,.$$

In the terminology of Section 7.3 [Eq. (7.38)], $f_0(r)=1$, $f_1(r)=2/r$, $f_2(r) = k^2 - l(l+1)r^2$. We saw in Section 7.3 that W satisfies the equation

$$\frac{dW}{dr'} = -\frac{f_1(r')}{f_0(r')}\,W\,.$$

Substituting for f_0 and f_1, we find that

$$\frac{dW}{dr'} = -\frac{2}{r'}W,$$

from which we conclude that $W(r') = C(1/r'^2)$.

Now we must determine C, which depends on the particular normalization of the functions. Since we know the asymptotic form of j_l and n_l at large distances, we can use this knowledge to determine $W(r')$ for large r', and hence C. Using Eqs. (7.87a) and (7.87b), we have

$$W(r') \xrightarrow[r'\to\infty]{} \frac{1}{kr'^2},$$

from which we conclude that $C = 1/k$. Of course, the identical result could be obtained by using the expressions for small r'. Thus, quite generally,

$$W(r') = 1/kr'^2,$$

and Eqs. (7.89a) and (7.89b) become

$$g_l(r, r') = b_1 j_l(kr) + b_2 n_l(kr) - kr'^2[j_l(kr)n_l(kr') - j_l(kr')n_l(kr)] \qquad (7.90a)$$

for $r > r'$, and

$$g_l(r, r') = b_1 j_l(kr) + b_2 n_l(kr) \qquad (7.90b)$$

for $r < r'$. Now we do not want the solutions to Schrödinger's equation to be singular at the origin, so for $r < r'$, $g_l(r, r')$ must have no singularity at $r = 0$. However, $n_l(kr)$ has a singularity of the type $(1/r)^{l+1}$ at $r = 0$, so we must exclude the term in n_l from the solution. Hence we take $b_2 = 0$. For large r, a possible boundary condition is to require that the solution have the form e^{ikr}/r, as we did in the last section. But this is the asymptotic form of the function $h_l^{(1)}$, so let us choose b_1 so as to ensure this behavior for $r > r'$. If we take

$$b_1 = -ikr'^2 h_l^{(1)}(kr'),$$

then for $r > r'$ we have, using Eq. (7.90a) with $b_2 = 0$ and Eq. (7.84a),

$$g_l(r, r') = -ikr'^2 h_l^{(1)}(kr) j_l(kr').$$

For $r < r'$, this value for b_1 leads to the result

$$g_l(r, r') = -ikr'^2 j_l(kr) h_l^{(1)}(kr').$$

This completes the specification of g_l. Note that $g_l(r, r')$ is continuous at $r = r'$, although, of course, its derivatives are not. One often introduces the notation $r_> \equiv$ larger of (r, r') and $r_< \equiv$ smaller of (r, r'). Using this convention, we have the following simple expression for g_l:

$$g_l(r, r') = -ikr'^2 j_l(kr_<) h_l^{(1)}(kr_>). \qquad (7.91)$$

Now, if we write Schrödinger's equation as

$$\frac{d^2\phi_l}{dr^2} + \frac{2}{r}\frac{d\phi_l}{dr} + \left[k^2 - \frac{l(l+1)}{r^2}\right]\phi_l(r) = \frac{2m}{\hbar^2}V(r)\phi_l(r),$$

then by our previous work we can immediately transform this into the integral equation

$$\phi_l(r) = Aj_l(kr) + \frac{2m}{\hbar^2}\int_0^\infty g_l(r, r')V(r')\phi_l(r')\,dr' . \tag{7.92}$$

Or, more explicitly,

$$\phi_l(r) = Aj_l(kr) - \frac{2mki}{\hbar^2}\int_0^\infty j_l(kr_<)h_l^{(1)}(kr_>)V(r')\phi_l(r')r'^2\,dr' . \tag{7.93}$$

We have added an arbitrary multiple of $j_l(kr)$, that solution of the homogeneous equation which is regular at the origin.

In our study of Schrödinger's equation, we have considered only the case $E \geq 0$, but it is easy to remedy this by defining

$$\kappa^2 \equiv -2mE/\hbar^2$$

and writing Schrödinger's equation as

$$\frac{d^2\phi_l}{dr^2} + \frac{2}{r}\frac{d\phi_l}{dr} + \left[-\kappa^2 - \frac{l(l+1)}{r^2}\right]\phi_l(r) = \frac{2m}{\hbar^2}V(r)\phi_l(r) .$$

If we now choose the square root of $-\kappa^2$ with *positive* imaginary part, then the solutions to the zero-order equation are $j_l(i\kappa r)$ and $n_l(i\kappa r)$. In this case, $h_l^{(1)}(i\kappa r)$ has just the type of behavior at large distances that we want in a bound state problem, namely, exponential decrease. All other solutions grow exponentially at large distances, so our integral equation becomes

$$\phi_l(r) = \frac{2m\kappa}{\hbar^2}\int_0^\infty j_l(i\kappa r_<)h_l^{(1)}(i\kappa r_>)V(r')\phi_l(r')r'^2\,dr' . \tag{7.94}$$

There is no inhomogeneous term present because there is no solution to the equation

$$\frac{d^2\phi_l}{dr^2} + \frac{2}{r}\frac{d\phi_l}{dr} + \left[-\kappa^2 - \frac{l(l+1)}{r^2}\right]\phi_l(r) = 0,$$

which satisfies the boundary conditions of regularity at $r = 0$ and exponential decrease at large distances.

It is interesting to compare these one-dimensional integral equations, for example Eq. (7.94), with the three-dimensional results of the last section. In the bound state case, we found in Section 7.4, Eq. (7.57), that

$$\psi(\mathbf{r}) = -\frac{m}{2\pi\hbar^2}\int \frac{e^{-\kappa|\mathbf{r}-\mathbf{r}'|}}{|\mathbf{r}-\mathbf{r}'|}V(\mathbf{r}')\psi(\mathbf{r}')\,d^3r' . \tag{7.57}$$

Since the Green's function depends only on $|\mathbf{r} - \mathbf{r}'|$, we can consider it as a function of r, r', and x, where x is the cosine of the angle between $\mathbf{r}$ and $\mathbf{r}'$. It is natural to expand it in a series of Legendre polynomials in x. To this end, we define

$$g_l(r, r') \equiv \frac{1}{2}\int_{-1}^{1} \frac{e^{-\kappa|\mathbf{r}-\mathbf{r}'|}}{|\mathbf{r}-\mathbf{r}'|}P_l(x)\,dx . \tag{7.95}$$

Then

$$\frac{e^{-\kappa|\mathbf{r}-\mathbf{r}'|}}{|\mathbf{r}-\mathbf{r}'|} = \sum_{l=0}^{\infty} (2l+1) g_l(r, r') P_l(x) , \tag{7.96}$$

where we have used the fact [Eq. (5.37), Theorem 5.3] that

$$\int_{-1}^{1} P_l(x) P_{l'}(x)\, dx = \frac{2}{2l+1} \delta_{ll'} .$$

We now insert Eq. (7.96) into Eq. (7.57) to get

$$\psi(\mathbf{r}) = -\frac{m}{2\pi\hbar^2} \int \sum_{l=0}^{\infty} (2l+1) g_l(r, r') P_l(x) V(r') \psi(\mathbf{r}')\, d^3r' .$$

Next we use the addition theorem for spherical harmonics, Eq. (5.83), which tells us that if x is the cosine of the angle between $\mathbf{r}$ and $\mathbf{r}'$ and if the orientation of $\mathbf{r}$ is specified by (θ, ϕ) and that of $\mathbf{r}'$ is specified by (θ', ϕ'), then

$$P_l(x) = \frac{4\pi}{2l+1} \sum_{m=-l}^{l} Y_{lm}(\theta, \phi) Y_{lm}^*(\theta', \phi') .$$

Thus

$$\psi(\mathbf{r}) = -\frac{2m}{\hbar^2} \int \sum_{l=0}^{\infty} \sum_{m=-l}^{l} g_l(r, r') Y_{lm}(\theta, \phi) Y_{lm}^*(\theta', \phi') V(r') \psi(\mathbf{r}')\, d^3r' . \tag{7.97}$$

We now look for solutions of the form

$$\psi(\mathbf{r}) = \phi_l(r) Y_{lm}(\theta, \phi) .$$

Inserting this into Eq. (7.97) and using the completeness and orthonormality of the spherical harmonics, we get

$$\phi_l(r) = -\frac{2m}{\hbar^2} \int_0^{\infty} g_l(r, r') V(r') \phi_l(r') r'^2\, dr' ,$$

since $V(r')$ depends only on the magnitude of $\mathbf{r}'$.

It is possible to show that $g_l(r, r')$, as defined by Eq. (7.95), is equal to $-\kappa j_l(i\kappa r_<) h_l^{(1)}(i\kappa r_>)$, so that the two approaches agree completely. This result follows from the relation

$$\frac{e^{ik|\mathbf{r}-\mathbf{r}'|}}{|\mathbf{r}-\mathbf{r}'|} = ik \sum_{l=0}^{\infty} (2l+1) j_l(kr_<) h_l^{(1)}(kr_>) P_l(x) \tag{7.98}$$

and Eq. (7.95). This is a very useful relationship in mathematical physics, but it would not be instructive to carry out an independent demonstration of it here. [One could, of course, regard the work of this section as a derivation of Eq. (7.98).] We content ourselves with showing that for $l = 0$, $g_l(r, r')$ as defined by Eq. (7.95) is indeed equal to $-\kappa j_l(i\kappa r_<) h_l^{(1)}(i\kappa r_>)$. To do this, we must evaluate

$$g_0(r, r') = \frac{1}{2} \int_{-1}^{1} \frac{e^{-\kappa|\mathbf{r}-\mathbf{r}'|}}{|\mathbf{r}-\mathbf{r}'|} P_0(x)\, dx .$$

We write

$$u = |\mathbf{r} - \mathbf{r}'| = (r^2+r'^2 - 2rr' \cos\theta)^{1/2} = (r^2 + r'^2 - 2rr'x)^{1/2}.$$

Thus

$$du = -\frac{rr'\,dx}{u}, \qquad dx = -\frac{u\,du}{rr'}.$$

Making use of this change of variable, we find that

$$g_0(r, r') = \frac{1}{2rr'}\int_{|r-r'|}^{|r+r'|} e^{-\kappa u}\,du = -\frac{1}{2rr'\kappa}\left[e^{-\kappa|r+r'|} - e^{-\kappa|r-r'|}\right].$$

In terms of the variables $r_>$ and $r_<$, this is

$$\begin{aligned} g_0(r, r') &= -\frac{1}{2\kappa}\,\frac{e^{-\kappa(r_>+r_<)} - e^{-\kappa(r_>-r_<)}}{r_> r_<} \\ &= \kappa\,\frac{e^{-\kappa r_>}}{\kappa r_>}\left[\frac{e^{\kappa r_<} - e^{-\kappa r_<}}{\kappa r_<}\right] \\ &= \kappa\left[\frac{e^{-\kappa r_>}}{\kappa r_>}\right]\left[\frac{\sinh(\kappa r_<)}{\kappa r_<}\right]. \end{aligned}$$

But by Eqs. (7.86a) and (7.86c),

$$j_0(i\kappa r_<) = \frac{\sin(i\kappa r_<)}{i\kappa r_<} \equiv \frac{\sinh(\kappa r_<)}{\kappa r_<}, \qquad h_0^{(1)}(i\kappa r_>) = -\frac{e^{-\kappa r_>}}{\kappa r_>},$$

so

$$g_0(r, r') = -\kappa j_0(i\kappa r_<)h_0^{(1)}(i\kappa r_>).$$

Thus there is a nice relationship between the techniques of Sections 7.3 and 7.4 whenever the potential in Schrödinger's equation is spherically symmetric. Depending on the type of problem, the three-dimensional approach or the one-dimensional approach may be more useful. Both of them have been used extensively in quantum theoretical studies of scattering processes.

At this juncture, it is perhaps worthwhile to emphasize again the importance of boundary conditions in determining the form of a Green's function. In the work of this section we have reduced ∇^2 to a one-dimensional operator by factoring out its angular dependence and have thereby seen how to apply in a straightforward way the boundary conditions appropriate to the scattering problem. The "$i\epsilon$-prescription" used in Section 7.4 is a very convenient shorthand device for achieving the same end. Its great practical utility more than justifies its use, but it is good to remember that it is in one sense a "trick." In considering a general three-dimensional operator it is often impossible to find such a prescription, and one is in trouble unless the operator in question can be brought to an effective one-dimensional form, where the methods of Section 7.3 can be used.

In this context, however, it should also be pointed out that whenever one can obtain an object which satisfies

$$\mathcal{O}G(x, x') = \delta(x - x'),$$

then if one also has the solutions to $\mathcal{O}\psi = 0$, the general solution to

$$\mathcal{O}\psi(x) = F(x)$$

can be written as

$$\psi(x) = \xi(x) + \int G(x, x')F(x')\,dx' .$$

However, there is an important difference between this formal solution and the very similar equations written in previous sections. There we imagined that the coefficients of the solutions to $\mathcal{O}\psi = 0$ appearing in $\xi(x)$ were *independent of* $F(x)$, the inhomogeneous term. In fact, strictly speaking, this is what we meant by a Green's function; it was a function which enabled us to solve a given operator equation in such a way that the coefficients occurring in $\xi(x)$ could be determined without having to think about the inhomogeneous term of the operator equation. However, if all we know about $G(x, x')$ is that it satisfies $\mathcal{O}G(x, x') = \delta(x - x')$, then it will, in general, be the case that the coefficients in $\xi(x)$ will depend on F. This is not surprising, since to any solution of $\mathcal{O}G(x, x') = \delta(x - x')$ we can add an arbitrary linear combination of solutions to $\mathcal{O}\psi(x) = 0$, so the equation $\mathcal{O}G(x, x') = \delta(x - x')$ does not uniquely specify the function $G(x, x')$; that is, one must specify definite boundary conditions for $G(x, x')$.

For example, suppose that for some reason the only function satisfying

$$(\nabla^2 + k^2)G(\mathbf{r}, \mathbf{r}') = \delta(\mathbf{r} - \mathbf{r}')$$

which was known to us was the function

$$G_-(\mathbf{r}, \mathbf{r}') = -\frac{1}{4\pi}\frac{e^{-ik|\mathbf{r}-\mathbf{r}'|}}{|\mathbf{r} - \mathbf{r}'|} . \tag{7.99a}$$

Then we could certainly write Schrödinger's equation,

$$(\nabla^2 + k^2)\psi(x) = \frac{2m}{\hbar^2}V(x)\psi(x) ,$$

as

$$\psi(\mathbf{r}) = \xi(\mathbf{r}) - \frac{m}{2\pi\hbar^2}\int G_-(\mathbf{r}, \mathbf{r}')\,V(\mathbf{r}')\psi(\mathbf{r}')d^3r' . \tag{7.99b}$$

But if we want the function $\psi(\mathbf{r})$ to behave at large values of $|\mathbf{r}|$ like

$$\psi(\mathbf{r}) \sim e^{i\mathbf{k}\cdot\mathbf{r}} + f(\theta, \phi)\frac{e^{+ikr}}{r} ,$$

that is, if we want the potential to produce only *outgoing* spherical waves, then the function $\xi(r)$ needed to satisfy the boundary conditions would not be simply $e^{i\mathbf{k}\cdot\mathbf{r}}$ (then we would have just *incoming* spherical waves), but rather $e^{i\mathbf{k}\cdot\mathbf{r}}$ *plus* a complicated linear combination of solutions to the homogeneous equation,

$$(\nabla^2 + k^2)\phi(\mathbf{r}) = 0 ,$$

whose coefficients would depend on $V(\mathbf{r})$. For this purpose, it would be convenient to take the solutions to be of the form

$$\phi_{lm}(\mathbf{r}) = j_l(kr)\, Y_{lm}(\theta, \phi) \,,$$

where we reject solutions depending on $n_l(kr)$ because they are singular at the origin. (Note that $e^{i\mathbf{k}\cdot\mathbf{r}}$ is a particular linear combination of these $\phi_{lm}(\mathbf{r})$. What are the coefficients of this linear combination? See Problem 7.14.) The reader should show (Problem 7.15) that if we take

$$\xi(\mathbf{r}) = \frac{A}{(2\pi)^{3/2}} e^{i\mathbf{k}\cdot\mathbf{r}} - \frac{4imk}{\hbar^2} \sum_{l=0}^{\infty} \sum_{m=-l}^{l} j_l(kr)\, Y_{lm}(\theta, \phi) \times \int Y_{lm}^*(\theta', \phi') j_l(kr')\, V(\mathbf{r}')\psi(\mathbf{r}')\, d\mathbf{r}' \,,$$

then Eq. (7.99b) reduces to the familiar result

$$\psi(\mathbf{r}) = \frac{A}{(2\pi)^{3/2}} e^{i\mathbf{k}\cdot\mathbf{r}} - \frac{m}{2\pi\hbar^2} \int \frac{e^{ik|\mathbf{r}-\mathbf{r}'|}}{|\mathbf{r}-\mathbf{r}'|} V(\mathbf{r}')\psi(\mathbf{r}')\, d\mathbf{r}' \,.$$

This is just the result of Eq. (7.68) with the proper sign in the exponential to guarantee *outgoing* spherical waves. (Naturally, we could also view the above situation in terms of adding an appropriate linear combination of terms to the "incoming spherical wave Green's function" of Eq. (7.99a) in order to turn it into the desired "outgoing spherical wave Green's function.")

Thus we see that even if one starts with a Green's function satisfying very inconvenient boundary conditions, it is possible to utilize the freedom in the choice of $\xi(r)$ (or equivalently the freedom to add solutions of the homogeneous equation to $G(x, x')$) to overcome this difficulty. It should be clear, however, that in three dimensions this procedure would, in general, present very serious practical difficulties. Fortunately, many of the most important operators of physics are sufficiently simple in form so that it is a relatively straightforward matter to obtain for them Green's functions which satisfy physically useful boundary conditions. However, it is also often the case that even for simple operators the boundary conditions in three (or even two) dimensions are so complicated that the determination of the Green's function becomes virtually impossible. The following section is devoted to such a problem.

7.6 AN APPLICATION TO THE THEORY OF DIFFRACTION

In many problems of practical interest in physics, the boundary conditions are quite complicated from a mathematical point of view, even though the physical situation may be comparatively simple; this means that the appropriate operator equation is made difficult by the form of the boundary condition, rather than by the complexity of the operator itself. This is typically the case in electromagnetic theory where the presence of conducting surfaces—often not even "perfectly conducting" surfaces—makes the solution of many realistic problems extremely difficult, even after the most convenient geometrical configurations have been chosen. However, because of the relatively simple form of the *operator*, it is often possible to solve the Green's function equation,

$$LG(\mathbf{r}, \boldsymbol{\rho}) = \delta^3(\mathbf{r} - \boldsymbol{\rho}) , \tag{7.100}$$

subject to some simple boundary condition. This would be the case in electromagnetic theory, for example, where $L = \nabla^2 + k^2$; if we then ask for the outgoing spherical wave solution, we get the Green's function of Eq. (7.66).

It is thus natural to ask the question: Can a solution to Eq. (7.100) help us obtain a solution to the equation,

$$L\psi = \phi , \tag{7.101}$$

where ψ must satisfy much more complicated boundary conditions than those imposed on $G(\mathbf{r}, \boldsymbol{\rho})$ in Eq. (7.100)?

To answer this question, let us first make the assumption that the operator L has the form

$$\nabla \cdot [f(\mathbf{r})\nabla] + g(\mathbf{r}) , \tag{7.102}$$

which is sufficiently general for many physical applications. Now, if ψ is any solution of Eq. (7.101), and if $G(\mathbf{r}, \boldsymbol{\rho})$ satisfies Eq. (7.100), then

$$\psi(\boldsymbol{\rho}) = \int_V \phi(\mathbf{r})G(\mathbf{r}, \boldsymbol{\rho})d^3\mathbf{r} + \int_V [\psi(\mathbf{r})LG(\mathbf{r}, \boldsymbol{\rho}) - G(\mathbf{r}, \boldsymbol{\rho})L\psi(\mathbf{r})]\, d^3\mathbf{r} , \tag{7.103}$$

where V is the volume throughout which we want the solution. Using the specific form of L [Eq. (7.102)], we can rewrite Eq. (7.103) as

$$\psi(\boldsymbol{\rho}) = \int_V \phi(\mathbf{r})G(\mathbf{r}, \boldsymbol{\rho})d^3\mathbf{r} + \int_V [\psi(\mathbf{r})\nabla \cdot f(\mathbf{r})\nabla G(\mathbf{r}, \boldsymbol{\rho}) - G(\mathbf{r}, \boldsymbol{\rho})\nabla \cdot f(\mathbf{r})\nabla\psi(\mathbf{r})]d^3\mathbf{r} .$$

From the properties of the del operator, we find that

$$\psi(\boldsymbol{\rho}) = \int_V \phi(\mathbf{r})G(\mathbf{r}, \boldsymbol{\rho})d^3\mathbf{r} + \int_V \nabla \cdot [\psi(\mathbf{r})f(\mathbf{r})\nabla G(\mathbf{r}, \boldsymbol{\rho}) - G(\mathbf{r}, \boldsymbol{\rho})f(\mathbf{r})\nabla\psi(\mathbf{r})]\, d^3\mathbf{r} .$$

By Gauss's theorem, this last expression can be written as an integral over the surface S which encloses V. Denoting the outward normal to S by $\mathbf{n}$, we have

$$\psi(\boldsymbol{\rho}) = \int_V \phi(\mathbf{r})G(\mathbf{r}, \boldsymbol{\rho})d^3\mathbf{r} + \int_S f(\mathbf{r})[\psi(\mathbf{r})\nabla G(\mathbf{r}, \boldsymbol{\rho}) - G(\mathbf{r}, \boldsymbol{\rho})\nabla\psi(\mathbf{r})] \cdot \mathbf{n}\, dS . \tag{7.104}$$

Since $\phi(\mathbf{r})$, $f(\mathbf{r})$ and $G(\mathbf{r}, \boldsymbol{\rho})$ are presumed known, all that is required for the evaluation of the integrals is the value of ψ and its normal derivative *on the surface S*.

It must be emphasized, however, that the situation is really more complicated than Eq. (7.104) suggests because $\psi(\mathbf{r})$ and its normal derivative are not *both* arbitrary on S. As a boundary condition we might have $\psi(\mathbf{r})$ specified on S *or* $\nabla\psi(\mathbf{r}) \cdot \mathbf{n}$ specified on S, or even a combination of $\psi(\mathbf{r})$ and $\nabla\psi(\mathbf{r}) \cdot \mathbf{n}$ specified, *but not both separately*. This is a familiar situation from one-dimensional problems, where for second-order equations we specify the value of the function at the endpoints, not both the value of the function and its first derivative at both endpoints.

In many circumstances, this difficulty can be overcome by making use of the freedom of choice of $G(\mathbf{r}, \boldsymbol{\rho})$ which arises from the fact that to any solution of Eq. (7.100) we can add an arbitrary multiple of any solution of $L\xi(\mathbf{r}) = 0$. If, for example, we can find a $G(\mathbf{r}, \boldsymbol{\rho})$ which satisfies Eq. (7.100) and also vanishes when $\mathbf{r}$ is on the surface, S, then Eq. (7.104) takes the form

$$\psi(\boldsymbol{\rho}) = \int_V \phi(\mathbf{r})G(\mathbf{r}, \boldsymbol{\rho})d^3\mathbf{r} + \int_S f(\mathbf{r})\psi(\mathbf{r})\nabla G(\mathbf{r}, \boldsymbol{\rho})\cdot\mathbf{n}\,dS\,.$$

Thus, if the boundary condition involves a specification of $\psi(\mathbf{r})$ on S the problem is formally solved. This is a very common problem in electrostatics [$f(\mathbf{r}) \equiv 1$, $g(\mathbf{r}) \equiv 0$ in Eq. (7.102)], where numerous special techniques are available for the solution of Eq. (7.100) subject to the requirement that $G(\mathbf{r}, \boldsymbol{\rho})$ vanish on S. For a discussion of some of these methods, the reader is referred to Jackson's excellent book, *Classical Electrodynamics*. Notice that Eq. (7.100) is, apart from factors of 4π, just Poisson's equation for the potential due to a point charge at $\boldsymbol{\rho}$. Thus the above technique applied to electrostatics boils down to just the replacement of Poisson's equation for an arbitrary charge distribution and some complicated restriction on the potential at certain boundaries by Poisson's equation for a point charge and a simple boundary condition. (Note that if the shape of S is complicated even the boundary condition that $G(\mathbf{r}, \boldsymbol{\rho})$ vanish for $\mathbf{r}$ on S may in fact not be so "simple"!)

However, situations can also arise in which even the *specification* of boundary conditions becomes difficult. Nevertheless, on the basis of physical considerations one may be able to get an idea of what to expect for $\psi(\mathbf{r})$ and its normal derivative on the boundary of some region.

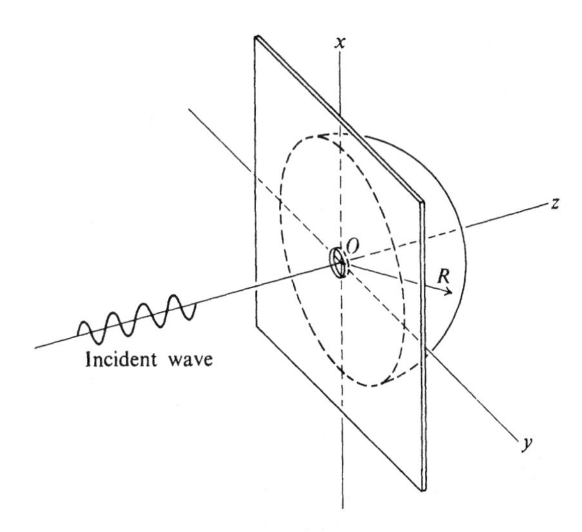

Fig. 7.8 A schematic illustration of the diffraction problem.

A beautiful example of such a situation, and one of the most important applications of Eq. (7.104), is the study of the diffraction of waves by a boundary. From a rigorous mathematical point of view, the theory of diffraction is one of the most complicated topics in classical physics. It involves solving Maxwell's equations in the presence of an awkward boundary, such as a barrier with a hole in it, and only under very special circumstances can an exact solution be obtained. However, in the freewheeling spirit of this chapter, let us try to use some physical ideas along with Eq. (7.104) to study this problem.

We envision a plane wave $e^{i\mathbf{k}_i\cdot\mathbf{r}}$ incident from the left on a flat boundary which contains an aperture, as in Fig. 7.8. The aperture is centered at the origin of the coordinate system; the screen lies in the xy-plane. Everywhere in space except on the screen the function describing the wave must satisfy the wave equation,

$$\nabla^2\psi + k_i^2\psi = 0\,.$$

We are interested in the solution to the right of the screen, so let us take as our volume a hemisphere of large radius centered at the origin, whose flat bottom lies infinitesimally close to the screen.

Now what can we say about the nature of the solution on the boundary? Certainly we would expect the solution at large distances from the scattering center (the aperture) to look like an outgoing spherical wave. Thus, near the curved part of the hemisphere, the wave should have the form

$$\psi(\mathbf{r}) \sim f(\theta, \phi)\frac{e^{ik_ir}}{r}\,. \tag{7.105}$$

Thus the appropriate Green's function for the wave equation in that region is

$$G(\mathbf{r}, \boldsymbol{\rho}) = -\frac{1}{4\pi}\frac{e^{ik_i|\mathbf{r}-\boldsymbol{\rho}|}}{|\mathbf{r}-\boldsymbol{\rho}|}\,, \tag{7.106}$$

which will, according to previous arguments, yield outgoing spherical waves in Eq. (7.104). What about the flat bottom surface of the hemisphere? Here we make the physically reasonable guess that on the back of the "opaque" screen (opposite from the side on which the incident wave is falling) the solution should be identically zero. In the aperture, we take the incoming wave itself as a first approximation:

$$\psi(\mathbf{r}) = e^{i\mathbf{k}_i\cdot\mathbf{r}}\,.$$

We now use these physical assumptions to write Eq. (7.104) in the form appropriate to this case. We have

$$\begin{aligned}\psi(\boldsymbol{\rho}) = &\int_{\text{ap}} [e^{i\mathbf{k}_i\cdot\mathbf{r}}\nabla G(\mathbf{r}, \boldsymbol{\rho}) - G(\mathbf{r}, \boldsymbol{\rho})\nabla e^{i\mathbf{k}_i\cdot\mathbf{r}}]\cdot\mathbf{n}\,dS \\ &+ \int f(\theta, \phi)\left[\frac{e^{ik_ir}}{r}\frac{\partial}{\partial r}G(\mathbf{r}, \boldsymbol{\rho}) - G(\mathbf{r}, \boldsymbol{\rho})\frac{\partial}{\partial r}\frac{e^{ik_ir}}{r}\right]_{r=R} R^2\,d\Omega\,. \end{aligned} \tag{7.107}$$

In the first integral, $\mathbf{n}$ is a unit vector in the negative z-direction (an *outward* normal), and the integral ranges over the entire aperture. The second integral is an angular integration over the curved surface of the hemisphere. The derivatives in the second integral are radial derivatives, since the outward normal is in the radial direction for a spherical surface. R is the radius of the hemisphere, and may be taken to be arbitrarily large. We want to show that this second term vanishes as $R \to \infty$. It is not sufficient to say that the solution drops off like $1/r$ at large distances, since there is a factor of r^2 in the volume element. However, using the value of $G(\mathbf{r}, \boldsymbol{\rho})$ given in Eq. (7.106), it is a simple matter to show that the second integral in Eq. (7.107) falls off like $1/R$. Thus this term vanishes as we make the radius of the hemisphere large, and

$$\psi(\boldsymbol{\rho}) = \int_{\mathrm{ap}} [e^{i\mathbf{k}_i\cdot\mathbf{r}}\nabla G(\mathbf{r}, \boldsymbol{\rho}) - G(\mathbf{r}, \boldsymbol{\rho})\nabla e^{i\mathbf{k}_i\cdot\mathbf{r}}]\cdot\mathbf{n}\, dS\,. \tag{7.108}$$

To simplify the work, let us assume that $\mathbf{k}_i$ is in the z-direction, which means that the wave is incident perpendicular to the screen. Now $\mathbf{n}$ is in the negative z-direction, and $\mathbf{k}_i\cdot\mathbf{r} = 0$, because the surface over which we integrate is in the xy-plane. Therefore, Eq. (7.108) reduces to

$$\psi(\boldsymbol{\rho}) = \int_{\mathrm{ap}} \left[ik_i G(\mathbf{r}, \boldsymbol{\rho}) - \frac{\partial}{\partial z} G(\mathbf{r}, \boldsymbol{\rho})\right] dS\,. \tag{7.109}$$

Even in this form the integration represents a formidable problem, so we shall simplify life further by considering $|\boldsymbol{\rho}|$ to be large compared to a typical screen dimension. Then we can use the asymptotic form of $G(\mathbf{r}, \boldsymbol{\rho})$ discussed in Section 7.4:

$$-\frac{1}{4\pi}\frac{e^{ik_i|\mathbf{r}-\boldsymbol{\rho}|}}{|\mathbf{r}-\boldsymbol{\rho}|} \xrightarrow[|\boldsymbol{\rho}|\to\infty]{} -\frac{1}{4\pi}\frac{e^{ik_i\rho}}{\rho}e^{-i\mathbf{k}_f\cdot\mathbf{r}}\,, \tag{7.110}$$

where $\mathbf{k}_f$ is a vector whose magnitude is k_i and whose direction is given by (θ, ϕ)—the same angles specifying the orientation of $\boldsymbol{\rho}$. Using the fact that

$$(\mathbf{k}_f)_z = |\mathbf{k}_f| \cos\theta = k_i \cos\theta\,,$$

and inserting Eq. (7.110) into Eq. (7.109), we get

$$\psi(\boldsymbol{\rho}) = -\frac{ik_i}{4\pi}\frac{e^{ik_i\rho}}{\rho}(1 + \cos\theta)\int_{\mathrm{ap}} e^{-i\mathbf{k}_f\cdot\mathbf{r}}\, dS\,. \tag{7.111}$$

This is the general expression for the diffraction of a wave by an aperture in the region whose distance from the origin is much larger than the size of the aperture. The precise form of the solution depends, of course, on the geometrical details. However, several important results follow immediately from the general result of Eq. (7.111).

First, it can be shown that if the wavelength λ of the radiation is small compared to the aperture dimensions, there is very intense radiation in the forward direction. In this direction, $\theta = 0$, and $\mathbf{k}_f\cdot\mathbf{r} = 0$, because $\mathbf{k}_f$ is perpendicular to the screen, on which $\theta = \pi/2$. Therefore

$$[\psi(\boldsymbol{\rho})]_{\text{forward}} = -\frac{ik_i A}{2\pi} \frac{e^{ik_i\rho}}{\rho}, \tag{7.112}$$

where A is the area of the aperture. Since $k_i = 2\pi/\lambda$, we see that

$$[\psi(\boldsymbol{\rho})]_{\text{forward}} = \psi(\rho, \theta = 0, \phi) = -i\frac{A}{\lambda}\frac{e^{ik_i\rho}}{\rho}.$$

If we denote the number of photons per square centimeter per second falling on the screen by I_0, then the number of photons per second at a distance $|\boldsymbol{\rho}| = R$ coming into the solid angle $d\Omega$ about some direction (θ, ϕ) is given by

$$I(\theta, \phi) = I_0|\psi(R, \theta, \phi)|^2 R^2 \, d\Omega .$$

For $\theta = 0$ this is

$$I(0, \phi) = (I_0 A)(A/\lambda^2) d\Omega ,$$

which is very large if $\lambda^2 \ll A$. Note that $I_0 A$ is just the number of photons per second falling on the aperture. If λ is much smaller than a typical aperture dimension then it is possible to evaluate the total intensity of diffracted radiation by integrating $I(\theta, \phi)$ over all solid angle, using Eq. (7.111) for $\psi(R, \theta, \phi)$. This integration yields precisely $I_0 A$, i.e., the number of diffracted photons per second arriving on the hemisphere is exactly equal to the number of photons per second falling on the screen, in the short wavelength limit.

It will also become apparent in the following examples that $\psi(\rho, \pi/2, \phi)$ —the solution on the screen—and its normal derivative are very small in absolute value compared to $|\psi(\rho, 0, \phi)|$. Thus the boundary condition on the screen is approximately, although not exactly, satisfied in this case.

Let us now see what $\psi(\boldsymbol{\rho})$ looks like for some simple aperture shapes. The easiest case is that of a rectangular aperture with sides a and b, centered about the origin. Then

$$\begin{aligned} \psi(\boldsymbol{\rho}) &= -\frac{ik_i}{4\pi}(1 + \cos\theta)\frac{e^{ik_i\rho}}{\rho}\int_{-a/2}^{a/2} e^{ik_x x}\, dx \int_{-b/2}^{b/2} e^{ik_y y}\, dy \\ &= -\frac{ik_i}{\pi}(1 + \cos\theta)\frac{\sin(k_x a/2)\sin(k_y b/2)}{k_x k_y}\frac{e^{ik_i\rho}}{\rho}. \end{aligned}$$

This has the form of Eq. (7.112) in the forward direction ($\theta = 0$) since

$$k_x = k_i \sin\theta \cos\phi$$

and

$$k_y = k_i \sin\theta \sin\phi.$$

It is also of interest to consider $\psi(\boldsymbol{\rho})$ on the screen (at $\theta = \pi/2$). We find that

$$\psi(\rho, \theta = \pi/2, \phi) = -\frac{i}{\pi k_i}\frac{\sin[(k_i a/2)\cos\phi]\sin[(k_i b/2)\sin\phi]}{\sin\phi\cos\phi}\frac{e^{ik_i\rho}}{\rho}.$$

If λ is small, k_i is large, so $\psi(\rho, \theta = \pi/2, \phi)$ is small on the screen, as we said above. The derivative normal to the screen is also very small when $\theta = \pi/2$. Thus the boundary conditions on the screen are approximately satisfied, at least for large ρ.

A particularly convenient arrangement is a circular aperture. Denoting a point in the aperture by $(r, \theta_r = \pi/2, \phi_r)$ and $\mathbf{k}_f$ by (k_i, θ, ϕ), we have

$$\psi(\boldsymbol{\rho}) = -\frac{ik_i}{4\pi}\frac{e^{ik_i\rho}}{\rho}(1 + \cos\theta)\int_0^R r\,dr\int_0^{2\pi} d\phi_r e^{-ik_i r \sin\theta \cos(\phi_r - \phi)} ,$$

where R is the radius of the aperture. Here we have used the fact that if α is the angle between two vectors whose directions are specified by (θ, ϕ) and (θ_r, ϕ_r), then

$$\cos\alpha = \cos\theta\cos\theta_r + \sin\theta\sin\theta_r\cos(\phi_r - \phi) .$$

Because of the rotational symmetry about the z-axis, the final result cannot depend on ϕ, so by changing variables to $x = k_i r \sin\theta$, we get

$$\psi(\boldsymbol{\rho}) = -\frac{i}{2k_i}\frac{e^{ik_i\rho}}{\rho}\frac{(1 + \cos\theta)}{\sin^2\theta}\int_0^{k_i R \sin\theta} x\,dx\left[\frac{1}{2\pi}\int_0^{2\pi} e^{-ix\cos\beta}\,d\beta\right] .$$

The term in brackets is easily transformed into the integral representation of the Bessel function of order zero, (Section 6.9). Thus

$$\psi(\boldsymbol{\rho}) = -\frac{i}{2k_i}\frac{(1 + \cos\theta)}{\sin^2\theta}\int_0^{k_i R\sin\theta} xJ_0(x)\,dx\,\frac{e^{ik_i\rho}}{\rho} .$$

Using Eq. (6.69), which gives an infinite series for the Bessel functions of integral order, the reader can show that

$$xJ_0(x) = \frac{d}{dx}[xJ_1(x)] ,$$

so finally

$$\psi(\boldsymbol{\rho}) = -\frac{iR}{2}\frac{(1 + \cos\theta)}{\sin\theta}J_1(k_i R\sin\theta)\frac{e^{ik_i\rho}}{\rho} . \tag{7.113}$$

This agrees with the general result of Eq. (7.112) when $\theta = 0$, for, as can be seen from Eq. (6.69), $J_1(x) \to x/2$ as $x \to 0$. Since $J_1(x) \sim 1/\sqrt{x}$ for large x, we can see that when $\theta = \pi/2$, $\psi(\boldsymbol{\rho})$ and its derivative normal to the screen are very small if the wavelength is small compared to R. Once again, the boundary conditions on the screen are approximately satisfied. The Bessel functions are oscillatory, so if we put a screen behind the circular aperture at a distance that is large compared to the radius of the aperture, the radiation falling on it will give rise to circular rings, starting with an intense central bright ring, then a dark ring as the Bessel function passes through its first zero, and so forth. Figure 7.9 shows the quantity

$$|\psi(\rho, \theta, \phi)|^2\rho^2 ,$$

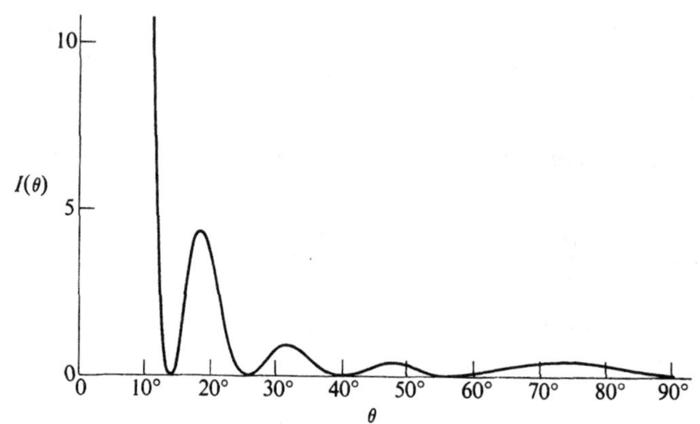

Fig. 7.9 The intensity of diffracted radiation in arbitrary units as a function of angle. The curve does not reach $\theta = 0$ until $I = 256$. This curve corresponds to a circular aperture with $kR = 16$ in Eq. (7.113). Essentially all the intensity is concentrated in the first peak.

which is proportional to the intensity; it is also independent of ϕ because of the symmetry about the z-axis. The intensity pattern of Fig. 7.9 has been verified experimentally.

It is interesting to ask the question, "What would we find if instead of having radiation fall on a screen containing a hole, we let it fall on a small screen exactly the size of the hole?" A screen of this type is called the screen *complementary* to the one we have been considering; a screen plus its complement form a completely opaque barrier. To solve this problem we use the configuration of Fig. 7.8 except that the "screen" is now transparent and the "aperture" is now opaque. In order that the boundary conditions in the xy-plane be consistent with our previous treatment, we take ψ_c to be given simply by $e^{i\mathbf{k}_i\cdot\mathbf{r}}$ in the transparent part, and identically equal to zero behind the opaque part. On the curved hemispherical surface we expect at large distances to find a scattered spherical wave superimposed on the incident plane wave:

$$\psi_c(\mathbf{r}) \sim e^{i\mathbf{k}_i\cdot\mathbf{r}} + F(\theta, \phi)e^{ik_i r}/r .$$

From Eq. (7.104) we have for the "complementary solution," $\psi_c(\boldsymbol{\rho})$,

$$\begin{aligned}\psi_c(\boldsymbol{\rho}) = &\int_{xy-\mathrm{ap}} [e^{i\mathbf{k}_i\cdot\mathbf{r}}\nabla G(\mathbf{r}, \boldsymbol{\rho}) - G(\mathbf{r}, \boldsymbol{\rho})\nabla e^{i\mathbf{k}_i\cdot\mathbf{r}}]\cdot\mathbf{n}\, dS \\ &+ \int_{\text{hemisphere}} [e^{i\mathbf{k}_i\cdot\mathbf{r}}\nabla G(\mathbf{r}, \boldsymbol{\rho}) - G(\mathbf{r}, \boldsymbol{\rho})\nabla e^{i\mathbf{k}_i\cdot\mathbf{r}}]\cdot\mathbf{n}\, dS \\ &+ \int_{\text{hemisphere}} F(\theta, \phi)\left[\frac{e^{ik_i r}}{r}\nabla G(\mathbf{r}, \boldsymbol{\rho}) - G(\mathbf{r}, \boldsymbol{\rho})\nabla\frac{e^{ik_i r}}{r}\right]\cdot\mathbf{n}\, dS\,, \quad (7.114)\end{aligned}$$

where we have used our assumptions about the behavior of ψ on the boundaries. In Eq. (7.114) the subscript $xy - \mathrm{ap}$ means that the integral goes over the entire xy-plane, minus the aperture. Now, we have seen before that the last term of

Eq. (7.114) vanishes when the radius of the hemisphere becomes very large. Thus

$$\psi_c(\boldsymbol{\rho}) = \int_{xy-\text{ap}} [e^{i\mathbf{k}_i\cdot\mathbf{r}}\nabla G(\mathbf{r}, \boldsymbol{\rho}) - G(\mathbf{r}, \boldsymbol{\rho})\nabla e^{i\mathbf{k}_i\cdot\mathbf{r}}]\cdot\mathbf{n}\, dS + \int_{\text{hemisphere}} [e^{i\mathbf{k}_i\cdot\mathbf{r}}\nabla G(\mathbf{r}, \boldsymbol{\rho}) - G(\mathbf{r}, \boldsymbol{\rho})\nabla e^{i\mathbf{k}_i\cdot\mathbf{r}}]\cdot\mathbf{n}\, dS\,. \tag{7.115}$$

If we now compare this result with Eq. (7.108) for $\psi(\boldsymbol{\rho})$, we see that the sum $\psi(\boldsymbol{\rho}) + \psi_c(\boldsymbol{\rho})$ obeys the simple relation

$$\psi(\boldsymbol{\rho}) + \psi_c(\boldsymbol{\rho}) = \int_S [e^{i\mathbf{k}_i\cdot\mathbf{r}}\nabla G(\mathbf{r}, \boldsymbol{\rho}) - G(\mathbf{r}, \boldsymbol{\rho})\nabla e^{i\mathbf{k}_i\cdot\mathbf{r}}]\cdot\mathbf{n}\, dS\,, \tag{7.116}$$

where S denotes, as before, the *entire* closed surface. Now using Green's theorem and remembering that $e^{i\mathbf{k}_i\cdot\mathbf{r}}$ is a solution of the wave equation, we see that

$$\psi(\boldsymbol{\rho}) + \psi_c(\boldsymbol{\rho}) = e^{i\mathbf{k}_i\cdot\boldsymbol{\rho}}\,. \tag{7.117}$$

This beautifully simple result is known in optics as *Babinet's principle* or *the principle of complementary screens.*

If we use Eq. (7.111) for $\psi(\boldsymbol{\rho})$ at distances that are large compared to the dimensions of the screen, then

$$\psi_c(\boldsymbol{\rho}) = e^{i\mathbf{k}_i\cdot\boldsymbol{\rho}} + f(\theta, \phi)\,\frac{e^{ik_i\rho}}{\rho}\,, \tag{7.118a}$$

where

$$f(\theta, \phi) = \frac{ik_i}{4\pi}(1 + \cos\theta)\int_{\text{ap}} e^{-i\mathbf{k}_f\cdot\mathbf{r}}\, dS\,. \tag{7.118b}$$

Equations (7.118a) and (7.118b) are of the same form as the results which we found in our study of the Lippmann-Schwinger equation. In the latter case, $f(\theta, \phi)$, the scattering amplitude, was given in terms of the potential and the wave function. Here we obtain $f(\theta, \phi)$ in terms of a boundary condition.

According to Eq. (7.113), the scattering of a plane wave by a circular, opaque obstacle is described by the function

$$f(\theta, \phi) = \frac{iR}{2}\,\frac{1 + \cos\theta}{\sin\theta}\, J_1(k_i R\sin\theta)\,, \tag{7.119}$$

where R is the radius of the obstacle and $k_i = 2\pi/\lambda$, as before. Several interesting attempts have been made to understand in terms of diffraction the quantum-mechanical scattering of short wavelength (high energy) nuclear projectiles (α-particles, for example) from nuclei. For reasonable values of the nuclear radius, scattering amplitudes given by Eq. (7.119) match the experimental results quite closely. They show a strong forward maximum and an oscillatory behavior away from the forward peak very similar to the pattern shown in Fig. 7-8. This provides a nice illustration of the wave aspect of particles.

7.7 TIME-DEPENDENT GREEN'S FUNCTIONS: FIRST ORDER

In the past few sections we have concentrated our attention of the application of Green's function techniques to the solution of certain partial differential equations in which the time did not play an essential role. In this section we begin looking at problems where the evolution of some system in time (as well as in space) is important.

Consider an equation such as

$$H\psi + \frac{\partial \psi}{\partial \tau} = 0 , \tag{7.120}$$

where τ is a time variable, and H is an operator, *independent of time* which is assumed to possess a complete set of eigenfunctions. There are numerous equations of this type, and their application extends into all domains of physics. A familiar example is the heat equation,

$$\nabla^2 T = \frac{c}{k} \frac{\partial T}{\partial t} ,$$

where T is the temperature, c is the specific heat per unit volume, and k is the heat conductivity. If we call $H = -\nabla^2$ and $\tau = kt/c$, then this has precisely the form of Eq. (7.120). Schrödinger's time-dependent equation is also of this form, namely,

$$\mathscr{H}\psi = i\hbar \frac{\partial \psi}{\partial t} ,$$

where $\mathscr{H}$ is the total Hamiltonian of the system described by the wave function ψ. For example, for a free particle,

$$\mathscr{H} = -\frac{\hbar^2}{2m} \nabla^2 ,$$

so Schrödinger's equation is

$$-\nabla^2 \psi = \frac{2mi}{\hbar} \frac{\partial \psi}{\partial t} .$$

Thus if we set $H = -\nabla^2$ and $\tau = (i\hbar/2m)t$, the equation has exactly the same form as the heat equation. Another equation of this type is the diffusion equation which reads

$$\nabla^2 \rho = \frac{1}{a^2} \frac{\partial \rho}{\partial t} ,$$

where ρ is the density of the diffusing material, and a is the diffusion constant. Here again the identification $H = -\nabla^2$ and $\tau = a^2 t$ brings the equation to the basic form of Eq. (7.120).

Since we assume that H has a complete orthonormal set of eigenfunctions,

$$H\phi_m = \lambda_m \phi_m ,$$

we can expand the solution to Eq. (7.120) in terms of them;

$$\psi(\mathbf{r}, \tau) = \sum_m A_m(\tau)\phi_m(\mathbf{r}) .$$

Since $\psi(\mathbf{r}, \tau)$ depends on τ, so will the functions $A_m(\tau)$. Now

$$\begin{aligned} H\psi = H\left[\sum_m A_m(\tau)\phi_m(\mathbf{r})\right] &= \sum_m A_m(\tau)H\phi_m(\mathbf{r}) \\ &= \sum_m \lambda_m A_m(\tau)\phi_m(\mathbf{r}) . \end{aligned}$$

Also,

$$\frac{\partial\psi}{\partial\tau} = \sum_m \frac{\partial A_m(\tau)}{\partial\tau}\phi_m(\mathbf{r}) .$$

Thus Eq. (7.120) becomes

$$\sum_m \left[\lambda_m A_m(\tau) + \frac{dA_m(\tau)}{d\tau}\right]\phi_m(\mathbf{r}) = 0 .$$

Since the $\phi_m(\mathbf{r})$ are complete and orthonormal, we conclude immediately that

$$\frac{dA_m(\tau)}{d\tau} + \lambda_m A_m(\tau) = 0$$

for all m. The solution for $A_m(\tau)$ is simply

$$A_m(\tau) = A_m(0)e^{-\lambda_m\tau} ,$$

and thus we obtain

$$\psi(\mathbf{r}, \tau) = \sum_m A_m(0)\phi_m(\mathbf{r})e^{-\lambda_m\tau} .$$

Now, to complete the solution to a partial differential equation which is of first order in the time derivative, we need to specify the boundary condition at $\tau = 0$. Assuming that we know $\psi(\mathbf{r}, 0)$ (for example, in the case of the heat equation, if the initial temperature distribution $T(\mathbf{r}, 0)$ is known), we may write

$$\psi(\mathbf{r}, 0) = \sum_m A_m(0)\phi_m(\mathbf{r}) .$$

Using the fact that the ϕ_m form a complete orthonormal set, we have

$$A_n(0) = \int \phi_n^*(\mathbf{r})\psi(\mathbf{r}, 0)\, d^3\mathbf{r} .$$

Thus

$$\psi(\mathbf{r}, \tau) = \sum_m e^{-\lambda_m\tau}\phi_m(\mathbf{r})\int \phi_m^*(\mathbf{r}')\psi(\mathbf{r}', 0)\, d^3\mathbf{r}' ,$$

or

$$\psi(\mathbf{r}, \tau) = \int G_1(\mathbf{r}, \mathbf{r}', \tau)\psi(\mathbf{r}', 0)\, d^3\mathbf{r}' , \tag{7.121a}$$

where

$$G_1(\mathbf{r}, \mathbf{r}', \tau) = \sum_m \phi_m(\mathbf{r})\phi_m^*(\mathbf{r}')e^{-\lambda_m \tau} . \tag{7.121b}$$

Note that

$$G_1(\mathbf{r}, \mathbf{r}', 0) = \sum_m \phi_m(\mathbf{r})\phi_m^*(\mathbf{r}') = \delta^3(\mathbf{r} - \mathbf{r}') , \tag{7.122}$$

as it must, of course, because of Eq. (7.121a). Also,

$$\left(H + \frac{\partial}{\partial \tau}\right) G_1(\mathbf{r}, \mathbf{r}', \tau) = 0 . \tag{7.123}$$

Thus $\psi(\mathbf{r}, \tau)$, as given by Eq. (7.121a), is indeed a solution which satisfies the prescribed boundary condition.

We have put a subscript 1 on our function $G_1(\mathbf{r}, \mathbf{r}', \tau)$ to denote that it arises from an equation which is first order in time. We will refer loosely to G_1 as a Green's function, although it does not satisfy

$$\left[H + \frac{\partial}{\partial \tau}\right] G(\mathbf{r}, \mathbf{r}', \tau) = \delta^3(\mathbf{r} - \mathbf{r}')\delta(\tau) . \tag{7.124}$$

We will soon see, however, that G_1 is closely related to a function which *is* a Green's function in this usual sense.

Since G_1 does not satisfy Eq. (7.124), but, rather, Eq. (7.123), G_1 is itself a solution to Eq. (7.120). In fact, if $\psi(\mathbf{r}, 0) = \delta(\mathbf{r} - \mathbf{r}_0)$, then Eq. (7.121a) yields immediately

$$\psi(\mathbf{r}, \tau) = G_1(\mathbf{r}, \mathbf{r}_0, \tau)$$

for $\tau \geq 0$. Thus G_1 is a solution of Eq. (7.120) for a very special initial condition: the so-called "point source" initial condition. This fact leads to a simple physical interpretation of the Green's function which we shall discuss shortly.

Needless to say, our choice of initial time was completely arbitrary. We could equally well have chosen the initial time to be τ', and then we would have

$$\psi(\mathbf{r}, \tau) = \int G_1(\mathbf{r}, \mathbf{r}', \tau, \tau')\psi(\mathbf{r}', \tau')\, d^3\mathbf{r}' , \tag{7.125a}$$

where

$$G_1(\mathbf{r}, \mathbf{r}', \tau, \tau') = \sum_m \phi_m(\mathbf{r})\phi_m^*(\mathbf{r}')e^{-\lambda_m(\tau - \tau')} . \tag{7.125b}$$

In other words, $G_1(\mathbf{r}, \mathbf{r}', \tau, \tau')$ *propagates* the function ψ in time from τ' to $\tau > \tau'$; for this reason, G_1 is often referred to as a *propagator*. This manner of speaking suggests that if we propagate ψ from τ_0 to τ_1 and then from τ_1 to τ_2, we should get the same result as if we propagate directly from τ_0 to τ_2.

Mathematically, we write

$$\psi(\mathbf{r}, \tau_1) = \int G_1(\mathbf{r}, \mathbf{r}'', \tau_1, \tau_0)\psi(\mathbf{r}'', \tau_0)\, d^3\mathbf{r}''$$

$$\begin{aligned}\psi(\mathbf{r}, \tau_2) &= \int G_1(\mathbf{r}, \mathbf{r}', \tau_2, \tau_1)\psi(\mathbf{r}', \tau_1)\, d^3\mathbf{r}' \\ &= \int\int G_1(\mathbf{r}, \mathbf{r}', \tau_2, \tau_1) G_1(\mathbf{r}', \mathbf{r}'', \tau_1, \tau_0)\psi(\mathbf{r}'', \tau_0)\, d^3\mathbf{r}'\, d^3\mathbf{r}'' .\end{aligned}$$

Using Eq. (7.125b), we find that

$$\begin{aligned}&\int G_1(\mathbf{r}, \mathbf{r}', \tau_2, \tau_1) G_1(\mathbf{r}', \mathbf{r}'', \tau_1, \tau_0)\, d^3\mathbf{r}' \\ &\quad = \int \sum_m \phi_m(\mathbf{r})\phi_m^*(\mathbf{r}') e^{-\lambda_m(\tau_2-\tau_1)} \sum_n \phi_n(\mathbf{r}')\phi_n^*(\mathbf{r}'') e^{-\lambda_n(\tau_1-\tau_0)}\, d^3\mathbf{r}' .\end{aligned}$$

Since

$$\int \phi_m^*(\mathbf{r}')\phi_n(\mathbf{r}')\, d^3\mathbf{r}' = \delta_{mn} ,$$

we obtain immediately

$$\begin{aligned}&\int G_1(\mathbf{r}, \mathbf{r}', \tau_2, \tau_1) G_1(\mathbf{r}', \mathbf{r}'', \tau_1, \tau_0)\, d^3\mathbf{r}' \\ &\quad = \sum_m \phi_m(\mathbf{r})\phi_m^*(\mathbf{r}'') e^{-\lambda_m(\tau_2-\tau_1)} e^{-\lambda_m(\tau_1-\tau_0)} \\ &\quad = \sum_m \phi_m(\mathbf{r})\phi_m^*(\mathbf{r}'') e^{-\lambda_m(\tau_2-\tau_0)} \\ &\quad = G_1(\mathbf{r}, \mathbf{r}'', \tau_2, \tau_0) .\end{aligned}$$

Thus, as we expected,

$$\psi(\mathbf{r}, \tau_2) = \int G_1(\mathbf{r}, \mathbf{r}'', \tau_2, \tau_0)\psi(\mathbf{r}'', \tau_0)\, d^3\mathbf{r}'' .$$

This idea of compounding propagations is a very important notion in physics; we restate the result below:

$$G_1(\mathbf{r}, \mathbf{r}'', \tau_2, \tau_0) = \int G_1(\mathbf{r}, \mathbf{r}', \tau_2, \tau_1) G_1(\mathbf{r}', \mathbf{r}'', \tau_1, \tau_0)\, d^3\mathbf{r}' .$$

Let us now determine the Green's function for an equation of the diffusion or heat conduction type. We work in one dimension for convenience, in the interval $-L/2 \leq x \leq L/2$, and impose periodic boundary conditions, $\psi(L/2, \tau) = \psi(-L/2, \tau)$. As mentioned above, the operator H is simply $-d^2/dx^2$ in this case, so we want to solve

$$-\frac{d^2\phi_m}{dx^2} = \lambda_m \phi_m(x) .$$

This equation has as its solution

$$\phi_m(x) = \frac{1}{\sqrt{L}} e^{i\sqrt{\lambda_m}x},$$

where $\lambda_m = (2\pi m/L)^2$, in order to satisfy the boundary condition, and m is any integer. Thus if we call $k_m = 2\pi m/L$, we get for $G_1(x, x', \tau)$, using Eq. (7.121b),

$$G_1(x, x', \tau) = \sum_{m=-\infty}^{\infty} \frac{1}{L} e^{ik_m(x-x')} e^{-k_m^2\tau}. \tag{7.126}$$

Now suppose that we want to obtain the Green's function on the *infinite* line. Let us employ the same type of limiting procedure used in making the transition from Fourier series to Fourier integrals in Section 5.7. In Eq. (7.126), as L becomes larger and larger, k_m changes only very slightly as we go from one value of m to the next in the sum. In fact, k_m changes by $2\pi/L$ each time m is incremented by one unit. Let us therefore set $2\pi/L = \Delta k_m$ and write

$$G_1(x, x', \tau) = \frac{1}{2\pi} \sum_{m=-\infty}^{\infty} \Delta k_m e^{ik_m(x-x')} e^{-k_m^2\tau}.$$

In this form we see that our expression is exactly what one finds when one discusses the Riemann integral of some function; we break the interval of integration in question up into small subintervals Δ_i and evaluate the function at some point x_i in each interval. Then

$$\sum_i \Delta_i f(x_i) \to \int f(x)\, dx.$$

In this spirit, as $L \to \infty$, we have for G_1 just

$$G_1(x, x', \tau) = \frac{1}{2\pi} \int_{-\infty}^{\infty} e^{ik(x-x')} e^{-k^2\tau}\, dk. \tag{7.127}$$

It is not difficult to evaluate this integral. We note that

$$ik(x - x') - k^2\tau = -\tau\left[k - \frac{i(x - x')}{2\tau}\right]^2 - \frac{(x - x')^2}{4\tau},$$

so that

$$G_1(x, x', \tau) = \frac{1}{2\pi} e^{-(x-x')^2/4\tau} \int_{-\infty}^{\infty} e^{-\tau(k-i\delta)^2}\, dk,$$

where $\delta = (x - x')/2\tau$. To evaluate this integral, we consider the integrand to be a function of a complex variable z, and investigate the contour integral along the curve C shown in Fig. 7.10. Thus

$$I = \int_C e^{-\tau(z-i\delta)^2}\, dz = 0$$

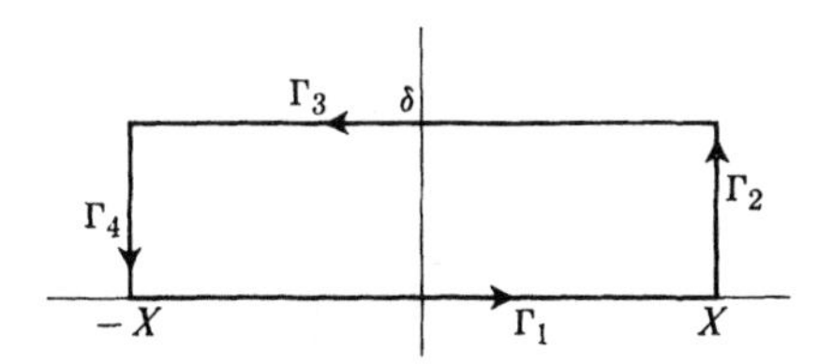

Fig. 7.10 The curve C used to evaluate the function $G_1(x, x', \tau)$. X is to be considered as being arbitrarily large.

since there are no singularities inside the contour. On Γ_1, $z = x$, and on Γ_3, $z = x + i\delta$. Thus

$$I = 0 = \int_{-X}^{X} e^{-\tau(x-i\delta)^2}\,dx + \int_{\Gamma_2} e^{-\tau(z-i\delta)^2}\,dz + \int_{X}^{-X} e^{-\tau x^2}\,dx + \int_{\Gamma_4} e^{-\tau(z-i\delta)^2}\,dz\,.$$

Now on Γ_2, $z = X + iy$, where X can become arbitrarily large. Hence

$$\int_{\Gamma_2} e^{-\tau(z-i\delta)^2}\,dz = i\int_0^{\delta} e^{-\tau[X+i(y-\delta)]^2}\,dy = ie^{-\tau X^2}\int_0^{\delta} e^{\tau(y-\delta)^2}e^{-2iX\tau(y-\delta)}\,dy\,.$$

Clearly, the integral on y yields a finite number since we integrate a bounded function over a finite interval. Thus as $X \to \infty$, the integral along Γ_2 vanishes because of the factor $e^{-X^2\tau}$. The same argument eliminates the integral along Γ_4. Thus

$$\int_{-\infty}^{\infty} e^{-\tau(x-i\delta)^2}\,dx = \int_{-\infty}^{\infty} e^{-\tau x^2}\,dx = \sqrt{\frac{\pi}{\tau}}\,,$$

where this last integral can be evaluated by real integral techniques, (or see Eq. 6.56). (It should be remarked that, although we have tacitly assumed that τ is real in this calculation, it is a simple matter [see Problem 31 (l) of Chapter 6] to show that the result is valid even when τ is pure imaginary.) Thus, finally,

$$G_1(x, x', \tau) = \frac{1}{\sqrt{4\pi\tau}}\,e^{-(x-x')^2/4\tau}\,. \tag{7.128}$$

Note that $G_1(x, x', \tau)$ is a symmetric function of x and x'. As τ tends to zero,

$$\lim_{\tau\to 0} G(x, x', \tau) = \lim_{\tau\to 0}\frac{1}{\sqrt{4\pi\tau}}\,e^{-(x-x')^2/4\tau}\,,$$

which is just one of the representations of the δ-function discussed in Section 5.3. Thus

$$\lim_{\tau\to 0} G(x, x', \tau) = \delta(x - x')\,,$$

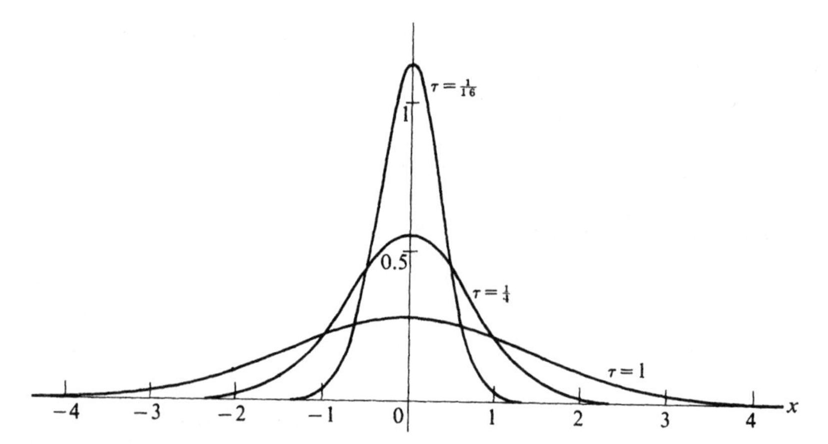

Fig. 7.11 The Green's function $G(x, x', \tau)$ as a function of x (centered at $x' = 0$ for convenience) for several values of τ.

in agreement with the general result of Eq. (7.122). As the time, τ, increases from zero, $G_1(x, x', \tau)$ looks like the familiar Gaussian, centered about $x = x'$ with half width equal to $\sqrt{2\tau}$. For small τ, the Gaussian is tall and narrow; as τ increases, the Gaussian broadens out and its height decreases, as shown in Fig. 7.11. However, the area under these Gaussians remains constant, since we see that

$$\int_{-\infty}^{\infty} G_1(x, x')\, dx = \frac{1}{\sqrt{4\pi\tau}} \int_{-\infty}^{\infty} e^{-(x-x')^2/4\tau}\, dx = 1\,,$$

independent of τ and x'. In terms of the diffusion equation, this means that the total quantity of diffusing gas (the integral of the density over all space) is constant, since

$$\psi(x, \tau) = \int_{-\infty}^{\infty} G_1(x, x', \tau)\psi(x', 0)\, dx'$$

implies that

$$\int_{-\infty}^{\infty} \psi(x, \tau)\, dx = \int_{-\infty}^{\infty} \psi(x', 0)\, dx'\,, \tag{7.129}$$

by the result above. In the diffusion equation, $\psi(x, \tau)$ is the density of the diffusing gas, so

$$\int_{-\infty}^{\infty} \psi(x, \tau)\, dx$$

is the total amount of diffusing gas. Equation (7.129) states that this quantity is constant in time.

In terms of this Green's function we can write the solution to the diffusion or heat equation at a later time in terms of the initial distribution:

$$\psi(x, \tau) = \frac{1}{\sqrt{4\pi\tau}} \int_{-\infty}^{\infty} e^{-(x-x')^2/4\tau} \psi(x', 0)\, dx' . \tag{7.130}$$

This equation has a simple physical interpretation: At $\tau = 0$, each element of space has an amount of diffusing material (or heat) equal to $\psi(x', 0)\, dx'$. The effect of this source at a later time at a point x is obtained by multiplying by the Green's function, which describes how the material (or heat) diffuses away from the point x'. The density (or temperature) at time τ at the point x is obtained by linear superposition of the effects at x due to all sources at the different points x'—that is, by integrating over the source.

Now suppose that we want to solve an equation of the form

$$H\psi + \frac{\partial \psi}{\partial \tau} = F(\mathbf{r}, \tau) . \tag{7.131}$$

We would like to find a Green's function such that

$$\psi(\mathbf{r}, \tau) = \psi_0(\mathbf{r}, \tau) + \int G(\mathbf{r}, \mathbf{r}', \tau, \tau') F(\mathbf{r}', \tau')\, d^3\mathbf{r}\, d\tau' , \tag{7.132}$$

where $\psi_0(\mathbf{r}, \tau)$ is a solution of the homogeneous equation. In the case of the heat equation or the diffusion equation, we might have a source of heat or a source of diffusing material which continuously varies in time described by the function $F(\mathbf{r}, \tau)$. In Schrödinger's equation, $F(\mathbf{r}, \tau)$ might represent a perturbation of some sort, such as an electromagnetic wave incident on a quantum mechanical system.

What is the relationship between the function $G_1(\mathbf{r}, \mathbf{r}', \tau, \tau')$ discussed above and a Green's function of the type required in Eq. (7.132)? First of all, the function $G(\mathbf{r}, \mathbf{r}', \tau, \tau')$ in Eq. (7.132) must satisfy

$$\left(H + \frac{\partial}{\partial \tau}\right) G(\mathbf{r}, \mathbf{r}', \tau, \tau') = \delta^3(\mathbf{r} - \mathbf{r}')\delta(\tau - \tau') , \tag{7.133}$$

whereas G_1 satisfies

$$\left(H + \frac{\partial}{\partial \tau}\right) G_1(\mathbf{r}, \mathbf{r}', \tau, \tau') = 0 . \tag{7.133'}$$

Now G_1 has one particularly interesting property which was emphasized above, namely,

$$\lim_{\tau \to \tau'} G_1(\mathbf{r}, \mathbf{r}', \tau, \tau') = \delta^3(\mathbf{r} - \mathbf{r}') , \tag{7.134}$$

according to Eq. (7.122). This suggests that if we can introduce a discontinuity in G_1 at $\tau = \tau'$, we may be able to satisfy Eq. (7.133). Since G must satisfy Eq. (7.134) everywhere except at $\mathbf{r} = \mathbf{r}'$, $\tau = \tau'$, it seems reasonable to try the modification

$$G(\mathbf{r}, \mathbf{r}', \tau, \tau') = G_1(\mathbf{r}, \mathbf{r}', \tau, \tau')\theta(\tau - \tau'); \tag{7.135}$$

that is, for $\tau > \tau'$, $G(\mathbf{r}, \mathbf{r}', \tau, \tau')$ is just equal to $G_1(\mathbf{r}, \mathbf{r}', \tau, \tau')$, but for $\tau < \tau'$, $G(\mathbf{r}, \mathbf{r}', \tau, \tau') = 0$. We now compute

$$\begin{aligned}
\left(H + \frac{\partial}{\partial\tau}\right) & G(\mathbf{r}, \mathbf{r}', \tau, \tau') \\
&= H[G_1(\mathbf{r}, \mathbf{r}', \tau, \tau')\theta(\tau - \tau'] + \frac{\partial}{\partial\tau}[G_1(\mathbf{r}, \mathbf{r}', \tau, \tau')\theta(\tau - \tau')] \\
&= \theta(\tau - \tau')HG_1(\mathbf{r}, \mathbf{r}', \tau, \tau') + \theta(\tau - \tau')\frac{\partial}{\partial\tau} G_1(\mathbf{r}, \mathbf{r}', \tau, \tau') \\
&\quad + G_1(\mathbf{r}, \mathbf{r}', \tau, \tau')\frac{\partial}{\partial\tau}\theta(\tau - \tau') \\
&= \theta(\tau - \tau')\left[H + \frac{\partial}{\partial\tau}\right] G_1(\mathbf{r}, \mathbf{r}', \tau, \tau') + G_1(\mathbf{r}, \mathbf{r}', \tau, \tau')\delta(\tau - \tau'),
\end{aligned}$$

since by Eq. (7.22),

$$\frac{\partial}{\partial\tau}\theta(\tau - \tau') = \delta(\tau - \tau').$$

Using Eq. (7.133′), we then have

$$\left(H + \frac{\partial}{\partial\tau}\right) G(\mathbf{r}, \mathbf{r}', \tau, \tau') = G_1(\mathbf{r}, \mathbf{r}', \tau, \tau')\delta(\tau - \tau').$$

But since $\delta(\tau - \tau')$ is zero except at $\tau = \tau'$, we are only interested in $G_1(\mathbf{r}, \mathbf{r}', \tau, \tau')$ at $\tau = \tau'$, which by Eq. (7.134) is just $\delta^3(\mathbf{r} - \mathbf{r}')$. Thus we have the desired result,

$$\left(H + \frac{\partial}{\partial\tau}\right) G(\mathbf{r}, \mathbf{r}', \tau, \tau') = \delta^3(\mathbf{r} - \mathbf{r}')\delta(\tau - \tau').$$

Using Eq. (7.135), we see that the solution to Eq. (7.131) is given by

$$\psi(\mathbf{r}, \tau) = \psi_0(\mathbf{r}, \tau) + \int d^3\mathbf{r}' \int_{-\infty}^{\tau} d\tau'\, G_1(\mathbf{r}, \mathbf{r}', \tau, \tau')F(\mathbf{r}', \tau'). \tag{7.136}$$

The object G_1 is seen to be very nearly a Green's function, in our original sense of the word. Since $F(\mathbf{r}, \tau)$ represents the effect of a known source of heat or diffusing material, once G_1 has been determined we need only evaluate the integrals in question.

A different situation arises in quantum mechanics, where Schrödinger's time-dependent equation has the form

$$\left(H + \frac{\partial}{\partial\tau}\right)\Psi(\mathbf{r}, \tau) = -\frac{2m}{\hbar^2}U(\mathbf{r}, \tau)\Psi(\mathbf{r}, \tau);$$

in the simplest case H might be given by $H = -\nabla^2$ and $\tau = i\hbar t/2m$. Thus we obtain from Eq. (7.136) Schrödinger's time-dependent integral equation

$$\Psi(\mathbf{r}, \tau) = \Psi_0(\mathbf{r}, \tau) - \frac{2m}{\hbar^2} \int d^3\mathbf{r}' \int_{-i\infty}^{\tau} d\tau' \, G(\mathbf{r}, \mathbf{r}', \tau, \tau') U(\mathbf{r}', \tau') \Psi(\mathbf{r}', \tau') \,, \tag{7.137}$$

where the integral is along the imaginary axis, since $\tau = i\hbar t/2m$ is pure imaginary.

When $U(\mathbf{r}, \tau)$ represents a perturbation like an oscillating electromagnetic field, Eq. (7.137) leads to the theory of transition probabilities and the "Fermi Golden Rule."

We can also use Eq. (7.137) to obtain a different perspective on the work of previous sections. Suppose that $U(\mathbf{r}, \tau)$ is a static (time-independent) potential. Then we would expect to see Eq. (7.137) reduce to the Lippmann-Schwinger equation (7.68). However, rather than simply replacing $U(\mathbf{r}, \tau)$ by $V(\mathbf{r})$, let us imagine that in the very remote past $U(\mathbf{r}, \tau)$ was "turned off" and that we slowly allowed $U(\mathbf{r}, \tau)$ to be "turned on." We may express this by writing

$$U(\mathbf{r}, \tau) = V(\mathbf{r}) e^{\epsilon' t} = V(\mathbf{r}) e^{-i\epsilon\tau}; \tag{7.138}$$

by making the real quantity $\epsilon' = \hbar\epsilon/2m$ small enough, we can approximate a static potential arbitrary well. Clearly, this should have no effect on any physically realistic scattering problem. If we take $H = -\nabla^2$, as we have done before in discussing scattering, then by direct analogy with Eq. (7.127)

$$\begin{aligned} G_1(\mathbf{r}, \mathbf{r}', \tau, \tau') = (1/(2\pi)^3) \int & dk_x \, dk_y \, dk_z \\ & \times \exp\{i[k_x(x - x') + k_y(y - y') + k_z(z - z')]\} \\ & \times \exp[-(k_x^2 + k_y^2 + k_z^2)(\tau - \tau')] \end{aligned}$$

where we imagine that we have put our system in a large rectangular box and afterwards let the sides become arbitrarily large. In spherical coordinates, this result appears as

$$G_1(\mathbf{r}, \mathbf{r}', \tau, \tau') = \frac{1}{(2\pi)^3} \int e^{i\mathbf{k}\cdot(\mathbf{r}-\mathbf{r}')} e^{-k^2(\tau-\tau')} \, d^3\mathbf{k} \,. \tag{7.139}$$

Putting Eq. (7.139) and Eq. (7.138) for $U(\mathbf{r}, \tau)$ into Eq. (7.137), we get

$$\begin{aligned} \Psi(\mathbf{r}, \tau) = \Psi_0(\mathbf{r}, \tau) & \\ & - (2m/\hbar^2(2\pi)^3) \int d^3\mathbf{r}' \int d^3\mathbf{k} \int_{-i\infty}^{\tau} d\tau' \, e^{i\mathbf{k}\cdot(\mathbf{r}-\mathbf{r}')} \\ & \times e^{-k^2(\tau-\tau')} V(\mathbf{r}') e^{-i\epsilon\tau'} \Psi(\mathbf{r}', \tau') \end{aligned} \tag{7.140}$$

with

$$\Psi_0(\mathbf{r}, \tau) = C e^{i\mathbf{k}_i\cdot\mathbf{r}} e^{-k_i^2\tau} \,,$$

where k_i is the wave number of the incident wave. Let us look for a solution of the form

$$\Psi(\mathbf{r}, \tau) = \psi(\mathbf{r})e^{-k_i^2\tau} .$$

With this substitution, Eq. (7.140) becomes

$$\begin{aligned}\psi(\mathbf{r})e^{-k_i^2\tau} &= Ce^{i\mathbf{k}_i\cdot\mathbf{r}}e^{-k_i^2\tau} \\ &\quad - \frac{2m}{\hbar^2}\int d^3\mathbf{r}' \frac{1}{(2\pi)^3}\int d^3\mathbf{k}e^{i\mathbf{k}\cdot(\mathbf{r}-\mathbf{r}')}V(\mathbf{r}')\psi(\mathbf{r}') \\ &\quad \times e^{-k^2\tau}\int_{-i\infty}^{\tau} e^{(k^2-k_i^2-i\epsilon)\tau'}\,d\tau' \end{aligned} \tag{7.141}$$

Now,

$$\int_{-i\infty}^{\tau} e^{(k^2-k_i^2-i\epsilon)\tau'}\,d\tau' = \frac{e^{(k^2-k_i^2-i\epsilon)\tau}}{(k^2 - k_i^2 - i\epsilon)} ; \tag{7.142}$$

the lower limit of integration gives a vanishing contribution because the perturbation was switched on adiabatically via the term $\exp(-i\epsilon\tau)$. Putting Eq. (7.142) into Eq. (7.141) and canceling a common factor of $\exp(-k_i^2\tau)$, we get

$$\begin{aligned}\psi(\mathbf{r}) &= Ce^{i\mathbf{k}_i\cdot\mathbf{r}} - \frac{2m}{\hbar^2}\int d^3\mathbf{r}' V(\mathbf{r}')\psi(\mathbf{r}')\frac{1}{(2\pi)^3} \\ &\quad \times\int \frac{e^{i\mathbf{k}\cdot(\mathbf{r}-\mathbf{r}')}}{(k^2 - k_i^2 - i\epsilon)} e^{-i\epsilon\tau}\,d^3\mathbf{k} .\end{aligned} \tag{7.143}$$

The integral

$$I(\mathbf{r}, \mathbf{r}') = -\frac{1}{(2\pi)^3}\int \frac{e^{i\mathbf{k}\cdot(\mathbf{r}-\mathbf{r}')}}{(k^2 - k_i^2 - i\epsilon)}\,d^3\mathbf{k}$$

is precisely the *outgoing spherical wave* Green's function $G_+(\mathbf{r}, \mathbf{r}')$ obtained for the operator $-\nabla^2 - k_i^2$ in Section 7.4 (Eq. 7.66). Thus

$$I(\mathbf{r}, \mathbf{r}') = -\frac{1}{4\pi}\frac{\exp[i(k_i + i\epsilon/2k_i)|\mathbf{r} - \mathbf{r}'|]}{|\mathbf{r} - \mathbf{r}'|} .$$

We now let $\epsilon \to 0$ and find that

$$\psi(\mathbf{r}) = Ce^{i\mathbf{k}_i\cdot\mathbf{r}} - \frac{m}{2\pi\hbar^2}\int \frac{e^{ik_i|\mathbf{r}-\mathbf{r}'|}}{|\mathbf{r} - \mathbf{r}'|} V(\mathbf{r}')\psi(\mathbf{r}')\,d^3\mathbf{r}' ,$$

which agrees with the Lippmann-Schwinger equation for ψ_+ which we derived earlier, Eq. (7.68). Thus in the limit of a static potential, the time-dependent formalism goes over to the form we should expect, with the prescription we used earlier for avoiding singularities being automatically provided. We now see that this prescription can be thought of as arising from the physically reasonable assumption that at sufficiently remote times in the past there is no interaction, just a free particle.

7.8 THE WAVE EQUATION

As a final example of the Green's function technique, we investigate the Lorentz-invariant, time-dependent wave equation. This is the basic equation of electrodynamics, and its properties have been exhaustively analyzed.

The homogeneous wave equation is

$$\Box^2\psi \equiv \nabla^2\psi - \frac{1}{c^2}\frac{\partial^2\psi}{\partial t^2} = 0\,. \tag{7.144}$$

For the sake of generality, we write this in the form

$$H\psi + \frac{\partial^2\psi}{\partial\tau^2} = 0\,, \tag{7.145}$$

where, in this case, $\tau = ct$ and $H = -\nabla^2$. For the moment, let us imagine that H is a general linear operator with a complete orthonormal set of eigenfunctions ϕ_n which satisfy

$$H\phi_n = \lambda_n\phi_n\,. \tag{7.146}$$

Then we look for a solution of Eq. (7.145) in the form

$$\psi(\mathbf{r}, \tau) = \sum_n A_n(\tau)\phi_n(\mathbf{r})\,. \tag{7.147}$$

Using Eq. (7.146) and Eq. (7.147), we find

$$H\psi + \frac{\partial^2\psi}{\partial\tau^2} = \sum_n (A_n\lambda_n + \ddot{A}_n)\phi_n = 0\,. \tag{7.148}$$

Since the eigenfunctions of H form a complete orthonormal set, we conclude from Eq. (7.148) that

$$\ddot{A}_n(\tau) + \lambda_n A_n(\tau) = 0\,.$$

Hence

$$A_n(\tau) = a_n e^{i\sqrt{\lambda_n}\tau} + b_n e^{-i\sqrt{\lambda_n}\tau}\,,$$

and

$$\psi(\mathbf{r}, \tau) = \sum_n (a_n e^{i\sqrt{\lambda_n}\tau} + b_n e^{-i\sqrt{\lambda_n}\tau})\phi_n(\mathbf{r})\,. \tag{7.149}$$

The constants a_n and b_n must be determined from the boundary conditions, which we assume to be of the familiar type; that is, $\psi(\mathbf{r}, 0)$ and $\dot{\psi}(\mathbf{r}, 0)$ are specified. Then

$$\psi(\mathbf{r}, 0) = \sum_n (a_n + b_n)\phi_n(\mathbf{r})\,,$$

$$\dot{\psi}(\mathbf{r}, 0) = i\sum_n \sqrt{\lambda_n}(a_n - b_n)\phi_n(\mathbf{r})\,.$$

We find that

$$a_n + b_n = \int \phi_n^*(\mathbf{r}')\psi(\mathbf{r}', 0)\, d^3\mathbf{r}' ,$$

$$a_n - b_n = -\frac{i}{\sqrt{\lambda_n}} \int \phi_n^*(\mathbf{r}')\dot{\psi}(\mathbf{r}', 0)\, d^3\mathbf{r}' ,$$

so

$$a_n = \frac{1}{2}\left[\int \phi_n^*(\mathbf{r}')\psi(\mathbf{r}', 0)\, d^3\mathbf{r}' + \frac{1}{i\sqrt{\lambda_n}} \int \phi_n^*(\mathbf{r}')\dot{\psi}(\mathbf{r}', 0)\, d^3\mathbf{r}'\right] ,$$

$$b_n = \frac{1}{2}\left[\int \phi_n^*(\mathbf{r}')\psi(\mathbf{r}', 0)\, d^3\mathbf{r}' - \frac{1}{i\sqrt{\lambda_n}} \int \phi_n^*(\mathbf{r}')\dot{\psi}(\mathbf{r}', 0)\, d^3\mathbf{r}'\right] .$$

Inserting these values for a_n and b_n into Eq. (7.149), we get

$$\psi(\mathbf{r}, \tau) = \int \sum_n \cos(\sqrt{\lambda_n}\tau)\phi_n(\mathbf{r})\phi_n^*(\mathbf{r}')\psi(\mathbf{r}', 0)\, d^3\mathbf{r}'$$
$$+ \int \sum_n \sin(\sqrt{\lambda_n}\tau)\frac{1}{\sqrt{\lambda_n}}\phi_n(\mathbf{r})\phi_n^*(\mathbf{r}')\dot{\psi}(\mathbf{r}', 0)\, d^3\mathbf{r}' .$$

In keeping with the notation of the previous section, let us write this as

$$\psi(\mathbf{r}, \tau) = \int G_2(\mathbf{r}, \mathbf{r}', \tau)\psi(\mathbf{r}', 0)\, d^3\mathbf{r}' + \int \tilde{G}_2(\mathbf{r}, \mathbf{r}', \tau)\dot{\psi}(\mathbf{r}', 0)\, d^3\mathbf{r}' , \tag{7.150}$$

where

$$G_2(\mathbf{r}, \mathbf{r}', \tau) = \sum_n \cos(\sqrt{\lambda_n}\tau)\phi_n(\mathbf{r})\phi_n^*(\mathbf{r}') , \tag{7.151a}$$

$$\tilde{G}_2(\mathbf{r}, \mathbf{r}', \tau) = \sum_n \frac{\sin\sqrt{\lambda_n}\tau}{\sqrt{\lambda_n}}\phi_n(\mathbf{r})\phi_n^*(\mathbf{r}') . \tag{7.151b}$$

Since the wave equation is second order in time, two boundary conditions have been required to determine the constants of integration. Hence we have obtained two Green's functions, one which acts on $\psi(\mathbf{r}, 0)$, the other which acts on $\dot{\psi}(\mathbf{r}, 0)$. For the first-order (in time) heat or diffusion equation, only one boundary condition is needed, and hence we found only one Green's function which acted on $\psi(\mathbf{r}, 0)$. Just as the Green's function obtained in the last section satisfied the heat equation, so both of our Green's functions, G_2 and $\tilde{G}_2$, satisfy the wave equation;

$$\Box^2 G_2(\mathbf{r}, \mathbf{r}', \tau) = 0 , \qquad \Box^2 \tilde{G}_2(\mathbf{r}, \mathbf{r}', \tau) = 0 .$$

Thus $\psi(\mathbf{r}, \tau)$ satisfies the wave equation. Note also that

$$G_2(\mathbf{r}, \mathbf{r}', \tau) = \frac{d}{d\tau}\tilde{G}_2(\mathbf{r}, \mathbf{r}'. \tau) ,$$

so that if we can calculate $\tilde{G}_2$, then G_2 can be obtained directly from it. The boundary conditions are clearly satisfied by the solution given in Eq. (7.150) since

$$G_2(\mathbf{r}, \mathbf{r}', 0) = \sum_n \phi_n(\mathbf{r})\phi_n^*(\mathbf{r}') = \delta^3(\mathbf{r} - \mathbf{r}') . \tag{7.152}$$

With these general results in mind, let us return to the specific case of the wave equation, and start by considering a region defined by a rectangular parallelepiped with sides L_1, L_2, and L_3 in the x, y- and z-directions, respectively. If we have periodic boundary conditions, the eigenfunctions of $H = -\nabla^2$ may be written as

$$\phi_{n_1n_2n_3}(\mathbf{r}) = \frac{1}{\sqrt{L_1L_2L_3}} e^{ik_xx}e^{ik_yy}e^{ik_zz} ,$$

where

$$k_x = \frac{2\pi n_1}{L_1} , \qquad k_y = \frac{2\pi n_2}{L_2} , \qquad k_z = \frac{2\pi n_3}{L_3} ,$$

and the eigenvalues are given simply by

$$\lambda_{n_1n_2n_3} = k_x^2 + k_y^2 + k_z^2 .$$

Let us now write $\tilde{G}_2$ in terms of these quantities. According to Eq. (7.151b),

$$\tilde{G}_2(\mathbf{r}, \mathbf{r}', \tau) = \frac{1}{L_1L_2L_3} \sum_{n_1, n_2, n_3}^{\infty} \frac{\sin \sqrt{k_x^2 + k_y^2 + k_z^2}\tau}{\sqrt{k_x^2 + k_y^2 + k_z^2}} e^{ik_x(x-x')}e^{ik_y(y-y')}e^{ik_z(z-z')} .$$

Now we can move to the limit in which the box becomes arbitrarily large, as we did before. Then

$$\frac{1}{L_1} \sum_{n_1=-\infty}^{\infty} \rightarrow \frac{1}{2\pi} \int_{-\infty}^{\infty} dk_x$$

and similarly for the sums on n_2 and n_3. Hence, in the limit of very large dimensions,

$$\tilde{G}_2(\mathbf{r}, \mathbf{r}', \tau) = \frac{1}{(2\pi)^3} \int_{-\infty}^{\infty} dk_x \int_{-\infty}^{\infty} dk_y \int_{-\infty}^{\infty} dk_z \frac{\sin k\tau}{k} e^{i\mathbf{k}\cdot(\mathbf{r}-\mathbf{r}')} ,$$

where $\mathbf{k} = k_x\mathbf{i} + k_y\mathbf{j} + k_z\mathbf{k}$ and $k^2 = k_x^2 + k_y^2 + k_z^2$. It will clearly be advantageous to evaluate this integral by using spherical coordinates. Setting $\mathbf{r} - \mathbf{r}' \equiv \boldsymbol{\rho}$, we get

$$\tilde{G}_2(\mathbf{r}, \mathbf{r}', \tau) = \frac{1}{(2\pi)^3} \int_0^{\infty} dk \cdot k^2 \sin k\tau \cdot \frac{1}{k} \int_0^{\pi} \sin \theta_k \, d\theta_k \int_0^{2\pi} d\phi_k e^{i\mathbf{k}\cdot\boldsymbol{\rho}} .$$

Since the integral is independent of the direction of $\boldsymbol{\rho}$, let us for convenience choose $\boldsymbol{\rho}$ to be in the z-direction. Then $\mathbf{k}\cdot\boldsymbol{\rho} = k\rho \cos \theta_k$, and making the change of variable $x = \cos\theta_k$, we have

$$\tilde{G}_2(\mathbf{r}, \mathbf{r}', \tau) = \frac{1}{(2\pi)^3}\int_0^\infty dk\cdot k\sin k\tau \int_{-1}^{1} dx \int_0^{2\pi} d\phi_k e^{ik\rho x}$$
$$= \frac{1}{2\pi^2\rho}\int_0^\infty \sin k\tau \sin k\rho \, dk$$
$$= \frac{1}{4\pi^2\rho}\int_{-\infty}^\infty \sin k\tau \sin k\rho \, dk$$
$$= \frac{1}{8\pi^2\rho}\int_{-\infty}^\infty [\cos k(\rho-\tau) - \cos k(\rho+\tau)]\, dk \, .$$

Since

$$\delta(x) = \frac{1}{2\pi}\int_{-\infty}^\infty \cos xy \, dy$$

(this follows immediately from Eq. 5.63), we get finally

$$\tilde{G}_2(\mathbf{r}, \mathbf{r}', \tau) = \frac{1}{4\pi\rho}[\delta(\rho-\tau) - \delta(\rho+\tau)] \, .$$

In terms of the original variables, this is

$$\tilde{G}_2(\mathbf{r}, \mathbf{r}', t) = \frac{1}{4\pi|\mathbf{r}-\mathbf{r}'|}[\delta(|\mathbf{r}-\mathbf{r}'| - ct) - \delta(|\mathbf{r}-\mathbf{r}'| + ct)] \, . \quad (7.153)$$

Clearly, our choice of $t = 0$ as the initial time was arbitrary. The reader can check that if we choose t' as the initial time, then

$$\psi(\mathbf{r}, t) = \int G_2(\mathbf{r}, \mathbf{r}', t, t')\psi(\mathbf{r}', t')\, d^3\mathbf{r}'$$
$$+ \int \tilde{G}_2(\mathbf{r}, \mathbf{r}', t, t')\dot{\psi}(\mathbf{r}', t')\, d^3\mathbf{r}' \, , \quad (7.154)$$

where

$$\tilde{G}_2(\mathbf{r}, \mathbf{r}', t, t') = \sum_n \frac{\sin[\sqrt{\lambda_n}c(t-t')]}{\sqrt{\lambda_n}}\phi_n(\mathbf{r})\phi_n^*(\mathbf{r}') \, , \quad (7.155a)$$

and

$$G_2(\mathbf{r}, \mathbf{r}', t, t') = \sum_n \cos[\sqrt{\lambda_n}c(t-t')]\phi_n(\mathbf{r})\phi_n^*(\mathbf{r}')$$
$$= \frac{1}{c}\frac{d}{dt}\tilde{G}_2(\mathbf{r}, \mathbf{r}', t, t') \, , \quad (7.155b)$$

For the case of the wave equation, $H = -\nabla^2$, $\tilde{G}_2$ is given by

$$\tilde{G}_2(\mathbf{r}, \mathbf{r}', t, t') = \frac{1}{4\pi|\mathbf{r}-\mathbf{r}'|}\{\delta[|\mathbf{r}-\mathbf{r}'| - c(t-t')]$$
$$- \delta[|\mathbf{r}-\mathbf{r}'| + c(t-t')]\} \, . \quad (7.156)$$

This gives the solution to the homogeneous wave equation, once the boundary conditions are specified.

However, just as in the case of the first-order equation studied in the last section, the inhomogeneous equation is very important. In classical electrodynamics, for example, Eq. (7.144) corresponds to a situation where no sources or currents are present. However, in general, one is interested in cases where there are currents and charges present. Thus, typically, we would like to be able to solve

$$\Box^2\psi \equiv \nabla^2\psi - \frac{1}{c^2}\frac{\partial^2\psi}{\partial t^2} = \frac{4\pi}{c}j(\mathbf{r}, t)\,, \tag{7.157}$$

or, in general,

$$H\psi + \frac{\partial^2\psi}{\partial\tau^2} = J(\mathbf{r}, \tau)\,. \tag{7.158}$$

Now, to solve Eq. (7.158), we need a Green's function $G(\mathbf{r}, \mathbf{r}', \tau, \tau')$ which satisfies

$$\left(H + \frac{\partial^2}{\partial\tau^2}\right)G(\mathbf{r}, \mathbf{r}', \tau, \tau') = \delta^3(\mathbf{r} - \mathbf{r}')\delta(\tau - \tau')\,.$$

Neither G_2 nor $\tilde{G}_2$ has this property; in fact, letting $\mathscr{G}$ stand for either G_2 or $\tilde{G}_2$,

$$\left(H + \frac{\partial^2}{\partial\tau^2}\right)\mathscr{G} = 0\,. \tag{7.159}$$

However, let us be guided by our experience in the previous section and look for G in the form

$$G_R(\mathbf{r}, \mathbf{r}', \tau, \tau') = \mathscr{G}(\mathbf{r}, \mathbf{r}', \tau, \tau')\theta(\tau - \tau')\,.$$

We have

$$\begin{aligned}\left(H + \frac{\partial^2}{\partial\tau^2}\right)&G_R(\mathbf{r}, \mathbf{r}', \tau, \tau')\\ &= \theta(\tau - \tau')H\mathscr{G} + \frac{\partial^2}{\partial\tau^2}[\mathscr{G}\theta(\tau - \tau')]\\ &= \theta(\tau - \tau')\left(H + \frac{\partial^2}{\partial\tau^2}\right)\mathscr{G} + 2\dot{\theta}(\tau - \tau')\dot{\mathscr{G}} + \mathscr{G}\ddot{\theta}(\tau - \tau')\,,\end{aligned}$$

where a dot indicates differentiation with respect to τ. By Eq. (7.159), this is just

$$\left(H + \frac{\partial^2}{\partial\tau^2}\right)G_R = 2\dot{\theta}(\tau - \tau')\dot{\mathscr{G}} + \mathscr{G}\ddot{\theta}(\tau - \tau')\,.$$

But $\dot{\theta}(\tau - \tau') = \delta(\tau - \tau')$, so

$$\begin{aligned}\left(H + \frac{\partial^2}{\partial\tau^2}\right)&G_R(\mathbf{r}, \mathbf{r}', \tau, \tau')\\ &= 2\delta(\tau - \tau')\dot{\mathscr{G}}(\mathbf{r}, \mathbf{r}', \tau, \tau') + \dot{\delta}(\tau - \tau')\mathscr{G}(\mathbf{r}, \mathbf{r}', \tau, \tau')\,,\end{aligned}$$

From the results of Section 7.2,

$$\delta(\tau - \tau')\, \dot{\mathscr{G}}(\mathbf{r}, \mathbf{r}', \tau, \tau') = [\, \dot{\mathscr{G}}(\mathbf{r}, \mathbf{r}', \tau, \tau')]_{\tau'=\tau}\delta(\tau - \tau')\,,$$
$$\dot{\delta}(\tau - \tau')\, \mathscr{G}(\mathbf{r}, \mathbf{r}', \tau, \tau') = -[\, \dot{\mathscr{G}}(\mathbf{r}, \mathbf{r}', \tau, \tau')]_{\tau'=\tau}\delta(\tau - \tau')\,.$$

Thus

$$\left(H + \frac{\partial^2}{\partial\tau^2}\right) G_R(\mathbf{r}, \mathbf{r}', \tau, \tau') = \delta(\tau - \tau')[\, \dot{\mathscr{G}}(\mathbf{r}, \mathbf{r}', \tau, \tau')\,]_{\tau'=\tau}\,. \tag{7.160}$$

If $\mathscr{G} = G_2$, the right-hand side of Eq. (7.160) vanishes by Eq. (7.155b), but if $\mathscr{G} = \tilde{G}_2$, then according to Eq. (7.155a) we get

$$\left(H + \frac{\partial^2}{\partial\tau^2}\right) G_R(\mathbf{r}, \mathbf{r}', \tau, \tau') = \delta(\tau - \tau')\delta^3(\mathbf{r} - \mathbf{r}')\,,$$

which is the desired form. Hence the Green's function for the homogeneous equation is

$$G_R(\mathbf{r}, \mathbf{r}', \tau, \tau') = \theta(\tau - \tau')\tilde{G}_2(\mathbf{r}, \mathbf{r}', \tau, \tau')\,, \tag{7.161}$$

and the solution to Eq. (7.158) can be written as

$$\psi_R(\mathbf{r}, \tau) = \psi_0(\mathbf{r}, \tau) + \int d^3\mathbf{r}' \int_{-\infty}^{\tau} d\tau'\, \tilde{G}_2(\mathbf{r}, \mathbf{r}', \tau, \tau')J(\mathbf{r}', \tau')\,, \tag{7.162}$$

where $\psi_0(\mathbf{r}, \tau)$ is a solution of the homogeneous equation.

Now for the special case of the wave equation, $\tilde{G}_2$ is given by Eq. (7.156); therefore, since G_R vanishes for $\tau < \tau'$, we may write

$$G_R(\mathbf{r}, \mathbf{r}', \tau, \tau') = \frac{1}{4\pi|\mathbf{r} - \mathbf{r}'|}\delta(|\mathbf{r} - \mathbf{r}'| - (\tau - \tau'))\theta(\tau - \tau')\,,$$

for when $\tau > \tau'$ the argument of the second δ-function in Eq. (7.156) is always nonzero. Since $\delta(|\mathbf{r} - \mathbf{r}'| - (\tau - \tau'))$ has zero argument only when $\tau > \tau'$, we may drop the factor $\theta(\tau - \tau')$ from G_R so that finally

$$G_R(\mathbf{r}, \mathbf{r}', \tau, \tau') = \frac{1}{4\pi|\mathbf{r} - \mathbf{r}'|}\delta(|\mathbf{r} - \mathbf{r}'| - (\tau - \tau'))\,. \tag{7.163}$$

Thus the solution to the wave equation is

$$\psi_R(\mathbf{r}, \tau) = \psi_0(\mathbf{r}, \tau) + \frac{1}{4\pi}\int d^3\mathbf{r}' \int_{-\infty}^{\infty} d\tau'\, \frac{1}{|\mathbf{r} - \mathbf{r}'|}\delta(|\mathbf{r} - \mathbf{r}'| - (\tau - \tau'))J(\mathbf{r}', \tau')\,.$$

The τ integration can be done immediately, giving

$$\psi_R(\mathbf{r}, \tau) = \psi_0(\mathbf{r}, \tau) + \frac{1}{4\pi}\int \frac{[J(\mathbf{r}', \tau')\,]_{\text{ret}}}{|\mathbf{r} - \mathbf{r}'|}\, d^3\mathbf{r}'$$

where the bracket $[\quad]_{\text{ret}}$ means that the time τ' is to be evaluated at the *retarded* time

$$\tau' = \tau - |\mathbf{r} - \mathbf{r}'| .$$

Writing this in terms of the time t, we have

$$\psi_R(\mathbf{r}, t) = \psi_0(\mathbf{r}, t) + \frac{1}{4\pi}\int \frac{[J(\mathbf{r}', t')]_{\text{ret}}}{|\mathbf{r} - \mathbf{r}'|} d^3\mathbf{r}' , \tag{7.164}$$

where in the bracket

$$t' = t - \frac{|\mathbf{r} - \mathbf{r}'|}{c} .$$

This is one of the most beautiful results in electromagnetic theory. It tells us that an observer at the field point $\mathbf{r}$ is affected at a given time t only by what a source at $\mathbf{r}'$ was doing at an earlier time t'; furthermore, this earlier time t' precedes the present time t by just the amount of time it takes a wave to travel from the source point $\mathbf{r}'$ to the field point $\mathbf{r}$, namely, $|\mathbf{r} - \mathbf{r}'|/c$. In general, an effect at one point in space-time is due to all disturbances throughout space which preceded the effect by exactly the amount of time necessary for the disturbances to propagate from their points of origin to the point where their effects are felt. ψ_R is referred to as the *retarded* solution to the inhomogeneous wave equation and G_R is called the *retarded* Green's function: hence the subscript R.

This is by no means the only Green's function that we could have defined. For example, we could write

$$G_A(\mathbf{r}, \mathbf{r}', \tau, \tau') = -\theta(\tau' - \tau)\tilde{G}_2(\mathbf{r}, \mathbf{r}', \tau, \tau') .$$

We leave it to the reader to show that this is also a Green's function for the inhomogeneous wave equation (called the *advanced* Green's function), and to show that the solution obtained from G_A corresponds to the effects of disturbances propagating back from the future to the present.

PROBLEMS

1. Show that the Green's functions of Eqs. (7.9′) and (7.11′) satisfy

$$\frac{d^2G(x, x')}{dx^2} = \delta(x - x') .$$

2. Show that

$$\frac{d\delta(x)}{dx} = -\delta(x)\frac{d}{dx}$$

when both elements of the equation are considered to operate on a function in an integrand; i.e., we really mean

$$\int_{-a}^{b} \frac{d\delta(x)}{dx} f(x)\, dx = -\int_{-a}^{b} \delta(x)\frac{df}{dx}\, dx$$

for any $f(x)$ with a continuous derivative.

3. Show that if H is a real, Hermitian operator (that is, $H = H^*$, $H = H^\dagger$) then the Green's function for the inhomogeneous equation

$$H\phi(x) = f(x)$$

is symmetric, that is, $G(x, x') = G(x', x)$.
[*Hint*: Show that the eigenfunctions of a real, Hermitian operator can always be chosen to be real.]

4. Show that the Green's function of Eq. (7.35) satisfies Eq. (7.36).

5. In Section 6.9 we discussed those solutions of Bessel's equation,

$$x^2 \frac{d^2 J_n}{dx^2} + x \frac{dJ_n}{dx} + (x^2 - n^2) J_n = 0 ,$$

which were finite at the origin. In particular, we chose a normalization such that for small x

$$J_n(x) \cong \frac{1}{n!}\left(\frac{x}{2}\right)^n .$$

There is, of course, another set of solutions, usually denoted by $N_n(x)$, which are irregular at the origin. They are conventionally normalized so that

$$N_0(x) \cong \frac{2}{\pi} \ln x , \qquad N_n(x) \sim -\frac{1}{\pi}\left(\frac{2}{x}\right)^n (n-1)!$$

(Note that for small values of x these $N_n(x)$ do indeed satisfy Bessel's equation.)
Consider the inhomogeneous Bessel's equation (for $n \geq 1$)

$$x^2 \frac{d^2 u}{dx^2} + x \frac{du}{dx} + (x^2 - n^2) u = f(x)$$

on the interval $[0, 1]$ with boundary condition $u(0) = 0 = u(1)$. Show that $u(x)$ is given by

$$u(x) = \int_0^1 G(x, x') f(x')\, dx ,$$

where

$$G(x, x') = -\frac{\pi}{2x' J_n(1)} J_n(x_<)[J_n(x_>) N_n(1) - N_n(x_>) J_n(1)] .$$

As usual, $x_>$ is the greater of x and x', and $x_<$ is the lesser of x and x'.

6. Prove that the Wronskian of a second-order linear differential operator never vanishes. Do this by first proving that $W(x)$ cannot be identically equal to zero (because if it were, u_1 and u_2 would be linearly dependent, contrary to assumption). Then, differentiating $W(x)$ and using the fact that u_1 and u_2 are solutions to

$$f_0(x) \frac{d^2 u}{dx^2} + f_1(x) \frac{du}{dx} + f_2(x) u = 0 ,$$

prove that

$$\frac{dW}{dx} = -\frac{f_1}{f_0} W .$$

Integrate this and show that if $f_0(x)$ and $f_1(x)$ are continuous and if $f_0(x)$ never vanishes, then $W(x)$ ***never*** vanishes.

7. Prove that for *any* linear differential operator of the form

$$f_0(x)\frac{d^n}{dx^n} + f_1(x)\frac{d^{n-1}}{dx^{n-1}} + \cdots + f_n(x),$$

we have

$$\frac{dW}{dx} = -\frac{f_1}{f_0}W,$$

where W is defined by

$$W(x) = \begin{vmatrix} u_1(x) & u_2(x) & \cdots & u_n(x) \\ u_1^{(1)}(x) & u_2^{(1)}(x) & \cdots & u_n^{(1)}(x) \\ u_1^{(n-1)}(x) & u_2^{(n-1)}(x) & \cdots & u_n^{(n-1)}(x) \end{vmatrix},$$

$u_1(x), u_2(x), \cdots, u_n(x)$ are the linearly independent solutions of the equations and $u^{(m)}$ denotes the mth derivative of u. Thus $W(x)$ is given by

$$W(x) = C\exp\left[-\int\frac{f_1(x)}{f_0(x)}dx\right].$$

8. Show that the Green's function for the differential operator

$$f_0(x)\frac{d^n}{dx^n} + f_1(x)\frac{d^{n-1}}{dx^{n-1}} + \cdots + f_n(x)$$

is given by

$$G(x, x') = \sum_{m=1}^{n} a_m(x')u_m(x) + \frac{N(x, x')}{f_0(x')W(x')}\theta(x - x'),$$

where $W(x')$ is defined as in the previous problem and

$$N(x, x') = \begin{vmatrix} u_1(x') & u_2(x') & \cdots & u_n(x') \\ u_1^{(1)}(x') & u_2^{(1)}(x') & & u_n^{(1)}(x') \\ \cdot & \cdot & & \cdot \\ \cdot & \cdot & & \cdot \\ \cdot & \cdot & & \cdot \\ u_1^{(n-2)}(x') & u_2^{(n-2)}(x') & & u_n^{(n-2)}(x') \\ u_1(x) & u_2(x) & \cdots & u_n(x) \end{vmatrix}.$$

The $a_m(x')$ are determined by the boundary condition which the Green's function must satisfy. As in Problem 7, $u_1, u_2, \cdots, u_n$ are the n linearly independent solutions of the homogeneous equation.

9. Using the results of the previous problem, find the Green's function for the general third-order linear differential operator,

$$\frac{d^3}{dx^3} + a_2\frac{d^2}{dx^2} + a_2\frac{d}{dx} + a_3$$

in terms of the roots, r_1, r_2 and r_3 of the characteristic polynomial

$$z^3 + a_1z^2 + a_2z + a_3 = 0.$$

Assume that the Green's function and its first two derivatives vanish at $x = 0$.

10. Schrödinger's equation in one dimension is

$$-\frac{h^2}{2m}\frac{d^2\psi}{dx^2} + V(x)\psi(x) = E\psi(x).$$

Using the methods discussed in Section 7.4, derive a Lippmann-Schwinger integral equation for $\psi(x)$,
a) for the case $E < 0$, b) for the case $E \geq 0$.
In case (b) you will find separate equations for $\psi_+(x)$ and $\psi_-(x)$ corresponding to different types of boundary conditions at infinity. Discuss the meaning of $\psi_+(x)$ and $\psi_-(x)$.

11. Solve Problem 10 by using the method of Section 7.3. What is the appropriate statement of boundary conditions to specify the Green's function completely in this case?

12. a) By making the change of variable $\phi_l(x) = f_l(x)/\sqrt{x}$ in Eq. (7.81), show that the following equation is obtained for $f_l(x)$:

$$x^2 f_l''(x) + x f_l'(x) + [x^2 - (l + 1/2)^2] f_l(x) = 0.$$

b) Using the results of Section 6.9, show that

$$j_l(x) \propto \frac{1}{\sqrt{x}} J_{l+1/2}(x) ,$$

where J denotes a Bessel function [see the discussion following Eq. (7.88b′)].

13. a) Show that the Green's function

$$G(t, t') = \frac{1}{2\omega_0 i} e^{i\omega_0|t-t'|}$$

satisfies

$$\left(\frac{d^2}{dt^2} + \omega_0^2\right) G(t, t') = \delta(t - t') .$$

This Green's function was used in Problem 7.10 to convert Schrödinger's differential equation into the Lippmann-Schwinger integral equation.

b) Show that a general solution to the equation

$$\ddot{X}(t) + \omega_0^2 X(t) = F(t)$$

can be written in terms of the function $G(t, t')$ above as

$$X(t) = A \cos \omega_0 t + B \sin \omega_0 t + \int_{-\infty}^{\infty} G(t, t') F(t') \, dt' ,$$

where A and B are to be determined in terms of the boundary conditions. For classical mechanical one-point boundary conditions, they will clearly depend on F. Show that if we require $x(0) = x_0$ and $\dot{x}(0) = 0$, then the equation reduces to the familiar result

$$X(t) = x_0 \cos \omega_0 t + \frac{1}{\omega_0} \int_0^t \sin \omega_0 (t - t') F(t') \, dt' .$$

Clearly, for one-point boundary conditions, it is easier to start with $(1/\omega_0) \times \theta(t - t') \sin \omega_0 (t - t')$ as Green's function, but it is not necessary to do so.

14. Show from Eq. (7.98) that

$$e^{i\mathbf{k}\cdot\mathbf{r}} = \sum_{l=0}^{\infty} i^l (2l + 1) j_l(kr) P_l (\cos \theta) ,$$

where θ is the angle between $\mathbf{k}$ and $\mathbf{r}$.
[*Hint*: Consider Eq. (7.98) in the limit $|\mathbf{r}'| \to \infty$.]

15. Making use of Eqs. (7.98), (7.86), (7.84a) and (5.83), show that if we choose

$$\xi(\mathbf{r}) = \frac{A}{(2\pi)^{3/2}} e^{i\mathbf{k}\cdot\mathbf{r}} - \frac{4imk}{\hbar^2} \sum_{l=0}^{\infty} \sum_{m=-\infty}^{\infty} j_l(kr) Y_{lm}(\theta,\phi) \int j_l(kr') Y_{lm}^*(\theta',\phi') V(\mathbf{r}')\psi(\mathbf{r}')\, d\mathbf{r}',$$

then the equation

$$\psi(\mathbf{r}) = \xi(\mathbf{r}) - \frac{m}{2\pi\hbar^2} \int \frac{e^{-ik|\mathbf{r}-\mathbf{r}'|}}{|\mathbf{r}-\mathbf{r}'|} V(\mathbf{r}')\psi(\mathbf{r}')\, d\mathbf{r}'$$

reduces to

$$\psi(\mathbf{r}) = \frac{A}{(2\pi)^{3/2}} e^{i\mathbf{k}\cdot\mathbf{r}} - \frac{m}{2\pi\hbar^2} \int \frac{e^{+ik|\mathbf{r}-\mathbf{r}'|}}{|\mathbf{r}-\mathbf{r}'|} V(\mathbf{r}')\psi(\mathbf{r}')\, d\mathbf{r}'.$$

Note that $\xi(\mathbf{r})$ is a linear superposition of solutions to $(\nabla^2 + k^2)\phi = 0$.

16. In obtaining the Lippmann-Schwinger bound-state equation by the Fourier transform method, it almost appears that we applied *no* boundary conditions at all (no "$i\epsilon$-prescription" was necessary) and yet we got a uniquely determined Green's function. Show that this is not actually the case, i.e., that we really did insert a boundary condition. How was this done? Relate your discussion to the work in Section 7.5.

17. Consider a thin ring of radius R. Initially, at $t = 0$, there is a temperature distribution $T(\phi, 0)$ around the ring, where ϕ is the appropriate angular coordinate. Using the Green's function method, find the temperature at later times. Explain any decisions you make concerning boundary conditions. Suppose that the ring is initially heated in a very small region around $\phi = \phi_0$, so that the form $T(\phi, 0) = T_0\delta(\phi - \phi_0)$ is a reasonable boundary condition. Find the temperature at later times, $t > 0$. What is the value of the uniform temperature which the ring finally attains? How is the rate of approach to equilibrium affected by increasing the radius of the ring?

18. a) Consider a one-dimensional rod of length L. Let one end, at $x = 0$, be held at zero temperature and the other end, at $x = L$, be held at temperature T_0. If the initial temperature distribution is $T(x, 0)$, show that the temperature at subsequent times is given by

$$T(x, t) = T_0 x/L + \int_0^L G_1(x, x', t)\, \tilde{T}(x', 0)\, dx',$$

where

$$\tilde{T}(x, 0) = T(x, 0) - T_0 x/L$$

and

$$G_1(x, x', t) = \frac{2}{L} \sum_{m=1}^{\infty} \sin\frac{m\pi x}{L} \sin\frac{m\pi x'}{L} e^{(-k/c)(m\pi/L)^2 t}.$$

[*Hint*: Note that $\tilde{T}(x, t) = T(x, t) - T_0 x/L$ also satisfies the heat equation. What are the boundary conditions satisfied by $\tilde{T}(x, t)$?

b) Using the result of part (*a*), find the Green's function which will enable one to solve the equation

$$\nabla^2 T = \frac{c}{k}\frac{\partial T}{\partial t}$$

on the infinite half-line $[0, \infty]$ when $T(x, 0)$ is given some prescribed functional value.

19. a) Using the results of the previous problem (or any other method), find in terms of a Green's function the solution to the heat equation for a one-dimensional rod whose ends at $x = -L/2$ and $x = L/2$ are kept at temperatures T_1 and T_2 respectively.

    b) Find an expression for the Green's function of part (a) in the limit $L \to \infty$, that is, for the infinite line. Compare this expression with the result found in Section 7.7 for periodic boundary conditions.

20. Using the eigenfunction expansion method, show that the potential energy in a two-dimensional, charge-free region bounded by circles of radii R_1 and R_2 is given by

$$V(r, \theta) = \int_0^{2\pi} G_1(\theta, \theta', r) V(R_1, \theta')\, d\theta' + \int_0^{2\pi} G_2(\theta, \theta', r) V(R_2, \theta')\, d\theta' ,$$

where $V(R_1, \theta)$ is the known value of the potential on the first surface as a function of θ, and $V(R_2, \theta)$ is its value on the second surface. Find the eigenfunction expansion of G_1 and G_2, using the fact that $\partial^2/\partial\theta^2$ is an Hermitian operator if one imposes periodic boundary conditions on $[0, 2\pi]$. Why do we require such conditions in this problem? Does the limit $V(R_1, \theta) = \text{const}$, $V(R_2, \theta) = \text{const}$ give the result you would expect? Show that $G_1(\theta, \theta', R_1) = \delta(\theta - \theta')$ and $G_2(\theta, \theta', R_2) = \delta(\theta - \theta')$.

21. a) Determine the Green's function, $G_1(x, y, x', y', t)$, for the homogeneous heat equation for the entire *xy*-plane (i.e., the two-dimensional Green's function for an infinite space).

    b) Assume that at $t = 0$ the temperature is a constant T_0 inside a square of side $2a$, centered at the origin, and zero outside this square. Determine the temperature distribution throughout the plane for all later times t in terms of the error function:

$$\text{erf}(x) = \frac{2}{\sqrt{\pi}}\int_0^x e^{-u^2}\, du .$$

22. Show that the Green's function for the two-dimensional Laplacian over the entire two-dimensional space is

$$G(x, y; x', y') = +\frac{1}{2\pi}\log[(x - x')^2 + (y - y')^2]^{1/2} .$$

[*Hint*: One can always add a constant to the Green's function and it will still satisfy $\nabla^2 G = \delta(x - x')\delta(y - y')$.]

23. Show that if the wave function ψ of a particle satisfies Schrödinger's equation,

$$H\psi = i\hbar\frac{\partial\psi}{\partial t} .$$

where H is a self-adjoint operator, and if

$$\int \psi^*(\mathbf{r}, 0)\psi(\mathbf{r}, 0)\, d^3r = 1 \tag{1}$$

at $t = 0$, then at any subsequent time t, we will also have

$$\int \psi^*(\mathbf{r}, t)\psi(\mathbf{r}, t) d^3r = 1\,. \tag{2}$$

Since in quantum mechanics $|\psi|^2$ is the probability density, $|\psi|^2 d^3r$ gives the probability of finding the particle in the volume element d^3r, and Eqs. (1) and (2) are a statement of the conservation of total probability. That is, if the total probability of finding the particle *someplace* is initially equal to unity [Eq. (1)], it remains equal to unity at all later times. Clearly, the interpretation of $|\psi|^2$ as a probability density would not make sense if this were not the case. Note that such an interpretation would not hold for solutions to the heat (or diffusion) equation.

24. Use Green's theorem to obtain Eq. (7.117) from Eq. (7.116).

25. a) Show that the function

$$G'(\mathbf{r}, \mathbf{r}', \tau, \tau') = -\, G_1(\mathbf{r}, \mathbf{r}', \tau, \tau')\theta(\tau' - \tau)$$

is a Green's function for Eq. (7.131), where G_1 satisfies Eqs. (7.134) and (7.135).

b) This means that we can write a time-dependent Schrödinger integral equation as

$$\psi(\mathbf{r}, t) = \psi_0(\mathbf{r}, t) - \frac{2m}{\hbar^2}\int d^3r' \int_\tau^{i\infty} G'(\mathbf{r}, \mathbf{r}', \tau, \tau')\, U(\mathbf{r}', \tau')\psi(\mathbf{r}', \tau')\, d\tau'$$

in the notation of Eq. (7.137). Show that U is a static potential which is "turned off" at remote *future* times, that is,

$$U(\mathbf{r}, \tau) = V(\mathbf{r})e^{-\epsilon' t} = V(\mathbf{r})e^{i\epsilon\tau}\,,$$

then we obtain in the limit $\epsilon \to 0$ the Lippmann-Schwinger equation with an *incoming* spherical wave boundary condition solution.

26. a) Note that Eq. (7.125b) does not make sense if $\tau' > \tau$ (assuming, as will usually be the case, that $\lambda_m \to +\infty$ as $m \to \infty$) and τ and τ' are real. This is also reflected in the discussion following Eq. (7.128), where we are tacitly assuming that $\tau \to 0$ from "the positive side of zero" (see Section 5.3). Thus the Green's function

$$G(\mathbf{r}, \mathbf{r}', \tau, \tau') = -\,\theta(\tau' - \tau)G_1(\mathbf{r}, \mathbf{r}', \tau, \tau')$$

obtained in the previous problem, although it formally seems to satisfy

$$\left(H + \frac{\partial}{\partial \tau}\right)G = \delta^3(\mathbf{r} - \mathbf{r}')\delta(\tau - \tau')\,,$$

is *not* an acceptable Green's function. Show that if one derives $G(\mathbf{r}, \mathbf{r}', \tau, \tau')$ for the case of the diffusion (or heat) equation by using the Fourier transform method, one obtains *only*

$$G(\mathbf{r}, \mathbf{r}', \tau, \tau') = \theta(\tau - \tau')G_1(\mathbf{r}, \mathbf{r}', \tau, \tau')\,,$$

where G_1 is given by Eq. (7.139).

b) For the case of Schrödinger's equation for a free particle, the expontial in Eq. (7.125b) is oscillatory because τ is imaginary. In this case, show that the Fourier transform method yields two Green's functions:

$$G_+(\mathbf{r}, \mathbf{r}', t, t') = \theta(t - t')G_1(\mathbf{r}, \mathbf{r}', t, t')$$

and

$$G_-(\mathbf{r}, \mathbf{r}', t, t') = -\theta(t' - t)G_1(\mathbf{r}, \mathbf{r}', t, t') ,$$

depending on whether one uses $+i\epsilon$ or $-i\epsilon$ to avoid the singularity arising in the Fourier transform method for this case. In the above equation for G_+ and G_-, G_1 is given by Eq. (7.139) with $\tau = i\hbar t/2m$.

27. a) Show that the function

$$G_A(\mathbf{r}, \mathbf{r}', \tau, \tau') = -\theta(\tau' - \tau)\tilde{G}_2(\mathbf{r}, \mathbf{r}', \tau, \tau')$$

is indeed a Green's function for the operator

$$\theta = H + \frac{\partial^2}{\partial \tau^2}$$

if $\tilde{G}_2$ satisfies Eqs. (7.152) and (7.157).

b) For the case of the wave equation, find an expression for $G_A(\mathbf{r}, \mathbf{r}', \tau, \tau')$ analogous to the one for G_R given in Eq. (7.163).

c) Find a solution to $\theta\psi(\mathbf{r}, \tau) = J(\mathbf{r}, \tau)$ in terms of G_A. Interpret the solution by analogy with Eq. (7.164).

d) Show that

$$G_A + G_R = \frac{1}{2\pi}\delta(s^2) ,$$

where s^2 is the Lorentz-invariant scalar product:

$$s^2 \equiv (\mathbf{r}-\mathbf{r}')^2 - (\tau - \tau')^2 = (x - x')^2 + (y - y')^2 + (z - z')^2 - c^2(t - t')^2.$$

28. Using the Fourier transform method (in *four dimensions*) obtain the Green's functions for the wave equation. Show that one obtains G_R [of Eq. (7.163)] or G_A (of the previous problem) depending on how one chooses to avoid a singularity which arises in the Fourier transform method (see a similar discussion in Section 7.4, (Eq. (7.63) *et seq.*)

29. Consider the time-dependent inhomogeneous Klein-Gordon equation:

$$\left[-\nabla^2 + \frac{1}{c^2}\frac{\partial^2}{\partial t^2} + \kappa^2\right]\psi(\mathbf{r}, t) = \rho(\mathbf{r}, t) .$$

a) Show that by using the Fourier transform method one obtains the following two integral expressions for the Green's functions G_R and G_A:

$$G_R(\mathbf{r}, t, \mathbf{r}', t') = \frac{c}{8\pi^2 Ri}\frac{d}{dR}\int_{-\infty}^{\infty}\frac{e^{i(R/c)\sqrt{q^2-\kappa^2c^2}}}{\sqrt{q^2 - \kappa^2c^2}}e^{-iq(t-t')}\,dq ,$$

$$G_A(\mathbf{r}, t, \mathbf{r}', t') = -\frac{c}{8\pi^2 Ri}\frac{d}{dR}\int_{-\infty}^{\infty}\frac{e^{-i(R/c)\sqrt{q^2-\kappa^2c^2}}}{\sqrt{q^2 - \kappa^2c^2}}e^{-iq(t-t')}\,dq ,$$

where $R = |\mathbf{r} - \mathbf{r}'|$. The two expressions correspond to two possible ways of avoiding a singularity which occurs in doing the Fourier transform integrals corresponding to the space variables. In the first integral, $\sqrt{q^2 - \kappa^2c^2}$ has *positive* imaginary part when $|q| < \kappa c$; in the second integral $\sqrt{q^2 - \kappa^2c^2}$ has *negative* imaginary part when $|q| < \kappa c$.

b) Show that $G_R = 0$ when $R/c > (t - t')$ and that $G_A = 0$ when $R/c + (t - t') > 0$. [*Hint*: If the integral for G_R is considered as an integral in the complex plane, then the plane is cut from $-\kappa c$ to $+\kappa c$ and the integral in question runs along the *top* of the cut. For G_A, the integral runs along the bottom of the cut.]

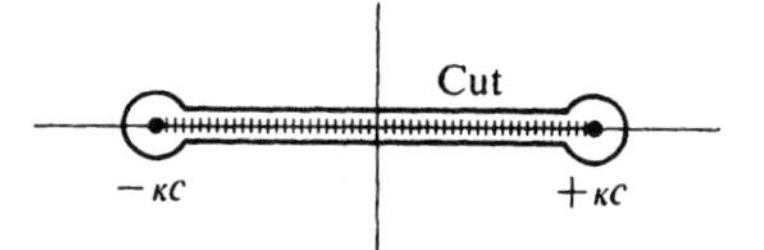

Fig. 7.12

c) When $R/c < (t - t')$, show that the integral occurring in G_R can be reduced to the contour integral shown in Fig. 7.12. Hence show that

$$G_R = -\frac{c}{4\pi^2 R}\frac{d}{dR}\left[\int_{-\kappa c}^{\kappa c}\frac{1}{\sqrt{\kappa^2c^2 - q^2}}\cosh\left[\frac{R}{c}\sqrt{\kappa^2c^2 - q^2}\right] \times e^{-iq(t-t')}\,dq\,\theta(T - R)\right],$$

and, similarly,

$$G_A = -\frac{c}{4\pi^2 R}\frac{d}{dR}\left[\int_{-\kappa c}^{\kappa c}\frac{1}{\sqrt{\kappa^2c^2 - q^2}}\cosh\left[\frac{R}{c}\sqrt{\kappa^2c^2 - q^2}\right] \times e^{-iq(t-t')}\,dq\,\theta(-T - R)\right],$$

where $\theta(x)$ equals 1 if $x > 0$ and equals zero if $x < 0$. Also, $T = c(t - t')$.

d) Making the change of variable $q = \kappa c \sin\theta$ and using some symmetry considerations, show that

$$G_R = -\frac{c}{8\pi^2 R}\frac{d}{dR}\left[\int_{-\pi}^{\pi}\cosh(\kappa R\cos\theta)e^{-i\kappa T\sin\theta}\,d\theta\,\theta(T - R)\right],$$

$$G_A = -\frac{c}{8\pi^2 R}\frac{d}{dR}\left[\int_{-\pi}^{\pi}\cosh(\kappa R\cos\theta)e^{-i\kappa T\sin\theta}\,d\theta\,\theta(-T - R)\right].$$

[*Hint*: Note that the limits of integration are $\pm\pi$, *not* $\pm\pi/2$.]

e) Show that this can be written as

$$G_R = -\frac{c}{8\pi^2 Ri}\frac{d}{dR}\left[\oint_C \frac{1}{z}\cosh\left[\frac{1}{2}\kappa R\left(z + \frac{1}{z}\right)\right]e^{-\kappa T(z-1/z)/2}dz\,\theta(T - R)\right],$$

with a similar expression for G_A. Here C is a counterclockwise contour around the *unit* circle.

f) By making two variable changes (in two different integrals), show that one obtains

$$G_R = -\frac{c}{8\pi^2 Ri}\frac{d}{dR}\left[\oint\frac{1}{\xi}\exp\left[-\frac{\kappa}{2}\sqrt{T^2 - R^2}\left(\xi - \frac{1}{\xi}\right)\right]d\xi\,\theta(T - R)\right].$$

Using the integral representation of the Bessel function [Eq. (6.65) and the sentence following this equation], show that this reduces to

$$G_R = -\frac{c}{4\pi R}\frac{d}{dR}[J_0(\kappa\sqrt{T^2 - R^2})\theta(T - R)].$$

Also,

$$G_A = -\frac{c}{4\pi R}\frac{d}{dR}[J_0(\kappa\sqrt{T^2 - R^2})\theta(-T - R)] .$$

[Here we have used the fact that $J_0(x) = J_0(-x)$, which is clear from Eq. (6.66)].

g) (We're almost done!) Carry out the differentiation using Eq. (7.22) to show that

$$G_R = \frac{c}{4\pi R}\delta(R - T) - \frac{\kappa c}{4\pi\sqrt{T^2 - R^2}}J_1(\kappa\sqrt{T^2 - R^2})\theta(T - R) ,$$

$$G_A = \frac{c}{4\pi R}\delta(R + T) - \frac{\kappa c}{4\pi\sqrt{T^2 - R^2}}J_1(\kappa\sqrt{T^2 - R^2})\theta(-T - R) .$$

Note that as $\kappa \to 0$, G_R goes over into the result of Eq. (7.163), apart from a factor of c which you should be able to explain (*Note*: $\delta(ax) = \delta(x)/|a|$). Also, note that $G_A + G_R$ can be written as

$$G_R + G_A = \frac{c}{2\pi}\delta(T^2 - R^2) - \frac{\kappa c}{4\pi\sqrt{T^2 - R^2}}J_1(\kappa\sqrt{T^2 - R^2})\theta(T^2 - R^2) ,$$

which is a function only of the Lorentz invariant $s^2 = R^2 - T^2 = (x - x')^2 + (y - y')^2 + (z - z')^2 - c^2(t - t')^2$.

FURTHER READINGS

CARSLAW, H. S., and J. C. JAEGER, *Conduction of Heat in Solids*, Second Edition. New York: Oxford University Press, 1959.

COURANT, R., and D. HILBERT, *Methods of Mathematical Physics* Vol. II. New York: Interscience, 1962.

DETTMAN, J., *Mathematical Methods in Physics and Engineering*. New York: McGraw-Hill, 1962.

FRIEDMAN, B., *Principles and Techniques of Applied Mathematics*. New York: Wiley, 1956.

JACKSON, J. D., *Classical Electrodynamics*. New York: Wiley, 1962.

LANCZOS, C., *Linear Differential Operations*. Princeton, N. J.: D. Van Nostrand, 1961.

MORSE, P. M. and H. FESHBACH, *Methods of Theoretical Physics*, Vol. I. New York: McGraw-Hill, 1953.

SOMMERFELD, A., *Partial Differential Equations in Physics*. New York: Academic Press, 1949.

CHAPTER 8

INTRODUCTION TO INTEGRAL EQUATIONS

INTRODUCTION

In the last chapter, we saw that by introducing the concept of a Green's function we were led in a natural way to reformulate certain problems in terms of integral equations rather than differential equations. In this chapter and the next, we shall study integral equations as a subject in their own right, independent of the particular physical origin of these equations. We shall find that the relevant mathematical setting for this study is in infinite-dimensional vector spaces; within this framework, we shall concentrate primarily, although not exclusively, on Hilbert space. A particularly interesting question, in light of the work in Chapters 3 and 4, is: What is the relationship between the results of finite-dimensional and infinite-dimensional vector space theory? We will see that there is a very close relationship and that in many important cases one can solve infinite-dimensional problems by finite-dimensional techniques. However, there are also important and interesting differences which have many consequences for physics.

This chapter is devoted to a study of several special types of integral equations which can be solved rather easily. We will also take this opportunity to introduce, in a more precise manner than in Chapter 5, some of the key concepts in the theory of infinite-dimensional vector spaces.

8.1 ITERATIVE TECHNIQUES—LINEAR INTEGRAL OPERATORS

In Chapter 7 we encountered several different integral equations of the form

$$f(x) = g(x) + \int_a^b k(x, y) f(y) \, dy , \tag{8.1}$$

where $g(x)$ and $k(x, y)$ are given functions. Equations of this form are known as *integral equations of the second kind.* If $g(x)$ is not identically zero, the equation is said to be inhomogeneous; if $g(x)$ vanishes everywhere, the equation is called homogeneous. If on the left-hand side of Eq. (8.1) we replace $f(x)$ by zero, the resulting equation is an *integral equation of the first kind.* An example of such an equation is

$$g(x) = \frac{1}{\sqrt{2\pi}} \int_{-\infty}^{\infty} e^{ixy} f(y) \, dy ,$$

where $g(x)$ is a known function and $f(y)$ is to be determined. For this particular case, we know from the theory of Fourier transforms that the solution is given by

$$f(y) = \frac{1}{\sqrt{2\pi}} \int_{-\infty}^{\infty} e^{-iyx} g(x)\, dx ,$$

but, in general, little is known about these equations and we will not discuss them further.

An important special case of Eq. (8.1) is the *Volterra* integral equation, which has the general form

$$f(x) = g(x) + \int_a^x \tilde{k}(x, y) f(y)\, dy . \tag{8.2}$$

Equation (8.1) reduces to this form if the function $k(x, y)$ vanishes when $y > x$. A generalization of Eq. (8.2), which one might encounter, for example, in Newtonian mechanics, has the form

$$f(x) = g(x) + \int_a^x h(x, y, f(y))\, dy . \tag{8.3}$$

Here the unknown function f may occur linearly, as in Eq. (8.2), or nonlinearly, depending on the form of $h(x, y, f(y))$. In this chapter we will discuss primarily linear integral equations, although in Section 8.4 we will look briefly at some of the simpler aspects of nonlinear equations.

Let us now rewrite Eq. (8.1) in the form

$$f = g + Kf , \tag{8.4}$$

where K is an operator defined by

$$Kf \equiv \int_a^b k(x, y) f(y)\, dy . \tag{8.5}$$

Evidently K is linear, since

$$\begin{aligned} K(f_1 + f_2) &= \int_a^b k(x, y)[f_1(y) + f_2(y)]\, dy \\ &= \int_a^b k(x, y) f_1(y)\, dy + \int_a^b k(x, y) f_2(y)\, dy \\ &= Kf_1 + Kf_2 . \end{aligned}$$

Therefore we may consider f and g as elements in some vector space, say V, and K as a linear operator which maps V into itself. The operator K, defined by Eq. (8.5), is called an *integral operator*, and $k(x, y)$ is called the *kernel* of the operator.

Equation (8.4) can now be stated as

$$(I - K)f = g , \tag{8.6}$$

where I is the identity operator. In this form the problem looks very much like the matrix equations discussed in Chapters 3 and 4. If we can compute $(I - K)^{-1}$, the problem is formally solved, since

$$f = (I - K)^{-1} g \tag{8.7}$$

(provided, of course, that the inverse operator is well defined). Note that although K may be an integral operator in Eq. (8.7), I is not an integral operator unless we want to use the δ-function:

$$If = \int_{-\infty}^{\infty} \delta(x - y) f(y) \, dy .$$

The most fortunate situation would be if K were in some sense a "small" operator, so that we could try to write

$$(I - K)^{-1} = I + K + K^2 + \cdots \tag{8.8}$$

by analogy with the power series for $(1 - x)^{-1}$ for small x. Since we have already required that K acting on *any* element of V produce another element of V, K^n is defined simply by successive applications of K: $K^2 = KK$, $K^3 = KK^2$, etc. This restriction on K is not trivial because it sometimes happens that reasonable-looking operators acting on V produce objects which are not in V. For example, consider the space $L_2[0, 1]$ of square-integrable functions of one variable defined on $[0, 1]$, and let the operator d/dx act on this space. Clearly, $f(x) = \sqrt{x}$ belongs to $L_2[0, 1]$, but

$$\frac{d}{dx}[f(x)] = \tfrac{1}{2}x^{-1/2}$$

does not. Thus d/dx does not transform every element of the space $L_2[0, 1]$ into another element of the same space, and cannot therefore be an admissible operator for our purposes. In fact, the assumption that a set of operators always transforms elements in a given space into elements of the *same* vector space has some strong consequences, as we shall point out later.

Suppose now that the right-hand side of Eq. (8.8) "converges." (We use quotation marks around "converges" because we have not yet given a careful definition of convergence for operators.) Then the operator to which it converges is the inverse of $(I - K)$, because multiplying $I + K + K^2 + K^3 + \cdots$ on either side by $(I - K)$ gives I. Thus we are led to speculate that the series

$$f = g + Kg + K^2g + K^3g + \cdots , \tag{8.9}$$

when it converges, is the solution to Eq. (8.4). The series in Eq. (8.9) is called the *Neumann series* by mathematicians, and the *Born series* by physicists, since it was Max Born who first utilized the basic iterative idea in quantum mechanics (although he was motivated by rather different considerations than the ones described here).

Let us now begin to investigate the circumstances under which the series of Eq. (8.9) converges. If K is an integral operator of the type defined by Eq. (8.5), then Eq. (8.9) can be written as

$$f(x) = g(x) + \int_a^b k(x, x')g(x') \, dx' + \int_a^b dx' \int_a^b dx'' \, k(x, x')k(x', x'')g(x'') + \cdots . \tag{8.10}$$

We are, of course, assuming that $g(x) \not\equiv 0$; otherwise the series of Eq. (8.9) or (8.10) degenerates into the result $f(x) \equiv 0$, which is always a possible (but uninteresting) solution when $g(x) \equiv 0$. We require that $|k(x, y)|$ be *bounded* when x and y lie in $[a, b]$:

$$\max_{x, y \in a, b} |k(x, y)| = M, \tag{8.11}$$

where M is some finite number. We further assume that the integral

$$\int_a^b |g(x)|\, dx \tag{8.12}$$

exists and is equal to a finite constant C. Then we may write

$$f(x) = g_0(x) + g_1(x) + g_2(x) + g_3(x) + \cdots, \tag{8.13}$$

where $g_0(x) \equiv g(x)$ and

$$g_n(x) = \int_a^b k(x, y) g_{n-1}(y)\, dy .$$

We next want to show that, with the given assumptions about $k(x, y)$ and $g(x)$,

$$|g_n(x)| \leq CM[M(b-a)]^{n-1} \tag{8.14}$$

for $n \geq 1$. For $n = 1$,

$$|g_1(x)| \leq \int_a^b |k(x, y) g(y)|\, dy \leq \int_a^b |k(x, y)| \cdot |g(y)|\, dy .$$

But $|k(x, y)| \leq M$, so

$$|g_1(x)| \leq M \int_a^b |g(y)|\, dy = MC .$$

Thus Eq. (8.14) is satisfied for $n = 1$. We now assume it to be true for $n = m - 1$ and then prove that it is true for $n = m$. This will establish the general result by induction. For $g_m(x)$, we have

$$\begin{aligned} |g_m(x)| &\leq \int_a^b |k(x, y)| \cdot |g_{m-1}(y)|\, dy \leq MC[M(b-a)]^{m-2} \int_a^b |k(x, y)|\, dy \\ &\leq MC[M(b-a)]^{m-1}, \end{aligned}$$

which is the desired result. Thus the mth term in the series of Eq. (8.13) is *dominated* by $MC[M(b-a)]^{m-1}$, that is, it is less in absolute magnitude than $MC[M(b-a)]^{m-1}$, for all x. Therefore

$$\begin{aligned} |f(x) - \sum_{n=0}^{N} g_n(x)| &\leq |g_{N+1}(x)| + |g_{N+2}(x)| + \cdots \\ &\leq MC \sum_{m=N}^{\infty} [M(b-a)]^m \\ &\leq MC[M(b-a)]^N \sum_{m=0}^{\infty} [M(b-a)]^m . \end{aligned}$$

If $M(b-a) < 1$, this last sum converges, so we can write

$$|f(x) - \sum_{n=0}^{N} g_n(x)| \leq \frac{MC[M(b-a)]^N}{1 - M(b-a)}$$

for *all* x. Since $M(b-a) < 1$, the right-hand side of this equation can be made as small as we please by choosing N sufficiently large. Thus as N tends to infinity, $\sum_{n=0}^{N} g_n(x)$ converges uniformly to a function $f(x)$. The reader can easily show that if $k(x, y)$ is continuous in x for all y and if $g(x)$ is continuous, then each $g_n(x)$ will be continuous, and the uniformly convergent sum will also be continuous. Since the series $f = g + Kg + K^2g + \cdots$ is uniformly convergent, we can calculate $(I - K)f$ by integrating term by term. We obtain

$$(I - K)f = g_0 + g_1 + g_2 + g_3 + \cdots - K(g_0 + g_1 + g_2 + \cdots) .$$

But $Kg_n = g_{n+1}$; therefore

$$(I - K)f = g_0 \equiv g ,$$

so the Neumann series does indeed give a solution to the problem.

We may summarize by saying that if $M < (b-a)^{-1}$, then the Neumann series of Eqs. (8.9) or (8.10) converges to the solution of Eq. (8.1). This in no way implies that M *must* be less than $(b-a)^{-1}$ for the Neumann series to converge. It is easy to construct kernels for which $M \geq (b-a)^{-1}$ but for which the Neumann series nevertheless converges. For example, the reader should calculate the Neumann series for

$$k(x, y) = \xi(x)\eta(y) ,$$

where

$$\int_a^b \xi(x)\eta(x)\, dx = 0 .$$

Another large class of kernels for which the restriction $M < (b-a)^{-1}$ is unnecessary is the Volterra class, i.e., equations of the form of Eq. (8.2). For this case, if we write

$$f(x) = \tilde{g}_0(x) + \tilde{g}_1(x) + \tilde{g}_2(x) + \cdots ,$$

where again $\tilde{g}_0(x) \equiv g(x)$ and

$$\tilde{g}_n(x) \equiv \int_a^x k(x, y)\tilde{g}_{n-1}(y)\, dy ,$$

then we can show that

$$|\tilde{g}_n(x)| \leq MC[M(x-a)]^{n-1}/(n-1)!$$

for $n \geq 1$, if $a \leq x \leq b$. Just as before, this is true for $n = 1$. Assuming it to be true for $n = m - 1$, we get

$$\begin{aligned} \tilde{g}_m(x) &\leq \int_a^x |k(x, y)| \cdot |\tilde{g}_{m-1}(y)|\, dy \\ &\leq MCM^{m-2} \int_a^x |k(x, y)|\, (y-a)^{m-2}\, dy/(m-2)! . \end{aligned}$$

Since $|k(x, y)| \leq M$, we have

$$|g_m(x)| \leq MC[M(x - a)]^{m-1}/(m - 1)!\,,$$

as required. Thus for $x \in [a, b]$, the nth term of the Neumann series is dominated by

$$MC[M(b - a)]^{n-1}/(n - 1)!\,,$$

so that the Neumann series is dominated by a series which converges for *all* values of $M(b - a)$. Hence the Neumann series *always* provides a solution to equations of the Volterra type.

8.2 NORMS OF OPERATORS

It may have occurred to the reader that the discussion of the previous section suffers from one serious limitation, namely, the requirement $M < (b - a)^{-1}$. In the case considered in Section 8.1, if the interval of integration is infinite, this requirement reduces to $M < 0$. But M is by definition positive, so no kernel can satisfy this requirement. However, to estimate a function by replacing it everywhere by its maximum value is a very crude approximation for infinite intervals. What we would like to be able to show is that we can replace

$$\max |k(x, y)| < (b - a)^{-1}$$

by a less restrictive requirement on $k(x, y)$. To do so we must first examine in some detail the problem of defining the magnitude of a linear *operator*.

In Hilbert space, we defined the norm of a *vector* f by

$$||f|| \equiv (f, f)^{1/2}\,,$$

where $(\ ,\ )$ is an inner product on the Hilbert space to which f belongs (see Definition 4.1). However, the concept of the norm of a vector is much more general; the existence of a norm is associated only with a definition of length, not of angle. There are many vector spaces on which one cannot define an inner product, but where nevertheless a length, or metric, can be defined. We can generalize the concept of norm to this wider class of spaces as follows.

Definition 8.1. A *norm* on a vector space, V, is a relation which assigns to any element, f, of V a real number denoted by $||f||$ such that

a) $||f + g|| \leq ||f|| + ||g||$, for all $f, g \in V$;
b) $||f|| = 0$ if and only if $f \equiv 0$;
c) $||af|| = |a| \cdot ||f||$ for any scalar a belonging to the field over which V is defined.

A vector space on which a norm can be defined is called a *normed vector space*. We have already studied one such space in some detail, namely, Hilbert space, where the norm of a vector can be derived from the existence of the inner product. Because of the axioms satisfied by the inner product (Definition 4.1), the norm defined in this way will have all the properties required by Definition 8.1.

Once a norm has been defined on a vector space, questions concerning the convergence of sequences of vectors can be considered.

Definition 8.2. A sequence of vectors $\{f_n\}$ in a normed vector space, V, is said to *converge in norm* to $f \in V$ if $\|f_n - f\| \to 0$ as $n \to \infty$.

This generalizes to an arbitrary normed space the notion of mean convergence introduced in Section 5.2 for Hilbert space.

Definition 8.3. A sequence of vectors $\{f_n\}$ in a normed vector space V is called a *Cauchy sequence* if, given any $\epsilon > 0$, there exists a number N such that $\|f_n - f_m\| < \epsilon$ whenever $m > N$ and $n > N$.

The most important property of a Cauchy sequence is that we need make no reference to a limit function. With this in mind, we make a final definition.

Definition 8.4. A normed vector space is said to be *complete* if every Cauchy sequence in the space tends to a vector which lies in the space.

A well-known vector space which is normed but not complete is the vector space of rational numbers with the usual Euclidean norm (absolute value). For example, the sequence $\{s_n\}$ where

$$s_n = \sum_{m=1}^{n} \frac{1}{m^2}$$

is a Cauchy sequence of rationals whose limit, $\pi^2/6$, is irrational.

Complete, normed vector spaces are called *Banach spaces,* after the Polish mathematician who played the key role in developing the theory of these spaces. A familiar Banach space is the space of continuous functions on the interval $[a, b]$, which we denote by $C[a, b]$. A norm for this space is

$$\|f(x)\| = \max_{x \in [a, b]} |f(x)| .$$

By the triangle inequality for real numbers,

$$|f(x) + g(x)| \leq |f(x)| + |g(x)|$$

for any x. Thus, in particular,

$$\max_{x \in [a, b]} |f(x) + g(x)| \leq \max_{x \in [a, b]} |f(x)| + \max_{x \in [a, b]} |g(x)| ,$$

or

$$\|f + g\| \leq \|f\| + \|g\| .$$

The other two requirements of Definition 8.1 are trivially satisfied. Clearly, $C[a, b]$ is a vector space, since the sum of two continuous functions is a continuous function. Finally, and most important, with a norm defined in this way $C[a, b]$ is a *complete* space, because the limit of a uniformly convergent sequence of continuous functions is continuous (see Problem 8.1). Thus $C[a, b]$ is a Banach space.

Another important family of Banach spaces are the so-called L_p spaces, defined to be the vector spaces of functions, $f(x)$, which satisfy

$$\int_D |f(x)|^p \, dx < \infty ,$$

where D is some subset (not necessarily finite) of the real line. A norm for this space is given by (see Problems 8.2 and 8.3)

$$||f|| = \left[\int_D |f(x)|^p \, dx\right]^{1/p}$$

and it can be shown (Generalized Riesz-Fisher theorem) that the space L_p is complete with this norm, and is thus a Banach space. We see, in particular, that the space, L_2, of square-integrable functions which we considered in Chapter 5 is a Banach space as well as being a Hilbert space. In fact, *any* Hilbert space is a Banach space since it is a complete inner-product space whose norm is automatically provided by the inner product.

Now that we have fixed the structure of Banach spaces, it is appropriate to consider how the space is changed by linear transformations. Throughout this chapter we shall limit the discussion to *bounded* linear transformations.

Definition 8.5. A transformation K on a normed linear vector space V is said to be a *bounded, linear transformation* if

a) $K(f + g) = Kf + Kg$ for all $f, g \in V$;
b) $K(af) = aKf$ for all $f \in V$ and all scalars a in the field over which V is defined;
c) there exists a constant M such that $||Kf|| \leq M||f||$ for all $f \in V$.

Definition 8.6. Let V be a normed vector space, and K a bounded transformation on V. The *smallest* M satisfying $||Kf|| \leq M||f||$ for all $f \in V$ is called the *norm* of K, which we will write as $||K||$. Note that we are using the same notation for the norm of an operator as for the norm of a function. To avoid confusion, vectors will always be written in lower case letters, operators in upper case letters.

It is not hard to see that $||K||$ as defined above is indeed a norm; for example

$$||(K_1 + K_2)f|| = ||K_1 f + K_2 f|| \leq ||K_1 f|| + ||K_2 f||$$

since $K_1 f$ and $K_2 f$ are elements of V. Applying Definition 8.6, we have, for all f,

$$\begin{aligned} ||(K_1 + K_2)f|| &\leq ||K_1|| \, ||f|| + ||K_2|| \, ||f|| \\ &\leq [||K_1|| + ||K_2||] \, ||f|| . \end{aligned}$$

But $||K_1 + K_2||$ is the smallest M satisfying $||(K_1 + K_2) f|| \leq M||f||$ for all f, so we have immediately

$$||K_1 + K_2|| \leq ||K_1|| + ||K_2|| .$$

The other properties of the norm of the operator follow similarly from the normed structure of the underlying vector space V. This leads to the conclusion that the set of all bounded linear transformations of a Banach space is a normed linear vector space—in fact, a Banach space, since the space can be shown to be complete. Now, what about the norm of the product of two bounded linear transformations on V?

$$||(K_1K_2)f|| = ||K_1(K_2f)|| \leq ||K_1|| \, ||K_2f|| \leq ||K_1|| \, ||K_2|| \, ||f||$$

for all f, so that $||K_1K_2|| \leq ||K_1|| \, ||K_2||$. In particular, this implies that

$$||K^n|| \leq ||K||^n . \tag{8.15}$$

The assumption that K is a bounded transformation has one particularly useful consequence. Consider a sequence of vectors, f_n, in V which converges in norm to some element $f \in V$, that is,

$$||f_n - f|| \to 0$$

as $n \to \infty$. It is natural to inquire whether or not the related sequence $\{Kf_n\}$ converges to Kf. To show that it does, consider the expression

$$||Kf_n - Kf|| = ||K(f_n - f)|| \leq ||K|| \, ||f_n - f|| \, ;$$

since $||K||$ is finite and $f_n \to f$ in norm, it follows that $Kf_n \to Kf$ in norm. This is called the *continuity property* of bounded linear transformations.

The introduction of an operator norm gives us a particularly powerful means of discussing the convergence of operator sequences.

Definition 8.7. We say that a sequence of operators, K_n, *converges in norm* to an operator, K, if $||K_n - K|| \to 0$ as $n \to \infty$.

There are, of course, other possible ways in which we could have defined the convergence of a sequence of operators. For example, we could say that $\{K_n\}$ converges to K if, for all $f \in B$,

$$||K_nf|| \to ||Kf|| . \tag{8.16}$$

This is certainly an acceptable way to define the convergence of operators, but it is *not* equivalent to convergence in norm.

To illustrate the difference between these two types of convergence in a Hilbert space, and also to introduce a particularly useful type of operator, we consider the so-called projection operators, P_n, which are defined for any orthonormal set of vectors, $\{\phi_i\}$, by

$$P_nf \equiv \sum_{i=1}^{n} (\phi_i, f)\phi_i . \tag{8.17}$$

We say that P_n is the projection operator onto the subspace of H spanned by the first n elements of the set $\{\phi_i\}$. It is clear from the definition of Eq. (8.17) that P_n is an Hermitian operator, since

$$(f, P_ng) = (P_nf, g) \tag{8.18}$$

for all $f, g \in H$. Also, P_n is idempotent:

$$P_n^2 = P_n . \tag{8.19}$$

(This last property is obvious geometrically, since after once projecting a function into a subspace of H, subsequent applications of the same projection leave the function unaffected.)

The operator norm of P_n satisfies

$$||P_n|| \leq 1 .$$

To show this, we consider

$$||P_n f|| = (P_n f, P_n f)^{1/2} = (f, P_n f)^{1/2} = [\sum_{i=1}^{n} (f, \phi_n)(\phi_n, f)]^{1/2} .$$

But by Bessel's inequality [Eq. (5.9)], we get immediately

$$||P_n f|| \leq ||f||$$

for all f. Hence $||P_n|| \leq 1$. The same proof may be applied to the operator $(I - P_n)$ to show that

$$||I - P_n|| \leq 1;$$

we now observe that both of the inequalities on the norms of P_n and $I - P_n$ can be converted into strict equalities by noting, for example, that there is always at least one f for which $||P_n f|| = ||f||$. In fact, whenever $f = \phi_1, \phi_2, \cdots, \phi_n$ we will have $||P_n f|| = ||f||$. Thus, $||P_n|| \geq 1$, and combining this with our previous result ($||P_n|| \leq 1$), we conclude that

$$||P_n|| = 1. \tag{8.20}$$

Similarly,

$$||I - P_n|| = 1. \tag{8.21}$$

The completeness of an orthonormal set $\{\phi_i\}$ as given in Definition 5.6 may be stated according to Eq. (5.10) in terms of projection operators as

$$\lim_{n \to \infty} (f, P_n f) = (f, f)$$

for all $f \in H$. Now using Eqs. (8.18) and (8.19), we have

$$||P_n f|| = (P_n f, P_n f)^{1/2} = (f, P_n^2 f)^{1/2} = (f, P_n f)^{1/2} ,$$

so

$$\lim_{n \to \infty} ||P_n f|| = (f, f)^{1/2} = ||If|| .$$

Thus P_n converges to I in the sense suggested by Eq. (8.16). Using a common terminology (albeit a somewhat misleading one), we may say that the sequence of operators $\{P_n\}$ *converges strongly* to I. If we think of the vectors as "points" in the vector space, then operators are analogous to functions; in this sense strong convergence of sequences of operators is analogous to the pointwise convergence of sequences of functions.

What about convergence in norm? For the sequence $\{P_n\}$ to converge to I in norm, given any ϵ, there must exist an N such that

$$||(P_n - I)f|| \leq \epsilon||f|| \tag{8.22}$$

whenever $n \geq N$, *for all* $f \in H$, that is, N is independent of f. Thus convergence in norm of operators is analogous to uniform convergence of a sequence of functions. But Eq. (8.22) cannot be satisfied, for, given any N, it is always possible to pick some function $h \in H$, in particular, $h = \phi_{N+1}$, such that

$$(P_N - I)h = (P_N - I)\phi_{N+1} = -\phi_{N+1} .$$

Then

$$||(P_N - I)h|| = ||h|| ,$$

so the sequence $\{P_n\}$ does *not* converge in norm to I. We see that convergence in norm is a *stronger* requirement than is strong convergence. In the sections that follow, when there can be no ambiguity, we shall often refer to convergence in norm simply as "convergence."

8.3 ITERATIVE TECHNIQUES IN A BANACH SPACE

Now that we have a method for specifying the magnitude of a linear operator, we can give a brief but complete account of the general iterative technique. Our discussion in this section will contain the results of Section 8.1 as a special case, but will also apply to infinite intervals and to operators more general than those of the integral type. We consider the operator equation

$$f = g + Kf$$

as an equation on a Banach space, B, with $g \in B$ and K a bounded linear transformation on B. We impose on the operator K, the restriction

$$||K|| < 1 , \tag{8.23}$$

which is analogous to the requirement $M(b - a) < 1$ of Section 8.1.

We want to show that the sequence of partial sums $\{f_n\}$, where

$$f_n = \sum_{m=0}^{n} K^m g ,$$

is a Cauchy sequence. Consider

$$||f_\nu - f_\mu|| = ||\sum_{m=\mu+1}^{\nu} K^m g|| \leq \sum_{m=\mu+1}^{\nu} ||K^m g|| \leq \sum_{m=\mu+1}^{\nu} ||K^m|| \, ||g|| ,$$

where we have taken $\nu > \mu$ for convenience. According to Eq. (8.15) $||K^m|| \leq ||K||^m$ so

$$||f_\nu - f_\mu|| \leq ||g|| \sum_{m=\mu+1}^{\nu} ||K||^m = ||g|| \, ||K||^{\mu+1} \sum_{m=0}^{\nu-\mu-1} ||K||^m .$$

But

$$\sum_{m=0}^{\nu-\mu-1} ||K||^m \leq \sum_{m=0}^{\infty} ||K||^m = (1 - ||K||)^{-1} ,$$

where we have summed the geometric series. Thus finally,

$$||f_\nu - f_\mu|| \le ||g|| \, ||K||^{\mu+1}(1 - ||K||)^{-1} . \tag{8.24}$$

Since $||K|| < 1$, $(1 - ||K||)^{-1}$ is a finite number. Hence Eq. (8.24) tells us that it is possible to choose μ large enough so that $||f_\nu - f_\mu||$ is smaller than any preassigned number. Therefore $\{f_n\}$ is a Cauchy sequence, and, since the space B is complete (by definition),

$$f_n \to f ,$$

where f is an element of B. Now, by the continuity property of bounded linear operators,

$$Kf_n \to Kf$$

in norm. But Kf_n is given by

$$Kf_n = K \sum_{m=0}^{n} K^m g = \sum_{m=1}^{n+1} K^m g = \sum_{m=0}^{n+1} K^m g - g = f_{n+1} - g .$$

Hence

$$(f_{n+1} - g) \to Kf$$

in norm. But $f_{n+1} \to f$ in norm, so

$$f = g + Kf .$$

Thus f is a solution of the basic equation. It is also the only solution, for suppose that

$$f_1 = g + Kf_1 , \qquad f_2 = g + Kf_2 .$$

Then

$$f_1 - f_2 = K(f_1 - f_2) ,$$

and therefore

$$||f_1 - f_2|| = ||K(f_1 - f_2)|| \le ||K|| \cdot ||f_1 - f_2|| .$$

Since $||K|| < 1$, this means that $||f_1 - f_2|| = 0$, so the solution we have found is unique.

Thus when $||K|| < 1$ it is always possible to solve Eq. (8.4). It is worth emphasizing that this does not mean that for $||K|| \ge 1$ the Neumann series must diverge. We have only shown that it *necessarily* converges for $||K|| < 1$. Clearly, the smaller $||K||$, the more rapidly the series will converge. If $||K||$ is sufficiently small, even one or two terms of the Neumann series may be adequate for the purpose at hand. This is often the case in physical problems.

A particularly important application of these results is to the space L_2 of square-integrable functions. Consider the integral operator K defined by

$$Kf = \int k(x, y) \, f(y) \, dy ,$$

acting on this space. If $k(x, y)$ is square-integrable as a function of two variables, so that

$$\int dx \int dy\, |k(x, y)|^2 < \infty\,, \tag{8.25}$$

then we can show that K is a transformation which maps every element of L_2 into an element of L_2. From Schwartz's inequality,

$$\left[\left|\int k(x, y)\, f(y)\, dy\right|\right]^2 \leq \int |k(x, y)|^2\, dy \cdot \int |f(y)|^2\, dy\,,$$

or

$$|Kf|^2 \leq \int |k(x, y)|^2\, dy \cdot ||f||^2\,.$$

Therefore

$$||Kf||^2 = \int |Kf|^2\, dx \leq ||f||^2 \int dx \int dy\, |k(x, y)|^2\,. \tag{8.26}$$

But this last integral is finite by Eq. (8.25), so $||Kf||$ is finite, as we asserted above.

Equation (8.26) also gives a useful estimate for the norm of K. From it we conclude that

$$||Kf|| \leq \left[\int dx \int dy\, |k(x, y)|^2\right]^{1/2} ||f||\,,$$

for all $f \in L_2$, and hence

$$||K|| \leq \left[\int dx \int dy\, |k(x, y)|^2\right]^{1/2}\,.$$

In particular, if

$$\int dx \int dy |k(x, y)|^2 < 1\,,$$

then $||K|| < 1$, and the above theory applies.

The integral form of Schrödinger's equation furnishes a good example of square-integrable kernels. According to Eqs. (7.57) and (7.68),

$$\psi(\mathbf{r}) = \xi(\mathbf{r}) + \int G_0(\mathbf{r}, \mathbf{r}', E)\, V(\mathbf{r}')\psi(\mathbf{r}')\, d\mathbf{r}'\,, \tag{8.27}$$

where

$$G_0(\mathbf{r}, \mathbf{r}', E) = -\frac{m}{2\pi\hbar^2} \frac{e^{iq|\mathbf{r}-\mathbf{r}'|}}{|\mathbf{r} - \mathbf{r}'|}\,; \tag{8.28}$$

here $\xi(\mathbf{r}) \equiv 0$ for the bound state case $(E < 0)$, and

$$\xi(\mathbf{r}) = \frac{A}{(2\pi)^{3/2}}\, e^{i\mathbf{q}\cdot\mathbf{r}}$$

for the scattering case $(E \geq 0)$. Also, q is defined by

$$q = |\mathbf{q}| = \sqrt{2mE/\hbar^2}\,,$$

and when $E < 0$ we take q with *positive* imaginary part. It will be convenient in what follows to consider q to be a complex number with positive imaginary part. Now the kernel of the above equation is

$$k(\mathbf{r}, \mathbf{r}', E) = G_0(\mathbf{r}, \mathbf{r}', E)\, V(\mathbf{r}')\,,$$

so

$$||K||^2 \leq \int d\mathbf{r} \int d\mathbf{r}'\, |k(\mathbf{r}, \mathbf{r}', E)|^2 = \int d\mathbf{r} \int d\mathbf{r}'\, |G_0(\mathbf{r}, \mathbf{r}', E)\, V(\mathbf{r}')|^2\,.$$

Using Eq. (8.28) for $G_0(\mathbf{r}, \mathbf{r}', E)$ and writing $q = \operatorname{Re} q + i \operatorname{Im} q$, we get

$$||K||^2 \leq \left(\frac{m}{2\pi\hbar^2}\right)^2 \int d\mathbf{r} \int d\mathbf{r}'\, \frac{e^{-2 \operatorname{Im} q |\mathbf{r}-\mathbf{r}'|}}{|\mathbf{r} - \mathbf{r}'|^2} |V(\mathbf{r}')|^2\,.$$

If we change variables by setting $\mathbf{r} = \boldsymbol{\rho} + \mathbf{r}'$, the integral can be handled easily. The result is

$$||K||^2 \leq \frac{m^2}{2\pi\hbar^4 \operatorname{Im} q} \int |V(\mathbf{r})|^2\, d\mathbf{r}\,, \tag{8.29}$$

which is finite *except* when $\operatorname{Im} q = 0$. But unfortunately this corresponds to the most interesting case, $E > 0$. This is when we are in the scattering region and would hope to be able to iterate Schrödinger's integral equation, at least under some conditions.

The reason that we encounter this difficulty is simply that we have failed to take adequate account of the physical situation. If $E > 0$, we are really interested in obtaining the scattering amplitude, $f(\theta, \phi)$, as was emphasized in Section 7.4. In terms of the wave function, $\psi(\mathbf{r})$, of Eq. (8.27),

$$f(\theta, \phi) = -\frac{m}{2\pi\hbar^2} \int e^{-i\mathbf{k}_f \cdot \mathbf{r}}\, V(\mathbf{r}) \psi(\mathbf{r})\, d\mathbf{r}\,.$$

From this expression, it is clear that we need only know $\psi(\mathbf{r})$ in those regions of space where the magnitude (strength) of $V(\mathbf{r})$ is significant, i.e., three-dimensional space should be weighted in importance by the function $V(\mathbf{r})$. With this in mind, we can write the above expression for $f(\theta, \phi)$ as

$$f(\theta, \phi) = -\frac{m}{2\pi\hbar^2} (\psi_{\text{free}}, \psi)_\rho\,,$$

where $\psi_{\text{free}} = e^{i\mathbf{k}_f \cdot \mathbf{r}}$ and the inner product $(\,,\,)_\rho$ is taken with respect to the weight function $V(\mathbf{r})$:

$$(f, g)_\rho \equiv \int f^*(\mathbf{r}) g(\mathbf{r})\, d\rho\, (\mathbf{r})\,, \tag{8.30a}$$

where $\rho(\mathbf{r})$ is defined by

$$d\rho\, (\mathbf{r}) \equiv V(\mathbf{r}) r^2 \sin\theta\, dr\, d\theta\, d\phi\,. \tag{8.30b}$$

Clearly, for this to be an inner product in the sense of Definition 4.1, $V(\mathbf{r})$ must not change sign as $\mathbf{r}$ varies. We will assume that this is the case; if it is not, the necessary modifications should be obvious to the reader. In this notation we may rewrite Eq. (8.27) as

$$\psi(\mathbf{r}) = \xi(\mathbf{r}) + \int G_0(\mathbf{r}, \mathbf{r}', E)\psi(\mathbf{r}')\, d\rho\,(\mathbf{r}')\,, \tag{8.31}$$

which we would like to consider as an integral equation on the Hilbert space whose inner product is given by Eq. (8.30a). This will be legitimate if $||\xi|| < \infty$, which will be the case if

$$\int d\rho\,(\mathbf{r}) = \int V(\mathbf{r})d\mathbf{r} < \infty\,,$$

since $|\xi(\mathbf{r})| = |Ae^{i\mathbf{k}_i\cdot\mathbf{r}}|$ is a constant. For most reasonable potentials $||\xi||$ is finite, although for the Coulomb potential $||\xi||$ is infinite. To be more precise, $||\xi||$ will be finite if $V(\mathbf{r})$ falls off faster than $1/r^3$ at large distances and is less singular than $1/r^3$ at the origin. For Eq. (8.31) to be an acceptable equation in Hilbert space, the norm of the integral operator in Eq. (8.31) must be finite. Denoting the integral operator whose kernel is $G_0(\mathbf{r}, \mathbf{r}', E)$ by K, we have

$$\begin{aligned} ||K||^2 &\leq \int d\rho\,(\mathbf{r}) \int d\rho\,(\mathbf{r}')\, |G_0(\mathbf{r}, \mathbf{r}', E)|^2 \\ &= \left(\frac{m}{2\pi\hbar^2}\right)^2 \int d\mathbf{r} \int d\mathbf{r}'\, V(\mathbf{r})\, V(\mathbf{r}')\, \frac{e^{-2\,\mathrm{Im}\, q|\mathbf{r}-\mathbf{r}'|}}{|\mathbf{r}-\mathbf{r}'|^2}\,. \end{aligned}$$

It is a relatively simple matter to evaluate this integral for the *central* potentials which typically occur in practice. The reason for this is that the integrand then depends only on $|\mathbf{r}|$, $|\mathbf{r}'|$, and $|\mathbf{r} - \mathbf{r}'|$, which form sides of a triangle. The integration over the angles which give the orientation of the triangle in space gives just $8\pi^2$. Writing $s = r + r'$, $t = r - r'$ and $u = |\mathbf{r} - \mathbf{r}'|$, one can easily show that for kernels which depend only on s, t, and u the volume element $d\mathbf{r}\, d\mathbf{r}'$ reduces to just $\pi^2(s^2 - t^2)u\, ds\, dt\, du$. The integration limits are defined by the inequalities $-u < t < u < s$. Thus

$$||K||^2 \leq \left(\frac{m}{2\hbar^2}\right)^2 \int_0^\infty ds \int_0^s du \int_{-u}^u dt\, (s^2 - t^2)\, \frac{e^{-2u\,\mathrm{Im}\, q}}{u}\, V(r)\, V(r')\,.$$

If, for example, $V(r)$ is the Yukawa potential, $V(r) = g^2e^{-\mu r}/r$, then since $s = r + r'$ and $rr' = (s^2 - t^2)/4$, we get

$$||K||^2 \leq \frac{m^2g^4}{\hbar^4} \int_0^\infty ds\, e^{-\mu s} \int_0^s \frac{du}{u}\, e^{-2u\,\mathrm{Im}\, q} \int_{-u}^u dt\,.$$

Doing the integrations, we find that

$$||K|| \leq \left[\frac{2}{1 + 2\,\mathrm{Im}\,(q/\mu)}\right]^{1/2} \frac{mg^2}{\mu\hbar^2}\,. \tag{8.32a}$$

Thus when $E > 0$ (that is, $\mathrm{Im}\, q = 0$),

$$||K|| \leq \sqrt{2}\, \frac{mg^2}{\mu\hbar^2}\,. \tag{8.32b}$$

Hence if the potential is weak enough, i.e., if g is small enough so $||K|| < 1$, we can iterate Schrödinger's integral equation. The iteration will converge in the weighted norm provided by Eq. (8.30a).

In the interest of not deceiving the reader, it should be emphasized that although the iteration is certain to converge if $V(\mathbf{r})$ is not too strong (and not too singular), it is, nevertheless, extremely difficult to perform the integrations involved in computing the successive terms of the series. If the reader has any doubts on this score, he is invited to try evaluating the first term in the iteration of Eq. (8.27) when $V(\mathbf{r})$ is given by some simple potential like the Yukawa potential. As we shall see, there are other methods of solution available which are often easier to employ than the method presented here. However, the Neumann series is a very important theoretical tool for developing other more practical methods of solution.

8.4 ITERATIVE TECHNIQUES FOR NONLINEAR EQUATIONS

Before leaving the subject of iterative solutions of integral equations, it is worthwhile contrasting the simple situation which prevails in the linear case with the more complicated features which arise in the nonlinear case. We shall see that the power of the iterative procedure is sharply reduced by nonlinearity, even when we restrict our attention to equations analogous to the Volterra equation discussed in Section 8.1.

Let us consider an equation of the form

$$f(x) = g(x) + \int_a^x h(x, y, f(y))\, dy\,, \qquad a \leq x \leq b\,. \tag{8.33}$$

We would like to develop an iterative procedure similar to the one we just employed to handle the linear case. However, since, in general, $f(y)$ does not occur linearly under the integration sign in Eq. (8.33), we shall proceed as follows. Set $f_0(x) \equiv g(x)$ and then by induction define

$$f_n(x) \equiv g(x) + \int_a^x h(x, y, f_{n-1}(y))\, dy\,. \tag{8.34}$$

The reader can easily verify that in the linear case, that is, $h(x, y, f(y)) = k(x, y)\, f(y)$, $f_n(x)$ is given by

$$f_n(x) = \sum_{m=0}^{n} K^m g\,,$$

where K is defined by Eq. (8.5). With this in mind, the basic idea is very simple. We start with $f_0(x) \equiv g(x)$ as a zeroth-order approximation to $f(x)$ and insert this approximation into Eq. (8.34) to obtain a first approximation, $f_1(x)$. Continuing this process, we would hope to generate a sequence which converges eventually to the solution to Eq. (8.33). One might expect that just as in the case of the Volterra linear integral equation, this process will always converge if $h(x, y, f(y))$ is bounded in the domain $a \leq x \leq b$, $a \leq y \leq b$. However, we must also remember that h depends on a third variable, namely, f itself.

Therefore we shall make the additional assumption that for (x, y) in the square $a \leq x \leq b,\ a \leq y \leq b$,

$$|h(x, y, f(y))| \leq M \tag{8.35}$$

whenever

$$|f(y) - g(y)| \leq \Delta\,, \tag{8.36}$$

where Δ is some given positive number and $y \in [a, b]$. These assumptions will at least enable us to start the iterative procedure.

Setting $f_0(x) = g(x)$, we obtain

$$\begin{aligned} f_1(x) &\equiv g(x) + \int_a^x h(x, y, f_0(y))\,dy \\ &= g(x) + \int_a^x h(x, y, g(y))\,dy\,, \end{aligned}$$

from which we may deduce that

$$|f_1(x) - g(x)| = |f_1(x) - f_0(x)| \leq M(x - a)\,, \tag{8.37}$$

since $|f_0(y) - g(y)| = 0 < \Delta$. But to continue one step further, we need to know that $|f_1(y) - g(y)| \leq \Delta$. Equation (8.37) tells us that this will be the case if $|y - a| \leq \Delta/M$. Thus, to get any further, we must restrict the variables so that

$$a \leq x \leq (\Delta/M) + a\,, \qquad a \leq y \leq (\Delta/M) + a\,. \tag{8.38}$$

Clearly, if $(\Delta/M) + a \geq b$, this is no restriction at all. Thus we can calculate

$$f_2(x) \equiv g(x) + \int_a^x h(x, y, f_1(y))\,dy\,.$$

Since $a \leq x \leq (\Delta/M) + a$, we have $|f_1(x) - g(x)| \leq \Delta$, and

$$|f_2(x) - g(x)| \leq M(x - a) \leq \Delta\,;$$

this will be true for all further iterations. However, we also hope to show, in analogy with the linear case, that f_n and f_{n-1} get very close to each other when n becomes sufficiently large. Hence we need to calculate the difference between successive approximations. Let us look first at $|f_2(x) - f_1(x)|$. We find that

$$|f_2(x) - f_1(x)| = \left|\int_a^x [h(x, y, f_1(y)) - h(x, y, f_0(y))]\,dy\right|\,. \tag{8.39}$$

For linear integral equations, it is relatively easy to estimate the difference occurring under the integral in Eq. (8.39), because

$$h(x, y, f(y)) = k(x, y)\,f(y)\,.$$

But in the nonlinear case it is necessary to put further restrictions on the function h in the region defined by Eq. (8.38). By analogy with the linear case, we require that there exist some finite number, N, such that

$$|h(x, y, \phi(y)) - h(x, y, \psi(y))| \leq N|\phi(y) - \psi(y)| \tag{8.40}$$

whenever $|\phi(y) - g(y)| \leq \Delta$ and $|\psi(y) - g(y)| \leq \Delta$. This requirement is called the *Lipschitz condition*.

Returning now to Eq. (8.39), we see that since $|f_1(y) - g(y)| \leq \Delta$ and $|f_0(y) - g(y)| = 0 < \Delta$, we can use the Lipschitz condition to get

$$|f_2(x) - f_1(x)| \leq N \int_a^x |f_1(y) - f_0(y)|\, dy\,.$$

But $|f_1(y) - f_0(y)| = |f_1(y) - g(y)| \leq \Delta$, so

$$|f_2(x) - f_1(x)| \leq N\Delta(x-a)\,.$$

Continuing to the next step,

$$|f_3(x) - f_2(x)| \leq \Delta N^2(x-a)^2/2!\,,$$

since $|f_2(x) - g(x)| \leq \Delta$. In general,

$$|f_n(x) - g(x)| \leq \Delta \tag{8.41}$$

and

$$|f_n(x) - f_{n-1}(x)| \leq \Delta N^{n-1}(x-a)^{n-1}/(n-1)!\,. \tag{8.42}$$

In view of Eq. (8.38), Eq. (8.42) can be written as

$$|f_n(x) - f_{n-1}(x)| \leq \Delta(\Delta N/M)^{n-1}/(n-1)!\,. \tag{8.43}$$

Thus for all $x \in [a, (\Delta/M) + a]$, $|f_n(x) - f_{n-1}(x)| \to 0$ uniformly as $n \to \infty$.

Just as in the linear case, we expect that $f_n(x)$ should tend to the solution of Eq. (8.33). To see this, we write $f_\nu(x)$ in the form

$$f_\nu(x) = f_0(x) + \sum_{n=1}^{\nu} [f_n(x) - f_{n-1}(x)]\,.$$

According to Eq. (8.43), each term of this sum is dominated by

$$\Delta(\Delta N/M)^{n-1}/(n-1)! \quad \text{for} \quad x \in [a, (\Delta/M) + a]\,.$$

Thus as ν goes to infinity, $f_\nu(x)$ tends uniformly toward some function, say, $f(x)$. We now show that $f(x)$ satisfies Eq. (8.33). We write

$$\begin{aligned} f_\nu(x) &= g(x) + \int_a^x h(x, y, f_{\nu-1}(y))\, dy \\ &= g(x) + \int_a^x h(x, y, f(y))\, dy + \int_a^x [h(x, y, f_{\nu-1}(y)) - h(x, y, f(y))]\, dy\,. \end{aligned}$$

Taking the limit $\nu \to \infty$ on both sides of this equation, we obtain

$$f(x) = g(x) + \int_a^x h(x, y, f(y))\, dy + \lim_{\nu\to\infty} R_\nu\,,$$

where

$$R_\nu = \int_a^x [h(x, y, f_{\nu-1}(y)) - h(x, y, f(y))]\, dy\,.$$

Since $|f(y) - g(y)| < \Delta$ because of Eqs. (8.41) and (8.43), we can use the Lipschitz condition to get

$$|R_\nu| \leq N \int_a^x |f_{\nu-1}(y) - f(y)|\, dy$$

$$\leq N\Delta \sum_{n=\nu}^{\infty} \frac{N^{n-1}}{(n-1)!} \int_a^x (y-a)^{n-1}\, dy$$

$$\leq \Delta \sum_{n=\nu}^{\infty} \frac{N^n (x-a)^n}{n!} \leq \Delta \sum_{n=\nu}^{\infty} \left(\frac{N\Delta}{M}\right)^n \frac{1}{n!},$$

which tends to zero as $\nu \to \infty$. Thus

$$f(x) = g(x) + \int_a^x h(x, y, f(y))\, dy,$$

and $f(x)$ is indeed the desired solution to Eq. (8.33).

We have therefore obtained an iterative solution in an interval, the length of which will generally depend on specific properties of $h(x, y, f(y))$. In the linear Volterra case, we were able to exhibit a convergent iterative series for *all* values of x, so we see that nonlinearity has a very serious effect on the iterative technique. The appropriate *region* of convergence is controlled by Δ/M, while the *rate* of convergence depends on the size of N, the constant in the Lipschitz condition [Eq. (8.40)].

As an example of nonlinear perturbation theory, we consider a problem from classical mechanics. Suppose that a particle is bound by a slightly anharmonic spring, with potential energy

$$V(x) = \tfrac{1}{2}kx^2 + \tfrac{1}{3}ax^3,$$

where a is a small constant. The equation of motion is

$$m\ddot{x} + kx + ax^2 = 0, \qquad \text{or} \qquad \ddot{x} + \omega_0^2 x = -\epsilon x^2,$$

where $\omega_0^2 = k/m$ and $\epsilon = a/m > 0$.

According to Eq. (7.35) (with $\gamma = 0$) the Green's function for the operator $d^2/dt^2 + \omega_0^2$ with one-point boundary conditions is

$$G(t, t') = \frac{1}{\omega_0} \sin[\omega_0(t-t')]\theta(t-t').$$

Assuming that the boundary condition is applied at $t = 0$ and has the form $x(0) = x_0 > 0$, $\dot{x}(0) = 0$, we get

$$x(t) = x_0 \cos\omega_0 t - \frac{\epsilon}{\omega_0} \int_0^t \sin\omega_0(t-t') x^2(t')\, dt'.$$

In keeping with the previous discussion, let us inquire under what conditions we can obtain an iterative solution for $x(t)$ if we require that

$$|x(t) - x_0 \cos\omega_0 t| < x_0. \tag{8.44}$$

For this particular problem, we have also

$$h(t, t', x(t')) = -(\epsilon/\omega_0) \sin \omega_0(t - t')x^2(t') .$$

Thus

$$|h(t, t', x(t'))| \leq (\epsilon/\omega_0)x^2(t') ,$$

which may be written as

$$\begin{aligned} |h(t, t', x(t'))| &\leq (\epsilon/\omega_0)[x(t') - x_0 \cos \omega_0 t' + x_0 \cos \omega_0 t']^2 \\ &\leq (\epsilon/\omega_0)[|x(t') - x_0 \cos \omega_0 t'| + x_0|\cos \omega_0 t'|]^2 . \end{aligned}$$

Using Eq. (8.44), we obtain immediately

$$|h(t, t', x(t'))| < 4\epsilon x_0^2/\omega_0 , \tag{8.45}$$

for all t and t' whenever $|x(t') - x_0 \cos \omega_0 t'| \leq x_0$.

Similarly, we may estimate the quantity

$$|h(t, t', x(t')) - h(t, t', \xi(t'))|$$

which appears in the Lipschitz condition. We have

$$|h(t, t', x(t')) - h(t, t', \xi(t'))| \leq (\epsilon/\omega_0)|x(t') + \xi(t')|\,|x(t') - \xi(t')| ,$$

which may be rewritten as

$$\begin{aligned} &|h(t, t', x(t')) - h(t, t', \xi(t'))| \\ &\quad \leq (\epsilon/\omega_0)|x(t') - x_0 \cos \omega_0 t' + \xi(t') - x_0 \cos \omega_0 t' + 2x_0 \cos \omega_0 t'| \cdot \\ &\qquad\qquad |x(t') - \xi(t')| . \end{aligned}$$

Using Eq. (8.44) again, we have

$$|h(t, t', x(t')) - h(t, t', \xi(t'))| \leq (4\epsilon x_0/\omega_0)|x(t') - \xi(t')| \tag{8.46}$$

for all t and t' whenever $|x(t') - x_0 \cos \omega_0 t'| \leq x_0$ and $|\xi(t') - x_0 \cos \omega_0 t'| \leq x_0$. Thus the quantities Δ, M, and N are given by

$$\Delta = x_0 , \tag{8.47a}$$

$$M = 4\epsilon x_0^2/\omega_0 , \tag{8.47b}$$

$$N = 4\epsilon x_0/\omega_0 . \tag{8.47c}$$

The above estimates involve no restrictions on t, but the previous discussion of the iterative method in the nonlinear case tells us that the iterative solution will be valid only in the interval $0 \leq t \leq \Delta/M$ or $0 \leq \omega_0 t \leq 1/(4\delta)$, where $\delta = \epsilon x_0/\omega_0^2$, and δ is the parameter in which we are attempting to expand $x(t)$. Hence we see that if δ is small, the expansion will converge for a large range of times.

With these results in hand, we begin the iterative procedure by setting

$$x_0(t) = x_0 \cos \omega_0 t . \tag{8.48}$$

By our previous work,

$$x_1(t) = x_0 \cos \omega_0 t - (\epsilon/\omega_0) \int_0^t \sin \omega_0(t - t')x_0^2(t')\, dt' .$$

Using Eq. (8.48) for $x_0(t')$, we get

$$x_1(t) = x_0\left[\cos\omega_0 t + \delta\left(\tfrac{1}{6}\cos 2\omega_0 t + \tfrac{1}{3}\cos\omega_0 t - \tfrac{1}{2}\right)\right]. \tag{8.49}$$

Note that $x_1(t)$ satisfies the boundary conditions at $t = 0$.

The main effect of the perturbation in first approximation is a slight change in the amplitude of the dominant harmonic ($\cos\omega_0 t$) and the addition of a second harmonic term ($\cos 2\omega_0 t$) with small amplitude (assuming that δ is small). By looking at $x_1(t)$ alone, it is not clear why the restriction $0 \leq \omega_0 t \leq 1/(4\delta)$ is necessary, since the size of the first-order correction is on the average independent of t. To see the reason for this restriction we look at $x_2(t)$. A few integrations show that

$$\begin{aligned} x_2(t) = x_0[&\cos\omega_0 t + \delta\left(\tfrac{1}{6}\cos 2\omega_0 t + \tfrac{1}{3}\cos\omega_0 t - \tfrac{1}{2}\right) \\ &+ \delta^2\left(\tfrac{1}{48}\cos 3\omega_0 t + \tfrac{1}{9}\cos 2\omega_0 t + \tfrac{29}{144}\cos\omega_0 t - \tfrac{1}{3}\right) \\ &+ \delta^3\left(\tfrac{1}{1080}\cos 4\omega_0 t + \tfrac{1}{144}\cos 3\omega_0 t - \tfrac{1}{27}\cos 2\omega_0 t + \tfrac{753}{2160}\cos\omega_0 t - \tfrac{23}{72}\right) \\ &+ \omega_0 t\left(\tfrac{5}{12}\delta^2\sin\omega_0 t + \tfrac{5}{36}\delta^3\sin\omega_0 t\right)]. \end{aligned} \tag{8.50}$$

This iterate has the same general form as $x_1(t)$ expect for the appearance of terms linear in $\omega_0 t$. Thus for t large, both $|x_2(t) - x_1(t)|$ and $|x_2(t) - x_0(t)|$ are dominated by terms proportional to t. These terms will remain bounded only if we restrict the range of t.

It should be noted that $x_2(t)$, although it contains terms of order δ^3, is complete only through terms of order δ^2; there are further terms of order δ^3 which will make their appearance in $x_3(t)$, along with terms of still higher order. (In general, there will be terms proportional to $\delta^{(2^n-1)}$ in the nth iterate.) It is also interesting to note that if we define

$$\omega_1 \equiv \omega_0\left[1 - \tfrac{5}{6}\delta^2\right]^{1/2}, \tag{8.51}$$

and replace ω_0 by ω_1 in $\cos\omega_0 t$, then through order δ^2, $x_2(t)$ can be written as in Eq. (8.50), but with the omission of the term linear in t and proportional to δ^2. This follows from the fact that through order δ^2

$$\cos\omega_0 t + \tfrac{5}{12}\delta^2\omega_0 t\sin\omega_0 t \simeq \cos\omega_1 t,$$

where ω_1 is given by Eq. (8.51). Thus, by redefining the frequency, it is possible to eliminate terms proportional to t, at least through order δ^2. It can be shown (see Problem 8.15) that a similar elimination can be achieved for all terms through order δ^3.

8.5 SEPARABLE KERNELS

We have now seen in some detail how to solve linear equations involving operators of small norm. However, it is clear, particularly from the example discussed at the end of Section 8.3, that there are many interesting problems which involve operators of finite, but not necessarily small, norm. We must therefore develop techniques of more general applicability.

We might hope that it will prove possible to treat operators on infinite-dimensional spaces in the same way that we treated operators on finite-dimensional spaces (Chapters 3 and 4). There we could simply take an operator, A, and turn it into a finite-dimensional matrix by computing its inner product between the finite number of vectors which constitute an orthonormal basis for the space. However, on infinite-dimensional spaces there must be an infinite number of vectors in any basis, so this procedure would lead us to infinite-dimensional matrices, and we have said nothing about such objects in previous chapters. (Even the problem of multiplying two infinite-dimensional matrices is not a trivial one because of convergence questions.)

On the other hand, we know that if we write a function, f, in terms of an orthonormal basis, $\{\phi_n\}$, of Hilbert space by computing the expansion coefficients, $a_n = (\phi_n, f)$, we can associate the infinite collection of a_n with an infinite column vector, and, more important, as n becomes large, say $n > N$, the elements a_n with $n > N$ become *small* (Bessel's inequality). Thus any function in Hilbert space can be approximated arbitrarily closely by a *finite* linear combination,

$$f \equiv \sum_{n=1}^{N} a_n \phi_n .$$

This encourages us to believe that perhaps the same situation will prevail for linear operators; namely, if we start computing the matrix elements of an operator, A, between the elements of an infinite orthonormal basis, then after we have computed enough matrix elements, A_{mn}, the elements will start to become small, just as did the a_n in the case of vectors. Thus we might hope that we can stop completely when we get to a large $N \times N$ array and just forget about the elements A_{mn}, with m or n greater than N. The value of N would, of course, have to depend on the degree of precision which one wanted in a given problem, but we imagine that any degree of precision can be obtained by this method.

If this were possible, the problem would come down to simple matrix diagonalization or inversion, which we can carry out by employing the methods of Chapters 3 and 4. We shall see that there is a rather large class of operators for which such an approach will work. It is the purpose of the subsequent sections of this chapter and also of the following chapter to present in some detail the theory of such operators.

It is best, however, to start with a type of operator for which no approximation is necessary in the reduction to a simple matrix. This is the class of operators of *finite rank*, so called because they are directly reducible to finite-dimensional matrices (and hence, by the discussion in Section 3.10, to matrices of finite rank).

To illustrate what we have in mind, let us consider a separable kernel, i.e., one which has the form

$$k_s(x, y) = \lambda \phi(x) \psi^*(y) , \tag{8.52}$$

where for the sake of simplicity we shall assume that ϕ and ψ are functions in Hilbert space. If

$$\int |k_s(x, y)|^2 \, dx \, dy < 1 , \tag{8.53}$$

then we know from our work in the previous section that the Neumann series converges. Let us look at this series. The iterative solution to the equation

$$f(x) = g(x) + \int k_s(x, y) f(y)\, dy \tag{8.54}$$

is

$$f(x) = g(x) + \int dy_1\, k_s(x, y_1)\, g(y_1) + \int dy_1 \int dy_2\, k_s(x, y_1) k_s(y_1, y_2)\, g(y_2) + \cdots$$

Since $k_s(x, y)$ is square integrable as a function of two variables, $\phi(x)$ and $\psi(y)$ must be square integrable as functions of a single variable. Thus if $g(x)$ is square integrable, then all the integrals in the iteration are well defined, and thus

$$\begin{aligned} f(x) = g(x) &+ \lambda \int dy_1\, \phi(x)\psi^*(y_1) g(y_1) \\ &+ \lambda^2 \int dy_1 \int dy_2\, \phi(x)\psi^*(y_1)\phi(y_1)\psi^*(y_2) g(y_2) + \cdots \end{aligned}$$

Let us set

$$\int \psi^*(y)\phi(y)\, dy = (\psi, \phi)\,.$$

Then $f(x)$ can be written as

$$\begin{aligned} f(x) &= g(x) + \lambda\phi(x)(\psi, g) + \lambda^2\phi(x)(\psi, \phi)(\psi, g) \\ &\quad + \lambda^3\phi(x)(\psi, \phi)^2(\psi, g) + \cdots \\ &= g(x) + \lambda\phi(x)(\psi, g) \sum_{n=0}^{\infty} \lambda^n (\psi, \phi)^n\,. \end{aligned} \tag{8.55}$$

If $|\lambda(\psi, \phi)| < 1$, the series of Eq. (8.55) converges. In fact, $|\lambda(\psi, \phi)|$ *is* less than unity. To prove this, note that Eq. (8.53) implies

$$|\lambda|^2 \int |\phi(x)|^2\, dx \int |\psi(y)|^2\, dy < 1\,. \tag{8.56}$$

But by Schwartz's inequality,

$$|(\psi, \phi)| \le [(\phi, \phi)(\psi, \psi)]^{1/2}\,.$$

Thus, by Eq. (8.56),

$$|\lambda(\psi, \phi)| \le [|\lambda|^2(\phi, \phi)(\psi, \psi)]^{1/2} < 1\,,$$

and we can sum the geometric series in Eq. (8.55) directly:

$$f(x) = g(x) + \lambda \frac{(\psi, g)}{1 - \lambda(\psi, \phi)} \phi(x)\,. \tag{8.57}$$

Although we derived this simple result in the framework of the Neumann series, its closed form suggests that it might be more general. That this is indeed the case may be seen by writing Eq. (8.54) as

$$\begin{aligned} f(x) &= g(x) + \lambda \int \phi(x)\psi^*(y) f(y)\, dy \\ &= g(x) + \lambda\phi(x)(\psi, f)\,. \end{aligned} \tag{8.58}$$

If we take the inner product of both sides of Eq. (8.58) with ψ, we get

$$(\psi, f) = (\psi, g) + \lambda(\psi, \phi)(\psi, f) . \tag{8.59}$$

Thus

$$(\psi, f) = \frac{(\psi, g)}{1 - \lambda(\psi, \phi)} , \tag{8.60}$$

and

$$f(x) = g(x) + \lambda \frac{(\psi, g)}{1 - \lambda(\psi, \phi)} \phi(x) \tag{8.61}$$

for all values of λ, *except* $\lambda = 1/(\psi, \phi)$. Thus right at the boundary of the domain of validity of the Neumann series, we seem to run into ambiguities. The reason for this may be most easily seen by considering the case where $g(x) = 0$. Then our equation becomes simply

$$f(x) = \int k_s(x, y) f(y) \, dy = \lambda \phi(x)(\psi, f) , \tag{8.62}$$

which is an eigenvalue problem of the usual type. Equation (8.62) implies that

$$(\psi, f) = \lambda(\psi, \phi)(\psi, f) ,$$

which has a nonzero solution for (ψ, f) if and only if

$$\lambda = 1/(\psi, \phi) ,$$

in which case (ψ, f) can be any constant. Thus the eigenvalue problem has as its solution

$$f(x) = A\phi(x) ,$$

where A is arbitrary. To fix A, a normalization condition is necessary, just as in the finite-dimensional case. The interesting result is that the condition for the existence of an eigensolution is the same condition under which we saw that a solution of the inhomogeneous equation given by Eq. (8.61) did not exist. But this result is familiar from finite-dimensional theory, where we know that if the homogeneous equation has a solution, then the inhomogeneous term of the inhomogeneous equation must be subject to certain conditions in order that a solution should exist. We see immediately from Eq. (8.61) that we must require $(\psi, g) = 0$ to compensate for the vanishing denominator. Thus in Eq. (8.59) we find that (ψ, f) is arbitrary, so the solution to Eq. (8.54), given by Eq. (8.58), becomes

$$f(x) = g(x) + B\phi(x) ,$$

where B is an arbitrary constant. Substituting this result into Eq. (8.54), we see that it is indeed a solution for any B when $\lambda(\psi, \phi) = 1$, that is, when $\phi(x)$ is an eigenvector of the integral operator K_s whose kernel is $k_s(x, y)$. This looks very much like the finite-dimensional Theorem 4.27; the resemblance is complete if we note that $(\psi, g) = 0$ means that g is orthogonal to the solution of

$$K_s^\dagger \psi = \lambda^* \int k_s(y, x)\, \psi(y) dy = \psi$$

when $\lambda = 1/(\psi, \phi)$.

Thus there is a close relation between this simple infinite-dimensional example and the finite-dimensional case. More interesting, however, is the hint that the divergence of the Neumann series is somehow related to the occurrence of a solution to the homogeneous equation.

Because separable kernels lead to integral equations with extremely simple closed-form solutions, physicists have naturally attempted to make use of them in practical problems. One example of their utility is found in the study of nuclear forces. For many years physicists have tried to determine the form of nucleon-nucleon potentials, and in doing so were forced to discard practically every sacred classical concept of what a potential should be. As mentioned in Chapter 1, the requirement of central potentials had to be abandoned, and it also had to be admitted that nuclear forces might depend on the intrinsic spin of the nuclear particles. It was natural that eventually even the requirement of *locality* would come into doubt.

By a *local potential*, we mean one whose action at a point depends only on what is happening at that point. Thus Schrödinger's equation is traditionally written with a local potential:

$$-\frac{\hbar^2}{2m} \nabla^2 \psi(\mathbf{r}) + V(\mathbf{r})\psi(\mathbf{r}) = E\psi(\mathbf{r}) \,. \tag{8.63}$$

However, it cannot be ruled out a priori that in quantum mechanics the potential may be an operator which acts in a nonlocal manner. For example, we could imagine writing Schrödinger's equation as

$$-\frac{\hbar^2}{2m} \nabla^2 \psi + \int U(\mathbf{r}, \mathbf{r}')\psi(\mathbf{r}')\, d\mathbf{r}' = E\psi(\mathbf{r}') \,. \tag{8.64}$$

We see from this equation that the effect of the potential operating on ψ does not depend just on the value of ψ at one point $\mathbf{r}$, as is the case in Eq. (8.63), but on the integral over some region of space weighted by the potential function $U(\mathbf{r}, \mathbf{r}')$. Note that if we want the Hamiltonian of Eq. (8.64) to be Hermitian, we must require that

$$U(\mathbf{r}, \mathbf{r}') = U(\mathbf{r}', \mathbf{r})^* \,, \tag{8.65}$$

The Green's function technique developed in Chapter 7 can readily be applied to Eq. (8.64). Proceeding in the manner described in Section 7.4, we find that $\psi(\mathbf{r})$ is given by the integral equation

$$\psi(\mathbf{r}) = \frac{A}{(2\pi)^{3/2}} e^{i\mathbf{q}_i \cdot \mathbf{r}} - \frac{m}{2\pi\hbar^2} \iint \frac{e^{iq_i|\mathbf{r}-\mathbf{r}'|}}{|\mathbf{r} - \mathbf{r}'|} U(\mathbf{r}', \mathbf{r}'')\psi(\mathbf{r}'')\, d\mathbf{r}'\, d\mathbf{r}'' \,, \tag{8.66}$$

where $\mathbf{q}_i$, m, $\hbar$, and A have the same significance as in Section 7.4. Note the similarity between this equation and Eq. (7.68). As was mentioned in Section

7.4, one of the most interesting physical quantities is the scattering amplitude $f^s(\mathbf{q}_f)$, which is defined in terms of the amplitude of the outgoing scattered wave. If we proceed as in Section 7.4, it is a simple matter to show that $f^s(\mathbf{q}_f)$ is given by

$$f^s(\mathbf{q}_f) = f(\mathbf{q}_f)\,, \tag{8.67a}$$

where f satisfies the integral equation

$$f(\mathbf{k}) = -\frac{4\pi^2 m}{\hbar^2}\hat{U}(\mathbf{k}, \mathbf{q}_i) - \frac{2m}{\hbar^2}\int \hat{U}(\mathbf{k}, \mathbf{q})\frac{1}{q^2 - (q_i + i\epsilon)^2} f(\mathbf{q})\, d\mathbf{q}\,, \tag{8.67b}$$

where $\hat{U}(\mathbf{k}, \mathbf{q})$ is the Fourier transform of $U(\mathbf{r}, \mathbf{r}')$, that is,

$$\hat{U}(\mathbf{k}, \mathbf{q}) = \frac{1}{(2\pi)^3}\iint e^{i\mathbf{q}\cdot\mathbf{r}'} e^{i\mathbf{k}\cdot\mathbf{r}}\, \hat{U}(\mathbf{r}, \mathbf{r}')\, d\mathbf{r}\, d\mathbf{r}'\,.$$

In Eqs. (8.67a) and (8.67b), $\mathbf{q}_i$ represents the incident direction of the particle and $\mathbf{q}_f$ represents its final direction. The magnitudes of $\mathbf{q}_i$ and $\mathbf{q}_f$ are equal and given by

$$|\mathbf{q}_i| = |\mathbf{q}_f| = \left[\frac{2mE}{\hbar^2}\right]^{1/2}.$$

Once we have $f(\mathbf{k})$, all the quantities of interest in the scattering process can be determined. For a general $U(\mathbf{k}, \mathbf{q})$ there is not much we can say about this function $f(\mathbf{k})$, but if we take U to be of the *separable* form, that is,

$$U(\mathbf{r}, \mathbf{r}') = -g^2 v(\mathbf{r}) v(\mathbf{r}')\,, \tag{8.68}$$

then we can make progress. Here $v(\mathbf{r})$ is a real-valued function, in order to insure that U is Hermitian. Using this form for U and denoting the Fourier transform of $v(\mathbf{r})$ by $\hat{v}(\mathbf{k})$, we bring Eq. (8.67b) to the form

$$f(\mathbf{k}) = \frac{4\pi^2 m g^2}{\hbar^2}\hat{v}(\mathbf{q}_i)\hat{v}(\mathbf{k}) + \frac{2mg^2}{\hbar^2}\int \hat{v}(\mathbf{k})\frac{\hat{v}(\mathbf{q})}{q^2 - (q_i + i\epsilon)^2} f(\mathbf{q})\, d\mathbf{q}\,,$$

or

$$f(\mathbf{k}) = \pi\tau\hat{v}(\mathbf{q}_i)\hat{v}(\mathbf{k}) + \frac{\tau}{2\pi}\int \hat{v}(\mathbf{k})\frac{\hat{v}(\mathbf{q})}{q^2 - (q_i + i\epsilon)^2} f(\mathbf{q})\, d\mathbf{q}\,, \tag{8.69}$$

where we have set $4\pi m g^2/\hbar^2 \equiv \tau$. But Eq. (8.69) has precisely the form of Eq. (8.54) if we make the identifications

$$x \to \mathbf{k}\,, \qquad y \to \mathbf{q}\,, \qquad g(x) \to \pi\tau\hat{v}(\mathbf{q}_i)\hat{v}(\mathbf{k})\,, \qquad \phi(x) \to \hat{v}(\mathbf{k})\,,$$

$$\psi^*(y) \to \frac{\hat{v}(\mathbf{q})}{q^2 - (q_i + i\epsilon)^2}\,,$$

and set $\lambda = \tau/2\pi$. Of course, the volume of integration is three-dimensional, but in Eq. (8.54) one need not be so literal-minded as to associate x and y with one-dimensional vectors and dy with a one-dimensional volume. The solution to Eq. (8.69) is given immediately by Eq. (8.61). We find that

$$f(\mathbf{k}) = \pi\tau\hat{v}(\mathbf{q}_i)\hat{v}(\mathbf{k}) + \lambda \frac{\displaystyle\int \frac{\hat{v}(\mathbf{q})}{q^2 - (q_i + i\epsilon)^2}[\pi\tau\hat{v}(\mathbf{q}_i)\hat{v}(\mathbf{q})]\, d\mathbf{q}}{\displaystyle 1 - \lambda \int \frac{\hat{v}(\mathbf{q})}{q^2 - (q_i + i\epsilon)^2}\hat{v}(\mathbf{q})\, d\mathbf{q}} \hat{v}(\mathbf{k}),$$

which can be written as

$$\begin{aligned} f(\mathbf{k}) &= \pi\tau\hat{v}(\mathbf{q}_i)\hat{v}(\mathbf{k})[1 + \lambda I/(1 - \lambda I)] \\ &= \pi\tau\hat{v}(\mathbf{q}_i)\hat{v}(\mathbf{k})/(1 - \lambda I), \end{aligned}$$

where we define

$$I \equiv \int \frac{\hat{v}(\mathbf{q})\hat{v}(\mathbf{q})}{q^2 - (q_i + i\epsilon)^2} d\mathbf{q}.$$

According to the definition of $f^s(\mathbf{q}_f)$, we see that

$$f^s(\mathbf{q}_f) = \frac{\pi\tau\hat{v}(\mathbf{q}_i)\hat{v}(\mathbf{q}_f)}{1 - \tau I/2\pi}. \tag{8.70}$$

Thus given any function $v(\mathbf{r})$ [or $v(\mathbf{q})$] the problem is solved. One obvious choice for $v(\mathbf{r})$ is a simple Yukawa potential, $v(\mathbf{r}) = e^{-\mu r}/r$, with the corresponding Fourier transform given by Eq. (7.58′):

$$\hat{v}(\mathbf{q}) = \left(\frac{2}{\pi}\right)^{1/2} \frac{1}{q^2 + \mu^2}. \tag{8.71}$$

Since this depends only on q^2, we see that $f^s(\mathbf{q}_f)$ will depend only on the magnitude of $\mathbf{q}_f$ (and of $\mathbf{q}_i$). We can therefore consider f^s to be a function of E alone, and not of the scattering angle. For this choice of $\hat{v}(\mathbf{q})$, the integral I can be readily computed by the contour integral techniques of Chapter 6. One finds that

$$I = -\frac{2\pi}{\mu}\frac{1}{(q_i + i\mu)^2}, \tag{8.72}$$

where we have let ϵ go to zero after performing the integration. Putting Eqs. (8.71) and (8.72) into Eq. (8.70), we find that

$$f^s(E) = \frac{2\tau}{(q_i^2 + \mu^2)^2 + \dfrac{\tau}{\mu}(q_i - i\mu)^2}. \tag{8.73}$$

If we make the definitions

$$a_0 \equiv \frac{2}{\mu}\left(1 - \frac{\mu^3}{\tau}\right)^{-1}, \qquad r_0 \equiv \frac{1}{\mu}\left(1 + \frac{2\mu^3}{\tau}\right),$$

then

$$f^s(E) = \left[-\frac{1}{a_0} + \frac{1}{2}r_0 q_i^2 + \frac{1}{2\tau}q_i^4 - iq_i\right]^{-1}. \tag{8.74}$$

We see that f^s has no angular dependence; i.e., it represents pure s-wave ($l = 0$) scattering.

It can be shown that for any well-behaved short-range *local* potential, the low energy ($\mathbf{q}_i$ small) behavior of the s-wave scattering amplitude is

$$f^s = \left[-\frac{1}{a_0} + \frac{1}{2} r_0 q_i^2 + \cdots - i q_i \right]^{-1} ,$$

where a_0 and r_0 can be calculated from the specific form of the local potential. (See, for example, Blatt and Weisskopf, *Theoretical Nuclear Physics*.) This is known as the effective range form of f^s; a_0 and r_0 are called the *scattering length* and the *effective range* respectively. Thus we see that it is possible to find a non-local potential, which can be handled in closed form, whose associated scattering amplitude looks just like the low energy scattering amplitude for a local potential. We leave it as an exercise for the reader to show that the above potential can also give rise to a bound state (i.e., a negative eigenvalue of Schrödinger's equation) if g is sufficiently large.

Because of the possibility of accommodating a bound deuteron and also because it gives rise to a scattering amplitude of the effective range type, the separable potential of Eq. (8.68) (and its generalization to spin-dependent forces) has often been used to analyze the low energy properties of the nucleon–nucleon system. In recent years, potentials of this type have also been used in theoretical studies of three-particle systems, where their simple form is very useful in obtaining tractable expressions for three-particle scattering amplitudes and bound state wave functions.

8.6 GENERAL KERNELS OF FINITE RANK

Now that we have seen how the separable kernel leads to a closed-form solution to an integral equation, it is not difficult to generalize to kernels of *finite rank*. The only complication is that instead of being able to reduce the vector space equation to a single scalar equation as we did for separable kernels, for kernels of finite rank we will reduce the vector space equation to a *system* of scalar equations. These equations can be written in the form of a finite-dimensional *matrix* equation, which is why such kernels are said to be of finite rank (see Section 3.10).

We define a *kernel of finite rank* to be of the form

$$K_F f \equiv \sum_{n=1}^{N} \phi_n(\psi_n, f) , \tag{8.75}$$

where f is any vector in the Banach space, B, on which K_F acts. The inner product used here need not be that appropriate to a Hilbert space; all we mean by (ψ_n, f) is a *linear functional*, denoted symbolically by (ψ_n, f), which takes any $f \in B$ into a scalar. The set of all linear functionals on a Banach space, B, is also a Banach space, often referred to as the *dual space* of B. In the case of a Hilbert space, this space of linear functionals is isomorphic to the Hilbert space itself. For the space L_p, the Banach space of linear functionals can be identified isomorphically with the space L_q, where $1/p + 1/q = 1$.

It can readily be shown (see Problem 8.21) that if $f \in L_p$, and $g \in L_q$ and if $1/p + 1/q = 1$, then

$$(g, f) = \int g^* f \, dx$$

exists, and it is in terms of this relation that we shall interpret Eq. (8.75). If the reader wishes, he can view Eq. (8.75) as an equation in Hilbert space, where the symbol $(\, ,)$ has its usual interpretation as an inner product. However, it is by no means necessary to do so; the results which we will obtain below are more general.

Since we have already seen that the finite-dimensional Theorem 4.27 holds for separable kernels, it will be interesting to see if it also holds in this more general case. We therefore want to consider the adjoint transformation, $K_F^\dagger$, which is defined by

$$K_F^\dagger \tilde{f} = \sum_{n=1}^{N} \psi_n(\phi_n, \tilde{f}) \,. \tag{8.76}$$

In the light of our previous discussion of Banach spaces, it is clear that if K_F acts on the space L_p, transforming $f \in L_p$ into $K_F f \in L_p$, then $K_F^\dagger$ acts on the space L_q $(1/p + 1/q = 1)$, transforming $\tilde{f} \in L_q$ into $K_F^\dagger \tilde{f} \in L_q$. We may remark at this point that seemingly more general forms for K_F such as, for example,

$$K_F' f \equiv \sum_{m,\, n=1}^{N} k_{mn} \phi_m (\psi_n, f) \,,$$

where k_{mn} is a collection of constants, are included in our definition of Eq. (8.75) since ϕ_m and ψ_n are not assumed to be normalized in any particular way and N, the range of the sum, is completely arbitrary.

We now consider, in the spirit of Theorem 4.27, the two equations

$$f - \lambda K f = g \tag{8.77}$$

and

$$\tilde{h} - \lambda^* K^\dagger \tilde{h} = 0 \,, \tag{8.78}$$

where f and g are vectors in B and $\tilde{h}$ is a vector in the dual space of B. Using the definitions of Eqs. (8.75) and (8.76) respectively, we get

$$f = g + \lambda \sum_{n=1}^{N} \phi_n(\psi_n, f) \tag{8.79}$$

and

$$\tilde{h} - \lambda^* \sum_{n=1}^{N} \psi_n(\phi_n, \tilde{h}) = 0 \,. \tag{8.80}$$

Equation (8.79) tells us that we need to determine only the collection of constants,

$$a_n \equiv (\psi_n, f) \,,$$

to obtain a solution. Using Eq. (8.79), we find that

$$(\psi_m, f) = (\psi_m, g) + \lambda \sum_{n=1}^{N} (\psi_m, \phi_n)(\psi_n, f) .$$

Setting

$$\alpha_{mn} \equiv (\psi_m, \phi_n) , \qquad b_m \equiv (\psi_m, g) , \tag{8.81}$$

this becomes

$$a_m - \lambda \sum_{n=1}^{N} \alpha_{mn} a_n = b_m .$$

It is convenient to write this in matrix notation as

$$(I - \lambda A)a = b , \tag{8.82}$$

where

$$[A]_{mn} = \alpha_{mn} , \qquad [a]_n = a_n , \qquad [b]_n = b_n .$$

Similarly, if we set $\tilde{a}_n = (\phi_n, \tilde{h})$, the homogeneous adjoint equation is

$$\tilde{a}_m - \lambda^* \sum_{n=1}^{N} (\phi_m, \psi_n)\tilde{a}_n = 0 ,$$

which, according to Eq. (8.81), may be written as

$$\tilde{a}_m - \lambda^* \sum_{n=1}^{N} \alpha_{nm}^* \tilde{a}_n = 0 .$$

In matrix notation, this is just

$$(I - \lambda A)^\dagger \tilde{a} = 0 , \tag{8.83}$$

where $[\tilde{a}]_n = \tilde{a}_n$. Thus our basic equations, Eqs. (8.77) and (8.78), have been transformed into the matrix equations, Eqs. (8.82) and (8.83).

But Eqs. (8.82) and (8.83) are in just the form required by Theorem 4.27. We conclude, therefore, that Eq. (8.82) has a solution if and only if

$$(b, \tilde{a}) = 0 = \sum_{n=1}^{N} b_n^* \tilde{a}_n \tag{8.84}$$

for all $\tilde{a}$ which are solutions of the homogeneous adjoint equation. Similarly, Eq. (8.77) will have solutions only under these conditions. Using the above definitions of b_n and $\tilde{a}_n$, this becomes

$$\sum_{n=1}^{N} (g, \psi_n)(\phi_n, \tilde{h}) = 0 .$$

But by our definition of $K^\dagger$ [Eq. (8.76)], this is just

$$(g, K^\dagger \tilde{h}) = 0 .$$

If $\tilde{h} \not\equiv 0$, $\lambda^* K^\dagger \tilde{h} = \tilde{h}$, by Eq. (8.78), so a solution to Eq. (8.77) exists if we require that

$$(g, \tilde{h}) = 0 \tag{8.85}$$

for all $\tilde{h}$ satisfying Eq. (8.78). Note that if Eq. (8.83) has no solution then $[I - \lambda A]^{-1}$ exists and Eq. (8.82) has a *unique* solution. Thus, if Eq. (8.78) has no solution, then Eq. (8.77) has a unique solution. We have, therefore, the following result:

Theorem 8.1. If K_F is a linear transformation of finite rank, then the equation

$$f = g + \lambda K_F f$$

has a solution if and only if $(g, \tilde{h}_i) = 0$ for all i, where $\tilde{h}_i$ satisfies the homogeneous adjoint equation

$$\lambda^* K_F^\dagger \tilde{h}_i = \tilde{h}_i \,.$$

If the homogeneous adjoint equation has no solutions then the solution to the inhomogeneous equation is unique.

By an identical argument, the same result holds for $K_F^\dagger$.

Corollary. If K_F is a linear operator of finite rank, then the equation

$$\tilde{f} = \tilde{g} + \lambda^* K_F^\dagger \tilde{f}$$

has a solution if and only if $(\tilde{g}, h_i) = 0$, where h_i is any solution of the homogeneous equation

$$\lambda K_F h_i = h_i \,.$$

If the homogeneous equation has no solution, then the solution to the inhomogeneous adjoint equation is unique.

The next corollary is an immediate consequence of the finite-dimensional Theorem 4.26.

Corollary. The equations

$$\lambda K_F h_i = h_i \qquad \text{and} \qquad \lambda^* K_F^\dagger \tilde{h}_i = \tilde{h}_i$$

have the same number of linearly independent solutions.

Theorem 8.1 and its two corollaries are often referred to as the Fredholm alternative ("alternative" because it expresses on option—*either* the inhomogeneous linear equation, $(I - \lambda K)f = g$, has a unique solution *or* the homogeneous adjoint equation possesses at least one solution). They represent an extension of the theorems of Section 4.8 to an infinite-dimensional space, but only for operators of a rather special type. However, in the next section we shall see that this result can be extended to a much broader class of operators.

Before extending these results, let us remark that if a unique solution to Eq. (8.77) exists, then we can find it explicitly by using Eqs. (8.79) and (8.82).

Equation (8.82) tells us that

$$a = (I - \lambda A)^{-1}b\,,$$

or, in component form

$$a_n \equiv (\psi_n, f) \equiv \sum_{m=1}^{N} [(I - \lambda A)^{-1}]_{nm} b_m$$
$$= \sum_{m=1}^{N} [(I - \lambda A)^{-1}]_{nm} (\psi_m, g)\,.$$

Putting this result into Eq. (8.79), we find that

$$f = g + \lambda \sum_{n=1}^{N} \sum_{m=1}^{N} \phi_n [(I - \lambda A)^{-1}]_{nm} (\psi_m, g)\,, \tag{8.86}$$

which gives us the desired solution if the equation

$$\lambda^* K_F^\dagger \tilde{h} = \tilde{h}$$

has no solutions, or, equivalently, if the equation

$$(I - \lambda A)^\dagger \tilde{a} = 0$$

has no solutions. Thus the problem of solving Eq. (8.77) reduces to calculating the inverse of a given matrix of scalars. Equation (8.86) may be thought of as defining a new operator, G_{K_F}, which is the inverse of $(I - \lambda K_F)$. According to Eq. (8.86), we write

$$G_{K_F} \equiv I + \lambda R_{K_F} = (I - \lambda K_F)^{-1}\,, \tag{8.87}$$

where

$$R_{K_F} f \equiv \sum_{n=1}^{N} \sum_{m=1}^{N} [(I - \lambda A)^{-1}]_{mn} \phi_m (\psi_n, f)\,; \tag{8.88}$$

R_{K_F} is called the *resolvent* operator corresponding to K_F. It is clearly also an operator of finite rank. It exists whenever $(I - \lambda A)^{-1}$ exists, that is, whenever

$$\det (I - \lambda A) \neq 0\,.$$

Since $\det (I - \lambda A)$ is a polynomial of degree N in λ, and therefore has N (not necessarily distinct) roots, we see that R_{K_F} is a *meromorphic function* of λ—that is, it is analytic everywhere except for a finite number of poles. At these poles, R_{K_F} fails to exist, and we have a solution to the equation

$$\lambda K_F h = h\,,$$

or, equivalently, to the equation

$$\lambda^* K_F^\dagger \tilde{h} = \tilde{h}\,.$$

It is a worthwhile exercise to check that

$$(I - \lambda K_F)(I + \lambda R_{K_F}) = (I + \lambda R_{K_F})(I - \lambda K_F) = I\,.$$

As an example of a kernel of finite rank, consider the equation

$$f(x) = x + \lambda \int_0^\pi \frac{\sin(x + x')}{\pi} f(x')\, dx'$$
$$= x + \lambda \frac{\sin x}{\sqrt{\pi}} \int_0^\pi \frac{\cos x'}{\sqrt{\pi}} f(x')\, dx' + \lambda \frac{\cos x}{\sqrt{\pi}} \int_0^\pi \frac{\sin x'}{\sqrt{\pi}} f(x')\, dx' . \quad (8.89)$$

Calling

$$a_1 = \frac{1}{\sqrt{\pi}} \int_0^\pi \cos x' f(x')\, dx' , \qquad a_2 = \frac{1}{\sqrt{\pi}} \int_0^\pi \sin x' f(x')\, dx' ,$$

and

$$b_1 = \frac{1}{\sqrt{\pi}} \int_0^\pi x \cos x\, dx , \qquad b_2 = \frac{1}{\sqrt{\pi}} \int_0^\pi x \sin x\, dx ,$$

We find that Eq. (8.89) may be transformed to

$$a_1 = b_1 + \lambda \frac{1}{\pi} \int_0^\pi \cos^2 x\, dx\, a_2 , \qquad a_2 = b_2 + \lambda \frac{1}{\pi} \int_0^\pi \sin^2 x\, dx\, a_1 .$$

Thus in matrix notation

$$a = b + \lambda M a \qquad \text{or} \qquad (I - \lambda M) a = b ,$$

where

$$M = \begin{bmatrix} 0 & \frac{1}{2} \\ \frac{1}{2} & 0 \end{bmatrix}, \qquad a = \begin{bmatrix} a_1 \\ a_2 \end{bmatrix}, \qquad b = \begin{bmatrix} b_1 \\ b_2 \end{bmatrix} = \frac{1}{\sqrt{\pi}} \begin{bmatrix} -2 \\ \pi \end{bmatrix}.$$

The inverse matrix $(I - \lambda M)^{-1}$ is

$$\begin{bmatrix} 1 & -\lambda/2 \\ -\lambda/2 & 1 \end{bmatrix}^{-1} = \frac{1}{1 - \lambda^2/4} \begin{bmatrix} 1 & \lambda/2 \\ \lambda/2 & 1 \end{bmatrix}.$$

Hence

$$a = (I - \lambda M)^{-1} b = \frac{1}{1 - \lambda^2/4} \begin{bmatrix} 1 & \lambda/2 \\ \lambda/2 & 1 \end{bmatrix} \frac{1}{\sqrt{\pi}} \begin{bmatrix} -2 \\ \pi \end{bmatrix},$$

so

$$a_1 = \frac{2}{\sqrt{\pi}} \frac{(\pi\lambda - 4)}{4 - \lambda^2} , \qquad a_2 = \frac{4}{\sqrt{\pi}} \frac{\pi - \lambda}{4 - \lambda^2} .$$

Thus

$$f(x) = x + 2\lambda \frac{(\lambda - 4/\pi)}{4 - \lambda^2} \sin x + 4\lambda \frac{(1 - \lambda/\pi)}{4 - \lambda^2} \cos x . \quad (8.90)$$

Clearly, when $\lambda = \pm 2$, this solution becomes infinite, corresponding to the existence of solutions to the homogeneous equation for these values.

The Neumann series for $f(x)$ is

$$f(x) = x + \frac{\lambda}{\pi}\int_0^\pi dx' \sin(x + x')x'$$
$$+ \frac{\lambda^2}{\pi^2}\int_0^\pi dx' \int_0^\pi dx'' \sin(x + x') \sin(x' + x'')x'' + \cdots .$$

According to our previously developed convergence criteria, this series will certainly converge when

$$\frac{|\lambda|}{\pi}\left[\int_0^\pi \int_0^\pi \sin^2(x + x')\,dx\,dx'\right]^{1/2} < 1 .$$

Performing the integrations, we find the requirement that $\lambda < \sqrt{2}$. The first few terms of the Neumann series, written out explicitly, are

$$f(x) = x + \lambda[\cos x - (2/\pi)\sin x] + \lambda^2[(1/2)\sin x - (1/\pi)\cos x] + \cdots \quad (8.91)$$

This result agrees with the first few terms in the power series expansion of Eq. (8.90). Since the power series converges whenever $\lambda < 2$, it is clear that the Neumann series must also converge if $\lambda < 2$. Thus we see that the estimate on the norm of an integral operator given by the integrated square of its kernel provides only a lower bound on the radius of convergence of the Neumann series. In fact, even if we knew the exact norm of the operator, the requirement $||K|| < 1$ only tells us that the Neumann series *will* converge if $||K|| < 1$, not that it will *fail* to converge if $||K|| > 1$.

When $\lambda = \pm 2$, the two eigenvectors of

$$f(x) = \lambda \int_0^\pi \frac{\sin(x + x')}{\pi} f(x')\,dx'$$

turn out to be $f_+(x) = A_+ (\sin x + \cos x)$ and $f_-(x) = A_- (\sin x - \cos x)$, corresponding respectively to $\lambda_+ = 2$ and $\lambda_- = -2$. If we normalize them to unity, we find that

$$f_+(x) = \sqrt{\frac{2}{\pi}}\sin\left(x + \frac{\pi}{4}\right), \qquad f_-(x) = \sqrt{\frac{2}{\pi}}\sin\left(x - \frac{\pi}{4}\right).$$

Note that these two solutions are orthogonal, as they should be, since $k(x, y) = (1/\pi)\sin(x + y)$ is a real symmetric kernel.

Figure 8.1 shows, for the case $\lambda = 1$, the exact solution compared to the iterates:

$$f_0(x) = x ,$$
$$f_1(x) = x + \lambda[\cos x - (2/\pi)\sin x] ,$$
$$f_2(x) = x + \lambda[\cos x - (2/\pi)\sin x] + \lambda^2[(1/2)\sin x - (1/\pi)\cos x] .$$

It is seen that $f_2(x)$ is a very good approximation to the exact solution, even though $||K_F|| \cong \frac{1}{2}$ when $\lambda = 1$.

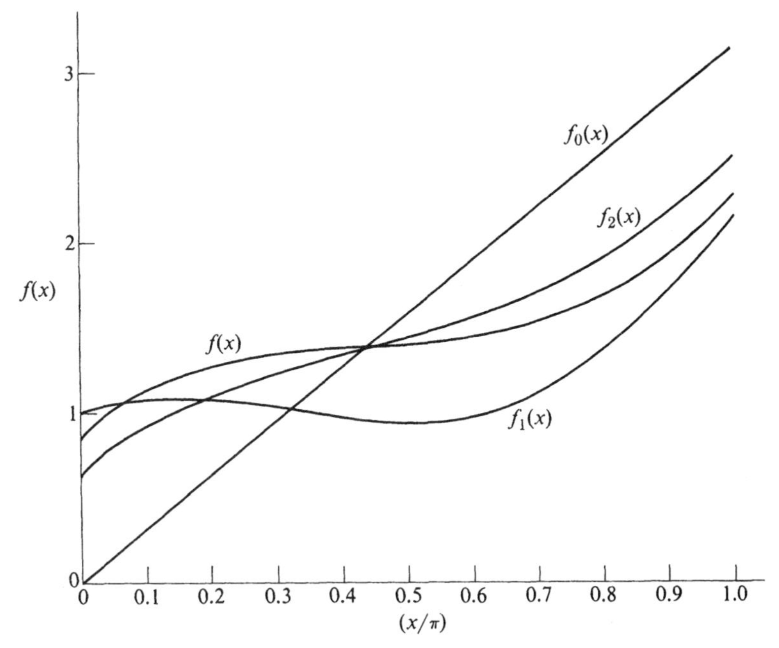

Fig. 8.1 The exact solution, $f(x)$, to Eq. (8.89) of the text, compared with the first three Neumann iterates, $f_0(x)$, $f_1(x)$, and $f_2(x)$.

8.7 COMPLETELY CONTINUOUS OPERATORS

Although the kernels of finite rank which we have discussed thus far lead to integral equations which can be solved in closed form, they are seldom *directly* relevant to physical problems. The integral equations which arise in physics (for example, in Schrödinger's integral equation and its relativistic generalizations) usually have kernels, $k(x, y)$, in which the x- and y-dependence cannot be unravelled into sums of the form of Eq. (8.75).

The reason that we have nevertheless spent so much time talking about these transformations of finite rank is explained by the following theorem.

> **Theorem 8.2.** Any linear transformation of Hilbert space whose kernel is continuous and square integrable as a function of two variables can be approximated arbitrarily closely in norm by a linear transformation of finite rank.

Proof. The simplest case is when the functions in the Hilbert space are defined on a finite interval. Then the two-dimensional Weierstrass theorem tells us that $k(x, y)$ can be approximated *uniformly* by a polynomial; in other words, given any δ there exists a polynomial $P_N(x, y)$, where N depends on δ such that

$$|k(x, y) - P_N(x, y)| < \delta$$

for all x, y. Note that a kernel in polynomial form is a kernel of finite rank since

$$P_N(x, y) = \sum_{m,\, n=0}^{N} C_{mn} x^m y^n$$

for all x, y. Hence

$$\int_a^b dx \int_a^b dy\, |k(x, y) - P_N(x, y)| \leq [\delta(b - a)]^2 .$$

Thus we can find an N which will provide a δ to make

$$||K - K_N|| < \epsilon$$

for any preassigned ϵ, where K_N is the transformation whose kernel is $P_N(x, y)$. Now if the Hilbert space is defined over an infinite interval, we need only truncate the kernel by setting it equal to zero for all x, y which lie outside some set, B, of finite area. Call this kernel k_B and the related linear transformation K_B. Then if we let $C(B)$ denote the complement of B in the xy-plane,

$$\int_{-\infty}^{\infty} dx \int_{-\infty}^{\infty} dy\, |k(x, y) - k_B(x, y)|^2 = \int_{C(B)} dx \int dy\, |k(x, y)|^2 .$$

But $k(x, y)$ is square integrable, so it must be possible to choose a set of points outside of which the integral of $|k|^2$ is arbitrarily small. Thus we can choose B so that

$$||K - K_B|| < \epsilon/2 .$$

But by previous reasoning, we can find a polynomial kernel $k_N(x, y)$ with related linear transformation K_N such that

$$||K_B - K_N|| < \epsilon/2 .$$

Thus

$$||K - K_N|| \leq ||K - K_B|| + ||K_B - K_N|| < \epsilon$$

for a suitable choice of N, which is what we wanted to show.

This theorem can easily be weakened to include the case of piecewise continuous kernels. Furthermore, the theorem can be extended to the class of square-integrable kernels with no continuity or boundedness requirements on $k(x, y)$ other than those implied by square integrability. The only difficulty in this more general case will be the fact that the kernel can become infinite at some points. For example, in the space $L_2[0, 1]$, the linear operator whose kernel is

$$k(x, y) = 1/\sqrt{x + y} \tag{8.92}$$

is an admissible operator because $k(x, y)$ is square integrable. However, at the point $x = 0$, $y = 0$ the kernel becomes infinite. This difficulty is easily overcome. We truncate the kernel by setting it equal to zero whenever $|k(x, y)| \geq$

L; where L is a number which may be made as large as we please. The truncated kernel, $k_L(x, y)$, can be approximated in the manner of Theorem 8.2. However, since $k(x, y)$ is square integrable, it must be possible to choose L sufficiently large so that we make negligible error in using $k_L(x, y)$ instead of $k(x, y)$; that is, the number

$$I_L = \int dx \int dy \, |k(x, y) - k_L(x, y)|^2$$

can be made as small as we please. I_L can be written as

$$I_L = \iint_{\mathscr{D}_L} |k(x, y)|^2 \, dx \, dy \, ,$$

where $\mathscr{D}_L$ is the set of points on which $|k(x, y)| \geq L$. Now clearly it must be possible to choose L sufficiently large so that the area of $\mathscr{D}_L$ is as small as we please [otherwise $k(x, y)$ would not be square integrable]. Also, it is clear that I_L must be finite for any $\mathscr{D}_L$ or again $k(x, y)$ would not be square integrable. There remains only the possibility that although the area of $\mathscr{D}_L$ is as small as we please, the integral of $|k(x, y)|^2$ over this arbitrarily small set is finite. Now it is possible to imagine objects which are singular in just such a way that their integral over an arbitrarily small set is finite; the δ-function, for example, has this property. However, in the Lebesgue theory of integration (and also in the Riemann theory) such objects are excluded from the class of admissible *functions*, and therefore we exclude them from the vector space of square-integrable functions. The kernel mentioned above [Eq. (8.92)], although singular at the origin, still has the property that the integral of $|k(x, y)|^2$ over a region of arbitrarily small area containing the origin tends toward zero as the area of the region tends toward zero. Thus *Theorem 8.2 can be extended to the space of square-integrable functions.*

Because of the practical importance of square-integrable kernels (see, for example, Section 8.3) this result is extremely useful, and in Chapter 9 we will develop its consequences in considerable detail. However, for many purposes it is best to concentrate not on the property of finite rank approximation, but rather on a more subtle property which can be inferred from it (and which is, in fact, equivalent to it). It is with this in mind that we introduce the concept of a *completely continuous* operator.

Definition 8.8. A linear operator on a normed vector space is said to be *completely continuous* if it transforms *any* bounded sequence of vectors (i.e., vectors whose norms are all less than some fixed constant) into a sequence containing a subsequence which converges in norm.

The use of the terminology *completely continuous* is fairly clear from the definition. An *ordinary* continuous (i.e., bounded) transformation maintains the status quo regarding convergent sequences; if $\{f_n\}$ converges, so does $\{Kf_n\}$ if K is continuous. But a *completely* continuous transform *forces* sequences to converge, even though they may not have converged originally. For the definition of complete continuity to be at all interesting, it must be possible to choose an infinite sequence of vectors which has a subsequence that does not converge. In a finite-dimensional space, it is impossible to find such a sequence because of the existence of the Bolzano-Weierstrass theorem* for real numbers. Thus the definition is empty on a finite-dimensional space; or if one likes, one may say that on a finite-dimensional vector space all operators are completely continuous.

Let us obtain some of the simpler properties of such transformations. Obviously, the sum of any finite number of completely continuous transformations is completely continuous. We can also show that the product of a completely continuous transformation and a bounded transformation is completely continuous. Suppose that K is completely continuous and let B be a bounded transformation. Consider KB acting on $\{f_n\}$, where $||f_n|| < C$ for all n. Then K acts on the sequence $\{Bf_n\}$, which is bounded because B is a bounded operator. Since K is completely continuous, there must be a subsequence of $\{KBf_n\}$ which converges in norm to an element of the space, so KB is completely continuous. Now consider BK acting on $\{f_n\}$. Since K is completely continuous, $\{Kf_n\}$ possesses a subsequence which converges in norm. Since B is bounded, it is continuous, so $\{BKf_n\}$ has a trivially related convergent subsequence.

We now show that the class of such transformations is not empty. Consider the space of square-integrable functions and take the separable transformation K_s with kernel

$$k_s(x, y) = \phi(x)\psi^*(y) ,$$

along with any sequence $\{f_n\}$, $||f_n|| < C$ for all n. Now

$$K_s f_n = \phi(x)(\psi, f_n) .$$

Since $|(\psi, f_n)| \leq ||\psi|| \cdot ||f_n||$, the sequence of complex numbers (ψ, f_n) is bounded in the complex plane and therefore possesses a convergent subsequence (Bolzano-Weierstrass theorem). Call the limit of this subsequence l, and consider

$$||K_s f_{n_k} - l\phi|| = ||\phi[(\psi, f_{n_k}) - l]|| \leq ||\phi|| \cdot |(\psi, f_{n_k}) - l| ,$$

where n_k $(k = 1, 2, 3, \cdots)$ defines the convergent subsequence. Since $|(\psi, f_{n_k}) - l|$ tends toward zero as n_k increases, we have

$$K_s f_{n_k} \to l\phi$$

in the norm of the space, so K_s is completely continuous. Hence any operator of finite rank is completely continuous, since it is a finite sum of separable operators.

Of course, there are also many operators which are not completely continuous. Consider $D_x = -id/dx$ acting on the space of square-integrable functions on [0, 1]. We know that $\{f_n = e^{inx}\}$ is a bounded sequence, but $\{Df_n =$

---

* For our purposes, the Bolzano-Weierstrass theorem may be stated as follows: From every bounded sequence of points in n-dimensional Euclidean space, it is possible to select a convergent subsequence.

$ne^{inx}\}$ does not contain a convergent subsequence. In fact, even the identity operator is not completely continuous.

A particularly interesting property of the class of completely continuous operators is that the limit of any sequence of completely continuous operators is itself a completely continuous operator.

Theorem 8.3. If $||K_i - K|| \to 0$ as $i \to \infty$, and if K_i is a completely continuous transformation for all i, then K is also completely continuous.

Proof. Let $\{f_n\}$ be any bounded sequence of functions. Then we can select a subsequence, $\{f_{n_1}^{(1)}\}$, such that $\{K_1 f_{n_1}^{(1)}\}$ converges in norm, since K_1 is completely continuous. Now consider the bounded sequence $\{f_{n_1}^{(1)}\}$. From this we can select a subsequence, $\{f_{n_2}^{(2)}\}$, such that $\{K_2 f_{n_2}^{(2)}\}$ converges in norm. Clearly, $\{K_1 f_{n_2}^{(2)}\}$ will still converge in norm. We can continue this process indefinitely, thus obtaining finally a sequence $\{\phi_m\}$ such that $\{K_i \phi_m\}$ converges in norm for all K_i. Now consider

$$K\phi_n - K\phi_m = K\phi_n - K_i\phi_n + K_i\phi_n - K_i\phi_m + K_i\phi_m - K\phi_m .$$

By the triangle inequality, we find that

$$\begin{aligned} ||K\phi_n - K\phi_m|| &\le ||K\phi_n - K_i\phi_n|| + ||K_i\phi_n - K_i\phi_m|| + ||K_i\phi_m - K\phi_m|| \\ &\le ||K - K_i|| \cdot ||\phi_n|| + ||K_i\phi_n - K_i\phi_m|| + ||K - K_i|| \cdot ||\phi_m|| . \end{aligned}$$

Since the ϕ_n are bounded,

$$||K\phi_n - K\phi_m|| \le C||K - K_i|| + ||K_i\phi_n - K_i\phi_m|| .$$

Because $||K - K_i|| \to 0$, we can pick some i sufficiently large so that

$$||K - K_i|| < \epsilon/(2C) .$$

Then when $m, n > N$ the second term on the right above can be made smaller than $\epsilon/2$. Thus for $m, n > N$, we have

$$||K\phi_n - K\phi_m|| < \epsilon .$$

Hence the sequence $\{K\phi_n\}$ is convergent, and the proof is complete.

As we remarked above, any transformation with a square-integrable kernel can be approximated arbitrarily closely in norm by a kernel of finite rank. Hence such transformations are completely continuous. This theorem underlines the fact that the projection operators, P_n, defined in Section 8.2 do not converge in norm to I, the identity operator. If they did, then since P_n is of finite rank and hence completely continuous, Theorem 8.3 would say that the identity operator is completely continuous, which it is not.

The above results show that any operator which can be approximated in norm by a sequence of operators of finite rank is completely continuous. We now want to show that the converse is also true, namely, any completely continuous operator can be approximated in norm arbitrarily closely by a sequence of operators of finite rank. In order that we may make use of a complete orthonormal basis, we will prove this result only in Hilbert space.

Patterning our work on the techniques of Chapter 5, let us imagine that we are dealing with an integral transformation whose kernel is $k(x, y)$. Let $\{\phi_n\}$ be an orthonormal basis in L_2. We write

$$k(x, y) = \sum_{n, m=1}^{\infty} (\phi_n \phi_m^*, k) \, \phi_n(x) \, \phi_m^*(y) \, ,$$

in the sense of mean convergence. Here we make use of the fact that the collection of all products $\phi_n(x)\phi_m(y)$ [or $\phi_n(x)\phi_m^*(y)$] forms a complete orthonormal basis for the Hilbert space of square-integrable functions of two variables to which $k(x, y)$ belongs. By truncating the sum at some large N, we obtain a kernel of finite rank

$$k_N(x, y) = \sum_{n, m=1}^{N} (\phi_n \phi_m^*, k) \, \phi_n(x) \, \phi_m^*(y) \, .$$

One would hope, on the basis of experience with Fourier series, that this will approximate $k(x, y)$ arbitrarily closely in norm for N sufficiently large. The related transformation K_N is defined by

$$K_N f = \sum_{n, m=1}^{N} \phi_n(x) \, (\phi_n \phi_m^*, k) \, (\phi_m, f) \, . \qquad (8.93)$$

We can abbreviate this by writing K_N as (see Problem 8.26)

$$K_N = P_N K P_N \, ,$$

where P_N is the projection operator defined by

$$P_N f \equiv \sum_{n=1}^{N} (\phi_n, f) \phi_n \, . \qquad (8.94)$$

With the above discussion as background, we now state the final theorem of this chapter in completely generality.

Theorem 8.4. Let $\{\phi_n\}$ be an orthonormal basis* for a Hilbert space H, and let K be a completely continuous operator on H. If we denote by P_N the operator that projects onto the subspace of H spanned by $\{\phi_1, \phi_2, \cdots, \phi_N\}$ and define $K_N \equiv P_N K P_N$, then the operator K_N tends in norm to K as N tends to infinity.

Proof. Consider

$$\begin{aligned} K - K_N &= K - P_N K P_N = K - P_N K + P_N K - P_N K P_N \\ &= (I - P_N) K + P_N K (I - P_N) \, . \end{aligned} \qquad (8.95)$$

---

* We are assuming here the existence of a *countable* basis for H. A Hilbert space with this property is called a *separable* Hilbert space; most, but not all, spaces which occur in physics are of this type. The proof of Theorem 8.4 can be extended to nonseparable Hilbert spaces.

If it were true that P_N tended toward I in operator norm, then the triangle inequality applied to Eq. (8.95) would complete the proof, which would then hold for *any bounded* operator. However, as we have already seen, this is *not* the case. The result will be proved by contradiction. The basic idea is that if the theorem is *not* true, then there must exist some real number a, such that for each n we can find a vector f_n, which satisfies

$$||(K - K_n)f_n|| > a||f_n||\,, \tag{8.96}$$

where a is independent of n and $a > 0$. If there were no such number a, then according to Definition 8.7, K_n would tend toward K in norm. If we divide both sides of Eq. (8.96) by $||f_n||$, we may say that if the theorem is *not* true then there exists a sequence of functions $\{f_n\colon ||f_n|| = 1\}$ such that

$$||(K - K_n)f_n|| > a \tag{8.97}$$

for all n. Assuming that K is completely continuous, we will show that Eq. (8.97) does not hold, so the theorem must be true. It is important to appreciate the fact that it is *not* enough to show that for any fixed $f \in H$, $||(K - K_n)f|| \to 0$. The triangle inequality applied to Eq. (8.95) would prove this.

We begin our proof by using the triangle inequality together with the relation $||P_N|| \leq 1$, to obtain from Eq. (8.95)

$$\begin{aligned} ||(K - K_N)f_N|| &\leq ||(I - P_N)Kf_N|| + ||P_N K(I - P_N)f_N|| \\ &\leq ||(I - P_N)Kf_N|| + ||K(I - P_N)f_N||\,. \end{aligned} \tag{8.98}$$

We must now show that both $||(I - P_N)Kf_N||$ and $||K(I - P_N)f_N||$ become small as N increases. It is at this point that the complete continuity of K becomes of central importance. It enables us to choose from the sequence $\{f_N\}$ a subsequence $\{\tilde{g}_n\}$ such that $\{K\tilde{g}_n\}$ converges to a limit, say $\xi \in H$. Consider also the sequence $\{(I - P_n)\tilde{g}_n\}$, which is a bounded sequence since $||I - P_n|| = 1$ and $||\tilde{g}_n|| = 1$. Because K is completely continuous, we can choose a subsequence $\{(I - P_m)g_m\}$, such that $\{K(I - P_m)g_m\}$ converges to a limit, say $\eta \in H$. Note that $\{Kg_m\}$ will converge to ξ since $\{K\tilde{g}_n\}$ converges to ξ and $\{g_m\}$ is a subsequence of $\{\tilde{g}_n\}$. In order to avoid confusion over notation, let us remind the reader that each g_m is equal to some element of our original collection $\{f_N\}$.

The meat of the proof is now behind us; only a few estimates of $||(I - P_m)Kg_m||$ and $||K(I - P_m)g_m||$ remain to be made. Consider first $||(I - P_m)Kg_m||$. We write

$$(I - P_m)Kg_m \equiv (I - P_m)(Kg_m - \xi) + (I - P_m)\xi\,,$$

where ξ is a fixed vector, and using the triangle inequality, we find that

$$\begin{aligned} ||(I - P_m)Kg_m|| &\leq ||(I - P_m)(Kg_m - \xi)|| + ||(I - P_m)\xi|| \\ &\leq ||Kg_m - \xi|| + ||\xi - P_m\xi||\,, \end{aligned}$$

since $||I - P_m|| = 1$. But by construction, Kg_m tends in norm toward ξ, and since ξ is fixed, $P_m\xi$ tends toward ξ in norm. Thus

$$||(I - P_m)Kg_m|| \to 0 \tag{8.99}$$

as $m \to \infty$.

Finally, we discuss $||K(I-P_m)g_m||$. By construction, as $m\to\infty$, $K(I-P_m)g_m$ tends in norm toward η. We will show that $||\eta||=0$. Consider

$$||\eta||^2 = (\eta, \eta) = \lim_{m\to\infty} (\eta, K(I-P_m)g_m) = \lim_{m\to\infty} ((I-P_m)K^\dagger\eta, g_m) ,$$

where we have used that fact that P_m is self-adjoint (regarding the second equality in the above equation; see Problem 8.21a). But

$$|((I-P_m)K^\dagger\eta, g_m)| \leq ||(I-P_m)K^\dagger\eta|| \, ||g_m|| ,$$

by Schwartz's inequality, so using $||g_m|| = 1$, we have

$$||\eta||^2 \leq \lim_{m\to\infty} ||(I-P_m)K^\dagger\eta|| .$$

Since $K^\dagger\eta$ is a fixed vector, $||(I-P_m)K^\dagger\eta||\to 0$ as $m\to\infty$, so $||\eta||=0$. Thus η is the zero vector, and since $K(I-P_m)g_m$ tends in norm toward η, we may write

$$||K(I-P_m)g_m|| \to 0 \tag{8.100}$$

as $m\to\infty$. Putting the results of Eqs. (8.99) and (8.100) into Eq. (8.98), we see that

$$||(K-K_m)g_m|| \to 0$$

as $m\to\infty$. Since the g_m form a subset of our original collection of f_N, *all* of which were assumed to satisfy $||(K-K_N)f_N|| > a$, where a is strictly positive, we see that we have reached the desired contradiction. Thus the theorem is proved.

This theorem suggests a very attractive method for solving equations involving completely continuous transformations. We just split any such transformation into a "small" operator plus an operator of finite rank. The former will yield to simple iterative techniques while the latter can be handled as an ordinary matrix, as we have seen before. In practice, many problems in physics and engineering must be handled in this way, since the operators are neither "small" enough for a simple Born approximation to be valid, nor simple enough to fall into the finite-rank category. The main problem that remains is to develop a method whereby the two processes, iteration and finite-rank approximation, can be combined. It is to this matter that Chapter 9 is primarily devoted.

PROBLEMS

1. Prove that the space C of continuous functions on $[a, b]$ with norm given by

$$||f|| = \max_{a\leq x\leq b} |f(x)|$$

for $f\in C$ is a Banach space, i. e., show that it is complete with respect to the above norm.

2. *Hölder's Inequality*

a) Show that the function $f(z) = z^\alpha - \alpha z - \beta$ takes on a maximum value at $z = 1$ if $\alpha < 1$ and that if $\beta = 1 - \alpha$, then this maximum value is zero. Thus $z^\alpha \leq \alpha z + \beta$ with $\alpha < 1$, $\beta = 1 - \alpha$.

b) By making use of the change of variables $z = x/y$, show that $x^\alpha y^\beta \leq \alpha x + \beta y$, where again $\alpha \leq 1$ and $\beta = 1 - \alpha$.

c) Suppose now that $|f(x)|^p$ and $|g(x)|^q$ are integrable functions with $1/p + 1/q = 1$. Use the result of part (b) to show that

$$|fg| \leq \frac{1}{p}|f|^p + \frac{1}{q}|g|^q,$$

so that $|fg|$ is also integrable.

d) Show that if $|f(x)|^p$ and $|g(x)|^q$ are integrable, then

$$\left|\int f(x)g(x)\,dx\right| \leq \left[\int |f(x)|^p\,dx\right]^{1/p}\left[\int |g(x)|^q\,dx\right]^{1/q}.$$

Hint: Apply part (*c*) to the functions

$$F(x) = f(x)\Big/\left[\int |f(x)|^p\,dx\right]^{1/p} \quad \text{and} \quad G(x) = g(x)\Big/\left[\int |g(x)|^q\,dx\right]^{1/q}.$$

This result generalizes Schwartz's inequality of Hilbert space and is known as Hölder's inequality.

3. *Minkowski's Inequality*

a) Show that $|f(x) + g(x)|^p \leq 2^p[|f(x)|^p + |g(x)|^p]$, and hence that if f and g belong to L_p (the space of p-integrable functions), then so does $f + g$. Thus L_p is indeed a vector space.

b) Show that for $p > 1$,

$$\int |f(x) + g(x)|^p\,dx \leq \int |f(x)|\,|f(x) + g(x)|^{p-1}\,dx + \int |g(x)|\,|f(x) + g(x)|^{p-1}\,dx,$$

and that $|f + g|^{p-1}$ belongs to L_q if f and g belong to L_p, where $1/p + 1/q = 1$.

c) Apply Hölder's inequality (see Problem 8.2) to each of the two integrals on the right-hand side of the inequality in part (*b*), and hence show that

$$\left[\int |f(x) + g(x)|^p\,dx\right]^{1/p} \leq \left[\int |f(x)|^p\,dx\right]^{1/p} + \left[\int |g(x)|^p\,dx\right]^{1/p}.$$

This is Minkowski's inequality. We see therefore that $[\int |f(x)|^p\,dx]^{1/p}$ does indeed define a norm for the space L_p; Minkowski's inequality is just a statement of the triangle inequality which is the essential ingredient in the definition of a norm.

4. Solve the equations

a) $f(x) = 1 + \lambda \int_0^\infty e^{-(x+y)} f(y)\,dy$ b) $f(x) = 1 + \lambda \int_0^x f(y)\,dy$

by iteration. For what values of λ will the iterations converge?

5. Multiplying both sides of Eq. (8.27) by $V^{1/2}(\mathbf{r})$ [assume $V(\mathbf{r})$ to be strictly positive], show that if we define

$$\tilde{\psi}(\mathbf{r}) = V^{1/2}(\mathbf{r})\psi(\mathbf{r}), \qquad \tilde{\xi}(\mathbf{r}) = V^{1/2}(\mathbf{r})\xi(\mathbf{r})$$

and

$$\tilde{k}(\mathbf{r}, \mathbf{r}', E) = V^{1/2}(\mathbf{r})G_0(\mathbf{r}, \mathbf{r}', E)V^{1/2}(\mathbf{r}') \,,$$

then

$$\tilde{\psi}(\mathbf{r}) = \tilde{\xi}(\mathbf{r}) + \int \tilde{k}(\mathbf{r}, \mathbf{r}', E)\tilde{\psi}(\mathbf{r}')\, d\mathbf{r}' \,.$$

Calling $\tilde{K}$ the integral operator whose kernel is $\tilde{k}(\mathbf{r}, \mathbf{r}', E)$, show that the norm of $\tilde{K}$ is the same as the norm of the kernel of Eq. (8.27) obtained by the weight-function method. Show also that if we iterate the equation for $\tilde{\psi}(\mathbf{r})$, the object $V^{1/2}(\mathbf{r})$ never actually appears in the calculation—the ordinary Born series for $\psi(\mathbf{r})$ is duplicated, apart from a multiplicative factor of $V^{1/2}(\mathbf{r})$.

6. Estimate the norm of the kernel of the Lippmann-Schwinger equation when $V(\mathbf{r}) = V_0 e^{-\mu r}$. For what values of V_0 will the Born series converge according to this estimate.

7. a) Suppose that we have an attractive δ-function potential in one dimension, $V(x) = -V_0\delta(x)$. Using the result of Problem 7.10, show that this potential will support one and only one bound state, no matter how large V_0 may be. Determine the binding energy and wave function of this state.

   b) Suppose that particles of positive energy E are incident on a δ-function potential, $V(x) = V_0\delta(x)$, in one dimension. What fraction of the particles is transmitted through the potential barrier and what fraction is reflected?

8. a) Using the result of Problem 7.10, show that if we have a potential $V(x)$ in one dimension which vanishes outside the interval $[-a/2, a/2]$, then, if we denote the wave function in the region $[-\infty, -a/2]$ by $\psi_1(x)$, the wave function in the region $[-a/2, a/2]$ by $\psi_2(x)$ and the wave function in $[a/2, \infty]$ by $\psi_3(x)$, then, calling $E = \hbar^2k^2/2m$ where $E > 0$ is the energy of the particle and m is its mass,

$$\psi_1(x) = e^{ikx} - \frac{im}{\hbar^2 k} e^{-ikx} \int_{-a/2}^{a/2} e^{ikx'} V(x')\psi_2(x')\, dx' \,,$$

$$\psi_3(x) = e^{ikx}\left[1 - \frac{im}{\hbar^2 k}\int_{-a/2}^{a/2} e^{-ikx'}\, V(x')\, \psi_2(x')\, dx'\right],$$

   and

$$\psi_2(x) = e^{ikx} - \frac{im}{\hbar^2 k}\int_{-a/2}^{a/2} e^{ik|x-x'|} V(x')\psi_2(x')\, dx' \,.$$

   Thus we see that in order to determine the wave function in the region where the potential doesn't act, we need only know the wave function in the region where the potential does act.

   b) In part (a) above, assume $V(x) = V_0 =$ constant in the region of interest. Then

$$\psi_2(x) = e^{ikx} - \frac{imV_0}{\hbar^2 k}\int_{-a/2}^{a/2} e^{ik|x-x'|}\psi_2(x')\, dx' \,.$$

   Show that the nth term in the iteration of this equation is less than $(mV_0a/\hbar^2k)^{n-1}$ in magnitude; so the iteration will converge if $(mV_0a/\hbar^2k) < 1$. Since $\hbar k =$

$p = mv$, this is equivalent to $V_0a/\hbar v < 1$. This result also holds for three-dimensional problems, although it is much more difficult to prove in that case. Note that the Hilbert space techniques used in the text give estimates for the convergence of the Born series in the three-dimensional case which are independent of the energy of the particle, whereas the result $V_0a/\hbar v < 1$ says that for *any* potential strength the Born series will converge if the energy of the particle is *sufficiently large*.

9. Using the weight-function method, show that in one dimension the iteration of the Lippmann-Schwinger equation will converge if

$$\frac{1}{\hbar v}\int_{-\infty}^{\infty} V(x)\, dx < 1 ,$$

where $V(x)$ is the potential. Note that if the range of $V(x)$ is a, then

$$\int_{-\infty}^{\infty} V(x)\, dx = \bar{V}a ,$$

where $\bar{V}$ is a mean potential strength. Thus this result is a simple generalization of the result of Problem 8(b). In that problem, the basic physical ideas are clearer, since there we can sharply delineate a region in which the potential acts and a region in which it does not act. The weight function becomes necessary because of the fact that most practical potentials have a nonvanishing magnitude at *all* points of space, although at *most* points the magnitude is small.

10. In Problem 8(a), assume that $V(x) = V_0 = \text{const}$, with V_0 small in the sense of Problem 8(b). Calculate ψ_1 and ψ_3 using the lowest-order approximation to ψ_2. What is the transmission probability in this approximation? What is the probability of reflection? Note that these two probabilities do not add up to 1. Why not? (For exact answers, see L. I. Schiff, *Quantum Mechanics*, McGraw-Hill, second edition, 1955, Section 17.)

11. Show that if $V(\mathbf{r})$ changes sign as $\mathbf{r}$ varies, then the appropriate modification of $\rho(\mathbf{r})$ in the weight-function method is to write $d\rho(\mathbf{r}) = |V(\mathbf{r})|d\mathbf{r}$. Obtain an estimate on the norm of the Lippmann-Schwinger integral operator using this weight function.

12. a) Making use of the definition $\psi(x) = e^{ikx}\phi(x)$ in the one-dimensional Lippmann-Schwinger equation, show that $\phi(x)$ satisfies the integral equation

$$\phi(x) = 1 - \frac{im}{h^2k}\int_{-\infty}^{x} V(x')\phi(x')\, dx' - \frac{im}{h^2k}e^{-ikx}\int_{x}^{\infty} e^{2ikx'}V(x')\phi(x')\, dx' .$$

b) Make some plausible arguments to suggest that the last term on the right-hand side of the above equation is negligible if k is large (i.e., if the energy of the incident particle is large).

c) Neglecting this term, show that the resulting $\phi(x)$, which we will call $\phi_E(x)$ satisfies the differential equation

$$\frac{d\phi_E(x)}{dx} = -\frac{im}{h^2k}V(x)\phi_E(x) ,$$

with the boundary condition $\phi_E(-\infty) = 1$. Hence show that

$$\phi_E(x) = \exp\left\{-\frac{im}{h^2k}\int_{-\infty}^{x} V(x')\, dx'\right\}.$$

This approximation is referred to as the *eikonal* approximation; its generalization to three dimensions has proved very useful in the analysis of certain high-energy scattering problems.

d) Returning to parts (a) and (b), show that for the case $V(x) = V_0$ for $x \in [-a/2, a/2]$, $V(x) = 0$ otherwise, the last term of the equation in part (a) really is small as $k \to \infty$. Do this by replacing $\phi(x)$ in the integral by the approximate $\phi_E(x)$ given by part (c). The resulting integral is simple; show that this term is less in magnitude than

$$\frac{1}{2}\left|\frac{E}{V_0} - \frac{1}{4}\right|^{-1},$$

so that for $E \gg V_0$ the term is small.

13. Consider the differential equation

$$\frac{d^2x(t)}{dt^2} + 2\gamma\frac{dx(t)}{dt} + \omega_0^2 x(t) = 0$$

for a damped harmonic oscillator.

a) Show that if $x(0) = x_0$ and $\dot{x}(0) = 0$, then $x(t)$ satisfies the integral equation

$$x(t) = x_0 \cos \omega_0 t + \frac{2x_0\gamma}{\omega_0} \sin \omega_0 t - 2\gamma \int_0^t \cos \omega_0(t - t')x(t')\, dt' .$$

b) Iterate this equation a few times and show that your result agrees with the exact result expanded to the appropriate order.

14. Show that $x_1(t)$ as given by Eq. (8.49) satisfies the requirement that energy be conserved *through order* δ where $\delta = \epsilon x_0/\omega_0^2 \equiv a x_0/k$. That is, show that

$$\tfrac{1}{2}m\dot{x}_1^2 + \tfrac{1}{2}kx_1^2 + \tfrac{1}{3}ax_1^3 = \tfrac{1}{2}kx_0^2 + \tfrac{1}{3}ax_0^3 + \mathcal{O}(\delta^2),$$

where $\mathcal{O}(\delta^2)$ denotes terms of order δ^2 and higher.

15. In solving the equation for the anharmonic oscillator in Section 8.4, we used the notation $x_0(t)$, $x_1(t)$, $x_2(t)$, etc., to denote the zeroth iterate, first iterate, second iterate, and so forth. Similarly, let us denote by $\bar{x}_0(t)$, $\bar{x}_1(t)$, $\bar{x}_2(t)$, etc., that part of the iterative solution containing all terms through order δ^0, order δ^1, order δ^2, and so forth. Clearly, $\bar{x}_0(t) = x_0(t)$ and $\bar{x}_1(t) = x_1(t)$, but $\bar{x}_2(t)$ contains only part of $x_2(t)$. Find $\bar{x}_3(t)$ [this will involve computing only a few of the terms in $x_3(t)$] and show that all terms in $\bar{x}_3(t)$ which are *linear* in t may be removed, *through order* δ^3, by replacing ω_0 by ω_1 in all the trigonometric functions. Write ω_1 as

$$\omega_1 = \omega_0 [1 + a\delta + b\delta^2 + c\delta^3]^{1/2} .$$

Find the values of a, b, and c which remove the terms linear in t. Using physical arguments, determine how large x_0 could be before the above expansion for ω_1 would diverge (we are imagining that the above redefinition of the frequency is carried out to all orders in δ).

16. a) Obtain an integral equation equivalent to the differential equation for the anharmonic oscillator whose motion is governed by

$$\ddot{x}(t) + \omega_0^2 x(t) = -ax^3(t),$$

by making use of the Green's function for the operator $d^2/dt^2 + \omega_0^2$. Assume $x(0) = x_0$, $\dot{x}(0) = 0$.

b) Iterate this equation once to obtain the first correction to the lowest-order solution, $x_0(t) = x_0 \cos \omega_0 t$.

17. For the system of Problem 8.16, find $\bar{x}_2(t)$ (we use the notation introduced in Problem 8.15). Show that if we write

$$\omega_1 = \omega_0 [1 + a\delta + b\delta^2]^{1/2},$$

then by appropriate choice of a and b, we can eliminate *through order* δ^2 all terms in $\bar{x}_2(t)$ proportional to t and t^2 by replacing ω_0 by ω_1 in all the trigonometric functions.

18. Solve the equations

a) $f(x) = x + \lambda \int_0^\infty e^{-(x+y)} f(y)\, dy$, b) $f(x) = x + \dfrac{\lambda}{\pi} \int_0^{\pi/2} \cos(x + y) f(y)\, dy$.

19. Find the unique bound-state solution to Schrödinger's integral equation for a separable potential of the form $-g^2 v(r) v(r')$ with $v(r) = e^{-\mu r}/r$. Show that a bound-state solution exists if and only if $g^2 > \hbar^2 \mu^3/(4\pi m)$, where m is the mass of the bound particle. Show also that the energy of the bound state is

$$E_B = -\frac{\hbar^2}{2m}\left[\left(\frac{4\pi m g^2}{\mu \hbar^2}\right)^{1/2} - \mu\right]^2.$$

What is the bound-state wave function?

20. Consider the quantity

$$f^S(E) = \frac{2\tau}{(q_i^2 + \mu^2)^2 + \dfrac{\tau}{\mu}(q_i - i\mu)^2},$$

where $\tau = 4\pi m g^2/\hbar^2$ and $q_i = (2mE/\hbar^2)^{1/2}$ [see Eq. (8.73)]. Show that if we consider E to be a complex variable, that is, if we do not restrict E to take on just physical real values, then $f^S(E)$ is a single-valued function of E on a two-sheeted Riemann surface cut along the positive real axis. Show that, as a function of E, $f^S(E)$ is analytic everywhere except for a pole of second order at $E = -\hbar^2\mu^2/2m$ on the first sheet, a simple pole at $E = -\hbar^2(\mu + \sqrt{\tau/\mu})^2/2m$ on the second sheet, and a so-called bound-state pole which may be either on the first or second sheet. If $\mu > \sqrt{\tau/\mu}$, then this pole is on the second sheet at $E = -\hbar^2(\sqrt{\tau/\mu} - \mu)^2/2m$; if $\sqrt{\tau/\mu} > \mu$, the pole is on the first sheet at $E = -\hbar^2(\sqrt{\tau/\mu} - \mu)^2/2m$. This value of E is precisely the energy of the bound state determined in the previous problem. We see, therefore, that as g^2 drops below the critical value necessary to produce a bound state, the effect of the bound state does not magically vanish from the scattering amplitude; the pole corresponding to the bound state merely moves from the first ("physical") sheet to the second sheet.

21. At the beginning of Section 8.6 in discussing kernels of finite rank, we introduced briefly the notion of a linear functional. The purpose of this problem and the two that follow it is to introduce some of the simpler aspects of linear functionals on a Banach space. We make the following *definition*. A *functional* F, on a Banach space $\boldsymbol{B}$, is an operation which assigns to every $g \in \boldsymbol{B}$ a complex number (assuming that $\boldsymbol{B}$ is defined over the field of complex numbers). The action of F on g is denoted by $F(g)$. F is said to be *linear* if $F(\alpha g_1 + \beta g_2) = \alpha F(g_1) + \beta F(g_2)$ for all complex numbers, α and β, and all $g_1, g_2 \in \boldsymbol{B}$. F is said to be *bounded* if there exists a real number M, such that $|F(g)| \leq M\|g\|$ for all $g \in \boldsymbol{B}$, M being independent of g. Let us now consider two aspects of this.

a) Let F be a bounded, linear functional on B. Suppose that we have in B a sequence $\{f_n\}$ which tends in norm to $f \in B$. Show that $F(f_n) \to F(f)$, that is, that F is continuous.

b) Show that if $f \in L_p$, then the functional F defined on L_q $(1/p + 1/q = 1)$ by

$$F(g) \equiv (f, g) \equiv \int f^*(x) g(x)\, dx\,,$$

where $g \in L_q$, is a bounded, linear functional. [*Hint*: Look back at Problem 8.2]. It can be shown that every bounded, linear functional on L_q can be written in this form, i. e., can be generated by an element of L_p.

22. Just as for a bounded, linear operator on B, we define the norm of a bounded, linear functional on B as the least uppper bound of the set of all M satisfying $|F(g)| \leq M||g||$ for all $g \in B$. As usual, we denote the norm of F by $||F||$.

a) Show that for the functional F, defined on L_q by

$$F(g) = (f, g)\,,$$

where $f \in L_p$ and $1/p + 1/q = 1$ (see the preceding problem), we have $||F|| \leq ||f||$. Here the norm of f is the norm appropriate to L_p, namely,

$$||f|| = \left[\int |f(x)|^p\, dx\right]^{1/p}.$$

b) Show that the function

$$g = \frac{|f|^{(p/q)+1}}{f^*}$$

belongs to L_q if $f \in L_p$ $(1/p + 1/q = 1)$ and therefore that $||F|| \geq ||f||$. Do this by showing that

i) $|F(g)| = |(f, g)| = ||f||^p$, ii) $||f||\,||g|| = ||f||^p$.

Hence $|F(g)| = ||f|| \cdot ||g||$, so $||F|| \geq ||f||$. Combining parts (a) and (b), we see that actually $||F|| = ||f||$.

23. Show that, by an obvious definition of addition and scalar multiplication, the bounded linear *functionals* on B form a Banach space in their own right. This space is called the *dual space* of B and is denoted by B^*. Since every linear functional on L_q is generated by an element of L_p $(1/p + 1/q = 1)$, we can identify L_q^* with L_p and say that L_p is the *dual space* of L_q. Note that the space L_2 of square-integrable functions is its own dual space.

There are three parts to this problem. First, you must show that your definition of addition and scalar multiplication is such that it gives rise to a linear vector space. Second, you must show that the definition of norm given in the previous problem really is a norm (as we did in Section 8.2 in discussing norms of operators). These two parts are easy. Not so easy is the problem of showing that the normed, linear vector space which you have constructed is *complete*.

[*Hint*: You must show that any Cauchy sequence of functionals $\{F_n\}$ has a limit which is also a bounded linear functional. Note that for any $f \in B$, $\{F_n(f)\}$ is a Cauchy sequence of complex numbers. Since the complex numbers are complete, you can define the limit of $\{F_n\}$ acting on any $f \in B$ in terms of the limit of $\{F_n(f)\}$, which is guaranteed to exist. Denote this limit by $F(f)$. It is then not difficult to show that F so defined is linear and bounded and that $||F_n - F|| \to 0$ as $n \to \infty$. The basic idea for all these proofs is contained in the following identity which would be relevant for the proof of the linearity of F:

$$F(\alpha g_1 + \beta g_2) - [\alpha F(g_1) + \beta F(g_2)] \equiv [F(\alpha g_1 + \beta g_2) - F_n(\alpha g_1 + \beta g_2)] - [\alpha F(g_1) - \alpha F_n(g_1)] - [\beta F(g_2) - \beta F_n(g_2)] .$$

Here we make use of the *known* linearity of F_n.]

24. Show that by defining addition and scalar multiplication in the obvious way, we can construct a Banach space out of the set of all bounded, linear *operators* on B. This proof can be patterned directly on the proof given in the previous problem for bounded linear functionals. Again, the completeness of the underlying space B, is the key ingredient in the proof.
25. Using Eqs. (8.75) and (8.88), show explicitly that
$$(I - \lambda K_F)(I + \lambda R_{K_F}) = I .$$
26. Show that if K_N is defined by Eq. (8.93) and P_N by Eq. (8.94), then $K_N = P_N K P_N$.

FURTHER READINGS

BLATT, J. M., and V. F. WEISSKOPF, *Theoretical Nuclear Physics*. New York: Wiley, 1952.

GOLDBERGER, M. L., and K. M. WATSON, *Collision Theory*. New York: Wiley, 1964.

LOVITT, W. V., *Linear Integral Equations*. New York: Dover, 1950.

MUSHELISHVILI, N. I., *Singular Integral Equations*. Groningen: P. Noordhoff, Ltd., 1953. This book gives a very useful and interesting account of integral equations with singular kernels.

NEWTON, R. G., *Scattering Theory of Waves and Particles*. New York: McGraw-Hill, 1966.

RIESZ, F., and B. SZ. NAGY, *Functional Analysis*. New York: Ungar, 1955.

CHAPTER 9

INTEGRAL EQUATIONS IN HILBERT SPACE

INTRODUCTION

In the last chapter we developed two of the basic techniques for the solution of integral equations, iteration and finite-rank approximation, and in this chapter we wish to exploit these methods to obtain some results which facilitate greatly the solution of practical problems. We shall begin by looking at the properties of completely continuous Hermitian operators in some detail; in particular, we shall obtain a completeness relation for the eigenfunctions of such operators and use these eigenfunctions to discuss perturbation theory and the solution of linear equations.

With the insights obtained from this special case, we will move on to consider general completely continuous operators and see how to solve linear equations by the method of finite-rank approximation. Finally, we will close the chapter by looking at the Fourier transform in the context of the theory of unitary operators on Hilbert space. This is one of the most important tools in mathematical physics, and the ease with which it can be handled in the context of infinite-dimensional vector spaces is a beautiful illustration of the power of the theory of linear operators.

9.1 COMPLETELY CONTINUOUS HERMITIAN OPERATORS

Before we attempt to deal with the most general type of completely continuous operators, we will investigate an important special case: the completely continuous, self-adjoint operators. This kind of operator is interesting in its own right; for example, the bound-state Schrödinger integral equations can be handled very deftly using the techniques we shall develop here. But in addition, this special case will give us some important clues as to how we can proceed to treat the general theory of completely continuous operators.

Just as in the finite-dimensional case, a completely continuous self-adjoint operator A satisfies

$$(x, Ay) = (Ax, y) \tag{9.1}$$

for all x, y in the vector space. Since A is completely continuous, it is also bounded, and therefore is defined on every element of the space. If A is an integral operator with kernel $a(t, t')$, then the Hermiticity requirement is equivalent to

$$a(t, t') = a^*(t', t)\ , \tag{9.2}$$

since

$$(x, Ay) = \int dt \int dt'\, x^*(t) a(t, t') y(t') ,$$

$$(Ax, y) = \int dt' \int dt\, a^*(t, t') x^*(t') y(t) = \int dt' \int dt\, x^*(t) a^*(t', t) y(t') .$$

Evidently Eq. (9.1) holds if Eq. (9.2) is obeyed. It should be said at this point that our change in notation, with the operator written as A and the vectors as x and y, is not capricious. We shall see in the next two sections that the theory of completely continuous, Hermitian operators is virtually identical to the finite-dimensional theory we discussed in Chapter 4. To emphasize this similarity, we shall observe the notational conventions of Chapter 4 in this section and the next. However, the reader should remember that the vectors x and y may be functions in this chapter, in which case we will write them as $x(t)$ and $y(t)$ if it is necessary to display their functional dependence.

We begin by recalling the definition of norm which we gave in Section 8.2, namely, $||A||$ is the smallest constant such that

$$||Ax|| \leq ||A|| \cdot ||x||$$

for all $x \in B$, where B is any normed vector space and A is a bounded operator. Now it is interesting to inquire whether or not there is actually some vector in the space for which one has

$$Ay = \pm ||A|| y . \tag{9.3}$$

Let H_1 be a Hilbert space, and let A be an operator with the property that there is always some vector in the space which satisfies Eq. (9.3). Suppose that we have found some vector x_1 which satisfies

$$Ax_1 = \pm ||A||_1 x_1 ,$$

where $||A||_1$ is the norm of A when it acts on the whole Hilbert space, H_1. The space of all vectors orthogonal to x_1, say, H_2, is also a Hilbert space; if A were bounded on H_1, then A would be necessarily bounded on H_2. Thus

$$||Ax|| \leq ||A||_2 \cdot ||x||$$

for all $x \in H_2$. Clearly, $||A||_2 \leq ||A||_1$. Now if A still has the property assumed above, then there is some $x_2 \in H_2$ such that

$$Ax_2 = \pm ||A||_2 x_2 ,$$

where $(x_1, x_2) = 0$. We can continue this process indefinitely, generating H_1, H_2, H_3, $\cdots$ and x_1, x_2, x_3, $\cdots$.

Several possibilities may now arise. It might be that $||A||_1 = ||A||_2 = ||A||_3 = \cdots$. This would be the case, for instance, if A were the identity operator. However, if A is completely continuous and $A \neq 0$, then this cannot happen, i.e., we cannot obtain some nonzero eigenvalue an infinite number of times. For if we did, we would have an infinite orthonormal sequence $\{x_n\}$, and $\{Ax_n\}$ would need a convergent subsequence since A is completely continuous. But if $Ax_n = ax_n$ for all the vectors of the sequence, then $\{ax_n\}$ must contain a convergent subsequence. However, $||x_m - x_n|| = \sqrt{2}$ for all m, n, so there can be no convergent subsequence, unless, of course, $a = 0$. Hence we can obtain a given, nonzero eigenvalue only a finite number of times.

However, it is possible to obtain a zero eigenvalue an infinite number of times. This, for example, would be true for an operator of finite rank on a Hilbert space. In any case, however, the eigenvalues λ_n must tend toward zero, because if they did not, x_n/λ_n would be a bounded sequence. But $Ax_n = \lambda_n x_n$, so $\{x_n\}$ must contain a convergent subsequence, which it cannot as noted above. This behavior of the eigenvalues could arise either because all but a finite number of eigenvalues are zero, or because there is an infinite collection of nonzero eigenvalues, λ_n, all of finite multiplicity, which become arbitrarily small for n sufficiently large. In this latter case, we might also have some zero eigenvalues (perhaps even an infinite number of them). Corresponding to these eigenvalues, we have an infinite number of eigenvectors, x_n, and we may hope that they will form a complete set, although the fact that there is an infinite number of x_n is only necessary for completeness, not sufficient.

In finite-dimensional spaces, the completeness property held for the eigenfunctions of self-adjoint, isometric, and normal transformations. In the case of completely continuous self-adjoint transformations, we shall now show that this same result still holds. Of course, basic properties such as reality of eigenvalues will be the same since the proofs of these properties given in Chapter 4 were in no way restricted to the finite-dimensional case. However, by way of emphasizing the difference between the finite-dimensional case and the completely continuous case, we mention that there are *no* completely continuous *isometric* transformations of an infinite-dimensional Hilbert space. Clearly, every isometry is *bounded* since

$$||Uy||^2 = (Uy, Uy) = (y, y) = ||y||^2$$

for any $y \in H$. Thus, $||U|| = 1$. However, consider any sequence $\{y_n\}$, which does not contain a convergent subsequence, and the related sequence $\{Uy_n\}$;

$$||Uy_n - Uy_m|| = ||U(y_n - y_m)|| = ||y_n - y_m|| ,$$

so U does not change the convergence properties of $\{y_n\}$ in any way. In particular, if $\{y_n\}$ does not contain a convergent subsequence, neither will $\{Uy_n\}$. Thus isometric transformations are not completely continuous.

The most important property of self-adjoint transformations is that they have *real* expectation values. This fact leads us to the following important result.

Theorem 9.1. Let A be a bounded, self-adjoint linear operator on the Hilbert space, H. Then if $||y|| = 1$,

$$\sup_{y \in H} |(y, Ay)| = ||A|| ,$$

where sup denotes the least upper bound.

Proof. By the Schwartz inequality, for all $y \in H$,

$$|(y, Ay)| \leq ||y|| \cdot ||Ay|| \leq ||A|| \cdot ||y||^2 . \tag{9.4}$$

Now, in analogy with our definition of the norm of A, we define the quantity $|A|$ as the smallest constant for which

$$|(y, Ay)| \leq |A| \cdot ||y||^2 , \tag{9.5}$$

for all $y \in H$. According to Eq. (9.4), $|A| \leq ||A||$. We now show that $|A| \geq ||A||$. To do this, we utilize the identity

$$||Ay||^2 = (Ay, Ay) \equiv \frac{1}{4}\left\{\left(\lambda y + \frac{1}{\lambda} Ay, A\left[\lambda y + \frac{1}{\lambda} Ay\right]\right) - \left(\lambda y - \frac{1}{\lambda} Ay, A\left[\lambda y - \frac{1}{\lambda} Ay\right]\right)\right\} ,$$

which holds for all real λ. Thus, using Eq. (9.5), we get

$$||Ay||^2 \leq \frac{1}{4} |A| \cdot \left[||\lambda y + \frac{1}{\lambda} Ay||^2 + ||\lambda y - \frac{1}{\lambda} Ay||^2 \right] ,$$

so that

$$||Ay||^2 \leq \frac{1}{2} |A| \cdot \left(\lambda^2 \cdot ||y||^2 + \frac{1}{\lambda^2} ||Ay||^2 \right) .$$

In particular, the inequality holds if $\lambda^2 = ||Ay||/||y||$. From this value, we find that

$$||Ay||^2 \leq |A| \cdot ||Ay|| \cdot ||y|| , \qquad \text{or} \qquad ||Ay|| \leq |A| \cdot ||y|| .$$

Thus $||A|| \leq |A|$. Since we have already shown that $||A|| \geq |A|$, we see that $||A|| = |A|$, as required. Note that this result holds for any bounded, self-adjoint operator, regardless of whether or not it is completely continuous. It might almost appear not to even depend on the Hermiticity of A. However, the reader can readily show that the identity in λ which is crucial to the proof is only true if A is self-adjoint.

With this result in hand, we can now carry out the program outlined at the beginning of this section.

Theorem 9.2. Let A be a completely continuous, self-adjoint transformation on a Hilbert space, H, and denote by C the set of all $y \in H$ such that $||y|| = 1$. The quantity $|(y, Ay)|$ attains a maximum value on this set, and the vector x, for which the maximum is attained is an eigenvector of A with eigenvalue $\lambda = \pm ||A||$.

Proof. Choose a sequence $\{y_n\}$, $y_n \in C$, such that (y_n, Ay_n) tends to λ_1, where $|\lambda_1| = ||A||$. The previous theorem tells us that this is always possible. Since we want to prove that λ_1 is an eigenvalue, let us consider

$$||Ay_n - \lambda_1 y_n||^2 = ||Ay_n||^2 - 2\lambda_1 (y_n, Ay_n) + \lambda_1^2 ||y_n||^2 .$$

Now $||Ay_n||^2 \leq ||A||^2 \cdot ||y_n||^2 = \lambda_1^2 ||y_n||^2$, so since $||y_n|| = 1$,

$$||Ay_n - \lambda_1 y_n||^2 \leq 2[\lambda_1^2 - \lambda_1(y_n, Ay_n)] .$$

But since $(y_n, Ay_n) \to \lambda_1$ by assumption,

$$||Ay_n - \lambda_1 y_n|| \to 0 . \tag{9.6}$$

Since A is completely continuous, the sequence $\{Ay_n\}$ contains a convergent subsequence, say, $\{A\eta_m\}$, where $\{\eta_m\}$ is a subset of $\{y_n\}$. For reasons which will soon be clear, we denote the vector to which $\{A\eta_m\}$ converges by $\lambda_1 x_1$:

$$||A\eta_m - \lambda_1 x_1|| \to 0 . \tag{9.7}$$

But

$$A\eta_m - \lambda_1\eta_m \equiv A\eta_m - \lambda_1 x_1 + \lambda_1 x_1 - \lambda_1 \eta_m , \tag{9.8}$$

so

$$||\lambda_1 x_1 - \lambda_1 \eta_m|| \leq ||A\eta_m - \lambda_1 x_1|| + ||A\eta_m - \lambda_1 \eta_m|| .$$

Thus by Eqs. (9.6) and (9.7),

$$||\lambda_1 x_1 - \lambda_1 \eta_m|| \to 0; \tag{9.9}$$

in other words, the sequence $\{\eta_m\}$ converges to x_1 (assuming, of course, that $|\lambda_1| = ||A|| \neq 0$). By the triangle inequality,

$$\begin{aligned} ||Ax_1 - \lambda_1 x_1|| &\leq ||Ax_1 - A\eta_m|| + ||A\eta_m - \lambda_1\eta_m|| + ||\lambda_1\eta_m - \lambda_1 x_1|| \\ &\leq 2||A|| \cdot ||x_1 - \eta_m|| + ||A\eta_m - \lambda_1\eta_m|| , \end{aligned}$$

so by Eqs. (9.6) and (9.9), we can make

$$||Ax_1 - \lambda_1 x_1|| < \epsilon$$

for any ϵ. Hence

$$Ax_1 = \lambda_1 x_1 ,$$

so that λ_1 is an eigenvalue of A corresponding to the eigenvector x_1. Also,

$$|(x_1, Ax_1)| = |(x_1, \lambda_1 x_1)| = |\lambda_1| = ||A|| .$$

This completes the proof of this theorem, which is the basic result on completely continuous, self-adjoint operators.

Now, let us consider again the Hilbert space H_2 of all vectors orthogonal to the vector x_1 found in the previous theorem. The operator A acting on this space is completely continuous and Hermitian. We write

$$||Ay|| \leq ||A||_2 \cdot ||y||$$

for all $y \in H_2$, where $||A||_2 \leq ||A||$. By repeating the procedure of the above theorem, we can find an x_2 such that

$$Ax_2 = \lambda_2 x_2 , \qquad |\lambda_2| = ||A||_2 .$$

We can continue the process indefinitely unless we arrive at a number, m, where the eigenvalue $\lambda_m = 0$ that is, $||A||_m = 0$. This means that after removing a number, m, of orthonormal vectors from our original space, H, we arrive at a space H_{m+1} such that A acting on this space produces only zero vectors. This means that H_{m+1} is the *null space* of A. This will clearly be the case for the operator of finite rank, A_F, defined by

$$A_F y \equiv \sum_{n=1}^{N} \alpha_n \phi_n (\phi_n, y) ,$$

where the ϕ_n form an orthonormal set. A_F has as eigenvectors the ϕ_n with corresponding eigenvalues, α_n. A_F also has as eigenvectors any vectors which are orthogonal to all the ϕ_n. Zero is the common eigenvalue for all these other eigenvectors. We may summarize the above discussion as follows:

Theorem 9.3. Let Y be the set of all vectors $y \in H$ such that $||y|| = 1$ and y is orthogonal to all the vectors $x_1, x_2, \cdots, x_n$ satisfying

$$Ax_n = \lambda_n x_n .$$

Then $|(y, Ay)|$ can always be maximized on this set if A is a completely continuous, self-adjoint operator. Furthermore, the vector x_{n+1} obtained in this way is an eigenvector of A with eigenvalue λ_{n+1}, where $|\lambda_{n+1}| = ||A||_{n+1}$ and $||A||_{n+1}$ is the norm of A in the subspace of H consisting of all vectors orthogonal to $x_1, x_2, \cdots, x_n$. This process generates an infinite sequence of vectors $\{x_n\}$ and a set of eigenvalues $\{\lambda_n\}$. Each nonzero eigenvalue is of finite multiplicity and $\lambda_n \to 0$ as $n \to \infty$.

Now we are in a position to obtain a completeness theorem for completely continuous, self-adjoint operators on a Hilbert space.

Theorem 9.4. The eigenvectors of the completely continuous, self-adjoint operator A form a complete, orthonormal set of vectors in the Hilbert space, H, on which A acts.

Proof. Let us assume at first that A has only nonzero eigenvalues.* It is clear that at the very least there is always a subspace of H in which this is true. For any $y \in H$, we consider

$$y_N = y - \sum_{n=1}^{N} (x_n, y) x_n ,$$

where the set $\{x_n\}$ consists of the eigenvectors of A. To establish completeness, we must show that as $N \to \infty$, $||y_N|| \to 0$. But y_N is obviously orthogonal to all x_n with $n \leq N$. Hence by our previous work

$$||Ay_N|| \leq |\lambda_{N+1}| \cdot ||y_N|| .$$

---

* We will also assume that A has an infinite number of eigenvectors belonging to nonzero eigenvalues. The case when A has only a finite number of nonzero eigenvalues is elementary and is left as a problem.

Since $||y_N|| \leq ||y||$, we have

$$||Ay_N|| \leq |\lambda_{N+1}| \cdot ||y|| .$$

But $\lambda_{N+1} \to 0$ as $N \to \infty$, so

$$||Ay_N|| \to 0$$

as $N \to \infty$. But it is easily shown that

$$Ay_N = Ay - \sum_{n=1}^{N} (x_n, y) Ax_n = Ay - \sum_{n=1}^{N} (x_n, Ay) x_n .$$

Thus we may say that

$$\sum_{n=1}^{N} (x_n, Ay) x_n$$

tends in norm to Ay as $N \to \infty$. We write this as

$$Ay \doteq \sum_{n=1}^{\infty} (x_n, Ay) x_n , \tag{9.10}$$

for any $y \in H$. Hence we have a completeness relation for all elements of the space of the form Ay.

If we knew that A were invertible, then we could immediately extend the completeness relation to all vectors in H, since then any vector in H could be written in the form Ay for some $y \in H$. However, more generally, we may argue as follows. Since H is a complete space,

$$\lim_{N \to \infty} \left[\sum_{n=1}^{N} (x_n, \eta) x_n \right]$$

converges in the mean to some vector, say $\tilde{\eta}$, because Bessel's inequality (Eq. 5.9) tells us that the sequence

$$\left\{ \eta_N = \sum_{n=1}^{N} (x_n, \eta) x_n \right\}$$

is a *Cauchy* sequence. We write

$$\tilde{\eta} \doteq \sum_{n=1}^{\infty} (x_n, \eta) x_n . \tag{9.11}$$

Since A is continuous,

$$A\tilde{\eta} \doteq \sum_{n=1}^{\infty} (x_n, \eta) Ax_n \doteq \sum_{n=1}^{\infty} (x_n, A\eta) x_n .$$

But by Eq. (9.10),

$$A\eta \doteq \sum_{n=1}^{\infty} (x_n, A\eta) x_n .$$

Hence

$$A\tilde{\eta} - A\eta = A(\tilde{\eta} - \eta) = 0 ,$$

except possibly on a set of measure zero. But we have assumed that A has no zero eigenvalues, so

$$\tilde{\eta} = \eta .$$

Thus Eq. (9.11) becomes

$$\eta \doteq \sum_{n=1}^{\infty} (x_n, \eta) x_n \tag{9.12}$$

for any $\eta \in H$, which is the familiar completeness relation. It should be remembered that the convergence of the infinite sum in Eq. (9.12) is defined in the sense of *mean* convergence.

The case when $Ax = 0$ has solutions is easily handled. The set of all such solutions forms a Hilbert space, and we may pick an orthonormal basis $\{y_n\}$ for it. (Note that according to Theorem 4.18, every y_n is orthogonal to every x_m belonging to a non-zero eigenvalue.) Each element of the basis satisfies $Ay_n = 0$. Now given any x, form

$$\tilde{x} = x - \sum_{n=1}^{\nu} (y_n, x) y_n ,$$

where ν is the number of elements in the set $\{y_n\}$. Clearly, $\tilde{x}$ lies in the orthogonal complement of the space spanned by $\{y_n\}$, so we can apply the above result to it:

$$\tilde{x} \doteq \sum_{m=1}^{\infty} (x_m, \tilde{x}) x_m \doteq \sum_{m=1}^{\infty} (x_m, x) x_m ,$$

where the $\{x_m\}$ are the eigenvectors of A when A is restricted to act on the orthogonal complement of the space spanned by the y_n. Hence

$$x \doteq \sum_{n=1}^{\nu} (y_n, x) y_n + \sum_{m=1}^{\infty} (x_m, x) x_m$$

for any $x \in H$. Therefore the set $\{x_n, y_m\}$ is a complete orthonormal basis in H, and we are done.

As a simple example of some of the previous results consider the equation

$$\int_0^1 t_<(1 - t_>) x_n(t')\, dt' = \lambda_n x_n(t), \tag{9.13}$$

which was shown in Section 7.1 to be equivalent to the differential eigenvalue problem

$$\frac{d^2 x_n(t)}{dt^2} + \mu_n x_n(t) = 0 , \qquad x_n(0) = 0 = x_n(1) ,$$

where $\mu_n = 1/\lambda_n$. The eigenvectors of Eq. (9.13) are therefore just

$$x_n(t) = \sqrt{2} \sin (n\pi t) ,$$

with eigenvalues

$$\lambda_n = 1/n^2\pi^2 .$$

This may also be verified by direct substitution in Eq. (9.13). Theorem 9.4 tells us that the vectors $x_n(t) = \sqrt{2} \sin (n\pi t)$ are a complete set of orthonormal functions on the space of square-integrable functions on [0, 1] which vanish at $t = 0$ and $t = 1$. From our previous discussion, we know that the norm of the integral operator defined by Eq. (9.13) must be equal to the largest eigenvalue, $1/\pi^2$. We have also seen that the norm of an integral kernel is less than or equal to the square root of the integrated square of the kernel [see, for example, Eq. (8.47) et. seq.]. In this case, therefore, we must have

$$\lambda_1^2 \leq \int_0^1 dt \int_0^1 dt' \, t_<^2(1 - t_>)^2 = 1/90 \qquad \text{or} \qquad |\lambda_1| \leq 1/9.49 = 0.1054 .$$

The actual value of λ_1 is $1/\pi^2$, that is, $\lambda_1 = 0.1013$, so we see that by approximating the norm of A by the square root of the integrated square of its kernel, we get an excellent estimate of the lowest eigenvalue. We can also estimate the second eigenvalue in this way by calculating in this approximate manner the norm of the operator *restricted* to the space of all vectors orthogonal to x_1. This means calculating the norm of

$$(I - P_1)A(I - P_1) ,$$

where P_1 is the projection onto the subspace spanned by x_1. In terms of the kernel of A, $a(t, t') = t_<(1 - t_>)$; this means that

$$|\lambda_2| \leq \left[\int_0^1 dt \int_0^1 dt'[a(t, t') - \lambda_1 x_1(t')]^2 \right]^{1/2} .$$

Since x_1 is an eigenvector of A, this integral reduces immediately to

$$|\lambda_2| \leq \left[\int_0^1 dt \int_0^1 dt' (a(t, t')^2 - \lambda_1^2) \right]^{1/2} .$$

Both quantities in the bracket are known, and we find that

$$|\lambda_2| \leq 0.029 ,$$

which is quite close to the exact value, $\lambda_2 = 1/4\pi^2 = 0.025$. For λ_1, the approximation was good to 4%, while for λ_2 the result is accurate to only 16%. Applying the some method to λ_3 yields

$$|\lambda_3| \leq 0.015 ,$$

compared with the exact value, $\lambda_3 = 1/9\pi^2 = 0.011$, which tells us that the estimate continues to worsen as we go to higher eigenvalues. However, such estimates are often useful in practical problems in getting an idea of the size of the lowest eigenvalue. Clearly, the method cannot be used to estimate higher eigenvalues unless the lower eigenvalues are known *exactly*.

These estimates can be understood more easily if we consider the following corollary of the previous theorem.

Corollary. If A is a self-adjoint integral operator whose square-integrable kernel is $a(t, t')$, then

$$a(t, t') \doteq \sum_{n=1}^{\infty} \lambda_n x_n(t) x_n^*(t') , \tag{9.14}$$

where $\{x_n\}$ and $\{\lambda_n\}$ are respectively the eigenvectors and eigenvalues of A, and the infinite sum is defined in the sense of mean convergence.

Proof. Denote the sum by

$$S(t, t') \doteq \sum_{n=1}^{\infty} \lambda_n x_n(t) x_n^*(t') .$$

We leave it for the reader to show that this sum is indeed convergent; this is just a matter of satisfying the Cauchy condition. Define the difference function

$$d(t, t') \equiv a(t, t') - S(t, t') ,$$

and consider the related integral operator, D:

$$Dy = \int d(t, t') y(t') \, dt' .$$

Let us assume that $||D||$ is *not* equal to zero. Now by Theorem 9.2, D has at least one nonzero eigenvalue, say λ, with associated eigenvector, $x(t)$. Let $x_n(t)$ $(n = 1, 2, \cdots)$ be any eigenvector of A [and hence of S, where S is the integral operator whose kernel is $S(t, t')$]. Consider

$$(x, x_n) \equiv \frac{1}{\lambda^*} (\lambda x, x_n) = \frac{1}{\lambda^*} (Dx, x_n) = \frac{1}{\lambda^*} (x, Dx_n) .$$

But

$$Dx_n = \int [a(t, t') - s(t, t')] x_n(t') \, dt' = (A - S) x_n = 0 .$$

Hence

$$(x, x_n) = 0 .$$

Thus, we have found a nonzero vector which cannot be expanded in terms of the supposedly complete set $\{x_n\}$, which contradicts the completeness theorem (Theorem 9.4). Hence our assumption that $||D|| \neq 0$ was incorrect, and therefore $||D|| = 0$. Thus, $d(t, t') = 0$, which proves the corollary.

This tells us that *every* symmetric, square-integrable function of two variables can be expanded in a series of the form of Eq. (9.14). We also have the result, which we leave to the reader to show, that

$$\int dt \int dt' \cdot |a(t, t')|^2 = \sum_{n=1}^{\infty} \lambda_n^2 , \tag{9.15}$$

where $\{\lambda_n\}$ is the set of eigenvalues of the integral operator A whose kernel is given by the symmetric function $a(t, t')$. Equation (9.15) is the basis of our estimates of $\lambda_1, \lambda_2, \cdots$ given above. It tells us that if there are no eigenvalues "near" λ_1, that is, $\lambda_1^2 \gg \lambda_2^2 > \lambda_3^2 > \cdots$, then

$$\int dt \int dt' \; |a(t, t')|^2 \cong \lambda_1^2 .$$

Until now we have ordered the eigenvalues according to decreasing magnitude of $|\lambda_n|$, $|\lambda_1|$ being the maximum value of $|(y, Ay)|$ which is attained when $y = x_1$, where

$$Ax_1 = \lambda_1 \cdot x_1 .$$

λ_1 may be positive or negative. The previous theorem enables us to make a similar ordering for the positive and negative eigenvalues separately.

Theorem 9.5. Suppose that A is a completely continuous Hermitian operator. Let $\lambda_1^+, \lambda_2^+, \cdots$ be the positive eigenvalues of A ordered so that $\lambda_1^+ \geq \lambda_2^+ \geq \cdots$, with corresponding eigenvectors $x_1^+, x_2^+, \cdots$. Similarly, let $\lambda_1^-, \lambda_2^-, \cdots$ be the negative eigenvalues of A with $\lambda_1^- \leq \lambda_2^-, \leq \cdots$, and denote the corresponding eigenvectors by $x_1^-, x_2^-, \cdots$. Then λ_{n+1}^+ is the *maximum* value of (y, Ay) when y varies over the Hilbert space, H, subject to $||y|| = 1$ and $(y, x_i^+) = 0$ for $i = 1, 2, \cdots, n$. Similarly, λ_{n+1}^- is the *minimum* value of (y, Ay) when y varies over H subject to $||y|| = 1$ and $(y, x_i^-) = 0$ for $i = 1, 2, \cdots, n$.

Proof. By Theorem 9.4, we have for any $y \in H$

$$Ay \doteq \sum_{n=1}^{\infty} (x_n, Ay)x_n \doteq \sum_{n=1}^{\infty} \lambda_n (x_n, y) x_n .$$

Therefore

$$(y, Ay) = \sum_{n=1}^{\infty} \lambda_n |(x_n, y)|^2 .$$

Using the notation just introduced, we can write this as

$$(y, Ay) = \sum_{\nu=1}^{N} \lambda_\nu^+ |(x_\nu^+, y)|^2 + \sum_{\mu=1}^{M} \lambda_\mu^- |(x_\mu^-, y)|^2 ,$$

where N is the number of positive eigenvalues and M is the number of negative eigenvalues (either M or N or both may be infinite). Suppose that $(y, x_i^+) = 0$ for $i = 1, 2, \cdots, n$. Then

$$(y, Ay) = \sum_{\nu=n+1}^{N} \lambda_\nu^+ |(x_\nu^+, y)|^2 + \sum_{\mu=1}^{M} \lambda_\mu^- |(x_\mu^-, y)|^2 .$$

Clearly, if $y = x_{n+1}^+$, then $(y, Ay) = \lambda_{n+1}^+$. Now consider any y and expand it as

$$y = \sum_{i=n+1}^{N} c_i^+ x_i^+ + \sum_{j=1}^{M} c_j^- x_j^- ,$$

where

$$\sum_{i=n+1}^{N} |c_i^+|^2 + \sum_{j=1}^{M} |c_j^-|^2 = 1; \tag{9.16}$$

y is orthogonal to $x_1^+, x_2^+, \cdots, x_n^+$ and is normalized to 1. For this y,

$$(y, Ay) = \sum_{\nu=n+1}^{N} \lambda_\nu^+ |c_\nu^+|^2 + \sum_{\mu=1}^{M} \lambda_\mu^- |c_\mu^-|^2 .$$

Since all λ_ν^+ with $\nu \geq n + 1$ satisfy $\lambda_\nu^+ \leq \lambda_{n+1}^+$, and since $\lambda_\mu^- \leq \lambda_{n+1}^+$ for all μ,

$$(y, Ay) \leq \lambda_{n+1}^+ \left[\sum_{\nu=n+1}^{N} |c_\nu^+|^2 + \sum_{\mu=1}^{M} |c_\mu^-|^2 \right] .$$

Thus according to Eq. (9.16)

$$(y, Ay) \leq \lambda_{n+1}^+ ,$$

which proves the result for positive eigenvalues. A similar argument gives the result for negative eigenvalues. We have *not* shown that $(y, Ay) = \lambda_{n+1}^+$ *only* for $y = x_{n+1}^+$; this would not be true, since λ_{n+1}^+ may have multiplicity $\mu > 1$. In our ordering of eigenvalues, we always list an eigenvalue and its corresponding eigenvector as many times as its multiplicity, μ, requires.

This theorem generalizes to infinite-dimensional spaces the results of Section 4.8 which were derived for finite-dimensional spaces. Thus we see that for completely continuous Hermitian operators, the results of finite-dimensional theory can be taken over to infinite-dimensional spaces virtually without alteration. Since we have obtained a complete, orthonormal set of eigenfunctions for A, it is also a straightforward matter to construct the projection operators, $E(\lambda)$, discussed in Section 5.11 and thereby write for any self-adjoint, completely continuous operator,

$$A = \int_{\lambda_{\min}}^{\lambda_{\max}} \lambda \, dE(\lambda) ,$$

where $E(\lambda)$ is the resolution of the identity belonging to A. Thus, for this special class of self-adjoint operators on Hilbert space we have essentially proved the Spectral Theorem. However, by no means are all self-adjoint operators on Hilbert space completely continuous. Even in the class of *bounded*, self-adjoint operators problems arise which prevent us from using the above methods to obtain the Spectral Theorem. Let us indicate briefly what one of the difficulties is.

Consider the bounded, Hermitian operator X, defined by

$$Xf(x) \equiv xf(x) ,$$

for all $f \in L_2[0, 1]$, where $L_2[0, 1]$ is the Hilbert space of square-integrable functions defined in the interval $[0, 1]$. Clearly, since $x \in [0, 1]$, the operator X is bounded. However, X is not completely continuous. Furthermore, the operator X has no eigenfunctions, since the equation

$$xf(x) = \lambda f(x)$$

cannot be solved to give any sensible sort of function. Of course, it is possible to find a nonsensible "eigenfunction," namely $\delta(x - x_0)$; that is, we could write

$$x\delta(x - x_0) = x_0\delta(x - x_0)$$

and say that $\delta(x - x_0)$ is an eigenfunction of X with eigenvalue x_0. This is true for any $x_0 \in [0, 1]$. Unfortunately, $\delta(x - x_0)$ is not a function in the usual sense of the word, and furthermore

$$\int_0^1 |\delta(x - x_0)|^2\, dx = \infty$$

if $x_0 \in [0, 1]$, so $\delta(x - x_0)$ is not even square-integrable. However, the δ-function does suggest the following possibility. Consider any point $x_0 \in [0, 1]$. We can construct a sequence of functions, which in the sense of Eq. (9.6) may be called "approximate" eigenfunctions, as follows. We define

$$\phi_{x_0}^{(N)}(x) = \begin{cases} \sqrt{\dfrac{N}{2}}, & \text{if} \quad 0 \le x_0 - \dfrac{1}{N} \le x \le x_0 + \dfrac{1}{N} \le 1 , \\ 0 & \text{otherwise} ; \end{cases}$$

clearly $(\phi_{x_0}^{(N)}, \phi_{x_0}^{(N)}) = 1$. The *square* of $\phi_{x_0}^{(N)}$ looks like a δ-function when N is large. Now a simple calculation shows that

$$||X\phi_{x_0}^{(N)} - x_0\phi_{x_0}^{(N)}|| = ||(x - x_0)\phi_{x_0}^{(N)}|| = \frac{1}{3N^2} ,$$

so

$$||X\phi_{x_0}^{(N)} - x_0\phi_{x_0}^{(N)}|| \to 0$$

as $N \to \infty$. Thus the set $\{\phi_{x_0}^{(N)}\}$ provides us with an "approximate" eigenfunction whose eigenvalue is x_0, because given an ϵ, no matter how small, we can find an N such that $\phi_{x_0}^{(N)}$ satisfies

$$||X\phi_{x_0}^{(N)} - x_0\phi_{x_0}^{(N)}|| < \epsilon .$$

But unfortunately the sequence $\{\phi_{x_0}^{(N)}\}$ is not a Cauchy sequence, since

$$||\phi_{x_0}^{(N)} - \phi_{x_0}^{(M)}|| = 2(1 - \sqrt{M/N})$$

(where we suppose that $N \ge M$), and hence the sequence does not converge to an element of $L_2[0, 1]$. We may summarize by saying that for every $x_0 \in [0, 1]$ we have an "approximate" eigenfunction with eigenvalue x_0, but there are *no* true eigenfunctions. For some operators it may even happen that there are both true and "approximate" eigenfunctions. It is because of difficulties like this that the concept of eigenfunctions and eigenvalues must be abandoned in order to obtain the Spectral Theorem for the most general type of self-adjoint operator.

9.2 LINEAR EQUATIONS AND PERTURBATION THEORY

Now that we have a completeness relation and a maximum-minimum principle available for completely continuous, self-adjoint operators, we can proceed to the important problems of perturbation theory and linear equations. To begin, let us consider the fundamental problem of solving a linear equation of the form

$$x = y + \lambda Ax , \tag{9.17}$$

where A is a completely continuous, self-adjoint operator. Let $\{x_i\}$ denote the complete set of eigenvectors of A, with associated eigenvalues λ_i. Taking the inner product of both sides of Eq. (9.17) with x_i, we get

$$\begin{aligned}(x_i, x) &= (x_i, y) + \lambda(x_i, Ax) \\ &= (x_i, y) + \lambda\lambda_i(x_i, x) ,\end{aligned}$$

and, therefore,

$$(1 - \lambda\lambda_i)(x_i, x) = (x_i, y) . \tag{9.18}$$

Thus if λ is not the reciprocal of any eigenvalue of A, then

$$(x_i, x) = \frac{(x_i, y)}{(1 - \lambda\lambda_i)} . \tag{9.19}$$

If $\lambda = \lambda_\nu^{-1}$, then Eq. (9.18) is well defined for $i = \nu$ only if $(x_\nu, y) = 0$, in which case the quantity (x_ν, x) is undetermined. If λ_ν has more than one associated eigenvector, a solution can exist only if y is orthogonal to all of them.

Now since

$$\phi \doteq \sum_{n=1}^{\infty} x_n(x_n, \phi)$$

in the sense of mean convergence, Eq. (9.19) tells us that

$$x \doteq \sum_{i=1}^{\infty} (1 - \lambda\lambda_i)^{-1}(x_i, y)x_i , \tag{9.20}$$

as long as λ is not the reciprocal of any λ_i. It is convenient to write this, using

$$(1 - \lambda\lambda_i)^{-1} = 1 + \lambda\lambda_i(1 - \lambda\lambda_i)^{-1} ,$$

as

$$x \doteq \sum_{i=1}^{\infty} (x_i, y)x_i + \lambda \sum_{i=1}^{\infty} \frac{\lambda_i}{1 - \lambda\lambda_i} (x_i, y)x_i \,.$$

Hence

$$x \doteq y + \lambda R_A y \,, \tag{9.21}$$

where for any $\xi \in H$, R_A is defined by

$$R_A \xi \equiv \sum_{i=1}^{\infty} \frac{\lambda_i}{1 - \lambda\lambda_i} (x_i, \xi)x_i ; \tag{9.22}$$

R_A is called the *resolvent* operator of A. Note that R_A is self-adjoint and that the infinite sum in Eq. (9.22) converges in the mean, since $\lambda_i/(1 - \lambda\lambda_i)$ is bounded ($\lambda_i \to 0$, $\lambda \neq 1/\lambda_i$ for any i), and the sequence of partial sums is therefore a Cauchy sequence. In fact, we can even say that the operator sequence of partial sums, $\{R_A^{(N)}\}$, where

$$R_A^{(N)} \xi \equiv \sum_{i=1}^{N} \frac{\lambda_i}{1 - \lambda\lambda_i} (x_i, \xi)x_i \,,$$

converges in operator norm to R_A, since

$$||(R_A - R_A^{(N)})\xi|| = \left[\sum_{i=N+1}^{\infty} \frac{\lambda_i^2}{(1 - \lambda\lambda_i)^2} |(x_i, \xi)|^2 \right]^{1/2} \leq \frac{|\lambda_{N+1}|}{\Delta_{\min}} \left[\sum_{i=N+1}^{\infty} |(x_i, \xi)|^2 \right]^{1/2} ;$$

$\Delta_{\min}$ is the minimum value of $|1 - \lambda\lambda_i|$ on the set of all eigenvalues of A, and we are ordering eigenvalues by decreasing absolute magnitude. Now

$$\sum_{i=N+1}^{\infty} |(x_i, \xi)|^2 \leq ||\xi|| \,,$$

so

$$||(R_A - R_A^{(N)})\xi|| \leq \frac{|\lambda_{N+1}|}{\Delta_{\min}} ||\xi|| \,.$$

But $\lambda_{N+1} \to 0$ as $N \to \infty$, $||R_A - R_A^{(N)}|| \to 0$ as $N \to \infty$. This means that since $R_A^{(N)}$ is an operator of finite rank, and hence completely continuous, then by Theorem 8.3, R_A is also completely continuous. We may summarize as follows:

Theorem 9.6. Let A be a completely continuous, self-adjoint operator on a Hilbert space, H. Then if λ is not equal to the inverse of any eigenvalue of A, $(I - \lambda A)^{-1}$ exists and is given by

$$(I - \lambda A)^{-1} = I + \lambda R_A \,, \tag{9.23}$$

where R_A is a completely continuous Hermitian operator defined for any $\xi \in H$ by

$$R_A \xi \equiv \sum_{i=1}^{\infty} \frac{\lambda_i}{1 - \lambda\lambda_i} (x_i, \xi)x_i \,. \tag{9.24}$$

The infinite sum converges in the mean; $\{x_i\}$ is the set of all eigenvectors of A.

Now suppose that $\lambda = 1/\lambda_\nu$. Then according to our previous work,

$$x = \tilde{x} + \sum_{i=1}^{\infty}{}' \frac{\lambda_i}{1 - \lambda\lambda_i} (x_i, y) x_i$$

is a solution to $x = y + \lambda Ax$ if and only if y is orthogonal to every eigenvector belonging to λ_ν. The prime on the summation means that eigenvectors belonging to λ_ν are to be omitted, and $\tilde{x}$ is *any* linear combination of eigenvectors belonging to λ_ν. Note that the statement that y be orthogonal to every eigenvector belonging to λ_ν is equivalent to the requirement that y be orthogonal to every linearly independent solution of the equation

$$A\xi = \lambda_\nu \xi ,$$

or, since A is self-adjoint,

$$A^\dagger \xi = \lambda_\nu \xi .$$

As in the case of operators of finite rank, we may combine all our results as follows:

Theorem 9.7. The Fredholm alternative holds for any completely continuous Hermitian operator.

This may be compared to Theorem 8.1 and its two corollaries proved for *arbitrary* transformations of finite rank.

Theorem 9.6 also gives us an example of a situation in which the method sketched nonrigorously in Section 7.2 can be carried out to obtain the Green's function of an operator, A. In the terminology of the last chapter, the Green's function of the operator $(I - \lambda A)$ is $(I + \lambda R_A)$; this may be seen by looking at Eq. (9.23). It should also be emphasized that R_A, as defined by Eq. (9.24), is a meromorphic function of λ, with poles located at $\lambda = 1/\lambda_i$.

Now, in most circumstances, when confronted with a linear equation of the type of Eq. (9.17), one will not be fortunate enough to know all the eigenvectors and eigenvalues of A, so that Eqs. (9.21) and (9.22) are often impractical. However, the result of Theorem 8.4 suggests that it may be possible to split A into two parts, $A = A_0 + A_1$, where A_1 is a "small" operator in the sense that $||A_1|| \ll ||A_0||$. It may then happen that one can deal with A_0 in a reasonable fashion. However, before we plunge into actual calculations, it is very important to be certain that the addition of a "small" A_1 to A_0 does not have a severe effect on the spectrum of A_0, since clearly the spectrum of A_0 (and A) will play a very important role in all considerations. It is reassuring to see that if A, A_0, and A_1 are completely continuous, then as $||A_1|| \to 0$, the eigenvalues of A_0 tend toward those of A.

Theorem 9.8. If $A = A_0 + A_1$ and A, A_0, and A_1 are completely continuous Hermitian operators, then we always have

$$\lambda_n^{(0)} - ||A_1|| \leq \lambda_n \leq \lambda_n^{(0)} + ||A_1|| ,$$

where $\lambda_n^{(0)}$ and λ_n are, respectively, the nth eigenvalues of A_0 and A.

Proof. First, consider $\lambda_n^{(0)+}$. Let C be the set of all $y \in H$ such that $||y|| = 1$ and $(y, x_i^{(0)+}) = 0$ for $i = 1, 2, \cdots, n-1$ where $x_i^{(0)+}$ denotes an eigenvector of A_0 with positive eigenvalue $\lambda_i^{(0)+}$. We have

$$\begin{aligned} \sup_{y \in C} (y, Ay) &= \sup_{y \in C} [(y, A_0 y) + (y, A_1 y)] \\ &\leq \sup_{y \in C} (y, A_0 y) + \sup_{y \in C} |(y, A_1 y)| . \end{aligned}$$

But by Theorems 9.1 and 9.5

$$\sup_{y \in C} (y, A_0 y) = \lambda_n^{(0)+} , \qquad \sup_{y \in C} (y, A_1 y) \leq ||A_1|| ,$$

so

$$\sup_{y \in C} (y, Ay) \leq \lambda_n^{(0)+} + ||A_1|| . \tag{9.25}$$

The term on the left of this equation looks very much like λ_n^+, but since y has to be orthogonal to $n-1$ eigenvectors of A_0 rather than $n-1$ eigenvectors of A, it is in fact *not* equal to λ_n^+. However, it is obviously greater than or equal to λ_n^+, since we can clearly find a normalized linear combination of the first n eigenvectors of A which lies in C (i.e., which is orthogonal to the first $n-1$ eigenvectors of A_0). For this vector, η, we will obviously have $(\eta, A\eta) \geq \lambda_n^+$, so

$$\sup_{y \in C} (y, Ay) \geq \lambda_n^+ .$$

Returning to Eq. (9.25), we have

$$\lambda_n^+ \leq \lambda_n^{(0)+} ||A_1|| .$$

For the second half of the inequality, we consider the set D of all y such that $||y|| = 1$ and $(y, x_i^+) = 0$ for $i = 1, 2, \cdots, n-1$. Then we have

$$\begin{aligned} \sup_{y \in D} (y, A_0 y) &= \sup_{y \in D} [(y, Ay) - (y, A_1 y)] \\ &\leq \sup_{y \in D} (y, Ay) + \sup_{y \in D} |(y, A_1 y)| . \end{aligned}$$

Now by the definition of D,

$$\sup_{y \in D} (y, Ay) = \lambda_n^+ , \qquad \text{so} \qquad \sup_{y \in D} (y, A_0 y) \leq \lambda_n^+ + ||A_1|| .$$

But by our previous argument,

$$\sup_{y \in D} (y, A_0 y) \geq \lambda_n^{(0)+} .$$

Hence

$$\lambda_n^{(0)+} \leq \lambda_n^+ + ||A_1|| , \qquad \text{or} \qquad \lambda_n^{(0)+} - ||A_1|| \leq \lambda_n^+ ,$$

which completes the proof for λ_n^+. The case of λ_n^- is handled in an identical fashion.

Now that we have confirmed the stability of the spectrum of a completely continuous Hermitian operator under small perturbations, let us investigate a perturbation approach to the solution of linear equations. We break A into two parts:

$$A = A_0 + A_1 ,$$

such that $||A_1||$ is very small. Then we write Eq. (9.17) as

$$x = y + \lambda A_0 x + \lambda A_1 x ,$$

or more suggestively,

$$(I - \lambda A_0)x = y + \lambda A_1 x .$$

Thus, according to Eq. (9.21), if λ^{-1} is not an eigenvalue of A_0,

$$x = y + \lambda A_1 x + \lambda R_{A_0}(y + \lambda A_1 x) ,$$

or

$$x = y + \lambda R_{A_0} y + \lambda(A_1 + \lambda R_{A_0} A_1)x . \tag{9.26}$$

This presents us with a new linear equation, but now the operator in question is one whose norm is small, since $||A_1||$ is assumed to be small. We have all the time assumed that λ^{-1} is not an eigenvalue of A_0. However, as $||A_1|| \to 0$, we know that the eigenvalues of A_0 get closer and closer to the eigenvalues of A. Thus, in this limit it is equivalent to assume that λ^{-1} is not an eigenvalue of A. As we make $||A_1||$ small, $||A_1 + \lambda R_{A_0} A_1||$ also becomes small; when $||A_1||$ becomes sufficiently small so that $|\lambda| \cdot ||A_1 + \lambda R_{A_0} A_1|| < 1$, then we can solve Eq. (9.26) by the Neumann series. Thus, in the limit $||A_1|| \to 0$,

$$x \to y + \lambda R_{A_0} y ,$$

if λ is not an eigenvalue of A. From Theorem 8.4, we know that it is always possible to approximate A in operator norm by a transformation of finite rank. Thus we can reduce the original problem to a matrix problem without knowing the eigenfunctions of A, by making use of the results of Section 8.6, in particular, Eq. (8.88). If A is an integral operator, the method of approximation will involve replacing the kernel by a series of step-functions, a method which we will discuss in more detail in the next section.

Now let us move on to the eigenvalue problem. For convenience, we write A in the form

$$A = A_0 + \epsilon A_1 ,$$

where we have put a constant, ϵ, in front of A_1 to emphasize the fact that this term will be small in norm. A, A_0, and A_1 are all completely continuous,

Hermitian operators. Consider the eigenvalue problem

$$Ax_n \equiv (A_0 + \epsilon A_1)x_n = \lambda_n x_n \,, \tag{9.27}$$

or, assuming that $\lambda_n \neq 0$,

$$\left(I - \frac{1}{\lambda_n} A_0\right) x_n = \frac{1}{\lambda_n} \epsilon A_1 x_n \,.$$

By our previous work, this linear equation with inhomogeneous term equal to $\epsilon A_1 x_n/\lambda_n$ can be solved by writing

$$x_n = \left(I - \frac{1}{\lambda_n} A_0\right)^{-1} \frac{1}{\lambda_n} \epsilon A_1 x_n \,.$$

By Eqs. (9.23) and (9.24), this is just

$$\begin{aligned} x_n &= \sum_{m=1}^{\infty} \left(1 - \frac{\lambda_m^{(0)}}{\lambda_n}\right)^{-1} \frac{1}{\lambda_n} \epsilon (x_m^{(0)}, A_1 x_n) x_m^{(0)} \\ &= \epsilon \sum_{m=1}^{\infty} (\lambda_n - \lambda_m^{(0)})^{-1} (x_m^{(0)}, A_1 x_n) x_m^{(0)} \,, \end{aligned} \tag{9.28}$$

where the $x_m^{(0)}$ are the eigenvectors of A_0 and $\lambda_m^{(0)}$ are the corresponding eigenvalues. We assume that λ_n is not equal to any of the eigenvalues of A_0. When $\epsilon||A_1||$ is sufficiently small, this certainly will be the case, *except* for the eigenvalue $\lambda_n^{(0)}$, which tends toward λ_n as $\epsilon \rightarrow 0$. We also assume for convenience that the spectrum of A_0 is nondegenerate. With the above difficulty in mind, let us isolate the possibly divergent denominator in Eq. (9.28) by writing

$$x_n = \frac{(x_n^{(0)}, \epsilon A_1 x_n)}{\lambda_n - \lambda_n^{(0)}} x_n^{(0)} + \epsilon \sum_{m=1}^{\infty}{}' \frac{(x_m^{(0)}, A_1 x_n)}{\lambda_n - \lambda_m^{(0)}} x_m^{(0)} \,, \tag{9.29}$$

where the prime on the summation means that the term with $m = n$ is omitted. Now Eq. (9.29) looks very much like an ordinary inhomogeneous linear equation. Before attempting to solve it by the methods we have developed, we should say something about the suspicious-looking constant multiplying $x_n^{(0)}$ in Eq. (9.29). Taking the inner product on both sides of Eq. (9.27) by $x_n^{(0)}$ and using the fact that

$$A_0 x_n^{(0)} = \lambda_n^{(0)} x_n^{(0)} \,,$$

we get

$$(x_n^{(0)}, \epsilon A_1 x_n) = (\lambda_n - \lambda_n^{(0)}) (x_n^{(0)}, x_n) \,. \tag{9.30}$$

This tells us that as $\lambda_n^{(0)} \rightarrow \lambda_n$, $(x_n^{(0)}, \epsilon A_1 x_n) \rightarrow 0$ in such a way that

$$\frac{(x_n^{(0)}, \epsilon A_1 x_n)}{\lambda_n - \lambda_n^{(0)}} = (x_n^{(0)}, x_n) < \infty \,.$$

This determines the constant, and allows us to write Eq. (9.29) as

$$x_n = (x_n^{(0)}, x_n) x_n^{(0)} + \epsilon \sum_{m=1}^{\infty}{}' \frac{(x_m^{(0)}, A_1 x_n)}{\lambda_n - \lambda_m^{(0)}} x_m^{(0)} \,. \tag{9.31}$$

Only under rather unusual circumstances would $(x_n^{(0)}, x_n) = 0$, and certainly not if ϵ is sufficiently small, for under this assumption we can generate a contradiction. Assuming that $(x_n^{(0)}, x_n) = 0$, Eq. (9.31) becomes

$$x_n = \epsilon \sum_{m=1}^{\infty}{}' \frac{(x_m^{(0)}, A_1 x_n)}{\lambda_n - \lambda_m^{(0)}} x_m^{(0)} .$$

It follows that

$$||x_n|| = \epsilon \left[\sum_{m=1}^{\infty}{}' \frac{|(x_m^{(0)}, A_1 x_n)|^2}{(\lambda_n - \lambda_m^{(0)})^2} \right]^{1/2} \leq \frac{\epsilon}{\Delta_n} \left[\sum_{m=1}^{\infty}{}' |(x_m^{(0)}, A_1 x_n)|^2 \right]^{1/2} ,$$

where Δ_n is the minimum of $|\lambda_n - \lambda_m^{(0)}|$ taken over all eigenvalues of A_0 except $\lambda_n^{(0)}$. For ϵ sufficiently small, Δ_n tends toward a number which is independent of ϵ. Thus, by completeness, we have

$$||x_n|| \leq \frac{\epsilon}{\Delta_n} (A_1 x_n, A_1 x_n)^{1/2} = \frac{\epsilon}{\Delta_n} ||A_1 x_n|| ,$$

so that

$$||x_n|| \leq \frac{\epsilon ||A_1||}{\Delta_n} ||x_n|| \qquad \text{or} \qquad 1 \leq \frac{\epsilon ||A_1||}{\Delta_n} .$$

For ϵ sufficiently small, we clearly have a contradiction.

Since the equation for x_n is homogeneous, x_n is arbitrary up to a multiplicative constant; setting $(x_n^{(0)}, x_n) = 1$ is a convenient way to fix that constant. With this choice, Eq. (9.31) takes the simple form

$$x_n = x_n^{(0)} + \epsilon \sum_{m=1}^{\infty}{}' \frac{(x_m^{(0)}, A_1 x_n)}{\lambda_n - \lambda_m^{(0)}} x_m^{(0)} . \tag{9.32}$$

Also, according to Eq. (9.30),

$$\lambda_n = \lambda_n^{(0)} + \epsilon (x_n^{(0)}, A_1 x_n) . \tag{9.33}$$

Equations. (9.32) and (9.33) are the starting points for studying perturbations in the spectra of completely continuous Hermitian operators.

Certainly the most promising attack on these equations for the case of $\epsilon ||A_1||$ small but finite would seem to be through iterative techniques. However, we shall have to proceed in a slightly different manner than in Sections 8.1 and 8.3, since the operator on the right-hand side of Eq. (9.32) contains an unknown quantity, λ_n. On the other hand, Eq. (9.33) is an equation for λ_n, but in terms of the unknown vector x_n. Thus we have a pair of coupled nonlinear equations for x_n and λ_n. Now we have seen that the norm of the operator on the right-hand side of Eq. (9.32) is equal to or less than $\epsilon ||A_1|| / \Delta_n$. Therefore, if this quantity is less than 1, we can iterate Eq. (9.32) to obtain

$$\begin{aligned} x_n = x_n^{(0)} &+ \epsilon \sum_{m=1}^{\infty}{}' \frac{(x_m^{(0)}, A_1 x_n^{(0)})}{\lambda_n - \lambda_m^{(0)}} x_m^{(0)} \\ &+ \epsilon^2 \sum_{l=1}^{\infty}{}' \sum_{m=1}^{\infty} \frac{(x_m^{(0)}, A_1 x_l^{(0)})(x_l^{(0)}, A_1 x_n^{(0)})}{(\lambda_n - \lambda_m^{(0)})(\lambda_n - \lambda_l^{(0)})} x_m^{(0)} + \cdots . \end{aligned} \tag{9.34}$$

When we have computed x_n to sufficient accuracy, we insert Eq. (9.34) into Eq. (9.33) and thus obtain a relation for λ_n. Of course, in general, this equation will be extremely complicated.

Another possibility would be to substitute Eq. (9.33) into Eq. (9.34) to obtain

$$x_n = x_n^{(0)} + \epsilon \sum_{m=1}^{\infty}{}' \frac{(x_m^{(0)}, A_1 x_n^{(0)}) x_m^{(0)}}{\lambda_n^{(0)} - \lambda_m^{(0)} + \epsilon(x_n^{(0)}, A_1 x_n)}$$
$$+ \epsilon^2 \sum_{m=1}^{\infty}{}' \sum_{l=1}^{\infty}{}' \frac{(x_m^{(0)}, A_1 x_l^{(0)})(x_l^{(0)}, A_1 x_n^{(0)})}{[\lambda_n^{(0)} - \lambda_m^{(0)} + \epsilon(x_n^{(0)}, A_1 x_n)][\lambda_n^{(0)} - \lambda_l^{(0)} + \epsilon(x_n^{(0)}, A_1 x_n)]} x_m^{(0)}$$
$$+ \cdots . \tag{9.35}$$

If $\epsilon \| A_1 \|$ is small enough that

$$\left| \frac{\epsilon(x_n^{(0)}, A_1 x_n)}{\lambda_n^{(0)} - \lambda_m^{(0)}} \right| < 1 , \tag{9.36}$$

for all $m \neq n$, then we can expand each of the denominators occurring in Eq. (9.35) in a power series in ϵ:

$$[\lambda_n^{(0)} - \lambda_m^{(0)} + \epsilon(x_n^{(0)}, A_1 x_n)]^{-1} = (\lambda_n^{(0)} - \lambda_m^{(0)})^{-1} \sum_{\mu=0}^{\infty} (-1)^\mu \epsilon^\mu \frac{(x_n^{(0)}, A_1 x_n)^\mu}{(\lambda_n^{(0)} - \lambda_m^{(0)})^\mu} . \tag{9.37}$$

Putting this result into Eq. (9.35), we obtain for the first few terms of the combined iteration and denominator expansion

$$x_n = x_n^{(0)} + \sum_{m=1}^{\infty}{}' \sum_{\mu=0}^{\infty} (-1)^\mu \epsilon^{\mu+1} \frac{(x_m^{(0)}, A_1 x_n^{(0)})}{[\lambda_n^{(0)} - \lambda_m^{(0)}]^{\mu+1}} (x_n^{(0)}, A_1 x_n)^\mu x_m^{(0)}$$
$$+ \sum_{m=1}^{\infty}{}' \sum_{l=1}^{\infty}{}' \sum_{\mu=0}^{\infty} \sum_{\nu=0}^{\infty} (-1)^{\mu+\nu} \epsilon^{\mu+\nu+2}$$
$$\times \frac{(x_m^{(0)}, A_1 x_l^{(0)})(x_l^{(0)}, A_1 x_n^{(0)})}{[\lambda_n^{(0)} - \lambda_m^{(0)}]^{\mu+1}[\lambda_n^{(0)} - \lambda_l^{(0)}]^{\nu+1}} (x_n^{(0)}, A_1 x_n)^{\mu+\nu} x_m^{(0)}$$
$$+ \cdots . \tag{9.38}$$

If ϵ is small enough, we obtain a convergent power series expansion for x_n. If we now write

$$x_n = \sum_{l=0}^{\infty} \epsilon^l f_n^{(l)} \tag{9.39}$$

and insert this expression into both sides of (9.38), we can calculate the vectors $f_n^{(l)}$ by equating the coefficients of equal powers of ϵ on either side of the equation. Doing this, we find for the first few $f_n^{(l)}$

$$f_n^{(0)} = x_n^{(0)} ,$$

$$f_n^{(1)} = \sum_{m=1}^{\infty}{}' \frac{(x_m^{(0)}, A_1 x_n^{(0)})}{\lambda_n^{(0)} - \lambda_m^{(0)}} x_m^{(0)} ,$$

and

$$f_n^{(2)} = \sum_{m=1}^{\infty}{}' \sum_{l=1}^{\infty}{}' \frac{(x_m^{(0)}, A_1 x_l^{(0)})(x_l^{(0)}, A_1 x_n^{(0)})}{(\lambda_n^{(0)} - \lambda_m^{(0)})(\lambda_n^{(0)} - \lambda_l^{(0)})} x_m^{(0)}$$

$$- (x_n^{(0)}, A_1 x_n^{(0)}) \sum_{m=1}^{\infty}{}' \frac{(x_m^{(0)}, A_1 x_n^{(0)})}{(\lambda_n^{(0)} - \lambda_m^{(0)})^2} x_m^{(0)} . \tag{9.40}$$

Putting these into Eq. (9.39), we obtain an expression for x_n through order ϵ^2. With this result, we then use Eq. (9.33) to calculate λ_n through third order in ϵ. The eigenvalue λ_n can also be written in a power series, the first term of which is $\lambda_n^{(0)}$. We have

$$\lambda_n = \sum_{l=0}^{\infty} \epsilon^l \lambda_n^{(l)} . \tag{9.41}$$

Using x_n as given by Eqs. (9.39) and (9.40), we find that

$$\lambda_n^{(1)} = (x_n^{(0)}, A_1 x_n^{(0)}), \qquad \lambda_n^{(2)} = \sum_{m=1}^{\infty}{}' \frac{|(x_m^{(0)}, A_1 x_n^{(0)})|^2}{\lambda_n^{(0)} - \lambda_m^{(0)}} \tag{9.42}$$

and similarly for higher-order terms.

These results agree exactly with those obtained in Rayleigh-Schrödinger perturbation theory (Section 4.11). Thus we have been able to give a derivation of Rayleigh-Schrödinger perturbation theory, at least in the nondegenerate case, and have shown that if ϵ is sufficiently small, the perturbation series for x_n will converge. It is difficult to give a more precise statement of convergence requirements, but it is fairly clear from the above discussion that if the difference between $\lambda_n^{(0)}$ and the nearest eigenvalue of A_0 is much greater in magnitude than $(\lambda_n - \lambda_n^{(0)})$, then the method has a good chance of converging. As a rule of thumb, this estimate is often helpful.

The results of Eqs. (9.39), (9.40), (9.41), and (9.42) suffer from the drawback that they require the knowlege of *all* the eigenvectors and eigenvalues of A_0. There are several ways to avoid this difficulty. One way is to make use of the proved *existence* of series expansions for λ_n and x_n and then obtain coupled equations for the vectors $f_n^{(l)}$ involving the $\lambda_n^{(l)}$ as was done in Section 4.11. In this way one obtains, for example,

$$(A_0 - \lambda_n^{(0)}) f_n^{(1)} = -(A_1 - \lambda_n^{(1)}) x_n^{(0)} ,$$
$$(A_0 - \lambda_n^{(0)}) f_n^{(2)} = -(A_1 - \lambda_n^{(1)}) f_n^{(1)} + \lambda_n^{(2)} x_n^{(0)} ,$$
$$(A_0 - \lambda_n^{(0)}) f_n^{(3)} = -(A_1 - \lambda_n^{(1)}) f_n^{(2)} + \lambda_n^{(2)} f_n^{(1)} + \lambda_n^{(3)} x_n^{(0)} , \tag{9.43}$$

and so forth, just as in Section 4.11. These equations involve only $x_n^{(0)}$ and $\lambda_n^{(0)}$ as their starting point, and one has just a system of inhomogeneous equations to solve. This system was discussed in some detail in Section 4.11.

Another approach is to make a direct attack on A itself, using the maximum and minimum principles of Theorem 9.5. Since we know that for any $y \in H$, $||y|| = 1$,

$$(y, Ay) \leq \lambda_1^+ ,$$

we can build "trial vectors," y_t, containing many variable parameters and then maximize (y, Ay) with respect to these parameters, subject to the constraint $||y_t|| = 1$. If the parameters occur linearly in y_t it can be shown (see Problem 5) that the problem reduces to a matrix problem, with the linear parameters in y_t being the unknowns. As we perform this process with more and more parameters, the successive values of (y_t, Ay_t) must increase toward, but never exceed, λ_1^+. If one chooses trial functions sensibly, this process can very quickly yield excellent approximations for λ_1^+ and x_1^+ (the function y_t which gives the good approximation to λ_1^+).

As an example of this method, we consider the equation

$$Ax = \int_0^1 t_<(1 - t_>)x(t')\,dt' = \lambda x(t) ,$$

where A is defined on the space of square-integrable function on $[0, 1]$. We already know (see Section 7.1) that this equation has the eigenfunctions

$$x_n(t) = \sqrt{2}\sin(n\pi t)$$

and eigenvalues

$$\lambda_n = 1/n^2\pi^2$$

for the boundary conditions $x_n(0) = 0 = x_n(1)$. Now the simplest normalized trial function which vanishes at $t = 0$ and $t = 1$ is

$$y_t(t) = \sqrt{30}\,t(1 - t) .$$

If we calculate (y_t, Ay_t), we find that

$$(y_t, Ay_t) = 17/168 = 0.10119 ,$$

which is less than $1/\pi^2 = 0.10132$, but not by much! Thus by making the most obvious choice, we get a result which is extremely close to the correct one. Needless to say, in most problems one is not quite so lucky. One can also calculate $||x_1^+ - y_t||$:

$$\begin{aligned} ||x_1^+ - y_t|| &= ||\sqrt{2}\sin(\pi t) - \sqrt{30}\,t(1 - t)|| \\ &= [2(1 - 4\sqrt{60}/\pi^3)]^{1/2} = 0.038 . \end{aligned}$$

Thus y_t approximates x_1^+ very closely in norm. If one wanted still greater accuracy, one would choose a trial function with several parameters, for example,

$$y_t = at(1 - t) + bt^2(1 - t)^2 ,$$

and then maximize (y_t, Ay_t) with respect to a and b subject to $||y_t|| = 1$. In practice, it is convenient to handle the constraint on the norm of y_t by using a Lagrange multiplier. Clearly, $\lambda_2^+, \lambda_3^+, \cdots$ can be computed in the same way by using more constraint equations to ensure that the proper orthogonality properties are satisfied.

9.3 FINITE-RANK TECHNIQUES FOR EIGENVALUE PROBLEMS

For most realistic physical problems, the chances of obtaining closed-form solutions are rather remote. This perhaps comes as a shock to the student who has been exposed to the beautiful exact solutions which can be obtained for the harmonic oscillator and the $1/r$ potential in both classical and quantum mechanics. However, problems like these are the exception rather than the rule in physics (although very important exceptions, since they often provide the insight necessary for the understanding of more complicated problems). In the case of the Coulomb potential, for example, all one need do is to consider the potential as arising from a charge distribution of finite extent (a proton) rather than a point charge, and one is immediately forced to look beyond exact solutions to some sort of approximation procedure.

However, in the problem just mentioned, the effect of the finite proton size can be thought of as a small perturbation so that the techniques discussed in Section 9.2 can be successfully applied. The type of problem with which we shall be concerned in this section is one in which the idea of a small perturbation of a well-understood physical system is inadequate. For such problems, we will make use of a different sort of approximation, namely, the notion of approximation by an operator of finite rank, which was discussed in Section 8.7.

What we want to do is to consider the eigenvalue problem of Eq. (9.27) in the case when $\epsilon||A_1||$ tends to zero, i.e., when we approximate A extremely closely by A_0, and in principle, arbitrarily closely (see Theorem 8.4). In this case, according to our work in the last section

$$x_n^{(0)} \to x_n$$

and

$$\lambda_n^{(0)} \to \lambda_n \,.$$

Let us focus our attention on a Hermitian operator whose kernel is $a(t, t')$ with t and t' in an interval $[a, b]$. We have already encountered such problems in Chapter 7 when discussing Schrödinger's integral equation. The basic idea which we will use is to divide $[a, b]$ into N small segments, each of length $(b - a)/N$, and define the set of orthogonal functions

$$\phi_n^{(N)}(t) = \begin{cases} 1\,, & a + \dfrac{n-1}{N}\,l \le t \le a + \dfrac{n}{N}\,l\,, \\ 0\,, & \text{otherwise}\,, \end{cases}$$

where $l = b - a$ and $n = 1, 2, \cdots, N$. If we wished, we could normalize $\phi_n^{(N)}(t)$ to 1 by multiplying it by $N^{1/2}$. Now we can approximate any kernel in L_2 arbitrarily closely by such functions; if the kernel is defined on an infinite range, then we truncate it by setting it equal to zero when t or t' is outside some large, but finite interval. It is usually possible to determine a sensible point of truncation by inspection of the kernel. With the $\phi_n^{(N)}$, we can approximate $a(t, t')$ arbitrarily closely by some $a_N(t, t')$, satisfying

$$||a(t, t') - a_N(t, t')|| < \epsilon$$

for any preassigned ϵ. If we call the operator associated with $a_N(t, t')$, A_N, then

$$||A - A_N|| < \epsilon .$$

Thus according to Theorem 9.8 the eigenvalues of A_N tend toward those of A, and according to Eq. (9.32) the eigenvectors of A_N tend in norm toward those of A. Hence the problem reduces to solving the equation

$$A_N x_n^{(0)}(t) = \lambda_n^{(0)} x_n^{(0)}(t) , \tag{9.44}$$

for some sufficiently large N. In terms of the $\phi_n^{(N)}$, the kernel of A_N can be written as

$$a_N(t, t') = \sum_{\mu,\nu=1}^{N} M_{\mu\nu}\phi_\mu^{(N)}(t)\phi_\nu^{(N)}(t') , \tag{9.45}$$

where

$$M_{\mu\nu} = (N/l)^2 \int dt \int dt' \; \phi_\mu^{(N)}(t)\phi_\nu^{(N)}(t')a(t, t') . \tag{9.46}$$

For convenience, we take $a(t, t')$ to be real. Note that $M_{\mu\nu}$ is simply the *average value* of the function $a(t, t')$ on the $\mu\nu$-square of the two-dimensional grid into which we have divided the tt'-plane. Using Eq. (9.45), we get for Eq. (9.44)

$$\sum_{\mu,\nu=1}^{N} \phi_\mu^{(N)}(t)M_{\mu\nu} \int \phi_\nu^{(N)}(t')x_n^{(0)}(t') \, dt' = \lambda_n^{(0)} x_n^{(0)}(t) .$$

If we write

$$C_\nu^{(n)} \equiv (N/l) \int \phi_\nu^{(N)}(t')x_n^{(0)}(t') \, dt'$$

for the average value of $x_n^{(0)}$ on the νth subinterval of $[a, b]$, then our equation becomes

$$\sum_{\mu,\nu=1}^{N} \phi_\mu^{(N)}(t)M_{\mu\nu}C_\nu^{(n)} \frac{l}{N} = \lambda_n^{(0)} x_n^{(0)}(t) . \tag{9.47}$$

Thus all we need to do is evaluate the scalars $C_\nu^{(n)}$ by using Eq. (9.47). Multiplying both sides by $\phi_{\nu'}^{(N)}(t)$ and integrating, we get

$$\frac{l}{N} \sum_{\nu=1}^{N} M_{\nu'\nu}C_\nu^{(n)} = \lambda_n^{(0)} C_{\nu'}^{(n)} \tag{9.48}$$

for $\nu' = 1, 2, \cdots, N$. Equation (9.48) thus reduces our problem to one of finding the eigenvalues and eigenvectors of a finite-dimensional matrix, $\boldsymbol{M}$. With Eq. (9.48) solved for $\lambda_n^{(0)}$ and $C_\nu^{(n)}$ ($\nu = 1, 2, \cdots, N$), then Eq. (9.47) tells us that

$$x_n^{(0)}(t) = \sum_{\nu=1}^{N} C_\nu^{(n)}\phi_\nu^{(N)}(t) . \tag{9.49}$$

A simple way of looking at these results is to consider the equation

$$Ax_n = \lambda_n x_n = \int_a^b a(t, t')x_n(t')\,dt'$$

from the viewpoint of simple Riemann integration theory. If we subdivide the interval of integration into N parts, each of length l/N, where l is the length of the interval, then our equation can be written as

$$\sum_{\nu=1}^{N} a(t, t_\nu)x_n(t_\nu)\left(\frac{l}{N}\right) = \lambda_n x_n(t) ,$$

where t_ν is some point in the νth interval. If N is large enough this will approximate the original integral arbitrarily closely, as long as the functions involved are continuous. Thus

$$\sum_{\nu=1}^{N} a(t_{\nu'}, t_\nu)x_n(t_\nu)\left(\frac{l}{N}\right) = \lambda_n x_n(t_{\nu'}) .$$

Defining

$$\tilde{M}_{\nu'\nu} \equiv a(t_{\nu'}, t_\nu) , \qquad \tilde{C}_\nu^{(n)} \equiv x_n(t_\nu) ,$$

we get simply

$$\frac{l}{N}\sum_{\nu=1}^{N} \tilde{M}_{\nu'\nu}\tilde{C}_\nu^{(n)} = \lambda_n \tilde{C}_{\nu'}^{(n)} , \tag{9.50}$$

which obviously bears a very close relation to Eq. (9.48). We see that the basic idea behind the practical solution of integral equations is no more complicated than the basic idea involved in evaluating the Riemann integral.

For $||A_1|| \ll 1$, the approximate eigenvalues of A are given by Eq. (9.48) and the related eigenvectors by Eqs. (9.48) and Eq. (9.49). To estimate the size of the error involved in the approximation, we can make use of the perturbation theory results of Eqs. (9.39), (9.40), (9.41), and (9.42). A particularly important point in this connection is the fact that if $x_m^{(0)}$ and $x_n^{(0)}$ are any approximate solutions of the form

$$x_m^{(0)} = \sum_{\nu=1}^{N} C_\nu^{(m)}\phi_\nu^{(N)} , \qquad x_n^{(0)} = \sum_{\nu=1}^{N} C_\nu^{(n)}\phi_\nu^{(N)} , \tag{9.51}$$

then

$$(x_m^{(0)}, A_1 x_n^{(0)}) = (x_m^{(0)}, (A - A_N)x_n^{(0)}) = 0 \tag{9.52}$$

so that the perturbation theory results simplify considerably. Indeed, at first glance it might seem that they simplify too much, namely, to

$$\lambda_n = \lambda_n^{(0)} \qquad \text{and} \qquad x_n = x_n^{(0)} ,$$

since all the terms contributing to $f_n^{(1)}, f_n^{(2)}$, etc., in Eq. (9.40) and to $\lambda_n^{(1)}$, $\lambda_n^{(2)}$, etc., in Eq. (9.42) appear to vanish. However, this is not the case. Vectors of the form of Eq. (9.51) are *not* the only eigenvectors of A_N. *Any* vector which is orthogonal to all N of the $\phi_n^{(N)}$ is an eigenvector of A_N with eigenvalue zero. There is an infinite number of such vectors, and we have not determined any of them. If we agree to label these starting with index $N+1$, that is, $x_{N+1}^{(0)}$, $x_{N+2}^{(0)}, \cdots$, then the corrections to x_n and λ_n can be put in the form

$$\begin{aligned} x_n &= x_n^{(0)} + \frac{1}{\lambda_n^{(0)}} \sum_{\nu=N+1}^{\infty} (x_\nu^{(0)}, A_1 x_n^{(0)}) x_\nu^{(0)} + \cdots, \\ \lambda_n &= \lambda_n^{(0)} + \frac{1}{\lambda_n^{(0)}} \sum_{\nu=N+1}^{\infty} |(x_\nu^{(0)}, A_1 x_n^{(0)})|^2 + \cdots. \end{aligned} \tag{9.53}$$

Since $A_0 x_\nu^{(0)} = 0$ for $\nu > N$, these can also be written as

$$\begin{aligned} x_n &= x_n^{(0)} + \frac{1}{\lambda_n^{(0)}} \sum_{\nu=N+1}^{\infty} (x_\nu^{(0)}, A x_n^{(0)}) x_\nu^{(0)} + \cdots, \\ \lambda_n &= \lambda_n^{(0)} + \frac{1}{\lambda_n^{(0)}} \sum_{\nu=N+1}^{\infty} |(x_\nu^{(0)}, A x_n^{(0)})|^2 + \cdots. \end{aligned} \tag{9.54}$$

In particular, note that the leading correction to $\lambda_n^{(0)}$, $(x_n^{(0)}, A_1 x_n^{(0)})$, vanishes. Equations (9.53) indicate that the first nonvanishing correction to $\lambda_n^{(0)}$ is always positive if $\lambda_n^{(0)}$ is positive, and that this correction is certainly less than or equal to $||A_1||^2/\lambda_n^{(0)}$. However, $||x_n - x_n^{(0)}|| \leq ||A_1||/\lambda_n^{(0)}$, so we have a better estimate on λ_n than on x_n. In either case, the corrections will grow steadily larger as we get to small eigenvalues.

As a typical example, we consider Schrödinger's integral equation for the bound states of a particle in a field given by the centrally symmetric potential $V(|\mathbf{r}|)$. From symmetry considerations we know that the angular part of the bound-state wave function is given by the spherical harmonic $Y_{lm}(\theta, \phi)$ where l is the angular momentum of the particle in question and m is the projection of the angular momentum on the z-axis. The radial wave functions satisfy Eq. (7.94) which we will reproduce here:

$$\phi_l(r) = \frac{2m\kappa}{\hbar^2} \int_0^\infty j_l(i\kappa r_<) h_l^{(1)}(i\kappa r_>) V(r') \phi_l(r') r'^2 \, dr' , \tag{9.55}$$

where κ is related to the binding energy of the particle by

$$\kappa = \sqrt{-2mE/\hbar^2}$$

and $r_>$ and $r_<$ are respectively the greater and lesser of r and r'. Now, as it stands, the kernel of Eq. (9.55) is not self-adjoint, because

$$k(r, r') \neq k^*(r', r) .$$

However, by using the symmetrization method discussed in Problem 8.5, we can easily bring it to this form. We define

$$\psi_l(r) \equiv r V^{1/2}(r) \phi_l(r) .$$

Then Eq. (9.55) becomes

$$\psi_l(r) = \frac{2m\kappa}{\hbar^2}\int_0^\infty rV^{1/2}(r)j_l(i\kappa r_<)h_l^{(1)}(i\kappa r_>)V^{1/2}(r')r'\psi_l(r')\,dr' . \tag{9.56}$$

The kernel is now the symmetric function

$$k_l(r, r') = \frac{2m\kappa}{\hbar^2}rV^{1/2}(r)j_l(i\kappa r_<)h_l^{(1)}(i\kappa r_>)V^{1/2}(r')r' .$$

This satisfies $k_l(r, r') = k_l^*(r', r)$ since, as the reader can easily show from Eqs. (7.86a) and (7.86c), the product $j_l(i\kappa r_<)h_l^{(1)}(i\kappa r_>)$ is always real.

As an illustration, we consider the familiar Coulomb potential,

$$V(r) = -Ze^2/r$$

with $l=0$ (although it is not difficult to work with other values of l). We choose the Coulomb potential in order to have exact results for comparison; actually, from the numerical point of view a potential like the Yukawa potential would be somewhat easier to handle because of its exponential cut-off.

Now, since by Eqs. (7.86a) and (7.86c),

$$j_0(i\kappa r_<) = \frac{\sinh(\kappa r_<)}{\kappa r_<}, \qquad h_0^{(1)}(i\kappa r_>) = -\frac{e^{-\kappa r_>}}{\kappa r_>},$$

we have

$$k_0(r, r') = \frac{2me^2Z}{\hbar^2\kappa}\left[\frac{\sinh(\kappa r_<)e^{-\kappa r_>}}{r_>^{1/2}r_<^{1/2}}\right], \tag{9.57}$$

and the integral equation becomes

$$\psi_0(r) = \frac{2Z}{\kappa a_0}\int_0^\infty \frac{1}{r_>^{1/2}r_<^{1/2}}e^{-\kappa r_>}\sinh(\kappa r_<)\psi_0(r')\,dr' .$$

The quantity $a_0 \equiv \hbar^2/me^2$ has the dimensions of length and is called the *Bohr radius*.

It is convenient to make the problem dimensionless by defining

$$\rho \equiv r/a_0 , \qquad q \equiv \kappa a_0 \equiv [-E/(\tfrac{1}{2}mc^2\alpha^2)]^{1/2} , \tag{9.58}$$

where $\alpha \equiv e^2/\hbar c$ is the dimensionless "fine structure" constant. In these units,

$$\psi_0(\rho) = \frac{2Z}{q}\int_0^\infty \frac{1}{\rho_>^{1/2}\rho_<^{1/2}}e^{-q\rho_>}\sinh(q\rho_<)\psi_0(\rho')\,d\rho' . \tag{9.59}$$

Before trying to solve this equation by finite-rank approximation, we should make certain that it is completely continuous by showing that the kernel is square-integrable. We have

$$I = \int_0^\infty d\rho\int_0^\infty d\rho'|k_0(\rho, \rho')|^2 = \frac{4Z^2}{q^2}\int_0^\infty d\rho\int_0^\infty d\rho'\,\frac{e^{-2q\rho_>}[\sinh(q\rho_<)]^2}{\rho_>\rho_<} .$$

Assuming that q is positive, the integrations yield

$$I = \frac{\pi^2 Z^2}{6q^2}, \tag{9.60}$$

so that the kernel is indeed square-integrable.

Now the problem posed by Eq. (9.59) is to find the eigenvalues (binding energies), q, for a given Z. However, the kernel appears to have a rather complicated, nonlinear dependence on q. Actually in the present case, this dependence is illusory, since the change of variables

$$x = q\rho, \qquad x' = q\rho', \tag{9.61}$$

leads to the eigenvalue equation

$$2Z \int_0^\infty \frac{e^{-x_>} \sinh(x_<)}{x_>^{1/2} x_<^{1/2}} \psi_0(x')\, dx' = q\psi_0(x). \tag{9.62}$$

Note that the change of variable of Eq. (9.61) supposes that q is positive, as we have already assumed in showing that the kernel of Eq. (9.59) is square-integrable. However, in most cases—for example, if $V(r) = g^2 e^{-\mu r}/r$—it will be impossible to remove the q-dependence in this way; under these circumstances it is often desirable to regard q as fixed and Z as the quantity to be determined. In other words, we ask what potential "strength" will produce a given binding energy.

We can illustrate the problem of obtaining solutions of an integral equation by solving Eq. (9.62) for the hydrogen atom, where $Z = 1$. If we truncate the kernel for $x > 7$, $x' > 7$, and divide the interval $[0, 7]$ into thirty-five equal parts, then the procedure discussed above in obtaining Eqs. (9.47) and (9.48) leads to the problem of calculating the eigenvalues of a 35×35 matrix. With modern computing machines this is an elementary problem. For the first few eigenvalues, we find that

$$q_1 = 0.994, \qquad q_2 = 0.495, \qquad q_3 = 0.326, \qquad q_4 = 0.233.$$

The exact values may be obtained by using the formula for the bound-state energies of hydrogen

$$E_n = -\tfrac{1}{2} mc^2\alpha^2/n^2.$$

According to Eq. (9.58), this means that

$$q_1 = 1 = 1.000, \quad q_2 = \tfrac{1}{2} = 0.500, \quad q_3 = \tfrac{1}{3} = 0.333, \quad q_4 = \tfrac{1}{4} = 0.250.$$

As suggested by Eq. (9.53), the approximate values are always below the exact values. As the eigenvalues get smaller, the agreement worsens, which is also in accord with Eq. (9.53). Note that

$$\sum_{n=1}^\infty q_n^2 = \sum_{n=1}^\infty \frac{1}{n^2} = \frac{\pi^2}{6},$$

which is equal to the integrated square of the kernel of Eq. (9.62), as may be seen by looking at Eq. (9.60). This is in agreement with the general result of Eq. (9.15). Since the wave functions of hydrogen are known, we can write the exact solutions to Eq. (9.62). Taking into account the changes of variable of Eqs. (9.58) and (9.61), and the definition of ψ_0 in terms of ϕ_0, we have for the first three eigenvectors (see, for example, L.I. Schiff, *Quantum Mechanics*. New York: McGraw-Hill, 1969)

$$\psi_{10} = 2\sqrt{x}\, e^{-x} ,$$

$$\psi_{20} = 2\sqrt{2x}(1 - x)e^{-x} ,$$

$$\psi_{30} = 2\sqrt{3x}(1 - 2x + \tfrac{2}{3}x^2)e^{-x} .$$

These are all orthogonal and normalized to 1 for convenience.

Figure 9.1 shows the exact ψ_{10} (solid curve) compared with the approximate ψ_{10} (dotted curve). The agreement is very close. Since, strictly speaking, the approximate ψ_{10} consists of a series of flat line segments in each subinterval, we have interpolated by replacing this by a point at the middle of the interval. This is an excellent approximation when the curve varies slowly, and is not so good where the curve varies more rapidly (for example, in Fig. 9.1, near the origin). In such regions of rapid variation, one should choose smaller subintervals to obtain higher accuracy. In the framework we have developed, this is easily done. Figures 9.2 and 9.3 show similar comparisons for the second and third eigenvectors. Note that the deviations from the exact result are a bit more noticeable in these cases, although the agreement is still good.

The calculation would have been simpler had we evaluated the matrix in the spirit of the Riemann integral scheme discussed above. There each matrix element is obtained simply by evaluating the kernel at a point in the xx'-plane. For the sake of comparison, we include in Fig. 9.3 the curve which was obtained by this method for ψ_{30}. The corresponding eigenvalue was $q_3 = 0.353$. It is

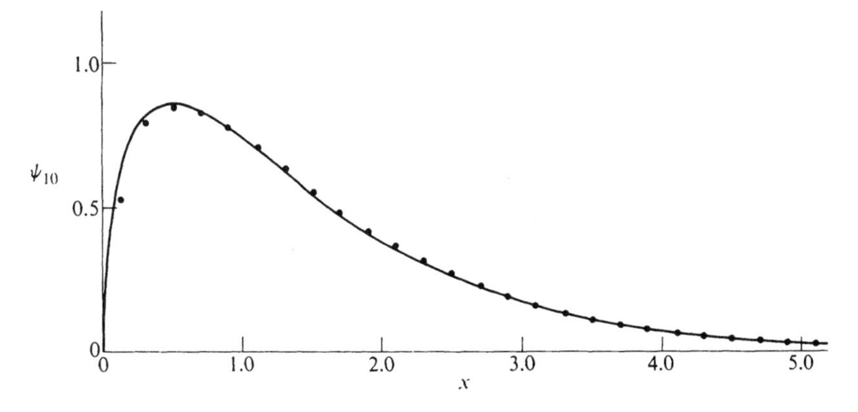

Fig. 9.1 The exact eigenvector, ψ_{10}, solid curve, and the approximate ψ_{10}, dotted curve.

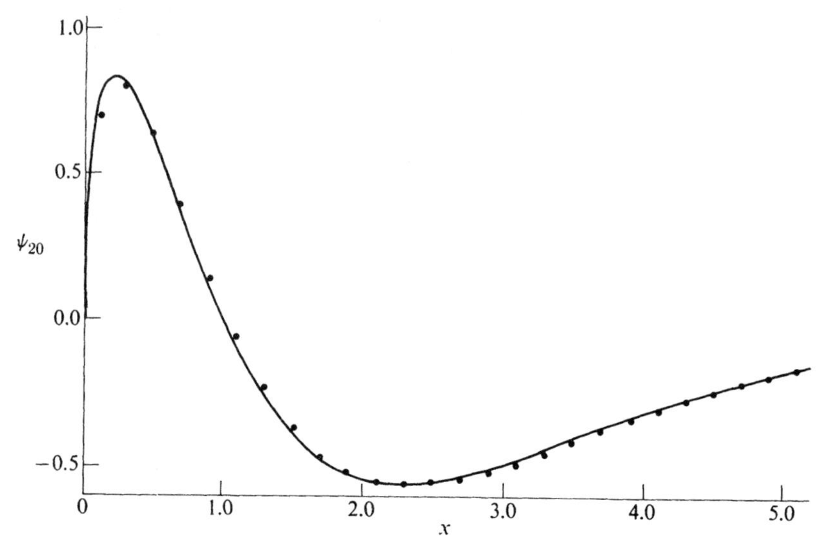

Fig. 9.2 The exact eigenvector, ψ_{20}, solid curve, and the approximate ψ_{20}, dotted curve.

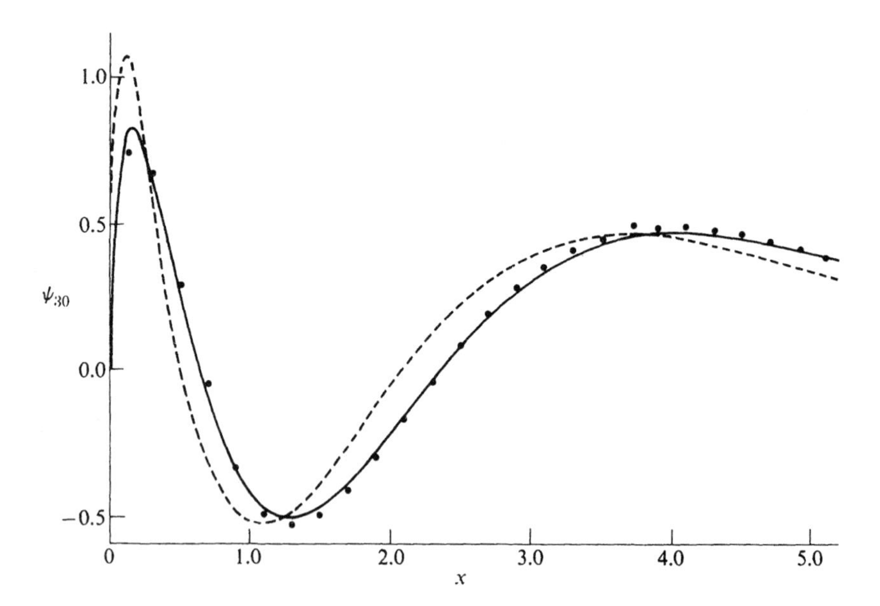

Fig. 9.3 The exact eigenvector, ψ_{30}, solid curve, compared with the approximate ψ_{30} as calculated from Eq. (9.48), dotted curve, and as calculated from Eq. (9.50), dashed curve.

clear that for a coarse subdivision such as the one employed here, the first method is more satisfactory; for sufficiently fine partitions of the interval of integration, the two approaches are equivalent.

9.4 THE FREDHOLM ALTERNATIVE FOR COMPLETELY CONTINUOUS OPERATORS

We now turn our attention to general completely continuous operators. We have less information to work with in this case than we did when considering Hermitian operators. For example, it is no longer possible to guarantee the existence of a complete set of eigenfunctions, and we have no maximum-minimum principles at our disposal. In dealing with non-Hermitian operators, however, we have already made some progress, because we know how to solve linear equations involving either operators with small norm or operators of finite rank. The link between these two types of operators is that any completely continuous transformation of Hilbert space can be written as

$$K = K_F + \Delta \,, \tag{9.63}$$

where K_F is an operator of finite rank and $||\Delta||$ can be made arbitrarily small. Using this basic decomposition, we will show in this section that the Fredholm alternative holds for an arbitrary completely continuous transformation. In doing so, we will also derive several useful results which will enable us to solve problems that arise in practice.

Suppose now that we want to solve

$$f = g + \lambda K f \,, \tag{9.64}$$

or, using Eq. (9.63),

$$(I - \lambda K_F - \lambda\Delta) f = g \,.$$

The problem is to calculate the inverse of $I - \lambda K_F - \lambda\Delta$. We already know how to calculate $(I - \lambda\Delta)^{-1}$ if $||\Delta|| < 1/\lambda$, and how to calculate $(I - \lambda K_F)^{-1}$ when K_F is any kernel of finite rank and λ is not a singular value of K_F. Therefore we would like to write $(I - \lambda K_F - \lambda\Delta)^{-1}$, assuming that it exists, in terms of these known inverses. To do this, we make use of the operator identity (see Problem 3.26)

$$(A - B)^{-1} \equiv A^{-1} + A^{-1}B(A - B)^{-1} \,, \tag{9.65}$$

which is valid if $(A - B)^{-1}$ and A^{-1} both exist. Now for this problem, we identify $(I - \lambda\Delta)$ with A and λK_F with B. Assuming that $||\Delta|| < 1/\lambda$, we know that $(I - \lambda\Delta)^{-1}$ exists and can be calculated by the Neumann series

$$(I - \lambda\Delta)^{-1} = I + \lambda\Delta + \lambda^2\Delta^2 + \lambda^3\Delta^3 + \cdots \,. \tag{9.66}$$

Thus if $(I - \lambda K_F - \lambda\Delta)^{-1}$ exists,

$$(I - \lambda K_F - \lambda\Delta)^{-1} = (I - \lambda\Delta)^{-1} + \lambda(I - \lambda\Delta)^{-1}K_F(I - \lambda K_F - \lambda\Delta)^{-1} \,. \tag{9.67}$$

But if K_F is an operator of finite rank, then so is $(I - \lambda\Delta)^{-1}K_F$, since if K_F is defined for all $f \in H$ by

$$K_F f \equiv \sum_{i=1}^{N} \phi_i (\psi_i, f) ,$$

then

$$(I - \lambda\Delta)^{-1} K_F f = \sum_{i=1}^{N} (I - \lambda\Delta)^{-1} \phi_i (\psi_i, f) \equiv \sum_{i=1}^{N} \phi_i^{(\Delta)} (\psi_i, f) . \tag{9.68}$$

We have made the definition

$$\phi_i^{(\Delta)} \equiv (I - \lambda\Delta)^{-1} \phi_i = \phi_i + \lambda\Delta\phi_i + \lambda^2\Delta^2\phi_i + \cdots \tag{9.69}$$

Let us define a new operator of finite rank, which depends explicitly on λ:

$$\tilde{K}_F(\lambda) \equiv (I - \lambda\Delta)^{-1} K_F . \tag{9.70}$$

Since $||\Delta|| < 1/\lambda$, the λ-dependence will be of the form of a convergent power series in λ. Thus, adopting the definitions,

$$G_K(\lambda) \equiv (I - \lambda K)^{-1} , \qquad G_\Delta(\lambda) \equiv (I - \lambda\Delta)^{-1} , \tag{9.71}$$

Eq. (9. 67) can be rewritten as

$$G_K(\lambda) = G_\Delta(\lambda) + \lambda \tilde{K}_F(\lambda) G_K(\lambda) . \tag{9.72}$$

In this notation, Eq. (9.70) becomes

$$\tilde{K}_F(\lambda) = G_\Delta(\lambda) K_F . \tag{9.73}$$

Equation (9.72) can be written in the familiar form

$$[I - \lambda \tilde{K}_F(\lambda)] G_K(\lambda) = G_\Delta(\lambda)$$

so all that remains to be done is to calculate the inverse of $[I - \lambda \tilde{K}_F(\lambda)]$, where $\tilde{K}_F$ is a transformation of finite rank. If the inverse exists, then

$$G_K(\lambda) = [I - \lambda \tilde{K}_F(\lambda)]^{-1} G_\Delta(\lambda) \tag{9.74}$$

and our original equation [Eq. (9.64)] is solved by

$$f = G_K(\lambda) g .$$

Equation (9.74) tells us that if $|\lambda| \cdot ||\Delta|| < 1$ (so that $G_\Delta(\lambda)$ exists), then $G_K(\lambda)$ exists if and only if $[I - \lambda \tilde{K}_F(\lambda)]^{-1}$ exists. From a practical point of view, the crucial fact is that Eq. (9.74) enables us to write $G_K(\lambda)$ as the product of two operators, each of which can, in principle, be computed by techniques already discussed: $G_\Delta(\lambda)$ is given by the straightforward iteration of Eq. (9.66), while $[I - \lambda \tilde{K}_F(\lambda)]^{-1}$ can be obtained by the methods of Section 8.6. According to Eqs. (9.68) and (9.70),

$$\tilde{K}_F(\lambda) f = \sum_{i=1}^{N} \phi_i^{(\Delta)} (\psi_i, f)$$

for any $f \in H$, where $\phi_i^{(\Delta)}$ may be written in terms of the original elements of the approximating kernel, K, as

$$\phi_i^{(\Delta)} = G_\Delta(\lambda)\phi_i \,.$$

Thus we can use Eqs. (8.87) and (8.88) to write

$$[I - \lambda \tilde{K}_F(\lambda)]^{-1} = I + \lambda R_{\tilde{K}_F}(\lambda) \,, \tag{9.75}$$

where

$$R_{\tilde{K}_F}(\lambda) f \equiv \sum_{m=1}^{N} \sum_{n=1}^{N} [(I - \lambda A)^{-1}]_{mn} \phi_m^{(\Delta)} (\psi_n, f) \,. \tag{9.76}$$

A is very simple; it is just an $N \times N$ matrix, the mn-element of which is given by

$$A_{mn} = (\psi_m, \phi_n^{(\Delta)}) = (\psi_m, G_\Delta \phi_n) \,. \tag{9.77}$$

According to Eq. (9.69), each element of A can be expressed as a power series in λ, which is convergent if $|\lambda| < ||\Delta||^{-1}$. Thus the original problem can be reduced to one of inverting a finite-dimensional matrix, so in principle we have solved the problem.

It is often possible to use a finite-rank transformation with only a few terms and thereby obtain a Δ whose norm is fairly small. Even with a simple separable kernel of the form

$$k_s(x, y) = \alpha(x)\beta(y) \,,$$

it is possible in many problems to get a Δ whose norm is much smaller than that of the original K. In specific cases this reduction may be very useful. However, it should be emphasized that no matter how small we make $||\Delta||$, for $|\lambda|$ sufficiently large $|\lambda| \cdot ||\Delta|| \geq 1$, and we can no longer be certain that $(I - \lambda\Delta)^{-1}$ exists.

However, if λ is an arbitrary complex number, then in any circle about the origin the complex λ-plane, there can be only a finite number of points for which $(I - \lambda K)^{-1}$ fails to exist. This is most apparent if we write everything in terms of resolvent operators:

$$G_K(\lambda) \equiv (I - \lambda K)^{-1} \equiv I + \lambda R_K(\lambda) \,,$$

$$G_\Delta(\lambda) \equiv (I - \lambda \Delta)^{-1} \equiv I + \lambda R_\Delta(\lambda) \,.$$

Using these relations along with Eq. (9.75) in Eq. (9.74), we obtain

$$I + \lambda R_K(\lambda) = (I + \lambda R_{\tilde{K}_F}(\lambda))(I + \lambda R_\Delta(\lambda)) \,,$$

so

$$R_K(\lambda) = R_\Delta(\lambda) + R_{\tilde{K}_F}(\lambda) \cdot (I + \lambda R_\Delta(\lambda)) \,. \tag{9.78}$$

Note that because of Eqs. (9.69) and (9.77), det $(I - \lambda A)$ is an analytic function of λ which is *not* identically zero (for example, det $(I - \lambda A) = 1$ when $\lambda = 0$). Therefore, if λ_0 is any point in the complex λ-plane for which det

$(I - \lambda_0 A) = 0$, it must be possible to find a neighborhood of λ_0 in which det $(I - \lambda A) \neq 0$ except at λ_0. Otherwise, all the derivatives of det $(I - \lambda A)$ would vanish at $\lambda = \lambda_0$, so the Taylor series about λ_0 would vanish identically. This means, according to the Bolzanno-Weierstrass theorem, that any finite circle in the complex λ-plane contains only a finite number of poles of $R_K(\lambda)$, since according to Eqs. (9.76) and (9.78) zeroes of det $(I - \lambda A)$ correspond to poles of $R_K(\lambda)$. Thus, $R_K(\lambda)$ is a *meromorphic* function of λ.

Another important by-product of breaking K into a transformation of finite rank, K_F, and a transformation with small norm, Δ, is the following result.

Theorem 9.9. The Fredholm alternative holds for any completely continuous transformation of Hilbert space.

Proof. The demonstration of this result is not difficult, but it is a bit long. We shall divide it into a series of lemmas.

Lemma 1. The equations

$$f = g + \lambda K f \tag{9.79}$$

and

$$\tilde{f} = \tilde{g} + \lambda^* K^\dagger \tilde{f}\,, \tag{9.80}$$

are equivalent, respectively, to the equations

$$f = g_\Delta + \lambda(I + \lambda R_\Delta) K_F f\,, \tag{9.81}$$

$$\tilde{f} = \tilde{g}_\Delta + \lambda^*(I + \lambda^* R_\Delta^\dagger) K_F^\dagger \tilde{f}\,, \tag{9.82}$$

where

$$g_\Delta = (I + \lambda R_\Delta) g\,, \qquad \tilde{g}_\Delta = (I + \lambda^* R_\Delta)\tilde{g}\,,$$

and R_Δ is the resolvent defined above. Δ is an operator which satisfies $||\Delta|| < 1/|\lambda|$, and K_F is an operator of finite rank.

Proof. The result follows immediately from our previous work. We write

$$f = g + \lambda K f = g + \lambda K_F f + \lambda \Delta f\,,$$

where Δ is such that $||\Delta|| < 1/|\lambda|$. By Theorem 8.4, there is always a K_F for which this will be true. Hence

$$(I - \lambda\Delta) f = g + \lambda K_F f\,,$$

or, since $(I - \lambda\Delta)^{-1}$ exists and since by definition

$$(I - \lambda\Delta)^{-1} \equiv I + \lambda R_\Delta\,,$$

we have

$$f = g_\Delta + \lambda(I + \lambda R_\Delta) K_F f$$

with

$$g_\Delta = (I + \lambda R_\Delta) g\,.$$

An identical argument applied to

$$\tilde{f} = \tilde{g} + \lambda^* K^\dagger \tilde{f}$$

leads to

$$\tilde{f} = \tilde{g}_\Delta + \lambda^*(I + \lambda^* R_\Delta^\dagger) K_F^\dagger \tilde{f} ,$$

as required. By reversing the steps, we immediately see that Eqs. (9.81) and (9.82) imply Eqs. (9,79) and (9.80), respectively, which completes the proof of equivalence.

Lemma 2. Every value of λ is either a regular value or a singular value of the operator K. If it is a regular value, then the equation

$$f = g + \lambda K f \tag{9.83}$$

has a unique solution given by

$$f = (I - \lambda K)^{-1} g ,$$

where $(I - \lambda K)^{-1}$ exists. If λ is a singular value, then Eq. (9.83) has a solution if and only if g is orthogonal to every solution of

$$\lambda^* K^\dagger \tilde{h} = \tilde{h} .$$

Proof. This result follows from Lemma 1 and Theorem 8.1. According to Lemma 1, Eq. (9.83) is equivalent to

$$f = g_\Delta + \lambda(I + \lambda R_\Delta) K_F f \tag{9.84}$$

for some appropriate Δ and K_F. But we have already shown

$$(I + \lambda R_\Delta) K_F$$

to be a transformation of finite rank, since K_F is a transformation of finite rank. Thus, by Theorem 8.1, Eq. (9.84) has a solution if and only if g_Δ is orthogonal to every solution of the homogeneous adjoint equation

$$\lambda^*[(I + \lambda R_\Delta) K_F]^\dagger \tilde{\phi} = \lambda^* K_F^\dagger (I + \lambda R_\Delta)^\dagger \tilde{\phi} = \tilde{\phi} . \tag{9.85}$$

If we call

$$\tilde{h} \equiv (I + \lambda R_\Delta)^\dagger \tilde{\phi} = (I + \lambda^* R_\Delta^\dagger) \tilde{\phi} ,$$

then the requirement that Eq. (9.84) have a solution becomes

$$(g_\Delta, \phi) = 0 = (g_\Delta, [I + \lambda^* R_\Delta^\dagger]^{-1} \tilde{h}) . \tag{9.86}$$

Since $g_\Delta = (I + \lambda R_\Delta) g$, this simply requires that

$$(g, \tilde{h}) = 0 .$$

Now $(I + \lambda^* R_\Delta^\dagger)^{-1}$ exists, and is by definition equal to $(I - \lambda^* \Delta^\dagger)$ (see Lemma 1). Moreover, $K^\dagger = K_F^\dagger + \Delta^\dagger$. Thus, according to Eq. (9.85), $\tilde{h}$ must obey the following equations:

$$\lambda^* K_F^\dagger \tilde{h} = \tilde{\phi} = (I + \lambda^* R_\Delta^\dagger)^{-1} \tilde{h} = (I - \lambda^* \Delta^\dagger) \tilde{h} = \tilde{h} + \lambda^* K_F^\dagger \tilde{h} - \lambda^* K^\dagger \tilde{h} ,$$

or

$$\lambda^* K^\dagger \tilde{h} = \tilde{h} .$$

Hence Eq. (9.84) has a solution if and only if g is orthogonal to all solutions of

$$\lambda^* K^\dagger \tilde{h} = \tilde{h} .$$

By Lemma 1, the same is true for Eq. (9.83), which completes the proof of this lemma.

Corollary. Every value of λ^* is either a regular value or a singular value of the operator $K^\dagger$. If it is a regular value, then the equation

$$\tilde{f} = \tilde{g} + \lambda^* K^\dagger \tilde{f} \tag{9.87}$$

has a unique solution given by

$$\tilde{f} = (I - \lambda^* K^\dagger)^{-1} \tilde{g} ,$$

where $(I - \lambda^* K^\dagger)^{-1}$ exists. If λ is a singular value, then Eq. (9.87) has a solution if and only if $\tilde{g}$ is orthogonal to every solution of $\lambda K h = h$.

Finally, we have

Lemma 3. The equations

$$\lambda K h_i = h_i , \qquad \lambda^* K^\dagger \tilde{h}_i = \tilde{h}_i$$

have the same number of linearly independent solutions.

Proof. According to Lemma 1, these two equations are equivalent respectively to

$$\lambda(I + \lambda R_\Delta) K_F h_i = h_i \tag{9.88}$$

and

$$\lambda^*(I + \lambda^* R_\Delta^\dagger) K_F^\dagger \tilde{h}_i = \tilde{h}_i \tag{9.89}$$

for some Δ and K_F. The last equation can be written as

$$\lambda^* K_F^\dagger (I + \lambda^* R_\Delta^\dagger)[(I + \lambda^* R_\Delta^\dagger)^{-1} \tilde{h}_i] = [(I + \lambda^* R_\Delta^\dagger)^{-1} \tilde{h}_i] , \tag{9.90}$$

since the inverse exists. If we define, as before,

$$\tilde{\phi}_i = (I + \lambda^* R_\Delta^\dagger)^{-1} \tilde{h}_i ,$$

then Eq. (9.90) becomes

$$\lambda^* K_F^\dagger (I + \lambda^* R_\Delta^\dagger) \tilde{\phi}_i = \tilde{\phi}_i ,$$

and Eq. (9.89) becomes

$$\lambda^*[(I + \lambda R_\Delta) K_F]^\dagger \tilde{\phi}_i = \tilde{\phi}_i . \tag{9.91}$$

Now according to the second corollary to Theorem 8.1, Eq. (9.88) and Eq. (9.91) have the same number of linearly independent solutions, $\tilde{h}_i$ and $\tilde{\phi}_i$. But $\tilde{h}_i$ is related to $\tilde{\phi}_i$ by

$$\tilde{h}_i = (I + \lambda^* R_\Delta^\dagger) \tilde{\phi}_i .$$

Since $(I + \lambda^* R_\Delta^\dagger)^{-1}$ exists, the $\tilde{h}_i$ will be linearly independent if the $\tilde{\phi}_i$ are.

Thus Eqs. (9.88) and (9.89) have the same number of linearly independent solutions, and hence the same is true of the equations

$$\lambda K h_i = h_i \quad \text{and} \quad \lambda^* K^\dagger \tilde{h}_i = \tilde{h}_i \, .$$

This completes the proof of Lemma 3 and also of Theorem 9.9.

Note that the Fredholm alternative is exactly the same for completely continuous transformations as for transformations of finite rank. We have tried to emphasize this by retaining the same notation in discussing the two cases (see Section 8.6).

9.5 THE NUMERICAL SOLUTION OF INTEGRAL EQUATIONS

Having assured ourselves that solutions to linear equations with completely continuous kernels exist, it is worthwhile to indicate how one can actually proceed to solve such equations. It is rare that a closed-form solution can be found, so we will concentrate in this section on an approximation scheme which can be refined in principle to arbitrary precision. To do this we will make use of the technique discussed in Section 9.3. However, before applying this method, we must be sure that the eigenvalues of the operators in question behave in a sensible way, i.e., that eigenvalues of approximate kernels come arbitrarily close to the eigenvalues of the kernel which they approximate as the approximation is made arbitrarily fine. For this purpose, let us write

$$(I - \lambda K)^{-1} \equiv (I - \lambda K_F - \lambda \Delta)^{-1}$$

as

$$(I - \lambda K)^{-1} = (I - \lambda K_F)^{-1} + \lambda (I - \lambda K_F)^{-1} \Delta (I - \lambda K)^{-1} \, . \tag{9.92}$$

When we have computed $(I - \lambda K)^{-1}$, the solution to

$$f = g + \lambda K f \tag{9.93}$$

is just

$$f = (I - \lambda K)^{-1} g \, . \tag{9.94}$$

Now we want to approximate K so closely by K_F that the first term in the iterative solution of Eq. (9.92) will provide [via Eq. (9.94)] a satisfactory solution to Eq. (9.93). First, we remark that Eq. (9.92) is valid if and only if $(I - \lambda K)^{-1}$ and $(I - \lambda K_F)^{-1}$ both exist. What we want to show is that if we make $||\Delta||$ sufficiently small, then the values of λ for which $(I - \lambda K)^{-1}$ fails to exist and the values of λ for which $(I - \lambda K_F)^{-1}$ fails to exist get close to each other. (For *Hermitian* operators we have Theorem 9.8 which gives us the desired result in a very precise form.)

For general completely continuous operators, we suppose first that λ is a singular value of K, but that no matter how small $||\Delta||$ becomes $(I - \lambda K_F)^{-1}$ remains bounded, say

$$||(I - \lambda K_F)^{-1}|| < B$$

for *all* Δ. But this assumption leads to a contradiction. For consider the equation

$$X = (I - \lambda K_F)^{-1} + \lambda(I - \lambda K_F)^{-1}\Delta X . \tag{9.95}$$

By familiar arguments, if $|\lambda| \cdot ||(I - \lambda K_F)^{-1}|| \cdot ||\Delta|| < 1$, this equation can be solved by iteration:

$$X = (I - \lambda K_F)^{-1} + \lambda(I - \lambda K_F)^{-1}\Delta(I - \lambda K_F)^{-1} + \cdots . \tag{9.96}$$

But $|\lambda| \cdot ||(I - \lambda K_F)^{-1}|| \cdot ||\Delta|| < 1$ means, by assumption, that $|\lambda| \cdot ||\Delta|| \cdot B < 1$. or

$$||\Delta|| < 1/(|\lambda| \cdot B) .$$

By Theorem 8.4, it is possible to obtain a Δ which satisfies this relation. The operator X is seen to have finite norm; in fact,

$$||X|| \leq \frac{B}{1 - \lambda B||\Delta||} < \infty .$$

But by multiplying the series of Eq. (9.96) on both sides by $(I - \lambda K)$, we find that

$$(I - \lambda K)X = X(I - \lambda K) = I .$$

Thus X is the bounded inverse of $I - \lambda K$. But this is impossible, since λ is assumed to be a singular value of K. Hence as $||\Delta|| \to 0$, $||(I - \lambda K_F)^{-1}|| \to \infty$, that is, λ comes closer and closer to being a singular value of K_F.

Since we know that any given singular value of K will come arbitrarily close to a singular value of K_F for $||\Delta||$ sufficiently small, it remains only to investigate the converse problem: Suppose that λ is *not* a singular value of K; can λ still be a singular value of K_F ? The answer is that if $||\Delta||$ is sufficiently small, then λ is also not a singular value of K_F. Of course, for some finite value of $||\Delta||$ it might happen that λ is a singular value of K_F, but if we keep refining our approximation ($||\Delta|| \to 0$), eventually we can be certain that λ is not a singular value of K_F. To see that this is true, note that Eq. (9.92) implies that if $(I - \lambda K)^{-1}$ exists, then $(I - \lambda K_F)^{-1}$ exists if

$$[I + \lambda\Delta(I - \lambda K)^{-1}]^{-1}$$

exists. In this case,

$$(I - \lambda K_F)^{-1} = (I - \lambda K)^{-1}[I + \lambda\Delta(I - \lambda K)^{-1}]^{-1} .$$

But if $||(I - \lambda K)^{-1}|| < C$, then when $||\Delta|| < 1/|\lambda| \cdot C$ the quantity $[I + \lambda\Delta \cdot (I - \lambda K)^{-1}]^{-1}$ can be computed by iteration. Thus if λ is not a singular value of K for $||\Delta||$ sufficiently small, it must also not be a singular value of K_F. We know, therefore, that if $||\Delta||$ is small enough, and if λ is not a singular value of K, then Eq. (9.92) is well defined and can be solved by iteration:

$$(I - \lambda K)^{-1} = (I - \lambda K_F)^{-1} + \lambda(I - \lambda K_F)^{-1}\Delta(I - \lambda K_F)^{-1} + \cdots . \tag{9.97}$$

For $||\Delta||$ very small compared to $[|\lambda| \cdot ||(I - \lambda K_F)^{-1}||]^{-1}$, we may neglect all but the first term in Eq. (9.97) and write

$$(I - \lambda K)^{-1} \cong (I - \lambda K_F)^{-1}$$

in the sense of operator norm. Then Eq. (9.94) can be written as

$$f = (I - \lambda K_F)^{-1} g + O(\Delta) \, ,$$

where $O(\Delta)$ refers to terms which become small in norm as $||\Delta|| \to 0$. Of course, what one means by $||\Delta||$ sufficiently small is clearly, according to Eq. (9.97), dependent on $||(I - \lambda K_F)^{-1}||$. If one is near a singular value of K, and hence of K_F, then $||(I - \lambda K)^{-1}||$ and $||(I - \lambda K_F)^{-1}||$ are large, and $||\Delta||$ must be correspondingly small.

We are now ready to solve linear equations in the same manner employed in the solution of eigenvalue problems involving self-adjoint transformations. Assuming that K is an integral operator, we truncate the kernel $k(x, y)$ in such a way as to ensure the accuracy we desire, and then divide the resulting region on the x(or y)-axis into small subregions. Defining the unnormalized functions $\phi_n^{(N)}(x)$ as in Section 9.3, we obtain a kernel of finite rank, $k_N(x, y)$, which approximates $k(x, y)$ as closely as we please. Proceeding in this way, we find that

$$f = \sum_{n=1}^{N} a_n \phi_n^{(N)} \, , \tag{9.98}$$

where a_n is given by

$$a_n = \frac{N}{l} \sum_{m=1}^{N} [(I - \lambda A)^{-1}]_{nm} b_m \tag{9.99}$$

and b_m is just

$$b_m = (\phi_m^{(N)}, g) \, . \tag{9.100}$$

The matrix A is given by

$$A_{\mu\nu} = \frac{N}{l} \int dx \int dy \, k(x, y) \phi_\mu^{(N)}(x) \phi_\nu^{(N)}(y) \, . \tag{9.101}$$

In these equations l is the length of the interval outside of which we truncate the kernel, and N is the number of subintervals into which we divide the main interval. In obtaining Eq. (9.101), we have assumed that the quantity

$$\left[||g||^2 - \frac{N}{l} \sum_{m=1}^{N} |b_m|^2 \right]^{1/2}$$

is small compared to $||f||$; if the $\phi_n^{(N)}$ formed a complete set, then this quantity would be exactly zero. The reader should derive these results using the method by which we obtained Eqs. (9.48) and (9.49); an expression for f can also be obtained using the Riemann integral approach that led to Eq. (9.50) (see Problem 9.25). With Eqs. (9.98) through (9.101) at our disposal, the problem of solving an integral equation with completely continuous kernel is reduced to a matter of matrix inversion.

As an example of a linear equation with completely continuous kernel, let us consider the Lippmann-Schwinger equation discussed in Chapter 7. Dropping the subscript ($\pm$) of Eq. (7.68), we write the equation as

$$\psi(\mathbf{r}) = \frac{A}{(2\pi)^{3/2}} e^{i\mathbf{q}_i\cdot\mathbf{r}} - \frac{m}{2\pi\hbar^2}\int \frac{e^{iq_i|\mathbf{r}-\mathbf{r}'|}}{|\mathbf{r}-\mathbf{r}'|} V(\mathbf{r}')\psi(\mathbf{r}')d^3\mathbf{r}' . \tag{9.102}$$

This is related to the scattering amplitude, $f(\theta, \phi)$, by

$$f^s(\mathbf{q}_f) = -\frac{(2\pi)^{1/2} m}{\hbar^2 A}\int e^{-i\mathbf{q}_f\cdot\mathbf{r}} V(\mathbf{r})\psi(\mathbf{r})\, d^3\mathbf{r} , \tag{9.103}$$

where $\mathbf{q}_f$ is a vector in the direction (θ, ϕ) which is equal in magnitude to $|\mathbf{q}_i|$. As remarked in Section 7.4, f^s is the quantity of real physical interest. Choosing the direction of the incident wave to be the z-direction and using $A = (2\pi)^{3/2}$ for convenience (f^s is clearly independent of A), we get

$$\psi(\mathbf{r}) = e^{iq_i z} - \frac{m}{2\pi\hbar^2}\int \frac{e^{iq_i|\mathbf{r}-\mathbf{r}'|}}{|\mathbf{r}-\mathbf{r}'|} V(\mathbf{r}')\psi(\mathbf{r}')\, d^3\mathbf{r}' ,$$

$$f^s(\mathbf{q}_f) = -\frac{m}{2\pi\hbar^2}\int e^{-i\mathbf{q}_f\cdot\mathbf{r}} V(\mathbf{r})\psi(\mathbf{r})\, d^3\mathbf{r} .$$

Just as in Section 7.5, we expand everything in terms of Legendre polynomials. We write

$$\psi(\mathbf{r}) = \sum_{l=0}^{\infty} (2l+1)\phi_l(r) P_l(\cos\theta) ,$$

$$f^s(\mathbf{q}_f) = \sum_{l=0}^{\infty} (2l+1) f_l^s(q_i) P_l(\cos\theta_f) ,$$

where we have used the fact that $|\mathbf{q}_f| = q_i$ and have written the scattering angle as θ_f. In this way, we readily obtain a set of equations for the ϕ_l and the f_l^s. We will use the case $l = 0$ as an example. In this case,

$$\phi_0(r) = \frac{\sin(q_i r)}{q_i r} - \frac{2m}{\hbar^2 q_i}\int_0^\infty \frac{\sin(q_i r_<) e^{iq_i r_>}}{r_< r_>} V(r')\phi_0(r') r'^2\, dr' \tag{9.104}$$

$$f_0^s(q_i) = -\frac{2m}{\hbar^2 q_i}\int_0^\infty \sin(q_i r) V(r)\phi_0(r) r\, dr . \tag{9.105}$$

In Eq. (9.104), $r_<$ and $r_>$ denote respectively the smaller of r and r' and the greater of r and r'.

To be specific, let us look at the case of an exponential potential,

$$V(r) = V_0 e^{-\mu r} .$$

Then Eqs. (9.104) and (9.105) become

$$\phi_0(r) = \frac{\sin(q_i r)}{q_i r} - \frac{2mV_0}{\hbar^2 q_i}\int_0^\infty \frac{1}{r}\sin(q_i r_<) e^{iq_i r_>} e^{-\mu r'} r'\phi_0(r')\, dr' , \tag{9.106}$$

$$f_0^s(q_i) = -\frac{2mV_0}{\hbar^2 q_i}\int_0^\infty \sin(q_i r) e^{-\mu r} r\phi_0(r)\, dr . \tag{9.107}$$

As it stands, neither the inhomogeneous term nor the kernel of Eq. (9.106) is square-integrable. However, the symmetrization process which we used in Section 9.3 will bring the equations to the desired form. We define

$$\xi_0(r) \equiv re^{-\mu r/2}\phi_0(r) .$$

Then our equations take the form

$$\xi_0(r) = \frac{1}{q_i} e^{-\mu r/2} \sin(q_i r) - \frac{2mV_0}{\hbar^2 q_i}\int_0^\infty e^{-\mu r/2} \sin(q_i r_<) e^{iq_i r_>} e^{-\mu r'/2}\xi_0(r')\, dr' , \tag{9.108}$$

$$f_0^s(q_i) = -\frac{2mV_0}{\hbar^2 q_i}\int_0^\infty \sin(q_i r) e^{-\mu r/2}\xi_0(r)\, dr . \tag{9.109}$$

It is easily verified that the inhomogeneous term of Eq. (9.108) is square-integrable. What about the kernel

$$k_0(r, r') = -\frac{2mV_0}{\hbar^2 q_i} e^{-\mu(r+r')/2} \sin(q_i r_<) e^{iq_i r_>} ?$$

We find that

$$I(q_i) = \int_0^\infty dr \int_0^\infty dr' |k_0(r, r')|^2 = \frac{2m^2V_0^2}{\hbar^4 \mu^2 (\mu^2 + q_i^2)} .$$

Then the norm of the operator, K_0, whose kernel is $k_0(r, r')$, satisfies

$$||K_0|| \leq \sqrt{2}\,\frac{mV_0}{\mu\hbar^2}(\mu^2 + q_i^2)^{-1/2} . \tag{9.110}$$

Thus the operator is completely continuous, and its norm tends toward zero as $q_i \to \infty$. Hence for q_i sufficiently large, we can always solve Eq. (9.108) by iteration. One might even hope that it will *always* be possible to solve Eq. (9.108) by iteration. However, for typical situations where an exponential potential is relevant (in nuclear physics, for example), the values of m, μ, and V_0 are such that the estimate on $||K_0||$ reads

$$||K_0|| \leq 2 , \tag{9.111}$$

when $q_i \ll \mu$. We have used

$$\begin{aligned} 1/\mu &= 1.4 \times 10^{-13}\,\text{cm} , \\ m &= 1.66 \times 10^{-24}\,\text{g} , \\ V_0 &= 30\,\text{MeV} = 4.8 \times 10^{-5}\,\text{erg} , \end{aligned}$$

where $1/\mu$ is approximately equal to the range of nuclear forces, m is the mass of the proton (or neutron), V_0 is a reasonable estimate of the strength of nuclear forces, and $\hbar$ is Planck's constant. For $q_i \cong 2\mu$, $||K_0||$ becomes less than 1, but nevertheless for an important range of values of q_i, $||K_0||$ is by no means negligible compared with 1. Of course, Eq. (9.111) is only an *inequality*; $||K_0||$

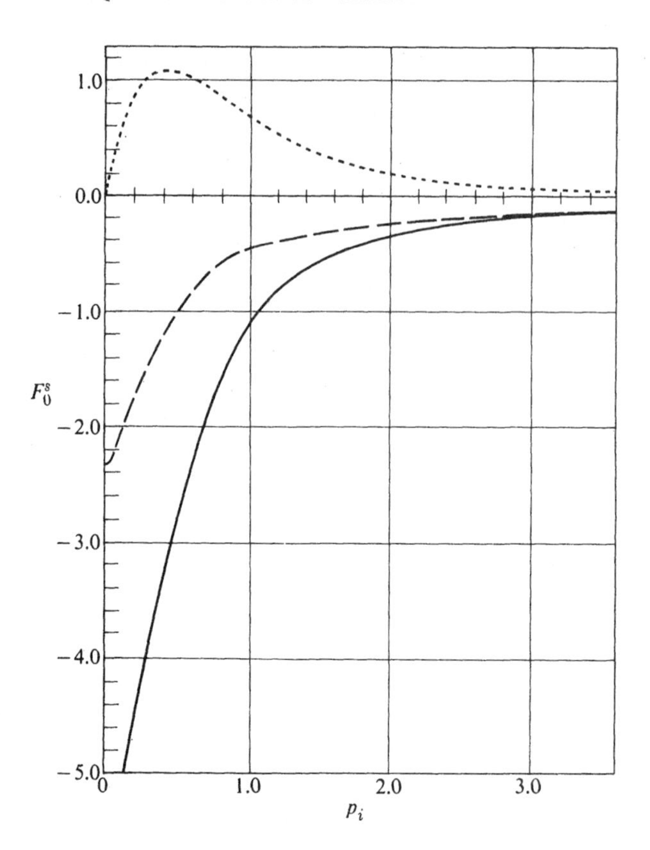

Fig. 9.4 The real part of F_0^s (dashed curve) and the imaginary part of F_0^s (dotted curve) for $l = 0$ scattering by an exponential potential. Included for comparison is the first iterative approximation, F_{0B}^s, which is purely real (solid curve). All the amplitudes are measured in units of the range of the potential $(1/\mu)$, and are plotted as functions of the dimensionless variable p_i, which is essentially the momentum of the scattered particle, $p_i = (2mE/\mu^2\hbar^2)^{1/2}$. The strength of the potential is such that the dimensionless parameter $(mV_0/\mu^2\hbar^2)$ appearing in Eqs. (9.112) and (9.113) is equal to 1.4, as discussed in the text.

might be less than 1, even for $q_i = 0$; however, for the iterative method to be of *practical* use, $||K_0||$ should be *much* less than 1, and the estimate provided by Eq. (9.111) makes that seem unlikely.

Thus we are left with the method which we used in the last section for solving the eigenvalue problem for a Hermitian operator. It is again convenient to scale the distances into appropriate units by defining

$$x \equiv \mu r\,, \qquad p_i = q_i/\mu = (2mE/\mu^2\hbar^2)^{1/2}\,,$$
$$\eta_0 = \mu\xi_0\,, \qquad F_0^s = \mu f_0^s\,.$$

With these changes, Eqs. (9.108) and (9.109) are transformed to

$$\eta_0(x) = \frac{1}{p_i} e^{-x/2} \sin (p_i x) - \frac{2mV_0}{\mu^2\hbar^2 p_i} \int_0^\infty e^{-x/2} \sin (p_i x_<) e^{ip_i x_>} e^{-x'/2} \eta_0(x')\, dx' , \tag{9.112}$$

$$F_0^s(p_i) = -\frac{2mV_0}{\mu^2\hbar^2 p_i} \int_0^\infty \sin (p_i x) e^{-x/2} \eta_0(x)\, dx . \tag{9.113}$$

To solve Eq. (9.112) numerically, we truncate the kernel when x or x' is greater than 7, and then divide the interval [0, 7] into 35 equal parts. Defining $\phi_n^{(N)}$ as usual, we can use Eqs. (9.98) through (9.101) to reduce the problem to one of matrix inversion. Figure 9.4 shows the results for both the real and imaginary parts of F_0^s as a function of p_i for $(mV_0/\mu^2\hbar^2) = 1.4$. This value is obtained from the physically relevant estimates of m, V_0, and μ listed above. For comparison, we also show in Fig. 9.4 the first iterative approximation (Born approximation) to F_0^s, obtained by writing

$$\eta_0^B(x) = \frac{1}{p_i} e^{-x/2} \sin (p_i x) ,$$

$$F_0^{sB} = -2\left(\frac{mV_0}{\mu^2\hbar^2}\right)\frac{1}{p_i}\int_0^\infty \sin (p_i x)\, e^{-x/2} \eta_0^B(x)\, dx = -4\left(\frac{mV_0}{\mu^2\hbar^2}\right)\frac{1}{4p_i^2+1} .$$

Just as we expect from Eq. (9.110), for $q_i \leq \mu$ and $p_i \leq 1$ the first iterative approximation to the amplitude differs considerably from the correct

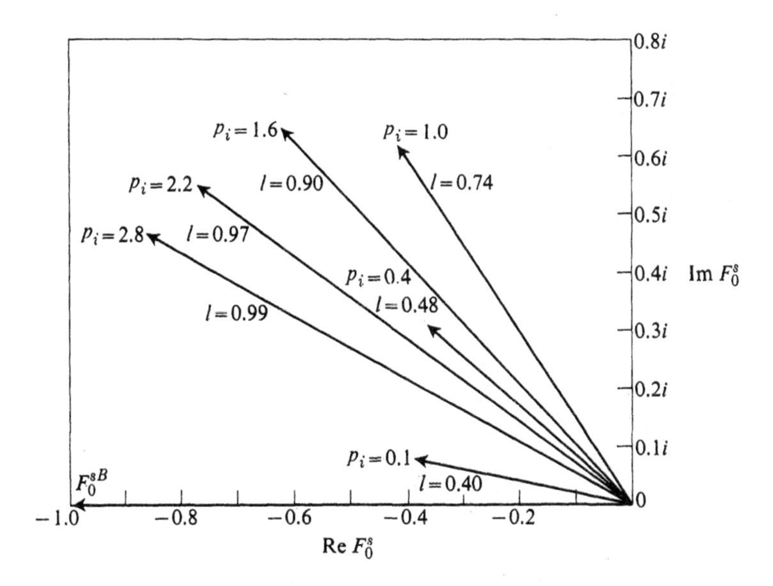

Fig. 9.5 Convergence of F_0^s to the first iterative approximation, as viewed in the complex plane.

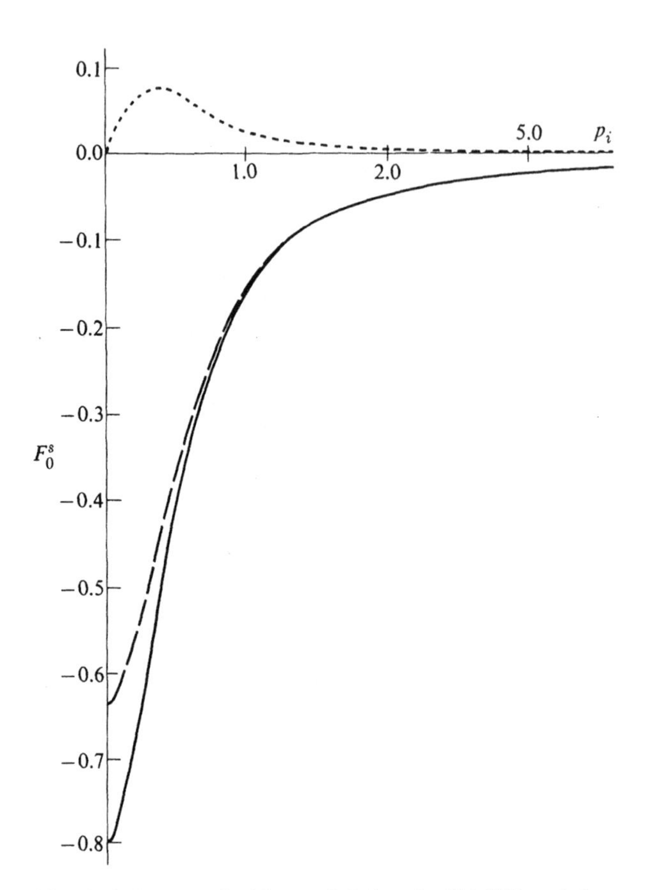

Fig. 9.6 Same as in Figure 9.4, but $(mV_0/\mu^2\hbar^2) = 0.2$.

amplitude. For $p_i = 1$, the real and imaginary parts of F_0^s are of the same order of magnitude, while F_0^{sB} is purely real. However, as p_i becomes larger, the imaginary part of F_0^s drops off more rapidly than the real part, and Fig. 9.4 shows that the real part of F_0^s tends toward F_0^{sB}. However, even for $p_i \cong 3$, the discrepancy is still striking if viewed from the vantage point of Fig. 9.5. In this figure we have divided all amplitudes by $|F_0^{sB}|$ to obtain a reasonable scale and have plotted F_0^s as a "vector" in the complex plane. The "vector" F_0^{sB} is now equal to -1 for all p_i, and we show it as a reference vector along the negative real axis. The tip of each vector is marked by the value of p_i to which it corresponds. We see that the vectors $F_0^s(p_i)$ are tending toward F_0^{sB} as p_i increases, but for $p_i = 2.8$ there is still much improvement to be desired. This is not surprising, for even at $p_i = 2.8$, Eq. (9.110) gives $||K_0|| = 0.66$, so although the relevant norm is less than unity, it is still not particularly small. Along each vector, we have written its length, l, and one is struck by the fact that the length of the complex "vectors" representing $F_0^s(p_i)$ tend very rapidly toward the length of F_0^{sB}, that is,

$$|F_0^s| \to |F_0^{sB}|$$

as $p_i \to \infty$, much faster than

$$|F_0^s - F_0^{sB}| \to 0 .$$

This means that the most important observable quantity, namely, the cross section, will be given deceptively well by the Born approximation, since it depends only on the *modulus* of F_0^s. However, quantities which depend on the phase of F_0^s will not be given so accurately by the Born approximation.

The iterative approach is much more successful for $(mV_0/\mu^2\hbar^2) = 0.2$. Figure 9.6 shows the real and imaginary part of F_0^s along with the first iterative approximation. Even when $p_i \leq 1$, the imaginary part of F is much smaller than the real part, and by the time $p_i = 1$, F_0^{sB} differs by only three percent from the real part of F_0^s. At $p_i = 3$, the imaginary part of F_0^s is smaller than the real part by more than an order of magnitude, and F_0^{sB} and the real part of F_0^s are virtually identical. Thus, in this case, the iterative approach is simple and accurate. But unfortunately, this will not be the case in many problems of practical interest, and more powerful techniques involving some form of finite-rank approximation must be used.

*9.6 UNITARY TRANSFORMATIONS

Up to now, most of our detailed work has been on completely continuous transformations of Hilbert space, although we managed to put the discussion of iterative techniques in the more general setting of *bounded* transformations. In this section we want to touch on unitary transformations of Hilbert spaces, which are important in mathematical physics because of their connection with the theory of group representations and quantum mechanics, even though they are never completely continuous (see Section 9.1).

At the outset we must make a distinction between unitary and isometric transformations which did not arise in finite-dimensional vector spaces. An isometry is defined in vector spaces of any dimension as a linear transformation U satisfying

$$(Uf, Ug) = (f, g)$$

for all f, g in the vector space. This implies that $U^\dagger$ is a left inverse of U:

$$U^\dagger U = I .$$

Now in a finite-dimensional space, the existence of a left inverse guarantees the existence of a right inverse, but on an infinite-dimensional space this is not always the case. *If* a right inverse also exists, U is said to be unitary. This right inverse must of course be equal to the left inverse, since if $BA = I = AC$, then $BAC = C$, so $B = C$. Thus for *unitary* transformations, we write

$$U^\dagger = U^{-1} .$$

To show that not all isometries are unitary consider a complete orthonormal set in a Hilbert space, H, and define U acting on any $f \in H$ by

$$Uf = \sum_{i=1}^{\infty} \phi_{i+1}(\phi_i, f) \, .$$

The adjoint transformation $U^\dagger$ is given by

$$U^\dagger g = \sum_{i=1}^{\infty} \phi_i(\phi_{i+1}, g) \, ,$$

for any $g \in H$. The completeness of the $\{\phi_i\}$ guarantees that

$$(Uf, Ug) = (f, g)$$

for any $f, g \in H$. Thus $U^\dagger$ is a left inverse. Suppose that it were also a right inverse; then we must have

$$(U^\dagger f, U^\dagger g) = (f, UU^\dagger g) = (f, g) \, ,$$

for any $f, g \in H$, so $U^\dagger$ must be an isometry. But for our $U^\dagger$,

$$(U^\dagger f, U^\dagger g) = \sum_{i=1}^{\infty} (f, \phi_{i+1})(\phi_{i+1}, g) \neq (f, g) \, ,$$

so $U^\dagger$ is not an isometry and U is therefore *not* unitary. The reader should show that this construction does not produce a nonunitary isometry on a *finite-dimensional* space.

On the space of square-integrable functions, L_2, it is possible to obtain an explicit characterization of all unitary operators. This will lead us to a neater presentation of the Fourier transform, which we have already discussed in an intuitive fashion in Section 5.7. We use the method of Bochner as presented by Riesz and Nagy.

Theorem 9.10. Let U be a unitary transformation of the space of square-integrable functions, L_2. Then for any $f \in L_2$ such that $g = Uf$, there exist two functions $k(\xi, x)$ and $h(\xi, x)$ which satisfy the relations

$$\int_0^\xi g(x)\, dx = \int k^*(\xi, x) f(x)\, dx \tag{9.114a}$$

and

$$\int_0^\xi f(x)\, dx = \int h^*(\xi, x) g(x)\, dx \, . \tag{9.114b}$$

Proof. We define the function $e_\xi(x)$ which is equal to the sign of x when x lies between zero and ξ and vanishes otherwise. Now for any unitary transformation, let

$$h(\xi, x) \equiv Ue_\xi(x) \, , \qquad k(\xi, x) \equiv U^{-1}e_\xi(x) = U^\dagger e_\xi(x) \, .$$

Consider the equation $g = Uf$. We have

$$(e_\xi, g) \equiv (e_\xi, Uf) = (U^\dagger e_\xi, f) = (k(\xi, x), f(x)) \, .$$

Writing this out explicitly, we get

$$\int_0^\xi g(x)dx = \int k^*(\xi, x)f(x)dx ,$$

making use of the fact that when $e_\xi(x) \not\equiv 0$,

$$e_\xi(x) = \operatorname{sgn}(x)$$

and *not* simply

$$e_\xi(x) = 1 .$$

Similarly,

$$(e_\xi, f) = (e_\xi, U^{-1}g) = (e_\xi, U^\dagger g) = (Ue_\xi, g) = (h(\xi, x), g(x)) ,$$

so that

$$\int_0^\xi f(x)\, dx = \int h^*(\xi, x) g(x)\, dx .$$

This completes the proof of the theorem.

The functions k and h have certain special properties. Suppose that in Eq. (9.114b) we set $f = U^{-1}e_\eta = k(\eta, x)$, and hence $g = Uf = e_\eta$. Thus

$$\int_0^\xi k(\eta, x)\, dx = \int_0^\eta h^*(\xi, x)\, dx . \tag{9.115a}$$

Doing the same in Eq. (9.114a), we find that

$$\int k^*(\xi, x)k(\eta, x)\, dx = \int_0^\xi e_\eta(x)\, dx .$$

Clearly, if ξ and η are of opposite sign, this last integral is zero. If $\xi > 0$, $\eta > 0$, then

$$\int k^*(\xi, x)k(\eta, x)\, dx = \min(\xi, \eta) .$$

If $\xi < 0$, $\eta < 0$, then

$$\begin{aligned}\int k^*(\xi, x)k(\eta, x)\, dx &= -\max(\xi, \eta) = \min(-\xi, -\eta) ,\\ &= \min(|\xi|, |\eta|) .\end{aligned}$$

Hence we have

$$\int k^*(\xi, x)k(\eta, x)\, dx = \begin{cases} \min(|\xi|, |\eta|) & \text{if } \xi\eta > 0 , \\ 0 & \text{if } \xi\eta \le 0 , \end{cases} \tag{9.115b}$$

and similarly

$$\int h^*(\xi, x)h(\eta, x)\, dx = \begin{cases} \min(|\xi|, |\eta|) & \text{if } \xi\eta > 0 , \\ 0 & \text{if } \xi\eta \le 0 . \end{cases} \tag{9.115c}$$

The converse of Theorem 9.10 is extremely important from the practical point of view.

Theorem 9.11. Suppose that we are given two functions $k(\xi, x)$ and $h(\xi, x)$ which belong to L_2 for all ξ and which satisfy Eqs. (9.115a), (9.115b), and (9.115c). Then the equations

$$\int_0^\xi g(x)\,dx = \int k^*(\xi, x) f(x)\,dx\,,$$

$$\int_0^\xi f(x)\,dx = \int h^*(\xi, x) g(x)\,dx\,,$$

define a unitary transformation of L_2, U, and its inverse such that $g = Uf$, $f = U^{-1}g$.

Proof. The proof of this result involves a mathematical technique which we have not employed before. The idea is to use the functions k and h to define operators V and U which act on the functions $e_\xi(x)$ introduced in proving Theorem 9.10. We make the definitions

$$Ue_\xi(x) \equiv h(\xi, x)\,, \qquad Ve_\xi(x) \equiv k(\xi, x)\,.$$

For any functions e_ξ, e_η belonging to this class, we have, according to Eq. (9.115a),

$$(Ve_\xi, e_\eta) = (e_\xi, Ue_\eta)\,.$$

Also, Eq. (9.115b) tells us that

$$(Ue_\xi, Ue_\eta) = (e_\xi, e_\eta)\,,$$

and Eq. (9.115c) yields

$$(Ve_\xi, Ve_\eta) = (e_\xi, e_\eta)\,.$$

We would now like to extend the definitions of U and V to a wider class of functions. The first obvious extension is to the class of all elements of L_2 which can be written as *linear combinations* of functions of the type $e_\xi(x)$. For example, we will have

$$U\left[\sum_{n=1}^N a_n e_{\xi_n}(x)\right] = \sum_{n=1}^N a_n h(\xi_n, x)\,,$$

$$V\left[\sum_{n=1}^N a_n e_{\xi_n}(x)\right] = \sum_{n=1}^N a_n k(\xi_n, x)\,.$$

By the additivity of the inner product, we have for any ϕ, ψ which belong to this more general class,

$$(V\phi, \psi) = (\phi, U\psi)\,, \qquad (V\phi, V\psi) = (\phi, \psi)\,, \qquad (U\phi, U\psi) = (\phi, \psi)\,.$$

Among the functions in this extended class are the step functions, which take on the values C_n $(n = 1, 2, \cdots, N)$ on the intervals I_n $(n = 1, 2, \cdots, N)$ and are zero elsewhere. But any $f \in L_2$ can be approximated in the mean arbitrarily closely by step functions; that is, we can find a sequence of step functions $\{\phi_n\}$ such that for any $f \in L_2$,

$$||\phi_n - f|| \to 0$$

as $n \to \infty$. Since U is unitary, at least on the class of step functions,

$$||U\phi_m - U\phi_n|| < \epsilon$$

when

$$||\phi_m - \phi_n|| < \epsilon .$$

Therefore, if the ϕ_n converge to some element, f, of L_2, so must the set $\{U\phi_n\}$. We define the function to which this set converges to be Uf. Thus we extend U to the entire space by continuity. It only remains to show that if $f, g \in L_2$, then we still have

$$(Uf, Ug) = (f, g) , \qquad (Vf, Vg) = (f, g) , \qquad (Vf, g) = (f, Ug) .$$

We prove the first of these results and leave the others to the reader. Consider

$$\begin{aligned}(Uf, Ug) - (f, g) \equiv (Uf, Ug) &- (U\phi_n, Ug) + (U\phi_n, Ug) \\ &- (U\phi_n, U\psi_m) + (U\phi_n, U\psi_m) \\ &- (\phi_n, g) + (\phi_n, g) - (f, g)\end{aligned}$$

where $\phi_n \to f$, $\psi_m \to g$. For step functions,

$$(U\phi_n, U\psi_m) = (\phi_n, \psi_m) ,$$

so

$$\begin{aligned}|(Uf, Ug) - (f, g)| &\leq |(Uf, Ug) - (U\phi_n, Ug)| \\ &\quad + |(U\phi_n, Ug) - (U\phi_n, U\psi_m)| \\ &\quad + |(\phi_n, \psi_m) - (\phi_n, g)| + |(\phi_n, g) - (f, g)| \\ &\leq |(Uf - U\phi_n, Ug)| + |(U\phi_n, Ug - U\psi_m)| \\ &\quad + |(\phi_n, \psi_m - g)| + |(\phi_n - f, g)| \\ &\leq ||Uf - U\phi_n|| \cdot ||Ug|| + ||Ug - U\psi_m|| \cdot ||U\phi_n|| \\ &\quad + ||\psi_m - g|| \cdot ||\phi_n|| \\ &\quad + ||\phi_n - f|| \cdot ||g|| .\end{aligned}$$

But since $\phi_n \to f$ and $\psi_m \to g$, and by definition $U\phi_n \to Uf$ and $U\psi_m \to Ug$, we can make the right-hand side of this equation as small as we please. Therefore

$$(Uf, Ug) = (f, g) .$$

Similarly,

$$(Vf, Vg) = (f, g) , \qquad (Vf, g) = (f, Ug) ,$$

so that we may conclude that

$$U^\dagger U = I , \qquad V^\dagger V = I , \qquad V = U^\dagger .$$

Hence U has a right inverse and a left inverse:

$$U^{-1} = U^\dagger = V ,$$

that is, U is a unitary transformation whose inverse is given by V. According to the previous theorem, U and V are represented by

$$\int_0^\xi g(x)\,dx = \int k^*(\xi, x) f(x)\,dx\,,$$

$$\int_0^\xi f(x)\,dx = \int h^*(\xi, x) g(x)\,dx\,,$$

if $g = Uf$. This completes the proof of the theorem.

Note that Eqs. (9.114a) and (9.114b) can be written more conveniently as

$$g(x) = \frac{d}{dx}\int k^*(x, y) f(y)\,dy\,, \tag{9.116a}$$

$$f(x) = \frac{d}{dx}\int h^*(x, y) g(y)\,dy\,. \tag{9.116b}$$

An extremely useful special case of this theorem results when we consider the functions

$$k(x, y) = \frac{1}{\sqrt{2\pi}}\,\frac{e^{-ixy} - 1}{-iy}$$

$$h(x, y) = k^*(x, y)\,.$$

We want to show that these functions satisfy the conditions of Theorem 9.11. Consider first

$$\begin{aligned}
\int k^*(\xi, x) k(\eta, x)\,dx &= \frac{1}{2\pi}\int_{-\infty}^{\infty} \frac{(e^{i\xi x} - 1)(e^{-i\eta x} - 1)}{x^2}\,dx \\
&= \frac{1}{2\pi}\int_{-\infty}^{\infty} \frac{1}{x^2}[e^{i(\xi-\eta)x} - e^{i\xi x} - e^{-i\eta x} + 1]\,dx \\
&= \frac{1}{2\pi}\int_{-\infty}^{\infty} \frac{1}{x^2}[\cos(\xi - \eta)x - \cos\xi x - \cos\eta x + 1]\,dx \\
&\quad + \frac{i}{2\pi}\int_{-\infty}^{\infty} \frac{1}{x^2}[\sin(\xi - \eta)x - \sin\xi x + \sin\eta x]\,dx\,.
\end{aligned}$$

The imaginary part vanishes since it is an odd function integrated over a symmetric domain. Note that the integral is well defined, because it converges at $x = \pm\infty$ and is *not* singular at $x = 0$. Thus

$$\begin{aligned}
\int k^*(\xi, x) k(\eta, x)\,dx &= \frac{1}{2\pi}\int_{-\infty}^{\infty} \frac{1}{x^2}[\cos(\xi - \eta)x - \cos\xi x - \cos\eta x + 1]\,dx \\
&= \frac{1}{\pi}\int_{-\infty}^{\infty} [\sin^2\tfrac{1}{2}|\xi|x + \sin^2\tfrac{1}{2}|\eta|x - \sin^2\tfrac{1}{2}|\xi - \eta|x]\frac{dx}{x^2}\,.
\end{aligned}$$

Making a change of variable in each term, we get

$$\int k^*(\xi, x) k(\eta, x)\,dx = \frac{1}{2\pi}[|\xi| + |\eta| - |\xi - \eta|]\int_{-\infty}^{\infty} \frac{\sin^2 z}{z^2}\,dz\,.$$

This integral can be readily evaluated by the complex variable methods of Chapter 6; it is equal to π. Thus

$$\int k^*(\xi, x)k(\eta, x)\,dx = \tfrac{1}{2}\left(|\xi| + |\eta| - |\xi - \eta|\right)$$
$$= \begin{cases} \min(|\xi|, |\eta|) & \text{if } \xi\eta > 0\,, \\ 0 & \text{if } \xi\eta \leq 0\,. \end{cases}$$

Likewise,

$$\int h^*(\xi, k)h(\eta, x)\,dx = \begin{cases} \min(|\xi|, |\eta|) & \text{if } \eta\xi > 0\,, \\ 0 & \text{if } \eta\xi \leq 0\,. \end{cases}$$

Finally

$$\int_0^\eta k(\xi, x)\,dx = \frac{1}{\sqrt{2\pi}}\int_0^\eta \frac{e^{-i\xi x} - 1}{-ix}\,dx = \frac{1}{\sqrt{2\pi}}\int_0^{\xi\eta} \frac{e^{-iz} - 1}{-iz}\,dz$$

and also

$$\int_0^\xi h^*(\eta, x)\,dx = \frac{1}{\sqrt{2\pi}}\int_0^\xi \frac{e^{-i\eta x} - 1}{-ix}\,dx = \frac{1}{\sqrt{2\pi}}\int_0^{\xi\eta} \frac{e^{-iz} - 1}{-iz}\,dz\,.$$

So

$$\int_0^\xi h^*(\eta, x)\,dx = \int_0^\eta k(\xi, x)\,dx\,.$$

Thus k and h fulfill all the hypotheses of Theorem 9.11, and we have the result:

Corollary. The formulas

$$g(x) = \frac{1}{\sqrt{2\pi}}\frac{d}{dx}\int_{-\infty}^{\infty} \frac{e^{ixy} - 1}{iy} f(y)\,dy \tag{9.117a}$$

and

$$f(x) = \frac{1}{\sqrt{2\pi}}\frac{d}{dx}\int_{-\infty}^{\infty} \frac{e^{-ixy} - 1}{-iy} g(y)\,dy \tag{9.117b}$$

define a unitary transform of L_2 and its inverse.

Note that if we interchange integration and differentiation, we obtain just the classical result of Fourier:

$$g(x) = \frac{1}{\sqrt{2\pi}}\int_{-\infty}^{\infty} e^{ixy}f(y)\,dy\,, \tag{9.118a}$$

$$f(x) = \frac{1}{\sqrt{2\pi}}\int_{-\infty}^{\infty} e^{-ixy}g(y)\,dy\,. \tag{9.118b}$$

Now we can easily justify this interchange if we interpret properly the infinite integration range. First note that any $f \in L_2$ can be approximated *in norm* arbitrarily closely by a function which vanishes outside a sufficiently large interval $[-L, L]$, namely, by the function f_L which equals $f(x)$ in $[-L, L]$ and is zero outside. Now let us consider the transform, g_L, of such an approximating function. We have

$$g_L(x) = \frac{1}{\sqrt{2\pi}} \frac{d}{dx} \int_{-L}^{L} \frac{e^{ixy} - 1}{iy} f(y)\, dy$$

$$= \frac{1}{\sqrt{2\pi}} \lim_{\epsilon \to 0} \int_{-L}^{L} \frac{e^{i(x+\epsilon)y} - e^{ixy}}{iy\epsilon} f(y)\, dy$$

$$= \frac{1}{\sqrt{2\pi}} \lim_{\epsilon \to 0} \int_{-L}^{L} \frac{\sin(\epsilon y/2)}{(\epsilon y/2)} e^{i\epsilon y/2} e^{ixy} f(y)\, dy\,. \qquad (9.119)$$

The function under the integral is less than $f(y)$ in absolute magnitude since

$$\left| \frac{\sin(\epsilon y/2)}{(\epsilon y/2)} e^{i\epsilon y/2} e^{ixy} \right| = \left| \frac{\sin(\epsilon y/2)}{(\epsilon y/2)} \right| \leq 1$$

for all values of ϵy. But on any finite interval, $|f(y)|$ is clearly integrable, because

$$\int_{-L}^{L} |f(y)|\, dy = \int_{-\infty}^{\infty} [e_L(y) - e_{-L}(y)]^* \cdot |f(y)|\, dy$$

$$= (e_L, |f|) - (e_{-L}, |f|) < \infty\,,$$

since e_L, e_{-L}, and $|f|$ all belong to L_2. Therefore, the integrand of Eq. (9.119) is bounded by an integrable function for all values of ϵ, and we can interchange the limit and the integration sign to get

$$g_L(x) = \frac{1}{\sqrt{2\pi}} \int_{-L}^{L} e^{ixy} f(y)\, dy = \frac{1}{\sqrt{2\pi}} \frac{d}{dx} \int_{-\infty}^{\infty} \frac{e^{ixy} - 1}{iy} f_L(y)\, dy\,.$$

Since this last transformation is unitary, $Uf_L \to Uf = g$ as $f_L \to f$ in norm. Thus we may write

$$g(x) = \frac{1}{\sqrt{2\pi}} \int_{-\infty}^{\infty} e^{ixy} f(y)\, dy\,,$$

if we mean by the integral from $-\infty$ to $+\infty$ the limit in norm of the integral from $-L$ to $+L$ as L tends toward infinity. The integration in the inverse transform,

$$f(x) = \frac{1}{\sqrt{2\pi}} \int_{-\infty}^{\infty} e^{-ixy} g(y)\, dy\,,$$

is to be interpreted in the same way. Thus the Fourier integral gives rise to a unitary transformation of Hilbert space. The transform and its inverse are given by the familiar formulas of Eqs. (9.118a) and (9.118b).

PROBLEMS

1. If A is a completely continuous Hermitian operator with only a finite number of nonzero eigenvalues, show that the eigenvectors of A can be chosen to form a complete orthonormal set.

2. Show that if we modify $k(x, y)$ on a set of measure zero in the plane, then the resulting kernel, $\tilde{k}(x, y)$ has essentially the same eigenvectors and eigenvalues as $k(x, y)$ and the solution to the equation

$$f(x) = g(x) + \lambda \int k(x, y) f(y)\, dy$$

is essentially the same as the solution to

$$\tilde{f}(x) = g(x) + \lambda \int \tilde{k}(x, y)\, \tilde{f}(y)\, dy\,.$$

What is meant by "essentially" in the above statement?

3. Show that if $\{\lambda_n\}$ and $\{x_n(t)\}$ are respectively the eigenvalues and eigenvectors of a completely continuous Hermitian integral operator, then the sum

$$S(t, t') = \sum_{n=1}^{\infty} \lambda_n x_n(t) x_n^*(t')$$

converges in the sense of mean convergence.

4. Prove that if $\{\lambda_n\}$ and $\{x_n(t)\}$ are respectively the eigenvalues and eigenvectors of the Hermitian integral operator, A, with square integrable kernel $a(t, t')$, then

$$\int dt \int dt'\, |a(t, t')|^2 = \sum_{n=1}^{\infty} \lambda_n^2\,.$$

5. Consider the Hermitian integral operator, A, whose square-integrable kernel is $a(t, t')$ and assume for convenience that all the eigenvalues of A are positive.

   a) Show that if we write a "trial eigenvector" of A as

$$\psi^t = \sum_{n=1}^{N} a_n f_n(t)\,,$$

   where $\{f_n(t) \colon n = 1, 2, \cdots, \nu\}$ is a given set of functions, then if we extremize $(\psi', A\psi')$ with respect to the coefficients a_n, subject to the constraint $(\psi', \psi') = 1$, the resulting equation for the a_n is

$$Mv = \lambda N v\,. \tag{1}$$

   In this equation λ is the Lagrange multiplier which is introduced to handle the constraint, and v is a column vector whose ith component is a_i. M and N are matrices whose ijth elements are

$$M_{ij} = \int f_i^*(t) a(t, t') f_j(t')\, dt\, dt'\,, \qquad i, j = 1, 2, \cdots, \nu\,,$$

$$N_{ij} = \int f_i^*(t) f_j(t)\, dt\,, \qquad i, j = 1, 2, \cdots, \nu\,.$$

   b) Show that M and N are Hermitian matrices.

   c) Show that the values of λ which satisfy Eq. (1) must be real.

   d) Will the eigenvectors corresponding to the various values of λ be orthogonal? Will they satisfy any relationship analogous to the orthogonality relationship?

   e) What is the relationship between the largest λ which solves Eq. (1) and the largest eigenvalue of A? We may imagine that if we choose the set $\{f_n(t)\}$ intelligently, it will be possible to arrive at a very good approximation to this largest eigenvalue.

6. a) Let us denote the space of square-integrable functions on the interval [0, 1] by $L_2[0, 1]$. Show that the operator P, defined on $L_2[0, 1]$ by

$$Pf(t) \equiv f(1 - t) ,$$

commutes with the integral operator A, acting on $L_2[0, 1]$, whose kernel is $a(t, t') = t_<(1 - t_>)$.

b) Show that P is self-adjoint and bounded. What is the norm of P?

c) What are the possible eigenvalues of P? Is P completely continuous? If so, why? If not, why not?

d) In accordance with Theorem 4.22 (assuming it to be applicable in this infinite-dimensional case), can you classify the eigenvectors of A [see part (a)] according to which eigenvalue of P they belong to?

7. a) Using the operator P of Problem 9.6, can you give an argument which suggests why the trial function

$$\phi_1^t = at(1 - t) + bt^2(1 - t)^2$$

would be a better function for obtaining a good value for the largest eigenvalue of A than the function

$$\phi_2^t = at(1 - t) + bt^2(1 - t) .$$

Here A is the Hermitian integral operator on $L_2[0, 1]$ whose kernel is $a(t, t') = t_<(1 - t_>)$. Note that we have already shown in Section 9.2 that $\phi_0^t = at(1 - t)$ gives an excellent starting point.

b) Explain why both ϕ_1^t and ϕ_2^t must certainly give a value for the largest eigenvalue of A which is at least as close to the true value as the value (computed in Section 9.2) given by ϕ_0^t.

c) Using the operator P, state directly, without further detailed computation (but making use of the computation in Section 9.2), what the function ϕ_2^t of part (a) would give as an estimate of the largest eigenvalue of A.

8. a) Show that the function $\phi^t = at(1 - t)(1 - 2t)$ is orthogonal to $\sin \pi t$ on the interval [0, 1] so it is an appropriate trial function to use in attempting to find the second largest eigenvalue (and corresponding eigenvector) of the operator A whose kernel is $a(t, t') = t_<(1 - t_>)$ on $L_2[0, 1]$.
[*Hint*: This can be done by brute force integration, but it is much simpler to use arguments involving the operator P of Problem 9.6.]

b) Using the function ϕ^t given in part (a), obtain an estimate of the second largest eigenvalue of A and compare with the exact value, $\lambda_2 = 1/(4\pi^2)$.

9. Consider the hydrogen atom, consisting of an electron bound to a proton. Neglecting the fact that electrons and protons have spin, one can show on the basis of the nonrelativistic Schrödinger equation,

$$H_0\phi_n = \lambda_n^{(0)}\phi_n ,$$

that there are an infinite number of bound states. Here H_0 is a Hermitian linear operator whose precise form is unimportant for our purpose. The state with lowest energy is a nondegenerate state with eigenvalue $\lambda_1^{(0)} = -13.6$ eV and eigenfunction ϕ_1. The next state with eigenvalue $\lambda_2^{(0)} = -3.4$ eV is fourfold degenerate. We can label the eigenvectors of these four states physically by $\phi_{2s}, \phi_{2p+}, \phi_{2po}$ and ϕ_{2p-}, where the 2 refers to $\lambda_2^{(0)}$, s and p refer to the fact that one state has angular momentum 0 (an "s-state") while the other has angular momentum $\hbar$ (a "p-state") and the subscripts $+$, 0, and $-$ refer to the components of angular momentum along the z-axis in the p-state, which can be either $\hbar$, 0 or $-\hbar$. In the first approximation these states are degenerate, but a closer look at the situation reveals that because of electromagnetic self-energy effects, the $2s$-state actually lies above the $2p$-state by an energy L whose value is roughly 4×10^{-6} eV. The states which lie nearest to this closely spaced $2s$-$2p$ cluster are the $3s$, $3p$, and $3d$ states (they are about 2 eV distant).

a) Imagine now that an electric field is applied to the hydrogen atom, so that the states of the system (the atom in the electric field) will be given by the eigenvectors which satisfy

$$H\psi_n = \lambda_n \psi_n \,,$$

where $H = H_0 + H_1$, H_0 being the Hermitian linear operator mentioned above, and H_1 being given by $H_1 = eEz$. E is the magnitude of the electric field and e is the electronic charge. We will assume the electric field to be of such a magnitude as to shift the energies of the hydrogen atom by a number of the order of magnitude of L. Assuming the perturbation theory of Section 9.2 to be valid for this problem, explain why in calculating the effect of the field on the $2s$-state we can forget about all other states except the $2p$-states without making a significant error.

b) Show that if we use Eq. (9.34) to calculate the function ψ_{2s} (this is the function which would reduce to ϕ_{2s} if the electric field were turned off) in terms of ϕ_{2s} and ϕ_{2p}, then only the first two terms in Eq. (9.34) are nonvanishing. Show that this would not be the case if we used the analogous Rayleigh-Schrödinger expression of Eq. (9.40). You may use the fact that

$$(\phi_{2s}, H_1\phi_{2p+}) = (\phi_{2s}, H_1\phi_{2p-}) = (\phi_{2p}, H_1\phi_{2p+}) = (\phi_{2p}, H_1\phi_{2p-}) = 0\,,$$

and set the only significant nonvanishing inner product equal to M, that is

$$(\phi_{2s}, H_1\phi_{2po}) = M\,.$$

c) Using the result of part (b) and utilizing Eqs. (9.33) and (9.34), find ψ_{2s} and λ_{2s} in terms of ϕ_{2s}, ϕ_{2po}, L, and M.

d) Holding E (that is M) fixed, let L become very small. Explain the simple result which you obtain for ψ_{2s} and λ_{2s} in this limit.

10. Consider the quantity

$$\Phi = (f,(A_0 - \lambda_1^{(0)})f) + 2(f(A_1 - \lambda_1^{(1)})x_1^{(0)})\,,$$

where $\lambda_1^{(0)}$ is the *largest* eigenvalue of A_0 with corresponding eigenvector $x_1^{(0)}$. A_1 is a small perturbation on A_0, and $\lambda_1^{(1)} = (x_1^{(0)}, A_1 x_1^{(0)})$. The spectrum of A_0 is assumed to be nondegenerate. A_0 and A_1 are both completely continuous Hermitian operators on a *real* Hilbert space, H. Show that for every $f \in H$,

$$\Phi \leq \lambda_1^{(2)}\,, \tag{1}$$

where $\lambda_1^{(2)}$ is given by Eq. (9.42). Show that the equality sign in Eq. (1) holds if and only if f is equal to $f_1^{(1)}$ as given by Eq. (9.40). Note that the assumption that $\lambda_1^{(0)}$ is the largest eigenvalue of A_0 is crucial, and that the assumption of a real Hilbert space is also important. How would one modify Φ if one wanted to extend this result to a complex Hilbert space?

Thus we have a maximum principle for $\lambda_1^{(2)}$ which will enable us to find $\lambda_1^{(2)}$ and $f_1^{(1)}$ by constructing trial functions in the manner indicated in Section 9.2.

Therefore the infinite summations in Eqs. (9.40) and (9.42) can be avoided, at least as far as $f_1^{(1)}$ and $\lambda_1^{(2)}$ are concerned. It can be shown that these sums can be avoided for all $f_i^{(n)}$ and $\lambda_i^{(n)}$ by the use of maximum principles.

11. a) By using Simpson's rule for the evaluation of real integrals, write a matrix expression to solve approximately the equation

$$f(x) = g(x) + \lambda \int_a^b k(x, y) f(y)\, dy\,,$$

where $g(x)$ and $k(x, y)$ are continuous functions. (For a statement and derivation of Simpson's rule, see any book on differential calculus, for example, R. Courant, *Differential and Integral Calculus*.) In particular, show that if the interval $[a, b]$ is broken up by $2n + 1$ equally spaced points $(n > 1)$, $a = x_1$, $x_2, \cdots, x_{2n}$, $b = x_{2n+1}$, the integral equation reduces to

$$a = (I - \lambda MD)^{-1} b\,.$$

Here $a_i = f(x_i)$, $b_i = g(x_i)$, $M_{ij} = k(x_i, x_j)$ and D is a diagonal matrix whose elements are $D_{11} = \frac{1}{3}$, $D_{22} = \frac{4}{3}$, $D_{33} = \frac{2}{3}$, $\cdots$, $D_{2n-1,\,2n-1} = \frac{2}{3}$, $D_{2n,\,2n} = \frac{4}{3}$, $D_{2n+1,\,2n+1} = \frac{1}{3}$.

b) Show how to use the same method to treat the homogeneous eigenvalue problem [that is, $g(x) \equiv 0$].

c) Show that if $k(x, y)$ is a real, symmetric kernel, then it is possible to convert the operator eigenvalue equation into a matrix eigenvalue equation involving a single real symmetric matrix.

12. Consider the integral equation

$$f(x) = 1 + 4 \int_0^{1/2} \sqrt{1 - xy}\, f(y)\, dy\,. \tag{1}$$

Compare the functions $f_1(x)$, $f_2(x)$, and $f_3(x)$ obtained by making the successive finite-rank approximations:

a) $\sqrt{1 - xy} \cong 1$, b) $\sqrt{1 - xy} \cong 1 - \frac{1}{2}xy$,
c) $\sqrt{1 - xy} \cong 1 - \frac{1}{2}xy - \frac{1}{8}x^2y^2$,

for the kernel of Eq. (1). These are obtained in an obvious manner from the power series for the square root.

13. The purpose of this problem (and the three that follow) is to develop further the Fredholm method of solving a linear integral equation whose kernel is $k(x, y)$. We begin by defining the determinant Δ, a function of $2n$ variables, by

$$\Delta \begin{bmatrix} x_1, x_2, \cdots, x_n \\ y_1, y_2, \cdots, y_n \end{bmatrix} \equiv \begin{vmatrix} k(x_1, y_1) & k(x_1, y_2) & \cdots & k(x_1, y_n) \\ k(x_2, y_1) & k(x_2, y_2) & & k(x_2, y_n) \\ \vdots & & & \vdots \\ k(x_n, y_1) & k(x_n, y_2) & \cdots & k(x_n, y_n) \end{vmatrix}.$$

In what follows, we will assume that $k(x, y)$ is a continuous function for $a \leq x \leq b$ and $a \leq y \leq b$ with $|k(x, y)| \leq M$. Generalizations from this special case to more general situations (e.g., square-integrable kernels on infinite intervals) is straightforward.

a) Consider the infinite series

$$D(\lambda) = 1 - \lambda \int_a^b \Delta\begin{bmatrix} x_1 \\ x_1 \end{bmatrix} dx_1 + \frac{\lambda^2}{2!}\int_a^b\int_a^b \Delta\begin{bmatrix} x_1, x_2 \\ x_1, x_2 \end{bmatrix} dx_1\, dx_2 + \cdots$$
$$+ (-1)^n \frac{\lambda^n}{n!}\int_a^b\int_a^b \cdots \int_a^b \Delta\begin{bmatrix} x_1, x_2, \cdots, x_n \\ x_1, x_2, \cdots, x_n \end{bmatrix} dx_1\, dx_2 \cdots dx_n + \cdots .$$

Using Hadamard's inequality (see Section 4.3), prove that the coefficient of $\lambda^n/n!$ is less in magnitude than $M^n n^{n/2}$, and hence show that the series $D(\lambda)$ converges for *all* λ. $D(\lambda)$ is called the *Fredholm determinant*.

b) Consider the infinite series

$$N(x, y, \lambda) = k(x, y) - \lambda \int_a^b \Delta\begin{bmatrix} x, x_1 \\ y, x_1 \end{bmatrix} dx_1 + \frac{\lambda^2}{2}\int_a^b\int_a^b \Delta\begin{bmatrix} x, x_1, x_2 \\ y, x_1, x_2 \end{bmatrix} dx_1\, dx_2 + \cdots$$
$$+ (-1)^n \frac{\lambda^n}{n!}\int_a^b\int_a^b \cdots \int_a^b \Delta\begin{bmatrix} x, x_1, x_2, \cdots, x_n \\ y, x_1, x_2, \cdots, x_n \end{bmatrix} dx_1\, dx_2 \cdots dx_n + \cdots .$$

Using the methods employed in part (a), prove that this series converges for all λ to a continuous function of x and y for $a \le x \le b$ and $a \le y \le b$. (For a discussion of what originally motivated the definition of N and D, see, for example, F. Smithies, *Integral Equations* (Chapter 5).

14. a) Expand the determinant

$$\Delta\begin{bmatrix} x, x_1, x_2, \cdots, x_n \\ y, x_1, x_2, \cdots, x_n \end{bmatrix}$$

by minors about its first row to show that

$$\Delta\begin{bmatrix} x, x_1, x_2, \cdots, x_n \\ y, x_1, x_2, \cdots, x_n \end{bmatrix} = k(x, y)\Delta\begin{bmatrix} x_1, x_2, \cdots, x_n \\ x_1, x_2, \cdots, x_n \end{bmatrix}$$
$$+ \sum_{i=1}^{n} k(x,x_i)\Delta\begin{bmatrix} x_i, x_1, x_2, \cdots, x_{i-1}, x_{i+1}, \cdots, x_n \\ y, x_1, x_2, \cdots, x_{i-1}, x_{i+1}, \cdots, x_n \end{bmatrix}$$

and therefore that

$$\int_a^b\int_a^b \cdots \int_a^b \Delta\begin{bmatrix} x, x_1, x_2, \cdots, x_n \\ y, x_1, x_2, \cdots, x_n \end{bmatrix} dx_1\, dx_2 \cdots dx_n$$
$$= k(x, y)\int_a^b\int_a^b \cdots \int_a^b \Delta\begin{bmatrix} x_1, x_2, \cdots, x_n \\ x_1, x_2, \cdots, x_n \end{bmatrix} dx_1\, dx_2 \cdots dx_n$$
$$- n\int_a^b\int_a^b \cdots \int_a^b k(x, z)\Delta\begin{bmatrix} z, x_1, x_2, \cdots, x_{n-1} \\ y, x_1, x_2, \cdots, x_{n-1} \end{bmatrix} dx_1\, dx_2 \cdots dx_{n-1}\, dz .$$

b) By expanding Δ about the first column, show in the manner of part (a) that

$$\int_a^b\int_a^b \cdots \int_a^b \Delta\begin{bmatrix} x, x_1, x_2, \cdots, x_n \\ y, x_1, x_2, \cdots, x_n \end{bmatrix} dx_1\, dx_2 \cdots dx_n$$
$$= k(x, y)\int_a^b\int_a^b \cdots \int_a^b \Delta\begin{bmatrix} x_1, x_2, \cdots, x_n \\ x_1, x_2, \cdots, x_n \end{bmatrix} dx_1\, dx_2 \cdots dx_n$$
$$- n\int_a^b\int_a^b \cdots \int_a^b \Delta\begin{bmatrix} x, x_1, x_2, \cdots x_{n-1} \\ z, x_1, x_2, \cdots x_{n-1} \end{bmatrix} k(z, y)\, dx_1\, dx_2 \cdots dx_{n-1}\, dz .$$

15. a) Using the result of Problem 9.14(a) in the expression of Problem 9.13(b) for $N(x, y, \lambda)$, show that

$$N(x, y, \lambda) = k(x, y)D(\lambda) + \lambda \int_a^b k(x, z)N(z, y, \lambda)\, dz .$$

Similarly, use the result of Problem 9.14(b) to show that

$$N(x, y, \lambda) = k(x, y)D(\lambda) + \lambda \int_a^b N(x, z, \lambda)k(z, y)\, dz .$$

b) Define

$$R(x, y, \lambda) \equiv \frac{N(x, y, \lambda)}{D(\lambda)} .$$

Show that the two equations of part (a) can be rewritten as

$$(I - \lambda K)(I + \lambda R(\lambda)) = I, \qquad \text{and} \qquad (I + \lambda R(\lambda))(I - \lambda K) = I ,$$

where $R(\lambda)$ is the integral operator whose kernel is $R(x, y, \lambda)$ and K is the integral operator whose kernel is $k(x, y)$. Thus we see that $I + \lambda R(\lambda)$ is the *inverse* of $I - \lambda K$, so that in the language of this chapter, $R(\lambda)$ is the *resolvent* of K. Therefore the equation

$$f = g + \lambda K f$$

can be solved for all values of λ for which $R(\lambda)$ exists by

$$f = g + \lambda R(\lambda) g .$$

In terms of the actual kernels, this may be written as

$$f(x) = g(x) + \lambda \int_a^b R(x, y, \lambda)g(y)\, dy = g(x) + \frac{\lambda}{D(\lambda)} \int_a^b N(x, y, \lambda)g(y)\, dy .$$

16. The only remaining question to settle is: What are the values of λ for which $R(\lambda)$ fails to exist? Clearly, since by Problem 9.13 $N(x, y, \lambda)$ and $D(\lambda)$ are analytic functions of λ, $R(\lambda)$ will fail to exist at any zero of $D(\lambda)$, unless that zero is canceled by a corresponding zero of $N(x, y, \lambda)$. In fact, such a cancellation cannot occur.

a) Using the results of Problem 9.13, show that

$$\frac{dD(\lambda)}{d\lambda} = - \int_a^b N(x, x, \lambda)\, dx .$$

b) Using the result of part (a), show that if $D(\lambda)$ has a zero at $\lambda = \lambda_0$, then $N(x, y, \lambda)$ *cannot* have a zero at this point. Prove this result by contradiction. First, write $D(\lambda) = (\lambda - \lambda_0)d(\lambda)$, where $d(\lambda)$ is analytic; next, assume that $N(x, y, \lambda)$ can also be written in this manner, i.e., as $N(x, y, \lambda) = (\lambda - \lambda_0)n(x, y, \lambda)$, where $n(x, y, \lambda)$ is an analytic function of λ; and finally show that under this assumption the equation of part (a) cannot be satisfied. Thus the zeroes of $D(\lambda)$ are precisely the values of λ for which $R(\lambda)$ does not exist. For these values, the homogeneous equation $f = \lambda K f$ will have a solution.

17. Let K be an integral operator with a continuous kernel and assume that the spectrum of K is nondegenerate. Using the results of Problems 9.15(a) and 9.16(a), show that the residue at a pole, λ_n, of $R(x, y, \lambda)$ is equal to

$$- \frac{\phi_n(x)\psi_n^*(y)}{\int \phi_n(x)\psi_n^*(x)\, dx} ,$$

where ϕ_n satisfies the equation $\lambda_n K\phi_n = \phi_n$, and ψ_n satisfies the adjoint equation $\lambda_n^* K^\dagger \psi_n = \psi_n$. Note that for Hermitian kernels this result is implicit in the eigenfunction expansion of the resolvent given by Eq. (9.22).

18. Show that if $k(x, y) = \phi(x)\psi^*(y)$, then the Fredholm method reduces to

$$D(\lambda) = 1 - \lambda \int \phi(x)\psi^*(x)\, dx\,, \qquad N(x, y, \lambda) = k(x, y) = \phi(x)\psi^*(y)\,,$$

that is, all higher terms in the Fredholm series for $D(\lambda)$ and $N(x, y, \lambda)$ vanish. Show that this leads to a solution of $f = g + \lambda K f$ in agreement with Eq. (8.57).

19. Assuming that λ is sufficiently small, expand $[D(\lambda)]^{-1}$ in powers of λ and show that through order λ^2, $R(x, y, \lambda)$ is equal to the Neumann series expansion of the resolvent through order λ^2. It can be shown that this equality holds to *all* orders, as we would expect.

20. Suppose that one wished to extend the Fredholm method to square-integrable kernels. The most serious obstacle to doing this is the appearance of terms like

$$I = \int_a^b k(x, x)\, dx$$

in the treatment of Problems 9.13 through 9.16. Even if $k(x, y)$ is square integrable, I may not exist, and hence the method fails. This happens, for example, in the case of kernels of the form $k(\mathbf{r}, \mathbf{r}') = G(\mathbf{r}, \mathbf{r}')V(\mathbf{r})$, where $G(\mathbf{r}, \mathbf{r}')$ is given by Eq. (7.66), even if one makes the appropriate weight-function modification in the manner of Section 8.3.

a) Give an argument which suggests that we can *always* modify $k(x, y)$ in such a way that $k(x, x) = 0$ without altering in any essential way the solutions to

$$f(x) = g(x) + \lambda \int k(x, y) f(y)\, dy\,, \tag{1}$$

where $k(x, y)$ is assumed to be square-integrable.
[*Hint*: See Problem 9.2.]

b) Part (a) seems to suggest that all difficulties with integrals like

$$\int_a^b k(x, x)\, \mathrm{d}x$$

can be removed simply by defining

$$\tilde{D}(\lambda) = 1 - \lambda \int \tilde{\Delta}\begin{bmatrix} x_1 \\ x_1 \end{bmatrix} dx_1 + \frac{\lambda^2}{2!} \int\int \tilde{\Delta}\begin{bmatrix} x_1, x_2 \\ x_1, x_2 \end{bmatrix} dx_1, dx_2 + \cdots$$

$$\tilde{N}(x, y, \lambda) = k(x, y) - \lambda \int \tilde{\Delta}\begin{bmatrix} x, x_1 \\ y, x_1 \end{bmatrix} dx_1 + \frac{\lambda^2}{2!} \int\int \tilde{\Delta}\begin{bmatrix} x, x_1, x_2 \\ y, x_1, x_2 \end{bmatrix} dx_1\, dx_2 + \cdots$$

where $\tilde{\Delta}$ is the same determinant as Δ *except* that whenever an entry of the form $k(x_i, x_i)$ appears in Δ, we replace it by zero to get the equivalent entry in $\tilde{\Delta}$. Thus

$$\Delta\begin{bmatrix} x_1 \\ x_1 \end{bmatrix} = k(x_1, x_1)\,, \qquad \tilde{\Delta}\begin{bmatrix} x_1 \\ x_1 \end{bmatrix} = 0\,,$$

$$\Delta\begin{bmatrix} x_1, x_2 \\ x_1, x_2 \end{bmatrix} = k(x_1, x_1)k(x_2, x_2) - k(x_1, x_2)k(x_2, x_1)\,,$$

$$\tilde{\Delta}\begin{bmatrix} x_1, x_2 \\ x_1, x_2 \end{bmatrix} = -k(x_1, x_2)k(x_2, x_1)\,,$$

and so forth. Letting

$$\tilde{R}(x, y, \lambda) \equiv \frac{\tilde{N}(x, y, \lambda)}{\tilde{D}(\lambda)},$$

and expanding $[\tilde{D}(\lambda)]^{-1}$ (assuming λ is sufficiently small), show that through order λ^2, $\tilde{R}(x, y, \lambda)$ is equal to the Neumann expansion of the resolvent of K. This equality can be shown to hold to all orders. In fact, the modification suggested by this problem is correct, namely, $\tilde{N}(x, y, \lambda)$, $\tilde{D}(\lambda)$, and $\tilde{R}(x, y, \lambda)$ as defined above do indeed provide us with a solution to Eq. (1).

21. Obtain the function $\tilde{D}(\lambda)$, defined in Problem 20, through order λ^2 for the kernel of the integral equation Eq. (9.102) if $V(\mathbf{r}) = g^2 e^{-\mu r}/r$. The variables s, t and u defined in Section 8.3 will be useful. Note that the integral which must be done is very similar to the one we encountered in the evaluation of the norm of this kernel in Section 8.3. Assuming q_i to be real, what is the difference between these two cases?
22. Using Eqs. (9.104) and (9.105), show that if we denote the second iterative approximation to the amplitude by $f_0^{s2B}(q_i)$, then for any real potential $V(r)$ we have

$$Im f_0^{s2B}(q_i) = q_i[f_0^{sB}(q_i)]^2 .$$

23. Calculate the second Born approximation to the s-wave scattering amplitude, $F_0^s(p_i)$, defined by Eqs. (9.112) and (9.113). Compare this result with the curves of Figs. 9.4, 9.5, and 9.6 and make any comments which seem appropriate.
24. It can be shown that the quantity $f_0^s(q_i)$ defined in Section 9.5 can be written as

$$f_0^s(q_i) = \frac{1}{q_i} e^{i\delta_0} \sin \delta_0 ,$$

where δ_0 is a function of q_i and is called the s-wave phase shift. Using Fig. 9.4, obtain a graph of $\delta_0(q_i)$ versus q_i. Are the results obtained for $\delta_0(q_i)$ from the two different curves in Fig. 9.4 consistent? To what extent is δ_0 determined uniquely?
25. a) Obtain Eqs. (9.98) through (9.101) using the methods employed in Section 9.3 to discuss the homogeneous problem.
    b) Derive an equation which "solves" the integral equation $f = g + Kf$ in the spirit of Riemann integration (see Eq. (9.50)).
    c) If you have access to a computer make up a few integral equations and solve them numerically using the results of part (a) or (b). For example, you might consider Eqs. (9.112) and (9.113), but with a potential of the opposite sign (i.e., $mV/\mu^2\hbar^2 = -1.4$). You will find that the scattering amplitude looks rather different than it did in the case of positive V and that there is a striking change in the behavior of δ_0 (see Problem 9.24) as a function of p_i. This change is caused by the existence of an s-wave bound state in the attractive potential (negative V) which is not present in a repulsive potential.
26. Suppose that $\{\phi_n\}$ and $\{\psi_m\}$ are sequences of step functions and that $\phi_n \to f$ and $\psi_m \to g$. Show that if for any step functions ϕ and ψ we have

$$(V\phi, \psi) = (\phi, U\psi) ,$$
$$(V\phi, V\psi) = (\phi, \psi) ,$$

and

$$(U\phi, U\psi) = (\phi, \psi) ,$$

then the sequences $\{V\phi_n\}$ and $\{U\psi_m\}$ converge to some limit in Hilbert space and that if we denote these limits by Vf and Ug respectively, then

$$(Vf, g) = (f, Ug) .$$

This completes the proof of Theorem 9.11.

27. a) Show that if we define

$$\mathscr{H}_n(x) \equiv H_n(x)e^{-x^2/2} ,$$

where $H_n(x)$ is the nth Hermite polynomial which satisfies Eq. (5.105), then the Hermite function $\mathscr{H}_n(x)$ satisfies

$$\mathscr{H}_n''(x) - x^2\mathscr{H}_n(x) + (2n + 1)\mathscr{H}_n(x) = 0 .$$

Note that according to Eq. (5.100), $\mathscr{H}_n(x)$ is orthogonal to $\mathscr{H}_m(x)$ on the interval $[-\infty, \infty]$.

b) Denoting the unitary Fourier transformation by U_F, show that the functions $\mathscr{H}_n(x)$ are eigenfunctions of U_F. Do this by proving that the Fourier transform of $\mathscr{H}_n(x)$, say $\hat{\mathscr{H}}_n(x)$, satisfies the same differential equation as does $\mathscr{H}_n(x)$, so $\hat{\mathscr{H}}_n(x)$ must equal (to within a multiplicative constant) $\mathscr{H}_n(x)$. In operator language, this is just

$$U_F\mathscr{H}_n = \lambda_n\mathscr{H}_n ,$$

where λ_n has not yet been determined.

[*Hint* $x^2e^{ixy} = -(d^2/dy^2)e^{ixy}$; also, integration by parts will be useful.]

c) Show that if $f(x)$ is any square-integrable function, then $U_F^2f(x) = f(-x)$. Hence show that the only possible eigenvalues of U_F are 1, -1, i, and $-i$. There is an infinite number of Hermite functions which belong to each of these four eigenvalues. Give a simple characterization of those Hermite functions which will have real eigenvalues and those which will have imaginary eigenvalues.

FURTHER READINGS

Dettman, J. W., *Mathematical Methods in Physics and Engineering*. New York: McGraw-Hill, 1962.

Goldberger, M. L., and K. M. Watson, *Collision Theory*. New York: Wiley, 1964.

Newton, R. G., *Scattering Theory of Waves and Particles*. New York: McGraw-Hill, 1966.

Riesz, F., and B. Sz. Nagy, *Functional Analysis*. New York: Ungar, 1955.

Smithies, F., *Integral Equations*. New York: Cambridge University Press, 1958.

CHAPTER 10

INTRODUCTION TO GROUP THEORY

INTRODUCTION

Thus far in this book we have discussed objects (vectors) which combine in a way that is essentially *additive*, i.e., they combine commutatively. In investigating the properties of these objects we were led in a natural way to consider *transformations* which changed one vector into another. We observed that these transformations themselves could be considered as constituting a vector space. However, since the transformations took one vector into another, they could also be combined in a way which was essentially *multiplicative* in nature; i.e., the commutative law did not in general hold. In this concluding chapter we want to introduce some of the techniques by which one can study collections of objects which are characterized in terms of their multiplicative properties. Such a collection is called a *group*.

10.1 AN INDUCTIVE APPROACH

Let us first recall the definition of a group given previously in Chapter 3:

Definition 10.1. A *group*, G, is a collection of objects which can be combined via a closed operation, which we will denote by a dot. By a closed operation we mean one such that if $a, b \in G$, then $a \cdot b \in G$. The operation must furthermore satisfy the following three axioms:

1) $a \cdot (b \cdot c) = (a \cdot b) \cdot c$ (associative law)
2) there exists an identity element, e, such that $a \cdot e = a$ for all $a \in G$.
3) for every $a \in G$, there exists an inverse element, a^{-1}, such that $a \cdot a^{-1} = e$.

Note that in this definition we have required that the identity be only a *right* identity and that the inverse be only a *right* inverse. However, because of the other group properties, it follows that e is also a *left* identity ($e \cdot a = a$) and a^{-1} is also a *left* inverse ($a^{-1} \cdot a = e$).

To show this we note that since $a^{-1} \in G$, it must have an inverse, which we denote by $(a^{-1})^{-1}$. According to axiom (3),

$$(a^{-1}) \cdot (a^{-1})^{-1} = e \,. \tag{10.1}$$

But by Axioms (1), (2), and (3) we may write Eq. (10.1) as

$$\begin{aligned} e = a^{-1} \cdot (a^{-1})^{-1} &= (a^{-1} \cdot e) \cdot (a^{-1})^{-1} = [a^{-1} \cdot (a \cdot a^{-1})] \cdot (a^{-1})^{-1} \\ &= [(a^{-1} \cdot a) \cdot a^{-1}] \cdot (a^{-1})^{-1} = (a^{-1} \cdot a) \cdot [a^{-1} \cdot (a^{-1})^{-1}] \,. \end{aligned}$$

Using Eq. (10.1) again, we have

$$e = (a^{-1} \cdot a) \cdot e = a^{-1} \cdot (a \cdot e) \, .$$

By Axiom (2), $a \cdot e = a$, so finally we obtain

$$e = a^{-1} \cdot a \, . \tag{10.2}$$

Hence a^{-1} is a left inverse as well as a right inverse. This result contrasts with the situation for general linear operators, where the existence of a right inverse does not necessarily imply the existence of a left inverse.

Using this result, we can immediately see that e must be a left identity. By Axiom (2), e is a right identity, that is,

$$a \cdot e = a \, ,$$

but according to Eq. (10.2), this is equivalent to

$$a \cdot (a^{-1} \cdot a) = a \, .$$

Using Axioms (1) and (3), we find that

$$a = a \cdot (a^{-1} \cdot a) = (a \cdot a^{-1}) \cdot a = e \cdot a \, ,$$

that is, we have

$$a = e \cdot a \, ,$$

so e is also a left identity.

Definition 10.2. A group, G is said to be *abelian* if $a \cdot b = b \cdot a$ for any $a, b \in G$.

Definition 10.3. A group with n elements is called a group of *order n*.

In the first three sections of this chapter, we shall consider only finite groups ($n < \infty$) in order to fix our thinking on the most basic aspects of group theory. In subsequent sections, we shall have some occasion to mention infinite groups; when we do, we shall always state explicitly that this is the case.

Let us now introduce a few groups, starting with the simplest ones.

Example 10.1. The most uncomplicated group which one can imagine is the one element group: $\{e\}$. It obviously satisfies all the requirements of Definition 10.1. It is not a particularly interesting group.

Example 10.2. Next we have the group with two distinct elements: $\{e, a\}$. Now $a \cdot a$ must belong to this group according to Definition 10.1. Thus we must have either $a \cdot a = a$ or $a \cdot a = e$. The former is immediately ruled out by the fact that since a has an inverse, $a \cdot a = a$ implies that $a = e$. This contradicts the assumption that the group has two distinct elements. Thus $a \cdot a = e$. The integers $\{1, -1\}$ form a group with this structure if they combine via ordinary arithmetical multiplication ($e = 1$, $a = -1$). Also, the integers $\{0, 1\}$ form a group with addition modulo two as the rule of combination: $0 + 0 = 0$, $0 + 1 = 1$, $1 + 0 = 1$, $1 + 1 = 0$. In this case, $e = 0$, $a = 1$. Another structure of this type is the group of permutations of two objects; e is the permutation which leaves the order of the objects unchanged, and a is the permutation which interchanges them.

This last example leads naturally to the following definition.

Definition 10.4. Two groups are said to be *isomorphic* if there exists a one-to-one operation-preserving correspondence between them. That is, suppose that G has elements $a, b, c, \cdots$, and G' has elements $a', b', c', \cdots$ If $(a \cdot b)' = (a' \times b')$ for all $a, b \in G$ and $a', b' \in G$, then the groups $\{G, \cdot\}$ and $\{G', \times\}$ are said to be isomorphic.

Evidently the three groups mentioned in Example 10.2 are isomorphic to each other: once we know everything about any *one* of them, we know everything about *all* of them.

Example 10.3. Let us now move on to groups with three distinct elements, which we denote by $\{e, a, b\}$. From the axioms we see immediately that $a \cdot b = e$, for if $a \cdot b = a$ or $a \cdot b = b$, then the existence of an inverse would imply $b = e$ in the first case and $a = e$ in the second case, which is contrary to our assumption that this group has three *distinct* elements. What about the product $a \cdot a$? Clearly, $a \cdot a = a$ implies that $a = e$, so either $a \cdot a = e$ or $a \cdot a = b$. But

$$a \cdot a = e \Longrightarrow a \cdot (a \cdot b) = b \Longrightarrow a = b\ ,$$

since $a \cdot b = e$. But this, too, is impossible because we have assumed three distinct elements. Thus $a \cdot a = b$. Similarly $b \cdot b = a$. Adopting the obvious notation for quantities like $a \cdot a$, we may write our three element group as $\{e, a, a^2\}$, with $a^3 = e$. This is the *only* possibility for the three-element group. The postulates of Definition 10.1 restrict us to just one form. A simple example of this group is the set of three cube roots of unity, under ordinary multiplication of complex numbers.

Clearly, it is always possible to construct a group of the form $\{e, a, a^2, a^3, \cdots, a^{n-1}\}$ for any n; the basic model of such a group is the set of n nth roots of unity. In general, consider an arbitrary group, $\{G, \cdot\}$ and let $a \in G$. Form the sequence $a^0 = e, a, a^2, \cdots, a^i, \cdots$, and let n be the smallest *nonzero* integer such that $a^n = e$. Then the element a is said to be of *order n* (if n is infinite, we say that a is of infinite order). In this case, all the elements $a^0, a^1, a^2, \cdots, a^{n-1}$ are distinct, for if $a^i = a^j$ $(i < n, j < n)$, then $a^{|i-j|} = e$. But since $0 \leq i < n$ and $0 \leq j < n$, $|i - j| < n$, which contradicts our assumption that n was the *smallest* nonzero integer for which $a^n = e$.

Definition 10.5. A group whose elements can be written as $\{e, a, a^2, \cdots, a^{n-1}\}$ is called a *cyclic* group (of order n).

Thus far, all our groups have been of the cyclic type, so it is natural to ask: Are all groups cyclic? The answer is provided by the following example.

Example 10.4. We apply the methods already used to the possible groups of order four, which we denote by $\{e, a, b, c\}$. Let us begin by assuming $a^2 = b$;

then we can write the corresponding group as $\{e, a, a^2, c\}$. By an argument used before, $a \cdot c$ must equal either e or a^2. If $a \cdot c = a^2$, we conclude that $a = c$, which contradicts the assumption of four distinct elements. Thus $a \cdot c = e$, so it begins to look like we are obtaining a cyclic group. We therefore look at a^3. Suppose that $a^3 = e$; then $a^3 = a \cdot c$, so $a^2 = c$, which is impossible. If $a^3 = a$, then $a^2 = e$, which again is impossible. Similarly, if $a^3 = a^2$, $a = e$, which is impossible. Thus $a^3 = c$, and we indeed have found a cyclic group of order four: $\{e, a, a^2, a^3\}$, with $a^4 = e$.

The same result is obtained if we assume that $a^2 = c$, except that it then follows by the same reasoning as above that $b = a^3$. As mentioned above, $a^2 = a$ is not permissible. However, there remains the possibility that $a^2 = e$. Clearly, either $a \cdot b = e$ or $a \cdot b = c$. If $a \cdot b = e$, then $a^2 \cdot b = a$; since $a^2 = e$, we have $a = b$, which is impossible. Thus $a \cdot b = c$. By the same process of elimination, we find that $b \cdot a = c$, $a \cdot c = b$ and $c \cdot a = b$. Since $a \cdot b = c$ and $b \cdot a = c$, we have

$$(b \cdot a) \cdot (a \cdot b) = c \cdot c \Longrightarrow b \cdot a^2 \cdot b = c^2 .$$

But $a^2 = e$, so we have immediately $b^2 = c^2$, which means that c^2 cannot equal b or c. Thus either $c^2 = e$ or $c^2 = a$. We first examine the case $c^2 = a$. Then $a \cdot c^2 = a^2 = e$, but $a \cdot c = b$ so we have $b \cdot c = e$, which gives us explicit expressions for the products of all elements of the group. We summarize by writing

$$\begin{gathered} a^2 = e , \qquad b^2 = c^2 = a , \\ b \cdot c = c \cdot b = e , \\ a \cdot b = b \cdot a = c , \qquad a \cdot c = c \cdot a = b . \end{gathered} \tag{10.3}$$

We now want to know if this group of order four is different from the cyclic group of order four which we have already found several times. A very convenient way of analyzing simple finite groups is by means of a *multiplication table*. We illustrate it by using the *cyclic group* of order four:

	e	a	b	c
e	e	a	b	c
a	a	b	c	e
b	b	c	e	a
c	c	e	a	b

(10.4)

The meaning of this table is simple: in the ij-box one writes the product of the ith element of the group times the jth element. The reader can check for himself that this table represents accurately the fourth-order cyclic group first discussed in Example 10.4. Now what about the group whose multiplicative properties are given by Eqs. (10.3)? We have for the multiplication table

	e	a	b	c
e	e	a	b	c
a	a	e	c	b
b	b	c	a	e
c	c	b	e	a

(10.5)

This looks completely different from the table for the fourth-order cyclic group. However, it may nevertheless be *isomorphic* to the fourth-order cyclic group. This will be the case if we can relabel the elements in such a way that the multiplication table is the same as that of the fourth-order cyclic group. If in Table (10.5) we redefine the elements according to

$$e \to e' , \qquad a \to b' ,$$
$$b \to a' , \qquad c \to c' ,$$

then Table (10.5) becomes

	e'	a'	b'	c'
e'	e'	a'	b'	c'
a'	a'	b'	c'	e'
b'	b'	c'	e'	a'
c'	c'	e'	a'	b'

(10.6)

Since Table (10.6) is the same as Table (10.4), we still have not found a group with a multiplicative structure different from that of the cyclic group of order four.

However, in the class of four-element groups, one possibility remains. We have been forced to conclude that $a^2 = e$, $a \cdot b = c$, $b \cdot a = c$, $a \cdot c = b$, $c \cdot a = b$, and $b^2 = c^2$. When we let $b^2 = c^2 = a$, we were led back to the cyclic group. There remains, however, the choice $b^2 = c^2 = e$. Then

$$a \cdot b = c \Longrightarrow a \cdot b^2 = c \cdot b \Longrightarrow c \cdot b = a ,$$

and similarly

$$b \cdot a = c \Longrightarrow b^2 \cdot a = b \cdot c \Longrightarrow b \cdot c = a .$$

This completes the multiplication table:

	e	a	b	c
e	e	a	b	c
a	a	e	c	b
b	b	c	e	a
c	c	b	a	e

(10.7)

The reader can check that there is no rearrangement of these elements which will give the Table (10.4). Thus we have finally found a noncyclic group. Note, however, that it is an abelian group; this is reflected in the fact that the array in Table (10.7) is symmetric with respect to the diagonal. This group is called the four-group. A simple realization of the four-group is the set of all transformations of a rectangle which leave the rectangle's orientation in space unchanged (remember that in a rectangle the right and left sides are indistinguishable, as are the top and bottom sides). If the rectangle sits on an xy-plane with its center at the origin, there are four such transformations: (i) the identity transformation, (ii) rotation through 180° about an axis perpendicular to the rectangle and passing through its center, (iii) reflection through the x-axis, and (iv) reflection through the y-axis. Note that we omit rotations through 360°, 540°, etc., since they just duplicate the transformations (i) and (ii) above. Figure 10.1 illustrates the configuration. We leave it to the reader to show that this is indeed a group and that its multiplication table is the same as that given in Table (10.7).

From Table (10.7) it also appears that the elements $\{e, a\}$ of the four-group form a group by themselves ($e \cdot e = e$, $e \cdot a = a$, $a \cdot e = a$, $a \cdot a = e$), as do $\{e, b\}$ and $\{e, c\}$. This leads to the following definition.

Definition 10.6. A set of elements H, which is contained in G, is said to be a *subgroup* of G if

i) the product of any pair of elements in H is in H,

ii) if $a \in H$, then $a^{-1} \in H$.

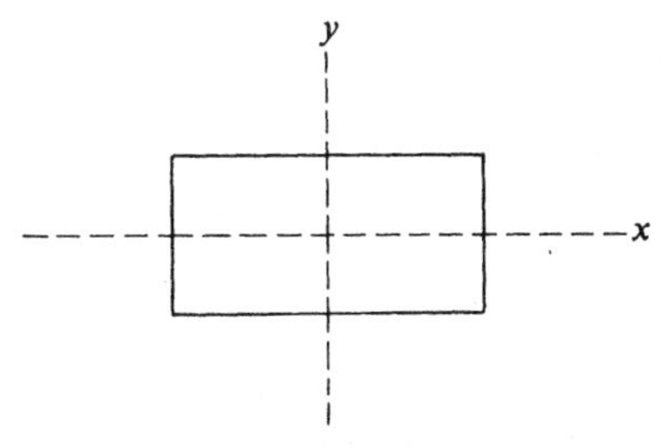

Fig. 10.1

Any subgroup is also a group because its elements are already in G, and hence obey the associative law; moreover, $a, a^{-1} \in H$ implies that $a \cdot a^{-1} = e$ also belongs to H. Clearly any group is a subgroup of itself, and the identity element is a subgroup of any group. As mentioned above, the four-group has $\{e, a\}$ as a subgroup, as well as isomorphic subgroups $\{e, b\}$ and $\{e, c\}$. Clearly, in any group, G, the collection $\{e, a, a^2, \cdots, a^{n-1}\}$, where n is the order of a and $a \in G$, is an abelian subgroup of G.

Another interesting feature of Table (10.7), which it shares with all the other multiplication tables written above, is the fact that each row (or column) contains each element of the group once and only once. This is not accidental; we have in fact the following result.

Theorem 10.1. If G is a group of order n, with elements $e, a_2, a_3, \cdots, a_n$, then every element of G occurs once and only once in the sequence

$$ea_i, a_2a_i, \cdots, a_na_i ,$$

for any i, and similarly for

$$a_ie, a_ia_2, \cdots, a_ia_n .$$

Note that in the statement of this theorem we have omitted the dot in denoting the product of group elements. We shall do this throughout the chapter, except when this omission might cause confusion.

Proof. If some element occurred twice, we would have $a_\mu a_i = a_\nu a_i$; this would imply $a_\mu = a_\nu$, contrary to our assumption of distinct group elements. Since there are n elements in the sequence and no element occurs twice, each element occurs once and only once. This explains the above mentioned structure of the group multiplication table. This very simple looking theorem is actually of central importance in the proof of most of the results of this chapter.

One could continue inductively, in the manner of this section, developing the groups of order five, order six, etc., but this would not be very instructive. We may remark, however, that not all groups are abelian like the ones which we have discussed so far; at order six one finds the first nonabelian group, the group of permutations of three objects. It should also be emphasized that not all groups have a finite number of elements. For example, the integers under addition ($e = 0$, $n^{-1} = -n$) form an (abelian) group of infinite order; the set of all unitary $n \times n$ matrices is an infinite nonabelian group.

10.2 THE SYMMETRIC GROUPS

By way of further illustrating group structure, we now discuss one of the most important groups of mathematics and physics, the group of permutations on n objects, called the symmetric group, S_n. The structure of the groups S_n is extraordinarily rich, and in this section we will consider only the simplest aspects of these groups.

There are $n!$ permutations of n objects, so S_n is of order $n!$. A typical one of these $n!$ elements will be denoted by

$$p = \begin{pmatrix} 1 & 2 & 3 & \cdots & n \\ m_1 & m_2 & m_3 & \cdots & m_n \end{pmatrix},$$

where the set $\{m_1, m_2, m_3, \cdots, m_n\}$ is some arrangement of the first n integers. This symbol means that 1 is replaced by m_1, 2 by m_2, etc. For example, the permutation

$$p = \begin{pmatrix} 1 & 2 & 3 & 4 & 5 & 6 \\ 6 & 5 & 4 & 3 & 2 & 1 \end{pmatrix},$$

acting on the arrangement $\{1\ 2\ 3\ 4\ 5\ 6\}$ produces the arrangement $\{6\ 5\ 4\ 3\ 2\ 1\}$; p acting on $\{2\ 5\ 3\ 4\ 6\ 1\}$ produces $\{5\ 2\ 4\ 3\ 1\ 6\}$. In this example, 1 and 6 go into each other in a closed manner: 1 is replaced by 6, and 6 is replaced by 1. Such a structure is called a *cycle* and will be written simply as (16). Similarly, the above example contains the cycles (25) and (34). We can thus write

$$\begin{pmatrix} 1 & 2 & 3 & 4 & 5 & 6 \\ 6 & 5 & 4 & 3 & 2 & 1 \end{pmatrix} = (16)(25)(34).$$

When we perform such a factorization, we say that we factor the given permutation into *disjoint* cycles, or cycles having no elements in common.

Of course, not all cycles contain only two elements. A cycle containing l elements is called an l-cycle, or a cycle of length l. For example, the permutation

$$\begin{pmatrix} 1 & 2 & 3 & 4 & 5 & 6 \\ 6 & 4 & 1 & 2 & 5 & 3 \end{pmatrix}$$

can be written as a disjoint product of a three-cycle, a two-cycle, and a one-cycle:

$$\begin{pmatrix} 1 & 2 & 3 & 4 & 5 & 6 \\ 6 & 4 & 1 & 2 & 5 & 3 \end{pmatrix} = (163)(24)(5).$$

The three-cycle, (163), is to be read as "1 is replaced by 6, 6 is replaced by 3 and 3 is replaced by 1." This structure closes on itself just like a two-cycle. Note that

$$(163) = (316) = (631),$$

but

$$(163) \neq (136).$$

Thus as the name "cycle" suggests, if one writes the elements of a cycle clockwise around a circle, one can start performing the sequence of replacements at *any* element and proceed clockwise back to the starting point. However, if one proceeds in a counterclockwise direction, one gets a different permutation.

The two-cycles, usually referred to as transpositions, are especially important because any cycle (and hence any permutation) can be written as a product of two-cycles (which are not, in general, disjoint). For example,

$$(163) = (13)(16)\ ,$$

since when (163) acts on say $\{136\}$, it produces $\{613\}$, and (16) acting on $\{136\}$ produces $\{631\}$, which in turn becomes $\{613\}$ when acted on by (13). Here we adopt the convention that when a string of cycles is written, the one on the right acts first, and the one on the left acts last. If the cycles are disjoint, it clearly makes no difference in which order they act, but if they are not disjoint, the order is crucial. For example,

$$(163)\{136\} = (13)(16)\{136\} = \{613\}\ ,$$

whereas

$$(16)(13)\{136\} = \{361\} = (136)\{136\}\ .$$

Also, the decomposition into two-cycles is *not* unique. In the above case,

$$(163) = (13)(16)$$

and

$$(163) = (316) = (36)(31)\ .$$

In general, a decomposition of an n-cycle into transpositions can be written as

$$(1\ 2\ 3\ \cdots\ n) = (1n)(1\ \ n-1)\ \cdots\ (13)(12)\ ,$$

as may be readily verified. The parity of a permutation is defined to be $(-1)^N$, where N is the number of transpositions in a given permutation. If N is odd, we speak of an odd permutation; if N is even, we speak of an even permutation.

If now by a *product* of permutations, we mean simply the two permutations carried out successively, then obviously the product of two permutations is another permutation. All the other group axioms are trivially satisfied so the collection of all permutations on n objects is a group, and this group will contain $n!$ elements. By our above discussion, it is easily seen that only S_1 and S_2 are abelian groups. S_n $(n \geq 3)$ is nonabelian. For example, (12) and (13) belong to S_3, and $(12)(13) \neq (13)(12)$.

Note that in our convention for writing a permutation, we have quite a bit of freedom in ordering; in the permutation symbol, only the vertical relationship matters, not the horizontal. Thus

$$\begin{pmatrix} 1 & 2 & 3 & 4 & 5 & 6 \\ 6 & 5 & 4 & 3 & 2 & 1 \end{pmatrix} = \begin{pmatrix} 6 & 5 & 4 & 3 & 2 & 1 \\ 1 & 2 & 3 & 4 & 5 & 6 \end{pmatrix} = \begin{pmatrix} 4 & 2 & 6 & 3 & 1 & 5 \\ 3 & 5 & 1 & 4 & 6 & 2 \end{pmatrix}.$$

Also, we have used numbers in our symbols merely for convenience (there is an arbitrarily large number of them). One could equally well use apple, pear, orange, $\cdots$, or Greek, Russian, French, $\cdots$, instead of one, two, three, $\cdots$

Example 10.5. The simplest symmetric groups are S_1, which contains the single element e, and S_2, which contains the elements e and (12). More interesting is S_3. Its six elements are

$$e, (12), (13), (23), (123), (321) .$$

As the reader can easily show, the multiplication table for S_3 is

	e	(12)	(13)	(23)	(123)	(321)
e	e	(12)	(13)	(23)	(123)	(321)
(12)	(12)	e	(321)	(123)	(23)	(13)
(13)	(13)	(123)	e	(321)	(12)	(23)
(23)	(23)	(321)	(123)	e	(13)	(12)
(123)	(123)	(13)	(23)	(12)	(321)	e
(321)	(321)	(23)	(12)	(13)	e	(123)

If we make the identifications $e = I$, $(12) = A$, $(13) = B$, $(23) = C$, $(123) = D$, $(321) = F$, then this table is identical to the one given in Table (3.4). Note that this group has quite a few subgroups. Clearly $\{e, (12)\}$ is a subgroup of order two, and $\{e, (13)\}$ and $\{e, (23)\}$ are isomorphic to it. $\{e, (123), (321)\}$ is a subgroup of order three, which is isomorphic to the cyclic group of order three discussed in Example 10.3. Inspection of the multiplication table shows that S_3 is a nonabelian group.

One of the most remarkable facts about S_n is embodied in the following theorem of Cayley.

Theorem 10.2. Every group G of order n is isomorphic to a subgroup of S_n.

Proof. The proof of this result is based on Theorem 10.1. Call the elements of G $a_1, a_2, a_3, \cdots, a_n$. Let a_i be any element of G. Then according to Theorem 10.1, the collection $\{a_ia_1, a_ia_2, \cdots, a_ia_n\}$ is a rearrangement of $\{a_1, a_2, \cdots, a_n\}$ in which every element occurs once and only once. Therefore, let us make the correspondence, which is clearly one-to-one,

$$a_i \rightarrow P_{a_i} = \begin{pmatrix} a_1 & a_2 & \cdots & a_n \\ a_ia_1 & a_ia_2 & \cdots & a_ia_n \end{pmatrix} .$$

Similarly, for $a_j \in G$,

$$a_j \rightarrow P_{a_j} = \begin{pmatrix} a_1 & a_2 & \cdots & a_n \\ a_ja_1 & a_ja_2 & \cdots & a_ja_n \end{pmatrix} .$$

For the element of G which is given by the product $a_i a_j$, we will have

$$a_i a_j \to P_{a_i a_j} = \begin{pmatrix} a_1 & a_2 & \cdots & a_n \\ a_i a_j a_1 & a_i a_j a_2 & \cdots & a_i a_j a_n \end{pmatrix}.$$

The crucial point in demonstrating that we have an isomorphism is to show that $P_{a_i} P_{a_j} = P_{a_i a_j}$. This result is easily obtained—it is merely an exercise in notation. The reader should keep this in mind as he plows through the manipulations which follow.

We have already remarked that the horizontal ordering of permutation symbols is unimportant. For example,

$$\begin{pmatrix} a_1 & a_2 & a_3 & \cdots & a_n \\ a_i a_1 & a_i a_2 & a_i a_3 & \cdots & a_i a_n \end{pmatrix} = \begin{pmatrix} a_3 & a_n & a_1 & \cdots & a_2 \\ a_i a_3 & a_i a_n & a_i a_1 & \cdots & a_i a_2 \end{pmatrix}.$$

In other words, we can rearrange the top row in any way we please provided we make a similar rearrangement of the bottom row. Therefore, since $\{a_j a_1, a_j a_2, \cdots, a_j a_n\}$ is a rearrangement of $\{a_1, a_2, \cdots, a_n\}$, we may write

$$P_{a_i} = \begin{pmatrix} a_1 & a_2 & \cdots & a_n \\ a_i a_1 & a_i a_2 & \cdots & a_i a_n \end{pmatrix} = \begin{pmatrix} a_j a_1 & a_j a_2 & \cdots & a_j a_n \\ a_i(a_j a_1) & a_i(a_j a_2) & \cdots & a_i(a_j a_n) \end{pmatrix}$$
$$= \begin{pmatrix} a_j a_1 & a_j a_2 & \cdots & a_j a_n \\ a_i a_j a_1 & a_i a_j a_2 & \cdots & a_i a_j a_n \end{pmatrix}.$$

Thus

$$P_{a_i} P_{a_j} = \begin{pmatrix} a_j a_1 & a_j a_2 & \cdots & a_j a_n \\ a_i a_j a_1 & a_i a_j a_2 & \cdots & a_i a_j a_n \end{pmatrix} \begin{pmatrix} a_1 & a_2 & \cdots & a_n \\ a_j a_1 & a_j a_2 & \cdots & a_j a_n \end{pmatrix}$$
$$= \begin{pmatrix} a_1 & a_2 & \cdots & a_n \\ a_i a_j a_1 & a_i a_j a_2 & \cdots & a_i a_j a_n \end{pmatrix},$$

since the multiplication can now be determined simply by inspection. Thus

$$P_{a_i} P_{a_j} = P_{a_i a_j},$$

and the proof of isomorphism is complete.

This result encompasses the remark made in Example 10.5, where it was noted that S_3 has a subgroup $\{e, (123), (321)\}$ which is isomorphic to the group of order three. Regarding the groups of order four, we take first the four-group. Using the multiplication table of Table (10.7), we find immediately the correspondence

$$P_e = \begin{pmatrix} e & a & b & c \\ e & a & b & c \end{pmatrix}, \qquad P_a = \begin{pmatrix} e & a & b & c \\ a & e & c & b \end{pmatrix},$$
$$P_b = \begin{pmatrix} e & a & b & c \\ b & c & e & a \end{pmatrix}, \qquad P_c = \begin{pmatrix} e & a & b & c \\ c & b & a & e \end{pmatrix},$$

by using the rule of correspondence of Cayley's theorem. In the compact cycle notation, these can be written as

$$P_e = e, \qquad P_a = (ea)(bc), \qquad P_b = (eb)(ac), \qquad P_c = (ec)(ab).$$

It is a simple matter to check that the multiplication table of Table (10.7) is obeyed.

In the case of the cyclic group of order four, we find, using Table (10.4),

$$P_e = \begin{pmatrix} e & a & b & c \\ e & a & b & c \end{pmatrix}, \quad P_a = \begin{pmatrix} e & a & b & c \\ a & b & c & e \end{pmatrix},$$

$$P_b = \begin{pmatrix} e & a & b & c \\ b & c & e & a \end{pmatrix}, \quad P_c = \begin{pmatrix} e & a & b & c \\ c & e & a & b \end{pmatrix}.$$

As above, this can be written as

$$P_e = e\,, \qquad P_a = (eabc)\,, \qquad P_b = (eb)(ac)\,, \qquad P_c = (ecba)\,.$$

This case gives us our first example of a four-cycle.

In looking at these two illustrations, one notes that except for the identity, all the permutations leave no symbol unchanged. A subgroup of S_n with this property is called a *regular subgroup* or a subgroup of regular permutations. It is clear from the construction used in proving Theorem 10.2 that every group of order n is isomorphic to a regular subgroup of S_n.

We close this section with a few results about the regular subgroups of S_n.

Lemma. In a regular subgroup, no two elements take a given symbol into the same symbol.

Proof. Suppose that p_1 and p_2 ($p_1 \neq p_2$) belong to a regular subgroup and that both p_1 and p_2 take a into b. But then $p_1 p_2^{-1}$ ($\neq e$) leaves b unchanged. However, $p_1 p_2^{-1}$ also belongs to the regular subgroup (since the subgroup is closed), which contradicts the assumption that the subgroup is regular and hence that every element (except the identity) changes *all* the symbols.

Lemma. In a regular subgroup, if we decompose a given permutation into disjoint cycles, then each cycle must have the same length. For example, in S_5, (12) (345) could not belong to a regular subgroup.

Proof. Suppose that p is an element of a regular subgroup which can be decomposed into two cycles of length l_1 and l_2 with $l_1 < l_2$. Now since a cycle of length l must satisfy $(a_1 a_2 \cdots a_l)^l = e$, p^{l_1} leaves all the symbols contained in the first cycle unchanged. However, p^{l_1} must change some of the symbols contained in the second cycle (all of them, in fact), since if we have a cycle of length $\lambda > l'$, then $(a_1 a_2 \cdots a_\lambda)^{l'} \neq e$. Hence p^{l_1}, which is a permutation belonging to the subgroup containing p, changes some symbols, but not others. This contradicts our assumption that p belongs to a regular subgroup.

This lemma is illustrated by the results just obtained for the regular subgroups of S_4 corresponding to the four-group and to the cyclic group of order four. Using this lemma, we can also prove the following very powerful theorem.

Theorem 10.3. Every group $\{G, \cdot\}$ of order n is cyclic if n is a prime number.

Proof. By the previous lemma, the subgroup of S_n to which G is isomorphic must contain elements which are either a product of n one-cycles or just one n-cycle, since n is prime. The identity is a product of n one-cycles, so G consists of an identity, plus $n - 1$ elements which correspond isomorphically to n-cycles. Call these elements $a_2, a_3, \cdots, a_n$. Now since a_i corresponds to an n-cycle, each element of the sequence $\{a_i, a_i^2, a_i^3, \cdots, a_i^{n-1}\}$ is distinct (note that $a_i^n = e$), since if the elements were not all distinct, say $a_i^\mu = a_i^\nu$, then we would have $a_i^m = e$ for $m < n$, which is impossible because a_i corresponds to an n-cycle. Thus the distinct elements $\{e, a_i, a_i^2, \cdots, a_i^{n-1}\}$ exhaust all elements of the group. That is, we have a cyclic group of order n. Clearly this argument works for any element a_i.

For example, the group of order three is isomorphic to the subgroup $\{e, (123), (321)\}$ of S_3. It is easily checked that

$$(123)^1 = (123), \qquad (123)^2 = (321), \qquad (123)^3 = e,$$

so we could write our group as

$$\{e, (123), (123)^2\}.$$

But also,

$$(321)^1 = (321), \qquad (321)^2 = (123), \qquad (321)^3 = e,$$

so we could equally well reorder the group and write it as

$$\{e, (321), (321)^2\}.$$

Theorem 10.3 tells us, among other things, that although we might be tempted to imagine that for n large the number of groups of order n is also large, this is in fact not the case. There is only one group of order 97, namely, the cyclic group.

10.3 COSETS, CLASSES, AND INVARIANT SUBGROUPS

Having discussed in some detail a special category of finite groups, let us now look at some of the most important general properties which are common to all groups.

> **Definition 10.7.** Let A be a subgroup of G, with elements $e, a_2, a_3, \cdots, a_m$ ($m \leq n$, where n is the order of G) and let $b \in G$, but $b \notin A$. Then the m distinct elements $be = b, ba_2, ba_3, \cdots, ba_m$ form a *left coset* of A, which we denote symbolically by bA. Similarly, $b, a_2b, a_3b, \cdots, a_mb$ form a right coset, Ab.

The notation bA is just a shorthand for the collection of objects formed by multiplying every element of A on the left by b. Note that a coset is *not* a subgroup, since it clearly cannot contain the identity element. If $ba_i = e$, then $b = a_i^{-1}$, which implies $b \in A$, contrary to the assumption in Definition 10.7. In fact, bA contains no elements of A at all, for if $ba_i = a_j$, then $b = a_j a_i^{-1}$, and hence $b \in A$, again contrary to assumption.

Lemma. Two left cosets of a subgroup A either contain all the same elements or else have no elements in common.

Proof. Let xA be one coset of A and yA be another. Suppose that $xa_i = ya_j$; then $y^{-1}x = a_j a_i^{-1}$, so $y^{-1}x \in A$. Thus $y^{-1}x$ applied to the group A must just be a rearrangement of A, according to Theorem 10.1. Hence A can be written as $\{y^{-1}xa_1, y^{-1}xa_2, \cdots, y^{-1}xa_m\}$, and the collection $\{y(y^{-1}xa_1), y(y^{-1}xa_2), \cdots, y(y^{-1}xa_m)\}$ must be identical with yA (the ordering of the elements of yA is immaterial). But this collection is just $\{xa_1, xa_2, \cdots, xa_m\}$, so it is equal to xA. Hence xA and yA are the same, and we conclude that if two cosets have one element in common, they have all elements in common. This is what we set out to prove.

Theorem 10.4 (Lagrange's theorem). The order, g, of the group G is an integral multiple of the order of any subgroup, A.

Proof. Consider all the distinct cosets of A, which we denote by $b_1A, b_2A, \cdots, b_{\mu-1}A$. (It is purely for convenience that we take the number of cosets to be $\mu - 1$.) Let h be the order of A; thus there are h elements in each coset. But every element of G must occur either in A or in one of its $\mu - 1$ distinct cosets, and no element can appear more than once. Hence $h + (\mu - 1)h = \mu h = g$, which is what was to be shown.

We call μ the *index* of the subgroup A. From this result it is apparent that the order of any element of a group must be an integral divisor of the order of the group, since each element, a, of order ν generates a subgroup $\{e, a, a^2, \cdots, a^{\nu-1}\}$ of order ν. This leads us to conclude again, as in Theorem 10.3, that the groups of prime order must be cyclic.

Example 10.6. Consider S_3, whose elements are e, (12), (13), (23), (123), and (321). The left cosets of the subgroup $\{e, (12)\} = A$ are:

$$\begin{aligned}
(13)A &= \{(13), (13)(12)\} &&= \{(13), (123)\}, \\
(23)A &= \{(23), (23)(12)\} &&= \{(23), (321)\}, \\
(123)A &= \{(123), (123)(12)\} &&= \{(123), (13)\}, \\
(321)A &= \{(321), (321)(12)\} &&= \{(321), (23)\}.
\end{aligned}$$

Thus there are two distinct left cosets of A:

$$(13)A = \{(13), (123)\}, \qquad (23)A = \{(23), (321)\}.$$

We may write symbolically

$$S_3 = A + (13)A + (23)A,$$

or equivalently,

$$S_3 = A + (123)A + (321)A.$$

Note that the right cosets of A are *not* the same as the left cosets:

$$\begin{aligned}
A(13) &= \{(13), (12)(13)\} = \{(13), (321)\}, \\
A(23) &= \{(23), (12)(23)\} = \{(23), (123)\},
\end{aligned}$$

and similarly for $A(123)$ and $A(321)$.

Another subgroup of S_3 is $B = \{e, (123), (321)\}$. Its left cosets are:

$$(12)B = \{(12), (12)(123), (12)(321)\}$$
$$= \{(12), (23), (13)\}\ ;$$

$(13)B$ and $(23)B$ must equal $(12)B$ according to the previous lemma. It also follows from that lemma that the right coset $B(12)$ is equal to $(12)B$. Thus there is only one distinct coset of B. In this case we may write symbolically

$$S_3 = B + (12)B\ .$$

An important relationship which can exist between group elements is that of conjugacy.

Definition 10.8. An element $b \in G$ is said to be *conjugate* to $a \in G$ if there exists some $x \in G$ such that

$$b = xax^{-1}\ .$$

This relationship of two group elements is analogous to that of similarity between matrices, which was discussed in Chapter 3. According to this definition, a is clearly conjugate to itself and if a is conjugate to b, then b is conjugate to a. We also have the following simple result.

Lemma. If a is conjugate to b and b is conjugate to c, then a is conjugate to c.

Proof. Since a is conjugate to b, there exists $x \in G$ such that $a = xbx^{-1}$; similarly, there exists $y \in G$ such that $b = ycy^{-1}$. Combining these two results, we find that

$$a = x(ycy^{-1})x^{-1} = xycy^{-1}x^{-1} = (xy)c(xy)^{-1}\ ,$$

since $(xy)^{-1} = y^{-1}x^{-1}$ as may be checked by direct multiplication. But $xy \in G$, so a is conjugate to c, as required.

Thus the relation of conjugacy is reflexive, symmetric, and transitive. This suggests the following definition.

Definition 10.9. All the elements of a group which are conjugate to each other form an *equivalence class*, referred to hereafter simply as a *class*.

According to this definition any two elements of a class must be of the same order. For suppose that a is of order n ($a^n = e$) and that b is conjugate to a ($b = xax^{-1}$ for some $x \in G$). Then $b^n = (xax^{-1})^n = xax^{-1}xax^{-1} \cdots xax^{-1} = xa^nx^{-1} = xex^{-1} = e$. We also have $b^m = xa^mx^{-1}$ for any $m < n$. Now if $b^m = e$, it follows that $a^m = e$, which is impossible since a is of order n and $m < n$. Thus n is the smallest integer for which $b^n = e$; that is, b is also of order n. Obviously, the identity forms a class by itself. For any other group element, a, we form the sequence

$$eae^{-1} = a,\ a_2aa_2^{-1},\ a_3aa_3^{-1},\ \cdots,\ a_naa_n^{-1}\ .$$

The elements of this sequence are all conjugate to each other (of course some elements may occur more than once), and hence form a class. In this manner, the elements of any group can be divided into classes. For abelian groups, this procedure is very simple: every element constitutes a class by itself, since all the elements commute.

Example 10.7. Consider once again the group S_3: $\{e, (12), (13), (23), (123), (321)\}$. The respective inverses are e, (12), (13), (23), (321) and (123). Of course, e constitutes a class by itself. Now let us conjugate (12) by all the elements of S_3:

i) $e(12)e^{-1} = (12)$,

ii) $(12)(12)(12)^{-1} = (12)$,

iii) $(13)(12)(13)^{-1} = (13)(12)(13) = (13)(132) = (13)(321)$
$= (13)(31)(32) = (23)$,

iv) $(23)(12)(23)^{-1} = (23)(12)(23) = (23)(21)(23) = (23)(231)$
$= (23)(312) = (23)(32)(31) = (13)$,

v) $(123)(12)(123)^{-1} = (123)(12)(321) = (13)(12)(12)(321)$
$= (13)(321) = (13)(31)(32) = (23)$,

vi) $(321)(12)(321)^{-1} = (321)(12)(123) = (321)(12)(231)$
$= (321)(12)(21)(23) = (321)(23) = (31)(32)(23)$
$= (13)$.

The above calculations illustrate some of the techniques that are useful in the manipulation of cycles. In particular, we have repeatedly made use of two important properties of cycles: (1) that the square of any two-cycle gives the identity element; and (2) that one can order a three-cycle (or generally, an n-cycle) in more ways than one to suit a given situation. Thus we conclude that (12), (13), and (23) form a class. This leaves only (123) and (321) to place in classes. We have

$$e(123)e^{-1} = (123)$$
$$(12)(123)(12)^{-1} = (12)(13)(12)(12) = (12)(13) = (132) = (321) .$$

We need go no further, since we see that (123) and (321) are conjugate to each other. They are therefore in the same class. Since this exhausts the elements of the group, we have determined all the classes of S_3. We summarize as follows:

$$\mathscr{C}_1 = e ,$$
$$\mathscr{C}_2 = \{(12), (13), (23)\} ,$$
$$\mathscr{C}_3 = \{(123), (321)\} .$$

We remark here without proof that in the case of the permutation groups, the division into classes corresponds always to the division according to cycle structure, just as in the above case. That means, for example, that in the case of

S_4 the $4! = 24$ elements fall into the classes:

$$\mathscr{C}_1 = e\,,$$
$$\mathscr{C}_2 = \{(12), (13), (14), (23), (24), (34)\}\,,$$
$$\mathscr{C}_3 = \{(123), (321), (124), (421), (134), (431), (234), (432)\}\,,$$
$$\mathscr{C}_4 = \{(12)(34), (13)(24), (14)(23)\}\,,$$
$$\mathscr{C}_5 = \{(1234), (1243), (1324), (1342), (1423), (1432)\}\,.$$

Now let H be a subgroup of G. The set $H' = aHa^{-1}$, where $a \in G$, is also a subgroup of G. For if $x \in H$ and $y \in H$, axa^{-1} and aya^{-1} are two elements of H'. But

$$(axa^{-1})(aya^{-1}) = a(xy)a^{-1}\,,$$

and since $xy \in H$, $a(xy)a^{-1} \in H'$. This proves that H' is closed; the other group properties of H' may readily be verified. Evidently, if $a \in H$, then H' is just a one-to-one mapping of H onto itself. $H' = aHa^{-1}$ is said to be a *conjugate subgroup* of H in G.

Definition 10.10. If for all $a \in G$, $aHa^{-1} = H$, H is said to be an *invariant* subgroup.

Under these circumstances, $aH = Ha$, so we arrive at an alternative formulation: H is an invariant subgroup if the left and right cosets formed with any $a \in G$ are the same. Thus the subgroup B of S_3 discussed in Example 10.6 is an invariant subgroup. The subgroup A of S_3 is not invariant. Clearly, in any group, the identity is an invariant subgroup, as is the whole group itself.

Definition 10.11. A group which has no invariant subgroups save the whole group and the identity is a *simple* group.

Lemma. If H is a subgroup of G, H is invariant if and only if it contains the elements of G in complete classes; that is, if H contains one element of a class, then it must contain them all.

Proof. Assume first that H is invariant; then for any $x \in H$ and $a \in G$, $axa^{-1} \in H$. Hence if $\mathscr{C}$ is a class which contains x, all members of $\mathscr{C}$ are also in H, by Definition 10.9. By the same argument, if $y \in \mathscr{C}'$ $(\mathscr{C}' \neq \mathscr{C})$, all members of $\mathscr{C}'$ must also be in H. Thus H contains only complete classes and may contain more than one class. In fact, since e always constitutes a class by itself, only the most trivial invariant subgroup (the group consisting of the identity alone) will contain just one class. Conversely, suppose that H contains only complete classes. Then every $x \in H$ belongs to some class $\mathscr{C}$; furthermore, every element of $\mathscr{C}$ is contained in H. This means that for *any* $a \in G$, $axa^{-1} \in H$. Since this argument holds for any x, H is invariant.

One of the most interesting results of group theory is that the *cosets* of an invariant subgroup themselves constitute a group!

Theorem 10.5. The collection consisting of an invariant subgroup H and all its distinct cosets is itself a group, called the factor group of G, usually denoted by G/H. (Remember that the left and right cosets of an *invariant* subgroup are identical.) Multiplication of two cosets aH and bH is defined as the set of all distinct products $z = xy$, with $x \in aH$ and $y \in bH$; the identity element of the factor group is the subgroup H itself.

Proof. Consider any coset aH. Then since $HH = H$,

$$(aH)H = aHH = aH\,; \qquad H(aH) = HaH = aHH = aH\,,$$

so H is indeed the identity. Here we have used $aH = Ha$, which follows from the fact that H is invariant. Now take any two cosets aH and bH. We have

$$(aH)(bH) = aHbH = abHH = abH\,,$$

so the product of two cosets yields another coset. Finally, the inverse of aH is clearly $a^{-1}H$, since H is the identity and

$$(aH)(a^{-1}H) = aa^{-1}HH = HH = H\,.$$

This completes the proof that G/H is indeed a group.

A simple illustration of this theorem is provided by S_3. We have already noted that $B = \{e, (123), (321)\}$ is an invariant subgroup. Its coset can be written as $(12)B$. Thus the factor group S_3/B consists of two elements:

$$E = B = \{e, (123), (321)\}\,, \qquad A = (12)B = \{(12), (13), (23)\}\,,$$

with $A^2 = E$. In discussing the factor group we will use capital letters to denote factor group elements. When multiplying two such elements we will denote group multiplication by a dot to avoid possible confusion with matrix multiplication.

From another point of view, the factor group is a mapping of one group G onto another group G'. This mapping preserves group products, but is not necessarily one to one. Such a map is called a homomorphism; it is to be contrasted with the isomorphism, which is a product-preserving, one-to-one map. We say that G is homomorphic to G'. In the example under discussion we can consider e, (123) and (321) as being mapped onto E, the identity; and (12), (13) and (23) as being mapped onto A ($A^2 = E$). According to the proof of Theorem 10.5, this mapping must preserve products. For example, (13)(12) is an element of the product of A by A. Since $(13)(12) = (123)$, this product belongs to E, as it should since $A \cdot A = E$. Similarly, $(123)(12) \in E \cdot A$, which is again as it should be since $(123)(12) = (13)$, which belongs to A. The reader can check that all the multiplications work out as they are supposed to according to Theorem 10.5. We also leave it to the reader to show that in *any* homomorphic mapping of G onto G', the elements of G which are mapped onto the identity of G' must form an invariant subgroup of G. The other elements of G' then must be the images of the cosets of the invariant subgroup.

Example 10.8. To illustrate further some of the ideas of this section and also to look at a slightly more complicated group than we have examined so far, let us analyze the group S_4. As mentioned earlier, S_4 can be divided into five classes according to cycle structure:

$$\mathscr{C}_1 = e\,,$$
$$\mathscr{C}_2 = (12),\ (13),\ (14),\ (23),\ (24),\ (34)\,,$$
$$\mathscr{C}_3 = (123),\ (124),\ (134),\ (234),\ (321),\ (421),\ (431),\ (432)\,,$$
$$\mathscr{C}_4 = (12)(34),\ (13)(24),\ (14)(23)\,,$$
$$\mathscr{C}_5 = (1234),\ (1243),\ (1324),\ (1342),\ (1423),\ (1432)\,.$$

We would like to determine whether there are any nontrivial invariant subgroups of S_4. Since we know that such an invariant subgroup can contain only complete classes and since furthermore the order of the subgroup must be an integral divisor of the order $(4! = 24)$ of S_4, there are, according to the previous lemma only two possibilities:

$$H = \mathscr{C}_1 + \mathscr{C}_4\,, \qquad H' = \mathscr{C}_1 + \mathscr{C}_3 + \mathscr{C}_4\,.$$

H is of order four and H' is of order twelve. Both these orders are integral divisors of 24; $H'' = \mathscr{C}_1 + \mathscr{C}_2$, for example, is excluded since $24/7 \neq$ integer. In addition, H'' is not even a subgroup. Now H' is just the subgroup of even permutations. Since any coset must have as many elements as the related subgroup, H' can have only one coset, the set $\mathscr{C}$ of all odd permutations (note that this is *not* a subgroup). In particular, the left and right cosets must be identical, so H' is indeed an invariant subgroup. The factor group S_4/H' (consisting of the two sets H' and $\mathscr{C}$) is isomorphic to the two-element cyclic group. It is obvious that the two-element cyclic group is a factor group of any S_n, since half the elements of S_n are even permutations and half are odd permutations.

That H is also a group is seen by noting that according to the discussion following the proof of Theorem 10.2, this collection of four elements is isomorphic to the four-group. Now what are the cosets of H? The computation is tedious but straightforward. We have

$$(12)H = \{(12),\ (34),\ (1324),\ (1423)\}\,,$$
$$(13)H = \{(13),\ (24),\ (1234),\ (1432)\}\,,$$
$$(23)H = \{(23),\ (14),\ (1243),\ (1342)\}\,.$$

According to the first lemma of this section, we must also have

$$(12)H = (34)H = (1324)H = (1423)H\,,$$
$$(13)H = (24)H = (1234)H = (1432)H\,,$$
$$(23)H = (14)H = (1243)H = (1342)H\,.$$

Now consider $(123)H$. We find that

$$(123)H = \{(123),\ (134),\ (432),\ (421)\}\,.$$

This contains half of $\mathscr{C}_3$, so we know (why?) that $(321)H$ must be the only remaining coset of H. In fact,

$$(321)H = \{(321), (234), (124), (431)\} .$$

This collection of six objects, $\{H, (12)H, (13)H, (23)H, (123)H, (321)H\}$, is isomorphic to S_3, and by inspection we may make the identifications:

$$H \rightarrow e , \qquad (12)H \rightarrow (12) , \qquad (13)H \rightarrow (13) ,$$
$$(23)H \rightarrow (23) , \qquad (123)H \rightarrow (123) , \qquad (321)H \rightarrow (321) .$$

For example, $(1324) \in (12)H$ and $(1234) \in (13)H$. A simple calculation gives

$$\begin{aligned}(1324)(1234) &= (4132)(1234) = (42)(43)(41)(14)(13)(12) \\ &= (42)(43)(13)(12) = (432)(123) = (324)(312) \\ &= (34)(32)(32)(31) = (34)(31) = (314) = (431) ,\end{aligned}$$

which belongs to $(321)H$. Since $(12)(13) = (132) = (321)$, we see that the isomorphism holds:

$$[(12)H][(13)H] = (321)H \qquad \text{and} \qquad (12)(13) = (321) .$$

10.4 SYMMETRY AND GROUP REPRESENTATIONS

The role of group theory in physics is intimately related to the symmetries of the world around us. The importance of such symmetries as translational and rotational invariance in giving rise to conservation of linear and angular momentum has long been known and is familiar to the student from classical mechanics. With the development of quantum mechanics, in which the physical world is separated from us by the intermediary of the "wave function," group theory became particularly significant. To see why this is the case, we must first say precisely what is meant by such phrases as "translational invariance" and "rotational invariance."

Suppose that we have a group, G, of operators, $U_1, U_2, U_3, \cdots$, which can act on elements, x, of a vector space, V. By a group of operators we simply mean a collection of operators on V which obeys the group axioms of Definition 10.1. The results of $U_1, U_2, U_3, \cdots$ acting on x will, as usual, be denoted by $U_1x, U_2x, U_3x, \cdots$ In three-dimensional space an example of such a group is the collection $\{T_{\mathbf{a}}\}$, where $\mathbf{a}$ is any three-dimensional vector, and $T_{\mathbf{a}}$ acts on vectors according to the rule

$$T_{\mathbf{a}}\mathbf{r} = \mathbf{r} - \mathbf{a} . \tag{10.8}$$

Thus $T_{\mathbf{a}}$ is a translation operator. Note that there is an infinitude of such operators, since $\mathbf{a}$ can be *any* vector. The entire collection is clearly a group; closure follows because the product of any two translations is again a translation. In fact, from the defining equation, we see that this group is abelian since

$$T_{\mathbf{a}}T_{\mathbf{b}} = T_{\mathbf{a}+\mathbf{b}} = T_{\mathbf{b}}T_{\mathbf{a}} .$$

Obviously, such operators can be generalized to any number of dimensions. Now we suppose further that on this vector space we have defined functions, $f, g, \cdots$, which assign to each vector some complex number. In three-dimensional Euclidean space, $f(\mathbf{r}) = x^4 + y^4 + z^4$ is an example of such a function.

In general, we denote the action of a function f on a vector x in V by $f(x)$. Let U_i be any element of the group of transformations written above. What can we say about the action of f on the transformed vector $U_i^{-1}x$? Let us define an operator $\mathscr{U}_i$ which acts on functions of x in such a manner that for all f

$$\mathscr{U}_i f(x) = f(U_i^{-1}x) .$$

Now consider the quantity $\mathscr{U}_i \mathscr{U}_j f(x)$. We have from the above equation

$$\mathscr{U}_i \mathscr{U}_j f(x) = \mathscr{U}_i f(U_j^{-1}x) = f(U_j^{-1}U_i^{-1}x) = f([U_iU_j]^{-1}x) .$$

Thus, if $U_iU_j = U_k$, then $\mathscr{U}_i \mathscr{U}_j = \mathscr{U}_k$, so that the elements $\mathscr{U}_1, \mathscr{U}_2, \cdots$ form a group, $\mathscr{G}$, which is isomorphic to G.

To illustrate what we have in mind, let us extend the example of the $\{T_\mathbf{a}\}$, defined in Eq. (10.8), a bit further. We ask: Is it possible to find an operator $\mathscr{T}_\mathbf{a}$ such that

$$\mathscr{T}_\mathbf{a} f(\mathbf{r}) = f(T_\mathbf{a}^{-1}\mathbf{r}) = f(\mathbf{r} + \mathbf{a}) \,?$$

In general, it will be necessary to say something about the functions $f(x)$, $g(x)$, $\cdots$ before we can obtain an explicit expression for $\mathscr{T}_\mathbf{a}$. For the sake of simplicity, suppose that we take our functions to be analytic. Then we can write a Taylor series for $f(\mathbf{r} + \mathbf{a})$:

$$f(\mathbf{r} + \mathbf{a}) = \sum_{n=0}^{\infty} \frac{1}{n!} [\mathbf{a} \cdot \nabla]^n f(\mathbf{r}) .$$

Symbolically, this can be written as

$$f(\mathbf{r} + \mathbf{a}) = e^{\mathbf{a} \cdot \nabla} f(\mathbf{r}) ,$$

so we see that

$$\mathscr{T}_\mathbf{a} = e^{\mathbf{a} \cdot \nabla} . \tag{10.9}$$

If one were quantum-mechanically minded, one would write this as

$$\mathscr{T}_\mathbf{a} = e^{i\mathbf{a} \cdot \mathbf{p}/\hbar} , \tag{10.10}$$

where $\mathbf{p} = -i\hbar\nabla$ is the quantum-mechanical momentum operator.

Similarly, if one consider rotations, R_α^z, of a coordinate system about the z-axis, so that $(r, \theta, \phi) \rightarrow (r, \theta, \phi - \alpha)$, then the corresponding operator would be

$$\mathscr{R}_\alpha^z = e^{\alpha (\partial/\partial\phi)} = e^{i\alpha L_z/\hbar} , \tag{10.11}$$

where $L_z = -i\hbar\partial/\partial\phi$ is the quantum-mechanical operator for the z-component of angular momentum. For rotation through angles β and γ about the y-axis and x-axis respectively, one has

$$\mathscr{R}^y_\beta = e^{i\beta L_y/\hbar}, \qquad \mathscr{R}^x_\gamma = e^{i\gamma L_x/\hbar}. \tag{10.12}$$

Groups of the type characterized by the collection $\{\mathscr{T}_\mathbf{a}\}$ or $\{\mathscr{R}^z_\alpha\}$ are called transformation groups. Both of these examples involve infinite groups. We have been able to parametrize the translation group by $\mathbf{a}$ $(= a_x, a_y, a_z)$, and the group of rotations about the z-axis by α. The parameters in question vary continuously, and the groups are therefore called *continuous* groups.

Now we make contact with our work in earlier chapters. Suppose that the functions which we have been discussing above belong to a vector space, H, and that acting on H we have some linear operator, A_x. Acting on any $f(x) \in H$, A_x gives another vector $g(x) \in H$:

$$A_x f(x) = g(x). \tag{10.13}$$

The subscript x on A serves to remind us that A_x is an operator on a space of functions whose action depends on the point at which the function is evaluated. For example, we might have

$$A_\mathbf{r} = \nabla^2 = \frac{\partial^2}{\partial x^2} + \frac{\partial^2}{\partial y^2} + \frac{\partial^2}{\partial z^2}.$$

Let $\mathscr{U} \in \mathscr{G}$ act on Eq. (10.13). Then we find that

$$\mathscr{U} A_x f(x) = \mathscr{U} g(x) = g(U^{-1}x).$$

Since $\mathscr{U}^{-1}\mathscr{U} = I$, we have equivalently,

$$\mathscr{U} A_x \mathscr{U}^{-1} \mathscr{U} f(x) = g(U^{-1}x),$$

or

$$\mathscr{U} A_x \mathscr{U}^{-1} f(U^{-1}x) = g(U^{-1}x).$$

But since $A_x f(x) = g(x)$, we have also

$$A_{U^{-1}x} f(U^{-1}x) = g(U^{-1}x),$$

so we conclude that

$$\mathscr{U} A_x \mathscr{U}^{-1} f(U^{-1}x) = A_{U^{-1}x} f(U^{-1}x),$$

for any $f \in H$, or

$$\mathscr{U} A_x \mathscr{U}^{-1} = A_{U^{-1}x}.$$

$\mathscr{U} A_x \mathscr{U}^{-1}$ is called the *transformed* operator; the above equation tells us that the transformed operator at the point x is equal to the untransformed operator at the point $U^{-1}x$. It is often simpler to find the transformed operator by evaluating $A_{x'}$, where $x' = U^{-1}x$, rather than $\mathscr{U} A_x \mathscr{U}^{-1}$. Now if it happens that the transformed operator at the point x equals the untransformed operator at the point x, that is, if

$$\mathscr{U} A_x \mathscr{U}^{-1} = A_x$$

for all $\mathscr{U} \in \mathscr{G}$, then we say that A_x is invariant under the action of the group $\mathscr{G}$. Similarly, if for all $\mathscr{U} \in \mathscr{G}$, $\mathscr{U} f(x) = f(x)$, then we say that f is invariant under $\mathscr{G}$. Note that the criterion for the invariance of A_x can also be written as

$$A_x \mathscr{U} = \mathscr{U} A_x .$$

In other words, if all $\mathscr{U} \in \mathscr{G}$ commute with A_x, then A_x is invariant under $\mathscr{G}$.

It is easily seen that the operator

$$A_{\mathbf{r}} = \nabla^2 = \frac{\partial^2}{\partial x^2} + \frac{\partial^2}{\partial y^2} + \frac{\partial^2}{\partial z^2}$$

is invariant under $\mathscr{T}_{\mathbf{a}}$ for all $\mathbf{a}$, where $\mathscr{T}_{\mathbf{a}}$ is defined by Eq. (10.9). This follows simply from the fact that the partial derivatives with respect to x, y and z commute with each other, and hence $\exp(\mathbf{a}\cdot\nabla)$ commutes with ∇^2. Thus we may say that the Laplacian is translation invariant. A similar situation prevails for the rotation operators $\mathscr{R}^z_\alpha$, $\mathscr{R}^y_\beta$ and $\mathscr{R}^x_\gamma$ as defined by Eqs. (10.11) and (10.12). According to Eq. (5.89) ∇^2 is simply related to $L^2 = L_x^2 + L_y^2 + L_z^2$, and since L_x, L_y and L_z all commute with L^2 (why?), we see that the rotation operators will also commute with L^2 and hence with ∇^2. Thus, ∇^2 is rotationally invariant.

As an example of a function which is rotationally invariant we mention $f(\mathbf{r}) = g(r)$, where g is any function and r is the magnitude of $\mathbf{r}$. This follows immediately from the fact that L_x, L_y and L_z depend only on the angles θ and ϕ which specify the orientation of $\mathbf{r}$ and not on the magnitude of $\mathbf{r}$ [see Eqs. (5.87)].

To get a feeling for how the above formalism works, it is useful to examine an operator which is *not* invariant under some group. For example, let us consider the operator X_x defined by $X_x f(x) \equiv x f(x)$. We shall investigate how this operator transforms under the x-translation operators $T_a (T_a x = x - a$, for all $x)$. We can compute the transform of X_x in two ways. The simplest method is to make use of the relation

$$\mathscr{T}_a X_x \mathscr{T}_a^{-1} = X_{T_a^{-1}x} .$$

Since $T_a^{-1}x = x + a$, we see immediately that

$$\mathscr{T}_a X_x \mathscr{T}_a^{-1} = X_x + a .$$

We can also proceed by letting $\mathscr{T}_a X_x \mathscr{T}_a^{-1}$ act on an arbitrary function:

$$\begin{aligned}\mathscr{T}_a X_x \mathscr{T}_a^{-1} f(x) &= \mathscr{T}_a X_x f(T_a x) = \mathscr{T}_a X_x f(x - a) \\ &= \mathscr{T}_a x f(x - a) = (x + a) f(x) \\ &= (X_x + a) f(x) .\end{aligned}$$

This is exactly the result obtained above; clearly the operator X_x is not invariant under the group of translations. For a slightly more complicated example of the transformation of operators, the reader should look at Problem 10.10.

Now let us look at the eigenvalue problem

$$A_x f(x) = \lambda f(x) . \tag{10.14}$$

We assume that we have a physical problem in mind, so that A_x is self-adjoint, and we also assume that A_x is invariant under some group of operators $\mathscr{G}$, a typical element of which we denote by $\mathscr{U}$. The question now arises: If the operator A_x is invariant under $\mathscr{G}$, are its eigenvectors invariant under $\mathscr{G}$? To answer this question we let $\mathscr{U}$ operate on both sides of the above equation to obtain

$$\mathscr{U} A_x f(x) = \lambda \mathscr{U} f(x) .$$

Since $\mathscr{U} A_x = A_x \mathscr{U}$,

$$A_x \mathscr{U} f(x) = \lambda \mathscr{U} f(x) .$$

This equation tells us that if $f(x)$ is an eigenfunction of A_x belonging to λ, then so is $\mathscr{U} f(x)$. This does *not* mean that $f(x) = \mathscr{U} f(x)$. However, if λ is a nondegenerate eigenvalue, then we conclude that $\mathscr{U} f(x)$ is just a simple multiple of $f(x)$:

$$\mathscr{U} f(x) = D(U) f(x) ,$$

where we write $D(U)$ to emphasize that the value of D may depend on which element of the group G is being used.

If, however, λ has multiplicity μ, then we must write

$$A_x f_i(x) = \lambda f_i(x) , \qquad i = 1, 2, \cdots, \mu ,$$
$$A_x \mathscr{U} f_i(x) = \lambda \mathscr{U} f_i(x) , \qquad i = 1, 2, \cdots, \mu .$$

In this case, we must conclude that $\mathscr{U} f_i(x)$ is a linear combination of the elements of the set $\{f_j(x)\}$:

$$\mathscr{U} f_i(x) = \sum_{j=1}^{\mu} D_{ji}(U) f_j(x) . \tag{10.15}$$

Now let us consider $\mathscr{U}_1 \mathscr{U}_2 f_i(x)$, where $\mathscr{U}_1, \mathscr{U}_2 \in \mathscr{G}$. By the previous equation we have

$$\mathscr{U}_1 \mathscr{U}_2 f_i(x) = \mathscr{U}_1 \sum_{j=1}^{\mu} D_{ji}(U_2) f_j(x) = \sum_{j=1}^{\mu} D_{ji}(U_2) \mathscr{U}_1 f_j(x) .$$

Since

$$\mathscr{U}_1 f_j(x) = \sum_{k=1}^{\mu} D_{kj}(U_1) f_k(x) ,$$

we have

$$\mathscr{U}_1 \mathscr{U}_2 f_i(x) = \sum_{j=1}^{\mu} \sum_{k=1}^{\mu} D_{kj}(U_1) D_{ji}(U_2) f_k(x) .$$

But also,

$$\mathscr{U}_1 \mathscr{U}_2 f_i(x) = \sum_{k=1}^{\mu} D_{ki}(U_1 U_2) f_k(x) .$$

Combining these last two equations gives the result

$$\sum_{k=1}^{\mu} \left[D_{ki}(U_1 U_2) - \sum_{j=1}^{\mu} D_{kj}(U_1) D_{ji}(U_2) \right] f_k(x) = 0 .$$

But the set $\{f_k(x):\ k = 1, 2, \cdots, \mu\}$ is linearly independent, so

$$D_{ki}(U_1U_2) = \sum_{j=1}^{\mu} D_{kj}(U_1)D_{ji}(U_2)$$

for all $i, k = 1, 2, \cdots, \mu$. Thus the matrices $D(U_1), D(U_2), \cdots$ form a group which is homomorphic to the group G and the group $\mathscr{G}$. Such a collection of matrices is called a *representation* of the group G. A representation of a group G is just a homomorphic mapping of G onto a collection of finite-dimensional matrices.

Thus we see that there is a very close relationship between symmetries and degeneracy in physical eigenvalue problems. For example, in quantum mechanics it is often the case that the Hamiltonian which describes some physical system is rotationally invariant. We would then expect the eigenfunctions corresponding to some degenerate energy level to transform among themselves under rotations and thus give rise to a representation of the rotation group. Since degeneracy tends to be the rule rather than the exception in the eigenvalue problems of physics, it is not surprising that the study of group representations has come to play a very important role in physics. It is to the study and classification of such representations that we now turn our attention.

10.5 IRREDUCIBLE REPRESENTATIONS

Let us begin with a few general inferences about the possibility of finding representations of groups. Clearly, for any group there is always one available homomorphism, namely, the mapping which assigns the one-dimensional matrix, 1, to every element of the group. All group properties are trivially satisfied, but this representation is not very useful. However, it is *always* a possible representation and cannot be neglected when we classify representations. Also, if we can find some n-dimensional representation, then a related one-dimensional representation may always be obtained by mapping each element of G into the determinant of its representative matrix. This follows immediately from the relation

$$\det(AB) = (\det A)(\det B)\ .$$

Another possibility arises when the group, G, has an invariant subgroup, H, whose associated factor group, G/H, has known representations. Since the factor group is homomorphic to G, we can find representations of G by assigning to each element $U \in G$ the matrix representative of that element of G/H onto which U is mapped.

A third interesting representation can be obtained from the group table itself. This is called the *regular* representation and we will make use of it later in this chapter. The basic idea is contained in Theorem 10.1: If we take some element of G, say U_ν, and multiply every group element by U_ν, we rearrange the group elements. Symbolically,

$$U_\nu U_j = U_i\ , \tag{10.16}$$

where i and j run through all the group elements. We translate this ordering into matrix form by saying that U_ν will be represented by a $g \times g$ matrix, $D_{ij}(U_\nu)$, where g is the order of the group. Each column will contain all zeroes except for one "1," the position of the "1" being determined from Eq. (10.16) as follows: If $U_\nu U_j = U_i$, then the ij-element of $D(U_\nu)$ is 1 and all other elements of the ith row and jth column are zero. This rule is consistent since if $U_\nu U_j = U_i$ we cannot also have $U_\nu U_k = U_i$ unless $j = k$ and, similarly, if $U_\nu U_j = U_i$ we cannot also have $U_\nu U_j = U_k$ unless $i = k$. In this representation the identity is given by the $g \times g$ unit matrix, and all the other elements of G are represented by matrices which have only zeroes on the diagonal.

We now show that this is indeed a product-preserving mapping. Consider the matrix equation

$$M = D(U_\nu)D(U_\mu) .$$

In components this is

$$M_{ij} = \sum_{k=1}^{g} D_{ik}(U_\nu)D_{kj}(U_\mu) , \qquad i, j = 1, 2, \cdots, g .$$

It follows directly from the nature of matrix multiplication that M must be the same type of matrix as $D(U_\nu)$ and $D(U_\mu)$; that is, each row and column of M can have only one nonzero element, which must equal 1. Therefore, we need only check that this one nonzero element is in the proper place. Hence suppose that $M_{ij} = 1$. Therefore, for some unique k,

$$D_{ik}(U_\nu) = 1, \qquad D_{kj}(U_\mu) = 1 .$$

These equations, in turn, mean that

$$U_\nu U_k = U_i , \qquad U_\mu U_j = U_k .$$

Hence

$$U_\nu U_\mu U_j = U_i ,$$

which says that the ij-element of $D(U_\nu U_\mu)$ is unity. Thus we can identify M_{ij} with $D_{ij}(U_\nu U_\mu)$, for *all* i and j. Hence

$$D(U_\nu U_\mu) = D(U_\nu)D(U_\mu) ,$$

so we have a representation, as claimed above.

As an example, we consider the cyclic group of order four, whose multiplication table is given in Table (10.4). We find that if we call $a_1 = e$, $a_2 = a$, $a_3 = b = a^2$, $a_4 = c = a^3$, then, for example,

$$\begin{aligned} aa_1 &= ae = a = a_2 , \\ aa_2 &= aa = b = a_3 , \\ aa_3 &= aa^2 = c = a_4 , \\ aa_4 &= aa^3 = e = a_1 , \end{aligned}$$

so that the 21-, 32-, 43- and 14-elements of $D(a)$ are equal to 1, and all the other elements vanish. A similar calculation gives $D(b)$ and $D(c)$. $D(e)$ is, of course, just the unit matrix. We list the results below:

$$D(e) = \begin{bmatrix} 1 & 0 & 0 & 0 \\ 0 & 1 & 0 & 0 \\ 0 & 0 & 1 & 0 \\ 0 & 0 & 0 & 1 \end{bmatrix}, \qquad D(a) = \begin{bmatrix} 0 & 0 & 0 & 1 \\ 1 & 0 & 0 & 0 \\ 0 & 1 & 0 & 0 \\ 0 & 0 & 1 & 0 \end{bmatrix},$$

$$D(b = a^2) = \begin{bmatrix} 0 & 0 & 1 & 0 \\ 0 & 0 & 0 & 1 \\ 1 & 0 & 0 & 0 \\ 0 & 1 & 0 & 0 \end{bmatrix}, \qquad D(c = a^3) = \begin{bmatrix} 0 & 1 & 0 & 0 \\ 0 & 0 & 1 & 0 \\ 0 & 0 & 0 & 1 \\ 1 & 0 & 0 & 0 \end{bmatrix}. \tag{10.17}$$

This demonstrates the existence of a representation which is not one-dimensional; the reader can easily prove that these four matrices satisfy Table (10.4). Moreover, since each matrix is distinct, the mapping from the cyclic group is an isomorphism; such a representation is said to be *faithful*. Unfortunately, several problems seem to arise from this example. For one thing, we can use this regular representation to generate many more representations by using similarity transformations, or by using the Kronecker (or direct) product introduced in Section 3.11.

The situation regarding similarity transforms is as follows: If $D(U_i)$ is a representation of G, then for any nonsingular matrix, S, of the same dimensionality as $D(U_i)$, we define

$$\tilde{D}(U_i) \equiv S^{-1}D(U_i)S,$$

where the tilde does not indicate the transpose of a matrix, but is merely a label. We then have

$$\tilde{D}(U_i)\tilde{D}(U_j) = S^{-1}D(U_i)SS^{-1}D(U_j)S = S^{-1}D(U_i)D(U_j)S = S^{-1}D(U_iU_j)S,$$

since $D(U_i)$ is a representation. Thus

$$\tilde{D}(U_i)\tilde{D}(U_j) = \tilde{D}(U_iU_j),$$

so $\tilde{D}(U_i)$ is also a representation.

In Section 3.11, we showed that if we denote the direct product of A and B by $A \otimes B$, then

$$(A_1 \otimes B_1)(A_2 \otimes B_2) = (A_1A_2) \otimes (B_1B_2), \tag{10.18}$$

if A_1 and A_2 (and B_1 and B_2) are of the same dimensionality. This suggests that given any two representations of a group G, say $D(U_i)$ and $D'(U_i)$, another representation can be obtained by the identification

$$U_i \longrightarrow D(U_i) \otimes D'(U_i).$$

This preserves products, since by Eq. (10.18)

$$\begin{aligned}[D(U_i) \otimes D'(U_i)][D(U_j) \otimes D'(U_j)] &= [D(U_i)D(U_j)] \otimes [D'(U_i)D'(U_j)] \\ &= D(U_iU_j) \otimes D'(U_iU_j).\end{aligned}$$

Thus we have indeed constructed another representation whose dimension will be the product of the dimensions of D and D'. Since we have already found one specific multidimensional representation, the regular representation, we can use direct products to generate an arbitrarily large number of other representations. Hence the problem of classifying representations would appear to be hopeless.

Before we accept such a pessimistic conclusion, however, it is a good idea to look at the problem from a different point of view. Let us start with some simple representations and then try to work up to more complicated ones. First, we must decide how to go about finding representations. In the previous section we have already seen that one way of constructing them is to find a collection of linearly independent functions which transform among themselves under the action of the group, that is, a collection of functions $\{f_i(x)\}$ which satisfy

$$\mathscr{U}f_i(x) = \sum_j D_{ji}(U)\, f_j(x)$$

for all $\mathscr{U} \in \mathscr{G}$.

Now the only problem is to decide how to find a set of functions which has this nice property of transforming into itself under the action of a group of operators. One possible way would be just to pick an arbitrary function $f(x)$ and operate on it with all the $\mathscr{U}_i \in \mathscr{G}$. In this way we find a set $\{f_i(x)\}$, where

$$f_i(x) \equiv \mathscr{U}_i f(x) \,.$$

By the group property, this set satisfies the basic requirement of closure:

$$\mathscr{U}_\nu f_j(x) = \mathscr{U}_\nu \mathscr{U}_j f(x) = \mathscr{U}_i f(x) = f_i(x) \,,$$

where i is determined by the multiplication table. If all elements of the set $\{f_i(x)\}$ are linearly independent, then we have succeeded in our aims. In this case,

$$\mathscr{U}_\nu f_j(x) = \sum_k D_{kj}(U_\nu)\, f_k(x) \,,$$

where

$$D_{kj}(U_\nu) = \delta_{ik} \,,$$

i being determined by $\mathscr{U}_\nu \mathscr{U}_j = \mathscr{U}_i$. Looking back a few paragraphs, we find that this is just the regular representation. However, this is by no means the only possibility. It might well happen that the $f_i(x)$ are not linearly independent. Then we must pick out a linearly independent subset which will, of course, also be invariant under the group operations (Why?). Carrying out the above process, we obtain in this case a representation of smaller dimension than the regular representation. Of course, it is often not necessary to do this, for one may well be able to spot a set of functions which transform among themselves without recourse to this unintuitive formalism.

Example 10.9. A particularly straightforward example is the group of rotations about the z-axis. This is an infinite, continuous group, but that is no reason to be daunted. It is, in fact, very easy to analyze this group.

Now when one talks about angles and rotations, the functions which come to mind are sines and cosines. Denoting a rotation through α by R_α ($R_\alpha\phi = \phi - \alpha$), we consider $f_1(\phi) = \cos\phi$ and $f_2(\phi) = \sin\phi$. Then

$$f_1(R_\alpha^{-1}\phi) = \cos(\phi+\alpha) = \cos\alpha f_1(\phi) - \sin\alpha f_2(\phi) ,$$
$$f_2(R_\alpha^{-1}\phi) = \sin(\phi+\alpha) = \sin\alpha f_1(\phi) + \cos\alpha f_2(\phi) .$$

From this we conclude, using Eq. (10.15), that

$$D^{(1)}(R_\alpha) = \begin{pmatrix} \cos\alpha & \sin\alpha \\ -\sin\alpha & \cos\alpha \end{pmatrix} .$$

(We use the superscript to distinguish between various representations of this group.) A simple multiplication of matrices shows that

$$D^{(1)}(R_\alpha)D^{(1)}(R_\beta) = D^{(1)}(R_{\alpha+\beta}) = D^{(1)}(R_\beta)D^{(1)}(R_\alpha) = D^{(1)}(R_\alpha R_\beta) ,$$

so we have found a two-dimensional representation.

Of course, $\cos\phi$ and $\sin\phi$ are not the only choice of functions we could have made. For instance, $e^{i\phi} = f(\phi)$ is such that

$$f(R_\alpha^{-1}\phi) = e^{i(\phi+\alpha)} = e^{i\alpha}e^{i\phi} = e^{i\alpha}f(\phi) .$$

Thus we obtain a one-dimensional representation:

$$D^{(2)}(R_\alpha) = e^{i\alpha} .$$

Similarly, if we choose $g(\phi) = e^{-i\phi}$, we find that

$$D^{(3)}(R_\alpha) = e^{-i\alpha} .$$

In fact it is clear that the function $f(\phi) = e^{im\phi}$ leads directly to a representation $D(R_\alpha) = e^{im\alpha}$. We choose m to be a positive or negative integer (or zero) so that $D(R_0) = D(R_{2\pi})$. Now since the functions $f_1(\phi) = e^{i\phi}$ and $f_2(\phi) = e^{-i\phi}$ are linearly independent we can use them together to form a set $\{f_i\}$. This leads to the two-dimensional representation

$$D^{(4)}(R_\alpha) = \begin{pmatrix} e^{i\alpha} & 0 \\ 0 & e^{-i\alpha} \end{pmatrix} .$$

This example brings us to the heart of the basic problem of representation theory. The two-dimensional representation $D^{(4)}(R_\alpha)$ is clearly just two one-dimensional representations joined together. In some sense, we would not want to call this a *basic* representation since it is compounded out of representations of lower dimensionality. The situation is complicated further if we look back at $D^{(1)}(R_\alpha)$. This does not, on the surface of things, look much like $D^{(4)}(R_\alpha)$, but if we diagonalize $D^{(1)}$, we find that

$$S^{-1}D^{(1)}(R_\alpha)S = \begin{pmatrix} e^{i\alpha} & 0 \\ 0 & e^{-i\alpha} \end{pmatrix} = D^{(4)}(R_\alpha) ,$$

where

$$S = \frac{1}{\sqrt{2}}\begin{pmatrix} 1 & i \\ i & 1 \end{pmatrix}$$

is a matrix independent of α. Thus we have a similarity transformation which will bring $D^{(1)}(R_\alpha)$ to exactly the form of $D^{(4)}(R_\alpha)$, and this can be done simultaneously for *all* R_α. Thus, even $D^{(1)}(R_\alpha)$ is essentially a sum of the two one-dimensional representations, $D^{(2)}$ and $D^{(3)}$.

This can also be the case for more complicated situations. For example, the four matrices composing the regular representation of the four-element cyclic group [Eq. (10.17)] can be diagonalized simultaneously by the matrix

$$S' = \begin{bmatrix} 1 & 1 & 1 & 1 \\ -1 & 1 & i & -i \\ 1 & 1 & -1 & -1 \\ -1 & 1 & -i & i \end{bmatrix}.$$

Therefore $D(e)$, $D(a)$, $D(b)$, and $D(c)$ are each sums of four one-dimensional representations.

We can generalize this as follows: If we find a representation $D(U_i)$ which can be brought to the form

$$\bar{D}(U_i) \equiv SD(U_i)S^{-1} = \begin{bmatrix} D^{(1)}(U_i) & & O \\ & U^{(2)}(D_i) & \\ O & & D^{(3)}(U_i) \end{bmatrix}, \qquad (10.19)$$

where S is the same for all U_i, then we will not consider it as a basic representation, but rather as a sum of representations.

In terms of the functions $f_i(x)$, this means that although we have found a linearly independent set $\{f_i\}$, the elements of which transform among themselves, there are actually smaller subsets which also transform into themselves under the group in question.

Representations of the type of Eq. (10.19) are called *reducible* representations; those which cannot be reduced in this manner are called *irreducible* representations. It is clearly the irreducible representations of groups that we should study, since other representations are simply built up out of them. However, even among irreducible representations we do not want to count the similarity transformation of a representation as a separate representation. Two representations which are related by a similarity transform are said to be *equivalent*. Thus, in Eq. (10.19), assuming $D^{(1)}$, $D^{(2)}$, and $D^{(3)}$ to be irreducible and inequivalent, we would write symbolically

$$D(U_i) = D^{(1)}(U_i) \oplus D^{(2)}(U_i) \oplus D^{(3)}(U_i)$$

for all $U_i \in G$. In practice, when we reduce a representation to this form, a given irreducible representation may occur more than once. Thus we would write generally

$$D(U_i) = a^{(1)}D^{(1)}(U_i) \oplus a^{(2)}D^{(2)}(U_i) \oplus \cdots \oplus a^{(N)}D^{(N)}(U_i), \qquad (10.20)$$

where $a^{(\nu)}$ is the number of times the irreducible representation $D^{(\nu)}(U_i)$, or any representation equivalent to it, appears in the decomposition.

A very useful characterization of a representation which is the same for all equivalent representations, is given by the *trace* of the representative matrices. The trace of $D(U)$ is denoted by $\chi(U)$ and is called the *character* of U in the representation in question. Since

$$\operatorname{tr}[C^{-1}D(U)C] = \operatorname{tr}[D(U)] \equiv \chi(U) ,$$

the character of U is the same for all equivalent representations. When we are dealing with an irreducible representation, $D^{(\nu)}(U)$, we will write for the corresponding character $\chi^{(\nu)}(U)$. Now within a given representation, many of the characters will be the same. In fact, consider a class, $\mathscr{C}_i$, of conjugate elements of the group G. Then if $U_\nu, U_\mu \in \mathscr{C}_i$, there exists some $U_\lambda \in G$ such that

$$U_\nu = U_\lambda U_\mu U_\lambda^{-1} .$$

This means that

$$D(U_\nu) = D(U_\lambda) D(U_\mu) D(U_\lambda^{-1}) ,$$

and hence

$$\begin{aligned} \chi(U_\nu) &\equiv \operatorname{tr}[D(U_\nu)] = \operatorname{tr}[D(U_\lambda)D(U_\mu)D(U_\lambda^{-1})] \\ &= \operatorname{tr}[D(U_\lambda^{-1})D(U_\lambda)D(U_\mu)] = \operatorname{tr}[D(U_\mu)] \equiv \chi(U_\mu) , \end{aligned}$$

that is, all elements of $\mathscr{C}_i$ have the same character. For this reason one can, with no ambiguity, refer to $\chi_i^{(\nu)}$ as the character of the ith class in the νth irreducible representation, thereby removing the necessity of referring explicitly to the group elements. Thus in a group with k classes, for a given representation, ν, there are k characteristic numbers,

$$\chi_1^{(\nu)}, \chi_2^{(\nu)}, \cdots, \chi_k^{(\nu)}$$

to be determined. These k constants, in fact, provide a surprisingly large amount of information about the group, as we will see in the following sections.

10.6 UNITARY REPRESENTATIONS, SCHUR'S LEMMAS, AND ORTHOGONALITY RELATIONS

In this section we will develop the main results of representation theory, which will enable us to determine irreducible representations. To begin, we prove a theorem which restricts tremendously the possible forms which a representation can take.

Theorem 10.6. Every representation of a finite group is equivalent to a unitary representation.

Proof. The first step is to construct an inner product in the n-dimensional vector space, V, of the representation and show that the representation is unitary with respect to this inner product. For $x, y \in V$, we denote the usual inner product of x and y by the familiar (x, y); in terms of this inner product, we define a new inner product as follows:

$$\{x, y\} = \sum_{U \in G} (D(U)x, D(U)y), \tag{10.21}$$

where D is any representation of G. It is easy to check that since $(\,,\,)$ is an inner product and since $D^{-1}(U)$ exists for all $U \in G$, $\{ \,,\, \}$ will also satisfy all the inner-product axioms. Consider now any $D(U')$. We want to show that

$$\{D(U')x, D(U')y\} = \{x, y\} .$$

From Eq. (10.21), we have

$$\begin{aligned}\{D(U')x, D(U')y\} &= \sum_{U \in G} (D(U)D(U')x, D(U)D(U')y) \\ &= \sum_{U \in G} (D(UU')x, D(UU')y) .\end{aligned} \tag{10.22}$$

But as U runs through all the elements of G, so does UU', according to Theorem 10.1. Of course, the ordering may be different, but that is immaterial. Hence

$$\sum_{U \in G} (D(UU')x, D(UU')y) = \sum_{\tilde{U} \in G} (D(\tilde{U})x, D(\tilde{U})y) = \{x, y\} . \tag{10.23}$$

Combining Eqs. (10.22) and (10.23), we get

$$\{D(U')x, D(U')y\} = \{x, y\}$$

for any $U' \in G$, as required.

But to say that $D(U')$ is unitary with respect to some special inner product is simply to say that we have written $D(U')$ in an inconvenient basis. By a similarity transformation we can easily correct this difficulty and make D unitary with respect to the original inner product. Let $\{\xi_i\}$ be a complete orthonormal basis with respect to the inner product $(\,,\,)$ and let $\{\eta_i\}$ be a complete orthonormal basis with respect to the special inner product $\{ \,,\, \}$. We define a linear operator S on V by

$$\eta_i = S\xi_i , \qquad i = 1, 2, \cdots, n . \tag{10.24}$$

Note that with this definition for any $x, y \in V$

$$\{Sx, Sy\} = (x, y) , \tag{10.25a}$$

or

$$(S^{-1}x, S^{-1}y) = \{x, y\} . \tag{10.25b}$$

This follows by expanding x and y in the $\{\xi_i\}$ basis:

$$x = \sum_{i=1}^{n} a_i\xi_i , \qquad y = \sum_{j=1}^{n} b_j\xi_j .$$

Then

$$\{Sx, Sy\} = \sum_{i,j=1}^{n} a_i^* b_j \{S\xi_i, S\xi_j\} = \sum_{i,j=1}^{n} a_i^* b_j \{\eta_i, \eta_j\} ,$$

by use of Eq. (10.24). Since the η_i form an orthonormal set with respect to $\{\ ,\ \}$, we have

$$\{Sx, Sy\} = \sum_{i,j=1}^{n} a_i^* b_j \delta_{ij} = \sum_{i=1}^{n} a_i^* b_i = (x, y)\ ,$$

which is just Eq. (10.25a). S is precisely the operator we need to transform the original representation.

For any $U \in G$, we define

$$\tilde{D}(U) \equiv S^{-1} D(U) S\ .$$

Consider

$$(\tilde{D}(U)x, \tilde{D}(U)y) = (S^{-1}D(U)Sx, S^{-1}D(U)Sy)\ .$$

By Eq. (10.25b), we have immediately

$$(\tilde{D}(U)x, \tilde{D}(U)y) = \{D(U)Sx, D(U)Sy\}\ .$$

But with respect to $\{\ ,\ \}$ $D(U)$ is unitary, so

$$(\tilde{D}(U)x, \tilde{D}(U)y) = \{Sx, Sy\}\ .$$

Finally, Eq. (10.25a) gives

$$(\tilde{D}(U)x, \tilde{D}(U)y) = (x, y)\ ,$$

so $\tilde{D}(U)$ is unitary with respect to the original inner product.

According to this theorem, we can restrict ourselves to unitary matrices in our search for irreducible representations. Note that although we have stated this theorem for finite groups, we have made very little use of the finite order of G, using it only in Eq. (10.21) to write the sum over all $U \in G$. This sum is finite since it contains only a finite number of terms. For many infinite groups, it is possible to make the transition from discrete sums to integrals over a continuous parameter; for example, in the case of continuous groups such a possibility is evident. However, even if it is possible to replace sums over group elements by integrals over group elements, it may be that the integral formulation of Eq. (10.21) will not be meaningful because the integrations run over an infinite range and are divergent. Nevertheless, there are many groups, such as the rotation groups, where this does not happen, and for them Theorem 10.6 is valid, the proof being identical to the one presented here, save for the presence of integrals instead of summations.

The fact that the matrices in a representation are equivalent to unitary representations is a very powerful result. From it follows the two main theorems of representation theory, the first and second lemmas of Schur. The proof of these two important group theoretical results hinges on the following lemma, which has nothing to do with group theory.

Lemma. If a matrix, M, commutes with a unitary matrix, U, and if we write $M = M_+ + iM_-$, where $M_+ = (M + M^\dagger)/2$ and $M_- = (M - M^\dagger)/2i$, then both M_+ and M_- commute with U.

Proof. If

$$UM = MU, \tag{10.26}$$

then

$$M^\dagger U^\dagger = U^\dagger M^\dagger .$$

Multiplying on the left by U, we get

$$UM^\dagger U^\dagger = UU^\dagger M^\dagger = M^\dagger ,$$

since U is unitary. Multiplying this last equation on the right by U, we find that

$$UM^\dagger U^\dagger U = M^\dagger U .$$

Thus, using the unitarity of U once more,

$$UM^\dagger = M^\dagger U . \tag{10.27}$$

Combining Eqs. (10.26) and (10.27), we readily obtain

$$U\left[\frac{M + M^\dagger}{2}\right] = \left[\frac{M + M^\dagger}{2}\right]U , \tag{10.28a}$$

$$U\left[\frac{M - M^\dagger}{2i}\right] = \left[\frac{M - M^\dagger}{2i}\right]U , \tag{10.28b}$$

or

$$UM_+ = M_+U , \qquad UM_- = M_-U ,$$

as required.

With this lemma at our disposal, we now prove the main result of this section.

> **Lemma (Schur).** If $D(U)$ is an element of an irreducible representation of G and if $D(U)M = MD(U)$ for all $U \in G$, then M must be a multiple of the unit matrix.

Proof. According to the previous lemma, if $MD(U) = D(U)M$ for all $U \in G$ and if we write

$$M = M_+ + iM_- , \tag{10.29}$$

then since the representation can be taken to be unitary,

$$M_+D(U) = D(U)M_+ \qquad \text{and} \qquad M_-D(U) = D(U)M_-$$

for all $U \in G$. $M_+ = (M + M^\dagger)/2$ and $M_- = (M - M^\dagger)/2i$ are both self-adjoint, so let us begin by considering the eigenvalue problem associated with M_+:

$$M_+x_n^{(i)} = \lambda_n x_n^{(i)} , \qquad i = 1, 2, \cdots, m_n ,$$

where m_n is the multiplicity of λ_n. Since M_+ is Hermitian, the set of all $x_n^{(i)}$ spans the N-dimensional space of the representation. For all $U \in G$, we have

$$D(U)M_+x_n^{(i)} = \lambda_n D(U)x_n^{(i)} .$$

But $[D(U), M_+] = 0$, so this is just

$$M_+[D(U)x_n^{(i)}] = \lambda_n[D(U)x_n^{(i)}] .$$

Since $D(U)x_n^{(i)}$ is an eigenvector of M_+ belonging to λ_n, it must be a linear combination of the $x_n^{(i)}$ $(i = 1, 2, \cdots, m_n)$:

$$D(U)x_n^{(i)} = \sum_{j=1}^{m_n} a_{ij}(U)x_n^{(j)} \tag{10.30}$$

for all $U \in G$. Let us compute the matrix D in the space spanned by the eigenvectors of M_+. We take $n = 1$ first. Using Eq. (10.30), we find that

$$(x_1^{(l)}, D(U)x_1^{(i)}) = \sum_{j=1}^{m_1} a_{ij}(U)(x_1^{(l)}, x_1^{(j)}) = a_{il}(U) , \qquad i, l = 1, 2, \cdots, m_1 , \tag{10.31a}$$

for all $U \in G$. For any $n \neq 1$, Eq. (10.30) tells us directly that

$$(x_n^{(l)}, D(U)x_1^{(i)}) = 0 ; \qquad l = 1, 2, \cdots, m_n , \quad i = 1, 2, \cdots, m_1 , \tag{10.31b}$$

for all $U \in G$, since the eigenvectors belonging to different eigenvalues of an Hermitian operator are orthogonal. Similarly, for $n \neq 1$, we also have from Eq. (10.30)

$$(x_1^{(l)}, D(U)x_n^{(i)}) = 0; \qquad l = 1, 2, \cdots, m_1 , \quad i = 1, 2, \cdots, m_n . \tag{10.31c}$$

Equations (10.31a), (10.31b), and (10.31c) mean that the matrix $D(U)$ looks like

$$D(U) = \left(\begin{array}{c|c} m_1 \times m_1 & 0 \\ \hline 0 & (N - m_1) \times (N - m_1) \end{array}\right),$$

for all $U \in G$, where N is the dimension of the representation. The $m_1 \times m_1$ matrix in the upper left-hand corner has elements which are given by the $a_{ij}(U)$ defined in Eq. (10.30). But the representation is assumed to be irreducible, so this can only be the case if $m_1 = N$; that is, the multiplicity of λ_1 must be equal to N, the dimensionality of the representation. The only Hermitian matrix having an eigenvalue whose multiplicity equals the dimension of the matrix is a real constant times the unit matrix. Thus $M_+ = c_+I$. Similarly, $M_- = c_-I$, so we obtain from Eq. (10.29) $M = (c_+ + ic_-)I$, and the proof is complete.

As an immediate corollary (really no more than a restatement) of this lemma, we have:

Corollary. If there exists a matrix which commutes with every element of a representation and is *not* a multiple of the unit matrix, then the representation is necessarily reducible.

At this point it is perhaps a good idea to say a word about a more abstract way of looking at irreducible representations which may clarify the situation, as well as boil down the preceding pages into a few words.

As a careful reading of the previous section shows, we can characterize the irreducible representations of a group, G, of order g in a more abstract manner as follows. Let

$$\mathscr{G} = \{D(U_i), \quad i = 1, 2, \cdots, g\}$$

be a group of linear operators on an n-dimensional vector space, V, and let this group of operators be homomorphic to G, that is, $\mathscr{G}$ is a representation of G. $\mathscr{G}$ is said to be an *irreducible* representation of G if no *proper* subspace of V is left invariant (i.e., mapped into itself) under all $D \in \mathscr{G}$. (The proper subspaces of V are all subspaces of V *except* for V itself.) It may of course happen that some *subset* of $\mathscr{G}$ leaves invariant some proper subspace of V, but if we have an irreducible representation, then $\mathscr{G}$ in its entirety must leave only V invariant. Obviously, if we look just at one unitary $D(U) \in \mathscr{G}$, then it will have n invariant subspaces, defined by the n orthonormal eigenvectors of $D(U)$.

In light of this, Schur's first lemma has a very simple meaning: If there is a matrix which commutes with $D(U)$ for all $U \in G$, the linearly independent eigenvectors of this matrix define a collection of subspaces of V which are invariant under $D(U)$ for all $U \in G$. The only admissible invariant subspace is V itself, and the only matrix which has every vector in V as an eigenvector is the unit matrix or a multiple thereof.

A simple, but very useful, consequence of Schur's first lemma is the fact that an abelian group can have only one-dimensional irreducible representations! This follows immediately from the statement that if G is abelian, then for any $U \in G$, $D(U)$ commutes with all elements of the representation D. Thus, by Schur's first lemma, for all $U \in G$,

$$D(U) = C(U)I,$$

where $C(U)$ is a constant which can depend on U. The unit matrix is irreducible only when it is the one-dimensional unit matrix. Thus, for an abelian group, all irreducible representations must be one-dimensional.

Schur's second lemma gives an analogous result for two *different* representations.

Lemma. Let $D(U)$ and $D'(U)$ be irreducible representations. If for all $U \in G$, $D(U)M = MD'(U)$, then either D and D' are equivalent or else $M = 0$.

Proof. In general, D and D' will have different dimensions, so that M need not be a square matrix. If D is $n \times n$ and D' is $m \times m$, then M must be $n \times m$. The proof starts out in the same direction as that of the first lemma. As before

$$D(U)M = MD'(U) \tag{10.32}$$

for all $U \in G$ implies that

$$M^\dagger D^\dagger(U) = D'^\dagger(U)M^\dagger$$

for all $U \in G$, where $M^\dagger$ is an $m \times n$ matrix, $(M^\dagger)_{ij} = M^*_{ji}$. Taking $D(U)$ to be unitary, we find that

$$M^\dagger D(U^{-1}) = D'(U^{-1})M^\dagger. \tag{10.33}$$

Multiplying Eq. (10.33) from the left by M, we get

$$MM^{\dagger}D(U^{-1}) = MD'(U^{-1})M^{\dagger}. \tag{10.34}$$

But by Eq. (10.32),

$$MD'(U^{-1})M^{\dagger} = D(U^{-1})MM^{\dagger},$$

so Eq. (10.34) becomes

$$MM^{\dagger}D(U^{-1}) = D(U^{-1})MM^{\dagger}$$

for all $U \in G$. The previous lemma (and the uniqueness of inverses) tells us straightaway that

$$MM^{\dagger} = cI.$$

We leave it to the reader to show in the same way that multiplying Eq. (10.33) on the right by M leads to

$$M^{\dagger}M = cI.$$

It should be observed that the matrices $MM^{\dagger}$ and $M^{\dagger}M$ have different dimensions, the former being $n \times n$, the latter $m \times m$.

We now consider three cases separately:

(i) $m = n$. Then if $c = 0$, we have $MM^{\dagger} = 0$, which means that for all i,

$$(MM^{\dagger})_{ii} = 0 = \sum_{j=1}^{n} M_{ij}(M^{\dagger})_{ji} = \sum_{j=1}^{n} |M_{ij}|^2.$$

Thus, for all i,

$$\sum_{j=1}^{n} |M_{ij}|^2 = 0,$$

so we conclude that $M_{ij} = 0$, for all i and j, and $M = 0$. (For a slightly stronger result see Problem 3.14) If, on the other hand, $c \neq 0$, then

$$\det(MM^{\dagger}) \neq 0,$$

so $\det M \neq 0$, and M is invertible. Thus according to Eq. (10.32),

$$D(U) = MD'(U)M^{-1}$$

for all $U \in G$, therefore D and D' are equivalent representations.

(ii) $n > m$. In this case, we make M into a square matrix by filling in $n - m$ columns of zeros. Calling the new matrix N, we have

$$N = \left(\begin{array}{c|c} \overset{\longleftarrow m \longrightarrow}{M} & \overset{\longleftarrow n-m \longrightarrow}{0} \end{array}\right) \Big\updownarrow n, \qquad N^{\dagger} = \left(\begin{array}{c} \overset{\longleftarrow n \longrightarrow}{M^{\dagger}} \\ \hdashline 0 \end{array}\right) \begin{array}{l} \updownarrow m \\ \updownarrow n-m \end{array}.$$

It is clear that $MM^{\dagger} = NN^{\dagger}$, so $MM^{\dagger} = cI$ implies that $NN^{\dagger} = cI$. Now

$$\det(NN^{\dagger}) = c^n = (\det N)(\det N^{\dagger}),$$

but from the form of N, we see that $\det N = 0$, so the constant, c, must vanish. Thus

$$NN^\dagger = 0 = MM^\dagger$$

and, just as in case (i), we conclude that $M = 0$.

(iii) $m > n$. In this case, we must add $m - n$ rows of zeros to M to turn it into a square matrix. Again denoting the new matrix by N, we have

$$N = \begin{pmatrix} M \\ \hdashline 0 \end{pmatrix} \begin{matrix} \updownarrow n \\ \updownarrow m-n \end{matrix} \quad (\text{width } m), \qquad N^\dagger = \left(\begin{array}{c:c} M^\dagger & 0 \end{array}\right) \updownarrow m \quad (\text{widths } n,\ m-n).$$

In this case, $NN^\dagger \neq MM^\dagger$, but we do have $N^\dagger N = M^\dagger M$. Thus, using the relation $M^\dagger M = cI$ instead of $MM^\dagger = cI$, we see that the new matrix N satisfies

$$N^\dagger N = cI\,;$$

as in case (i), we conclude that $N = 0$. Hence $M = 0$ as well, and the proof of the lemma is complete.

Note that in proving these lemmas, we have not made direct reference to the number of elements in the group G. We have actually proved these results for any finite-dimensional *unitary* representation. Of course, Theorem 10.6 tells us that for a finite group the representations can always be chosen to be unitary. But Schur's two lemmas hold also for the important class of infinite groups which have finite-dimensional unitary representations.

The reason that these two lemmas are so important is that it turns out to be rather easy to find matrices, M, which satisfy the conditions of the lemmas. These matrices in turn provide us with a vast amount of information about the irreducible representations and their characters. One simple example of a matrix which satisfies the conditions of Schur's first lemma is defined by

$$M_i^{(\nu)} = \sum_{U \in \mathscr{C}_i} D^{(\nu)}(U)\,. \tag{10.35}$$

$M_i^{(\nu)}$ is just the sum of all representative matrices of the elements of the class $\mathscr{C}_i$ in the νth irreducible representation. Now let U' be any element of G and consider the quantity

$$D^{(\nu)}(U')M_i^{(\nu)}D^{(\nu)}(U')^{-1}\,.$$

We have

$$\begin{aligned} D^{(\nu)}(U')M_i^{(\nu)}D^{(\nu)}(U')^{-1} &= \sum_{U \in \mathscr{C}_i} D^{(\nu)}(U')D^{(\nu)}(U)D^{(\nu)}(U'^{-1}) \\ &= \sum_{U \in \mathscr{C}_i} D^{(\nu)}(U'UU'^{-1})\,. \end{aligned}$$

But by the definition of a class, if $U \in \mathscr{C}_i$, then $U'UU'^{-1} \in \mathscr{C}_i$. Since a similarity transformation just rearranges the elements of $\mathscr{C}_i$, we have from the previous equation

$$D^{(\nu)}(U')M_i^{(\nu)}D^{(\nu)}(U')^{-1} = \sum_{\tilde{U}\in\mathscr{C}_i} D^{(\nu)}(\tilde{U}) = M_i^{(\nu)} .$$

Thus, for all $U' \in G$,

$$D^{(\nu)}(U')M_i^{(\nu)} = M_i^{(\nu)}D^{(\nu)}(U') ,$$

so $M_i^{(\nu)}$ satisfies the requirements of Schur's first lemma. We conclude that

$$M_i^{(\nu)} = c_i^{(\nu)}I .$$

The constant $c_i^{(\nu)}$ can be readily evaluated. We have, calling the dimension of the νth representation n_ν,

$$\operatorname{tr} M_i^{(\nu)} = c_i^{(\nu)}n_\nu . \tag{10.36}$$

But

$$\operatorname{tr} M_i^{(\nu)} = \sum_{U\in\mathscr{C}_i} \operatorname{tr} D^{(\nu)}(U) = \sum_{U\in\mathscr{C}_i} \chi^{(\nu)}(U) .$$

The trace of all the representative matrices in a given class is the same, as was pointed out at the end of the last section. Hence using the notation established there, we have

$$\operatorname{tr} M_i^{(\nu)} = g_i\chi_i^{(\nu)} , \tag{10.37}$$

where g_i is the number of elements in the class $\mathscr{C}_i$. Combining Eqs. (10.36) and (10.37), we obtain

$$c_i^{(\nu)} = \frac{g_i}{n_\nu}\chi_i^{(\nu)} .$$

Therefore

$$M_i^{(\nu)} = \frac{g_i}{n_\nu}\chi_i^{(\nu)}I , \tag{10.38}$$

that is, the sum of all the representative matrices in a given class is a simple multiple of the unit matrix. We shall soon have occasion to make use of this result.

A more fruitful example of a matrix which fulfills all the requirements of Schur's lemmas is the following:

$$M = \sum_{U\in G} D^{(\nu)}(U)XD^{(\mu)}(U^{-1}) = \sum_{U\in G} D^{(\nu)}(U)XD^{(\mu)}(U)^{-1} ,$$

where $D^{(\nu)}$ and $D^{(\mu)}$ are irreducible representations and X is an arbitrary matrix, although in order for this expression to have any meaning, X must have n_ν rows and n_μ columns. Multiplying on the left by $D^{(\nu)}(U')$ and on the right by $D^{(\mu)}(U')^{-1}$, we find that

$$\begin{aligned} D^{(\nu)}(U')MD^{(\mu)}(U')^{-1} &= \sum_{U\in G} D^{(\nu)}(U')D^{(\nu)}(U)XD^{(\mu)}(U)^{-1}D^{(\mu)}(U')^{-1} \\ &= \sum_{U\in G} D^{(\nu)}(U'U)XD^{(\mu)}(U^{-1}U'^{-1}) . \end{aligned}$$

Calling $\tilde{U} = U'U$ and remembering that $\tilde{U}^{-1} = U^{-1}U'^{-1}$, we have

$$D^{(\nu)}(U')MD^{(\mu)}(U')^{-1} = \sum_{U\in G} D^{(\nu)}(\tilde{U})XD^{(\mu)}(\tilde{U})^{-1} . \qquad (10.39)$$

But as U runs through the entire group, so does $\tilde{U} = U'U$, although in a different order. Thus the right-hand side of Eq. (10.39) is just equal to M, and

$$D^{(\nu)}(U')M = MD^{(\mu)}(U') \qquad (10.40)$$

for all $U' \in G$. Assuming that $D^{(\nu)}$ is not equivalent to $D^{(\mu)}$ (in the case $n_\nu = n_\mu$), we have by Schur's lemmas

$$M = c(X)I\delta_{\nu\mu} .$$

In other words, $M = 0$ if the representations have different dimensionality (Schur's second lemma) and $M = cI$ if they have the same dimensionality (Schur's first lemma). The value of the constant depends, of course, on the choice of X. Thus we arrive at the result

$$\sum_{U\in G} D^{(\nu)}(U)XD^{(\mu)}(U)^{-1} = c(X)I\delta_{\mu\nu} . \qquad (10.41)$$

If $D^{(\nu)}$ and $D^{(\mu)}$ are chosen to be unitary, this can also be written as

$$\sum_{U\in G} D^{(\nu)}(U)XD^{(\mu)\dagger}(U) = c(X)I\delta_{\mu\nu} . \qquad (10.42)$$

To exploit this equation, let us write it in component form:

$$\sum_{U\in G}\sum_{i',l'} D^{(\nu)}_{ii'}(U)\ X_{i'l'}\ [D^{(\mu)\dagger}(U)]_{l'l} = c(X)\delta_{il}\delta_{\mu\nu} .$$

Now suppose that we choose the matrix X so that all its elements are zero except for the jk-element. Denoting the related $c(X)$ by c_{jk}, we have

$$\sum_{U\in G} D^{(\nu)}_{ij}(U)[D^{(\mu)\dagger}(U)]_{kl} = c_{jk}\delta_{il}\delta_{\mu\nu} . \qquad (10.43)$$

Since

$$(A^\dagger)_{kl} = A^*_{lk} ,$$

we obtain from Eq. (10.43)

$$\sum_{U\in G} D^{(\mu)*}_{lk}(U)D^{(\nu)}_{ij}(U) = c_{jk}\delta_{il}\delta_{\mu\nu} , \qquad (10.44)$$

for all i, j, k, l, μ, and ν.

All that remains is to determine c_{jk}. To do this, we set $\mu = \nu$, $l = i$ and sum on i. Thus

$$\sum_{U\in G}\sum_{i} D^{(\nu)*}_{ik}(U)D^{(\nu)}_{ij}(U) = c_{jk}\sum_i \delta_{ii} = c_{jk}n_\nu .$$

But since $D^{(\nu)}$ is unitary, the sum on i on the left-hand side of this equation is just equal to δ_{kj}. Hence

$$c_{jk}n_\nu = \sum_{U\in G} \delta_{jk} = g\delta_{jk} ,$$

where g is the order of the group. Therefore

$$c_{jk} = \frac{g}{n_\nu}\delta_{jk}\,,$$

and we have finally from Eq. (10.44) the remarkable orthogonality theorem,

$$\sum_{U\in G} D_{lk}^{(\mu)*}(U)D_{ij}^{(\nu)}(U) = \frac{g}{n_\nu}\delta_{jk}\delta_{il}\delta_{\mu\nu}\,, \tag{10.45}$$

for all i, j, k, l, μ, and ν.

Now let us see what this means in simple terms. The sum on the left-hand side of Eq. (10.45) looks very much like an inner product of two vectors, except that the sum is over the elements of a group rather than a conventional subscript. However, there is no law against indexing the components of a vector by using the elements of a group. We may say, therefore, that Eq. (10.45) is an orthogonality relationship for the inner product of two g-dimensional vectors, each vector being identified by three labels. Equation (10.45) says that if, from any irreducible representation we pick the ij-element of the representative matrices, then this gives us a g-dimensional vector which is orthogonal to the g-dimensional vector obtained from any other element, say the $i'j'$-element, and *also* to any vector obtained in this way from any *other* representation. Thus in the g-dimensional space we have $n_1^2 + n_2^2 + \cdots + n_N^2$ orthogonal vectors, where N is the number of inequivalent irreducible representations. Since the number of orthogonal vectors in a g-dimensional space cannot exceed g, we must have

$$\sum_{\nu=1}^{N} n_\nu^2 \leq g\,. \tag{10.46}$$

Thus we arrive at the important result that a finite group can have only a finite number of inequivalent irreducible representations, all of which must have dimension less than $\sqrt{g}$.

From Eq. (10.45) it is a simple matter to obtain an orthogonality relation for the characters of a representation. In Eq. (10.45), if one sets $k = l$ and sums on l, one finds that

$$\sum_{U\in G}\sum_{l=1}^{n_\mu} D_{ll}^{(\mu)*}(U)D_{ij}^{(\nu)}(U) = \frac{g}{n_\nu}\delta_{\mu\nu}\sum_{l=1}^{n_\mu}\delta_{il}\delta_{lj} = \frac{g}{n_\nu}\delta_{\mu\nu}\delta_{ij}\,.$$

But

$$\sum_{l} D_{ll}^{(\mu)*}(U) = \operatorname{tr} D^{(\mu)*}(U) = \chi^{(\mu)*}(U)\,, \tag{10.47}$$

so

$$\sum_{U\in G}\chi^{(\mu)*}(U)D_{ij}^{(\nu)}(U) = \frac{g}{n_\nu}\delta_{\mu\nu}\delta_{ij}\,.$$

Finally, setting $i = j$ and summing on j, we find that

$$\sum_{U\in G}\chi^{(\mu)*}(U)\sum_{j=1}^{n_\nu} D_{jj}^{(\nu)}(U) = \frac{g}{n_\nu}\delta_{\mu\nu}\sum_{j=1}^{n_\nu}\delta_{jj} = g\delta_{\mu\nu}\,.$$

Using Eq. (10.47) once more, we arrive at

$$\sum_{U\in G} \chi^{(\mu)*}(U)\chi^{(\nu)}(U) = g\delta_{\mu\nu} . \tag{10.48}$$

Since all the members of a given class have the same character, we can convert the sum on $U \in G$ in Eq. (10.48) into a sum on the classes. If g_i equals the number of elements in $\mathscr{C}_i$, and if k denotes the number of classes, Eq. (10.48) becomes

$$\sum_{i=1}^{k} g_i \chi_i^{(\mu)*}\chi_i^{(\nu)} = g\delta_{\mu\nu} . \tag{10.49}$$

Equation (10.49) tells us that the N nonequivalent irreducible representations provide us with N k-dimensional orthogonal vectors via the class characters. Since in a k-dimensional space there cannot be more than k orthogonal vectors, we conclude that

$$N \leq k . \tag{10.50}$$

In the case of abelian groups, Eq. (10.50) has the same content as Eq. (10.46). Since abelian groups have only one-dimensional representations, we have

$$\sum_{\nu=1}^{N} n_\nu^2 = \sum_{\nu=1}^{N} 1 = N .$$

So Eq. (10.46) becomes

$$N \leq g .$$

On the other hand, for an abelian group every element forms a separate class. Thus $k = g$, and Eq. (10.50) becomes

$$N \leq g .$$

Hence both Eqs. (10.46) and (10.50) lead to the same inequality for an abelian group.

It is also worth noting that in deriving the orthogonality relations of Eqs. (10.45) and (10.49), we have been writing sums on group elements quite freely. However, there are, as mentioned before, cases involving infinite groups where the sum can be replaced by an integration. Furthermore, for certain groups, integrals over the entire group will be convergent, and the proofs of this section will hold virtually unchanged. As a simple example, we look briefly at the group of rotations about the z-axis. We have already found the one-dimensional representations

$$D^{(m)}(\alpha) = e^{im\alpha} . \tag{10.51}$$

The logical candidate for integration variable for this group is clearly α, with infinitesimal element $d\alpha$. For this case, we expect $\sum_{U\in G}$ to be replaced by $\int_0^{2\pi} d\alpha$. In particular, since

$$g = \sum_{U\in G} 1 ,$$

we expect g to be replaced by

$$\int_0^{2\pi} 1 d\alpha = 2\pi$$

in this case. For groups of this type we refer to g as the volume of the group. According to the above speculations, we would expect Eq. (10.45) to read

$$\int_0^{2\pi} D^{(m)*}(\alpha) D^{(n)}(\alpha) d\alpha = 2\pi\, \delta_{mn}, \tag{10.52}$$

since the dimension of all representations is 1. We see that if $D^{(m)}(\alpha)$ is given by Eq. (10.51), then Eq. (10.52) is obeyed. Similar considerations can be applied to other continuous groups, such as the rotation group and the unitary groups. However, there are exceptions to this simple-minded type of extension of the theory of finite groups, the most significant one being the Lorentz group.

10.7 THE DETERMINATION OF GROUP REPRESENTATIONS

In this section we want to apply the major theorems proved thus far to the problem of actually finding group representations. In doing so, we shall also be able to sharpen slightly the orthogonality relations obtained in the previous section so that they become completeness relations.

First, we note that the characters of a representation, D, which is *not* irreducible can be simply written in terms of those of the irreducible representations. If D is written in the form of Eq. (10.19)—as it always can be by use of a similarity transformation—then the character, χ_i, of a given class will have the form

$$\chi_i = a^{(1)}\chi_i^{(1)} + a^{(2)}\chi_i^{(2)} + \cdots + a^{(N)}\chi_i^{(N)} \equiv \sum_{\nu=1}^{N} a^{(\nu)}\chi_i^{(\nu)}. \tag{10.53}$$

This follows directly from the definition of the characters in terms of the trace of the representative matrices. $a^{(\nu)}$ is just the number of times the νth irreducible representation occurs in the decomposition of D given by Eq. (10.20). Multiplying Eq. (10.53) on both sides by $g_i\chi_i^{(\mu)*}$, summing on i, and using the orthogonality relation Eq. (10.49), we get

$$\sum_{i=1}^{k} g_i\chi_i^{(\mu)*}\chi_i = \sum_{\nu=1}^{N} a^{(\nu)} \sum_{i=1}^{k} g_i\chi_i^{(\mu)*}\chi_i^{(\nu)} = \sum_{\nu=1}^{N} g a^{(\nu)}\delta_{\mu\nu} = g a^{(\mu)}.$$

Hence

$$a^{(\mu)} = \frac{1}{g}\sum_{i=1}^{k} g_i\chi_i^{(\mu)*}\chi_i, \tag{10.54}$$

so the number of times a given irreducible representation occurs in an arbitrary representation is readily determined from the characters of that representation. Thus if two representations have the same characters they must be equivalent, since the $a^{(\mu)}$ will then be the same for all of them.

As an example, let us apply Eqs. (10.53) and (10.54) to the regular representation discussed in Section 10.5. This special case will lead to several useful results. The characters of the regular representation are easily determined. Since its dimensionality is g (the number of elements of G), the character χ_1 of the identity class will be equal to the trace of the g-dimensional unit matrix:

$$\chi_1 = g . \tag{10.55}$$

All other characters vanish, since all other representative matrices have only zeros on the diagonal, as was pointed out in Section 10.5. Hence

$$\chi_i = 0 , \qquad i \neq 1 . \tag{10.56}$$

Thus Eq. (10.53) becomes, for $i = 1$,

$$g = \sum_{\nu=1}^{N} a^{(\nu)} \chi_1^{(\nu)} .$$

But in any representation of dimension n_ν, the character of the identity class is just n_ν, the trace of the n_ν-dimensional unit matrix. Thus

$$g = \sum_{\nu=1}^{N} a^{(\nu)} n_\nu . \tag{10.57}$$

On the other hand, using Eqs. (10.55) and (10.56) together with the fact that $g_1 = 1$, we find from Eq. (10.54) that

$$a^{(\mu)} = \frac{1}{g} g_1 \chi_1^{(\mu)*} \chi_1 = \chi_1^{(\mu)*} = n_\mu , \tag{10.58}$$

so Eq. (10.57) becomes

$$g = \sum_{\nu=1}^{N} n_\nu^2 . \tag{10.59}$$

Equation (10.58) says that in the regular representation *every* irreducible representation occurs precisely as many times as its dimensionality. Equation (10.59) is a much stronger result than Eq. (10.46). It tells us that the $D_{ij}^{(\nu)}(U)$ provide us with exactly g orthogonal vectors (because $g = \sum n_\nu^2$) in the g-dimensional vector space generated by the g group elements via their irreducible representations. Thus the $D_{ij}^{(\nu)}(U)$ $(U \in G)$ are a *complete* set of vectors for this space. Normalizing in the manner required by Eq. (10.45), we may say that the "representation vectors" whose g components are

$$\sqrt{\frac{n_\nu}{g}}\, D_{ij}^{(\nu)}(U_1),\ \sqrt{\frac{n_\nu}{g}}\, D_{ij}^{(\nu)}(U_2),\ \cdots,\ \sqrt{\frac{n_\nu}{g}}\, D_{ij}^{(\nu)}(U_g) \tag{10.60}$$

for $\nu = 1, 2, \cdots, N$; $i, j = 1, 2, \cdots, n_\nu$ form a complete orthonormal set in the g-dimensional complex vector space. Using a cumbersome, but minimal notation we may express this completeness properly as

$$\sum_{\nu=1}^{N} \sum_{i=1}^{n_\nu} \sum_{j=1}^{n_\nu} \sqrt{\frac{n_\nu}{g}}\, D_{ij}^{(\nu)}(U) \sqrt{\frac{n_\nu}{g}}\, D_{ij}^{(\nu)*}(U') = \delta_{UU'} ,$$

where $\delta_{UU'}$ is a Kronecker δ-symbol which is equal to 1 if $U = U'$ and is zero otherwise. This may be written more simply as

$$\sum_{\nu=1}^{N} \sum_{i=1}^{n_\nu} \sum_{j=1}^{n_\nu} \frac{n_\nu}{g} D_{ij}^{(\nu)}(U) D_{ij}^{(\nu)*}(U') = \delta_{UU'} . \tag{10.61}$$

Having strengthened the orthogonality properties of the D's in this manner, we are led naturally to conjecture that a similar strengthening can be achieved for the characters. To see that this is indeed the case, let us sum both sides of Eq. (10.61) over all $U \in \mathscr{C}_l$ and all $U' \in \mathscr{C}_m$. We find that

$$\sum_{\nu=1}^{N} \sum_{i=1}^{n_\nu} \sum_{j=1}^{n_\nu} n_\nu \Big(\sum_{U \in \mathscr{C}_l} D_{ij}^{(\nu)}(U) \Big) \Big(\sum_{U' \in \mathscr{C}_m} D_{ij}^{(\nu)*}(U') \Big) = g \sum_{U \in \mathscr{C}_l} \sum_{U' \in \mathscr{C}_m} \delta_{UU'} . \tag{10.62}$$

The right-hand side of this equation vanishes unless $\mathscr{C}_l = \mathscr{C}_m$, for otherwise $U \neq U'$. If $\mathscr{C}_l = \mathscr{C}_m$, then each term with $U = U'$ will contribute 1 to the sum and all other terms will contribute zero. There are g_l contributing terms if g_l is the number of elements in $\mathscr{C}_l$, so in the case $\mathscr{C}_l = \mathscr{C}_m$, the right-hand side of Eq. (10.62) equals gg_l. But according to the definition of Eq. (10.35),

$$\sum_{U \in \mathscr{C}_l} D^{(\nu)}(U) = M_l^{(\nu)} .$$

Hence Eq. (10.62) becomes

$$\sum_{\nu=1}^{N} \sum_{i=1}^{n_\nu} \sum_{j=1}^{n_\nu} n_\nu \, [M_l^{(\nu)}]_{ij} [M_m^{(\nu)}]_{ij} = gg_l \delta_{lm} . \tag{10.63}$$

Now by Eq. (10.38),

$$[M_l^{(\nu)}]_{ij} = \frac{g_l}{n_\nu} \chi_l^{(\nu)} \delta_{ij} , \qquad [M_m^{(\nu)}]_{ij} = \frac{g_m}{n_\nu} \chi_m^{(\nu)} \delta_{ij} ,$$

so Eq. (10.63) reduces to

$$\sum_{\nu=1}^{N} \sum_{i=1}^{n_\nu} \sum_{j=1}^{n_\nu} \frac{1}{n_\nu} \chi_l^{(\nu)} \chi_m^{(\nu)*} \delta_{ij} = \frac{g}{g_m} \delta_{lm} ,$$

where we have made use of the fact that $\delta_{ij}^2 = \delta_{ij}$. Carrying out the sum on i and j, we obtain finally

$$\sum_{\nu=1}^{N} \chi_l^{(\nu)} \chi_m^{(\nu)*} = \frac{g}{g_m} \delta_{lm} . \tag{10.64}$$

Thus if we consider the N constants obtained from the characters of the lth class in each irreducible representation as components of a vector in N-dimensional space, then this vector is orthogonal to the vector obtained in a similar way by using the mth class. Therefore, we have k orthogonal vectors in an N-dimensional space, and we conclude that

$$k \leq N .$$

Combining this result with Eq. (10.50), we get

$$k = N . \tag{10.65}$$

Hence the number of irreducible representations is equal to the number of classes in the group! Combining Eqs. (10.49) and (10.64), we may say that the "character vectors" whose k elements are

$$\sqrt{\frac{g_1}{g}}\chi_1^{(\nu)}, \sqrt{\frac{g_2}{g}}\chi_2^{(\nu)}, \cdots, \sqrt{\frac{g_k}{g}}\chi_k^{(\nu)} \tag{10.66}$$

for $\nu = 1, 2, \cdots, N$ form a *complete* orthonormal set in k-dimensional space.

We now summarize all the basic results in their final form:

a) $$\sum_{\nu=1}^{N} n_\nu^2 = g , \tag{10.59}$$

b) $$k = N , \tag{10.65}$$

c) $$\sum_{U \in G} \sqrt{\frac{n_\mu}{g}} D_{lk}^{(\mu)*}(U) \sqrt{\frac{n_\nu}{g}} D_{ij}^{(\nu)}(U) = \delta_{il}\delta_{kj}\delta_{\mu\nu} , \tag{10.45}$$

d) $$\sum_{\nu=1}^{N} \sum_{i=1}^{n_\nu} \sum_{j=1}^{n_\nu} \sqrt{\frac{n_\nu}{g}} D_{ij}^{(\nu)}(U) \sqrt{\frac{n_\nu}{g}} D_{ij}^{(\nu)*}(U') = \delta_{UU'} , \tag{10.61}$$

e) $$\sum_{i=1}^{N} \sqrt{\frac{g_i}{g}} \chi_i^{(\mu)*} \sqrt{\frac{g_i}{g}} \chi_i^{(\nu)} = \delta_{\mu\nu} , \tag{10.49}$$

f) $$\sum_{\nu=1}^{N} \sqrt{\frac{g_i}{g}} \chi_i^{(\nu)} \sqrt{\frac{g_j}{g}} \chi_j^{(\nu)*} = \delta_{ij} . \tag{10.64}$$

Here g is the order of the group, g_i is the number of elements in the ith class, k is the number of classes, N is the number of irreducible representations and n_ν is the dimensionality of the νth representation.

With this arsenal at our disposal, we can easily determine the characters and representations of some of the basic finite groups. We start with the simplest group, namely, the one-element group consisting of the identity. Since there is only one class, there is only one representation which is one-dimensional. Thus the problem is solved:

$$D^{(1)}(e) = 1 .$$

Consider now the two element group, $\{e, a\}$. The classes are

$$\mathscr{C}_1 = e , \qquad \mathscr{C}_2 = a ,$$

so there are two representations. Since the group is abelian, these must be one-dimensional. Thus we expect to find

$$\sum_{\nu=1}^{2} n_\nu^2 = 2 .$$

Since $n_\nu = 1$ for both $\nu = 1$ and $\nu = 2$, this relation is obeyed. For one-dimensional representations, the characters are the same as the representations, so we can concentrate on the characters. It is useful to make a table of the form

$\nu \backslash i$	1	2
1	1	α
2	1	β

(10.67)

where i labels the class and ν labels the representation. The $i\nu$th element of the square array is just $\chi_i^{(\nu)}$. Since the character of the identity class is equal to the dimensionality of the representation, we have filled in the first column in Eq. (10.67) accordingly (in what follows we shall always label the identity class by $i = 1$). Since in the case of one-dimensional representations the characters are equal to the representation matrices, they must obey the multiplication table of the group. Since $a^2 = e$, we have in this case $\alpha^2 = 1$, $\beta^2 = 1$, $\alpha = \pm 1$, $\beta = \pm 1$. Thus the two distinct representations form the character table

$\nu \backslash i$	1	2
1	1	1
2	1	-1

(10.68)

Note that the "character vector" (1, 1) is orthogonal to the "character vector" (1, -1), as it should be according to Eq. (10.49).

The three element group, $\{e, a, a^2\}$ with $a^3 = e$, has three classes

$$\mathscr{C}_1 = e\,, \qquad \mathscr{C}_2 = a\,, \qquad \mathscr{C}_3 = a^2\,,$$

and therefore three one-dimensional representations. Note that Eq. (10.59) is satisfied in this case. Using the multiplication table for the group, which says that the square of element two is equal to element three, we can write immediately for the character table

$\nu \backslash i$	1	2	3
1	1	α	α^2
2	1	β	β^2
3	1	γ	γ^2

and $\alpha^3 = \beta^3 = \gamma^3 = 1$. Thus α, β, and γ can take on three possible values, 1, $e^{2\pi i/3}$ and $e^{4\pi i/3}$. The three distinct representations are therefore

$\nu \backslash i$	1	2	3
1	1	1	1
2	1	ϵ	ϵ^2
3	1	ϵ^2	ϵ

(10.69)

with $\epsilon = {}^{2\pi i/3}$. The three resulting "character vectors," $(1, 1, 1)$, $(1, \epsilon, \epsilon^2)$ and $(1, \epsilon^2, \epsilon)$ are mutually orthogonal, as the reader may readily verify.

At the fourth order, we find two groups awaiting our attention, the cyclic group of order four and the four-group. For the cyclic group, we shall just remark that by our treatment of the previous three cyclic groups, it should be clear to the reader that the characters of the cyclic group of order n are generated in an obvious way by the n nth roots of unity. Thus the character table for the fourth-order cyclic group looks like

$\nu \backslash i$	1	2	3	4
1	1	1	1	1
2	1	δ	δ^2	δ^3
3	1	δ^2	1	δ^2
4	1	δ^3	δ^2	δ

where $\delta = e^{2\pi i/4} = i$ and we have used the fact that $\delta^4 = 1$ to simplify the array. The generalization to the cyclic group of order n is obvious.

For the four-group, if we denote the elements as usual by $\{e, a, b, c\}$, then the discussion in Section 10.3 makes it clear that $\{e, a\}$ is an invariant subgroup. The factor group is

$$E = \{e, a\}\,, \qquad A = \{b, c\}\,.$$

Thus representations of $\{E, A\}$ [see Eq. (10.68)] can be transferred to the four-group by assigning the representative of E to both e and a and the representative of A to both b and c. The character assignment $\chi(E) = 1$, $\chi(A) = 1$ gives $\chi(e) = 1$, $\chi(a) = 1$, $\chi(b) = 1$ and $\chi(c) = 1$. The assignment $\chi(E) = 1$, $\chi(A) = -1$ gives $\chi(e) = 1$, $\chi(a) = 1$, $\chi(b) = -1$ and $\chi(c) = -1$. Similarly, the invariant subgroup $\{e, b\}$ gives rise to two assignments for the four-group. One of these is the trivial representation (all characters equal unity) which we have already obtained from $\{e, a\}$; the other is $\chi(e) = 1$, $\chi(a) = -1$, $\chi(b) = 1$ and $\chi(c) =$

-1. Finally, the invariant subgroup $\{e, c\}$ gives rise to one new set of characters, $\chi(e) = 1$, $\chi(a) = -1$, $\chi(b) = -1$, and $\chi(c) = 1$. Thus the character table of the four-group looks like

$\nu \backslash i$	1	2	3	4
1	1	1	1	1
2	1	1	-1	-1
3	1	-1	1	-1
4	1	-1	-1	1

(10.70)

Once again, the orthogonality relations are all satisfied.

Since five is a prime number, the only group of order five is the cyclic group which is dealt with by extension of our discussion of the cyclic group of order four. The same remark applies to the cyclic group of order six, which brings us to S_3, the six-element group of permutations on three objects. This is the smallest nonabelian group. We know that it has three classes,

$$\mathscr{C}_1 = e\,, \qquad \mathscr{C}_2 = \{(12), (13), (23)\}\,, \qquad \mathscr{C}_3 = \{(123), (321)\}\,,$$

with corresponding characters χ_1, χ_2, and χ_3. The number of elements in each class is $g_1 = 1$, $g_2 = 3$, and $g_3 = 2$. Since there are three classes, there must be three irreducible representations. Hence by Eq. (10.59)

$$\sum_{\nu=1}^{3} n_\nu^2 = 6\,.$$

The only way to add three squares to obtain the value six is if $n_1 = 1$, $n_2 = 1$, and $n_3 = 2$. Thus for the first time we will have a two-dimensional representation. Using the facts above, we can start our character table as follows:

$\nu \backslash i$	1	2	3
1	1	1	1
2	1	a	b
3	2	c	d

We have put in the trivial representation directly and have used the fact that $\chi_1^{(\nu)} = n_\nu$. Now by Eq. (10.49)

$$\sum_{i=1}^{3} g_i \chi_i^{(1)*} \chi_i^{(2)} = 0\,,$$

that is

$$1 + 3a + 2b = 0 . \tag{10.71}$$

Similarly, from

$$\sum_{i=1}^{3} g_i \chi_i^{(1)*} \chi_i^{(3)} = 0 ,$$

we conclude that

$$3c + 2d + 2 = 0 . \tag{10.72}$$

Using Eqs. (10.71) and (10.72), we simplify the character table to

$\nu \backslash i$	1	2	3
1	1	1	1
2	1	a	$-\frac{3a+1}{2}$
3	2	c	$-\frac{3c+2}{2}$

We can now use other orthogonality properties of the characters [Eq. (10.61)] to write

$$\sum_{\nu=1}^{3} \chi_1^{(\nu)*} \chi_2^{(\nu)} = 0 ,$$

$$\sum_{\nu=1}^{3} \chi_1^{(\nu)*} \chi_3^{(\nu)} = 0 .$$

These both lead directly to

$$a + 2c + 1 = 0 ,$$

so the character table now contains only one unknown:

$\nu \backslash i$	1	2	3
1	1	1	1
2	1	$-1 - 2c$	$1 + 3c$
3	2	c	$-1 - \frac{3}{2}c$

A simple way of tying down the value of c is to note that $(123)^3 = e$ and $(123)^2 = (321)$. Therefore in the one-dimensional representation $\chi_3^3 = 1$ and also $\chi_3^2 = \chi_3$, since (123) and (321) both belong to the same class. Thus $\chi_3 = 1$, and this implies that $1 + 3c = 1$, or $c = 0$. The character table now takes its final form:

ν \ i	1	2	3
1	1	1	1
2	1	-1	1
3	2	0	-1

The second irreducible representation is the so-called antisymmetric representation, obtained by representing odd permutations by -1 and even ones by $+1$. It clearly exists for all S_n. The reader should check that the six normalization conditions implied by Eqs. (10.49) and (10.61),

$$\sum_{i=1}^{3} g_i \chi_i^{(\nu)*} \chi_i^{(\nu)} = 6\,, \qquad \nu = 1, 2, 3\,,$$

and

$$\sum_{\nu=1}^{3} \chi_i^{(\nu)*} \chi_i^{(\nu)} = 6/g_i\,, \qquad i = 1, 2, 3\,,$$

are also satisfied. It should be emphasized that this is by no means the fastest way of obtaining the character table for S_3. For example, the fact that $\mathscr{C}_1$ and $\mathscr{C}_3$ combine to form an invariant subgroup whose factor group is isomorphic to $\{e, a\}$ enables one to write the two one-dimensional representations immediately from Eq. (10.68). Then the character vector of the two-dimensional representation is uniquely determined by orthogonality requirements, as the reader can show.

The characters of the two one-dimensional representations are, of course, exactly equal to the corresponding one-dimensional representation matrices:

$$D^{(1)}(U) = 1\,,$$

for all $U \in S_3$, and

$$D^{(2)}(e) = D^{(2)}(123) = D^{(2)}(321) = 1\,,$$
$$D^{(2)}(12) = D^{(2)}(13) = D^{(2)}(23) = -1\,.$$

However, the two-dimensional representation requires some additional work to obtain the representative matrices. The identity element is no problem. We have

$$D^{(3)}(e) = \begin{pmatrix} 1 & 0 \\ 0 & 1 \end{pmatrix}. \tag{10.74}$$

As for the matrix $D^{(3)}(12)$, let us choose it to be diagonal, since one can always pick one element of a representation to be diagonal in addition to the unit matrix (why?). Thus

$$D^{(3)}(12) = \begin{pmatrix} a & 0 \\ 0 & b \end{pmatrix}.$$

But $(12) \in \mathscr{C}_2$, so by the character table, $\chi_2^{(3)} = 0$, that is, the trace of $D^{(3)}(12)$ is zero. Hence $b = -a$. Also, since $(12)^2 = e$, we must have $a^2 = 1$. Thus choosing $a = 1$,

$$D^{(3)}(12) = \begin{pmatrix} 1 & 0 \\ 0 & -1 \end{pmatrix}. \tag{10.75}$$

Note that if we had taken $a = -1$, we would have found a different matrix, but one which is related to the matrix of Eq. (10.75) by a similarity transformation. Now what about $D^{(3)}(13)$ and $D^{(3)}(23)$? Clearly, since $(13)^2 = e$,

$$D^{(3)}(13) = D^{(3)}(13)^{-1}. \tag{10.76}$$

But since we need consider only unitary representations,

$$D^{(3)}(13)^\dagger = D^{(3)}(13)^{-1},$$

and combining this with Eq. (10.76), we get

$$D^{(3)}(13) = D^{(3)}(13)^\dagger,$$

that is, $D^{(3)}(13)$ is Hermitian. Similarly, $D^{(3)}(23)$ is Hermitian. Thus, using the fact that the trace of both $D^{(3)}(13)$ and $D^{(3)}(23)$ is zero, we may write in complete generality

$$D^{(3)}(13) = \begin{pmatrix} \alpha & \beta \\ \beta^* & -\alpha \end{pmatrix}, \qquad D^{(3)}(23) = \begin{pmatrix} \gamma & \delta \\ \delta^* & -\gamma \end{pmatrix} \tag{10.77}$$

where α and γ are real. But by Eqs. (10.35) and (10.38),

$$M_2^{(3)} = D^{(3)}(12) + D^{(3)}(13) + D^{(3)}(23) = \frac{g_2}{n_3}\chi_2^{(3)} I = 0,$$

since $\chi_2^{(3)} = 0$. Thus using Eqs. (10.75) and (10.77), we have

$$\begin{pmatrix} 1+\alpha+\gamma & \beta+\delta \\ \beta^*+\delta^* & -1-\alpha-\gamma \end{pmatrix} = \begin{pmatrix} 0 & 0 \\ 0 & 0 \end{pmatrix},$$

so that the matrices $D^{(3)}(13)$ and $D^{(3)}(23)$ take the form

$$D^{(3)}(13) = \begin{pmatrix} \alpha & \beta \\ \beta^* & -\alpha \end{pmatrix}, \qquad D^{(3)}(23) = \begin{pmatrix} -(\alpha+1) & -\beta \\ -\beta^* & (\alpha+1) \end{pmatrix}.$$

Finally, we use the unitarity conditions:

$$D^{(3)}(13)^\dagger D^{(3)}(13) = \begin{pmatrix} \alpha^2+|\beta|^2 & 0 \\ 0 & \alpha^2+|\beta|^2 \end{pmatrix} = \begin{pmatrix} 1 & 0 \\ 0 & 1 \end{pmatrix},$$

$$D^{(3)}(23)^\dagger D^{(3)}(23) = \begin{pmatrix} (\alpha+1)^2+|\beta|^2 & 0 \\ 0 & (\alpha+1)^2+|\beta|^2 \end{pmatrix} = \begin{pmatrix} 1 & 0 \\ 0 & 1 \end{pmatrix}.$$

These yield the two equations

$$\alpha^2 + |\beta|^2 = 1\,, \qquad (\alpha + 1)^2 + |\beta|^2 = 1\,,$$

which are readily solved to give $\alpha = -\frac{1}{2}$, $\beta = (\sqrt{3}/2)e^{i\phi}$, where ϕ is arbitrary. Thus

$$D^{(3)}(13) = \begin{pmatrix} -\frac{1}{2} & \frac{\sqrt{3}}{2}e^{i\phi} \\ \frac{\sqrt{3}}{2}e^{-i\phi} & \frac{1}{2} \end{pmatrix}, \tag{10.78}$$

$$D^{(3)}(23) = \begin{pmatrix} -\frac{1}{2} & -\frac{\sqrt{3}}{2}e^{i\phi} \\ -\frac{\sqrt{3}}{2}e^{-i\phi} & \frac{1}{2} \end{pmatrix}. \tag{10.79}$$

Now $(123) = (13)(12)$, so

$$\begin{aligned} D^{(3)}(123) = D^{(3)}(13)D^{(3)}(12) &= \begin{pmatrix} -\frac{1}{2} & \frac{\sqrt{3}}{2}e^{i\phi} \\ \frac{\sqrt{3}}{2}e^{-i\phi} & \frac{1}{2} \end{pmatrix}\begin{pmatrix} 1 & 0 \\ 0 & -1 \end{pmatrix} \\ &= \begin{pmatrix} -\frac{1}{2} & -\frac{\sqrt{3}}{2}e^{i\phi} \\ \frac{\sqrt{3}}{2}e^{-i\phi} & -\frac{1}{2} \end{pmatrix}. \end{aligned} \tag{10.80}$$

Also, $(321) = (123)^{-1}$, so

$$D^{(3)}(321) = D^{(3)}(123)^{-1} = D^{(3)}(123)^{\dagger} = \begin{pmatrix} -\frac{1}{2} & \frac{\sqrt{3}}{2}e^{i\phi} \\ -\frac{\sqrt{3}}{2}e^{-i\phi} & -\frac{1}{2} \end{pmatrix}, \tag{10.81}$$

which completes the determination of the representative matrices except for ϕ. However, it is always possible, without loss of generality, to choose $\phi = 0$, since a simple multiplication shows that any matrix of the form

$$\begin{pmatrix} a & be^{i\phi} \\ ce^{-i\phi} & d \end{pmatrix}$$

is reduced to the form

$$\begin{pmatrix} a & b \\ c & d \end{pmatrix}$$

when similarity transformed by the unitary matrix

$$\begin{pmatrix} e^{i\phi} & 0 \\ 0 & 1 \end{pmatrix}.$$

This fact, along with Eqs. (10.74), (10.75), (10.78), (10.79), (10.80), and (10.81), enables us to write finally

$$\begin{gathered} D^{(3)}(e) = \begin{pmatrix} 1 & 0 \\ 0 & 1 \end{pmatrix}, \quad D^{(3)}(12) = \begin{pmatrix} 1 & 0 \\ 0 & -1 \end{pmatrix}, \quad D^{(3)}(13) = \begin{pmatrix} -\frac{1}{2} & \frac{\sqrt{3}}{2} \\ \frac{\sqrt{3}}{2} & \frac{1}{2} \end{pmatrix}, \\ D^{(3)}(23) = \begin{pmatrix} -\frac{1}{2} & -\frac{\sqrt{3}}{2} \\ -\frac{\sqrt{3}}{2} & \frac{1}{2} \end{pmatrix}, \quad D^{(3)}(123) = \begin{pmatrix} -\frac{1}{2} & -\frac{\sqrt{3}}{2} \\ \frac{\sqrt{3}}{2} & -\frac{1}{2} \end{pmatrix}, \\ D^{(3)}(321) = \begin{pmatrix} -\frac{1}{2} & \frac{\sqrt{3}}{2} \\ -\frac{\sqrt{3}}{2} & -\frac{1}{2} \end{pmatrix}. \end{gathered} \tag{10.82}$$

We have already written these matrices in Eq. (3.5), where we pointed out that they were isomorphic to the group of permutations of three objects. Here we have started with S_3 and relentlessly derived these isomorphic matrices.

It is interesting to note that in this case, when the representation matrices are not all equal to characters, Eq. (10.45) is not equivalent to Eq. (10.49) as it is for purely one-dimensional cases. For example, the $i = 1$, $j = 1$ element of all the matrices of the two-dimensional representation [see Eq. (10.82)] gives a "representation vector" which is, according to Eq. (10.60),

$$\frac{1}{\sqrt{3}}\left(1,\, 1,\, -\frac{1}{2},\, -\frac{1}{2},\, -\frac{1}{2},\, -\frac{1}{2}\right).$$

On the other hand, the $i = 1$, $j = 2$ element gives

$$\frac{1}{\sqrt{3}}\left(0,\, 0,\, \frac{\sqrt{3}}{2},\, -\frac{\sqrt{3}}{2},\, -\frac{\sqrt{3}}{2},\, \frac{\sqrt{3}}{2}\right).$$

These two vectors are normalized to unity and are mutually orthogonal, as they should be according to Eq. (10.45). The nontrivial one-dimensional representation gives the "representation vector"

$$\frac{1}{\sqrt{6}}\,(1,\, -1,\, -1,\, -1,\, 1,\, 1)\,,$$

which is normalized correctly to unity and is orthogonal to each of the two "representation vectors" written above. The reader may want to check the remaining orthonormality relations implied by Eqs. (10.45) and (10.61).

10.8 GROUP THEORY IN PHYSICAL PROBLEMS

Now that we know how to obtain the characters and representations of groups, we can indicate some of the ways in which group theory can be used to increase our understanding of certain basic physical processes.

As a particularly simple illustration of what we have in mind, consider the problem of a quantum-mechanical system which has various eigenstates which are degenerate because of the existence of some symmetry of Schrödinger's equation. As we mentioned in Section 10.4, the collection of eigenfunctions belonging to some degenerate eigenvalue will transform into itself under any transformation belonging to the symmetry group in question. According to Eq. (10.15), if $\mathscr{U} \in \mathscr{G}$, and $U \in G$, then

$$\mathscr{U}\psi_m(\mathbf{r}) = \sum_{m'} D_{m'm}(U)\psi_{m'}(\mathbf{r})\,,$$

where $D(U)$ $(U \in G)$ is a representation of the group G. Now suppose that we apply to the system a weak perturbation having a smaller symmetry group than the original system. For example, we might have an atom, which in its natural environment in free space is a rotationally invariant system, but which when put in, say, the lattice of a cubic symmetric crystal feels a potential which has the symmetry of the cube. The symmetry of the cube is obviously much smaller than that of complete rotational invariance; in fact, the infinite-element symmetry group of all rotations in three dimensions is replaced by the 24-element symmetry group of the cube (this group is isomorphic to the group, S_4, of permutations on four objects).

When the atom finds itself in this new cubic potential, the energy levels will be modified in such a way that the degenerate eigenvectors belonging to the new energy levels will transform among themselves according to the representations of the group, $\mathcal{O}$, of symmetries of the cube. In other words, a μ-fold degenerate level of the rotationally invariant system must break up into M sets of degenerate levels belonging to $\mathcal{O}$, with multiplicities $\mu_1, \mu_2, \cdots, \mu_M$. Clearly, $\mu_1 + \mu_2 + \cdots + \mu_M = \mu$.

The proper framework for a discussion of these questions is provided by degenerate perturbation theory (Section 4.12). If the reader will look back at that section, he will see that the important objects in the discussion of perturbation theory are the so-called matrix elements of the perturbing operator between states of the unperturbed system. To be precise, let us suppose as in Section 4.12, that A_0 is the linear operator governing the unperturbed system:

$$A_0\phi_{n,\nu,i} = \lambda_{n,\nu}\,\phi_{n,\nu,i}\,.$$

Here we have chosen the indices to conform to the group theoretical facts which we have learned in this chapter. ν labels the irreducible representation which tells us how the eigenfunction transforms under the symmetry group, G_0, of A_0, and i is the column of the representation to which $\phi_{n,\nu,i}$ belongs. That is, in standard terminology,

$$\mathscr{V}\phi_{n,\nu,i} = \sum_{j=1}^{n_\nu} D_{ji}^{(\nu)}(V)\phi_{n,\nu,j}\,; \tag{10.83a}$$

V is any element of G_0 ($\mathscr{V}$ is the corresponding element of $\mathscr{G}_0$) and n denotes any additional labels which may be necessary to specify uniquely the eigenvalue under consideration. For example, the νth representation may be of relevance more than once as we run through the entire spectrum of A_0.

We now imagine that the perturbation A_1 is invariant under each element, U, of the symmetry group G_1 and that $\psi_{n,\nu,i}$ is a set of functions which transform according to the νth representation of G_1:

$$\mathscr{U}\psi_{n,\nu,i} = \sum_{j=1}^{n_\nu} D_{ji}^{(\nu)}(U)\psi_{n,\nu,j} \tag{10.83b}$$

where $\mathscr{U} \in \mathscr{G}_1$. We wish to consider matrix elements of the form

$$M_{n,\nu,i;n',\mu,j} \equiv (\psi_{n,\nu,i},\, A_1\psi_{n',\mu,j})\,. \tag{10.84}$$

Since $\mathscr{U}^\dagger\mathscr{U} = \mathscr{U}\mathscr{U}^\dagger = I$, we may write Eq. (10.84) as

$$\begin{aligned} M_{n,\nu,i;n',\mu,j} &= (\psi_{n,\nu,i},\, \mathscr{U}^\dagger\mathscr{U}A_1\mathscr{U}^\dagger\mathscr{U}\psi_{n',\mu,j}) \\ &= (\mathscr{U}\psi_{n,\nu,i},\, \mathscr{U}A_1\mathscr{U}^{-1}\mathscr{U}\psi_{n,\mu,j})\,. \end{aligned}$$

Using the transformation properties of the $\psi_{n,\nu,i}$ we may write this as

$$M_{n,\nu,i;n',\mu,j} = \left(\sum_{k=1}^{n_\nu} D_{ki}^{(\nu)}(U)\psi_{n,\nu,k}, [\mathscr{U} A_1 \mathscr{U}^{-1}] \sum_{l=1}^{n_\mu} D_{lj}^{(\mu)}(U)\psi_{n',\mu,l} \right),$$

or

$$M_{n,\nu,i;\,n',\mu,j} = \sum_{k=1}^{n_\nu} \sum_{l=1}^{n_\mu} D_{ki}^{(\nu)*}(U) D_{lj}^{(\mu)}(U) (\psi_{n,\nu,k}, [\mathscr{U} A_1 \mathscr{U}^{-1}]\psi_{n',\mu,l}) \,. \tag{10.85}$$

But since we have assumed A_1 to be invariant under G_1,

$$\mathscr{U} A_1 \mathscr{U}^{-1} = A_1 \,,$$

so Eq. (10.85) becomes

$$M_{n,\nu,i;\,n',\mu,j} = \sum_{k=1}^{n_\nu} \sum_{l=1}^{n_\mu} D_{ki}^{(\nu)*}(U) D_{lj}^{(\mu)}(U) (\psi_{n,\nu,k}, A_1\psi_{n',\mu,l}) \,. \tag{10.86}$$

Now the crucial step is to sum both sides of Eq. (10.86) over all $U \in G_1$, so we will be able to take advantage of the orthogonality relation of Eq. (10.45) on the right-hand side. On the left-hand side, since M does *not* depend on U [see Eq. (10.84)], we get just gM, where g is the order of G_1. Thus

$$gM_{n,\nu,i;\,n',\mu,j} = \sum_{k=1}^{n_\nu} \sum_{l=1}^{n_\mu} \sum_{U \in G_1} D_{ki}^{(\nu)*}(U) D_{lj}^{(\mu)}(U) M_{n,\nu,k;\,n',\mu,l} \,. \tag{10.87}$$

But the main part of this sum involving the representation elements can be evaluated immediately by using the orthogonality theorem [Eq. (10.45)] which tells us directly that

$$\sum_{U \in G_1} D_{ki}^{(\nu)*}(U) D_{lj}^{(\mu)}(U) = \frac{g}{n_\nu} \delta_{kl}\delta_{ij}\delta_{\mu\nu} \,.$$

Inserting this into Eq. (10.87), we obtain

$$M_{n,\nu,i;\,n',\mu,j} = \frac{1}{n_\nu} \delta_{\mu\nu}\delta_{ij} \sum_{l=1}^{n_\nu} M_{n,\nu,l;\,n',\nu,l} \,. \tag{10.88}$$

Thus we see that matrix elements of this type vanish unless the eigenfunctions belong to the same representation and, furthermore, to the same column of this representation. This result can often be helpful in perturbation calculations by severely limiting the number of eigenfunctions which can occur in the perturbation-theory sums [see, e.g., Eq. (4.60)].

On the most elementary level, Eq. (10.88) tells us that a perturbation, A_1, with the *same* symmetry group as the dominant operator, A_0, cannot split degeneracies in first order, since, in computing the matrix of A_1 between the degenerate states belonging to some eigenvalue of A_0, we will find only diagonal elements [$i = j$ in Eq. (10.88)], and all the diagonal elements are equal. Hence by Eq. (4.65b), all the first-order energies will be equal, and there is no splitting of the degeneracy. This is not surprising since if A_1 has the same symmetry as A_0, then the operator $A = A_0 + \epsilon A_1$ will also have the same symmetry as A_0.

Therefore, according to the general arguments of Section 10.4, the degenerate levels of A_0 belonging to some irreducible representation of G_0 can never split since these new levels of A would have a smaller multiplicity than the original level. This in turn would mean that there exist several linear combinations of eigenfunctions of the original degenerate level which transform among themselves under G_0. The original level must therefore give rise to a reducible representation of G_0, contrary to assumption. Thus no splitting can occur.

A more useful application of Eq. (10.88) is found by considering the case where A_1 has a symmetry lower than that of A_0 and where the group G_1, under which A_1 is invariant is a subgroup of the larger group G_0, under which A_0 is invariant. The key fact is that in such a case the representations $D^{(\nu)}(V)$ $(V \in G_0)$ of G_0 will give rise to representations of G_1 by associating $D^{(\nu)}(V)$ with the $U \in G_1$ which corresponds to $V \in G_0$. In general, this representation will be reducible. For example, the group $\{e, (12)\}$ is a subgroup of $S_3 = \{e, (12), (13), (23), (123), (321)\}$. Now, in the two-dimensional representation of S_3, e is represented by the unit matrix while (12) is represented by

$$\begin{pmatrix} 1 & 0 \\ 0 & -1 \end{pmatrix}.$$

Thus for the group $\{e, (12)\}$ the representation

$$e \rightarrow \begin{pmatrix} 1 & 0 \\ 0 & 1 \end{pmatrix}, \qquad (12) \rightarrow \begin{pmatrix} 1 & 0 \\ 0 & -1 \end{pmatrix}$$

is perfectly acceptable, but it is obviously reducible into the sum of two one-dimensional representations. In a more complicated case, where the decomposition is not so obvious, the number of times an *irreducible* representation of G_1 is contained in a representation of G_1 *induced* by G_0 can be calculated simply by knowing the characters of the irreducible representations of G_1 and using Eq. (10.54).

With this point of view clearly in mind, it is easy to see what happens when we split a degeneracy by applying a perturbation A_1, which is invariant under G_1, to a system A_0, which is invariant under G_0. A given degenerate level of A_0 belongs to an *irreducible* representation $D^{(\nu)}$ of G_0, i.e., the eigenfunctions of this level transform among themselves *under the action of* G_0 according to $D^{(\nu)}$. However, we can also consider this level as belonging to a *reducible* representation of G_1; that is, certain sets of linear combinations of wave-functions belonging to this level transform into themselves *under the action of* G_1.

We can see this as follows. Suppose $D(V)$ is some *irreducible* representation of G_0. If we restrict ourselves to the subset of V's which are also elements of G_1 and denote a typical element of this subset by U, then $D(U)$ is in general a *reducible* representation of G_1. Therefore there exists some matrix S (independent of U) such that for all $U \in G_1$,

$$S^{-1}D(U)S = D'(U) \qquad \text{or} \qquad D(U) = SD'(U)S^{-1}.$$

$D'(U)$ has the same form as the matrix shown in Eq. (10.19), that is, it has square matrices (*irreducible* representations of G_1) on its diagonal, with all other elements equal to zero. Suppose that $\{f_i\}$ is a set of functions which transform among themselves under G_0 according to the irreducible representation $D(V)$. Then if we restrict our attention to those elements, U, which also belong to G_1, we may write

$$\begin{aligned}\mathscr{U} f_i &= \sum_j D_{ji}(U)\, f_j \\ &= \sum_j [SD'(U)S^{-1}]_{ji}\, f_j \\ &= \sum_{j,\,k,\,m} S_{jk} D'_{km}(U)[S^{-1}]_{mi}\, f_j .\end{aligned}$$

This last relation can be written as

$$\mathscr{U} \sum_i S_{il} f_i = \sum_{j,\,k} S_{jk} D'_{kl}(U) f_j \,.$$

If we write

$$g_l = \sum_i S_{il} f_i \qquad \text{then} \qquad \mathscr{U} g_l = \sum_k D'_{kl}(U) g_k \,.$$

Looking at the form of $D'(U)$ [see Eq. (10.19)] we can pick out by inspection the sets of g's which transform among themselves according to the irreducible representations of G_1 on the diagonal of $D'(U)$.

If we now apply degenerate perturbation theory to the perturbing operator A_1, using the linear combinations of degenerate eigenfunctions appropriate to G_1, then according to Eq. (10.88) there will be no nonzero matrix elements of A_1 which connect different irreducible representations of G_1, and within an irreducible representation of G, the diagonal elements will give the only nonvanishing contributions. These diagonal elements are all equal within a given irreducible representation. Thus the matrix of A_1, calculated between appropriately chosen linear combinations of eigenfunctions belonging to a degenerate level of A_0, might appear as follows:

$$\begin{pmatrix}
a & & & & & & & & & & & & \\
 & a & & & & & & & & & & & \\
 & & a & & & & & & & & & & \\
 & & & a & & & & & & \text{O} & & & \\
 & & & & b & & & & & & & & \\
 & & & & & b & & & & & & & \\
 & & & & & & b & & & & & & \\
 & & & & & & & c & & & & & \\
 & & & & & & & & c & & & & \\
 & & \text{O} & & & & & & & c & & & \\
 & & & & & & & & & & c & & \\
 & & & & & & & & & & & c & \\
 & & & & & & & & & & & & c
\end{pmatrix}$$

In this case, a thirteenfold degenerate level of A_0, belonging to a thirteen-dimensional representation of G_0, has been split into three levels—fourfold, threefold, and sixfold degenerate—belonging to three different irreducible representations of G_1. (The reader should note that we are assuming that a given irreducible representation of G_1 occurs only once in the decomposition. What would the matrix of A_1 look like if some irreducible representation occurred twice?) Of course, one would not often know in advance the correct linear combination of wave-functions, so the above diagonal array would in practice look like a complicated 13×13 matrix, which would take the above form after being similarity transformed (diagonalized). The similarity transform would also give the correct linear combinations of the original wave-functions in the manner explained in Section 4.12.

To illustrate the ideas contained in the above discussion, we consider the example mentioned at the beginning of this section, namely, a rotationally invariant system (an atom in free space) placed in a weak, cubic-symmetric potential (in a crystal with cubic symmetry). Before we can solve the problem, we must first obtain the characters of the irreducible representations in question. Let us take the rotation group first. This is a continuous group, and therefore strictly speaking lies outside the framework developed in the preceding sections. However, if we are interested only in characters, we may proceed as follows. We know from elementary considerations (see Section 5.8) that the $(2l + 1)$ spherical harmonics $Y_{lm}(\theta, \phi)$ $(m = -l, -l + 1, \cdots, l - 1, l)$ are eigenfunctions of the rotationally invariant operator L^2 and all belong to the single eigenvalue $l(l + 1)$ $(l = 0, 1, 2, \cdots)$. These degenerate functions must therefore transform among themselves according to a representation of the rotation group (which can be shown to be irreducible).

We know that a general rotation can be brought by a unitary transformation to the form of a simple rotation about an axis. Let us call the angle of rotation about this axis Φ. Since all rotations through an angle Φ are unitarily equivalent, such rotations form a class which can be labeled by Φ. To obtain the character of such a class, we need consider only a simple rotation through Φ about the z-axis. Such a rotation takes $Y_{lm}(\theta, \phi)$ into $e^{im\Phi} Y_{lm}(\theta, \phi)$ (see Section 5.8). Thus the matrix representing this rotation will be

$$\begin{pmatrix} e^{-il\Phi} & & & & \text{O} \\ & e^{-i(l-1)\Phi} & & & \\ & & \ddots & & \\ & & & e^{i(l-1)\Phi} & \\ \text{O} & & & & e^{il\Phi} \end{pmatrix}.$$

The trace is readily computed; we find that

$$\chi^{(l)}(\Phi) = \sum_{m=-l}^{l} e^{im\Phi} = \frac{\sin(l + \frac{1}{2})\Phi}{\sin\left(\frac{\Phi}{2}\right)}, \tag{10.89}$$

where we have evaluated the sum by using the formula for summing a geometric series. Note that the dimension, n_l, of the lth representation is

$$\lim_{\Phi\to 0} \frac{\sin (l + \frac{1}{2})\Phi}{\sin (\frac{\Phi}{2})} = 2l + 1 ,$$

as we should expect since $\Phi = 0$ corresponds to the identity transformation. When $l = 0$, we find the trivial one-dimensional transformation in which every rotation is represented by 1. When $l = 1$, we obtain the three-dimensional representation which should be familiar from classical mechanics; it is given, as a function of the Euler angles, by Eq. (1.25).

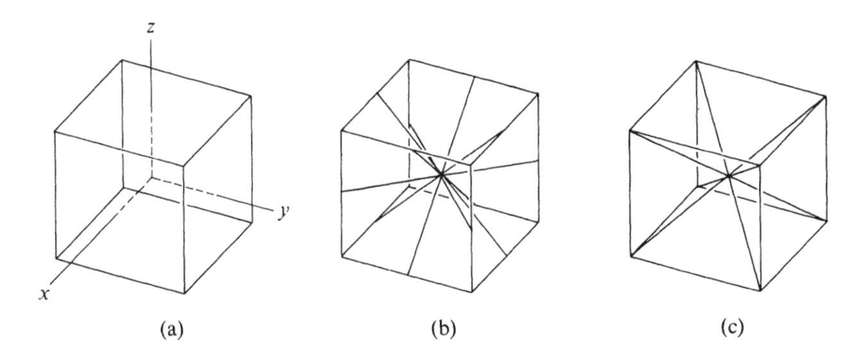

Fig. 10.2 The axes of symmetry of a cube.

Having found the class characters for all the representations of the group of three-dimensional rotations, let us now move to more familiar ground and find the representations and characters of the group $\mathcal{O}$ of symmetries of the cube. This group contains 24 elements in five classes, being isomorphic to S_4. The class $\mathscr{C}_1$ is the identity class. Figure 10.2(a) shows the axes of rotation which are relevant for the classes $\mathscr{C}_2$ and $\mathscr{C}_4$. $\mathscr{C}_2$ consists of the six rotations through $\pm\pi/2$ radians about each of the x, y, and z axes; the class $\mathscr{C}_4$ consists of the three elements corresponding to rotation through π radians about the x-, y-, and z-axes. Figure 10.2(b) shows the axes relevant to the class $\mathscr{C}_5$, which consists of rotations through π radians about each of these six axes. Finally, the class $\mathscr{C}_3$ consists of rotations through $\pm 2\pi/3$ radians about each of the four axes shown in Fig. 10.2(c). We have labeled these classes in this particular manner so they will correspond to the class assignments for S_4 given in Section 10.3. Just as with S_4, the classes $\mathscr{C}_1$ and $\mathscr{C}_4$ combine together to give an invariant subgroup V, the factor group of which is isomorphic to S_3. Thus, in the manner illustrated in the previous section, we can lift representations of S_3 onto S_4 (and hence onto $\mathcal{O}$). Since $\mathcal{O}$ has 24 elements and five classes, we must have

$$n_1^2 + n_2^2 + n_3^2 + n_4^2 + n_5^2 = 24 ,$$

where n_ν is the dimension of the νth representation. The only possible collection of n_ν which can satisfy this relation is $n_1 = 1$, $n_2 = 1$, $n_3 = 2$, $n_4 = 3$, and $n_5 = 3$.

According to our previous remarks, the first, second, and third representations (with $n_1 = 1$, $n_2 = 1$, and $n_3 = 2$) will correspond to those of $S_3 = S_4/V$. Thus, in usual manner, we can fill in a large part of the character table for $\mathcal{O}$:

$\nu \backslash i$	1	2	3	4	5
1	1	1	1	1	1
2	1	-1	1	1	-1
3	2	0	-1	2	0
4	3	a	b	c	d
5	3	α	β	γ	δ

(10.90)

Since $g_1 = 1$, $g_2 = 6$, $g_3 = 8$, $g_4 = 3$, and $g_5 = 6$, we can use Eq. (10.49) to obtain

$$6a + 8b + 3c + 6d = -3\,,$$
$$-6a + 8b + 3c - 6d = -3\,,$$
$$-8b + 6c = -6\,.$$

The equations for α, β, γ, and δ are identical to the above three. These equations are readily solved to give $b = 0$, $c = -1$, $d = -a$; similarly $\beta = 0$, $\gamma = -1$, $\delta = -\alpha$. Thus Table (10.90) becomes

$\nu \backslash i$	1	2	3	4	5
1	1	1	1	1	1
2	1	-1	1	1	-1
3	2	0	-1	2	0
4	3	a	0	-1	$-a$
5	3	α	0	-1	$-\alpha$

The normalization condition implied by Eq. (10.49) gives $a^2 = 1 = \alpha^2$, that is, $a = \pm 1$ and $\alpha = \pm 1$. The two distinct choices are $a = 1$, $\alpha = -1$ or $a =$

-1, $\alpha = 1$. These lead to the same character table except for an interchange of the last two rows. Thus the character table takes its final form:

$\nu \backslash i$	1	2	3	4	5
1	1	1	1	1	1
2	1	-1	1	1	-1
3	2	0	-1	2	0
4	3	1	0	-1	-1
5	3	-1	0	-1	1

(10.91)

In this case, the representation matrices can also be readily obtained. For the first three representations, the matrices are just those previously obtained for S_3. The three-dimensional representations are obtained in a simple manner from the three-dimensional representation of the rotation group. Since we will not need these matrices here, we will omit a detailed analysis of them.

With all the routine work out of the way, we can see what happens when a rotationally invariant atomic system is put into a weak cubic-symmetric environment. First, we note that the various representations of the rotation group induce representations of $\mathcal{O}$. Consider, for example, the $l = 1$ representation of the rotation group. The classes of the cubic group correspond to the following angles of rotations (in radians):

$$\mathscr{C}_1 \to 0\,, \qquad \mathscr{C}_2 \to \pm\pi/2\,, \qquad \mathscr{C}_3 \to \pm 2\pi/3\,, \qquad \mathscr{C}_4 \to \pi\,, \qquad \mathscr{C}_5 \to \pi\,.$$

The characters corresponding to these rotations are obtained from Eq. (10.89). We find that

$$\chi_1 = 3\,, \qquad \chi_2 = 1\,, \qquad \chi_3 = 0\,, \qquad \chi_4 = -1\,, \qquad \chi_5 = -1\,.$$

By looking at the character table of $\mathcal{O}$ [Table (10.91)], we see that this has given us an *irreducible* representation of $\mathcal{O}$, corresponding to the fourth row of the character table. Thus an eigenvalue of the rotationally invariant system belonging to the $l = 1$ representation of the rotation group will not be split when subjected to a perturbation of cubic symmetry.

Moving to the $l = 2$ representation and using Eq. (10.89), we find that

$$\chi_1 = 5\,, \qquad \chi_2 = -1\,, \qquad \chi_3 = -1\,, \qquad \chi_4 = 1\,, \qquad \chi_5 = 1$$

for the characters of the representation of $\mathcal{O}$ induced by the rotation group. Now, to determine the number of times a given *irreducible* representation of $\mathcal{O}$

is contained in this reducible representation, we use Eq. (10.54). We find that

$$a^{(1)} = \frac{1}{24}\sum_{i=1}^{5} g_i\chi_i^{(1)}\chi_i = \frac{1}{24}[5 - 6 - 8 + 3 + 6] = 0\,,$$

$$a^{(2)} = \frac{1}{24}\sum_{i=1}^{5} g_i\chi_i^{(2)}\chi_i = \frac{1}{24}[5 + 6 - 8 + 3 - 6] = 0\,,$$

$$a^{(3)} = \frac{1}{24}\sum_{i=1}^{5} g_i\chi_i^{(3)}\chi_i = \frac{1}{24}[10 + 0 + 8 + 6 + 0] = 1\,,$$

$$a^{(4)} = \frac{1}{24}\sum_{i=1}^{5} g_i\chi_i^{(4)}\chi_i = \frac{1}{24}[15 - 6 + 0 - 3 - 6] = 0\,,$$

$$a^{(5)} = \frac{1}{24}\sum_{i=1}^{5} g_i\chi_i^{(5)}\chi_i = \frac{1}{24}[15 + 6 + 0 - 3 + 6] = 1\,.$$

Thus the third and fifth *irreducible* representations of $\mathcal{O}$ occur in the *reducible* representation induced by the $l = 2$ representation of the rotation group. Therefore the fivefold degenerate state corresponding to an $l = 2$ representation of the rotation group will split in a cubic symmetric field into a twofold degenerate state and a threefold degenerate state.

The lifting of degeneracy in the problem follows *completely* from the symmetry of the problem without any knowledge of the explicit form of the interactions! Such effects have been verified in innumerable experiments on the spectra of atoms in solids, providing one of the most beautiful examples of the power of group theoretical techniques in physics. In recent years, similar group theoretical methods have been used in elementary particle physics to obtain information about the relationship between the masses of various elementary particles, using only certain assumptions about the symmetry of the "unperturbed" elementary particle Hamiltonian and about the nature of the perturbation term which breaks the symmetry.

As a final illustration of the utility of group theoretical techniques in physics, we shall modify Eq. (10.85) slightly to obtain an interesting result concerning the transitions between quantum-mechanical states. We have already considered the case where A_1 is invariant under all $\mathcal{U} \in \mathcal{G}_1$. Suppose now that A_1 transforms under $\mathcal{G}_1$ according to some irreducible representation of $\mathcal{G}_1$. This would be the case if A_1 were one of a collection of operators, which we might write as $\{A_1^{(\lambda,m)}, m = 1, 2, \cdots, n_\lambda\}$, where λ indicates the irreducible representation under which the set of $A_1^{(\lambda,m)}$ transform; i.e.,

$$\mathcal{U} A_1^{(\lambda,m)} \mathcal{U}^{-1} = \sum_{m'=1}^{n_\lambda} D_{m'm}^{(\lambda)}(U) A_1^{(\lambda,m')}\,.$$

Thus we would rewrite Eq. (10.85) as

$$M_{n,\nu,i;\,n',\mu,j}^{(\lambda,m)} = \sum_{k=1}^{n_\nu}\sum_{l=1}^{n_\mu}\sum_{m'=1}^{n_\lambda} D_{ki}^{(\nu)*}(U) D_{lj}^{(\mu)}(U) D_{m'm}^{(\lambda)}(U)\,(\psi_{n,\nu,k},\, A_1^{(\lambda,m')}\psi_{n',\mu,l})\,, \quad (10.92)$$

where we have defined

$$M^{(\lambda,m)}_{n,\nu,i;\,n',\mu,j} \equiv (\psi_{n,\nu,i},\, A_1^{(\lambda,m)}\psi_{n',\mu,j}) \;.$$

Thus Eq. (10.92) takes the form

$$M^{(\lambda,m)}_{n,\nu,i;\,n',\mu,j} = \sum_{k=1}^{n_\nu}\sum_{l=1}^{n_\mu}\sum_{m'=1}^{n_\lambda} D^{(\nu)}_{ki}{}^*(U)D^{(\mu)}_{lj}(U)D^{(\lambda)}_{m'm}(U)M^{(\lambda,m')}_{n,\nu,k;\,n',\mu,l} \;. \tag{10.93}$$

What can we say about the product $D^{(\mu)}_{lj}(U)D^{(\lambda)}_{m'm}(U)$? According to Eq. (3.60) this is equal to the lm', jm-element of the direct product of $D^{(\mu)}$ and $D^{(\lambda)}$:

$$[D^{(\mu)}(U) \otimes D^{(\lambda)}(U)]_{lm',jm} = D^{(\mu)}_{lj}(U)D^{(\lambda)}_{m'm}(U) \;.$$

Thus Eq. (10.93) becomes

$$M^{(\lambda,m)}_{n,\nu,i;\,n',\mu,j} = \sum_{k=1}^{n_\nu}\sum_{l=1}^{n_\mu}\sum_{m'=1}^{n_\lambda} D^{(\nu)}_{ki}{}^*(U)[D^{(\mu)}(U) \otimes D^{(\lambda)}(U)]_{lm',jm}M^{(\lambda,m')}_{n,\nu,k;\,n',\mu,l} \;. \tag{10.94}$$

Now since the direct product of two representations is again a representation, the quantity in square brackets in Eq. (10.94) will be a representation matrix element, but even though $D^{(\lambda)}$ and $D^{(\mu)}$ are *irreducible*, their direct product will usually be *reducible*. The number of times that an irreducible representation of G occurs in the direct product is easily determined [by Eq. (10.54)] if we know the characters of the irreducible representations, since the character of U corresponding to $D^{(\mu)}(U) \otimes D^{(\lambda)}(U)$ is just $\chi(U) = \chi^{(\mu)}(U)\chi^{(\lambda)}(U)$, according to Eq. (3.61). Let us label the irreducible representations occurring in $D^{(\mu)} \otimes D^{(\lambda)}$ by the elements ρ, of a set $R_{\lambda\mu}$ of integers. Then we can write

$$[D^{(\mu)}(U) \otimes D^{(\lambda)}(U)]_{lm',jm} = \sum_{\rho\in R_{\lambda\mu}}\sum_{\sigma,\tau=1}^{n_\rho} a^{(\lambda\mu\rho)}_{lm',jm;\,\sigma\tau}D^{(\rho)}_{\sigma\tau}(U) \;. \tag{10.95}$$

Putting Eq. (10.95) into Eq. (10.94) and, as before, summing both sides of Eq. (10.94) over all $U \in G$, we find that

$$M^{(\lambda,m)}_{n,\nu,i;\,n',\mu,j} = \frac{1}{n_\nu}\sum_{\rho\in R_{\mu\lambda}} \delta_{\nu\rho} \sum_{k,l,m'} a^{(\lambda\mu\nu)}_{lm',jm;\,ik}M^{(\lambda,m')}_{n,\nu,k;\,n',\mu,l} \;. \tag{10.96}$$

Without any further work, we can see that unless $\nu \in R_{\lambda\mu}$ the expression in Eq. (10.96) vanishes. In other words, unless the direct product of $D^{(\lambda)}$ and $D^{(\mu)}$ contains the irreducible representation $D^{(\nu)}$, the quantity $M^{(\lambda,m)}_{n,\nu,i,n',\mu,j}$ must equal zero for any choice of i, j, and m.

The usefulness of this result can most easily be seen by considering the rate of transitions between two quantum-mechanical states of a system which is invariant under some symmetry group. Suppose the levels belong respectively to the νth and μth representations of the symmetry group. It is known from quantum mechanics that if A_1 is the operator which induces the transition, then the transition rate is proportional to

$$|(\psi_{n,\nu,i},\, A_1\psi_{n',\mu,j})|^2 \,,$$

where we label the wave functions in the usual manner. If A_1 transforms according to some representation λ of the symmetry group in question, we see that unless the νth representation is contained in the direct product of the μth representation times the λth representation, the transition cannot occur. This restriction follows purely from the symmetry of the problem.

The theory of selection rules of quantum mechanics follow from these considerations. For example, a rotationally invariant atomic (or nuclear) system has energy levels which, as we have seen, can be classified according to the irreducible representations of the rotation group. Now the operator which induces electric dipole transitions between atomic levels is proportional to the electronic position vector $\mathbf{r}$ which transforms according to the $l = 1$ representation of the rotation group, as we saw in Chapter 1. Thus for a transition to take place from a level labeled by l to a level labeled by l', the l' representation of the rotation group must occur in the direct product of the l-representation times the 1-representation. The study of the decomposition of the direct product of representations is thus very important for the applications of group theory to physics, since such a study leads to a direct determination of the selection rules which play a central role in atomic, nuclear, and elementary particle physics.

PROBLEMS

1. Consider the plane of Fig. 10.3, where the objects at A, B, and C are fixed to the circular loop. This figure is clearly invariant under rotation through 120° and 240° and under reflection through the lines a, b, and c. Show that these five operations plus the identity form a group. Write out the group multiplication table. What are the classes into which the various group elements fall?

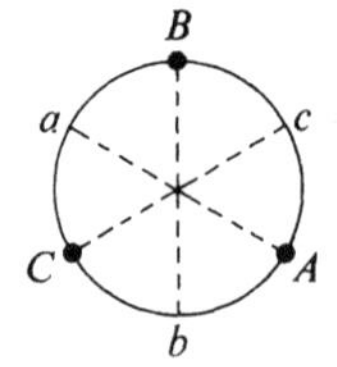

Fig. 10.3

2. Consider the group built up from the unit matrix and the Pauli spin matrices. The elements of this group are $\pm I$, $\pm i\sigma_x$, $\pm i\sigma_y$, and $\pm i\sigma_z$, where

$$\sigma_x = \begin{pmatrix} 0 & 1 \\ 1 & 0 \end{pmatrix}, \quad \sigma_y = \begin{pmatrix} 0 & -i \\ i & 0 \end{pmatrix}, \quad \sigma_z = \begin{pmatrix} 1 & 0 \\ 0 & -1 \end{pmatrix}.$$

What is the multiplication table? Find all subgroups and classes. Are there any invariant subgroups?

3. Find the symmetry group, including only rotations, of a square, rectangular parallelepiped (i.e., square on the top and bottom and rectangular on the sides). Show that this group is isomorphic to the symmetry group of the square which

includes both rotations and reflection. What are the classes? Show that there is an invariant subgroup. Find the factor group associated with this invariant subgroup.

4. Determine all the elements of the group of rotational symmetries of the regular tetrahedron (i.e., a tetrahedron whose sides are equilateral triangles). Put these elements into the classes to which they belong. Find all the subgroups. Are there any invariant subgroups?
5. Consider an arbitrary group, G, with a subgroup, H. Show that if H is of index two, then H is invariant. Remember that the index of a subgroup is the ratio of the order of the group to the order of the subgroup.
6. Find the subgroup of S_5 to which the cyclic group of order five is isomorphic. Demonstrate the isomorphism explicitly by performing all the relevant multiplications of the elements of S_5.
7. Let $(123\cdots l)$ be a cycle of length l. Prove that if $\lambda < l$ then $(123\cdots l)^\lambda \neq e$, while $(123\cdots l)^l = e$.
8. Let G and G' be two groups. Show that in a homomorphism of G onto G' the elements of G which are mapped into the identity of G' form an invariant subgroup, H, of G.
9. Show that the group G' of the previous problem is isomorphic to the factor group G/H, where H is as defined in the previous problem.
10. a) By direct calculation, show that

$$\nabla^2 = \frac{\partial^2}{\partial x^2} + \frac{\partial^2}{\partial y^2} + \frac{\partial^2}{\partial z^2}$$

is invariant under rotation through the Euler angles α, β and γ. To refresh your memory about Euler angles, see Section 1.4.
[*Hint*: Note that the most general rotation is built up from three simpler rotations.]

b) Determine how the three components of the quantum mechanical angular momentum [see Eqs. (5.86)] transform under the group of rotations about the z-axis.

11. Show that the set $P_n = [1, x, x^2, \cdots, x^{n-1}]$ transforms into itself under the action of the translation group in one dimension. Denoting a translation through a distance a by T_a (that is, $T_a x = x - a$)· find the n-dimensional representation of the group of all one-dimensional translations which is provided by the set of functions P_n. Is this representation equivalent to a unitary representation? Show explicitly for the case of the three-dimensional representation that $T_a T_b = T_{a+b}$.
12. Show that the following three sets of functions, [$\cos 3\theta$], [$\sin 3\theta$] and [$\cos 4\theta$, $\sin 4\theta$], transform among themselves under the elements of the group discussed in Problem 10.1. What are the representations and characters which are obtained from these three sets of functions? Are they irreducible representations? What can you say about the representation provided by [$\cos 5\theta$, $\sin 5\theta$]?
13. Show that if C_i is a class containing $S_1, S_2, \cdots, S_{g_i}$ and $C_{i'}$ is the collection of elements consisting of $S_1^{-1}, S_2^{-1}, \cdots, S_{g_i}^{-1}$, then $C_{i'}$ is also a class. Show that for a general irreducible representation, i.e., one which is not necessarily unitary,

$$\sum_i g_i \chi_i^{(\mu)} \chi_{i'}^{(\nu)} = g\delta_{\mu\nu}, \qquad a_\mu = \frac{1}{g} \sum_i g_i \chi_i \chi_{i'}^{(\mu)},$$

where a_μ is the number of times the μth irreducible representation occurs in the representation (not necessarily irreducible) whose characters are χ_i.

14. Show that any 2×2 matrix which commutes with the three Pauli matrices given in Problem 10.2 must be a multiple of the unit matrix.
[*Hint*: Show that any 2×2 matrix can be written as a linear combination of the unit matrix and the Pauli matrices.]

15. Show that the most general 2×2 unitary matrix whose eigenvalues are 1 and -1 can be written as

$$U = \begin{pmatrix} \cos\theta & \sin\theta e^{i\phi} \\ \sin\theta e^{-i\phi} & -\cos\theta \end{pmatrix}.$$

16. Find the characters and irreducible representations of the group discussed in Problem 10.3.
[*Hint*: For the two-dimensional representation, the result of Problem 10.15 may be useful.]

17. Using the three-dimensional representation of the rotation group, given, for example, by Eq. (1.25), find the matrices for a three-dimensional representation of the symmetry group, $\mathcal{O}$, of the cube. We saw in Section 10.8 that there were *two* three-dimensional representations of $\mathcal{O}$. Give a simple prescription for finding the matrices belonging to the second three-dimensional representation (i.e., the one not given by the rotation group).

18. Using the results of Problem 10.4, find the characters and representations of the group of rotational symmetries of the regular tetrahedron.

19. Suppose that we have a linear operator with cubic symmetry, and we apply a small perturbation having the symmetry of a rectangular parallelepiped. Making use of the results of Problem 10.16 and the work of Section 10.8, find how the various possible degenerate states of cubic symmetry break up when this perturbation is applied.

20. Consider a linear operator, A_0, with rotational symmetry. Suppose that a small perturbation having the symmetry of a regular tetrahedron is applied. Using the results of Problem 10.18 and of Section 10.8, find how the four states of A_0 with the lowest degeneracies (onefold, threefold, fivefold, and sevenfold) break up when this perturbation is applied.

21. Consider all possible representations of the cubic group, $\mathcal{O}$, formed by taking the direct product of the various irreducible representations of $\mathcal{O}$. Find out how many times the irreducible representations of $\mathcal{O}$ occur in these product representations.

22. In quantum mechanics one defines an electron spin wave function using α for spin "up" and β for spin "down"; α and β are orthonormal. For a system of three electrons we can form product wave functions of the form $\alpha(1)\alpha(2)\alpha(3)$, $\alpha(1)\alpha(2)\beta(3)$, etc., there being $2^3 = 8$ such products. Clearly, these eight functions transform among themselves under the action of the elements of S_3, and they are mutually orthogonal. They therefore form a set of basis functions for a representation of S_3. Find how many times the various irreducible representations of S_3 occur in this representation. Is any representation missing? Why? Can you relate some of the objects you're dealing with to physics?

23. Suppose that all the eight product spin functions of the previous problem are degenerate eigenfunctions of some linear operator, A_0. Let A_0 be perturbed by an operator of the form $A_1 = AP_{12} + BP_{13} + CP_{23}$, where P_{ij} transforms coordinate i into coordinate j and vice versa. For example,

$$P_{12}\alpha(1)\beta(2)\alpha(3) = \alpha(2)\beta(1)\alpha(3) .$$

Show that the matrix which must be diagonalized to obtain the first-order corrections to the eigenvalue of A_0 to which the product functions belong can be written in the form

$$\begin{bmatrix} E & 0 & 0 & 0 & 0 & 0 & 0 & 0 \\ 0 & E & 0 & 0 & 0 & 0 & 0 & 0 \\ 0 & 0 & E & 0 & 0 & 0 & 0 & 0 \\ 0 & 0 & 0 & E & 0 & 0 & 0 & 0 \\ 0 & 0 & 0 & 0 & E_{11} & E_{12} & 0 & 0 \\ 0 & 0 & 0 & 0 & E_{12} & E_{22} & 0 & 0 \\ 0 & 0 & 0 & 0 & 0 & 0 & E_{11} & E_{12} \\ 0 & 0 & 0 & 0 & 0 & 0 & E_{12} & E_{22} \end{bmatrix},$$

where $E = A + B + C$, $E_{11} = A - (B + C)/2$, $E_{12} = \sqrt{3}(B - C)/2$ and $E_{22} = -E_{11}$, if we use the representation of S_3 given in Section 10.7.

24. If in the previous problem we had used a different (but unitarily equivalent) two-dimensional representation of S_3, we would have found different values for E_{11}, E_{12}, and E_{22}. Would the eigenvalues of the matrix of Problem 10.23 be different? Why?

FURTHER READINGS

DIXON, J. D., *Problems in Group Theory*. Waltham, Mass.: Blaisdell, 1967. Contains solutions.

LEDERMANN, W., *The Theory of Finite Groups*. New York: Interscience, 1964.

HAMERMESH, M., *Group Theory*. Reading, Mass.: Addison-Wesley, 1962.

TINKHAM, M., *Group Theory and Quantum Mechanics*. New York: McGraw-Hill, 1964.

WEYL, H., *Theory of Groups and Quantum Mechanics*. London: Methuen, 1931

WIGNER, E. P., *Group Theory*. New York: Academic Press, 1959.

GENERAL BIBLIOGRAPHY

MATHEMATICAL METHODS OF PHYSICS

COURANT, R., and D. HILBERT, *Methods of Mathematical Physics* (2 Volumes). New York: Wiley-Interscience, 1961.

DETTMAN, J. W., *Mathematical Methods in Physics and Engineering*. New York: McGraw-Hill, 1962.

IRVING, J., and N. MULLINEUX, *Mathematics in Physics and Engineering*. New York: Academic Press, 1959.

MARGENAU, H., and G. M. MURPHY, *The Mathematics of Physics and Chemistry*. Princeton, N. J.: D. Van Nostrand, 1961.

MATHEWS, J., and R. L. WALKER, *Mathematical Methods of Physics*. New York: W. A. Benjamin, 1964.

MORSE, P. M., and H. FESHBACK, *Methods of Theoretical Physics* (2 Volumes). New York: McGraw-Hill, 1953.

SAGAN, H., *Boundary and Eigenvalue Problems in Mathematical Physics*. New York: Wiley, 1961.

SCHWARTZ, L., *Mathematics for the Physical Sciences*. Reading, Mass.: Addison-Wesley, 1966.

WHITTAKER, E. T., and G. N. WATSON, *A Course of Modern Analysis*. New York: Cambridge University Press, 1958.

WILF, H. S., *Mathematics for the Physical Sciences*. New York: Wiley, 1962.

WYLIE, C., *Advanced Engineering Mathematics*, Second edition. New York: McGraw-Hill, 1960.

MATHEMATICS BOOKS

APOSTOL, T. M., *Mathematical Analysis*. Reading, Mass.: Addison-Wesley, 1957.

BERBERIAN, S. K., *Introduction to Hilbert Space*. New York: Oxford University Press, 1961.

DUNFORD, N., and J. T. SCHWARTZ, *Linear Operators*, Part I, II. New York: Wiley-Interscience, 1966.

HALMOS, P. R., *Introduction to Hilbert Space*. New York: Chelsea, 1951.

KOLMOGOROV, A. N., and S. V. FOMIN, *Functional Analysis*, Volumes I, II. Groyloch, 1957.

RAINVILLE, E., *Special Functions*. New York: Macmillan, 1960.

RIESZ, F., and B. SZ. NAGY, *Functional Analysis*. New York: Ungar, 1955.

RUDIN, W., *Principles of Mathematical Analysis*, Second edition. New York: McGraw-Hill, 1964.

SAKS, S., and A. ZYGMUND, *Analytic Functions*. New York: Hafner, 1952.

SEELEY, R., *An Introduction to Fourier Series and Integrals.* New York: W. A. Benjamin, 1966.

SNEDDON, I. N., *Fourier Transforms.* New York: McGraw-Hill, 1951.

TITCHMARSH, E. C., *The Theory of Functions.* New York: Oxford University Press, 1939.

WATSON, G. N., *A Treatise on the Theory of Bessel Functions,* Second edition. New York: Cambridge University Press, 1944.

WIDDER, D. V., *Advanced Calculus,* Second edition. Englewood Cliffs, N. J.: Prentice Hall, 1961.

BOOKS OF SOLVED PROBLEMS IN MATHEMATICAL PHYSICS

CHOQUET-BRUHAT, Y., *Problems and Solutions in Mathematical Physics.* San Francisco: Holden-Day. 1967.

LEBEDEV, N. N., I. P. SKOLSKAYA, and Y. S. UFLYAND, *Problems of Mathematical Physics.* Englewood Cliffs, N. J.: Prentice Hall, 1965.

MISYURKEYEV, I. V., *Problems in Mathematical Physics with Solutions.* New York: McGraw-Hill, 1966.

USEFUL MATHEMATICAL TABLES

BATEMAN, H., *Higher Transcendental Functions* (3 Volumes), based in part on notes left by H. Bateman and compiled by the staff of the Bateman Manuscript Project. New York: McGraw-Hill, 1953–1955.

BIERENS DE HAAN, D., *Nouvelles Tables d'Integrales Definies,* Edition of 1867, corrected, with an English translation of the introduction by J. F. Ritt. New York: Hafner, 1957.

FLETCHER, A., J. C. P. MILLER, L. ROSENHEAD, and L. J. COMRIE, *An Index of Mathematical Tables* (2 Volumes). Reading, Mass.: Addison-Wesley, 1962.

JAHNKE-EMDE-LOSCH, *Tables of Higher Functions.* Sixth edition. New York: McGraw-Hill, 1960.

JOLLEY, L. B. W., *Summation of Series,* Second edition. New York: Dover Publications, 1961.

MAGNUS, W., and F. OBERHETTINGER, *Formulas and Theorems for the Special Functions of Mathematical Physics,* translated from the German by John Wermer. New York: Chelsea, 1949.

MARCUS, M., *Basic Theorems in Matrix Theory,* Applied Mathematics Series, Pamphlet No. 57, U. S. Department of Commerce.

INDEX TO VOLUME ONE

INDEX TO VOLUME TWO

$$F = -\frac{d}{dx} U$$

$$P = F/A \qquad F = mg \qquad \frac{d}{dx}\frac{1}{2}m\dot{x}^2 = m\ddot{x}$$

$$P = F/dA \qquad \frac{d}{dx} U = F$$

$$dP = mg/dA$$

$$\int dP = mg\int \frac{1}{dA} \qquad \frac{d}{dx} mg x = mg$$

$$P - P_0 \; mg \ln A$$